STRONGLY CORRELATED SYSTEMS, COHERENCE AND ENTANGLEMENT

STRONGLY CORRELATED SYSTEMS, COHERENCE AND ENTANGLEMENT

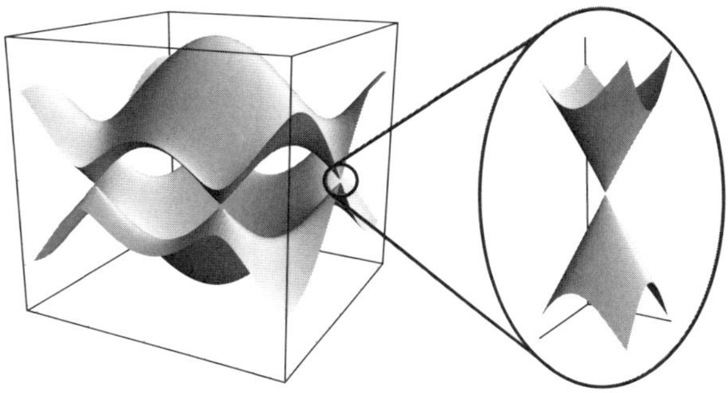

Editors

J. M. P. Carmelo
Universidade do Minho, Portugal

P. D. Sacramento
Instituto Superior Técnico, Portugal

J. M. B. Lopes dos Santos
Universidade do Porto, Portugal

V. Rocha Vieira
Instituto Superior Técnico, Portugal

World Scientific

NEW JERSEY • LONDON • SINGAPORE • BEIJING • SHANGHAI • HONG KONG • TAIPEI • CHENNAI

Published by

World Scientific Publishing Co. Pte. Ltd.

5 Toh Tuck Link, Singapore 596224

USA office: 27 Warren Street, Suite 401-402, Hackensack, NJ 07601

UK office: 57 Shelton Street, Covent Garden, London WC2H 9HE

British Library Cataloguing-in-Publication Data
A catalogue record for this book is available from the British Library.

STRONGLY CORRELATED SYSTEMS, COHERENCE AND ENTANGLEMENT

ISBN-13 978-981-270-572-3
ISBN-10 981-270-572-4

Printed in Singapore by B & JO Enterprise

Preface

This book presents a collection of review papers on recent work in the related areas of strongly correlated systems, the effects of coherence on macroscopic systems and entanglement in quantum systems. These areas have attracted considerable interest not just due to their inherent complexity and unexpected nontrivial phenomena, but also due to their potential applications in various fields, from material science to information technology. We review topics in the fields of A) Strong Correlations, Transport and Dynamics in Complex Materials, B) Strongly Correlated Magnetic Systems, C) Quantum Coherent Systems and D) Quantum Entanglement.

It is well known that low dimensionality enhances quantum fluctuations and electron electron correlations. The first two chapters review the Luttinger Liquid paradigm in the context of the one dimensional Hubbard model, and some recent results on finite energy transport and spectral properties of this same model, based on the exact Bethe Ansatz solution. These have found striking confirmation in photoemission studies. Next, we review aspects of strong correlations in the cuprates and their manifestation in photoemission experiments, again, a major source of information for these materials. Following the discovery by the Manchester group of Andre Geim that two-dimensional carbon sheets (atomic monolayers of graphite) can be prepared in planar form and are remarkably stable, there has been a surge of theoretical and experimental activity on this remarkable material, graphene. Its properties are fascinating, even at the one-electron level, and the potential for applications (electronics, sensors, hydrogen storage) is enormous. We include an introduction to the basic physics of this most unusual material. The field of the quantum Hall effect is reviewed, in particular new and less familiar topics like the anomalous Hall effect. We conclude this part with two chapters on aspects of non-equilibrium phenomena such as in spin systems and in the context of the Falicov-Kimball model.

In the second part, spin systems in low-dimensional materials and the importance of the effects of disorder in spin systems are considered. Spin disorder is a particularly important ingredient in the transport and optical properties of manganites and hexaborides, the focus of Chapters 10 and 11. The field of transport properties in magnetic materials and their important potential applications in sensor technology are reviewed, paying special attention to the presence of magnetic domain walls and their effects on transport with a view on the growing field of spintronics.

In a third set of chapters we focus our attention on topics related to quantum coherence in fields such of Bose-condensation and superconductivity. We gather articles on atomic correlations in ultra cold quantum gases, and discuss developments on the theory of Bose condensation and its kinetic and dynamic properties. Also, several articles are dedicated to aspects of superconductors like the effect of singular density of states on the superconducting properties and effects of disorder. The importance of the interdisciplinarity of the concepts that emerged to understand superconductors and its relation to other branches of physics like chiral symmetry breaking is also addressed. The experimental capability of controlling cold atoms confined in small spatial regions, for instance with the use of optical lattices, has enabled the experimental observation of Bose condensation, the control of atomic configurations,and even of interactions, providing a very clean way to simulate various physical systems, like traditional correlated condensed matter systems. It also became possible to control, in a systematic way, quantum states, with possible applications in information technology. This establishes an interface between traditional condensed matter or atomic physics problems with the fourth part of this book dedicated to entanglement and its possible applications.

In this last part of the book we take a look at some of the recent developments in quantum information and computation and their relation to condensed matter physics. Entanglement was recognized, very early in the development of quantum mechanics, as the characteristic feature which enforced its entire departure from classical physics, but it remained, for a long time, as something strange or at least mysterious. Later, after the introduction of Bell's inequalities and their experimental verification, it started to be considered as a useful resource for quantum information and computation, for example, increasing our understanding of quantum mechanics, and even of nature. More recent advances not just in quantum optics but also in macroscopic quantum coherent systems, such as those discussed in part 3, have led to a reanalysis of several known results in condensed matter physics and to the development of new methods and techniques, in a fruitful cooperation between quantum information and computation theory and condensed matter physics.

J.M.P. Carmelo, J.M.B. Lopes dos Santos, V. Rocha Vieira and P.D. Sacramento

Contents

PART 1

Strong Correlations, Transport and Dynamics in Complex Materials

Chapter 1

Correlation effects in one-dimensional systems

J.M.P. Carmelo

GCEP-Center of Physics, Universidade do Minho, Campus Gualtar, P-4710-057 Braga, Portugal

P.D. Sacramento

Departamento de Física and CFIF, Instituto Superior Técnico, Av. Rovisco Pais, 1049-001 Lisboa, Portugal

D. Bozi

Instituto de Ciencia de Materiales, CSIC, Cantoblanco, E-28949 Madrid, Spain

L.M. Martelo

Departamento de Física, Faculdade de Engenharia, Universidade do Porto, P-4200-465 Porto, Portugal

We review developments concerning the effect of correlations on the electronic properties of one-dimensional systems, focusing our analysis on the one-dimensional Hubbard model. We consider methods used to describe the exotic properties of these systems, ranging from bosonization associated with the Tomonaga and Luttinger liquid behavior, to the Bethe ansatz solution, referring to all energy scales of solvable quantum problems and the pseudoparticle description. We use that description to study the model energy spectrum and the low-energy quantities. In the ensuing companion chapter we discuss the relation of the electronic operators to these quantum objects.

Contents

1.1. Effects of correlations

1.1.1. *Introduction*

This chapter is complementary to the ensuing companion chapter, Ref. 1, where some of the concepts and theoretical tools summarized here are applied to the study of low-dimensional correlated systems and a pseudofermion theory for the study of the finite-energy dynamical properties is reviewed.

The main point concerning the issues studied here and in the next chapter is that, when the electronic movements are restricted to low-dimensional geometries, the effects of the electron-electron interactions become non perturbative, and thus conventional Fermi liquid theory[2] does not apply. Instead, the low-energy physics of such interacting problems shows some basic similarities with that of the Tomonaga and Luttinger models.[3,4] The concept of a Tomonaga-Luttinger liquid[5] follows such similarities and refers to interacting low-dimensional models whose low-energy behavior belongs to the universality class of those models.

The one-dimensional Hubbard model[6] is one of the interacting electronic models whose low-energy physics belongs to such an universality class. Its importance is that it is the simplest lattice model which describes the effects of electronic correlations in low-dimensional complex materials. Indeed, the exotic non-Fermi-liquid behavior associated with the concept of a Tomonaga-Luttinger liquid is observed in some of such materials, see Refs. 7-11. One of the techniques used in the study of the low-energy physics of interacting models belonging to that universality class is bosonization, see Refs. 12-15. Some of these models have exact solutions which combined with their global symmetries provide the spectrum of all energy eigenstates. For instance, the global symmetry of the Hubbard model was recently shown to be $[SO(4) \times U(1)]/Z_2$.[16] Its exact energy spectrum and spectral properties can be calculated by combining symmetry with Bethe-ansatz techniques, see Refs. 17-31.

As further discussed in the ensuing chapter, Ref. 1, finite-energy spectral functions of the one-dimensional Hubbard model can be evaluated by expressing the generators of its energy eigenstates, associated with the model Bethe-ansatz solution and global symmetry, in terms of suitable rotated-electron operators.[28] Such operators are related to the electronic creation and annihilation operators by a unitary transformation.[28,32] Moreover, the combination of the rotated-electron basis with the information provided by the Bethe-ansatz solution and global symmetry reveals that the energy eigenstates correspond to simple occupancy configurations of exotic quantum objects which are closely related to the rotated electrons. The statistics of such objects can be defined in terms of a generalized Pauli principle.[33]

In addition to reviewing the several concepts and theoretical tools involved in the description of one-dimensional correlated electronic problems, below we also consider the specific case of the one-dimensional Hubbard model. Both here and in the following chapter we summarize how the low-energy and finite-energy physics of the model is described in terms of the above exotic objects. This provides important information about the non-perturbative microscopic processes which control the unconventional properties observed in many low-dimensional complex materials, which are described in Refs. 7-11.

1.1.2. *Fermi liquid theory*

In many three-dimensional systems the effect of interactions between fermionic particles is taken into account using the so-called Fermi liquid theory. In this theory[2] the excitation spectrum is fermionic but i) the quasiparticle parameters are renormalized with respect to the free system (like the effective mass), ii) the thermodynamic quantities have a behavior similar to the free electron case but also with renormalized parameters, iii) the lifetime of the quasiparticles is finite, except at the Fermi surface where it diverges like $\tau \sim (\epsilon - \epsilon_F)^{-2}$, (therefore the quasiparticles are well defined quantities for energies close to the Fermi surface) and iv) new collective modes emerge in the system. The lifetime of the quasiparticles is a consequence of the interactions and the energy of the excitations is expressed as,

$$\Delta E = \sum_{\vec{k}} \left(\epsilon_0(\vec{k}) - \mu \right) \Delta N(\vec{k}) + \frac{1}{2} \sum_{\vec{k},\vec{k}'} \Delta N(\vec{k}) f(\vec{k},\vec{k}') \Delta N(\vec{k}') + \cdots \,, \tag{1.1}$$

where $\Delta N(\vec{k})$ is the deviation of the quasiparticle distribution with respect to the equilibrium distribution and $f(\vec{k},\vec{k}')$ results from the residual interactions between the quasiparticles. The interactive term is of the same order of magnitude as the free term since the deviation of the energies, with respect to the chemical potential, is also small in the regime where the quasiparticles have a long lifetime. The fermionic nature of the quasiparticles implies that the excitation spectrum remains similar to the free electron case. The spectrum is a continuum of low energy excitations due to the excitations of particles and holes of arbitrarily small energies and momenta around the Fermi surface, in addition to the plasma mode at finite frequency.

1.1.3. *One-dimensional systems*

The one-dimensional case is special due to the fact that the Fermi surface has only two points. An important consequence is that any instability of momentum $2k_F$ couples the two states at the Fermi surface and may lead to a gap in the spectrum (Peierls instability). The resulting spectrum is qualitatively different from the spectrum of the Fermi gas (which has no gap) showing that the interactions have an important role in the one-dimensional case. In general, in systems of higher dimension, any instability that couples two points of the Fermi surface has a null measure and therefore is not relevant to the behavior of the system (with the exception of nesting). The Peierls instability in one dimension suggests that the excitations of the system may have a different nature and can be described in terms of collective bosonic excitations.

Two models of one-dimensional conductors are normally considered in the literature. The first is a continuum model where one considers electrons with weak interactions and where the electrons occupy states that are extended. The other model is suitable in the opposite limit, where the electrons have wave functions which are more localized, typically with a strong atomic character. The model is immersed in a lattice and describes situations of narrow bands where the interactions (or at least the correlations) between the electrons are typically strong. The continuum model was originally considered by Tomonaga[3] and Luttinger[4] and the lattice model was introduced by Hubbard.[6] Actually, the Tomonaga-Luttinger model also constitutes a good starting point in situations where the interactions between the electrons are not weak, as long as one is interested in the low-energy and small-momentum properties, where the lattice details are not important.

1.1.4. *Tomonaga and Luttinger models*

The dispersion relation of the free electron gas is such that only the electrons close to the Fermi surface are important. It is therefore usual to linearize the dispersion relation in the form $\epsilon_r(k) = v_F(rk - k_F)$ introducing two branches ($r = \pm 1$) around the two Fermi points $\pm v_F$. It is then necessary to introduce a cut-off k_0. Such a procedure leads to the Tomonaga model.[3] It is also possible (and convenient) to consider a dispersion relation without cut-off, extending the bands to $\pm\infty$ (note that considering only the regime of small energies these additional states are not expected to affect the behavior). This choice corresponds to the Luttinger model.[4] This model describes therefore two types of fermions corresponding to the two branches $r = \pm 1$, whose densities interact via two interactions: one term corresponds to interactions between electrons in the same branch (g_2) and the other between electrons in different branches (g_4). The Luttinger model is defined by $H = H_0 + H_2 + H_4$ where,

$$H_0 = \sum_{r,k,s} v_F(rk - k_F) : c^\dagger_{rks} c_{rks} : ,$$
$$H_2 = \frac{1}{L} \sum_{p,s,s'} \left(g_{2,||}(p)\delta_{s,s'} + g_{2,\perp}(p)\delta_{s,-s'} \right) \rho_{+,s}(p)\rho_{-,s'}(-p) ,$$
$$H_4 = \frac{1}{2L} \sum_{r,p,s,s'} \left(g_{4,||}(p)\delta_{s,s'} + g_{4,\perp}(p)\delta_{s,-s'} \right) : \rho_{r,s}(p)\rho_{r,s'}(-p) : . \quad (1.2)$$

The first term is the usual kinetic operator and in the interacting terms the spin, s, dependence on the interactions is considered in the forms $||$ or $\perp$. The operators have to be normal ordered to eliminate the infinite number of states introduced in the model, by subtracting the average value in the ground state.

Besides these terms that only include low-momentum scattering, the Luttinger model may be extended considering additional terms that take into account the possibility of finite momentum excitations, where the two branches are coupled, allowing backscattering or considering Umklapp processes. The model without these additional terms is exactly solvable.[5] The Hamiltonian conserves the total charge, the spin, and also these quantities separately in each branch. Therefore the charge and spin currents are also conserved.

1.1.5. *Bosonization*

One way to solve the Luttinger model results from the property that the commutator of the density operators is given by,

$$[\rho_{r,s}(p), \rho_{r',s'}(-p')] = -\delta_{r,r'}\delta_{s,s'}\delta_{p,p'} \frac{rpL}{2\pi} , \quad (1.3)$$

where L is the system length. Therefore the density operators satisfy bosonic commutation relations. Defining charge and spin density operators in each branch $\rho_r(p) = (\rho_{r,\uparrow}(p) + \rho_{r,\downarrow}(p)) / \sqrt{2}$, $\sigma_r(p) = (\rho_{r,\uparrow}(p) - \rho_{r,\downarrow}(p)) / \sqrt{2}$, and using Kronig's identity, where the kinetic term of the Hamiltonian can be written in a bilinear form in the density operators, one obtains that the Luttinger model can be diagonalized via a Bogoliubov-Valatin transformation leading to a separation of the charge and spin degrees of freedom $\tilde{H} = \tilde{H}_\rho + \tilde{H}_\sigma + \tilde{H}_c$,

$$\tilde{H}_\nu = \frac{2\pi v_\nu}{L} \sum_{p\neq 0} : (\tilde{\nu}_+(p)\tilde{\nu}_+(-p) + \tilde{\nu}_-(p)\tilde{\nu}_-(p)) : ,$$

$$\tilde{H}_c = \frac{\pi}{2L} \sum_{\nu=\rho,\sigma} \left(\frac{v_\nu}{K_\nu}(N_{+,\nu} + N_{-,\nu})^2 + v_\nu K_\nu (N_{+,\nu} - N_{-,\nu})^2 \right) , \quad (1.4)$$

($\nu = \rho, \sigma$) where the charge and spin velocities are given by,

$$v_\nu = \sqrt{\left(v_F + \frac{1}{2\pi}(g_{4\|} \pm g_{4\perp})\right)^2 - \left(\frac{1}{2\pi}(g_{2\|} \pm g_{2\perp})\right)^2} , \quad (1.5)$$

($+, -$ corresponds to ρ, σ respectively), $N_{r,\nu} = \rho_{r,\nu}(p = 0)$, and the parameters K_ν are given by,

$$K_\nu = \sqrt{\frac{\pi v_F + \frac{1}{2}(g_{4\|} \pm g_{4\perp}) - \frac{1}{2}(g_{2\|} \pm g_{2\perp})}{\pi v_F + \frac{1}{2}(g_{4\|} \pm g_{4\perp}) + \frac{1}{2}(g_{2\|} \pm g_{2\perp})}} . \quad (1.6)$$

The operators of the charge and spin degrees of freedom commute among themselves and are separately conserved, as said above. Moreover, they propagate with different velocities leading to an effective separation in real space.

The non-existence of fermionic excitations implies that there are no quasiparticles at the Fermi surface. The residue of the pole of the Green function is zero. The density of states vanishes at the Fermi surface and therefore we find a spectral weight reduction near that surface. There is ample experimental evidence for these unusual properties. Using photoemission experiments it was found[7] that there is no Fermi edge in the dispersion relation in the two quasi-one-dimensional compounds $K_{0.3}MoO_3$ and $(TaSe_4)_2I$. Also, spin-charge separation was observed[8] in the material $SrCuO_2$. Further evidence for spin-charge separation was found in the organic conductor TTF-TCNQ[9] and it was further established that the experimental results are not compatible with standard band theory and that the interaction/correlation effects are determinant. Further unconventional behavior was observed in the Mott insulators $SrCuO_2$ and Sr_2CuO_3 whose dispersion relations are not consistent with band theory.[10,11]

The diagonalization of the Luttinger model shows that the Hamiltonian and its excitations are described by bosonic modes. However, the calculation of arbitrary correlation functions requires products of fermionic operators. The full solution of the model including all its correlation functions involves the representation of the fermionic operators in terms of the bosonic operators. This procedure is called bosonization.[12]

The charge density and the current, $j_s(x)$, satisfy the following commutator,

$$[\rho_s(x), j_s(x')] = -\frac{i}{\pi}\frac{\partial}{\partial x}\delta(x - x'), \tag{1.7}$$

where $j_s(x) =: \rho_{+,s}(x) - \rho_{-,s}(x)$:. Let us then consider two conjugate fields, $\Phi_s(x)$ and $\Pi_s(x)$, satisfying the canonical commutation relation $[\Phi_s(x), \Pi_s(x')] = i\delta(x - x')$. Identifying the operators $\rho_s(x) = -\frac{1}{\pi}\frac{\partial}{\partial x}\Phi_s(x)$ and $j_s(x) = \Pi_s(x)$, satisfies the commutation relation. This result suggests that we may represent the density by a bosonic field of the form $\frac{\partial \Phi_s(x)}{\partial x} = -\pi\rho_s(x)$. The introduction of a fermion at site x creates a kink (soliton) of amplitude π in the bosonic field. The introduction of one particle at point x implies that the rest of the particles have to adjust to accept the new particle at x. We may therefore express the fermionic operator in the form of a translation operator (plus a phase needed to yield the anti-commutation relations),[13]

$$\psi_{r,s}(x) \sim \lim_{\alpha \to 0} \frac{1}{2\pi\alpha} e^{irk_F x - ir\Phi_{r,s}(x) + i\pi\int_{-\infty}^{x} dz \Pi_{r,s}(z)} . \tag{1.8}$$

The Luttinger model may also be expressed as $H = H_\rho + H_\sigma$, where,

$$H_\nu = \int dx \left(\frac{\pi u_\nu K_\nu}{2}\Pi_\nu^2 + \frac{u_\nu}{2\pi K_\nu}(\partial_x \phi_\nu)^2 \right), \tag{1.9}$$

with $\nu = \rho, \sigma$. Such a transformation allows the calculation of all the correlation functions of the Luttinger model (gaussian model). The correlation functions are characterized by critical exponents which are non-universal. The calculation of the correlation functions is reduced in this context to averages over a gaussian distribution. For instance, within bosonization it can be shown that,[14]

$$< n(x)n(0) >= \frac{K_\rho}{(\pi x)^2} + A_1 \frac{\cos(2k_F x)}{x^{1+K_\rho}} \frac{1}{\ln^{-3/2}(x)} + A_2 \frac{\cos(4k_F x)}{x^{4K_\rho}}, \tag{1.10}$$

and

$$< \vec{S}(x) \cdot \vec{S}(0) >= \frac{1}{(\pi x)^2} + B_1 \frac{\cos(2k_F x)}{x^{1+K_\rho}} \frac{1}{\ln^{1/2}(x)} . \tag{1.11}$$

Also, the quantity K_ρ determines the singularity of the momentum distribution,

$$n_k \sim \frac{1}{2} - \frac{k - k_F}{|k - k_F|}|k - k_F|^\alpha, \tag{1.12}$$

where $\alpha = (K_\rho + 1/K_\rho - 2)/4$ is the exponent of the single-particle density of states $N(\omega) \sim |\omega|^\alpha$ as well.

One needs to determine the parameter K_ρ for each specific model. We may take several routes. A possible way is to note that the coefficient u_ρ/K_ρ in the Hamiltonian is proportional to the variation of the ground state energy with respect to the particle number, since the gradient of the phase field ϕ_ρ is proportional to the density,

$$\frac{1}{L}\frac{\partial^2 E_0(n)}{\partial n^2} = \frac{\pi}{2}\frac{u_\rho}{K_\rho} . \tag{1.13}$$

Once the ground state energy for instance for the Hubbard model may be obtained from the exact solution via the Bethe ansatz, we may calculate the parameter K_ρ and therefore the critical exponents. It turns out that $1/2 < K_\rho < 1$. For large on-site repulsion, U, $K_\rho = 1/2$ and $\alpha = 1/8$. We will return to this point ahead. Very similar results can be obtained for the closely related Tomonaga model.

1.1.6. *Tomonaga-Luttinger liquids*

Even though the Tomonaga and Luttinger models are very simplified, they describe in a qualitatively correct way the low energy properties of many interacting one-dimensional systems. A broadly used nomenclature classifying interacting one-dimensional systems whose low-energy behavior falls in the universality class of the Tomonaga and Luttinger models is that of a Tomonaga-Luttinger liquid. One of the consequences of this universality is that the critical behavior of the Tomonaga-Luttinger liquids is determined by the critical exponents of the Tomonaga and Luttinger models (however, the values of the exponents depend on the parameters of each specific model). Another class of one-dimensional systems is the Luther-Emery class[15] which groups systems with gaps in the spectrum. In particular, for instance the addition of the backscattering term to the Luttinger model leads in some regimes (for an attractive interaction) to a gap in the spin excitations. The Umklapp term leads to a gap in the charge excitations.

1.2. Hubbard model

A model whose low-energy physics is in the class of the Tomonaga-Luttinger liquids is the repulsive Hubbard model, away from half-filling. On the other hand, the attractive Hubbard model has a gap in the spin excitations and is in the class of the Luther-Emery model. The model is described by the Hamiltonian,

$$\hat{H} = -t\sum_{j,\sigma}\left(c^\dagger_{j,\sigma}c_{j+1,\sigma} + c^\dagger_{j+1,\sigma}c_{j,\sigma}\right) + U\hat{D} - \frac{U}{2}N + \frac{U}{4}N_a ,$$
$$\hat{D} = \sum_j c^\dagger_{j,\uparrow}c_{j,\uparrow}c^\dagger_{j,\downarrow}c_{j,\downarrow} , \tag{1.14}$$

describing a tight-binding model for N electrons with nearest-neighbor amplitude t where electrons of opposite spins interact with each other via a local repulsive

Coulomb potential, U, on a lattice with N_a sites. In this model only states with $l = 0$ are taken into account and therefore in each lattice site the maximal number of electrons is 2. $\hat{D}$ is the double occupancy operator. In general, we use units of Planck constant $\hbar$ and lattice constant a such that $\hbar = a = 1$. The model has $[SO(4) \times U(1)]/Z_2$ global symmetry and (if N_a is even) commutes with the six generators of the spin and eta-spin $SU(2)$ algebras and the generator of a hidden symmetry which is half of the number operator of sites singly occupied by "rotated electrons".[16] The spin generators and the eta-spin generators are given by $\hat{S}_s^z = -\frac{1}{2}[\hat{N}_\uparrow - \hat{N}_\downarrow]$, $\hat{S}_s^\dagger = \sum_j c_{j,\downarrow}^\dagger c_{j,\uparrow}$ and $\hat{S}_s = \sum_j c_{j,\uparrow}^\dagger c_{j,\downarrow}$; $\hat{S}_c^z = -\frac{1}{2}[N_a - \hat{N}]$, $\hat{S}_c^\dagger = \sum_j (-1)^j c_{j,\downarrow}^\dagger c_{j,\uparrow}^\dagger$ and $\hat{S}_c = \sum_j (-1)^j c_{j,\uparrow} c_{j,\downarrow}$. We call S_c (and S_s) the η-spin (and spin) value of an energy eigenstate and S_c^z (and S_s^z) its η-spin (and spin) projection.

1.2.1. *Bethe ansatz solution*

The Hubbard model has been solved exactly via the Bethe ansatz.[17] That solution refers to a subspace spanned by the lowest-weight states (LWSs) of both the η-spin and spin algebras. The latter states are such that $S_\alpha = -S_\alpha^z$ where $\alpha = c, s$. Within the thermodynamic limit the solution involves degrees of freedom that correspond to different rapidity branches. In addition to a $c0$ charge-momentum rapidity, there are sets of $\alpha\nu$ rapidities. The general rapidity branch label $\alpha\nu$ is such that $\alpha = c, s$ and $\nu = 0, 1, 2, ...$ for $\alpha = c$ and $\nu = 1, 2, ...$ for $\alpha = s$. The $c\nu$ and $s\nu$ rapidities are associated with the charge and spin degrees of freedom, respectively. For $\nu > 0$ the $c\nu$ and $s\nu$ rapidities may be associated with charge 2ν-holon and spin 2ν-spinon composite objects, respectively, where holons and spinons are elementary "particles" which carry η-spin $1/2$ and spin $1/2$, respectively.[28]

For electronic densities $n \leq 1$, the ground state has finite occupancies for the charge $c0$ and spin $s1$ branches only, reinforcing the idea that in that system there is a separation of degrees of freedom. However, the corresponding quantum objects called pseudoparticles in Refs. 27,28 are not independent and have residual interactions. In the $U \to \infty$ limit the equations that determine the residual interactions between those pseudoparticles and corresponding degrees of freedom decouple (but the spin degrees of freedom affect the charge degrees of freedom through a boundary condition term).

The Bethe ansatz equations are given by,

$$q_j = q(k_j) = k_j + \frac{2}{N_a} \sum_{\alpha\nu\neq c0} \sum_{j'=1}^{N^*_{\alpha\nu}} N^R_{\alpha\nu}(\Lambda_{\alpha\nu,\,j'}) \arctan\Big(\frac{\sin(k_j) - \Lambda_{\alpha\nu,\,j'}}{\nu U/4t}\Big)\,; \quad j = 1, 2, ..., N_a\,, \tag{1.15}$$

$$\begin{aligned} q_j &= q(\Lambda_{\alpha\nu,\,j}) \\ &= k_{\alpha\nu,\,j} - (\delta_{\alpha,c} - \delta_{\alpha,s})\frac{2}{N_a}\sum_{j'=1}^{N_a} N^R_{c0}(k_{j'}) \arctan\Big(\frac{\Lambda_{\alpha\nu,\,j} - \sin(k_{j'})}{\nu U/4t}\Big) \\ &- \frac{1}{N_a}\sum_{\nu'=1}^{N_a/2}\sum_{j'=1}^{N^*_{\alpha\nu'}} N^R_{\alpha\nu'}(\Lambda_{\alpha\nu',\,j'})\Theta_{\nu,\,\nu'}\Big(\frac{\Lambda_{\alpha\nu,\,j} - \Lambda_{\alpha\nu',\,j'}}{U/4t}\Big)\,; \\ & j = 1, 2, ..., N^*_{\alpha\nu}\,; \qquad \alpha\nu \neq c0\,, \end{aligned} \tag{1.16}$$

where $k_{\alpha\nu,\,j} = \delta_{\alpha,c} 2\,\mathrm{Re}\,\{\arcsin(\Lambda_{c\nu,\,j} + i\nu U/4t)\}$ with $j = 1, 2, ..., N^*_{\alpha\nu}$ for $\alpha\nu \neq c0$ and the value of the number $N^*_{\alpha\nu} \leq N_a$ is defined by Eqs. (B.6) and (B.7) of Ref. 28. The occupied and unoccupied values k_j of the charge-momentum rapidity and the occupied and unoccupied values $\Lambda_{\alpha\nu,\,j}$ of the $\alpha\nu$ rapidities of a given energy eigenstate are determined by these equations which are valid for large values of N_a and N and were first introduced by Takahashi.[19] Here we wrote them in functional form in terms of the distribution functions $N^R_{c0}(k_j)$ and $N^R_{\alpha\nu}(\Lambda_{\alpha\nu,\,j})$, whose occupancies are well defined for each state. The function $\Theta_{\nu,\,\nu'}(x)$ is given in Eq. (B.5) of Ref. 28. The equations (1.15) and (1.16) include the discrete bare-momentum values q_j of the form $q_j = [2\pi/N_a]\,I^{c0}_j$ and $q_j = [2\pi/N_a]\,I^{\alpha\nu}_j$ for $\alpha\nu \neq c0$ where the numbers I^{c0}_j and $I^{\alpha\nu}_j$ with $j = 1, 2, ..., N_a$ and $j = 1, 2, ..., N^*_{\alpha\nu}$, respectively, are the quantum numbers whose occupancy configurations describe the energy eigenstates. The latter numbers are integers or half-odd integers as a result of the following boundary conditions,

$$e^{iq_j\,N_a} = (e^{i\pi})^{[\sum_{\alpha=c,\,s}\sum_{\nu=1}^{\infty} N_{\alpha\nu}]}\,, \tag{1.17}$$

in the case of the $c0$ branch and,

$$e^{iq_j\,N_a} = (e^{i\pi})^{[1+N^*_{\alpha\nu}]} = (e^{i\pi})^{[1+N_{c0}+N_{\alpha\nu}]}\,; \qquad \alpha = c, s\,, \qquad \nu = 1, 2, ...\,, \tag{1.18}$$

for the $\alpha\nu \neq c0$ branches. Thus, for $\alpha\nu \neq c0$ the quantum numbers $I^{\alpha\nu}_j$ are integers (half-odd integers), if $N^*_{\alpha\nu}$ is odd (even). On the other hand, the quantum numbers I^{c0}_j are integers (half-odd integers), if $\frac{N_a}{2} - \sum_{\alpha\nu\neq c0} N_{\alpha\nu}$ is odd (even). There is for the $c0$ branch (and the $\alpha\nu \neq c0$ branches) of all energy eigenstates a one-to-one correspondence between the discrete bare-momentum value q_j and the discrete charge-momentum rapidity value k_j (and discrete $\alpha\nu$ rapidity value $\Lambda_{\alpha\nu,\,j}$) with the same value for the index j such that $j = 1, ..., N_a$ (and $j = 1, ..., N^*_{\alpha\nu}$). That correspondence is such that there is no level crossing between the set of N_a $c0$ pseudoparticle discrete bare-momentum values $\{q_j\}$ (and $N^*_{\alpha\nu}$ $\alpha\nu \neq c0$ pseudoparticle discrete bare-momentum values $\{q_j\}$) and the set of charge momentum rapidity values $\{k_j\}$ (and $\alpha\nu$ rapidity values $\{\Lambda_{\alpha\nu,\,j}\}$). This means that if $q_j > q_{j'}$ for the $c0$ branch (and for a $\alpha\nu \neq c0$ branch), then $k_j > k_{j'}$ (and $\Lambda_{\alpha\nu,\,j} > \Lambda_{\alpha\nu,\,j'}$) for the same values of j and j', respectively.

The occupancies of the bare-momentum values q_j obey a Pauli principle, *i.e.* a discrete bare-momentum value q_j can either be unoccupied or singly occupied. Such occupancies can be described by bare-momentum distribution functions $N_{\alpha\nu}(q_j)$. Moreover, there is also a one-to-one correspondence between the occupied discrete bare-momentum values q_j and the occupied discrete charge-momentum rapidity values k_j or discrete $\alpha\nu$ rapidity values $\Lambda_{\alpha\nu,\,j}$, such that $j = 1, ..., N_{c0}$ or $j = 1, ..., N_{\alpha\nu}$, respectively. That correspondence is behind the equalities $N_{c0}(q_j) = N^R_{c0}(k_j)$ and $N_{\alpha\nu}(q_j) = N^R_{\alpha\nu}(\Lambda_{\alpha\nu,\,j})$ for the same values of the index j. The bare-momentum distribution functions read $N_{\alpha\nu}(q_j) = 1$ for occupied discrete bare-momentum values q_j and $N_{\alpha\nu}(q_j) = 0$ for unoccupied discrete bare-momentum values q_j. The pseudoparticle representation of Ref. 28 corresponds to the description of the energy eigenstates in terms of the discrete bare-momentum q_j occupancy configurations, instead of the charge-momentum rapidity k_j and $\alpha\nu$ rapidity $\Lambda_{\alpha\nu,\,j}$ occupancy configurations. Thus, the above distributions $N_{\alpha\nu}(q_j)$ are the $\alpha\nu$ bare-momentum pseudoparticle distribution functions. These functions are for all energy eigenstates the eigenvalues of the following pseudoparticle bare-momentum distribution function operators,

$$\hat{N}_{\alpha\nu}(q_j) = b^\dagger_{q_j,\,\alpha\nu}\, b_{q_j,\,\alpha\nu}\,. \tag{1.19}$$

Here the operator $b^\dagger_{q_j,\,\alpha\nu}$ (and $b_{q_j,\,\alpha\nu}$) creates (and annihilates) a $\alpha\nu$ pseudoparticle of bare-momentum q_j. Each LWS of both the η-spin and spin algebras is uniquely specified by the values of $[N_a - N]$, $[N_\uparrow - N_\downarrow]$, and the set of bare-momentum distribution functions $\{N_{\alpha\nu}(q_j)\}$ such that $\nu = 0, 1, 2, ...$ for $\alpha = c$ and $\nu = 1, 2, ...$ for $\alpha = s$ and $j = 1, ..., N^*_{\alpha\nu}$.

One finds by straightforward manipulation of the Bethe-ansatz equations (1.15) and (1.16) that the spacings $[k_{j+1} - k_j]$, $[k_{c\nu,\,j+1} - k_{c\nu,\,j}]$, and $[\Lambda_{\alpha\nu,\,j+1} - \Lambda_{\alpha\nu,\,j}]$ depend on the value of j and are given by,

$$k_{j+1} - k_j = \frac{2\pi}{L}\frac{1}{2\pi\rho(k_j)}\,; \qquad k_{c\nu,\,j+1} - k_{c\nu,\,j} = \frac{2\pi}{L}\frac{1}{2\pi\rho_{c\nu}(\Lambda_{c\nu,\,j})}\,;$$

$$\Lambda_{\alpha\nu,\, j+1} - \Lambda_{\alpha\nu,\, j} = \frac{2\pi}{L}\frac{1}{2\pi\sigma_{\alpha\nu}(\Lambda_{\alpha\nu,\, j})}\,. \tag{1.20}$$

Here $2\pi\rho_{c\nu}(\Lambda) = \frac{1}{2}\, 2\pi\sigma_{c\nu}(\Lambda)\,\mathrm{Re}\,\sqrt{1-(\Lambda+i\nu U/4t)^2}$ and the functionals $2\pi\rho(k)$ and $2\pi\sigma_{\alpha\nu}(\Lambda)$ are the solutions of well defined coupled integral equations. In contrast, the bare-momentum spacing is independent of j and given by $[q_{j+1} - q_j] = 2\pi/L$ and hence one can replace q_j by a continuous bare-momentum q so that the charge-momentum rapidity $k_j = k(q_j)$ and the $\alpha\nu$ rapidity $\Lambda_{\alpha\nu,\, j} = \Lambda_{\alpha\nu}(q_j)$ are for each energy eigenstate described by functions of q, $k(q)$ and $\Lambda_{\alpha\nu}(q)$, respectively. The Takahashi's equations (1.15) and (1.16) are expressed in terms of the corresponding bare-momentum pseudoparticle distribution functions as follows,

$$q = k(q) + \frac{1}{\pi}\sum_{\alpha\nu\neq c0}\int_{-q_{\alpha\nu}}^{q_{\alpha\nu}} dq'\, N_{\alpha\nu}(q')\arctan\Big(\frac{\sin(k(q)) - \Lambda_{\alpha\nu}(q')}{\nu U/4t}\Big)\,, \tag{1.21}$$

$$\begin{aligned} q = k_{\alpha\nu}(q) &- (\delta_{\alpha,c} - \delta_{\alpha,s})\frac{1}{\pi}\int_{q_c^-}^{q_c^+} dq'\, N_{c0}(q')\arctan\Big(\frac{\Lambda_{\alpha\nu}(q) - \sin(k(q'))}{\nu U/4t}\Big) \\ &- \frac{1}{2\pi}\sum_{\nu'=1}^{N_a/2}\int_{-q_{\alpha\nu}}^{q_{\alpha\nu}} dq'\, N_{\alpha\nu'}(q')\Theta_{\nu,\,\nu'}\Big(\frac{\Lambda_{\alpha\nu}(q) - \Lambda_{\alpha\nu'}(q')}{U/4t}\Big)\,; \qquad \alpha\nu \neq c0\,. \end{aligned} \tag{1.22}$$

Here $k_{\alpha\nu}(q) = \delta_{\alpha,c}2\,\mathrm{Re}\,\{\arcsin(\Lambda_{c\nu}(q) + i\nu U/4t)\}$ for $\alpha\nu \neq c0$, the function $\Theta_{\nu,\,\nu'}(x)$ is defined in Eq. (B.5) of Ref. 28, and the limiting bare-momentum values $q_{\alpha\nu}$ and $q_c^{\pm}$ are defined by Eqs. (B.14) and (B.16)-(B.17), respectively, of that reference. For a given energy eigenstate specified by the set of bare-momentum distribution functions $\{N_{\alpha\nu}(q)\}$, the solution of Eqs. (1.21) and (1.22) uniquely defines the set of occupied and unoccupied values of the charge rapidity momentum function $k(q)$ and set of $\alpha\nu$ rapidity functions $\Lambda_{\alpha\nu}(q)$ associated with that state.

For LWSs the energy and the momentum spectra are given by,

$$\begin{aligned} E = &-\frac{t\,N_a}{\pi}\int_{q_c^-}^{q_c^+} dq\, N_{c0}(q)\cos(k(q)) - \frac{U}{2}N + \frac{U}{4}N_a \\ &+ \frac{2t\,N_a}{\pi}\sum_{\nu=1}^{N_a/2}\int_{-q_{c\nu}}^{q_{c\nu}} dq\, N_{c\nu}(q)\,\mathrm{Re}\Big\{\sqrt{1-(\Lambda_{c\nu}(q) + i\nu U/4t)^2}\Big\} \end{aligned} \tag{1.23}$$

and

$$P = \frac{L}{2\pi}\Big\{\int_{q_{c0}^-}^{q_{c0}^+} dq\, N_c(q)\, q + \sum_{\nu=1}^{\infty}\int_{-q_{s\nu}}^{q_{s\nu}} dq\, N_{s\nu}(q)\, q$$
$$+ \sum_{\nu=1}^{\infty}\int_{-q_{c\nu}}^{q_{c\nu}} dq\, N_{c\nu}(q)\,[\frac{\pi}{a}(1+\nu) - q]\Big\}, \qquad (1.24)$$

respectively, where when $|P| > \pi$ the value of the momentum should be brought to the first Brillouin zone. The states occupation numbers are not independent. For instance, they have to obey the sum rules,

$$N = N_{c0} + 2\sum_{c\nu \neq c0} \nu N_{c\nu}\,,$$
$$N_\uparrow - N_\downarrow = N_{c0} - 2\sum_{s\nu} \nu N_{s\nu}\,, \qquad (1.25)$$

and the values of the quantum numbers I_j^{c0} and $I_j^{\alpha\nu}$ are contained in intervals such that the corresponding bare-momentum values $q_j = [2\pi/N_a]\, I_j^{c0}$ and $q_j = [2\pi/N_a]\, I_j^{\alpha\nu}$ belong to the ranges $q_c^- \leq q_j \leq q_c^+$ and $-q_{\alpha\nu} \leq q_j \leq q_{\alpha\nu}$, respectively. For the low-energy subspace spanned by states with vanishing occupancies for the sets of numbers $\{N_{s\nu}\} = 0$ and $\{N_{c\nu}\} = 0$ for $\nu > 1$ and $\nu > 0$, respectively, the magnetization provided in Eq. (1.25) simplifies to $N_\uparrow - N_\downarrow = N_{c0} - 2N_{s1}$.

The generators of the LWSs onto the electronic vacuum can be expressed as products of the pseudoparticle operators $b^\dagger_{q,\alpha\nu}$ and all energy eigenstates are also eigenstates of the operators (1.19) whose eigenvalues are the pseudoparticle numbers. The pseudoparticles do not obey fermionic statistics (except for the $c0$ pseudoparticles) but their statistics can be classified according the generalized Pauli principle of Haldane.[33] The pseudoparticle operator anticommutation relations are given by,

$$\{b^\dagger_{q_j,\,\alpha\nu},\, b_{q_{j'},\,\alpha'\nu'}\} = \delta_{\alpha\nu,\,\alpha'\nu'}\, F(q_j,\, q_{j'})\,;$$
$$\{b^\dagger_{q_j,\,\alpha\nu},\, b^\dagger_{q_{j'},\,\alpha'\nu'}\} = \{b_{q_j,\,\alpha\nu},\, b_{q_{j'},\,\alpha'\nu'}\} = 0\,, \qquad (1.26)$$

where

$$F(q_j,\, q_{j'}) = \delta_{q_j,\, q_{j'}}\,, \qquad (1.27)$$

when for $\alpha\nu = \alpha'\nu'$ both the $I_j^{\alpha\nu}$ and $I_{j'}^{\alpha\nu}$ numbers such that $q_j = [2\pi/L]\, I_j^{\alpha\nu}$ and $q_j = [2\pi/L]\, I_{j'}^{\alpha\nu}$, respectively, are integers or half-odd integers and,

$$F(q_j,\, q_{j'}) = \frac{i}{L}\frac{1}{e^{+i(q_j - q_{j'})/2}\,\sin([q_j - q_{j'}]/2)}\,, \qquad (1.28)$$

when for $\alpha\nu = \alpha'\nu'$ the above $I_j^{\alpha\nu}$ numbers are integers (or half-odd integers) and the $I_{j'}^{\alpha\nu}$ numbers are half-odd integers (or integers).

The momentum dependent creation and annihilation operators can be formally defined locally on an effective $\alpha\nu$ lattice, whose lattice constant $a_{\alpha\nu}$ is defined so that the length of such a lattice is $\alpha\nu$ independent and equal to L: $a_{\alpha\nu} = a\frac{N_a}{N^*_{\alpha\nu}}$. Hence the above numbers $N^*_{\alpha\nu}$ are also the number of $\alpha\nu$ lattice sites. The numbers $N^*_{\alpha\nu} \leq N_a$ correspond to the upper and lower bounds on the quantum numbers $I_j^{\alpha\nu}$ for $\alpha\nu \neq c0$ such that $j = 1, ..., N^*_{\alpha\nu}$ of the above Bethe-ansatz equations. Such equations are valid for $N_a >> 1$ within the so called Takahashi string hypothesis[19] and provide naturally the values of the number $N^*_{\alpha\nu} \leq N_a$, which are given in Eqs. (B.6) and (B.7) of Ref. 28. The corresponding numbers $\pm q_{\alpha\nu}$ refer to the largest possible absolute bare-momentum value (the boundaries of the $\alpha\nu$ bare-momentum Brillouin zone). Only for one branch ($c0$-pseudoparticles), does the total number of allowed discrete momenta equals the number of "real" lattice sites N_a.

In the standard Bethe ansatz literature one often uses the charge $c0$ and spin $s1$ rapidity density functions $2\pi\rho(k)$ and $2\pi\sigma_{s1}(\Lambda)$, respectively, appearing in Eq. (1.20),[17] which are the only relevant ones for the above low-energy subspace. For that subspace, they obey the simplified integral equations,

$$2\pi\rho(k) = 1 + \frac{U}{4\pi t}\cos k \int_{-B}^{B} d\Lambda' \frac{2\pi\sigma_{s1}(\Lambda')}{[U/4t]^2 + [\sin k - \Lambda']^2},$$

$$2\pi\sigma_{s1}(\Lambda) = \frac{U}{4\pi t}\int_{-Q}^{Q} dk' \frac{2\pi\rho(k')}{[U/4t]^2 + [\sin k' - \Lambda]^2} - \frac{U}{8\pi t}\int_{-B}^{B} d\Lambda' \frac{2\pi\sigma_{s1}(\Lambda')}{[U/4t]^2 + [(\Lambda - \Lambda')/2]^2}.$$

In these equations the cutoff parameters Q and B are defined by,

$$\int_0^Q dk' 2\pi\rho(k') = \pi n = 2k_F,$$

$$\int_0^B d\Lambda' 2\pi\sigma_{s1}(\Lambda') = \pi n_\downarrow = k_{F\downarrow}.$$

The solution of the problem reveals that in general the state of the system is metallic, except at half-filling ($n = N/N_a = 1$) where it constitutes a Mott-Hubbard insulator for any finite value of $U > 0$.[17] The thermodynamics of the model was solved and leads to a low temperature specific heat that is linear in the temperature and to a susceptibility that is finite (as in a Luttinger liquid and Fermi liquid). Except for the one-electron properties, this result suggests an alternative description of the system in a form closer to that of the Landau theory.

From the thermodynamics point of view, the system has similar properties to those of a Fermi liquid but the correlation functions are qualitatively different. In particular, the correlation function of a single particle is qualitatively different, as seen above for the Luttinger liquid. The analysis of the correlation functions of the Hubbard model will reveal that at low energies the model is indeed of the Tomonaga-Luttinger liquid class. However, at finite energies a new description, reviewed in Ref. 1, is necessary.

Below, we consider often the above-mentioned low-energy subspace where the limiting bare-momentum values q_{s1} and $q_c^{\pm}$ defined by Eqs. (B.14) and (B.16)-(B.17) of Ref. 28, respectively, simplify and except for corrections of order $1/L$ can be written as,

$$q_{s1} = k_{F\uparrow}\,; \quad q_c^{\pm} = \pm\pi\,; \quad q_{Fs1} = k_{F\downarrow}\,; \quad q_{Fc0} = 2k_F\,. \tag{1.29}$$

In this equation we also provided the values of the $c0$ and $s1$ Fermi momenta which appear in the ground-state bare-momentum distributions used below.

1.2.2. *Landau liquid description*

As discussed above, the Bethe ansatz solution can be described in terms of a pseudoparticle representation associated with the bare momenta $q_j = [2\pi/N_a]\, I_j^{c0}$ and $q_j = [2\pi/N_a]\, I_j^{\alpha\nu}$ and corresponding quantum numbers I_j^{c0} and $I_j^{\alpha\nu}$ which have a regular distribution, similar to that of the discrete momenta of usual non-interacting fermionic systems. For example, the ground state of the system is obtained considering a symmetrical distribution of the numbers I_j^{c0} around the origin, filling the acessible numbers until a value such that the maximal occupied number is according to Eq. (1.29), $|q| = q_{Fc0} = 2k_F = \pi n$ (the maximal value of $|q|$ is π), defining the Fermi surface of the $c0$ band. In the same way, the numbers I_j^{s1} are distributed in a symmetrical way, such that the maximal occupied number corresponds to the value $|q| = q_{Fs1} = k_{F\downarrow}$ given in Eq. (1.29) (the maximal value is also provided in that equation and reads $q_{s1} = k_{F\uparrow}$ where $k_{F\sigma} = N_\sigma/N$, with $\sigma =\uparrow,\downarrow$). At zero magnetization $q_{Fs1} = q_{s1} = k_F$, the $s1$ bare-momentum band is full and at half-filling the $c0$ band is full as well.

Away from half-filling and at finite magnetization the low-energy excitations around the ground state lead to small deviations relative to the equilibrium distributions of the quantum numbers I_j^{c0}, I_j^{s1}. Those excitations are just particle-hole excitations in the $c0$ and $s1$ bare-momentum bands, like in the usual description of a Fermi liquid, except that the $s1$ pseudoparticles are not strictly fermions and both the $c0$ and $s1$ pseudoparticles do not have a one-to-one correspondence to the electrons, as $U/t \to 0$. Their occupancy configurations correspond to exact energy eigenstates of the many-body system obtained from the Bethe ansatz solution. In the standard rapidity description the excitations are obtained by introducing "holes" or "particles" in the distribution functions of the related charge-momentum rapidities k_j and $\alpha\nu$ rapidities $\Lambda_{\alpha\nu,\,j}$.[18]

Let us now focus our attention on the ground state distributions which, except for corrections of the order of $1/L$, can be written as $N^0_{c0}(q) = \Theta(q_{Fc0} - |q|) = \Theta(2k_F - |q|)$ and $N^0_{s1}(q) = \Theta(q_{Fs1} - |q|) = \Theta(k_{F\downarrow} - |q|)$. Here q is the above pseudoparticle bare momentum. At low energies the excitations are characterized by deviations from the ground-state distributions. Then one may introduce general distributions $N_{c0}(q) = N^0_{c0}(q) + \Delta N_{c0}(q)$ and $N_{s1}(q) = N^0_{s1}(q) + \Delta N_{s1}(q)$. In the limit when the deviations are small the energy of the system may be expanded around the ground-state distributions as follows,

$$\begin{aligned}
E &= E_0 + E_1 + E_2\,,\\
E_1 &= \frac{L}{2\pi}\left\{\int_{-\pi}^{\pi} dq\Delta N_{c0}(q)\epsilon_{c0}(q) + \int_{-k_{F\uparrow}}^{k_{F\uparrow}} dq\Delta N_{s1}(q)\epsilon_{s1}(q)\right\}\,,\\
E_2 &= \frac{L}{(2\pi)^2}\{\int_{-\pi}^{\pi} dq\int_{-\pi}^{\pi} dq'\Delta N_{c0}(q)\frac{f_{c0\,c0}(q,q')}{2}\Delta N_{c0}(q')\\
&+ \int_{-k_{F\uparrow}}^{k_{F\uparrow}} dq\int_{-k_{F\uparrow}}^{k_{F\uparrow}} dq'\Delta N_{s1}(q)\frac{f_{s1\,s1}(q,q')}{2}\Delta N_{s1}(q')\\
&+ \int_{-\pi}^{\pi} dq\int_{-k_{F\uparrow}}^{k_{F\uparrow}} dq'\Delta N_{c0}(q)f_{c0\,s1}(q,q')\Delta N_{s1}(q)\}\,,
\end{aligned} \tag{1.30}$$

in a way analogous to a Fermi liquid (to simplify, at this stage we only consider the lower energy excitations). This reformulation of the problem has the advantage of a standard band-like interpretation of the excitation spectrum. The energies $\epsilon_{c0}(q)$ and $\epsilon_{s1}(q)$ are the charge $c0$ and spin $s1$ bands and the parameters $f_{c0\,c0}$, $f_{s1\,s1}$, and $f_{c0\,s1} = f_{s1\,c0}$ describe the residual interactions between the pseudoparticles. Even though the formulation is similar to that of a Fermi liquid, the pseudoparticles refer to energy eigenstates that do not decay in time.

It is shown in Ref. 28 that for $\nu > 0$ the $c\nu$ and $s\nu$ pseudoparticles are composite objects: the $c\nu$ (and $s\nu$) pseudoparticles are η-spin singlet 2ν-holon (and spin-singlet 2ν-spinon) composite objects of η-spin $1/2$ ν holons of η-spin projection $1/2$ and ν holons of η-spin projection $-1/2$ (and spin $1/2$ ν spinons of spin projection $1/2$ and ν spinons of spin projection $-1/2$). The η-spin projection $1/2$ (and $-1/2$) holons correspond to rotated-electron unoccupied sites (and doubly-occupied sites). The spinons of spin projection $\pm 1/2$ refer to the spins of the rotated electrons which singly occupied sites.The original non-perturbative electronic problem (the spectral function of the 1D Hubbard model is fully incoherent) becomes "perturbative" in the pseudoparticle basis.[23]

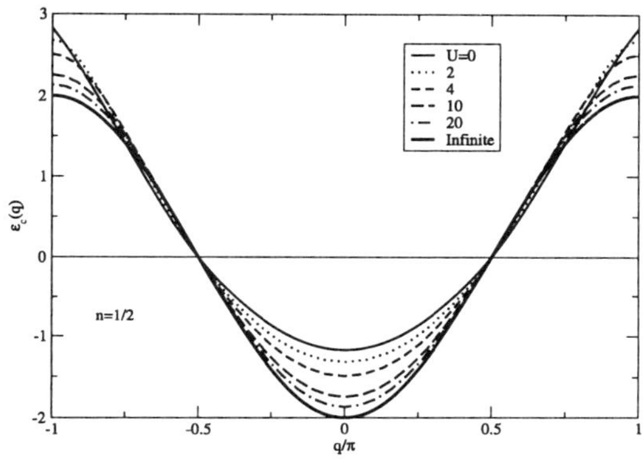

Fig. 1.1. The pseudoparticle energy band $\epsilon_{c0}(q)$ in units of t for density $n = 1/2$ and various values of U/t. Reproduced with permission of the American Physical Society from Ref. 27.

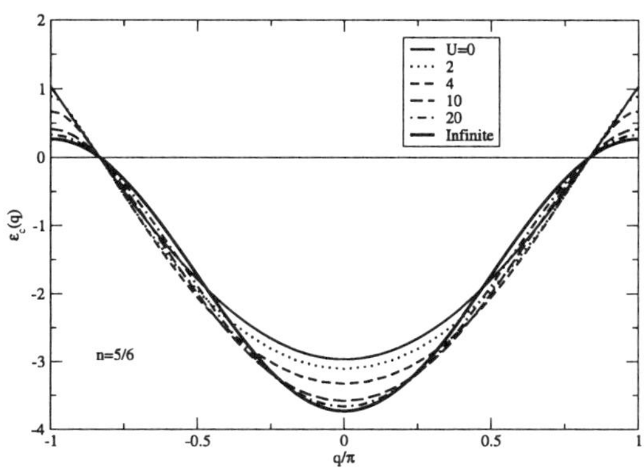

Fig. 1.2. The pseudoparticle energy band $\epsilon_{c0}(q)$ in units of t for density $n = 5/6$ and various values of U/t. Reproduced with permission of the American Physical Society from Ref. 27.

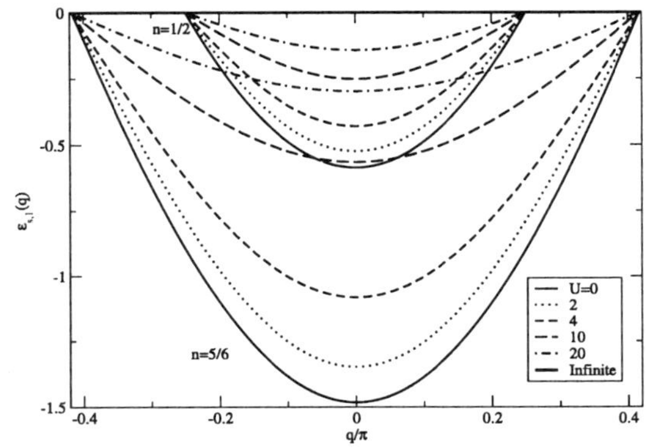

Fig. 1.3. The pseudoparticle energy band $\epsilon_{s1}(q)$ in units of t for density $n = 5/6$ and various values of U/t. Reproduced with permission of the American Physical Society from Ref. 27.

The pseudoparticle band expressions can be expressed in terms of the following integrals,[28]

$$\epsilon_{c0}(q) = \int_Q^{k^0(q)} dk' 2t\eta(k') \, ,$$
$$\epsilon_{s1}(q) = \int_B^{\Lambda_{s1}^0(q)} d\Lambda' 2t\eta_{s1}(\Lambda') \, ,$$

where the functions in the upper limits are such that their inverse functions are given by the following integrals,

$$q = \int_0^{k^0(q)} dk' 2\pi\rho(k') \, ,$$
$$q = \int_0^{\Lambda_{s1}^0(q)} d\Lambda' 2\pi\sigma_{s1}(\Lambda') \, . \tag{1.31}$$

At the $c0$ and $s1$ Fermi momenta these functions read $k^0(2k_F) = Q$ and $\Lambda_{s1}^0(k_{F\downarrow}) = B$, respectively. The other distributions involved in the above band expressions obey the integral equations,

$$2t\eta(k) = 2t\sin k + \frac{U}{4\pi t}\cos k\int_{-B}^{B} d\Lambda' \frac{2t\eta_{s1}(\Lambda')}{[U/4t]^2 + [\sin k - \Lambda']^2},$$

$$2t\eta_{s1}(\Lambda) = \frac{U}{4\pi t}\int_{-Q}^{Q} dk' \frac{2t\eta(k')}{[U/4t]^2 + [\sin k' - \Lambda]^2}$$
$$- \frac{U}{8\pi t}\int_{-B}^{B} d\Lambda' \frac{2t\eta_{s1}(\Lambda')}{[U/4t]^2 + [(\Lambda - \Lambda')/2]^2}.$$

The velocities associated with the energy bands are given by $v_{c0}(q) = d\epsilon_{c0}(q)/dq$ and $v_{s1}(q) = d\epsilon_{s1}(q)/dq$. The bands are such that $\epsilon_{c0}(2k_F) = 0$ and $\epsilon_{s1}(k_{F\downarrow}) = 0$. The energy bands are shown in Figs. 1.1,1.2, and 1.3.

The pseudoparticle f-functions which describe their residual interactions read,

$$f_{c0\,c0}(q,q') = 2\pi v_{c0}(q)\Phi_{c0\,c0}(q,q') + 2\pi v_{c0}(q')\Phi_{c0\,c0}(q',q)$$
$$+ [2\pi v_{c0}]\sum_{j=\pm1}\Phi_{c0\,c0}(2k_F j,q)\Phi_{c0\,c0}(2k_F j,q')$$
$$+ [2\pi v_{s1}]\sum_{j=\pm1}\Phi_{s1\,c0}(k_{F\downarrow}j,q)\Phi_{s1\,c0}(k_{F\downarrow}j,q'), \quad (1.32)$$

$$f_{s1\,s1}(q,q') = 2\pi v_{s1}(q)\Phi_{s1\,s1}(q,q') + 2\pi v_{s1}(q')\Phi_{s1\,s1}(q',q)$$
$$+ [2\pi v_{s1}]\sum_{j=\pm1}\Phi_{s1\,s1}(k_{F\downarrow}j,q)\Phi_{s1\,s1}(k_{F\downarrow}j,q')$$
$$+ [2\pi v_{c0}]\sum_{j=\pm1}\Phi_{c0\,s1}(2k_F j,q)\Phi_{c0\,s1}(2k_F j,q'), \quad (1.33)$$

$$f_{c0\,s1}(q,q') = 2\pi v_{c0}(q)\Phi_{c0\,s1}(q,q') + 2\pi v_{s1}(q')\Phi_{s1\,c0}(q',q)$$
$$+ [2\pi v_{c0}]\sum_{j=\pm1}\Phi_{c0\,c0}(2k_F j,q)\Phi_{c0\,s1}(2k_F j,q')$$
$$+ [2\pi v_{s1}]\sum_{j=\pm1}\Phi_{s1\,s1}(k_{F\downarrow}j,q)\Phi_{s1\,c0}(k_{F\downarrow}j,q'), \quad (1.34)$$

where $v_{c0} = v_{c0}(2k_F)$ and $v_{s1} = v_{s1}(k_{F\downarrow})$. While the f functions are associated with the residual interactions of the pseudoparticles, the functions Φ are the phase shifts, in units of π, of the collisions between the corresponding pseudofermions. The latter objects are introduced in the following chapter, Ref. 1. The phase shifts appearing in the above f-function expressions are functions of the two momentum values. Alternatively, one can define phase shifts which depend on the corresponding two rapidity values. The two types of phase shifts are related according to

$$\begin{aligned}
\Phi_{c0\,c0}(q,q') &= \bar{\Phi}_{c0\,c0}(4t\sin k^0(q)/U, 4t\sin k^0(q')/U),\\
\Phi_{c0\,s1}(q,q') &= \bar{\Phi}_{c0\,s1}(4t\sin k^0(q)/U, 4t\Lambda^0_{s1}(q')/U),\\
\Phi_{s1\,s1}(q,q') &= \bar{\Phi}_{s1\,s1}(4t\Lambda^0_{s1}(q)/U, 4t\Lambda^0_{s1}(q')/U),\\
\Phi_{s1\,c0}(q,q') &= \bar{\Phi}_{s1\,c0}(4t\Lambda^0_{s1}(q)/U, 4t\sin k^0(q')/U).
\end{aligned}$$

The rapidity phase shifts satisfy the integral equations,

$$\begin{aligned}
\bar{\Phi}_{c0\,c0}(r,r') &= \frac{1}{\pi}\int_{-y_0}^{y_0} dr''\,\frac{\bar{\Phi}_{s1\,c0}(r,r')}{1+(r-r'')^2},\\
\bar{\Phi}_{c0\,s1}(r,r') &= -\frac{1}{\pi}\text{arc }\tan(r-r') + \frac{1}{\pi}\int_{-y_0}^{y_0} dr''\,\frac{\bar{\Phi}_{s1\,s1}(r'',r')}{1+(r-r'')^2},\\
\bar{\Phi}_{s1\,c0}(r,r') &= -\frac{1}{\pi}\text{arc }\tan(r-r') + \frac{1}{\pi}\int_{-y_0}^{y_0} dr''\,G(r,r'')\bar{\Phi}_{s1\,c0}(r,r')\,,\\
\bar{\Phi}_{s1\,s1}(r,r') &= \frac{1}{\pi}\text{arc }\tan(\frac{r-r'}{2}) - \frac{1}{\pi^2}\int_{-x_0}^{x_0} dr''\,\frac{\text{arc }\tan(r''-r')}{1+(r-r'')^2}\\
&\quad + \int_{-y_0}^{y_0} dr''G(r,r'')\bar{\Phi}_{s1\,s1}(r'',r')\,. \qquad (1.35)
\end{aligned}$$

Here $x_0 = 4t\sin Q/U$, $y_0 = 4tB/U$, and the kernel $G(r,r')$ is given by,

$$G(r,r') = -\frac{1}{2\pi}\left[\frac{1}{1+((r-r')/2)^2}\right]\left[1-\frac{1}{2}\left(t(r)+t(r')+\frac{l(r)-l(r')}{r-r'}\right)\right], \qquad (1.36)$$

where

$$\begin{aligned}
t(r) &= \frac{1}{\pi}\left[\text{arc }\tan(r+x_0) - \text{arc }\tan(r-x_0)\right],\\
l(r) &= \frac{1}{\pi}\left[\ln(1+(r+x_0)^2) - \ln(1+(r-x_0)^2)\right].
\end{aligned}$$

The following phase-shift parameters play an important role in the quantum-liquid physics,

$$\begin{aligned}
\zeta^i_{c0\,c0} &= 1 + \bar{\Phi}_{c0\,c0}(x_0,x_0) + (-1)^i\bar{\Phi}_{c0\,c0}(x_0,-x_0)\,,\\
\zeta^i_{c0\,s1} &= \bar{\Phi}_{c0\,s1}(x_0,y_0) + (-1)^i\bar{\Phi}_{c0\,s1}(x_0,-y_0)\,,\\
\zeta^i_{s1\,c0} &= \bar{\Phi}_{s1\,c0}(y_0,x_0) + (-1)^i\bar{\Phi}_{s1\,c0}(y_0,-x_0)\,,\\
\zeta^i_{s1\,s1} &= 1 + \bar{\Phi}_{s1\,s1}(y_0,y_0) + (-1)^i\bar{\Phi}_{s1\,s1}(y_0,-y_0)\,, \quad i=0,1\,.
\end{aligned}$$

These parameters can be written as $\zeta^i_{c0\,c0} = \zeta^i_{c0\,c0}(x_0)$, $\zeta^i_{c0\,s1} = \zeta^i_{c0\,s1}(x_0)$, $\zeta^i_{s1\,c0} = \zeta^i_{s1\,c0}(y_0)$, and $\zeta^i_{s1\,s1} = \zeta^i_{s1\,s1}(y_0)$ where the functions on the right-hand side of these equations are defined as follows,

$$\zeta^1_{c0\,c0}(r) = 1 + \frac{1}{\pi}\int_{-y_0}^{y_0} dr'' \frac{\zeta^1_{s1\,c0}(r'')}{1+(r-r'')^2}\,,$$
$$\zeta^1_{c0\,s1}(r) = \frac{1}{\pi}\int_{-y_0}^{y_0} dr'' \frac{\zeta^1_{s1\,s1}(r'')}{1+(r-r'')^2}\,,$$
$$\zeta^1_{s1\,c0}(r) = t(r) + \int_{-y_0}^{y_0} dr'' G(r,r'')\zeta^1_{s1\,c0}(r'')\,,$$
$$\zeta^1_{s1\,s1}(r) = 1 + \int_{-y_0}^{y_0} dr'' G(r,r'')\zeta^1_{s1\,s1}(r'')\,. \tag{1.37}$$

The parameters corresponding to the symetrical linear combination of the phase shifts are obtained as the inverse of the transpose of the matrix whose entries are the antisymmetrical parameters given here. The point is that the above phase-shift parameters are the elementary pieces of other quantities which play the same role as the Landau parameters of Fermi liquid theory. Such pseudoparticle Landau parameters are given by,

$$v_{c0} + F^i_{c0\,c0} = v_{c0}[\zeta^i_{c0\,c0}]^2 + v_{s1}[\zeta^i_{s1\,c0}]^2\,,$$
$$v_{s1} + F^i_{s1\,s1} = v_{s1}[\zeta^i_{s1\,s1}]^2 + v_{c0}[\zeta^i_{c0\,s1}]^2\,,$$
$$F^i_{c0\,s1} = F^i_{s1\,c0} = v_{c0}\zeta^i_{c0\,c0}\zeta^i_{c0\,s1} + v_{s1}\zeta^i_{s1\,s1}\zeta^i_{s1\,c0}\,, \quad i = 0,1\,. \tag{1.38}$$

The pseudoparticle Landau parameters can be defined in a way similar to that of the Fermi-liquid theory quasiparticles,

$$F^i_{c0\,c0} = \frac{1}{2\pi}\sum_{j=\pm1}(j)^i f_{c0\,c0}(2k_F, j2k_F)\,,$$
$$F^i_{s1\,s1} = \frac{1}{2\pi}\sum_{j=\pm1}(j)^i f_{s1\,s1}(k_{F\downarrow}, jk_{F\downarrow})\,,$$
$$F^i_{c0\,s1} = F^i_{s1\,c0} = \frac{1}{2\pi}\sum_{j=\pm1}(j)^i f_{c0\,s1}(2k_F, jk_{F\downarrow})$$
$$= \frac{1}{2\pi}\sum_{j=\pm1}(j)^i f_{s1\,c0}(k_{F\downarrow}, j2k_F)\,. \tag{1.39}$$

1.2.3. *Low-temperature thermodynamics*

Many low-energy quantities of the one-dimensional Hubbard model can be expressed in terms of the phase-shift parameters and related pseudoparticle Landau parameters. As in a Fermi liquid the low-temperature specific heat does not depend on such parameters and only involves the c0 and s1 pseudoparticle Fermi velocities. For electronic densities in the range $n < 1$ and spin densities $m > 0$ the specific heat reads,[22]

$$\frac{c_V}{N_a} = \left(\frac{k_B^2 \pi}{3}\right) \left(\frac{1}{v_{c0}} + \frac{1}{v_{s1}}\right) T \, . \tag{1.40}$$

This result is obtained considering the first-order momentum distribution deviation contributions to the energy when expressed in terms of low-temperature Fermi-Dirac distributions, for both the $c0$ pseudoparticles and $s1$ pseudoparticles. The energy deviation is expressed in terms of the deviations,[22]

$$\begin{aligned} \Delta N_{c0}(q) &= N_{c0}(q) - \Theta(2k_F - |q|) \, , \\ \Delta N_{s1}(q) &= N_{s1}(q) - \Theta(k_{F,\downarrow} - |q|) \, , \end{aligned} \tag{1.41}$$

where $N_{c0}(q)$ and $N_{s1}(q)$ are the Fermi-Dirac distributions.

Also, the static charge and spin susceptibilities may be obtained in a way similar to that of a Fermi liquid.[24] The magnetic susceptibility at zero temperature and spin density $m = 0$ was obtained first by Shiba.[20] Here we follow the procedure of Ref. 24 and present the expressions derived in that reference for $m > 0$. For most cases the $m \to 0$ limit of the obtained expressions provides the corresponding $m = 0$ expression. The basic procedure corresponds to using expressions for the chemical potential and the magnetic field given by,[24]

$$\begin{aligned} \mu(n) &= \frac{U}{2} - \epsilon_{c0}^0(2k_F) - \frac{1}{2}\epsilon_{s1}^0(k_{F\downarrow}) \, , \\ H(m) &= -\frac{\epsilon_{s1}^0(k_{F\downarrow})}{2} \, , \end{aligned} \tag{1.42}$$

where $\epsilon_{c0}(q) = \epsilon_{c0}^0(q) + \mu - \mu_0 H$ and $\epsilon_{s1}(q) = \epsilon_{s1}^0(q) + 2\mu_0 H$. The charge susceptibility is then expressed as,

$$\chi_c|_{H,m} = -\frac{1}{n^2} \frac{1}{\partial\mu(n)/\partial n|_{H,m}} \, , \tag{1.43}$$

and the spin susceptibility may be expressed as,

$$\chi_s|_{\mu,n} = \frac{2\mu_0}{\partial H(m)/\partial m|_{\mu,n}} \, . \tag{1.44}$$

It was obtained that these quantities can be written in terms of the above phase-shift parameters as follows,[24]

$$
\begin{aligned}
\chi_c|_H &= \frac{1}{\pi n^2}\left(\frac{(\zeta^1_{c0\,c0})^2}{v_{c0}} + \frac{(\zeta^1_{s1\,c0})^2}{v_{s1}}\right),\\
\chi_s|_\mu &= \frac{\mu_0^2}{\pi}\left(\frac{(\zeta^1_{c0\,c0} - 2\zeta^1_{c0\,s1})^2}{v_{c0}} + \frac{(\zeta^1_{s1\,c0} - 2\zeta^1_{s1\,s1})^2}{v_{s1}}\right),\\
\chi_c|_m &= \frac{1}{\pi n^2}\left(\frac{1}{v_{c0}(\zeta^0_{c0\,c0} + \zeta^0_{c0\,s1}/2)^2 + v_{s1}(\zeta^0_{s1\,c0} + \zeta^0_{s1\,s1}/2)^2}\right),\\
\chi_s|_n &= \frac{\mu_0^2}{\pi}\left(\frac{1}{v_{c0}(\zeta^0_{c0\,s1}/2)^2 + v_{s1}(\zeta^0_{s1\,s1}/2)^2}\right).
\end{aligned}
\tag{1.45}
$$

The dependence of these thermodynamic quantities on the various parameters is discussed in Ref. 24.

Alternatively, the above given charge and spin susceptibilities can be expressed in terms of the Landau parameters provided in Eqs. (1.38) and (1.39), as in Fermi liquid theory.

1.3. Summary

In this chapter we have briefly reviewed several schemes used in the description of the unusual properties of low-dimensional correlated systems. A hint on these properties is provided by the Tomonaga and Luttinger models where bosonization techniques allow the solution at low energies.

We have devoted most of our attention to the one-dimensional Hubbard model whose low-energy physics can, in spite of the lack of Fermi liquid behavior, be described by a functional theory in terms of pseudoparticle bare-momentum distributions, which resembles that of Fermi liquid theory. Except that in the limit of zero interaction the pseudoparticles do not map onto electrons, and for $U > 0$ the one-electron spectral function is fully incoherent, the low-temperature thermodynamics and the low-energy charge and spin susceptibilities can be derived as in a Fermi liquid. However, it has proven exceedingly difficult in the past to obtain information on correlation functions via the exact Bethe ansatz solution, if we do not restrict to the asymptotic regime in space and time. The calculation of correlation functions at general momentum and frequency is a complex problem that has only been solved recently, as shown in the following chapter.

In the ensuing companion chapter we review a transformation which maps the pseudoparticles considered here onto non-interacting pseudofermions. That enables the evaluation of matrix elements between energy eigenstates and the construction of a pseudofermion dynamical theory. Such a theory provides expressions for finite-energy correlation and spectral functions.

Acknowledgments

We thank the support of FCT under the grant POCTI/FIS/58133/2004 and that of the ESF Science Programme INSTANS 2005-2010.

References

1. J. M. P. Carmelo, P. D. Sacramento, D. Bozi, and L. M. Martelo, Dynamical and spectral properties of low dimensional materials, in "Strongly correlated systems, coherence and entanglement", edited by J. M. P. Carmelo, J. M. B. Lopes dos Santos, V. Rocha Vieira, and P. D. Sacramento, World Scientific, Singapore (2007), page 29.
2. L. D. Landau, The theory of a Fermi liquid, *Sov. Phys. JETP*. **3**, 920 (1957); L. D. Landau, Oscillations in a Fermi liquid, *Sov. Phys. JETP*. **5**, 101 (1957); D. Pines, P. Noziéres, *The theory of Fermi liquids I*, (Benjamin, New York, 1966).
3. S. Tomonaga, Remarks on Bloch's method on sound waves applied to many-fermion problems, *Prog. Theor. Phys.*. **5**, 544 (1950).
4. J. M. Luttinger, An exactly soluble model of a many-fermion system, *J. Math. Phys.*. **4**, 1154 (1963).
5. J. Voit, One-dimensional Fermi liquids, *Rep. Prog. in Phys.*. **58**, 977 (1995).
6. J. Hubbard, Electron correlations in narrow energy bands, *Proc. R. Soc. A*. **276**, 238 (1963).
7. B. Dardel, D. Malterre, M. Grioni, P. Weibel, Y. Baer and F. Lévy, Unusual photoemission spectral-function of quasi-one-dimensional metals, *Phys. Rev. Lett.*. **67**, 3144 (1991).
8. C. Kim, A. Y. Matsuura, Z. X. Shen, N. Motoyama, H. Eisaki, S. Uchida, T. Tohyama, S. Maekawa, Observation of Spin-Charge Separation in One-Dimensional SrCuO2, *Phys. Rev. Lett* **77**, 4054 (1996).
9. R. Claessen, M. Sing, U. Schwingenschlögl, P. Blaha, M. Dressel, C. S. Jacobsen, Spectroscopic Signatures of Spin-Charge Separation in the Quasi-One-Dimensional Organic Conductor TTF-TCNQ, *Phys. Rev. Lett.*. **88**, 096402-1 (2002).
10. M.Z. Hasan *et al*, Momentum-Resolved Charge Excitations in a Prototype One-Dimensional Mott Insulator, *Phys. Rev. Lett.*. **88**, 177403 (2002).
11. A. Koitzsch *et al*, Current spinon-holon description of the one-dimensional charge-transfer insulator SrCuO2: Angle-resolved photoemission measurements, *Phys. Rev. B*. **73**, 201101 (2006).
12. A. Luther and I. Peschel, Single-particle states, Kohn anomaly, and pairing fluctuations in one dimension, *Phys. Rev. B*. **9**, 2911 (1974); D. C. Mattis, New wave-operator identity applied to study of persistent currents in 1D, *J. Math. Phys.*. **15**, 609 (1974);
13. F. D. M. Haldane, Effective Harmonic-Fluid Approach to Low-Energy Properties of One-Dimensional Quantum Fluids, *Phys. Rev. Lett.*. **47**, 1840 (1981); F. D. M. Haldane, Luttinger liquid theory of one-dimensional quantum fluids. 1. Properties of the Luttinger model and their extension to the general 1D interacting spinless Fermi gas, *J. Phys. C*. **14**, 2585 (1981).
14. H. J. Schulz, Correlation exponents and the metal-insulator transition in the one-dimensional Hubbard model, *Phys. Rev. Lett.*. **64**, 2831 (1990).

15. A. Luther and V. J. Emery, Backward Scattering in the One-Dimensional Electron Gas, *Phys. Rev. Lett.*. **33**, 589 (1974); P. A. Lee, Comments on a Solution of a One-Dimensional Fermi-Gas Model, *Phys. Rev. Lett.*. **34**, 1247 (1975).
16. J. M. P. Carmelo and M. J. Sampaio, private communication.
17. E. H. Lieb, F. Y. Wu, Absence of Mott Transition in an Exact Solution of the Short-Range, One-Band Model in One Dimension, *Phys. Rev. Lett.*. **20**, 1445 (1968).
18. A. A. Ovchinnikov, Excitation spectrum in one-dimensional Hubbard model, *Zh. Eksp. Teor. Fiz.*. **57**, 2137 (1969) [*Sov. Phys. JETP*. **30**, 1160 (1970)]
19. M. Takahashi, One-dimensional Hubbard model at finite temperature, *Prog. Theor. Phys.*. **47**, 69 (1972).
20. H. Shiba, Magnetic Susceptibility at Zero Temperature for the One-Dimensional Hubbard Model, *Phys. Rev. B*. **6**, 930 (1972).
21. M. Ogata, H. Shiba, Bethe-ansatz wave function, momentum distribution, and spin correlation in the one-dimensional strongly correlated Hubbard model, *Phys. Rev. B*. **41**, 2326 (1990).
22. J. M. P. Carmelo, P. Horsch, P. A. Bares, A. A. Ovchinnikov, Renormalized pseudoparticle description of the one-dimensional Hubbard model thermodynamics, *Phys. Rev. B*. **44**, 9967 (1991).
23. J. M. P. Carmelo, A. A. Ovchinnikov, Generalization of the Landau liquid concept: example of the Luttinger liquids, *J. Phys. Cond. Mat.*. **3**, 757 (1991).
24. J. M. P. Carmelo, P. Horsh, A. A. Ovchinnikov, Static properties of one-dimensional generalized Landau liquids, *Phys. Rev. B*. **45**, 7899 (1992).
25. K. Penc, F. Mila, H. Shiba, Spectral Function of the 1D Hubbard Model in the U infinite limit, *Phys. Rev. Lett.*. **75**, 894 (1995).
26. J. M. P. Carmelo, N. M. R. Peres, Complete pseudohole and heavy-pseudoparticle operator representation for the Hubbard chain, *Phys. Rev. B*. **56**, 3717 (1997).
27. J. M. P. Carmelo and P. D. Sacramento, Finite-energy Landau liquid theory for the one-dimensional Hubbard model: Pseudoparticle energy bands and degree of localization/delocalization, *Phys. Rev. B*. **68**, 085104 (2003).
28. J. M. P. Carmelo, J. M. Román, K. Penc, Charge and spin quantum fluids generated by many-electron interactions, *Nucl. Phys. B*. **683**, 387 (2004).
29. J. M. P. Carmelo, K. Penc, D. Bozi, Finite-energy spectral-weight distributions of a 1D correlated metal, *Nucl. Phys. B*. **725**, 421 (2005); *Nucl. Phys. B*. **737** 351 (2006), *Erratum*.
30. J. M. P. Carmelo and K. Penc, General spectral function expressions of a 1D correlated model, *Eur. Phys. J. B*. **51**, 477 (2006).
31. J. M. P. Carmelo, L. M. Martelo, K. Penc, The low-energy limiting behavior of the pseudofermion dynamical theory, *Nucl. Phys. B*. **737**, 237 (2006); J. M. P. Carmelo, K. Penc, Correlation-function asymptotic expansions: Universality of prefactors of the one-dimensional Hubbard model, *Phys. Rev. B*. **73**, 113112 (2006).
32. J. Stein, Flow equations and the strong-coupling expansion for the Hubbard model. *J. Stat. Phys.* **88**, 487 (1997).
33. F. D. M. Haldane, Fractional statistics in arbitrary dimensions: A generalization of the Pauli principle, *Phys. Rev. Lett.*. **67**, 937 (1991).

Chapter 2

Dynamical and spectral properties of low dimensional materials

J.M.P. Carmelo

GCEP-Center of Physics, Universidade do Minho, Campus Gualtar, P-4710-057 Braga, Portugal

P.D. Sacramento

Departamento de Física and CFIF, Instituto Superior Técnico, Av. Rovisco Pais, 1049-001 Lisboa, Portugal

D. Bozi

Instituto de Ciencia de Materiales, CSIC, Cantoblanco, E-28949 Madrid, Spain

L.M. Martelo

Departamento de Física, Faculdade de Engenharia, Universidade do Porto, P-4200-465 Porto, Portugal

This chapter follows its companion, chapter 1. Here we review different methods based on the Bethe ansatz solution of the one-dimensional Hubbard model, in order to study quantities related to charge transport and the momentum dependent conductivity. Moreover, we report recent developments on finite-energy dynamical properties. This is achieved by introducing new entities called pseudofermions which are basically free, in the sense that their energies are additive, and where the effect of the interactions appears through phase shifts that are absorbed by their discrete momentum values. The resulting pseudofermion dynamical theory enables the evaluation of matrix elements between energy eigenstates and hence the derivation of finite energy expressions for the one- and two-electron correlation and spectral functions. Comparison with experimental results is also discussed.

Contents

2.1. Introduction

In the previous companion chapter, Ref. 1, we reviewed the physics of low-dimensional correlated models of interest for the description of the unusual properties observed in low-dimensional complex materials.[2,3] Our analysis has focused on the one-dimensional Hubbard model and its Bethe-ansatz solution, Refs. 4-8. Here we start by reviewing the application of the concepts introduced in Ref. 9, concerning the frequency-dependent conductivity of metals, to that correlated model, discussed in Refs. 10-16. Moreover, in this chapter we also report results involving the combination of the conformal invariance of the model low-energy spectrum with its Bethe-ansatz solution to derive low-energy correlation-function expressions as discussed in Refs. 17-23.

However, while the low-energy physics of the one-dimensional Hubbard model is well understood, new methods to study its finite-energy physics were introduced only recently. For instance, the investigations of Ref. 24 reveal that in the vicinity of the upper-Hubbard bands lower limit, the finite-energy physics can be mapped onto a low-energy conformal field theory so that one can derive expressions for the corresponding finite-energy one- and two-electron spectral functions. Nevertheless, the study of general expressions for the finite-energy spectral functions of the one-dimensional Hubbard model requires the use of more complex methods. As discussed in this chapter, such expressions can be evaluated by expressing the generators of the model energy eigenstates associated with its Bethe-ansatz solution and global symmetry, in terms of suitable quantum-object operators.[25] These objects are easier to relate to rotated electrons than to electrons. In turn, the rotated-electron operators are connected to the electronic creation and annihilation operators by a unitary transformation.[25,26]

The above procedure leads to a dynamical theory for finite values of the on-site repulsion U, which is a generalization of the technique introduced in Refs. 27 and 28 for $U \rightarrow \infty$. The general method to deal with the problem is presented in Refs. 29-31, where it is shown that the suitable description of the exact energy eigenstates to derive matrix elements between such states involves the introduction of quantum objects called pseudofermions, which as mentioned above are closely related to the rotated electrons. The pseudofermion dynamical theory introduced in Refs. 29-31 is used in the description of the finite energy properties observed in low-dimensional complex materials in Refs. 32-36.

Here we summarize and review such recent developments which enable a transparent description of the exact energy eigenstates of a non-perturbative many-body system in terms of the pseudofermions having in mind the calculation of one- and two-electron spectral-weight distributions. The operators associated with those objects obey anti-commutation relations that are close to fermionic, but which reflect a system with twisted boundary conditions due to phase shifts arising from zero-momentum forward-scattering collisions between them. Therefore, the momenta of these pseudofermions are affected by these phase shifts but their energies are simply additive and described by the same energy bands as the corresponding pseudoparticles, which were studied in the companion chapter.

As mentioned above, we start by reviewing several studies on the transport of charge and corresponding conductivity in the one-dimensional Hubbard model by use of techniques which profit from the use of its Bethe-ansatz solution. The first part of the chapter includes the discussion of various correlation functions like the conductivity (both the Drude peak and the regular part of the optical conductivity) and study of the instabilities of the model at small energies, where the Tomonaga-Luttinger liquid like description described in the previous chapter applies. In such low-energy studies we rely on the use of conformal field theory. In the second part of the chapter we summarize the pseudofermion dynamical theory and some of its applications. Two methods to calculate finite-energy expressions for correlation and dynamical functions are shortly reviewed, growing up in complexity and culminating in a nearly full description of the dependence of dynamical correlation functions on finite momentum and energy. We consider specifically the one-electron spectral function and discuss the relation of the theoretical predictions with recent photoemission experimental results. One reaches a quite good agreement between theory and experiments, as illustrated in this chapter.

2.2. Correlation functions

2.2.1. *Conductivity: Drude peak and regular part*

The optical conductivity is given generally by,

$$\sigma(\omega) = 2\pi D\delta(\omega) + \sigma^{reg}(\omega) . \tag{2.1}$$

The first term is the Drude peak and the second term is the the regular part of the conductivity. The quantity D is the charge stiffness and characterizes the response of the system to a static electric field, within linear response theory. The regular part describes the absorption of light of finite frequency by the system. The second term can be obtained in different ways. For instance it can be expressed as,

$$\sigma^{reg}(\omega) = \frac{\pi}{N_a} \sum_{\nu \neq 0} \frac{| < \nu | J | 0 > |^2}{\omega_{\nu,0}} \delta(\omega - \omega_{\nu,0})$$
$$\sim \lim_{k \to 0} \frac{\omega \Im \chi_\rho(k,\omega)}{k^2} , \tag{2.2}$$

where the current operator reads,

$$J = -it \sum_{j,\sigma} [c^\dagger_{j,\sigma} c_{j+1,\sigma} - c^\dagger_{j+1,\sigma} c_{j,\sigma}] , \tag{2.3}$$

and the summation runs over energy eigenstates, $\omega_{\nu,0} = E_\nu - E_0$ is the excitation energy,

$$\chi_\rho(k,\omega) = - \sum_{\nu \neq 0} \frac{| < \nu | n(k) | o > |^2 2\omega_{\nu,0}}{\omega^2_{\nu,0} - (\omega + i\delta)^2} \tag{2.4}$$

is the charge-charge response function, and

$$n(k) = \sum_{k',\sigma} c^\dagger_{k'+k,\sigma} c_{k',\sigma} . \tag{2.5}$$

The Drude peak may be obtained in several ways. For instance, in terms of the response of the energy eigenvalues (in particular of the ground state if the temperature is zero) to an external flux, ϕ, piercing the one-dimensional system (forming a closed circle). The zero-temperature charge stiffness may then be obtained using,

$$D = \frac{1}{2} \frac{d^2 E_0}{d(\phi/L)^2} |_{\phi=0} . \tag{2.6}$$

The Drude peak has a weight given by $2\pi D = 2u_\rho K_\rho$.[11] The total weight is proportional to the kinetic energy and is given by,[10]

$$\sigma_{tot} = \int_{-\infty}^{\infty} \sigma(\omega) d\omega = -\frac{\pi}{L} < H_{kin} > . \tag{2.7}$$

Except very close to half-filling the weight is almost entirely in the Drude peak. As we approach half-filling the weight shifts to the regular part of the conductivity.[11] At half-filling the dc conductivity vanishes due to the insulating behavior.

The Drude peak has recently attracted interest in particular with respect to its value at finite temperatures. Kohn proposed[9] that the value of $D(T = 0)$

can be used to distinguish between an ideal insulator, $D(0) = 0$, and an ideal conductor, $D(0) \neq 0$. At finite temperatures the definition was generalized[12] in the following way: i) if $D(T) > 0$, the system behaves as an ideal conductor, ii) if $D(T) = 0$ and $\sigma^{reg}(\omega \to 0, T) = 0$, the system behaves as an ideal insulator and iii) if $D(T) = 0$ and $\sigma^{reg}(\omega \to 0, T) > 0$ the system behaves as a normal conductor. An integrable system is expected to behave as either an ideal conductor or insulator, at least at zero temperature. For instance, for the Hubbard model the system is an ideal insulator at half-filling and any value of $U \neq 0$, and is an ideal conductor otherwise. The question then arises at finite temperature. It was conjectured using results from exact diagonalizations of small systems that integrable systems retain these properties at finite temperatures:[12] that is $D(T)$ should be zero if $D(0)$ is zero but finite otherwise. Several authors using different methods confirmed and disagreed with these results and it was argued that some non-integrable systems may have similar properties (for references see[13,14]).

Using the Bethe ansatz solution the problem may be addressed computing explicitly the effect of a flux on the energy levels. This was carried out explicitly for a model of one-dimensional spinless fermions (related to the Heisenberg model through a Jordan-Wigner transformation)[13] and for the Hubbard model.[14] For instance, for the Hubbard model the influence of the flux is easilly introduced in the Bethe ansatz equations considering a Peierls substitution at the Hamiltonian level. Its effect on the Bethe ansatz equations is simply the replacements $2\pi I_j^{c0} \to 2\pi I_j^{c0} + \phi_\uparrow$, $2\pi I_j^{c\nu} \to 2\pi I_j^{c\nu} - \nu(\phi_\uparrow + \phi_\downarrow)$, and $2\pi I_j^{s\nu} \to 2\pi I_j^{s\nu} + \nu(\phi_\uparrow - \phi_\downarrow)$, where we distinguish the flux that the $\uparrow$ and $\downarrow$ electrons feel (this is useful to study spin transport as well as charge transport). For $U/t \gg 1$ the results obtained for the two models reveal i) the conjecture is correct, ii) away from half-filling the energy levels are flux dependent and $D(T) \neq 0$. At half-filling the energy levels are flux independent and $D(T) = D(0) = 0$. In general the flux felt by the lowest energy excitations is renormalized by the higher energy levels. It turns out that at half-filling the renormalization of this effective flux is such that it vanishes in a non-trivial way. However, these results were only obtained in an expansion around large U and the exact corresponding behaviors remain an open question for $U > 0$.

The presence of an external flux also induces currents in the system (the stiffness is the first derivative of these currents to the external flux). It is also possible to study these currents in linear response theory.[16] Such an analysis shows that the coupling of the system to the external probes is through the pseudoparticles considered in the companion chapter, which describe the modes of the system. Also, due to interacting nature of the system, the effective charge and spin carried by the pseudoparticles are renormalized with respect to the free charge and spin of an electron. The Yang holons considered in Ref. 25 have a charge $-2e$ but do not carry current due to their infinite mass.

The finite energy part of the conductivity involves the calculation of the matrix elements of the current operator between the ground state and excited states or the calculation of a charge-charge correlation function. This is a much harder problem. Ideally one must calculate matrix elements, and this is indeed possible. However, we may follow a simpler route first.

2.2.2. *Critical exponents and conformal field theory*

For a conformal invariant system, there exists a set of correlation functions that behave near the conformal critical point, in the long-range limit, as

$$< 0|\phi(z,z^*)\phi(z',(z')^*)|0 >\propto \frac{1}{(z-z')^{2\Delta^+}}\frac{1}{(z^*-(z')^*)^{2\Delta^-}}\,, \tag{2.8}$$

where $z = x + iv\tau$ and $z^* = x - iv\tau$, v is the "light" velocity, and τ the Euclidian time ($\tau = -it$). The exponents (Δ^+, Δ^-) are the conformal dimensions. These can be obtained considering a conformal transformation that maps an infinite system into a system placed on a cylinder (periodic boundary conditions). One obtains that the energies and momenta of the finite system with respect to the ground state are given to leading order by,

$$E^L - E_0^L = \frac{2\pi}{L}v(\Delta^+ + \Delta^-)\,,$$
$$P^L - P_0^L = \frac{2\pi}{L}(\Delta^+ - \Delta^-)\,.$$

The above conformal-field theory analysis refers to a system where there is only one type of excitation that becomes critical. However, for several models such as the one-dimensional Hubbard model, there are two or more critical excitations. Specifically, at low energies both the charge ($c0$ pseudoparticles) and spin ($s1$ pseudoparticles) excitations considered in the first companion chapter are gapless away from half-filling and at finite spin density. The theory can then be generalized considering that there are two types of conformal fields which have different "light" velocities and that both become critical.[17] The relation between the finite-size energy correction and the conformal dimensions generalizes straightforwardly,

$$E = E_0 + \frac{2\pi}{N_a}\left[v_{c0}(\Delta_{c0}^+ + \Delta_{c0}^-) + v_{s1}(\Delta_{s1}^+ + \Delta_{s1}^-)\right]\,, \tag{2.9}$$

(specifying for the Hubbard model) where $v_{c0} = v_{c0}(2k_F)$, $v_{s1} = v_{s1}(k_{f\downarrow})$, and the velocities $v_{c0}(q)$ and $v_{s1}(q)$ are those considered in chapter 1.

For correlation functions involving physical fields, conformal field theory provides the asymptotic expansions but not the value of the corresponding prefactors. The problem was solved recently by use of the pseudofermion dynamical theory considered in the ensuing section. As shown in the second paper of Ref. 31, the suitable use of that theory provides the universal part of the prefactors of the correlation-function asymptotic expansions of the one-dimensional Hubbard model.

A certain correlation-function expression involves a well-defined set of excited states. The procedure then implies calculating the energy differences for a finite system with respect to the ground state. The energies are calculated for instance using the Bethe ansatz.[18,19,21] Typically one considers i) small deviations in the number of electrons and down-spin electrons $\Delta N = (N_a/\pi)\Delta 2k_F = (N_a/\pi)\Delta q_{Fc0}$ and $\Delta N_\downarrow = (N_a/\pi)\Delta k_{F\downarrow} = (N_a/\pi)\Delta q_{Fs1}$ associated with small changes in the above $c0$ and $s1$ Fermi momenta given in Eq. (1.29) of the companion chapter; ii) we may also consider finite-momentum $c0$ and $s1$ band particle-hole processes from or to bare momenta in the vicinity of $-2k_F$ (and $-k_{F\downarrow}$) to or from bare momenta in the neighborhood of $2k_F$ (and $k_{F\downarrow}$). This originates small changes Δq_{Fc0} and Δq_{Fs1} in the pseudoparticle Fermi momenta. The numbers of $c0$ and $s1$ pseudoparticles transferred, D_{c0} and D_{s1}, respectively, are given by $D_{c0} = (N_a/2\pi)\,\Delta[q^+_{Fc0}-q^-_{Fc0}]$ and $D_{s1} = (N_a/2\pi)\,\Delta[q^+_{Fs1}-q^-_{Fs1}]$. Finally iii) there is a second type of elementary $c0$ and $s1$ band particle-hole processes which involve small momentum $\pm 2\pi/N_a$ and occur around the same Fermi point whose number we denote by $N^\pm_{c0}$ and $N^\pm_{s1}$. Let $q^\pm_p$ and $q^\pm_h$ denote the momenta of the "particles" and "holes" around $q^\pm_{Fc0} = \pm 2k_F$, respectively, and $p^\pm_p$ and $p^\pm_h$ denote the momenta of the particles and holes around $q^\pm_{Fs1} = \pm k_{F\downarrow}$ up to corrections of order $1/N_a$, respectively. The appropriate distributions to describe these processes are given by,[7]

$$
\begin{aligned}
N_{c0}(q) &= \Theta(2k_F + \Delta 2k_F + (\operatorname{sgn} q)\frac{2\pi}{N_a}D_{c0} - |q|) \\
&+ \frac{2\pi}{N_a}\left(\sum_p [\delta(q-q_p^+) + \delta(q-q_p^-)] - \sum_h [\delta(q-q_h^+) + \delta(q-q_h^-)]\right),
\end{aligned}
$$

$$
\begin{aligned}
N_{s1}(q) &= \Theta(k_{F\downarrow} + \Delta k_{F\downarrow} + (\operatorname{sgn} q)\frac{2\pi}{N_a}D_{s1} - |q|) \\
&+ \frac{2\pi}{N_a}\left(\sum_q [\delta(q-q_p^+) + \delta(q-q_p^-)] - \sum_h [\delta(q-q_h^+) + \delta(q-q_h^-)]\right).
\end{aligned}
$$

It is then possible to obtain the energy and momentum deviations and the conformal dimensions which are given by,[6,7]

$$\Delta_{c0}^{\pm} = \frac{1}{2}\left(\zeta^1_{c0\,c0}D_{c0} + \zeta^1_{c0\,s1}D_{s1} \pm \zeta^0_{c0\,c0}\frac{\Delta N}{2} \pm \zeta^0_{c0\,s1}\frac{\Delta N_\downarrow}{2}\right)^2 + N_{c0}^{\pm}\,,$$
$$\Delta_{s1}^{\pm} = \frac{1}{2}\left(\zeta^1_{s1\,c0}D_{c0} + \zeta^1_{s1\,s1}D_{s1} \pm \zeta^0_{s1\,c0}\frac{\Delta N}{2} \pm \zeta^0_{s1\,s1}\frac{\Delta N_\downarrow}{2}\right)^2 + N_{s1}^{\pm}\,,$$

where the phase-shift parameters ζ^0 and ζ^1 are given in the companion chapter, Ref. 1.

The physical momentum deviation spectrum reads,

$$\Delta P = \frac{2\pi}{N_a}\left(\Delta_{c0}^{+} - \Delta_{c0}^{-} + \Delta_{s1}^{+} - \Delta_{s1}^{-}\right) + 2D_{c0}2k_F + 2D_{s1}k_{F\downarrow}\,. \qquad (2.10)$$

These results hold at low energies. The one-dimensional Hubbard model maps into a conformal invariant field theory at low-energy. Strict low-energy conformal invariance requires for the finite system that the value of N_{c0} must be even and that of N_{s1} odd.[18] In the gapless regimes for both excitations (charge and spin) the critical behavior is described by two sets of conformal fields with central charge 1, provided that the $c0$ and $s1$ velocities at the corresponding Fermi points are different. However, we should note that in some special cases the two velocities are equal. Then the system reduces to one set of conformal fields with central charge 2.[18] The relation and consistency between the alternative bosonization techniques and those based on the conformal invariance of the low-energy spectrum of the Hubbard model as obtained by use of the Bethe ansatz solution to study its low-energy Tomonaga-Luttinger liquid behavior was investigated in Ref. 8.

Until now our analysis referred to the asymptotic behavior of the correlation functions in space-time. We may as well consider their behavior in momentum-frequency space. In this case the theory holds in the small-momentum and low-energy regime. The exponents were first obtained for the singular behavior detected for specific lines of the (k,ω) plane defined by the following relations between the excitation momentum k and excitation energy/frequency ω, $\omega = \pm v_{c0}(k-k_0)$ or $\omega = \pm v_{s1}(k-k_0)$ (characteristic of the dispersion relations of the $c0$ charge and $s1$ spin branches of excitations), where k_0 refers to characteristic momenta where a singularity in the correlation function occurs.[21] The correlation function expressions in the vicinity of $\omega \sim \pm v_{c0}(k-k_0)$ or $\omega \sim \pm v_{s1}(k-k_0)$ are obtained by Fourier transform of the corresponding space and time correlation functions and are of the form,

$$g(k,\omega) \sim C[\omega = \mp v_{c0}(k-k_0)]^{2(\Delta_{s1}^{+}+\Delta_{s1}^{-}+\Delta_{c0}^{\pm})-1}\,, \quad \omega \sim \pm v_{c0}(k-k_0)\,,$$
$$g(k,\omega) \sim C[\omega = \mp v_{s1}(k-k_0)]^{2(\Delta_{c0}^{+}+\Delta_{c0}^{-}+\Delta_{s1}^{\pm})-1}\,, \quad \omega \sim \pm v_{s1}(k-k_0)\,. \qquad (2.11)$$

This formula is applicable only if all the conformal dimensions are nonzero. If one of them vanishes the corresponding singularity disappears, that is the constant C is zero. In turn, for all remaining directions of the (k,ω) plane in the vicinity of the above points $(\pm k_0, 0)$ corresponding to regions with finite spectral weight, the correlation functions expressions are instead given by,[22,23]

$$g(k,\omega) \sim C[\omega]^{2(\Delta_{s1}^{+}+\Delta_{s1}^{-}+\Delta_{c0}^{+}+\Delta_{c0}^{-})-2}\,, \quad \omega \sim \pm v(k-k_0)\,,$$
$$v > \min\{v_{s1}, v_{c0}\} > 0\,, \quad v \neq \max\{v_{s1}, v_{c0}\}\,. \tag{2.12}$$

Below we will use the latter expression, which refers to the most general situation, whereas the former expressions correspond to a regime associated with a small measure and hence difficult to detect in experiments on low-dimensional materials.

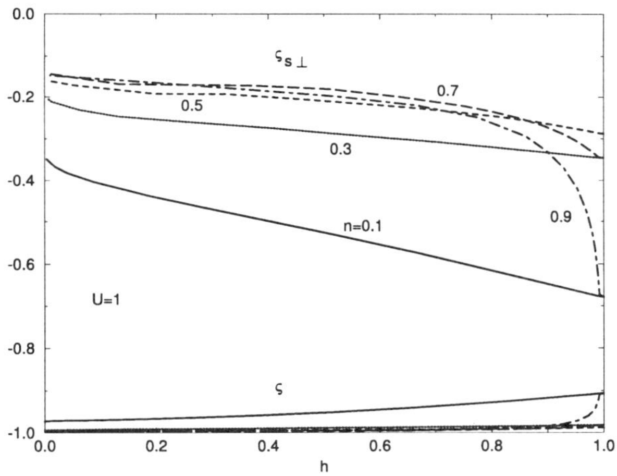

Fig. 2.1. The exponents $\zeta_{s\perp}$ and $\zeta_{\uparrow}$ for various densities and $U/t = 1$ as a function of the magnetic field normalized to the field that fully polarizes the electrons. Reproduced with permission of the American Physical Society from Ref. 23.

Calculations of the critical exponents by use of the latter general expression were performed in Refs. 22,23. Several possible instabilities were considered for various correlation functions such as the single-particle Green function, the charge-charge and spin-spin correlation functions, the spin-transverse function, and the singlet and triplet superconductivity functions. These are obtained considering physical fields of the form $\Phi_{1p}(k) = c^{\dagger}_{k\uparrow}$, $\Phi_{\rho}(k) = n_{k,\sigma} = \sum_{k'}\sum_{\sigma} c^{\dagger}_{k',\sigma}c_{k+k',\sigma}$, $\Phi_{\sigma}(k) = n_{k,\uparrow} - n_{k,\downarrow}$, $\Phi_{ss}(k) = \sum_{k'} c^{\dagger}_{k',\downarrow}c^{\dagger}_{k-k',\uparrow}$, $\Phi_{\sigma\perp}(k) = \sum_{k'} c^{\dagger}_{k',\downarrow}c_{k-k',\uparrow}$. The leading instability for the transverse spin func-

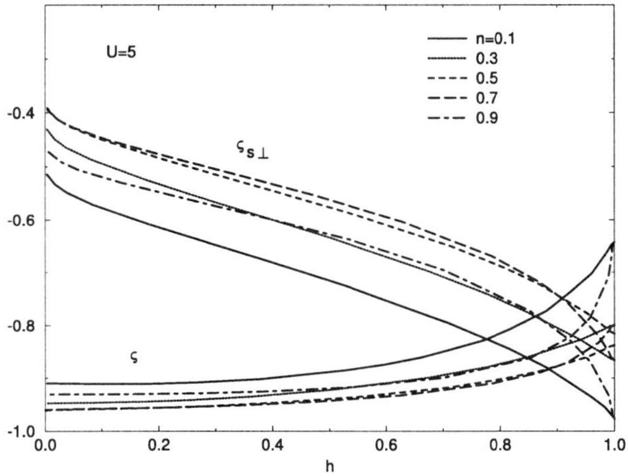

Fig. 2.2. The exponents $\zeta_{s\perp}$ and $\zeta_{\uparrow}$ for various densities and $U/t = 5$ as a function of the magnetic field normalized to the field that fully polarizes the electrons. Reproduced with permission of the American Physical Society from Ref. 23.

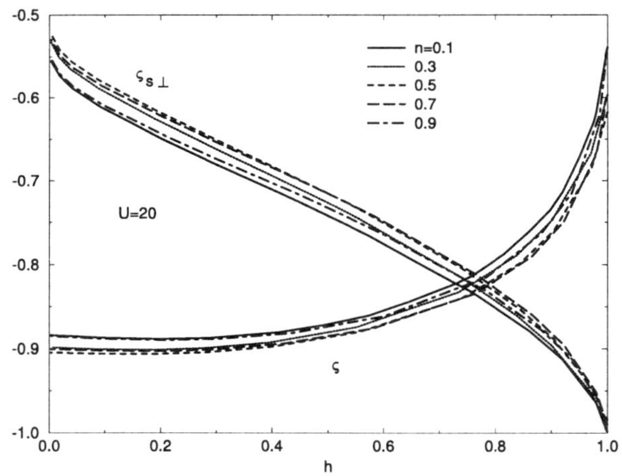

Fig. 2.3. The exponents $\zeta_{s\perp}$ and $\zeta_{\uparrow}$ for various densities and $U/t = 20$ as a function of the magnetic field normalized to the field that fully polarizes the electrons. Reproduced with permission of the American Physical Society from Ref. 23.

tion occurs at $k_0 = \pm 2k_F$ and is of the form,

$$\Re\chi_{\sigma\perp}(\pm 2k_F, \omega) \sim \omega^{\zeta_{s\perp}}$$

$$\zeta_{s\perp} = -2 + 2 \sum_{\alpha=c0,s1} \left(\left(\frac{\zeta^1_{\alpha\, c0}}{2} \right)^2 + \left(\frac{\zeta^1_{\alpha\, s1}}{2} \right)^2 \right), \quad (2.13)$$

where here and below the phase-shift parameters ζ^1 (and ζ^0) are those introduced in the previous chapter.

For spin density $m > 0$ there are singularities in the charge-charge and spin-spin functions at $k_0 = \pm 2k_{F\sigma}$ where $\sigma = \uparrow, \downarrow$ and the following leading behavior was found,[22]

$$\begin{aligned} \Re\chi_\rho(\pm 2k_{F\sigma}, \omega) &\sim \omega^{\zeta_{cs\sigma_z}} \,, \\ \Re\chi_{\sigma_z}(\pm 2k_{F\sigma}, \omega) &\sim \omega^{\zeta_{cs\sigma_z}} \,, \\ \zeta_{cs\downarrow} &= -2 + 2 \sum_{\alpha=c0,s1} \left(\zeta^1_{\alpha\, s1}\right)^2 \,, \\ \zeta_{cs\uparrow} &= -2 + 2 \sum_{\alpha=c0,s1} \left(\zeta^1_{\alpha\, c0} - \zeta^1_{\alpha\, s1}\right)^2 \,. \end{aligned}$$

The singlet superconductivity function has singularities at $k_0 = \pm 2k_F$ and $k_0 = \pm[k_{F,\uparrow} - k_{F\downarrow}]$ given by,

$$\begin{aligned} \Re\chi_{ss}(\pm 2k_F, \omega) &\sim \omega^{\zeta_{ss+}} \,, \\ \Re\chi_{ss}(\pm[k_{F,\uparrow} - k_{F\downarrow}], \omega) &\sim \omega^{\zeta_{ss-}} \,, \end{aligned} \qquad (2.14)$$

with

$$\begin{aligned} \zeta_{ss+} &= -2 + 2 \sum_{\alpha=c0,s1} \left(\left(\frac{\zeta^1_{\alpha\, c0}}{2}\right)^2 + \left(\zeta^0_{\alpha,c} + \frac{\zeta^0_{\alpha,s}}{2}\right)^2 \right) , \\ \zeta_{ss-} &= -2 + 2 \sum_{\alpha=c0,s1} \left(\left(\frac{\zeta^1_{\alpha\, c0}}{2} - \zeta^1_{\alpha\, s1}\right)^2 + \left(\zeta^0_{\alpha\, c0} + \frac{\zeta^0_{\alpha\, s1}}{2}\right)^2 \right) . \end{aligned}$$

For $U > 0$ the superconductivity exponents are always positive and therefore there is no instability. On the contrary, $-1 < \zeta_{s\perp} < 0$, $-1/2 < \zeta_{cs\downarrow} < 0$, and $-1/2 < \zeta_{cs\uparrow} < 2$ and therefore there are instabilities in the corresponding correlation functions. In turn, the triplet superconductivity function has no low-energy singularities.

The σ one electron Green function has singularities at the Fermi momenta $k_0 = \pm k_{F\sigma}$ such that,

$$\begin{aligned} \Re G_\sigma(\pm k_{F\sigma}, \omega) &\sim \omega^{\zeta_{\sigma_z}} \\ \zeta_\uparrow &= -2 + \frac{1}{2} \sum_{\alpha=c0,s1} \left(\left(\zeta^1_{\alpha\, c0} - \zeta^1_{\alpha\, s1}\right)^2 + \left(\zeta^0_{\alpha\, c0}\right)^2 \right) , \\ \zeta_\downarrow &= -2 + \frac{1}{2} \sum_{\alpha=c0,s1} \left(\left(\zeta^1_{\alpha\, s1}\right)^2 + \left(\zeta^0_{\alpha\, c0} + \zeta^0_{\alpha\, s1}\right)^2 \right) . \end{aligned} \qquad (2.15)$$

The excitations that are described by the single-particle Green function change the electron number but are interesting in the context of the spectral function probed by photoemission or inverse photoemission or when chains are coupled and electrons are transferred from one chain to another. We will return to this point later. The results show that while removing or adding single electrons is dominant at zero magnetic field, the presence of the field brings about a dominance for the transverse $\pm 2k_F$ spin density wave over all the remaining instabilities for a large domain of U/t and density values. The above exponents which control the leading instabilities are plotted in Figs. 2.1,2.2, and 2.3. Note that when the exponent vanishes the singularities become logarithmic in the frequency.

In this low energy regime the exponents are non-universal, consistent with the Tomonaga-Luttinger liquid behavior addressed in the companion chapter, and for $m = 0$ can be expressed in terms of a single phase-shift parameter. We call it ζ_0. It is obtained as the limit of the phase-shift parameter $\zeta^1_{c0\,c0}$ as $m \to 0$. It is related to the charge stiffness as $\zeta_0 = \sqrt{2K_\rho}$. The exponent of the momentum distribution was obtained via the Bethe ansatz[20] and reads,

$$\alpha = \frac{1}{16\alpha_c}(\alpha_c - 4)^2 \, , \tag{2.16}$$

where $\alpha_c/4$ is the ratio of the charge susceptibility and the charge contribution to the specific heat coefficient.

2.2.3. *Finite-energy problems which can be mapped onto a low-energy conformal field theory*

The studies of Ref. 24 report a particular type of finite-energy subspace of the Hubbard model where due to symmetry, the model can be mapped onto a low-energy conformal-invariant quantum problem. The excited states which span such a subspace control the physics in the vicinity of the upper-Hubbard band lower limit. In particular the one-electron addition spectral function, the dynamical structure factor, and the spin singlet Cooper-pair addition spectral function were studied in that reference. Defining the threshold for the various Hubbard bands as ME_u, where $M = \sum_\nu N_{c\nu} + L_{c,-1/2}$, the above subspace is spanned by states such that either $M = L_{c,-1/2}$ or the $c\nu$ pseudoparticles are created for $\nu > 0$ at $q = \pm q_{c\nu}$. Here $q_{c\nu}$ is the limiting momentum of the $c\nu$ band given in Eq. (1.29) of Ref. 1, $N_{c\nu}$ is the number of the $c\nu$ pseudoparticles considered in the first chapter, and $L_{c,-1/2}$ denotes the number of Yang holons of eta-spin projection $-1/2$ defined below in subsection 2.3.1. Such excited states lead to the dominant contributions to the finite-energy correlation functions in the vicinity of some upper-Hubbard band lower limits. In contrast, for general finite-energy excitations the problem cannot be mapped onto a low-energy conformal field theory and is instead handled by the pseudofermion dynamical theory introduced in the ensuing section.

When the finite-energy correlation-function behavior is controlled by transitions to the subspace considered in Ref. 24, one obtains power-law expressions similar to those given above for low energy of the form $\Im\chi(k^l_M, \omega) \sim (\omega - ME_u)^\varsigma$ where k^l_M is a momentum which plays the role of the above momentum k_0 and $E_u = \mu$, where μ is the chemical potential, which for $n \to 1$ equals half of the Mott-Hubbard gap. This general expression refers to a regime where $(\omega - ME_u) \sim \pm v(k - k^l_M)$ with $v > \min\{v_{s1}, v_{c0}\} > 0$ and $v \neq \max\{v_{s1}, v_{c0}\}$. In turn, for singularities along lines such that $(\omega - ME_u) \sim \pm v_{c0}(k - k^l_M)$ or $(\omega - ME_u) \sim \pm v_{s1}(k - k^l_M)$ the exponents are different and of the form of those given in Ref. 21. In either case for $m = 0$ the exponents only involve the above phase-shift parameter $\zeta_0 = \sqrt{2K_\rho}$ where K_ρ is the parameter characteristic of the Tomonaga-Luttinger liquid low-energy behavior, as discussed in Ref. 1. The leading term is controlled by excitations involving the creation of Yang holons or $c\nu$ pseudoparticles at $q = \pm q_{c\nu}$. Since these objects have a non-interacting character, the dependence of the exponents on U/t occurs only through K_ρ. In contrast, for other finite-energy singularities the expressions involve the momentum-dependent phase shifts, which cannot be obtained by use of methods relying on bosonization or conformal field theory and a new procedure summarized in the ensuing section is required.

The results obtained in the limit of zero magnetization yield for the one-electron addition upper-Hubbard band,[24]

$$\Im\chi_{1p}(\pi - lk_F, \omega) \sim (\omega - E_u)^{\zeta^l_{1p}}, l = \pm 1, \pm 3, \pm 5, \cdots,$$
$$\zeta^l_{1p} = -\frac{3}{2} + \frac{1}{2}\left([\frac{1}{\sqrt{2K_\rho}}]^2 + [\frac{l\sqrt{2K_\rho}}{2}]^2\right). \quad (2.17)$$

Only for $l = 1$ is the exponent negative and hence singular behavior occurs.

In the case of the charge dynamical structure factor it was obtained that,

$$\Im\chi_\rho(\pi \mp 2k_F, \omega) \sim (\omega - E_u)^{\zeta_\rho},$$
$$\zeta_\rho = -2\left(1 - [\frac{\sqrt{2K_\rho}}{2}]^2 - [\frac{1}{\sqrt{2K_\rho}}]^2\right). \quad (2.18)$$

In general the momentum values $k^l_M = [\pi \mp 2k_F]$ of this expression are finite and thus in the metallic phase this correlation function is not related to the zero-momentum frequency-dependent optical conductivity. However, in the half-filling limit $n \rightarrow 1$ these momentum values vanish. Thus, for the particular case of the Mott-Hubbard insulator one can use the following relation between $\Im\chi_\rho(k,\omega)$ and the regular part of the frequency dependent optical conductivity $\sigma_{reg}(\omega)$,

$$\Re\sigma_{reg}(\omega) = \lim_{k\rightarrow 0} \frac{\omega\Im\chi_\rho(k,\omega)}{k^2} . \quad (2.19)$$

By use of such a relation one obtains for any finite value of U/t that,[15,24]

$$\Re\sigma_{reg}(\omega) \sim (\omega - E_{MH})^{1/2} , \quad (2.20)$$

where E_{MH} is the half-filling Mott-Hubbard gap.

The metallic regime of the optical conductivity was studied in Ref. 15 by a preliminary version of the method introduced in the ensuing section. In such a regime there are two contributions referring to the Drude peak and the regular part, respectively, as discussed above. In that reference, the result obtained by exact diagonalizations that there is an apparent pseudogap between the Drude peak and the threshold for the upper Hubbard band, was clarified. Indeed, recalling that the regular part involves a matrix element of the current operator between the ground state and excited states, it was shown that the spectral weight for frequencies smaller than the optical gap is very small since the dominant terms vanish due to a parity selection rule once the corresponding excited states have the same parity as the ground state and the current changes the parity. These terms result from transitions that leave the N_{c1} number unchanged. Other transitions change that number by 1. While the first group of transitions correspond to low energy values the second group correspond to finite energies, and are relevant for the upper-Hubbard band(s). The studies of Ref. 15 arrived to an expression for the frequency-dependent conductivity, which at half filling coincides with that given in Eq. (2.20).

Experiments performed on (TMTSF)$_2X$ salts have shown the presence of a Drude peak and a pseudogap. In the case X=ClO$_4$,[2,3] it was possible to determine that the system is very close to half-filling, $n \sim 0.995$. In this case the spectral weight of the Drude peak is decreased and the weight of the upper band allows a comparison of the theory for the threshold exponent. This was carried out[15] and a reasonable fit was obtained.

Finally, we note that no singularities were found in the superconductivity correlation functions and therefore we do not present the results here.[24] We note however that there is an instability due to the creation of a $c1$ pseudoparticle along a line in the upper-Hubbard band defined by the dispersion relation $\epsilon_{c1}(q)$ considered in the previous chapter. This instability becomes dominant[32] close to half-filling and at small values of U/t. This may originate a superconducting instability due to chain coupling to nearby chains as discussed[32] in the context of the phase diagram of the $(TMTTF)_2X$ salts.

2.3. Rotated electrons and the pseudofermion dynamical theory

2.3.1. *Pseudoparticles, rotated electrons, and pseudofermions*

To complete the bosonization of the Luttinger model it was important, not only to diagonalize the Hamiltonian, but also to represent the fermionic operators in terms of the bosonic operators, as discussed in Ref. 1. Similarly, the Bethe-ansatz solution which diagonalizes the one-dimensional Hubbard model leads to objects whose occupancy configurations describe the exact energy eigenstates, but to evaluate correlation functions (in particular dynamic), it is necessary to represent the fermionic operators of the Hubbard model (electrons) in terms of the operators for such objects. Such a construction has been fulfilled in part recently in Refs. 25,29–31, allowing the derivation of general finite-energy expressions for one- and two-electron correlation and spectral functions.

Electron double occupancy is a good quantum number (it is conserved) in the limit of $U \to \infty$ but for finite values of U/t it is not conserved. However, it is possible to define new fermionic operators, associated with fermionic objects called rotated electrons, through a canonical transformation, $\hat{V}$, such that the double occupancy of these rotated electrons is a good quantum number for all finite values of U/t.[25,26] Hence the operator $\tilde{D} = \hat{V}^\dagger \hat{D} \hat{V}$ commutes with the Hubbard Hamiltonian $[\hat{H}, \tilde{D}] = 0$. The rotated electrons may be viewed as dressed electrons, carrying the same charge and spin as the electrons. For $U/t > 0$, the complete set of $4^{N_a^2}$ energy eigenstates $\{|\Psi_{U/t}\rangle\}$ generated by combining the Bethe-ansatz solution with the global symmetries of the model, can be written as $|\Psi_{U/t}\rangle = \hat{V}^\dagger |\Psi_\infty\rangle$ where $\{|\Psi_\infty\rangle\}$ is a suitably chosen set of $U/t \to \infty$ energy eigenstates and $\hat{V}$ is the corresponding electron - rotated-electron unitary operator. We note however that there are infinite choices for the unitary operator $\hat{V}$ and corresponding rotated-electron operators such that rotated-electron double occupancy is a good quantum number for $U/t > 0$. Most of the corresponding rotated electrons refer to choices of $U/t \to \infty$ sets of energy eigenstates $\{|\Psi_\infty\rangle\}$ such that the states $\{\hat{V}^\dagger |\Psi_\infty\rangle\}$ correspond to a complete basis yet are not in general energy eigenstates for $U/t > 0$.[26] Here we consider a unitary operator $\hat{V}$ which corresponds to a choice of the states $\{|\Psi_\infty\rangle\}$ whose corresponding states $|\Psi_{U/t}\rangle = \hat{V}^\dagger |\Psi_\infty\rangle$ are energy eigenstates for U/t finite. That operator can be written as $\hat{V} = e^{-\hat{S}}$ and hence $\hat{V}^\dagger = e^{\hat{S}}$. Moreover, the operator $\hat{S}$ can be written as $\hat{S} = \hat{S}(\infty) + \Delta\hat{S}$, where $\hat{S}(\infty)$ corresponds to the operator $S(l)$ for $l = \infty$

defined in Eq. (61) of Ref. 26 and $\Delta\hat{S}$ has the general form provided in its Eq. (64). We emphasize that the unitary operator $\hat{V}$ which generates the set of energy eigenstates associated with the Bethe-ansatz solution and the global symmetry of the model considered here corresponds to one and only one choice for the numbers $D^{(k)}(\mathbf{m})$ on Eq. (64) of Ref. 26, where $k = 1, 2, ...$ refers to the number of rotated electron doubly occupied sites.

The operator $\hat{V}$ transforms any operator $\hat{O}$ into $\hat{O}_r = \hat{V}^\dagger\hat{O}\,\hat{V}$. Let us consider the operators $\hat{R}_{c,-1}$, $\hat{R}_{c,+1}$, $\hat{R}_{s,-1}$, and $\hat{R}_{s,+1}$, that count the number of electron doubly occupied sites, unoccupied sites, and spin $\downarrow$ and spin $\uparrow$ singly occupied sites, respectively. Importantly, the operators $\hat{R}_{c,-1\,r}$, $\hat{R}_{c,+1,r}$, $\hat{R}_{s,-1\,r}$, and $\hat{R}_{s,+1\,r}$ that count the corresponding numbers for the rotated electrons commute with the Hubbard model. The energy eigenstates are also eigenstates of such four operators. Following the studies of Ref. 25, let us denote the latter four operators by $\hat{M}_{c,-1/2}$, $\hat{M}_{c,1/2}$, $\hat{M}_{s,-1/2}$, and $\hat{M}_{s,1/2}$, respectively. Indeed, in that reference it is shown that all energy eigenstates can be described by occupancy configurations of three elementary objects only: the holons of η-spin $1/2$ and zero spin, the η-spinless spinons of spin $1/2$, and the spinless and η-spinless c or $c0$ pseudoparticles. The operator $\hat{M}_{c,\pm 1/2}$ counts the number of holons with η-spin projection $\pm 1/2$ and the operator $\hat{M}_{s,\pm 1/2}$ counts the number of spinons with spin projection $\pm 1/2$. The c or $c0$ pseudoparticles are the objects introduced in the companion chapter . The $c0$ band is populated by N_{c0} $c0$ pseudoparticles and $N^h_{c0} = [N_a - N_{c0}]$ $c0$ pseudoparticle holes. The corresponding number operators $\hat{N}_{c0}$ and $\hat{N}^h_{c0}$ are fully defined by the above operators. Indeed, they are given by $\hat{N}_{c0} = \hat{M}_{s,1/2} + \hat{M}_{s,-1/2}$ and $\hat{N}^h_{c0} = \hat{M}_{c,1/2} + \hat{M}_{c,-1/2}$. The spinons refer to the spins of the rotated electrons which singly occupy sites whereas the charge degrees of freedom of the same rotated electrons are described by the $c0$ pseudoparticles. This justifies why the number of the latter objects equals that of the spinons. On the other hand, the $-1/2$ and $1/2$ holons describe the rotated-electron pairs and rotated-hole pairs of the doubly occupied and unoccupied sites, respectively.

In turn, the $c\nu$ (and $s\nu$) pseudoparticles defined in chapter 1 in terms of the Bethe-ansatz quantum numbers are η-spin singlet (and spin singlet) composite 2ν-holon (and of 2ν-spinon) objects. Both these composite objects, as well as the $c0$ pseudoparticles, are not invariant under the unitary operator $\hat{V}$. An interesting symmetry is that the holons and spinons which are not part of composite pseudoparticles remain invariant under that operator. We call them $\pm 1/2$ Yang holons and $\pm 1/2$ Heilmann and Lieb (HL) spinons and their numbers $L_{c,\,\pm 1/2}$ and $L_{s,\,\pm 1/2}$, respectively. Those are fully determined by the values S_c, S^z_c, S_s, and S^z_s of each energy eigenstate as $L_{c,\,\pm 1/2} = [S_c \mp S^z_c]$ and $L_{s,\,\pm 1/2} = [S_s \mp S^z_s]$.

The relations between the various entities is summarized in Fig. 2.4.

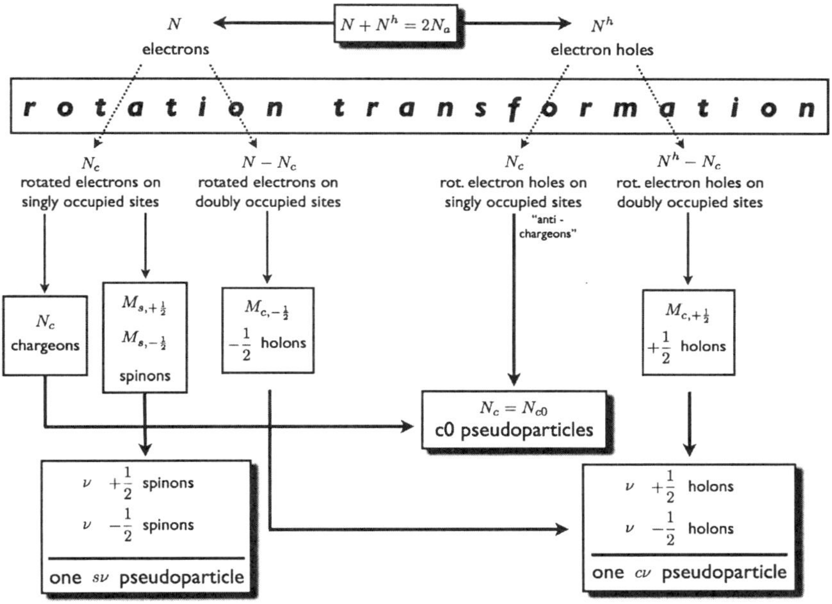

Fig. 2.4. Flow chart describing how the electrons and the holes, due to the unitary transformation generated by $\hat{V}(U/t)$ are described in terms of rotated electrons and rotated holes. These, in turn, are closely related to the holons and spinons. Except for the c or $c0$ pseudoparticles, the $c\nu$ and $s\nu$ pseudoparticles considered in the companion chapter, Ref. 1, are composite objects of holons and spinons, respectively. (The chargeons correspond to the charge part of the rotated electrons which singly occupy sites[25]).

In the previous chapter we showed that any energy eigenstate is described by a charge momentum rapidity function $k(q)$ and a set of $\alpha\nu$ rapidity functions $\{\Lambda_{\alpha\nu}(q)\}$ for each $\alpha\nu$ branch with finite occupancy in the state, where $\alpha = c, s$ and $\nu = 1, 2, \ldots$. Let us now call $k^0(q)$ the initial ground state charge momentum rapidity function. Consider an excited energy eigenstate which is generated from the initial ground state by changing the occupancy configurations of a finite number of objects. The charge momentum rapidity function $k(q)$ of that state is related to $k^0(q)$ as follows,

$$k(q) = k^0(q) + \Delta k(q) = k^0(q) + \frac{dk^0(q)}{dq}\Delta Q_{c0}(q)\,, \tag{2.21}$$

and equivalently for the rapidity functions $\Lambda_{\alpha\nu}(q)$ of the $\alpha\nu$ branches with finite occupancy in the excited state. For simplicity, in the following we use as example the $c0$ pseudoparticle branch, but the treatment is valid for the composite pseudoparticle branches as well (by letting $k^0(q) \to \Lambda^0_{\alpha\nu}(q)$ whenever $\alpha\nu \neq c0$). A normal Taylor expansion of $k^0(q)$ yields,

$$k^0(q + \delta(q)) = k^0(q) + \frac{dk^0(q)}{dq}\delta(q) + \dots \,, \tag{2.22}$$

where $|\delta(q)|$ is a small number. By defining $Q^{\Phi}_{c0}(q) = L\Delta Q_{co}(q)$, we find by use of the Bethe ansatz equations provided in Ref. 1 that the two expansions become equal provided that we define $\delta(q)$ as,

$$\delta(q) = \Delta Q_{co}(q) = \frac{Q^{\Phi}_{c0}(q)}{L} \,. \tag{2.23}$$

Up to first order in the small bare-momentum deviations, we can thus write the following relationship between the excited-state and ground-state charge momentum rapidity functions (and $\alpha\nu$ rapidity functions),

$$k(q) = k^0\left(q + \frac{Q^{\Phi}_{c0}(q)}{L}\right) \,,$$

$$\Lambda_{\alpha\nu}(q) = \Lambda^0_{\alpha\nu}\left(q + \frac{Q^{\Phi}_{\alpha\nu}(q)}{L}\right) \,, \qquad \alpha = c, s \,, \qquad \nu = 1, 2, \dots \,. \tag{2.24}$$

This is quite remarkable since it states that the excited-state charge momentum rapidity function and $\alpha\nu$ rapidity functions can be expressed in terms of the corresponding initial ground-state functions, provided that we shift the bare momenta by an amount $Q^{\Phi}_{\alpha\nu}(q)/L$. We see that the charge momentum rapidity function and $\alpha\nu$ rapidity functions of all excited states that we may be interested in can thus be expressed in terms of the ground-state corresponding functions, provided that we use a slightly shifted value for the momenta.

Following the studies of Refs. 29–31, let us now define new objects, called pseudofermions, which are generated from the corresponding pseudoparticles and carry discrete momentum values given by,

$$\bar{q}_j = \bar{q}(q_j) = q_j + \frac{Q_{\alpha\nu}(q_j)}{L} = \frac{2\pi}{L} I^{\alpha\nu}_j + \frac{Q_{\alpha\nu}(q_j)}{L} \,. \tag{2.25}$$

Here $Q_{\alpha\nu}(q_j)/L$ is the following *canonical-momentum functional*,

$$\frac{Q_{\alpha\nu}(q_j)}{L} = \frac{2\pi}{L} \sum_{\alpha'\nu'} \sum_{j'=1}^{N^*_{\alpha'\nu'}} \Phi_{\alpha\nu,\alpha'\nu'}(q_j, q_{j'}) \Delta N_{\alpha'\nu'}(q_{j'}) \,. \tag{2.26}$$

This expression involves the phase shifts introduced in the preceding chapter and can be obtained from the Bethe ansatz equations given in that paper. The pseudofermions refer to a normal-ordered description relative to the initial ground state.

The point is that the excitation energy relative to that state is additive in the energy contributions of the pseudofermions involved in the transition to the excited states. The pseudofermions are well defined in a subspace spanned by the initial ground state and all energy eigenstates generated from it by application of one- and two-electron operators. Fortunately, such a subspace refers to all energy scales and the suitable use of the pseudofermion description enables the evaluation of finite-energy expressions for the one- and two-electron correlation and spectral functions. Indeed, it turns out that the information recorded in the pseudoparticle f-functions considered in the companion chapter is transferred over to the canonical momentum within the functional $Q_{\alpha\nu}(q_j)/L$. Moreover, for the one- and two-electron subspace where the pseudofermions are defined, only the energy contributions of first order in the momentum deviations have physical significance. It follows that these new objects do not have energy residual interactions between themselves, in contrast to the corresponding pseudoparticles. The total energy of the system only contains the terms of the energy bands and the interacting term is absorbed in the momentum values, which are now shifted from their original, bare, values through the inclusion of the phase shifts $Q_{\alpha\nu}(q_j)$ in the pseudofermion momenta.

Except for the discrete momentum values, the pseudofermions have the same properties as the corresponding pseudoparticles. Hence we are able to relate such objects to the rotated electrons which are connected themselves to the original electrons by a unitary transformation. On the other hand, the pseudofermion canonical momentum $\bar{q}_j = q_j + Q_{\alpha\nu}(q_j)/L$ is fully determined by the quantum numbers $q_j = [2\pi/L]I_j^{\alpha\nu}$ obtained from the Bethe-ansatz solution in the first chapter. Along this path the relation of the original electrons to the Bethe-ansatz quantum numbers has been clarified. Clearly this procedure may be extended to other models solvable by the Bethe ansatz.

Recalling that in the large U/t limit the rotated electrons are equal to the electrons, we can now take advantage of the large U/t treatments to reach insight on the finite U/t problem. In the limit $U/t \rightarrow \infty$, the energy eigenstates of the Hubbard model can be described by a product of a spinless fermion wave function with a squeezed spin wave function.[5] The spin wave function refers to the spins of the singly occupied sites. The spin momentum affects the boundary condition of the spinless-fermion discrete momentum values and the two systems are coupled. Since the boundary conditions are affected, the discrete momentum values of the spinless fermions are also affected so that the anti-commutation relations, while fermionic in real space, are altered in momentum space.[27] A similar situation occurs for the $c0$ pseudofermions for $U/t > 0$. We recall that the U/t-finite energy eigenstates $|\Psi_{U/t}\rangle$ can be generated from the corresponding $U/t \rightarrow \infty$

energy eigenstates, according to $|\Psi_{U/t}\rangle = \hat{V}^{\dagger}|\Psi_{\infty}\rangle$. Hence due to the canonical character of the operator $\hat{V}^{\dagger}$, which maps the spinless fermions of the $U/t \to \infty$ case onto the $c0$ pseudofermions for $U/t > 0$, their operators also anti-commute in real space but since their discrete momentum values are affected by the extra term $Q_{c0}(q_j)/L$, resulting from the phase shifts generated by the ground-state - excited-state transition, the treatment is formally similar to that of the altered boundary conditions. Therefore, their anti-commutation relations are given by,

$$\{f^{\dagger}_{\bar{q},\alpha\nu}, f_{\bar{q}',\alpha'\nu'}\} = \delta_{\alpha\nu,\alpha'\nu'} \frac{1}{N^*_{\alpha\nu}} e^{-i(\bar{q}-\bar{q}')/2} e^{iQ_{\alpha\nu}(q)/2} \frac{\sin(Q_{\alpha\nu}(q)/2)}{\sin([\bar{q}-\bar{q}']/2)} . \quad (2.27)$$

Such a relation can be shown to be valid for all remaining $\alpha\nu$ pseudofermion branches.

In summary, for the Hubbard model defined in the subspace generated by a given initial ground state and excited states generated from it by application of one- and two-electron operators, we have a pseudofermion description without energy interactions. Since we are able to establish a relation, for any excited energy eigenstate belonging to that subspace, between the pseudofermions and the original electrons, we are able to calculate matrix elements between the ground state and these states.[30]

Once the $c0$ pseudofermions are spinless and η-spinless objects and the composite $c\nu$ and $s\nu$ pseudofermions are η-spin-zero and spin-zero objects, respectively, their scattering matrix has dimension 1×1, i.e. it is just a complex number, in contrast to the representation of Ref. 37, whose scatterers are η-spin $\frac{1}{2}$ and spin $\frac{1}{2}$ objects.

2.3.2. *The pseudofermion dynamical theory*

The above properties of the pseudofermions enables the construction of a pseudofermion dynamical theory which provides finite-energy expressions for correlation and spectral functions. In order to obtain zero-temperature correlation functions we need to evaluate matrix elements between the ground state and the set of excited states coupled by the one- or two-electron operator under consideration. (At finite temperature the procedure is more complex because we have to consider matrix elements between the excited states as well). The exact energy eigenstates obtained from the Bethe ansatz solution and the model global symmetry are not easily expressed in terms of electronic occupancy configurations. However, as we have seen above, there is a general route to evaluate such matrix elements. Typically, we are interested in correlation functions involving one- and two-electron operators. The first step is to rewrite the electronic operators in terms of rotated electrons.[29,30] For $U/t > 0$ the above operator $\hat{S}$ can be expanded in a series of t/U and the corresponding first-order term has an universal form given in Eq. (41) of Ref. 26. The use of that first-order expansion leads to the following expression of the electronic annihilation operator in terms of rotated-electron operators,

$$\begin{aligned} c_{i\sigma} &= \hat{V}(U/t)\tilde{c}_{i\sigma}\hat{V}^{\dagger}(U/t) \\ &= \tilde{c}_{i\sigma} - \frac{t}{U}\sum_{\delta=\pm1}[\tilde{c}_{i+\delta,\sigma}(\tilde{n}_{i+\delta,-\sigma} - \tilde{n}_{i,-\sigma}) - \tilde{c}^{\dagger}_{i,-\sigma}\tilde{c}_{i+\delta,-\sigma}\tilde{c}_{i,\sigma} \\ &\quad + \tilde{c}^{\dagger}_{i+\delta,-\sigma}\tilde{c}_{i,-\sigma}\tilde{c}_{i,\sigma}] + O\left(\frac{t^2}{U^2}\right). \end{aligned} \tag{2.28}$$

There is an identification between the rotated electrons as defined above and the $c0$ pseudofermions, holons and spinons. Each one- or two-electron correlation function corresponds to a well defined subspace spanned by the initial ground state and the excited states which couple to it. This is the above pseudofermion subspace for the correlation function under consideration. Thus, there is a specific pseudofermion description for each initial ground state and one- or two-electron operator. For each choice of the latter operator, the excited-state permitted values for the numbers of up and down spins, rotated-electron doubly occupied sites ($-1/2$ holons), unoccupied sites ($+1/2$ holons), and spin σ singly occupied sites (σ spinons) are well defined. Taking this into account we may express the electron operators in terms of rotated electron operators and identify the relation of the latter operators to the pseudofermion operators. At the end of such a procedure we have replaced a matrix element in terms of electron operators by a matrix element in terms of pseudofermion operators, which is much simpler to calculate.[30] The procedure is similar to the treatment in the limit $U \to \infty$ where the matrix element is replaced by the determinant of the spinless fermions anti-commutators.[27,28] However, for $U/t > 0$ one has such determinants for both $c0$ and $s1$ pseudofermion operators.[30] The anti-commutators involve the phase shifts defined in the companion chapter, which are two-pseudofermion phase shifts. Such phase shifts have two contributions: a collision-less contribution which results from the shifts in the quantum numbers of the Bethe-asatz solution from integers to half integers or vice versa, plus the $\alpha\nu$ shift given in Eq. (2.26). The latter shift results from the collisions between all pseudofermions (scatterers) with those created under the ground-state - excited-state transition (scattering centers). This is consistent with the form of the momentum shift $Q_{\alpha\nu}(q_j)/L$ given in that equation.

As it turns out, the substitution $c^l_{i\sigma} \to \tilde{c}^l_{i\sigma}$ in the operators appearing in the matrix elements is usually enough to a high degree of accuracy. For instance, for the spectral function it accounts for over 99% of the total spectral weight.[29,30] This property greatly simplifies the calculations. Actually, the canonical transformation is not known exactly and one needs to consider expansions in powers of t/U which work well at large values of U/t, but as U decreases become increasingly difficult. However, that is only needed for contributions beyond the above substitution.

As examples let us illustrate the following local elementary processes:

1. Creation of one spin-projection $\pm 1/2$ rotated electron at the unnocupied site j is equivalent to annihilation of a $+1/2$ Yang holon and creation of a local $c0$ pseudofermion and a local $\pm 1/2$ spinon at the same site.

2. Creation of one spin-projection $\sigma = \pm 1/2$ rotated electron at a spin-projection $\mp 1/2$ rotated-electron singly occupied site j is equivalent to annihilation of a local $\mp 1/2$ spinon and a local $c0$ pseudofermion and creation of a local $-1/2$ holon at such a site.

3. Creation of two rotated electrons of opposite spin projection onto the unnocupied site j is equivalent to the annihilation of a $+1/2$ Yang holon and creation of a local $-1/2$ holon at such a site.

4. Annihilation of one spin-projection $\sigma = \pm 1/2$ rotated electron and creation of one-spin projection $\sigma = \mp 1/2$ rotated electron at the singly-occupied site j is equivalent to the annihilation of one local $\pm 1/2$ spinon and creation of one local $\mp 1/2$ spinon.

The leading contributions in terms of the pseudofermion operators which correspond to most spectral weight are as follows:[30]

$$
\begin{aligned}
\tilde{c}^{\dagger}_{j,\downarrow}(1-\tilde{n}_{j,\uparrow}) &= \frac{1}{C_J}\, f^{\dagger}_{x_{j'},s1} f^{\dagger}_{x_j,c0}\,,\\
\tilde{c}^{\dagger}_{j,\uparrow}(1-\tilde{n}_{j,\downarrow}) &= \frac{1}{C_J}\, f^{\dagger}_{x_j,c0}\,,\\
\tilde{c}^{\dagger}_{j,\uparrow}\tilde{n}_{j,\downarrow} &= \frac{1}{C_J}\, f^{\dagger}_{x_{j''},c1} f_{x_j,c0} f_{x_{j'},s1}\,,\\
\tilde{c}^{\dagger}_{j,\downarrow}\tilde{n}_{j,\uparrow} &= \frac{1}{C_J}\, f^{\dagger}_{x_{j'},c1} f_{x_j,c0}\,,\\
\tilde{c}^{\dagger}_{j,\downarrow}\tilde{c}^{\dagger}_{j,\uparrow} &= \frac{1}{C_J}\, f^{\dagger}_{x_{j''},c1}\,,\\
\tilde{c}^{\dagger}_{j,\downarrow}\tilde{c}_{j,\uparrow} &= \frac{1}{C_J}\, f^{\dagger}_{x_{j'},s1}\,,
\end{aligned}
\tag{2.29}
$$

where we have omitted higher-order contributions involving products of an increasing number of pseudofermion operators. Here the constant C_J is defined in Eq. (46) of the first paper of Ref. 30 and has in general a different value for each of those expressions and the sites coordinates x_j refer to the rotated-electron lattice, which coincides with that of the original electrons. The $c0$ effective lattice equals that of the electrons and rotated electrons and hence the indices j of the $c0$ pseudofermion operators sites x_j which appear in the expressions of Eq. (2.29) equal those of the corresponding rotated electrons. In contrast, for a certain site x_j corresponding to a specific index j of the rotated-electron lattice, the corresponding site $x_{j'}$ of the $s1$ effective lattice is such that the index j' of the $s1$ pseudofermion operators appearing in the same expressions equals the integer number closest to $j\, n_\downarrow$ whereas for the site $x_{j''}$ of the $c1$ effective lattice j'' is in those expressions the integer number closest to $j\,(1-n)$.[30]

2.3.3. *Application: the one-electron spectral function*

An important application of the pseudofermion dynamical theory is the calculation of the one-electron spectral function which can be measured by photoemission and inverse photoemisison experiments on low-dimensional materials. The one-electron spectral function $B^l(k,\,\omega)$ such that $l=-1$ (and $l=+1$) for electron removal (and addition) reads,

$$B^{-1}(k,\,\omega)=\sum_{\sigma,\,\gamma}|\langle\gamma|\,c_{k,\,\sigma}|\,GS\rangle|^2\,\delta(\omega+\Delta E_\gamma^{N-1})\,, \tag{2.30}$$

and

$$B^{+1}(k,\,\omega)=\sum_{\sigma,\,\gamma'}|\langle\gamma'|\,c_{k,\,\sigma}^\dagger|\,GS\rangle|^2\,\delta(\omega-\Delta E_{\gamma'}^{N+1})\,. \tag{2.31}$$

Here $c_{k,\,\sigma}$ and $c_{k,\,\sigma}^\dagger$ are electron operators of momentum k and $|GS\rangle$ denotes the initial N-electron ground state. The γ and γ' summations run over the $N-1$ and $N+1$-electron excited states, respectively, and ΔE_γ^{N-1} and $\Delta E_{\gamma'}^{N+1}$ are the corresponding excitation energies. For $n>0$ and $m=0$, there are in the initial ground state no $-1/2$ Yang holons and HL spinons, no composite $c\nu$ pseudofermions, and no $s\nu$ pseudofermions belonging to $\nu>1$ branches, the $s1$ pseudofermions band is full, and the $c0$ pseudofermions occupy $0\leq|\bar{q}|\leq 2k_F$ (leaving $2k_F<|\bar{q}|\leq\pi$ empty).

The ground state and excited energy eigenstates can be expressed in terms of occupancy configurations of pseudofermions. For electron removal, the dominant processes involve creation of one hole both in $\epsilon_{c0}(\bar{q})$ and $\epsilon_{s1}(\bar{q})$. (The $c0$ and $s1$ energy bands are those studied in the previous chapter with q replaced by $\bar{q}$.) For electron addition, these dominant processes lead to two structures: A *lower-Hubbard band* (LHB) generated by creation of one particle in $\epsilon_{c0}(\bar{q})$ and one hole in $\epsilon_{s1}(\bar{q})$; A UHB generated by creation of one hole both in $\epsilon_{c0}(\bar{q})$ and $\epsilon_{s1}(\bar{q})$ and either one particle in $\epsilon_{c1}^0(\bar{q})$ for $n<1$ or one $-1/2$ Yang holon for

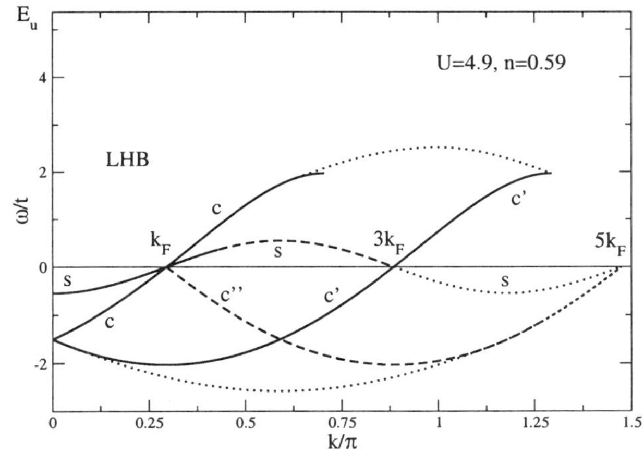

Fig. 2.5. The solid and dashed lines denoted by the letters c, c', c'' and s are singular and edge branch lines, respectively. Electron removal (LHB addition) corresponds to $\omega < 0$ (and $\omega > 0$) and $\omega = E_u$ marks the UHB lower limit. Reproduced with permission of EDP Sciences from Ref. 33.

$n \rightarrow 1$. According to the pseudofermion dynamical theory,[29,30] both the one-electron spectral-weight singularities and edges are located on pseudofermion branch lines.[33,35] Such lines are generated by processes where a specific pseudofermion is created or annihilated for the available values of momentum $\bar{q}$ and the remaining quantum objects are created or annihilated at their *Fermi points*. The weight distribution shape of the singular (and edge) branch lines is controlled by negative (and positive) exponents smaller than zero (and one). The electron removal ($\omega < 0$) and LHB addition ($\omega > 0$) singular and edge branch lines are represented in the Fig. 2.5 by solid and dashed lines, respectively. The solid lines correspond to the regimes where there is singular behavior. For simplicity, the figure does not represent the $\omega > E_u$ UHB region. For instance, the s (c) branch line is obtained allowing a hole to scan the $s1$ ($c0$) band. When drawing these lines one has to consider the two Fermi points for each pseudofermion band and the quantum shake-up effect resulting from the eventual changes of integer to half-integer Bethe-ansatz quantum numbers. The dashed-dotted lines and some of the branch lines of Fig. 2.5 are border lines for the $\omega < E_u$ domain of the (k, ω)-plane whose spectral weight is generated by dominant processes. (There is a region limited above by the s line for $k_F < k < 3k_F$ and below by the c'' and c' lines for $k_F < k < 2k_F$ and $2k_F < k < 3k_F$, respectively, which does not belong to that domain.) The dominant processes also include particle-hole pseudofermion processes which lead to spectral weight both inside and outside but in the close vicinity of that domain.

The calculation of the matrix elements for the entire allowed regions of the (k, ω) plane is technically complex. However, a full treatment for the whole (k, ω) plane is possible.[29,30] This involves consideration of four different cases (where in the following *"P"* stands for those *P*seudofermions created or annihilated away from the Fermi points):

(1) *2P contribution:* Neither the $c0$ nor the $s1$ pseudofermion or pseudofermion hole are created at any of the Fermi points. This contribution will lead to the overall "background" of the spectral-function weight distribution, since both pseudofermions or pseudofermion holes are dispersive, leading to contributions over nearly the whole range of allowed k and ω values.
(2) *s-branch (1P):* The $c0$ pseudofermion or pseudofermion hole is created at the left or the right $c0$ Fermi point and the $s1$ pseudofermion hole is created away from any of the $s1$ Fermi points. This will lead to a line in the (k, ω) plane, following the dispersion of the $s1$ pseudofermion hole.
(3) *c-branch (1P):* The $s1$ pseudofermion hole is created at the left or the right $s1$ Fermi point and the $c0$ pseudofermion hole is created away from any of the $c0$ Fermi points. This will lead to a line in the (k, ω) plane, following the dispersion of the $c0$ pseudofermion or pseudofermion hole.
(4) *Fermi contribution (0P):* Both pseudofermions or pseudofermion holes are created at their left or right Fermi points, respectively. This contribution leads to a spectral weight distribution in the vicinity of certain points in the (k, ω) plane.

Let us first consider the behavior in the vicinity of the branch lines, where the spectral weight distribution may display singular peaks and the weight is in general larger. We start by considering the spin s branch line for $0 < k < 3k_F$, the charge c branch line for $0 < k < \pi - k_F$, and the charge c' branch line for $0 < k < \pi + k_F$ (see Fig. 2.5). The parametric equations that define these branch lines read $\omega(k) = \epsilon_s(q)$ for the s line where $q = q(k) = (1+l)\, k_F - l\, k$ for $(1+l)k_F/2 < k < k_F + (1+l)k_F$ and $\omega(k) = \epsilon_\alpha(q)$ for the $\alpha = c$ line ($\iota' = +1$) and $\alpha = c'$ line ($\iota' = -1$) where $q = q(k) = k + \iota'\, k_F$ for $0 < k < \pi - \iota'\, k_F$. Here, $l = \pm 1$ and $\epsilon_{c'}(q) \equiv \epsilon_c(q)$. The following expression describes the weight distribution in the vicinity of the $\alpha = s, c, c'$ branch lines for ω values such that $(\epsilon_\alpha(q) + l\, \omega)$ is small and positive,

$$B^l(k,\, \omega) = C^l_\alpha(k)\Big(\epsilon_\alpha(q) + l\, \omega\Big)^{\zeta_\alpha(k)}, \quad \alpha = s, c, c'\,. \tag{2.32}$$

The k-dependence of the prefactor $C^l_\alpha(k)$, such that $C^l_\alpha(k) > 0$ for $U/t > 0$, is in general involved and can be studied numerically. For $U/t \to 0$, $C^l_\alpha(k)$ behaves as $C_s^l(k) \rightarrow \delta_{l,-1}\, C_s$, $C_c^l(k) \rightarrow \delta_{l,+1}\, C_c$, and $C_{c'}^l(k) \rightarrow 0$, where C_s and C_c are independent of k. The expression of the exponent of Eq. (2.32) involves the two-pseudofermion phases shifts discussed in Ref. 1. For simplicity here we denote the corresponding branch indices $c0$ and $s1$ by c and s respectively. Such expressions read,

$$\zeta_s(k) = -1 + \sum_{\iota=\pm1} [\Big\{\frac{1}{2\sqrt{2}} - \iota\,\Phi_{s\,s}(\iota\,k_F,\,q)\Big\}^2 + \Big\{\frac{\iota}{2\xi_0} + \frac{(1+l)\,\xi_0}{4} - l\,\Phi_{c\,s}(\iota\,2k_F,\,q)\Big\}^2],$$
$$\zeta_\alpha(k) = -1 + \sum_{\iota=\pm1} [\Big\{\frac{1}{2\sqrt{2}}[1 - l(1+\iota\iota')] + \iota\,\Phi_{s\,c}(\iota\,k_F,\,q)\Big\}^2 + \Big\{\iota'\frac{\xi_0}{4} - \Phi_{c\,c}(\iota\,2k_F,\,q)\Big\}^2], \quad (2.33)$$

where $\alpha = c$ for $\iota' = +1$ and $\alpha = c'$ for $\iota' = -1$.

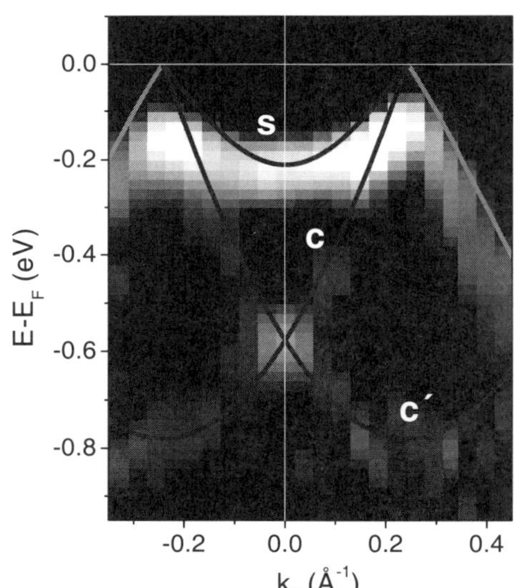

Fig. 2.6. Angle-resolved photoemission spectra of TTF-TCNQ measured along the easy transport axis and matching theoretical branch lines. Reproduced with permission of EDP Sciences from Ref. 33.

The weight shape of the UHB singular branch lines is controlled by the same exponents as the corresponding electron-removal branch lines. There are also branch lines in the upper Hubbard band. Details of the various contributions to the spectral weight may be found in Refs. 33,35.

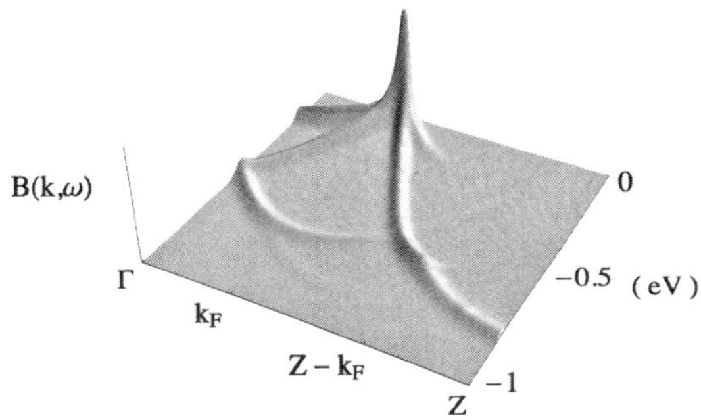

Fig. 2.7. Theoretical spectral weight for TTF-TCNQ.[36]

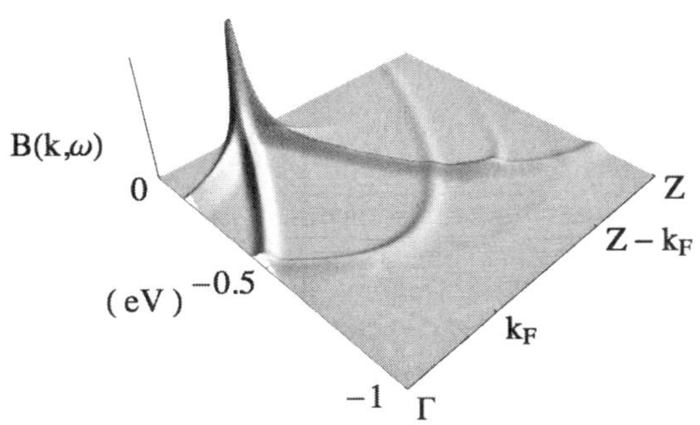

Fig. 2.8. Theoretical spectral weight for TTF-TCNQ, from a different angle.[36]

The experimental dispersions for the compound TTF-TCNQ in the electron removal spectrum of this quasi-1D conductor as measured by ARPES are shown in Fig. 2.6 where the theoretical results are superimposed.[33,35] A qualitative description of the same results and a discussion of the importance of correlation effects versus band theory were presented in Ref. 34. A more detailed study of the singular features of the TCNQ part is presented in Ref. 35. The agreement between the theoretical predictions and the observed experimental results is quite good. Not only the theory predicts correctly the location of the observed features, but it is in qualitative agreement with their relative weights. Importantly, the observed features coincide with the locations of the branch lines, as expected. The results confirm the utility of the Hubbard model for the understanding of these systems, particularly at high energies. At low energies other modes interfere, like phonons. Moreover the results show that the separation of the charge and spin degrees of freedom is a phenomenon which occurs at all energy scales and not only at low energy.

We should note that the expressions obtained for the exponents along the branch lines are not valid as the dispersive pseudofermion enters the linear region of its spectrum. This issue is discussed in detail in Ref. 35. Such a linear energy dispersion corresponds to the regime of validity of the low-energy Tomonaga-Luttinger liquid description associated with expressions (2.11) and (2.12). When coming into such a linear region the momentum and energy dependence of the spectral weight is in the vicinity of a α-branch line different. In this region the created pseudofermion or pseudofermion hole, and the small-momentum pseudofermion particle-hole excitations, share the same velocity. As explained in Ref. 35, the low-energy expression (2.11) arises from a "velocity resonance effect".

Finally we present in Figs. 2.7, and 2.8 results for the spectral function in the full (k, ω) plane after calculating explicitly all the relevant matrix elements.[36] This is a numerically quite involved calculation. The results confirm that most of the spectral weight is located at the branch lines. However, this property refers to the one-electron spectral function only. For the finite-energy expressions of two-electron correlation and spectral functions the above considered surface contribution is instead the dominant one.

2.4. Conclusions

In this chapter we reviewed different issues regarding the Drude peak and the regular part of the one-dimensional Hubbard model frequency dependent conductivity and the use of conformal field theory to study the low-energy behavior of correlation functions of that solvable model. In addition, we also considered the finite-energy expressions of correlation and spectral functions and reviewed a transformation which maps the pseudoparticles considered in the companion chapter, Ref. 1, onto non-interacting pseudofermions. That enables the evaluation of matrix elements between energy eigenstates and the construction of a pseudofermion dynamical theory which provides the above finite-energy expressions.

The studies of this chapter considered mainly the one-dimensional Hubbard model, which describes successfully some of the exotic properties observed in low-dimensional materials. The pseudofermion dynamical theory and other approaches reviewed here also apply to related integrable interacting problems and therefore have wide applicability.

Acknowledgments

We thank the support of FCT under the grant POCTI/FIS/58133/2004 and that of the ESF Science Programme INSTANS 2005-2010.

References

1. J. M. P. Carmelo, P. D. Sacramento, D. Bozi, and L. M. Martelo, Correlation effects in one-dimensional systems, in "Strongly correlated systems, coherence and entanglement", edited by J. M. P. Carmelo, J. M. B. Lopes dos Santos, V. Rocha Vieira, and P. D. Sacramento, World Scientific, Singapore (2007), page 3.
2. V. Vescoli *et al*, Dimensionality-driven insulator-to-metal transition in the Bechgaard salts, *Science*. **281**, 1181 (1998).
3. A. Schwartz *et al*, On-chain electrodynamics of metallic (TMTSF)2X salts: Observation of Tomonaga-Luttinger liquid response, *Phys. Rev. B*. **58**, 1261 (1998).
4. E. H. Lieb, F. Y. Wu, Absence of Mott transition in an exact solution of the short-range, one-band model in one dimension, *Phys. Rev. Lett.*. **20**, 1445 (1968).
5. M. Ogata, H. Shiba, Bethe-ansatz wave function, momentum distribution, and spin correlation in the one-dimensional strongly correlated Hubbard model, *Phys. Rev. B*. **41**, 2326 (1990).
6. J. M. P. Carmelo, A. A. Ovchinnikov, Generalization of the Landau liquid concept: example of the Luttinger liquids, *J. Phys. Cond. Mat.*. **3**, 757 (1991).
7. J. M. P. Carmelo, P. Horsh, A. A. Ovchinnikov, Static properties of one-dimensional generalized Landau liquids, *Phys. Rev. B*. **45**, 7899 (1992).
8. K. Penc and J. Sòlyom, One-dimensional Hubbard model in a magnetic field and the multicomponent Tomonaga-Luttinger model, *Phys. Rev. B*. **47**, 6273 (1993).
9. W. Kohn, Theory of the Insulating State, *Phys. Rev.*. **133**, A171 (1964).
10. D. Baeriswyl, J. Carmelo and A. Luther, Correlation effects on the oscillator strength of optical absorption: Sum rule for the one-dimensional Hubbard model, *Phys. Rev. B*. **33**, 7247 (1986).
11. H. J. Schulz, Correlation exponents and the metal-insulator transition in the one-dimensional Hubbard model, *Phys. Rev. Lett.*. **64**, 2831 (1990).
12. X. Zotos and P. Prelovsek, Evidence for ideal insulating or conducting state in a one-dimensional integrable system, *Phys. Rev. B*. **53**, 983 (1996).
13. N. M. R. Peres, P. D. Sacramento, D. K. Campbell and J. M. P. Carmelo, Curvature of levels and charge stiffness of one-dimensional spinless fermions, *Phys. Rev. B*. **59**, 7382 (1999).

14. N. M. R. Peres, R. G. Dias, P. D. Sacramento and J. M. P. Carmelo, Finite-temperature transport in finite-size Hubbard rings in the strong-coupling limit, *Phys. Rev. B.* **61**, 5169 (2000).
15. J. M. P. Carmelo, N. M. R. Peres and P. D. Sacramento, Finite-Frequency Optical Absorption in 1D Conductors and Mott-Hubbard Insulators, *Phys. Rev. Lett.*. **84**, 4673 (2000).
16. N. M. R. Peres, P. D. Sacramento and J. M. P. Carmelo, Charge and spin transport in the one-dimensional Hubbard model, *J. Phys. Cond. Matt.*. **13**, 5135 (2001).
17. A. G. Izergin, V. E. Korepin and N. Yu. Reshetikhin, Conformal dimensions in Bethe ansatz solvable models, *J. Phys. A.* **22**, 2615 (1989).
18. F. Woynarovich, Finite-size effects in a non-half-filled Hubbard chain, *J. Phys. A.* **22**, 4243 (1989).
19. H. Frahm, V. E. Korepin, Critical exponents for the one-dimensional Hubbard model, *Phys. Rev. B.* **42**, 10553 (1990).
20. N. Kawakami and S.-K. Yang, Luttinger anomaly exponent of momentum distribution in the Hubbard chain, *Phys. Lett. A.* **148**, 359 (1990).
21. H. Frahm, V. E. Korepin, Correlation functions of the one-dimensional Hubbard model in a magnetic field, *Phys. Rev. B.* **43**, 5653 (1991).
22. J. M. P. Carmelo, P. Horsch, D. K. Campbell, A. H. Castro Neto, Magnetic effects, dynamical form factors, and electronic instabilities in the Hubbard chain, *Phys. Rev. B.* **48**, 4200 (1993).
23. J. M. P. Carmelo, F. Guinea and P. D. Sacramento, Instabilities of the Hubbard chain in a magnetic field, *Phys. Rev. B.* **55**, 7565 (1997).
24. J. M. P. Carmelo, L. M. Martelo and P. D. Sacramento, One- and two-electron spectral function expressions in the vicinity of the upper-Hubbard bands lower limit, *J. Phys. Cond. Matt.*. **16**, 1375 (2004).
25. J. M. P. Carmelo, J. M. Román, K. Penc, Charge and spin quantum fluids generated by many-electron interactions, *Nucl. Phys. B.* **683**, 387 (2004).
26. J. Stein, Flow equations and the strong-coupling expansion for the Hubbard model. *J. Stat. Phys.* **88**, 487 (1997).
27. K. Penc, F. Mila, H. Shiba, Spectral Function of the 1D Hubbard Model in the U infinite limit, *Phys. Rev. Lett.*. **75**, 894 (1995).
28. K. Penc, K. Hallberg, F. Mila, H. Shiba, Spectral functions of the one-dimensional Hubbard model in the U infinite limit: How to use the factorized wave function, *Phys. Rev. B.* **55**, 15475 (1997).
29. J. M. P. Carmelo, K. Penc, D. Bozi, Finite-energy spectral-weight distributions of a 1D correlated metal, *Nucl. Phys. B.* **725**, 421 (2005); *Nucl. Phys. B.* **737** 351 (2006), *Erratum.*
30. J. M. P. Carmelo and K. Penc, General spectral function expressions of a 1D correlated model, *Eur. Phys. J. B.* **51**, 477 (2006); J. M. P. Carmelo and K. Penc, Spectral microscopic mechanisms and quantum phase transition in a 1D correlated problem, *J. Phys.: Cond. Matt.* **18**, 2881 (2006).
31. J. M. P. Carmelo, L. M. Martelo, K. Penc, The low-energy limiting behavior of the pseudofermion dynamical theory, *Nucl. Phys. B.* **737**, 237 (2006); J. M. P. Carmelo, K. Penc, Correlation-function asymptotic expansions: Universality of prefactors of the one-dimensional Hubbard model, *Phys. Rev. B.* **73**, 113112 (2006).

32. J. M. P. Carmelo, F. Guinea, K. Penc and P. D. Sacramento, Superconductivity driven by chain coupling and electronic correlations, *Eur. Phys. Lett.* **68**, 839 (2004).
33. J. M. P. Carmelo, K. Penc, L. M. Martelo, P. D. Sacramento, J. M. B. Lopes dos Santos, R. Claessen, M. Sing, U. Schwingenschlögl, One-electron singular branch lines of the Hubbard chain, *Europhys. Lett.*. **67**, 233 (2004).
34. M. Sing, U. Schwingenschlögl, R. Claessen, P. Blaha, J. M. P. Carmelo, L. M. Martelo, P. D. Sacramento, M. Dressel, C. S. Jacobsen, Electronic structure of the quasi-one-dimensional organic conductor TTF-TCNQ, *Phys. Rev. B*. **68**, 125111 (2003).
35. J. M. P. Carmelo, K. Penc, P. D. Sacramento, M. Sing and R. Claessen, The Hubbard model description of the TCNQ related singular features in photoemission of TTF-TCNQ, *J. Phys. Cond. Matt.*. **18**, 5191 (2006).
36. D. Bozi, Ph.D. thesis (unpublished), cond-mat0606380, (2006).
37. F. H. L. Eßler, V. E. Korepin, Scattering matrix and excitation spectrum of the Hubbard model, *Phys. Rev. Lett.* **72**, 908 (1994).

Chapter 3

Electron spectral function of high-temperature cuprate superconductors

Tiago C. Ribeiro

Global Modelling and Analytics Group, Credit Suisse, One Cabot Square, London E14 4QJ, United Kingdom

Xiao-Gang Wen

Department of Physics, Massachusetts Institute of Technology, Cambridge, Massachusetts 02139, USA

We address the doping evolution of the low energy electronic structure of high-temperature superconducting copper-oxide compounds, as described by the $tt't''J$ model. Following experimental evidence for well defined quasiparticles in the normal state of these doped Mott insulators, we use a new slave-particle basis that includes electron-like operators, namely, the doped-carrier basis, and extensively discuss the mean-field electron spectral function of the $tt't''J$ model. We show that the above mean-field theory reproduces many aspects of the non-trivial microscopic single electron dynamics probed by angle-resolved photoemission experiments in hole and electron doped cuprates; these include: the emergence of spectral peaks inside the Mott gap upon doping away from half-filling; the differentiation between the nodal and antinodal regions of momentum space, which displays distinct properties in the hole and electron doped regimes; the low energy spectral weight arcs, whose length increases with doping; the nodal dispersion kink, which is sharper in the underdoped regime; the strong dispersion renormalization, which renders the dispersion close to $(0,\pi)$ and $(\pi,0)$ surprisingly flat. We further argue that measured angle-resolved photoemission spectral dispersions, together with the associated spectral weight intensity, impose strong constraints on the character of coexisting short-range correlations. The agreement between our results and experimental data supports that the two predominant local spin correlations in cuprate superconductors are: (i) d-wave singlet pairing correlations, and (ii) staggered moment correlations.

Contents

3.1. Introduction

High-temperature superconductivity, together with a large body of novel phenomenology, occurs in the copper-oxide compounds in the regime intermediate to the Néel state and the Fermi liquid metal, where electrons evolve from being strongly correlated local moments to weakly correlated itinerant entities.[1–3] This regime, which constitutes a major and long-standing problem in the field of condensed matter physics,[4–6] is the focus of this paper. Specifically, we study the doping evolution of the microscopic single electron dynamics, as probed by angle-resolved photoemission spectroscopy (ARPES) experiments in hole and electron doped cuprates. Our emphasis lies on the so called underdoped regime, where electron occupancy per unit cell is close to unity, since it exhibits a most striking phenomenology.[3,7]

Even though the d-wave superconducting (dSC) phase that intervenes between the undoped Mott insulator and the overdoped metal is believed to be conventional, as attested by flux quantization experiments[8–10] and the evidence for long-lived Bogoliubov nodal excitations,[11–14] quasiparticle properties in these materials deviate in many ways from the BCS pairing state of otherwise effectively free electrons.[15,16] These deviations are particularly notable in ARPES data, which reveals strong band renormalization effects, such as the nodal dispersion kink[17–21] and the flat dispersion around $(0, \pi)$ and $(\pi, 0)$,[21–23] throughout the various cuprate families. In addition, this experimental probe unveils a peculiar redistribution of spectral intensity along the observed dispersive features, leading to the formation of, for instance, low energy spectral weight arcs around the nodal direction in hole doped samples.[17,18,24,25] ARPES also provides clear evidence for the asymmetry between the hole and electron doped regimes, as in the latter low energy spectral weight predominates close to the Brillouin zone boundary instead.[26] In this paper, we propose a microscopic description of the above deviations in the correlated superconducting state, as well as other phenomena which we detail below. Interestingly, reproducing the deviations from BCS mean-field theory observed in the low temperature superconducting state provides valuable information concerning the high-energy local physics of cuprate superconductors, specifically, we find that the single electron dynamics follows from the interplay between coexisting d-wave singlet pairing correlations and short-range antiferromagnetic (AF) correlations between local moments.

The aforementioned novel phenomenology, which includes the nodal dispersion kink and the spectral weight arcs, is more pronounced near the Mott insulating state,[19,24,27] which is well described by the AF Heisenberg model.[28] Hence, below, we consider the two-dimensional generalized-tJ Hamiltonian in order to model the physics of underdoped copper-oxide layers.[1–3] Since we want to describe single electron properties of the dSC state, whose sole gapless excitations are well defined Bogoliubov quasiparticles, in this paper we resort to mean-field theory. Different mean-field approaches to the generalized-tJ model have been used in the literature and, notably so, a variety of slave-particle mean-field theories have been the subject of thorough study.[29–38] These theories are formally equivalent once all fluctuations around the mean-field saddle-point are properly considered, otherwise, they strictly correspond to different approximations. Therefore, a judicious choice of the mean-field slave-particle operators is required. We follow recent ARPES data showing sharp spectral peaks in the normal state,[39] and choose to use the doped-carrier slave-particle framework of the generalized-tJ model[37,38] since it decouples the electron operator into spinon and dopon operators, where the latter are electron-like operators (by this we mean that dopons are charge-e and spin-1/2 fermionic operators, which thus define sharp quasiparticle excitations).

The structure of this paper is as follows. In Sec. 3.2 we review ARPES data concerning the doping evolution of the electron spectral function in both hole and electron doped cuprates. We also compare this data to $tt't''J$ model numerical results, thus motivating the use of the $tt't''J$ Hamiltonian as a model of cuprate compounds. The doped-carrier framework is reviewed in Sec. 3.3 and, in Sec. 3.4, we extensively compare the corresponding MF theory electron spectral function both to cuprate experimental data and to $tt't''J$ model numerical results. Naturally, our mean-field theory results cannot be but a rudimentary representation of a strongly correlated superconductor's spectral function; for one thing, the mean-field spectral function reduces to a collection of δ-peaks, while spectral features of strongly correlated systems are remarkably broad.[27] Even though our mean-field theory overlooks the spectral linewidth, in Sec. 3.4 we show that it reproduces: (i) the dispersion of observed spectral features; (ii) the total spectral intensity associated with these features. We remark that both (i) and (ii) encode crucial information concerning the local dynamics of single electrons. In Sec. 3.5 we discuss the simple theoretical/conceptual picture that stems from the aforementioned agreement between mean-field results and experimental observations. Finally, in Sec. 3.6 we summarize our main results.

3.2. ARPES and the cuprates

In the course of the last 20 years both undoped, hole doped, and electron doped cuprates have been heavily studied using a variety of experimental probes. Given the layered nature of these materials, ARPES has proved to be a highly valuable technique which provides both momentum and frequency resolved information concerning the microscopic electronic structure of the CuO_2 planes.[27] ARPES measures the single electron spectral properties, from which the quasiparticle dispersion, quasiparticle lifetime and quasiparticle weight can be extracted. These features unveil not only the electronic character of excitations, but also the local correlations that determine the dynamics of electrons and holes. As an example, take the role played by ARPES in establishing the validity of local moment physics in the undoped AF compounds[40] and the d-wave symmetry of electron pairs in the dSC state.[41] Due to the recent improvement of experimental resolution and in the quality of deeply underdoped samples, ARPES has also addressed the evolution of the electronic structure between the AF and dSC phases.[18,24,25,42] The resulting data provides crucial insights to the understanding of the short-range correlations that develop in these materials. We consider that such an understanding lies at the heart of the cuprate underdoped regime problem and, as such, the central point of this paper is to propose a theoretical interpretation of the aforementioned experiments, which we briefly review in the remainder of this section.

3.2.1. *Undoped compounds*

ARPES experiments have been conducted in several half-filled cuprates, like La_2CuO_4,[43] $Sr_2CuO_2Cl_2$,[44] $Ca_2CuO_2Cl_2$,[45] Nd_2CuO_4[26] and Sm_2CuO_4,[46] all of which show similar results. These experiments remove a photo-electron from the copper-oxide plane and, therefore, probe the single hole dynamics described by the effective three-band Hubbard Hamiltonian that models the CuO_2 layers.[47,48] Both theoretical arguments[49] and experimental evidence[50,51] show that the hole doped in the oxygen valence band forms a (Zhang-Rice) singlet bond with a copper atom hole. The effective model describing the hopping of the Zhang-Rice singlet in the two-dimensional square lattice with a spin-1/2 in all occupied lattice sites is the generalized-tJ Hamiltonian.[49] Notably, as we detail below, the dispersive feature that appears in half-filled cuprates' ARPES data right below the chemical potential is well fit by the $tt't''J$ Hamiltonian with model parameters independently determined from band calculations and other experimental probes.[52–55]

Using the nearest-neighbor (NN) hopping parameter $t \approx 400$meV (obtained from band calculations[56,57]), and the NN spin exchange constant $J \approx 130$meV (determined both from band calculations[56] and from fitting Raman scattering[58] and neutron scattering[59] experiments), the tJ model reproduces the spectral dispersion of half-filled cuprates along $(0,0) - (\pm\pi, \pm\pi)$.[2,52] Since $t \approx 3J$, holes move faster than the underlying spin background and, thus, the tJ model band width is renormalized down to $\approx 2.2J \approx 300$meV, in accordance with experiments.[18,25,27,40,42,44,60] We emphasize that the above agreement between the (pure) tJ model and half-filled cuprates only applies to the dispersion along $(0,0) - (\pm\pi, \pm\pi)$. In fact, if $t' = t'' = 0$ the tJ model dispersion along $(0,\pi) - (\pi, 0)$ is almost flat,[2,40,52] quite unlike the experimental band dispersion along $(0,\pi) - (\pi, 0)$, whose width is of order J.[27,40] Remarkably, band calculations also estimate $t' \approx -2t'' \approx -J$,[57] in which case the corresponding $tt't''J$ model reproduces the dispersion width both along $(0,0) - (\pm\pi, \pm\pi)$ and $(0,\pi) - (\pi, 0)$. Interestingly, the values of t' and t'' are predicted by band theory to differ between different cuprate families[57] in consonance with the observed differences in the $(0,\pi) - (\pi, 0)$ dispersion width between these same families.[60] The above facts offer strong support to the applicability of the $tt't''J$ model within the present context.

3.2.2. *Hole doped compounds*

We now focus on the evolution of the aforementioned dispersive features observed in undoped compounds as these materials are (hole-)doped away from half-filling. Since the dispersion of the undoped parent compounds resembles a gapped $d_{x^2-y^2}$-wave symmetry dispersion [in the sense that its energy is higher at $(0,\pi)$ and $(\pi, 0)$ than at $(\frac{\pi}{2}, \frac{\pi}{2})$], one could naively expect that, upon doping, the AF state's single hole dispersion continuously evolves into the SC quasiparticles' $d_{x^2-y^2}$-wave dispersion. This scenario turns out to be countered by experimental evidence concerning both data close to the Brillouin zone boundary and data around the nodal direction, as we detail below.

First, we refer to data close to the Brillouin zone boundary. In particular, we focus on the broad spectral line shape at $(0, \pi)$ which, in undoped samples, peaks around 300meV. The energy at which the maximum of this spectral hump is observed decreases continuously as materials are doped away from half-filling into the SC state.[27,60] Still, in the SC state, the above mentioned peak's energy does not match the SC gap energy as defined by the sharp SC quasiparticle peak at $(0, \pi)$. Instead, the latter peak develops at a lower energy, namely, the leading edge midpoint energy of the aforementioned spectral hump.[61,62] Hence, the $(0, \pi)$ spectrum naturally defines two different energy scales which became known as the high energy pseudogap (the energy at the hump maximum) and the low energy pseudogap (the leading edge midpoint energy).[63] As shown in Fig. 62 of Ref. 27, these energy scales differ by a factor of $\times 4$, indicating that the dispersion from the AF insulators does not evolve into that of the SC quasiparticles. We can even argue that the high energy pseudogap scale is unrelated to superconductivity. Indeed, experiments in different materials show that this energy scale changes with the values of t' and t''[60] (as estimated by band theory arguments) in a way that is consistent with plain tight-binding arguments. Additional evidence for the connection between the aforementioned high energy band and short-range AF correlations comes from ARPES data showing similar high energy dispersions along the perpendicular directions $(0, \pi) - (0, 0)$ and $(0, \pi) - (\pi, \pi)$.[64]

The above evidence for two separate energy scales is further complemented by data concerning the nodal direction spectral lineshapes. Notably, the U-shaped dispersion observed along the diagonal direction in undoped materials is also resolved by experiments in SC hole doped compounds.[18,42] As shown, for instance, in Fig. 4 of Ref. 18 and in Fig. 4(a) of Ref. 42, this dispersion evolves smoothly with doping, always remaining far below the Fermi energy. In addition, a low energy band that disperses linearly across the Fermi level along the nodal direction emerges in doped compounds, as depicted in Fig. 4(a) of Ref. 42. The latter dispersion, whose spectral peaks develop above the AF band as the CuO_2 layers are doped away from half-filling, is characteristic of $d_{x^2-y^2}$-wave superconductors. The above experimental evidence is observed both in NaCCOC[18,42] and LSCO,[24,25] and suggests the existence of states inside the Mott gap that lose spectral weight as the insulating composition is approached.

From all the above, the spectral function of hole underdoped samples suggests that the AF and $d_{x^2-y^2}$-wave dispersions coexist at different energies: the former, defined by the high-energy spectral hump, fades away as the density of doped holes increases; the latter, defined by the Dirac-like nodal quasiparticles, loses its spectral intensity as the doping level vanishes. The existence of these two dispersive features argues in favor of the presence of both AF and dSC short-range correlations in cuprate superconductors (as described by the doped-carrier MF theory which we discuss in this paper).

The above evidence for two spectral dispersions is identified throughout the Brillouin zone[21] and, thus, it defines a new energy scale, namely, the one that separates the two spectral features. This is the dip energy that intervenes between the low energy quasiparticle peaks and the high energy spectral hump and, as seen in Fig. 4 of Ref. 21, it occurs at $\approx$ 70meV $\approx \frac{J}{2}$. As documented by experiments, a kink appears in the momentum distribution curve derived dispersion at the dip energy scale.[17,18,20] In the literature, this kink is considered to be a manifestation of a bosonic mode whose nature has been the topic of a vivid debate in recent years.[19–21,27,65] In the context of the present work, the kink derives from the coexistence of two distinct short-range correlations, each manifested in separate energy ranges.

An additional remarkable feature displayed by ARPES measurements in underdoped samples has to do with the strong momentum dependence of the spectral weight distribution.[17,18,24,25,66] Above we refer to experimental data showing the development of peaks close to the Fermi level upon hole doping. The same experiments also show that these peaks appear first in the nodal direction and are only observed close to the antinodes at higher hole densities (see Fig. 3 of Ref. 24). As a result, the low energy spectral weight forms arcs around the nodal direction, as observed in various materials.[18,24,25] This anisotropy in momentum space, which consists in the robustness of nodal quasiparticles almost into the insulator regime and in the rapid loss of antinodal quasiparticle coherence upon decreasing hole doping, became known as the nodal-antinodal dichotomy, and is consistent with a diversity of experimental observations: while tunneling evidence for quasiparticle interference effects,[67] thermal conductivity[14] and c-axis penetration depth measurements[68] support the survival of nodal excitations in the deeply underdoped regime, the absence of coherent peaks in tunneling data[69–72] of underdoped samples indicates that quasiparticles close to the BZ boundary are incoherent in the same limit; Raman scattering,[73] optical transient grating spectroscopy,[74] c-axis penetration depth,[75] optical conductivity[76] and c-axis resistivity[77] measurements are all consistent with scattering processes which are anisotropic throughout the Brillouin zone; the disparity between different critical magnetic fields in the cuprate materials suggests that the coupling between nodal (antinodal) quasiparticles and a magnetic field is orbital-like (Zeeman-like) in character.[78,79] Furthermore, all the experimental evidence supports that the nodal-antinodal dichotomy is sharper in more underdoped samples.

3.2.3. *Electron doped compounds*

Despite less studied, electron doped cuprates like Ce doped Nd_2CuO_4,[26,80,81] $PrLaCuO_4$,[82] and Sm_2CuO_4,[46] have also been the subject of ARPES studies. Since ARPES only probes occupied electronic states, in electron doped materials these experiments identify both the lower Hubbard band, which is approximately 1.3eV below the Fermi energy,[26] and the bottom of the upper Hubbard band, which is intercepted by the chemical potential near $(0, \pi)$ and $(\pi, 0)$.[26] This state of affairs is to be contrasted with that of hole underdoped materials, in which case spectral signatures of doped holes appear in the nodal region. This particle-hole asymmetry is well understood within the context of the $tt't''J$ model:[53,54,83] a particle-hole transformation changes the sign of t' and t'', which control the dispersion along $(0, \pi) - (\pi, 0)$; changing the sign of these parameters switches the dispersion maximum and minimum between the wave-vectors $(\frac{\pi}{2}, \frac{\pi}{2})$ and $(0, \pi)$. The presence of electron pockets in these materials is consistent with the rigid filling of the upper Hubbard band in the AF state and is reproduced by numerical calculations of the $tt't''J$ model.[53,84] Since AF long-range order survives up to $x \approx 0.10 - 0.14$, these spectral weight pockets are observed in a wide doping range.[26,81]

Given that the chemical potential shifts to the bottom of the upper Hubbard band upon electron doping, ARPES has access to the full correlation gap and detects how the spectral weight is transfered from the lower Hubbard band into the mid-gap region close to the Fermi level. In particular, ARPES data displays the build up of spectral intensity around the nodal direction as the electron concentration is increased and these materials approach the SC state.[26,46,80–82] This spectral weight approaches the Fermi level upon doping and, in the SC phase, it defines the gapless Dirac dispersion characteristic of $d_{x^2-y^2}$-wave superconductors. Hence, the electron spectral function in electron doped SC samples consists of two different contributions at low energy: it displays the spectral features close to the Brillouin zone boundary that evolve continuously from the pockets at $(0, \pi)$ and $(\pi, 0)$ in the AF phase;[26,80,81] in addition, it shows the spectral weight that arises from the nodal Bogoliubov quasiparticles. These two separate contributions provide evidence for both AF and dSC local correlations which, in the electron doped regime, manifest themselves in distinct momentum space regions.

3.3. Doped-carrier approach of the $tt't''J$ model

In this paper, we recur to the doped-carrier basis of the $tt't''J$ model to address the doping dependence of the single electron spectral function of doped Mott insulators in general, with a particular emphasis on the cuprate underdoped regime's phenomenology. The use of the $tt't''J$ model is partly justified by the evidence discussed in Sec. 3.2. For a more thorough discussion of this model and the subsequent comparison to the cuprate's phenomenology, we defer the reader to the extensive reviews in Refs. 2 and 3. Ultimately, we want to handle the above model analytically, hence, we make use of the slave-particle machinery, namely, we resort to the doped-carrier framework of the generalized-tJ model. This framework was originally introduced in Ref. 37, and it is extensively discussed in a number of subsequent papers.[38,85–87] Unlike other slave-particle approaches,[32,34,36] the above formalism leads to a fully fermionic mean-field theory and, therefore, its central advantage is the ease to describe fermionic quasiparticles. In addition, and to our best knowledge, this approach is the most successful of all slave-particle approaches in describing ARPES experiments (see Ref. 37 as well as the remaining of this paper), scanning tunneling microscopy (STM) experiments,[85] and penetration depth experiments[87] in the cuprate superconductors. Furthermore, the doped-carrier approach is also semi-quantitatively consistent with the observed phase diagram of both hole and electron doped cuprates.[38]

3.3.1. *Doped-carrier framework*

In what follows, we consider the two-dimensional $tt't''J$ Hamiltonian

$$H_{tJ} = J \sum_{\langle ij\rangle \in NN} \left(\boldsymbol{S}_i.\boldsymbol{S}_j - \frac{1}{4}\mathcal{P}n_i n_j \mathcal{P} \right) - \\ - \sum_{\langle ij\rangle} t_{ij} \mathcal{P} \left(c_i^\dagger c_j + c_j^\dagger c_i \right) \mathcal{P} \tag{3.1}$$

where $t_{ij} = t, t', t''$ for first, second and third nearest neighbor (NN) sites respectively. $c_i^\dagger = [c_{i,\uparrow}^\dagger c_{i,\downarrow}^\dagger]$ is the electron creation spinor operator, $n_i = c_i^\dagger c_i$ and $\boldsymbol{S}_i = c_i^\dagger \boldsymbol{\sigma} c_i$ are the electron number and spin operators and $\boldsymbol{\sigma}$ are the Pauli matrices. The operator $\mathcal{P}$ projects out states with on-site double electron occupancy and, therefore, the $tt't''J$ model Hilbert space consists of states where every site has either a spin-1/2 or a vacancy [a].

In the doped-carrier formulation of the generalized-tJ model, the projected

[a] The above applies to the hole doped case. In the electron doped case, the operator $\mathcal{P}$ rather projects out states with vacant sites, in which case the corresponding $tt't''J$ model Hilbert space consists of states where every site either has a spin-1/2 or is doubly occupied.

electron operators are recast as [b]

$$\mathcal{P}c^{\dagger}_{i,\sigma}\mathcal{P} = s_{\sigma}\frac{1}{\sqrt{2}}\widetilde{\mathcal{P}}\left[\left(\frac{1}{2} + s_{\sigma}\widetilde{S}^{z}_{i}\right)d_{i,-\sigma} - \widetilde{S}^{s_{\sigma}}_{i}d_{i,\sigma}\right]\widetilde{\mathcal{P}} \tag{3.2}$$

where $s_{\sigma} = (+1), (-1)$ for $\sigma = \uparrow, \downarrow$. In Eq. (3.2) we introduce the lattice spin-1/2 operators $\widetilde{\boldsymbol{S}}_i$ and the fermionic *dopon* creation spinor operators $d^{\dagger}_i = [d^{\dagger}_{i,\uparrow} d^{\dagger}_{i,\downarrow}]$. The latter carry unit electric charge and spin-1/2, and thus describe electron-like excitations of the strongly correlated system. We also introduce the projection operator $\widetilde{\mathcal{P}} = \prod_i(1 - d^{\dagger}_{i,\uparrow}d_{i,\uparrow}d^{\dagger}_{i,\downarrow}d_{i,\downarrow})$ which enforces the no-double-occupancy constraint for dopons. We remark that the density of carriers doped away from half-filling x is equal to the dopon density $x = \frac{1}{N}\sum_i \widetilde{\mathcal{P}}d^{\dagger}_i d_i\widetilde{\mathcal{P}}$, where N is the number of sites. Since we are interested in the low doping regime, the dopon density is small and we relax the above no-double-occupancy constraint for dopons when we construct the mean-field theory in Sec. 3.3.2.[37,38]

The $tt't''J$ model Hamiltonian can be written in terms of the lattice spin operators operators $\widetilde{\boldsymbol{S}}_i$ and the dopon operators $d^{\dagger}_i$ and d_i upon replacing Eq. (3.2) in Eq. (3.1).[37,38] The resulting expression is $H_{tJ} = H_J + H_t$, where the Heisenberg interaction is

$$H_J = J\sum_{\langle ij\rangle\in NN}\left(\widetilde{\boldsymbol{S}}_i.\widetilde{\boldsymbol{S}}_j - \frac{1}{4}\right)\widetilde{\mathcal{P}}\left(1 - d^{\dagger}_i d_i\right)\left(1 - d^{\dagger}_j d_j\right)\widetilde{\mathcal{P}} \tag{3.3}$$

and the hopping term is [c]

$$H_t = \sum_{\langle ij\rangle}\frac{t_{ij}}{2}\widetilde{\mathcal{P}}\left[\left(d^{\dagger}_i\boldsymbol{\sigma}d_j\right)\cdot\left(i\widetilde{\boldsymbol{S}}_i\times\widetilde{\boldsymbol{S}}_j - \frac{\widetilde{\boldsymbol{S}}_i + \widetilde{\boldsymbol{S}}_j}{2}\right) + \right.$$
$$\left. + \frac{1}{4}d^{\dagger}_i d_j + d^{\dagger}_i d_j\widetilde{\boldsymbol{S}}_i.\widetilde{\boldsymbol{S}}_j + h.c.\right]\widetilde{\mathcal{P}} \tag{3.4}$$

3.3.2. *Doped-carrier mean-field theory*

A doped-carrier mean-field theory is constructed upon applying a decoupling scheme to Eq. (3.3) and Eq. (3.4) that yields a mean-field quadratic Hamiltonian in the end. Here, we follow the scheme developed in Refs. 37 and 38, which overview in this section [d].

[b]In the electron doped case, the projected electron operator that acts on the Hilbert space that consists of states with no empty sites can be recast as $\mathcal{P}c^{\dagger}_{i,\sigma}\mathcal{P} = \frac{1}{\sqrt{2}}\widetilde{\mathcal{P}}\left[\left(\frac{1}{2} - s_{\sigma}\widetilde{S}^{z}_{i}\right)d^{\dagger}_{i,\sigma} - \widetilde{S}^{s_{\sigma}}_{i}d^{\dagger}_{i,-\sigma}\right]\widetilde{\mathcal{P}}$.

[c]In the electron doped case we rather use the projected electron operator that acts on the Hilbert space that consists of states with no empty sites. As a result, we obtain the same expression for H_t, apart from an overall $(-)$ sign.

[d]Ref. 86 extends this scheme to handle both hole and electron doped systems, as well as interactions between hole and electron doped systems.

In order to obtain the fully fermionic mean-field theory we recast the lattice spin operators $\widetilde{\boldsymbol{S}}_i$ in terms of chargeless spin-1/2 fermionic spinon operators $f_i^\dagger = [f_{i,\uparrow}^\dagger f_{i,\downarrow}^\dagger]$, as $\widetilde{\boldsymbol{S}}_i \equiv \frac{1}{2} f_i^\dagger \boldsymbol{\sigma} f_i$. It is also convenient to introduce the Nambu representation of the above fermionic operators, namely, $\psi_i^\dagger = [\psi_{i,1}^\dagger \psi_{i,2}^\dagger] = [f_{i,\uparrow}^\dagger f_{i,\downarrow}]$ and $\eta_i^\dagger = [\eta_{i,1}^\dagger \eta_{i,2}^\dagger] = [d_{i,\uparrow}^\dagger d_{i,\downarrow}]$. Given the above notation, we now introduce the expectation values used in our decoupling scheme. At low enough doping the tJ model develops AF order,[28] which yields non-zero staggered magnetization order parameters

$$
\begin{aligned}
m &= \frac{1}{2}\langle \psi_i^\dagger \psi_i - 1\rangle \\
n &= -\langle \sum_{\nu=2,3} t_\nu \sum_{\hat{u}\in\nu\, NN} \eta_i^\dagger \eta_{i+\hat{u}} + h.c.\rangle
\end{aligned}
\tag{3.5}
$$

where $\hat{u} = \pm\hat{x} \pm \hat{y}$ for $\nu = 2$ and $\hat{u} = \pm 2\hat{x}, \pm 2\hat{y}$ for $\nu = 3$, and t_2 and t_3 are effective hopping parameters whose values we address further below. In addition to the AF order parameter, we also introduce the singlet bond order parameters which capture the spin liquid correlations, namely,

$$
\chi = \langle \psi_i^\dagger \sigma_z \psi_j\rangle \; ; \;\; \Delta = (-)^{i_y - j_y} \langle \psi_i^\dagger \sigma_x \psi_j\rangle \tag{3.6}
$$

with $\langle ij\rangle \in NN$. Finally, the Bose amplitudes which describe the quantum coherence between the doped carriers and the spin background are given by

$$
b_0 = \langle f_i^\dagger d_i\rangle \; ; \;\; b_1 = \langle \frac{3}{16} \sum_{\nu=1,2,3} t_\nu \sum_{\hat{u}\in\nu\, NN} f_i^\dagger d_{i+\hat{u}}\rangle \tag{3.7}
$$

where $t_1 = t$ and $\hat{u} = \pm\hat{x}, \pm\hat{y}$ for $\nu = 1$.

In terms of these order parameters, the doped-carrier mean-field Hamiltonian is given by[37,38]

$$
\begin{aligned}
H_{tJ}^{MF} = &\sum_{\boldsymbol{k}} \begin{bmatrix} \psi_{\boldsymbol{k}}^\dagger & \eta_{\boldsymbol{k}}^\dagger \end{bmatrix} \begin{bmatrix} \alpha_{\boldsymbol{k}}^z \sigma_z + \alpha_{\boldsymbol{k}}^x \sigma_x & \beta_{\boldsymbol{k}} \sigma_z \\ \beta_{\boldsymbol{k}} \sigma_z & \gamma_{\boldsymbol{k}} \sigma_z \end{bmatrix} \begin{bmatrix} \psi_{\boldsymbol{k}} \\ \eta_{\boldsymbol{k}} \end{bmatrix} + \\
&+ \sum_{\boldsymbol{k}} \left(\nu^\psi \psi_{\boldsymbol{k}+\boldsymbol{Q}}^\dagger \psi_{\boldsymbol{k}} + \nu_{\boldsymbol{k}}^\eta \eta_{\boldsymbol{k}+\boldsymbol{Q}}^\dagger \eta_{\boldsymbol{k}} \right) - E_0
\end{aligned}
\tag{3.8}
$$

In the above equation, $\alpha_{\boldsymbol{k}}^z = -(\frac{3\tilde{J}}{4}\chi - t_1 x)(\cos k_x + \cos k_y) + a_0$ where a_0 is the Lagrange multiplier that ensures $\langle f_i^\dagger f_i\rangle = 1$, and $\tilde{J} = (1-x)^2 J$. In addition, $\alpha_{\boldsymbol{k}}^x = -\frac{3\tilde{J}}{4}\Delta(\cos k_x - \cos k_y)$, $\beta_{\boldsymbol{k}} = \frac{3b_0}{8}[t_1(\cos k_x + \cos k_y) + 2t_2 \cos k_x \cos k_y + t_3(\cos 2k_x + \cos 2k_y)] + b_1$, $\gamma_{\boldsymbol{k}} = t_2 \cos k_x \cos k_y + \frac{t_3}{2}(\cos 2k_x + \cos 2k_y) - \mu_d$, $\nu^\psi = 2(n - 0.34\tilde{J}m)$, and $\nu_{\boldsymbol{k}}^\eta = -2m(\gamma_{\boldsymbol{k}} + \mu_d)$. Furthermore, $E_0 = -[\frac{3\tilde{J}N}{4}(\chi^2 + \Delta^2) + 0.68\tilde{J}Nm^2 - 4Nmn - 2Nb_0 b_1 - N\mu_d]$, where we introduce the dopon chemical potential μ_d that sets the doping level $\langle d_i^\dagger d_i\rangle = x$.

Even though the $tt't''J$ model is intrinsically a one-band model, the above mean-field approach contains two different families of spin-1/2 fermions, namely spinons and dopons, and thus presents a two-band description of the same model. As a result, H_{tJ}^{MF} has a total of four fermionic bands described by the eigenenergies

$$\begin{aligned} \epsilon_{1,\boldsymbol{k}}^{\pm} &= \pm\sqrt{\rho_{\boldsymbol{k}} - \sqrt{\delta_{\boldsymbol{k}}}} \\ \epsilon_{2,\boldsymbol{k}}^{\pm} &= \pm\sqrt{\rho_{\boldsymbol{k}} + \sqrt{\delta_{\boldsymbol{k}}}} \end{aligned} \tag{3.9}$$

where

$$\begin{aligned} \delta_{\boldsymbol{k}} &= \beta_{\boldsymbol{k}}^2 \left[(\gamma_{\boldsymbol{k}} + \alpha_{\boldsymbol{k}}^z)^2 + (\alpha_{\boldsymbol{k}}^x)^2 \right] + \frac{1}{4} \left[\gamma_{\boldsymbol{k}}^2 - (\alpha_{\boldsymbol{k}}^x)^2 - (\alpha_{\boldsymbol{k}}^z)^2 \right]^2 \\ \rho_{\boldsymbol{k}} &= \beta_{\boldsymbol{k}}^2 + \frac{1}{2} \left[\gamma_{\boldsymbol{k}}^2 + (\alpha_{\boldsymbol{k}}^x)^2 + (\alpha_{\boldsymbol{k}}^z)^2 \right] \end{aligned} \tag{3.10}$$

In the absence of spinon-dopon mixing, *i.e.* when $b_0, b_1 = 0$, the bands $\epsilon_{1,\boldsymbol{k}}^{\pm}$ and $\epsilon_{2,\boldsymbol{k}}^{\pm}$ describe the spinon d-wave dispersion [which amounts to $\pm\sqrt{(\alpha_{\boldsymbol{k}}^x)^2 + (\alpha_{\boldsymbol{k}}^z)^2}$], as well as the dispersion of a hole surrounded by staggered local moments (note that the latter is given by $\gamma_{\boldsymbol{k}}$, which includes only intra-sublattice hopping processes, as appropriate in the one-hole limit of the $tt't''J$ model[2,88,89]). Upon the hybridization of spinons and dopons the eigenbands $\epsilon_{1,\boldsymbol{k}}^{\pm}$ and $\epsilon_{2,\boldsymbol{k}}^{\pm}$ differ from the bare spinon and dopon bands by a term of order $b_0^2, b_1^2, b_0 b_1 \sim x$. In particular, the lowest energy bands $\epsilon_{1,\boldsymbol{k}}^{\pm}$ are d-wave-like with nodal points along the $(0,0) - (\pm\pi, \pm\pi)$ directions and describe electronic excitations that coherently hop between NN sites. The highest energy bands $\epsilon_{2,\boldsymbol{k}}^{\pm}$ are mostly derived from the bare dopon bands and, therefore, describe excitations with reduced NN hopping.

Finally, we note that the explicit form of the above mean-field Hamiltonian depends on the values of t_2 and t_3, which are determined phenomenologically upon fitting the high energy dispersion $\epsilon_{2,\boldsymbol{k}}$ to both numerical results and cuprate ARPES data [e]. Following Ref. 38, we assume these parameters depend on the doping level x, being given by

$$\begin{aligned} t_2 &= t_J + r(x)t' \\ t_3 &= \frac{t_J}{2} + r(x)t'' \end{aligned} \tag{3.11}$$

[e] We remark that, even though t_2 and t_3 are used to control the high energy dispersion $\epsilon_{2,\boldsymbol{k}}$ in consonance with numerical and experimental data, there is no such direct numerical/experimental input on the low energy band $\epsilon_{1,\boldsymbol{k}}$, and all its properties result from the theory. Interestingly, Refs. 37 and 85, together with the present paper, find that a variety of low energy spectral properties associated with the $\epsilon_{1,\boldsymbol{k}}$ bands are consistent with ARPES and STM experiments in both hole and electron doped cuprates.

where we take $t_J = J$ and $r(x) = \left(1 - \frac{x}{0.3}\right)$. In the remaining of the paper, we also use $t' \approx -2t'' \approx -J$[40,57] [f].

3.4. Doped-carrier mean-field electron spectral function

As Sec. 3.3 details, the doped-carrier mean-field theory describes the physical degrees of freedom of doped Mott insulators in terms of spinon and dopon operators. The former are spin-1/2 excitations of the spin background, while the latter correspond to doped carriers in the deeply underdoped limit (the specific physical picture being that of vacancies surrounded by staggered local moments). The above mean-field theory introduces an additional object, namely, the spin-singlet pair of a spinon and a dopon, which accounts for vacancies surrounded by a singlet spin configuration that enhances the vacancy mobility. Therefore, the doped-carrier mean-field theory captures the interplay between different local spin correlations and the single hole dynamics which, ultimately, translates into the properties of the electron spectral function. The purpose of this section is to compute the doped-carrier mean-field electron spectral function, therefore addressing how AF and dSC short-range correlations are reflected in the spectral properties of superconducting electrons close to the Mott insulating state. We also compare our results to ARPES data from hole and electron doped cuprates, thus proposing a specific theoretical interpretation of experimental results.

3.4.1. *Mean-field electron operator*

In order to proceed with the computation of the doped-carrier mean-field theory single-electron spectral function, we simplify the expression for the projected electron operators. Specifically, we drop the projection operators $\widetilde{\mathcal{P}}$ and introduce appropriate fermionic averages to approximately recast Eq. (3.2) in terms of a single fermionic operator. Hence, we introduce the mean-field electron annihilation operator [g]

$$c^{MF}_{i,\sigma} \equiv \frac{1}{\sqrt{2}} \left(d^{\dagger}_{i,-\sigma} + b_0 f^{\dagger}_{i,-\sigma} \right) \tag{3.12}$$

which we use to compute the coherent contribution to the single-electron spectral function (note that to obtain the incoherent contribution, various convolution integrals neglected in this paper must be dealt with).

[f] If we use the projected electron operators defined above for the electron doped regime, t' and t'' effectively change sign.[53,90]

[g] In the electron doped regime, Eq. (3.12) rather represents an electron creation operator.

As a sanity check on the form of $c_{i,\sigma}^{MF}$, we show that it is consistent with important single-electron spectral weight sum rules. In fact, if we use the relations

$$\begin{aligned}\langle d_{i,-\sigma}^\dagger d_{i,-\sigma}\rangle &= 1-\langle d_{i,-\sigma} d_{i,-\sigma}^\dagger\rangle = \frac{x}{2}\\ \langle f_{i,-\sigma}^\dagger d_{i,-\sigma}\rangle &= \langle d_{i,-\sigma}^\dagger f_{i,-\sigma}\rangle = b_0\\ \langle f_{i,-\sigma}^\dagger f_{i,-\sigma}\rangle &= \langle f_{i,-\sigma} f_{i,-\sigma}^\dagger\rangle = \frac{1}{2}\end{aligned} \tag{3.13}$$

it is a trivial matter to show that

$$\begin{aligned}\langle c_{i,\sigma}^{MF}\left(c_{i,\sigma}^{MF}\right)^\dagger\rangle &= \frac{x+3b_0^2}{4}\\ \langle\left(c_{i,\sigma}^{MF}\right)^\dagger c_{i,\sigma}^{MF}\rangle &= \frac{2-x-b_0^2}{4}\end{aligned} \tag{3.14}$$

In the instance of full coherence, $b_0 = \sqrt{x}$, and the expressions in Eq. (3.14) reduce to $\langle c_{i,\sigma}^{MF}\left(c_{i,\sigma}^{MF}\right)^\dagger\rangle = x$ and $\langle\left(c_{i,\sigma}^{MF}\right)^\dagger c_{i,\sigma}^{MF}\rangle = \frac{1-x}{2}$. These are the correct $tt't''J$ model sum rules.[91] (It is relevant to note that the above sum rules are broken in other slave-particle mean-field schemes.)

3.4.2. *Mean-field single-electron spectral function*

By definition, the single-electron spectral function is

$$A(\boldsymbol{k},\omega) = -\frac{1}{\pi}\mathrm{Im}G_c(\boldsymbol{k},\omega+i0^+) \tag{3.15}$$

where the one-electron Green's function (in time and real space variables) is given by

$$G_c(i,t) = -\langle T_t\left[c_{i,\sigma}(t)c_{0,\sigma}^\dagger(0)\right]\rangle \tag{3.16}$$

Using the mean-field approximation for the electron operator in Eq. (3.12), the $T=0$ the mean-field spectral function reduces to a collection of δ-peaks and can be written as

$$A(\boldsymbol{k},\omega) = \sum_{\substack{l\in\{1,2\}\\ s\in\{+,-\}}} Z_{l,\boldsymbol{k}}^s\delta\left(\omega-\epsilon_{l,-\boldsymbol{k}}^s\right) \tag{3.17}$$

in the hole doped case and

$$A(\boldsymbol{k}+(\pi,\pi),\omega) = \sum_{\substack{l\in\{1,2\}\\ s\in\{+,-\}}} Z_{l,\boldsymbol{k}}^s\delta\left(\omega+\epsilon_{l,-\boldsymbol{k}}^s\right) \tag{3.18}$$

in the electron doped case [h]. The spectral weight in each peak is given by

$$Z^{s}_{l,\boldsymbol{k}} = \left| {}^{N}\langle 0| \left[\frac{1}{\sqrt{2}} \left(d^{\dagger}_{\boldsymbol{k},\downarrow} + b_0 f^{\dagger}_{\boldsymbol{k},\downarrow} \right) \right] \left(\phi^{s}_{l,-\boldsymbol{k}} \right)^{\dagger} |0\rangle^{N} \right|^{2} \tag{3.19}$$

where the notation in Eq. (3.19) is such that $(\phi^{s}_{l,\boldsymbol{k}})^{\dagger}$ is the appropriate linear combination of the Nambu operators $\psi_{\boldsymbol{k},\alpha}$ and $\eta_{\boldsymbol{k},\beta}$ that creates the mean-field quasiparticle eigenstate with momentum $\boldsymbol{k}$ and eigenenergy $\epsilon^{s}_{l,\boldsymbol{k}}$, where $l \in \{1,2\}$ and $s \in \{+,-\}$ [i]. Eq. (3.19) can be shown to reduce to

$$\begin{aligned} \frac{1+b_0^2}{2\,Z^{s}_{l,\boldsymbol{k}}} &= 1 + \frac{\left(\gamma_{\boldsymbol{k}} + \beta_{\boldsymbol{k}} b_0 + \epsilon^{s}_{l,\boldsymbol{k}}\right)^2}{\left(\beta_{\boldsymbol{k}} - \gamma_{\boldsymbol{k}} b_0 - \epsilon^{s}_{l,\boldsymbol{k}} b_0\right)^2} + \\ &+ \frac{(1+b_0)^2 \left[\left(\epsilon^{s}_{l,\boldsymbol{k}} + \alpha^{z}_{\boldsymbol{k}}\right)\left(\epsilon^{s}_{l,\boldsymbol{k}} + \gamma_{\boldsymbol{k}}\right) - \beta^2_{\boldsymbol{k}}\right]^2 \left[\left(\epsilon^{s}_{l,\boldsymbol{k}} - \gamma_{\boldsymbol{k}}\right)^2 + \beta^2_{\boldsymbol{k}}\right]}{\left(\alpha^{x}_{\boldsymbol{k}}\right)^2 \left(\epsilon^{s}_{l,\boldsymbol{k}} - \gamma_{\boldsymbol{k}}\right)^2 \left[\beta_{\boldsymbol{k}} - b_0\left(\epsilon^{s}_{l,\boldsymbol{k}} + \gamma_{\boldsymbol{k}}\right)\right]^2} \end{aligned} \tag{3.20}$$

As it follows from Eqs. (3.17) and (3.18), in this paper we approximate the spectral function of doped Mott insulators by a collection of δ-peaks and, thus, we completely miss the spectral broadening that characterizes strongly interacting systems, such as the cuprate compounds. In this context, we emphasize that in the present work we are not concerned with the broadening of the electron spectral function and, instead, we address: (i) the dispersion in energy-momentum space of spectral features; and (ii) the distribution of spectral intensity along these same spectral features. We note that the former can be addressed by the location of the δ-peaks in Eqs. (3.17) and (3.18), whereas the latter is addressed by the magnitude of the coherence factors given by Eq. (3.20). The physical meaning and significance of this approximation is discussed in Sec. 3.5.3.

3.4.3. *Hole doped case*

In what follows, we use Eq. (3.17) to list and discuss a variety of features of the doped-carrier electron spectral function, as of relevance to the cuprates' hole

---

[h] The wave vector (π,π) is added in (3.18) to make up for the fact that under a particle-hole transformation the hopping parameters change sign and the dispersions are shifted in momentum space by (π,π).

[i] Note that in order to obtain Eq. (3.19) we make use of the fact that the $T=0$ mean-field groundstate is $|GS\rangle_{MF} = \Pi_{\boldsymbol{k}} \left(\phi^{-}_{1,\boldsymbol{k}}\right)^{\dagger} \left(\phi^{-}_{2,\boldsymbol{k}}\right)^{\dagger} |0\rangle^{N}$, where $|0\rangle^{N}$ is the vacuum for the Nambu operators $\psi_{\boldsymbol{k},\alpha}$ and $\eta_{\boldsymbol{k},\beta}$.

(under)doped regime. The $tt't''J$ model parameters we use are $t = 3J$, and $t' = -2t'' = -J$,[40,57] which imply

$$t_2^{HD} = J - J\left(1 - \frac{x}{0.3}\right)$$
$$t_3^{HD} = \frac{J}{2} + \frac{J}{2}\left(1 - \frac{x}{0.3}\right) \tag{3.21}$$

3.4.3.1. *Dispersive features*

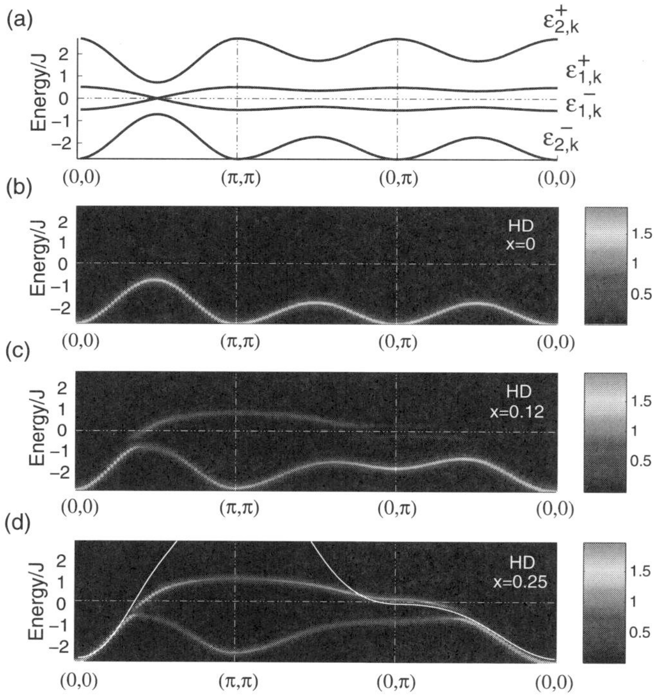

Fig. 3.1. Energy dispersions along $(0,0)-(\pi,\pi)-(0,\pi)-(0,0)$ at different hole doping levels. (a) Mean-field dispersions $\epsilon^{\pm}_{1,\boldsymbol{k}}$ and $\epsilon^{\pm}_{2,\boldsymbol{k}}$ for $x = 0$. In this case, $\epsilon^{\pm}_{1,\boldsymbol{k}}$ stand for the spinon bands and $\epsilon^{\pm}_{2,\boldsymbol{k}}$ for the dopon bands. (b)-(d) show the mean-field electron spectral function (3.17) with Lorentzian broadening $\Gamma = J/10$ for $x = 0$, $x = 0.12$ and $x = 0.25$ respectively. The solid white line in (d) plots the tight-binding dispersion $-2[t(\cos k_x + \cos k_y) + 2t' \cos k_x \cos k_y + t''(\cos 2k_x + \cos 2k_y) + 2t'''(\cos 2k_x \cos k_y + \cos 2k_y \cos k_x) + 2t'''' \cos 2k_x \cos 2k_y] - \mu$ with $t = 148.8$meV, $t' = -40.9$meV, $t'' = 13.0$meV, $t''' = 14.0$meV and $t'''' = -12.7$meV as obtained from the fitting to the experimental energy dispersions in Ref. 92. We chose μ to obtain $x = 0.25$ and use $J = 130$meV.

In Fig. 3.1(a) we plot the four mean-field bands for $x = 0$ along high symmetry directions. These are the d-wave paired spinon bands $\epsilon^{\pm}_{1,\boldsymbol{k}}$ and the dopon bands $\epsilon^{\pm}_{2,\boldsymbol{k}}$. For $x = 0$, the electron spectral function (3.17) vanishes for all bands but $\epsilon^{-}_{2,\boldsymbol{k}}$ [see Fig. 3.1(b)], which is the lower Hubbard band, and whose dispersion agrees with the self-consistent Born approximation in the undoped limit.[52,54,55] [The upper Hubbard band is not visible since these results apply to the $tt't''J$ model, where $U \to +\infty$.]

Away from half-filling spinons and dopons mix and spectral weight is transfered from the $\epsilon^{-}_{2,\boldsymbol{k}}$ band to the lower energy $\epsilon^{\pm}_{1,\boldsymbol{k}}$ bands [Figs. 3.1(c) and 3.1(d)]. The spinon-dopon hybridization leads to the appearance of spectral peaks above the insulating valence band (which is $\epsilon^{-}_{2,\boldsymbol{k}}$), hence inside the Mott-Hubbard gap. This result is consistent with the experimental observation of in-gap spectral features whose weight vanishes in the parent insulator compound.[18,24,25,42] Figs. 3.1(b)-3.1(d) also show that the low energy spectral weight in bands $\epsilon^{\pm}_{1,\boldsymbol{k}}$, which increases with x, is mostly transfered from the region around (π, π) and not from the region around $(0, 0)$. Physically, this fact reflects the weakening of local AF correlations with increasing hole concentration.

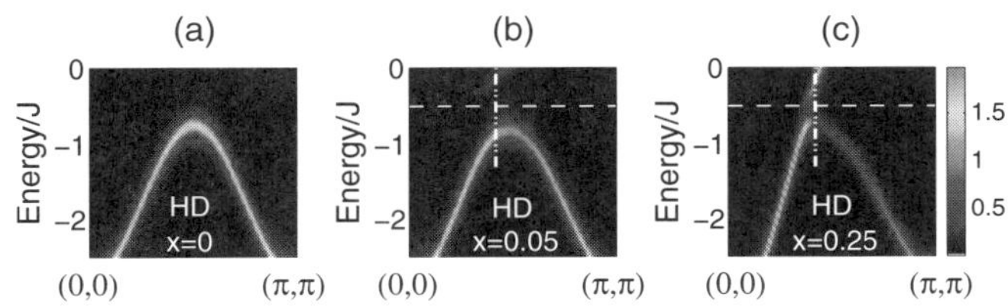

Fig. 3.2. Nodal direction electron spectral function (3.17) with Lorentzian broadening $\Gamma = J/10$ for doped hole density (a) $x = 0$, (b) $x = 0.05$ and (c) $x = 0.25$. The presence of two spectral features leads to a peak-dip-hump spectral structure along the vertical white dash-dotted line in (b) and (c). The horizontal white dashed line in (b) and (c) indicates the energy $\omega = J/2$ which lies at the dip energy and at which the nodal dispersion kink is observed (see main text).

Figs. 3.2(a)-3.2(c) focus on the evolution of the spectral function along the nodal direction between the undoped insulator ($x = 0$) and the overdoped Fermi liquid with a large Fermi surface ($x = 0.25$). As holes are introduced in the undoped system the dispersion characteristic of the AF insulator persists at high energy, while a second spectral feature emerges at low energy. The former disperses back in energy for momenta beyond $(\frac{\pi}{2}, \frac{\pi}{2})$, whereas the latter disperses linearly across the Fermi level ($\omega = 0$) close to $(\frac{\pi}{2}, \frac{\pi}{2})$, as seen in ARPES experiments.[42] Fig. 3.5(a) shows that the nodal point deviates from $(\frac{\pi}{2}, \frac{\pi}{2})$ toward $(0, 0)$. At low doping the Fermi point evolves linearly with x and, for $x > 0.10$ it remains around $k_x, k_y = 0.45\pi$. A similar pattern is observed experimentally in different cuprate families.[17,22,42,93]

In doped band insulators the chemical potential falls on top of the valence band and moves with doping in accordance with conventional rigid band filling of hole pockets. This picture is to be contrasted with the non-trivial behavior that follows the hybridization of spinons and dopons. In this case, the chemical potential remains in the insulating gap and, consequently, a new energy scale $\omega \approx \frac{J}{2}$ marked by the horizontal white dashed line in Figs. 3.2(b) and 3.2(c) appears. It separates the low energy dispersion reminiscent of the underlying SC correlations from the high energy spectral feature that reflects the local AF correlations which survive even in the absence of long-range AF order. We remark that such an energy scale is observed throughout the Brillouin zone[21] not only in underdoped cuprate samples[20,25,42] but also in overdoped materials.[94]

3.4.3.2. *Low energy spectral arcs*

The absence of hole Fermi pockets with area x raises the question of where in momentum space the hole density comes from. To answer this, we depict the doping evolution of the electron momentum distribution $n_{\boldsymbol{k}} = \int_{-\infty}^{0} d\omega A(\boldsymbol{k}, \omega)$ in Figs. 3.3(a) - 3.3(c). At low doping, most of the spectral weight comes from a small region around $(\frac{\pi}{2}, \frac{\pi}{2})$ and, in that sense, resembles the hole pocket scenario. More generally, the doping induced changes in $n_{\boldsymbol{k}}$ occur beyond the minimum gap locus, denoted by the black dashed line in Figs. 3.3(a) - 3.3(c), toward (π, π). By definition, this locus identifies the momentum vector $\boldsymbol{k}$ where the d-wave gap is lower along the direction defined by θ in Fig. 3.3(d). The variation of $n_{\boldsymbol{k}}$ across the minimum gap locus is particularly sharp in the overdoped regime [see Fig. 3.3(c) where $x = 0.20$] as expected from the proximity to the Fermi liquid regime.

The electron momentum distribution integrates the spectral function in energy space and, as such, misses the momentum dependence of the spectral weight transfered from the high energy band $\epsilon_{2,\boldsymbol{k}}^{-}$ to the low energy band $\epsilon_{1,\boldsymbol{k}}^{-}$. In order to address this dependence, in Figs. 3.3(d) - 3.3(f) we plot $Z_{1,\boldsymbol{k}}^{-}$ (which stands for the spectral weight of states in the low energy band $\epsilon_{1,\boldsymbol{k}}^{-}$). We find that it forms arcs of low energy spectral weight around the nodal direction, as observed in LSCO[24] and NaCCOC[18,25] experimental data. As depicted in these figures, the total spectral weight at low energy increases with x and, consequently, both the arc length and the spectral intensity in the arc region increase upon doping.

Figs. 3.4(a) - 3.4(c) show how the spectral weight of states in band $\epsilon_{1,\boldsymbol{k}}^{-}$ changes along the minimum gap locus for different doping values. Clearly, at low

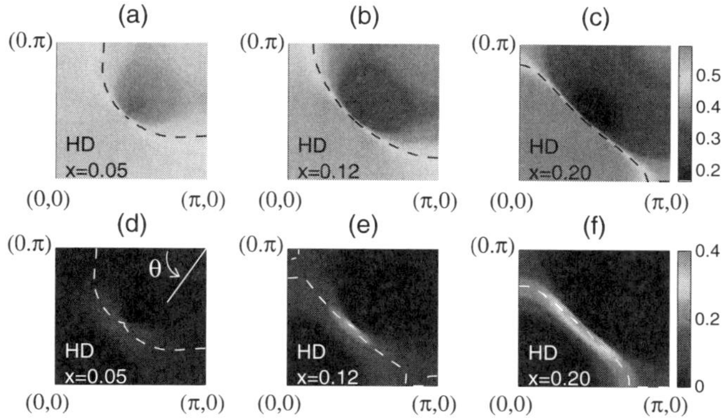

Fig. 3.3. (a)-(c) Electron momentum distribution $n_{\boldsymbol{k}} = Z^-_{1,\boldsymbol{k}} + Z^-_{2,\boldsymbol{k}}$ (top color scale). The black dashed line depicts the minimum gap locus. (d)-(f) Electron spectral weight $Z^-_{1,\boldsymbol{k}}$ of the $\epsilon^-_{1,\boldsymbol{k}}$ band states (bottom color scale). The white dashed line represents the maximum spectral weight locus. Doped hole density is $x = 0.05$ for (a) and (d), $x = 0.12$ for (b) and (e), and $x = 0.20$ for (c) and (f).

energy, states along the diagonal direction $[(0,0) - (\pm\pi, \pm\pi)]$ have more spectral weight than those close to the Brillouin zone boundary. This fact is consistent with the experimental nodal-antinodal dichotomy, a term motivated by the observation that underdoped cuprates display stronger quasiparticle features around the nodes than in the antinodal region.[17,18,25,66]

The mean-field spectral function is singular at the Dirac point and, in particular, the $Z^-_{1,\boldsymbol{k}}$ plots along the minimum gap locus are discontinuous at $\theta = \frac{\pi}{4}$. Indeed, in Figs. 3.4(a) - 3.4(c) the spectral weight at the nodal point is denoted by the symbol $(\times)$ above the curve defined by the spectral weight at other $\boldsymbol{k}$ vectors along the minimum gap locus. In order to avoid this discontinuity, we define the maximum spectral weight locus as the set of vectors $\boldsymbol{k}$ that maximize $Z^-_{1,\boldsymbol{k}}$ for each θ. This locus is plotted as the white dashed line in Figs. 3.3(d) - 3.3(f) and differs slightly from the the minimum gap locus [plotted in Figs. 3.3(a) - 3.3(c)]. The $Z^-_{1,\boldsymbol{k}}$ plot along the maximum spectral weight locus is continuous across $\theta = \frac{\pi}{4}$ and also displays the aforementioned differentiation between the nodal and antinodal regions [Figs. 3.4(d) - 3.4(f)].

An interesting question has to do with the spectral arcs' doping evolution. In order to quantify the doping dependence of the arc length ($\Delta\theta_{arc}$), we consider the full angular width at half maximum of the $Z^-_{1,\boldsymbol{k}}$ plots along both the minimum gap locus and the maximum spectral weight locus. In Fig. 3.5(b) we see that both definitions of arc length are consistent with two facts: (i) the arc length does

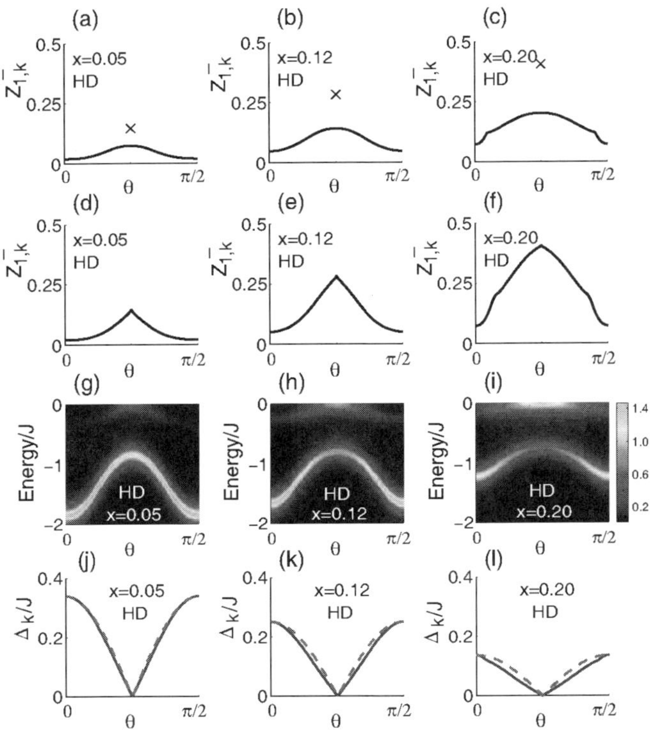

Fig. 3.4. Mean-field electron spectral function results for hole doping level $x = 0.05$ [(a),(d),(g) and (j)], $x = 0.12$ [(b),(e),(h) and (k)] and $x = 0.20$ [(c),(f),(i) and (l)]. (a)-(c) Electron spectral weight $Z^-_{1,\boldsymbol{k}}$ along minimum gap locus. The spectral weight is singular at $\theta = \pi/4$, and $Z^-_{1,\boldsymbol{k}}$ at the nodal point is depicted by the $\times$ symbol. (d)-(f) Electron spectral weight $Z^-_{1,\boldsymbol{k}}$ along maximum spectral weight locus. (g)-(i) Electron spectral function (3.17) along minimum gap locus. A Lorentzian broadening with $\Gamma = J/10$ is used. (j)-(l) $\epsilon^+_{1,\boldsymbol{k}}$ dispersion gap (solid line) and pure $d_{x^2-y^2}$-wave dispersion [$\Delta_{\boldsymbol{k}} = \Delta_0 \cos 2\theta_{\boldsymbol{k}}$] with same gap magnitude (dashed line) along minimum gap locus. Note that the pure $d_{x^2-y^2}$-wave dispersion lies above $\epsilon^+_{1,\boldsymbol{k}}$ implying $B < 1$ (see main text). The angular variable θ used to parameterize the points along the minimum gap locus and the maximum spectral weight locus is defined in Fig. 3.3(d).

not vanish in the zero doping limit; and (ii) the low energy spectral arcs elongate away from the nodal direction with increasing hole density. We also find that the quantitative estimate of the spectral arc length depends sensitively on whether we chose to measure it along the minimum gap locus or the maximum spectral weight locus. Interestingly, in Fig. 3.5(b) we compare both theoretical estimates with the arc length values deduced from Fig. 2(c) of Ref. 24 to find quantitative agreement between our mean-field results and both the doping dependence and the spread in

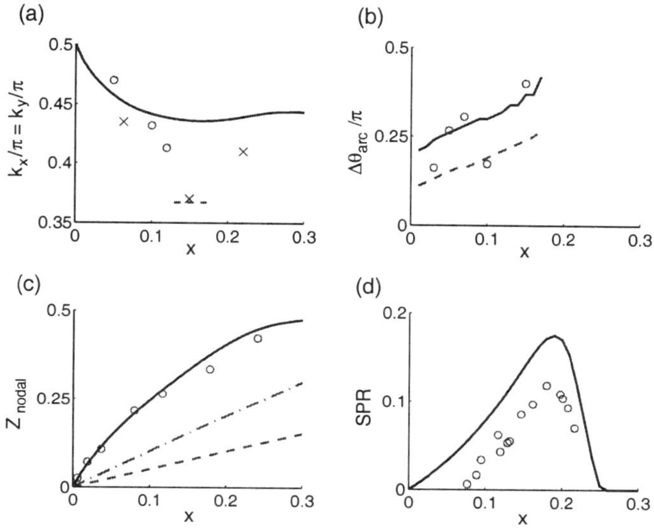

Fig. 3.5. (a) Mean-field theory results for the hole doping dependence of the nodal point location along the $(0, 0) - (\pi, \pi)$ direction (solid line). Comparison to BSCCO[93] (dashed line), LSCO[17,22] ($\times$) and NaCCOC[42] ($\circ$) results. (b) Low energy spectral weight arc length $\Delta\theta_{arc}$ as measured by the full angular width at half spectral weight maximum along minimum gap locus (solid line) and maximum spectral weight locus (dashed line). Comparison to LSCO results[24] ($\circ$). We only plot results up to $x = 0.17$, at which point the minimum gap locus changes from hole to electron topology. (c) Hole doping dependence of nodal spectral weight Z_{nodal} in the doped-carrier mean-field theory (solid line), the $U(1)$ slave-boson mean-field theory (dash-dotted line) and the $SU(2)$ slave-boson mean-field theory (dashed line). Comparison to variational Monte Carlo results for the tJ model with $t = 3J$[95] ($\circ$). (d) SC peak ratio (SPR), *i.e.* the relative intensity of the spectral function peak at $(0, \pi)$, as given by the theoretical ratio $Z^-_{1,(0,\pi)}/(Z^-_{1,(0,\pi)} + Z^-_{2,(0,\pi)})$ (solid line) and as measured in BSCCO[94,96] ($\circ$).

experimental data.

Figs. 3.3(d) - 3.3(f) and 3.4(a) - 3.4(f) demonstrate that the spectral weight intensity also increases with doping level. We now compare this doping dependence to that expected from other theoretical methods. Specifically, in Fig. 3.5(c) we plot the doping evolution of the nodal spectral weight as determined by variational Monte Carlo calculations, as well as by three different mean-field theories, namely, the doped-carrier, the $U(1)$ slave-boson, and the $SU(2)$ slave-boson mean-field theories. When compared to the slave-boson approaches, the doped-carrier mean-field results are seen to deviate from the pure linear behavior away from the $x \to 0$ limit. Furthermore, nodal states in the doped-carrier mean-field theory show enhanced quasiparticle behavior when compared to the results from slave-boson mean-field theory. Fig. 3.5(c) also depicts the almost exact quantitative agreement between the doped-carrier mean-field results and those obtained in the variational Monte Carlo approach to the tJ model with $t = 3J$[95] (note that the latter approach implements the projection onto the tJ model Hilbert space exactly). Even though this quantitative agreement has to be taken with a grain of salt (after all the two approaches use different values of t' and t''), we find that it embodies at least two significant facts: (i) the two curves display the same qualitative shape, which vanishes linearly with x and saturates at higher x values; and (ii) the doped-carrier mean-field approach captures the enhancement of nodal quasiparticle properties due to projection.

One final remark is due since, above, we refer to arcs of low energy spectral weight in the d-wave gaped SC state and, therefore, these arcs are not true, ungaped, Fermi arcs. Such ungaped Fermi arcs appear in the doped-carrier mean-field theory when Δ is zero and the magnitude of the spinon-dopon mixing mean-field order parameters b_0 and b_1 is non-vanishing, as it applies to the crossover between the Nernst and strange metal regimes (see the phase-diagrams in Refs. 37 and 38).

3.4.3.3. *Two spectral gaps*

In Figs. 3.4(g) - 3.4(i) we plot the spectral function along the minimum gap locus. Following the aforementioned nodal-antinodal dichotomy, in the deeply underdoped regime the SC peaks (*i.e.* those in the band $\epsilon^-_{1,\mathbf{k}}$) are faint in the antinodal region, and the spectrum near $(0.\pi)$ mostly reflects the high energy dispersion $\epsilon^-_{2,\mathbf{k}}$ [see Fig. 3.4(g) with $x = 0.05$]. This result is consistent with ARPES data in various cuprate families.[17,18,24,60,96] In this case, as we move away from the nodal direction the spectral function displays two gap features, of two different energy scales: (i) close to the diagonal direction the low energy d-wave gap dispersion is easily identified; and (ii) near the Brillouin zone boundary only the high energy gap associated with the motion of a hole in the presence of strong local AF correlations is resolved. Evidence for such a two gap structure, of two different energy scales, has been reported for NaCCOC[18] compounds.

3.4.3.4. *Peak-dip-hump spectral structure*

As doping is varied toward optimal doping the $\epsilon^-_{1,\mathbf{k}}$ band's spectral intensity close to $(0, \pi)$ increases [Figs. 3.4(h) and 3.4(i)].[17,96,97] The spectral function in this momentum space region then shows two different spectral features and, thus, it is consistent with the experimentally observed peak-dip-hump structure.[21,27,64]

The peak and hump energy scales have become known as the low energy and the high energy pseudogap respectively.[27] Therefore, in the doped-carrier theoretical framework the energy $\epsilon^-_{1,\boldsymbol{k}}$ at $(0,\pi)$ corresponds to the low energy pseudogap and the energy $\epsilon^-_{2,\boldsymbol{k}}$ at $(0,\pi)$ corresponds the high energy pseudogap. In agreement with the experimental evidence alluded to in Sec. 3.2, the low energy pseudogap $\epsilon^-_{1,(0,\pi)}$ coincides with the SC gap energy scale and the high energy pseudogap $\epsilon^-_{2,(0,\pi)}$ continuously evolves upon underdoping to match the large energy scale at $(0,\pi)$ observed in the half-filled insulators.

The existence of two spectral dispersions and two energy scales implies that similar peak-dip-hump structures are present in other regions of momentum space[21] and, indeed, in Figs. 3.2(b) and 3.2(c) we see a nodal peak-dip-hump line shape (indicated by the vertical white dash-dotted line), in agreement with ARPES results for LSCO,[17] NaCCOC,[18] Bi2201,[20] and Bi2212[21] compounds. We remark that the spectral peak-dip-hump is observed in single layer compounds and, thus, it is not a bilayer artifact.

ARPES experiments also assess the doping evolution of the spectral intensity of the SC peak at $(0,\pi)$. Specifically, the SC peak ratio (SPR), defined to be the intensity of the peak at $(0,\pi)$ normalized by the intensity of the entire $(0,\pi)$ spectral line, is studied.[94,96] We plot the ARPES data in Fig. 3.5(d) together with the corresponding theoretical ratio $\frac{Z^-_{1,(0,\pi)}}{Z^-_{1,(0,\pi)}+Z^-_{2,(0,\pi)}}$ and find that both curves are qualitatively similar: (i) they increases in the underdoped regime; and (ii) around optimal doping a downturn is observed.

3.4.3.5. *Spectral weight transfer to low energy: the role of t' and t''*

The spectral intensity in the low energy band $\epsilon^-_{1,\boldsymbol{k}}$ is stronger in the momentum space region where the high energy band $\epsilon^-_{2,\boldsymbol{k}}$ is closer to $\epsilon^-_{1,\boldsymbol{k}}$. This fact, that is easily appreciated in Figs. 3.4(g) - 3.4(i), is consistent with experiments. Indeed, the SC peak in the cuprates' ARPES spectrum is stronger when the energy difference between the Fermi level and the high energy hump decreases below $\approx J$, while it dies out when the hump disperses to higher energy.[94] The above observation underlies two facts, namely:

(i) the location of the low energy spectral weight arcs. In fact, at low doping, the band $\epsilon^-_{2,\boldsymbol{k}}$ disperses strongly along $(0,\pi)-(\pi,0)$, being closer to the Fermi level around $(\frac{\pi}{2},\frac{\pi}{2})$ and, thus, dopons hybridize more strongly with spinons in the nodal region, where spectral weight arcs appear at low energy.

(ii) the doping dependence of the low energy spectral weight arcs. To understand this, recall that t_2 and t_3 are explicitly chosen to change with doping so that we reproduce the doping induced renormalization of t' and t'', which flattens the $\epsilon^-_{2,\boldsymbol{k}}$ dispersion along $(0,\pi)-(\pi,0)$ in accordance with experiments[27,60] [see the expression for $\gamma_{\boldsymbol{k}}$ and the form of the effective parameters t_2 and t_3 given by Eq. (3.11)]. This doping dependence decreases the high energy pseudogap, bringing the band $\epsilon^-_{2,\boldsymbol{k}}$ closer to the band $\epsilon^-_{1,\boldsymbol{k}}$ around $(0,\pi)$, thus enhancing the spinon-dopon mixing in this part of momentum space [Figs. 3.4(g) - 3.4(i)] and increasing of the low energy spectral weight arc's length [Fig. 3.5(b)].

Naturally, the above interpretation of experimental data implies that, in case the high energy dispersion is strongly modified (say, as when we move from the hole doped to the electron doped regime), the low energy properties of the spectral function change accordingly. Indeed, in Sec. 3.4.4 we find that, upon changing the signs of t' and t'' so that $\epsilon^{\pm}_{2,\boldsymbol{k}}$ fits electron doped materials' data, the momentum distribution of low energy spectral weight in band $\epsilon^-_{1,\boldsymbol{k}}$ is strongly modified in agreement with experiments. The above picture also suggests that, in the absence of t' and t'', the high energy dispersion along $(0,\pi)-(\pi,0)$ is flat (as it follows from having $t_2 = 2t_3$) and the nodal-antinodal dichotomy is reduced, in accordance with variational Monte Carlo results for the tJ model,[95] which yield a nearly constant spectral intensity along $(0,\pi)-(\pi,0)$.

3.4.3.6. *Flat dispersion around* $(0,\pi)$ *and* $(\pi,0)$

Above, we show that the mean-field electron spectral function displays two dispersive features whose momentum and energy dependent spectral intensity evolves with doping in a manner that is consistent with experiments. In what follows, we discuss the shape of these two spectral features and show how strong correlations also induce a non-trivial renormalization of the quasiparticle dispersion.

Remarkably, all cuprate families exhibit an extreme flattening of the dispersion in an extended region around $(0,\pi)$.[21–23] This ubiquitous phenomenon has motivated interpretations involving anomalous scattering mechanisms in this region of momentum space, such as the nearly AF Fermi liquid and Van Hove singularity scenarios.[98–101] As noted in the literature,[63,85,89] the coexistence of the observed nodal dispersion's large energy scale and the small energy scale of the dispersion in the antinodal region cannot be explained with the bare hopping parameters estimated from band calculations and requires a careful fine-tuning in order to be reproduced. In fact, both Refs. 92 and 102 use effective dispersions with five hopping parameters to capture the different energy scales in the experimental dispersion. In Fig. 3.1(d) we use the parameters from Ref. 92 to compare the general features of the experimental normal state dispersion (solid white line) to that of the doped-carrier mean-field theory. We find that the mean-field dispersion reproduces both the dispersion between $(0,0)$ and the nodal point, as well as the flat dispersion around $(0,\pi)$. ARPES does not probe the dispersion above the Fermi energy and the aforementioned experimental fit uses band calculation results to fix the energy at (π,π),[92] whence the mismatch between the doped-carrier mean-field dispersion, obtained within the $tt't''J$ model context, and this fit above the Fermi level. Such a mismatch is supported by exact diagonalization calculations of the $tt't''J$ model which indicate that the dispersion above the Fermi level is less dispersive than expected from bare hopping parameters.[53]

3.4.3.7. *Dispersion kink*

The kink observed in the dispersions derived from the ARPES' momentum distribution curves of several cuprate families in both the underdoped and overdoped regimes, both below and above T_c, has been heavily discussed in the literature,[17,18,27] where both phonon[19,20] and magnetic modes[21,65,103] were proposed as the underlying cause for the observed effect. Experiments support that this feature is associated with the aforementioned peak-dip-hump spectral structure and, in particular, in NaCCOC,[18] LSCO,[17] Bi2201,[20] and Bi2212[21] the kink was shown to appear at the same energy as the dip. The horizontal white dashed line in Figs. 3.2(b) and 3.2(c) separates the SC peak from the high energy hump and, thus, defines the dip energy. Following the presence of two bands with different dispersions, a kink naturally appears at the dip energy, whose value $\omega \approx \frac{J}{2}$ is in good agreement with the experimental kink energy.[19] Notably, the kink is sharper in Fig. 3.2(b) than in Fig. 3.2(c) and, therefore, the dispersion kink obtained in the doped-carrier theory is found to smoothen out with increasing hole density, which is consistent with experiments.[19]

3.4.3.8. *Minimum gap locus topology*

Figs. 3.3(a) - 3.3(c) illustrate that the topology of the minimum gap locus changes from hole to electron-like as the doped hole density is increased. This transition occurs around optimal doping, more specifically at $x = 0.17$ [j]. Such a topology change has been observed in LSCO,[22,104] Bi2201,[105] and in the anti-bonding sheet of Bi2212.[106,107] Variational Monte Carlo calculations have also demonstrated a change in the underlying Fermi surface topology around optimal doping.[108]

Interestingly, both the minimum gap locus in Fig. 3.3(a) and the maximum spectral weight locus in Fig. 3.3(d) show long straight segments close to the Bril-

[j] This value differs from that found in Ref. 37, namely $x = 0.20$, since in this paper we take smaller values of t'/t and t''/t and, consequently, the minimum gap locus has a smaller curvature, thus crossing $(0,\pi)$ at higher electron density.

louin zone boundary. A similar observation in deeply underdoped LSCO samples led the authors of Ref. 17 to propose that the resulting nesting enhances scattering processes in the antinodal region, thus destroying quasiparticle features in this part of momentum space.[17] In addition, this scattering mechanism has also been suggested to be connected to the appearance of charge ordered states.[70,72,109] We remark that, here, the straight segments close to the Brillouin zone boundary for low values of x result exclusively from the interaction between spinons and dopons. The nodal-antinodal dichotomy in the present theoretical scenario is not a consequence of such a minimum gap locus structure. However, the peculiar form of the renormalized d-wave band $\epsilon^-_{1,\boldsymbol{k}}$ may indeed induce specific ordered structures in real space once fluctuations around the mean-field saddle point are properly included. This is a problem to be addressed in the future.

3.4.3.9. *d-wave gap renormalization*

Finally, we discuss the renormalization of the d-wave gap which, experiments show, tends to deviate from the pure d-wave dispersion $\Delta_{\boldsymbol{k}} = \Delta_0 \cos 2\theta_{\boldsymbol{k}}$.[110] A similar effect is present in the above mean-field approach, where the spinon d-wave dispersion is flattened close to the nodal points due to the hybridization with dopons. Since the spinon-dopon mixing increases for larger x, the deviation from the pure d-wave form increases with hole density, as depicted in Figs. 3.4(j) - 3.4(l). We fit the gap dispersion along the minimum gap locus to the functional form $\Delta_{\boldsymbol{k}} = \Delta_0[B\cos 2\theta_{\boldsymbol{k}} + (1-B)\cos 6\theta_{\boldsymbol{k}}]$ and find that B equals 0.97, 0.93 and 0.89 for $x = 0.05$, $x = 0.12$ and $x = 0.20$ respectively. Experiments generally find $B \sim 0.9$,[110] in agreement with the MF theory estimate in the $x \approx 0.10 - 0.20$ range. ARPES data also suggests that, around optimal doping, B increases with hole concentration as the system goes from the underdoped to the overdoped regime.[110] This doping dependence differs from that of our mean-field theory, which is intrinsically a low doping approach and, as such, should be firstly compared to phenomenology in the underdoped region of the phase diagram. As explained above, due to the nodal-antinodal dichotomy ARPES experiments cannot resolve the full SC d-wave dispersion in the deeply underdoped regime and, then, cannot address the doping dependence of B for low x. Since other experiments seem to indicate the reduction of arc length[24,111] and the increase of gap velocity at the nodes[14] with decreasing doping, we naively expect the SC gap dispersion to be less renormalized in the underdoped limit.

3.4.4. *Electron doped case*

In the previous section, we discuss the mean-field spectral function for the hole doped SC state and compare it to hole doped cuprates' ARPES data. Now, we shift gears to address the properties of the electron spectral function in Eq. (3.18), as of interest to the cuprates' electron doped regime. The $tt't''J$ model parameters we use are $t = 3J$, and $t' = -2t'' = J$,[53,90] which imply the use of the effective hopping parameters

$$t_2^{ED} = J + J\left(1 - \frac{x}{0.3}\right)$$
$$t_3^{ED} = \frac{J}{2} - \frac{J}{2}\left(1 - \frac{x}{0.3}\right) \tag{3.22}$$

in the mean-field Hamiltonian in Eq. (3.8).

3.4.4.1. *Hole doped versus electron doped*

Before listing and discussing the properties of the mean-field spectral function in the parameter regime defined above, we refer to a few differences between the hole and electron doped regimes of cuprate materials. We also comment on a couple of limitations in our calculation of the electron doped regime's spectral function.

Firstly, we recap that ARPES data of hole doped materials displays two dispersive features, one inherited from the undoped parent compound (the lower Hubbard band) and another whose spectral intensity increases as the electron concentration is reduced from half-filling (the mid-gap band). Since the chemical potential shifts to higher energy upon electron doping, the ARPES data of electron doped compounds shows three dispersive features instead, namely: (i) the lower Hubbard band, which is observed approximately 1.3eV below the Fermi energy;[26] (ii) the mid-gap spectral features, which emerge as spectral weight is transfered from the lower Hubbard band to the mid-gap region once the electron density becomes larger than $n = 1$;[26] and (iii) the upper Hubbard band, which is displayed by ARPES data since it is crossed by the chemical potential in the electron underdoped regime. Since in this paper we use the $tt't''J$ model, we cannot reproduce the above three spectral features shown by ARPES experiments in electron doped materials. This model sets the correlation gap $U \to +\infty$ and completely disregards the presence of the lower Hubbard band. As depicted in Fig. 3.6, the mean-field spectral function captures both the mid-gap spectral weight and the upper Hubbard band. The former leads to a somewhat flat dispersion below the Fermi level (which lies at $\omega = 0$), while the latter strongly disperses above this energy, in accordance with cluster perturbation theory results for the Hubbard model.[112–114]

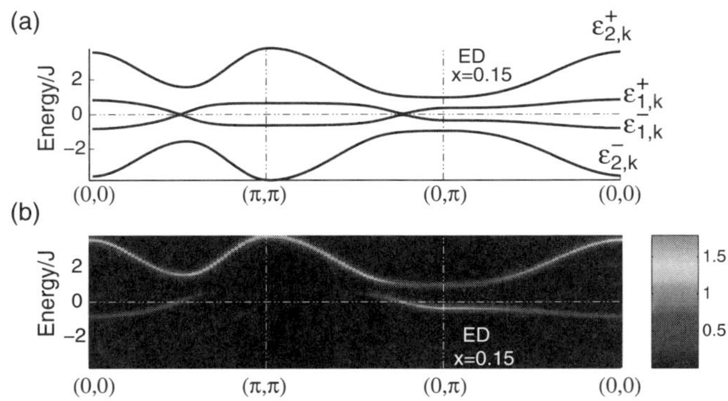

Fig. 3.6. Energy dispersion along $(0,0)-(\pi,\pi)-(0,\pi)-(0,0)$ for the electron doped $x = 0.15$ SC state. (a) mean-field dispersions $\epsilon^{\pm}_{1,\boldsymbol{k}}$ and $\epsilon^{\pm}_{2,\boldsymbol{k}}$. (b) mean-field electron spectral function (3.18) with Lorentzian broadening given by $\Gamma = J/10$.

Secondly, note that the hopping parameters t' and t'' have different signs in the hole and electron doped regimes. This sign change enhances (reduces) the robustness of AF (dSC) correlations in the electron doped case[38,90] and, consequently, long-range AF order survives up to larger doping values in the electron doped regime. Hence, while hole-underdoped cuprate samples display long-range dSC order coexisting with strong short-range AF correlations, deeply electron-underdoped cuprate samples have long-range AF order coexisting with strong short-range dSC correlations. Even though the doped-carrier mean-field Hamiltonian in Eq. (3.8) describes states with long-range AF order (if $m, n \neq 0$), it is particularly suited to describe states with long-range dSC order (due to the particular representation of lattice spin operators in terms of fermionic spinon operators). Still, due to its particular choice of operators to describe the generalized-tJ model (namely, spinons and dopons), the doped-carrier framework can account for both AF and dSC correlations at short length scales. Therefore, even in the cuprates' electron doped regime, the doped-carrier approach can be employed to address phenomena determined by short-range physics, such as the distribution of spectral weight throughout momentum space. Thus, below, we first study the doped-carrier spectral function in the AF state without accounting for dSC correlations (Sec. 3.4.4.2). This allows us to identify the specific signatures of AF correlations in the electron doped regime. After that, in Sec. 3.4.4.3 and in Sec. 3.4.4.4, we ignore the possible occurrence of AF long-range order, and focus on the properties of the dSC state with strong *local* AF correlations.

3.4.4.2. *Electron pockets*

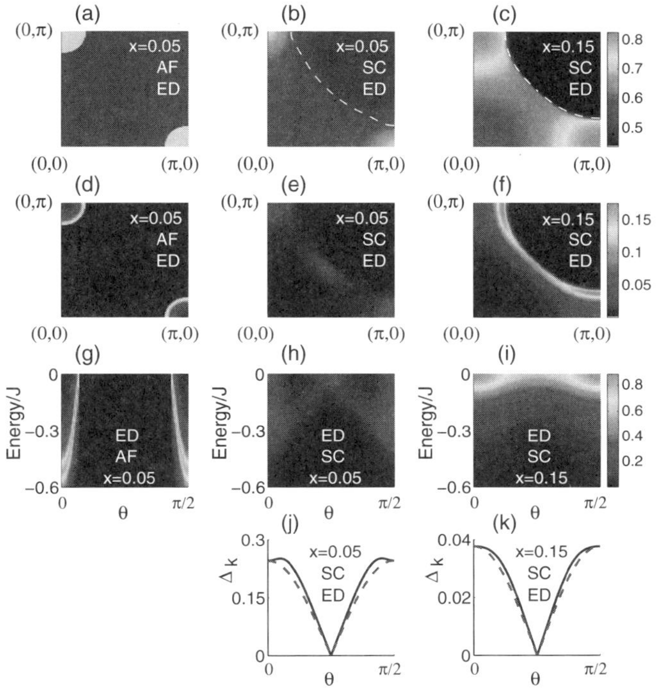

Fig. 3.7. Mean-field electron spectral function results for the electron doped $x = 0.05$ AF state [(a),(d) and (g)], the $x = 0.05$ dSC state [(b),(e),(h) and (j)] and the $x = 0.15$ dSC state [(c),(f),(i) and (k)]. (a)-(c) Electron momentum distribution $n_{\boldsymbol{k}} = 1 - Z^{-}_{1,\boldsymbol{k}} - Z^{-}_{2,\boldsymbol{k}}$ (top color scale). The white dashed line in (b) and (c) depicts the minimum gap locus. (d)-(f) Integrated electron spectral weight around the Fermi level (middle color scale). The energy window $[-0.15J, 0.15J]$ is used. (g)-(i) Electron spectral function (3.18) along minimum gap locus (bottom color scale). A Lorentzian broadening with $\Gamma = J/10$ is used in (d)-(i). (j) and (k) $\epsilon^{+}_{1,\boldsymbol{k}}$ dispersion gap (solid line) and pure d-wave dispersion ($\Delta_{\boldsymbol{k}} = \Delta_0 \cos 2\theta_{\boldsymbol{k}}$) with same gap magnitude (dashed line) along minimum gap locus. Note that the pure d-wave dispersion lies below $\epsilon^{+}_{1,\boldsymbol{k}}$ implying $B > 1$ (see main text). The energy scales in (j) and (k) are different. The angular variable θ used to parameterize the points along the minimum gap locus is defined in Fig. 3.3(d).

In the hole doped regime, doped carriers strongly frustrate the AF spin background and change the nature of the surrounding spin correlations, as reflected in the absence of hole pockets around $(\frac{\pi}{2}, \frac{\pi}{2})$. AF correlations are more robust in the electron doped case due to the sign change of t' and t''[90] and, as a result, the chemical potential crosses the bottom of the upper Hubbard band and gives rise to electron pockets. The sign of t' and t'' also implies that such pockets appear around $(0, \pi)$ and $(\pi, 0)$.[26,53,83] Figs. 3.7(a), 3.7(d), and 3.7(g) depict the corresponding mean-field results for the electron underdoped AF state (where $m, n \neq 0$ and $b_0, b_1 = 0$). These results are consistent with the aforementioned phenomenology. The first two figures show the electron momentum distribution function and the integrated spectral weight in an energy window around the Fermi level, respectively, both of which display sharp pockets. The third figure depicts the spectral function along the $(0, \pi) - (\pi, 0)$ line, which shows the corresponding dopon pocket dispersion.

3.4.4.3. *Nodal spectral weight as the signature of superconducting correlations*

In Sec. 3.4.4.2 we see that, in the electron doped regime, AF correlations lead to the buildup of electron spectral weight around $(0, \pi)$ and $(\pi, 0)$. As discussed above, we now focus on the paramagnetic SC solutions for $x = 0.05$ and $x = 0.15$ to find that the spectral weight transfer from the antinodes to the nodal region underlies the emergence of SC quasiparticles near $(\frac{\pi}{2}, \frac{\pi}{2})$.

In Figs. 3.7(b), 3.7(e), and 3.7(h), we plot mean-field results regarding the paramagnetic dSC phase at doping level $x = 0.05$ [k] and in Figs. 3.7(c), 3.7(f), and 3.7(i), we plot the corresponding results at doping level $x = 0.15$. Specifically, Figs. 3.7(b) and 3.7(c) depict the electron momentum distribution function and, we find, doped electrons spread between the minimum gap locus and $(0, 0)$. This distribution is uneven along the region surrounding the minimum gap locus, since the areas around $(0, \pi)$ and $(\pi, 0)$ carry most of the spectral intensity. Comparing to Fig.3.7(a) supports that the momentum space anisotropy in Figs. 3.7(b) and 3.7(c) follows from the presence of strong local AF correlations in the paramagnetic doped Mott insulator state.

Further looking at Figs. 3.7(e), 3.7(f), 3.7(h), and 3.7(i), elucidates the interesting interplay between AF and dSC correlations in the electron doped case, which plays an important role in the doping evolution of the spectral intensity around the Fermi level. Specifically:

(i) as the insulating parent compound is doped with electrons away from half-filling, dSC correlations grow in relevance and spectral weight is transfered from

[k] We remark that, for $x = 0.05$, both the doped-carrier mean-field theory of the $t = 3J$, $t' = -2t'' = J$ model and the electron doped cuprate compounds show long-range AF order. Hence, the results hereby shown for the $x = 0.05$ dSC paramagnetic state cannot be directly compared to experiments, since spinons are gapped along the nodal direction in the $x = 0.05$ AF state. Therefore, the spectral weight that appears in Figs. 3.7(e) and 3.7(h) close to $(\frac{\pi}{2}, \frac{\pi}{2})$ at the Fermi level should be at higher energy in the AF phase with local dSC correlations.

the pockets at $(0, \pi)$ and $(\pi, 0)$ to the nodal region [Figs. 3.7(e) and 3.7(h)]. Interestingly, experiments show that, even in the AF phase, there is a buildup of spectral weight around $(\frac{\pi}{2}, \frac{\pi}{2})$ inside the Mott-Hubbard gap which forms the d-wave nodal dispersion in the SC state.[26,81,82] Therefore, we associate the experimentally observed spectral intensity along the diagonal direction to the increasing relevance of dSC correlations.

(ii) at higher values of doping, the spectrum close to the Fermi energy displays a large, d-wave gapped, "Fermi surface" [Figs. 3.7(f) and 3.7(i)]. This behavior is also reminiscent of that observed in ARPES data.[26,46,80,81]

Notably, as a result of the t' and t'' sign change, the direction of the nodal-antinodal dichotomy shifts when we go from the hole doped to the electron doped regime.[115] In the former, both AF and SC correlations contribute to the presence of quasiparticles around $(\frac{\pi}{2}, \frac{\pi}{2})$ and lead to low energy spectral weight arcs [Figs. 3.3(d) - 3.3(f)]. In the latter, AF correlations enhance spectral weight close to $(0, \pi)$ and $(\pi, 0)$ while SC correlations favor low energy quasiparticles along the diagonal direction.

3.4.4.4. *d-wave gap renormalization*

In Sec. 3.4.3 we discuss how local AF correlations renormalize the d-wave SC dispersion along the minimum gap locus in the hole doped regime [Figs. 3.4(j) - 3.4(l)]. In particular, we find that the d-wave dispersion is flattened around the diagonal direction, which implies that $B < 1$ if the functional form $\Delta_{\boldsymbol{k}} = \Delta_0[B \cos 2\theta_{\boldsymbol{k}} + (1 - B) \cos 6\theta_{\boldsymbol{k}}]$ is used to fit the experimental dispersion.

In the electron doped case, short-range AF correlations are relevant close to the antinodes and, consequently, the d-wave dispersion deviates from the pure d-wave form close to the Brillouin zone boundary instead. Recurring to the above functional form to fit the dispersions in Figs. 3.7(j) and 3.7(k) we obtain $B = 1.04$. A value $B > 1$ indeed implies that the d-wave dispersion is flattened or, for large enough B, displays a downturn away from the nodal direction. This fact finds experimental support in ARPES[82] and Raman scattering[116] data, both of which provide evidence for a non-monotonic d-wave SC gap. In particular, the ARPES dispersion suggests $B = 1.43$,[82] which is much larger than the one obtained from Figs. 3.7(j) and 3.7(k). Even though we can fine-tune the mean-field parameters to closely reproduce the non-monotonic d-wave dispersion at mean-field level, we believe that better agreement with experiments may be reached, for instance, by introducing an extended d-wave pairing gap[117] or by accounting for fluctuations around the mean-field saddle-point.

Still, we highlight the fact that the doped-carrier mean-field theory correctly reproduces the sign of $(B - 1)$ in the hole and electron doped regimes. Hence, we propose that the experimentally observed deviation from the pure d-wave dispersion reflects the effect of local AF correlations in the dSC state.

3.5. Discussion

The purpose of this section is to clarify the simple physical picture that stems from our results and to argue that many aspects of the phenomenology of underdoped cuprates can be understood in terms of the interplay between *local* spin correlations and the doped carrier dynamics. We first comment on the role short-range spin correlations play in determining the hole dynamics (Sec. 3.5.1) and on how these correlations are captured by the doped-carrier multi-band description (Sec. 3.5.2). We then discuss some of the experimental evidence favoring the coexistence of short-range dSC and AF correlations (Sec. 3.5.3) and argue that the doping induced renormalization of t' and t'' follows from the growing relevance of dSC correlations over AF correlations (Sec. 3.5.4). Note that the discussion below focuses on the underlying intuitive physical picture that motivates the doped-carrier mean-field framework rather than on formally rigorous arguments.

3.5.1. *The role of short-range correlations*

In this paper, we are interested in single electron spectral properties of doped Mott insulators, with particular emphasis in the SC state near the Mott insulator transition, as observed in ARPES experiments of the high-T_c cuprate materials. ARPES is a most relevant experimental probe which inserts a hole in the system and resolves its dynamics as a function of momentum. In the limit of conventional uncorrelated metals, quasiparticles are dressed electrons whose dispersive features are largely determined by an effective external potential landscape that does not depend on the exact position of all other electrons. In strongly correlated systems, however, the dynamics of a hole is deeply intertwined with its surroundings. Given the large Coulomb gap, near half-filling the hole is encircled by localized spins, in which case the hole dynamics is closely related to nearby spin correlations. To illustrate how short-range spin correlations affect the hole dynamics and the related hole spectral features we consider two opposite limits, namely, that of staggered local moments and and that of resonating singlet bonds.

In the presence of staggered local moments, inter-sublattice hopping is strongly renormalized while intra-sublattice hopping is only mildly frustrated. The simple minded argument is that when a vacancy moves to the NN site it leaves a local moment misplaced in the staggered spin arrangement. Consequently, the electron dispersion in an AF background is symmetric across the $(0,\pi)-(\pi,0)$ line. In addition, the dispersion along this line is controlled by t' and t'' as determined by the corresponding bare hopping processes.[52–55]

Interestingly, in the presence of resonating singlet bonds exactly the opposite *may* apply. These states can be thought of as a quantum superposition of dimer coverings of the lattice. In this case, the motion of a vacancy between two sites requires the rearrangement of the dimers and it effectively hybridizes different coverings. Depending on the actual coverings that constitute the spin state, as well as on the relative phases, the vacancy hopping between two sites can be either favored or frustrated. Therefore, the effective renormalization of t, t' and t'' depends on the nature of the underlying spin liquid correlations. As it turns out, for those spin liquid states believed to be of interest to the $tt't''J$ model, t' and t'' are reduced by a factor $\sim x$ and coherent hole hopping between 2nd and 3rd NN is strongly frustrated. [In this case, the NN neighbor hopping integral is only renormalized down to the spin energy scale J.] This fact is easily appreciated in the slave-boson formulation of the generalized-tJ model.[118]

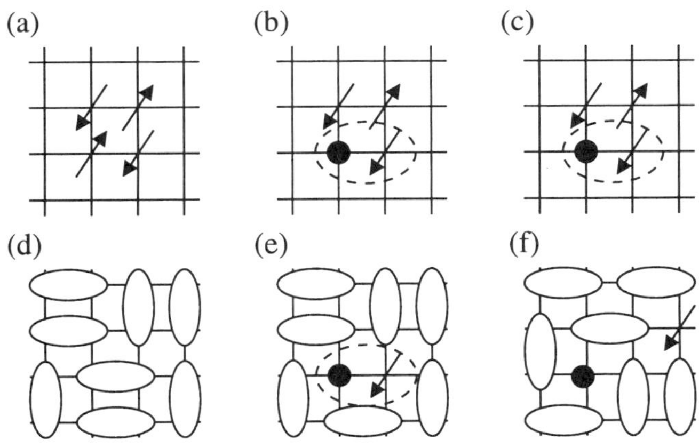

Fig. 3.8. Schematic pictures for the introduction and evolution of a hole in two different spin backgrounds. In the case of staggered local moments (a) removing an electron with spin $S_z=+\frac{1}{2}$ leaves a vacancy and a nearby spin $S_z=-\frac{1}{2}$ which, together, carry the same quantum numbers as a hole (b). Due to the rigidity of the staggered moment spin configuration the vacancy and the extra $S_z=-\frac{1}{2}$ spin remain bound to each other (c). Removing an electron from a resonating singlet bond spin configuration (d) leaves the same local signature of a hole (e). However, the liquid nature of the spin background takes the extra spin $S_z=-\frac{1}{2}$ away from the vacancy (f).

The above simple argument naturally explains why different spin correlations induce distinct electron dispersions. Interestingly, different spin correlations can also underlie different spectral intensities associated with the corresponding dispersions. To motivate this point, note that removing one electron introduces an additional spin-1/2 in the system, which lies around the electrically charged vacancy site [Figs. 3.8(b) and 3.8(e)]. As the system evolves after the electron removal, two distinct behaviors can be observed: (i) in the presence of staggered moments the extra spin-1/2 stays close to the vacancy [Fig. 3.8(c)], whereas (ii) the liquid nature of resonating valence bonds screens the doped spin away from the vacancy [Fig. 3.8(f)]. As a result, in the underdoped limit, the former state has larger electron spectral weight than the latter one.

Notably, the above discussion is closely connected to numerical work in the $tt't''J$ model. In fact, the results reported in Ref. 115 show that one hole states in the 2D $tt't''J$ model are the superposition of two states where: (i) the vacancy is surrounded by a staggered spin pattern and (ii) the vacancy is surrounded by a uniform spin pattern that screens the hole spin-1/2 away. This work also shows that the former has large quasiparticle spectral weight and strongly renormalizes t, while the latter has vanishing spectral intensity and strongly renormalizes t' and t''. Therefore, the above numerical study suggests that both staggered moment and singlet bond states are relevant to the $tt't''J$ model. These states are also of relevance to the cuprate materials as they are naturally related to the neighboring AF and dSC phases in the underdoped sector of the cuprate phase diagram.

Since ARPES probes the dispersion renormalization and the distribution of spectral weight in energy-momentum space as a function of doping, it is a highly valuable source of information about the evolution of short-range correlations with varying electron density. Based on our mean-field results, below we argue that ARPES data suggests t, t' and t'' are renormalized in consonance with the presence of staggered AF correlations in the undoped compounds and with the growing relevance of singlet bond correlations as materials are doped away from half-filling.

3.5.2. *Two-band description of the local energetics*

Quantum Monte Carlo simulations of the tJ and Hubbard models[119–122] show that the electron spectral function has a four band structure. In Refs. 121 and 122, the two bands below the Fermi level were interpreted in terms of two different states, namely: (i) holes on top of an otherwise unperturbed spin background and (ii) holes dressed by spin excitations. This interpretation is consistent with the aforementioned numerical work of Ref. 115 supporting a two-band picture of single hole states in the $tt't''J$ model and translates into the fact that there exist two relevant spin configurations around the vacancy.

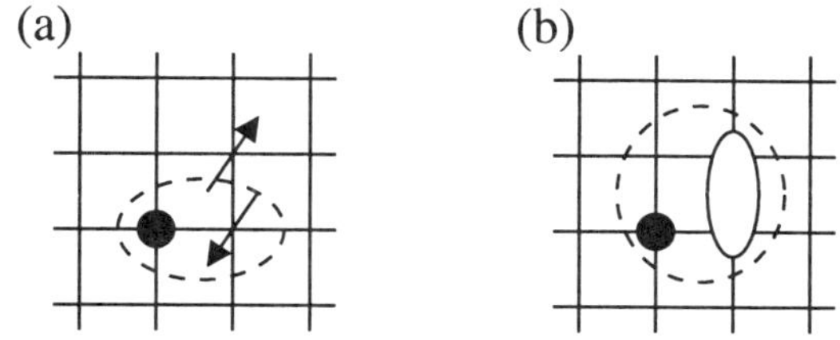

Fig. 3.9. Schematic representation of the dopon (a) and the holon (b). The dopon is obtained by removing a lattice spin and corresponds to the vacancy plus the nearby spin introduced by doping. The holon is obtained when the spin in the dopon configuration (a) is absorbed by the spin background and, thus, is the composite object made of a vacancy and its encircling spin singlet configuration.

The above results concern the properties of the single hole $tt't''J$ model, which describes a single vacancy hopping in a spin background. Since in the low doping limit vacancies are dilute, the doped-carrier framework borrows from the local problem of a single vacancy surrounded by spins. Hence, in the doped-carrier formulation of the $tt't''J$ model, the vacancy can also be screened by two different spin structures. The one-dopon state corresponds to the hole obtained immediately upon removing a lattice spin from the system. Pictorially, this hole is the composite object made of a vacancy and a nearby spin [see Fig. 3.9(a)]. Since local AF correlations predominate close to half-filling, we take the one-dopon state to correspond to a vacancy encircled by staggered local moments. In addition, the dopon spin can be screened away by the spin background, thus creating a bosonic charge-e and spinless spinon-dopon pair which, in the slave-particle jargon, is called a "holon". Hence, within the doped-carrier framework, a holon is the object that represents a vacancy encircled by a local spin singlet configuration.

To understand the rationale underlying the above two-band picture, note that hole motion, as described by the hopping term, is frustrated by and frustrates the AF spin correlations driven by the Heisenberg term. Hence, in the underdoped regime, the $tt't''J$ model local physics is that of the competition between the hole kinetic energy and the spin exchange energy. The two-band description intrinsic to the doped-carrier formulation of the $tt't''J$ model, which is intrinsically a one-band model, captures the above competition: (i) the one-*dopon* state preserves the local AF correlations favored by the exchange term but does not hop coherently between NN, thus frustrating the kinetic energy term; (ii) the one-*holon* state, which is the singlet pair of a dopon and a spinon, hops coherently between NN but distorts the spin background and frustrates the exchange energy. The evidence for coherent NN hopping following the spinon-dopon hybridization comes, for instance, from the nodal spectral functions in Figs. 3.2(b) and 3.2(c), which show a linear quasiparticle dispersion across the Fermi point near $(\frac{\pi}{2}, \frac{\pi}{2})$.

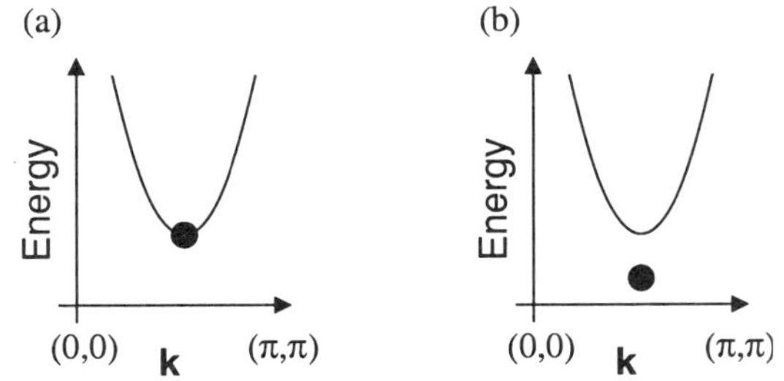

Fig. 3.10. Schematic band diagram for a hole introduced into a spin system with AF spin correlations. If the hole does not change the background correlations it occupies the bottom of the coherent band (a). If, instead, the moving hole distorts the surrounding spin configuration and decreases its kinetic energy, it lies below the aforementioned dispersion (b).

The above description provides a simple picture for the fact that, upon doping the insulating parent compound, the chemical potential does not fall on top of the valence band, thus preventing the formation of hole pockets with size x.[18,24,25,42] If short-range spin correlations are only of the staggered moment type, the hole inserted in the system sits at the bottom of a band that disperses like a single hole in an AF background [Fig. 3.10(a)]. However, if the hole distorts the surrounding spins in a manner that decreases its kinetic energy, it can appear below the AF band bottom and, hence, inside the gap [Fig. 3.10(b)]. Therefore, the absence of hole pockets in hole doped cuprates results from a change of spin correlations that enhances NN hopping. Note that the resulting kinetic energy gain stabilizes the formation of spinon-dopon pairs and, thus, it stabilizes the dSC phase as well.

3.5.3. *Interplay between dSC and AF correlations*

In the dSC phase AF correlations are relegated to high energy while, at low energy, d-wave singlet bond correlations underlie long-range dSC order. Even though at mean-field level the dopon dispersion is determined by $delta$-peaks in the spectral function, a vacancy surrounded by local Néel spins is not a true coherent state in the spin liquid background. Hence, a less approximate theoretical approach would lead to a broadened dopon dispersion.

The claim that both AF and dSC correlations are important to determine the dynamics of holes and, consequently, to fit experiments, means that, at least at short time scales, both staggered moment and d-wave spin singlet spin configurations exist around the vacancy. In addition, if both short-range correlations leave fingerprints in the electron spectral function, the above time scales are long enough to affect the vacancy dynamics. Since we find that, at zero temperature, the spinon-dopon mixing order parameter $b_0 \approx \sqrt{x}$, dopons spend most of the time paired to a spinon. The pairing between spinons and dopons is local and, therefore, the typical distance between these two entities can be denoted by na, where a is the lattice constant and n is a number of order unity. The relevance of AF correlations is tied to the time scale that measures how long the dopon and the spinon spend far apart, which is set by $\frac{na}{v} \sim \frac{\hbar}{\Gamma}$, where v is the characteristic velocity of spinons and dopons, and Γ is the dopon dispersion's broadening energy scale. Given that the spinon and dopon bandwidths are set by J, we expect $\Gamma \sim J$, which is of the order of the dopon band width.

The above argument is consistent with the remarkably broad hump, with a width of order J, observed by ARPES in the cuprate materials.[18,20,21,24,27,42,64,103] Previous work has identified the role of phonons and of self-trapping polaron effects in broadening the spectral feature that disperses as a hole in an AF background.[123] Here, we propose that the resilience of the role played by local AF correlations in the dSC spin liquid state also implies a broad high energy hump.

Γ is related to the inverse of the distance over which staggered moments might appear as the spin configuration surrounding the vacancy. Despite the broad hump measured by experiments, as long as the above distance is large enough to allow for the local physics encoded in our mean-field approximation, it is physically relevant to consider the sharp dopon mean-field dispersion. Ultimately, the relevance of the two-band approach must be decided upon comparison to experiments. Thus, below, we summarize the evidence for two different spectral dispersions in ARPES data, after what we comment on other *non-trivial* properties of the electron spectral function which can be understood as the result of a dopon band at high energy.

Cuprates' ARPES experiments identify two different dispersions. At lower energy, a dispersion linear across the Fermi level close to $(\frac{\pi}{2}, \frac{\pi}{2})$ has spectral intensity vanishing with underdoping. These two signatures, namely the linear dispersion and small spectral weight, are consistent with resonating singlet bond spin correlations around the vacancies as discussed in Sec. 3.5.1. The higher energy dispersion is reminiscent of a hole with strongly frustrated NN hopping and with 2nd and 3rd NN hopping consistent with band calculation expectations. Besides, it carries most of the electron spectral weight. These constitute strong evidence for vacancies surrounded by staggered spin configurations (Sec. 3.5.1). Even though the high energy dispersion inherited from the undoped parent compound gets broader and fainter upon doping,[42] as expected given the doping induced frustration of AF correlations, there exists experimental support for the presence of two pseudogap energy scales at $(0, \pi)$ all the way into the overdoped regime.[27] Therefore, experiments suggest that short-range AF and dSC correlations coexist throughout a vast range of the high-T_c phase diagram.

As we extensively discuss above, the doped-carrier mean-field theory captures the aforementioned two spectral dispersions. In order to further motivate the relevance of the resulting two-band mean-field description, we note that the dopon dynamics enclosed in the high energy dispersion affects spectral properties at low energy in consistency with experimental observations. Specifically, the impact of the dopon dispersion on the electron spectral function is two-fold (see Sec. 3.4): (i) it renormalizes different dispersive features, such as the nodal dispersion which displays a kink at the same energy as the dip of the peak-dip-hump structure,[17–21,27,103] and the d-wave gap along the minimum gap locus;[110] (ii) it determines the transfer of spectral weight to low energy, as well as its momentum space distribution, leading to the formation of low energy spectral intensity arcs around the nodal direction[18,24,25] and to the evolution of the peak intensity at $(0, \pi)$ that resembles experiments.[94,96]

The above deviations from plain BCS theory are more striking at low doping density, namely, the kink is sharper [Figs. 3.2(b) and 3.2(c)] and the arc length shorter [Fig. 3.5(b)], as observed by experiments.[19,24] This fact supports that the nodal peak-dip-hump structure, the nodal dispersion "kink" and the spectral weight arcs are fingerprints of AF correlations in the metallic state, which grow as doping is reduced. It also supports the physical relevance of the dopon to understand the hole dynamics in (under)doped Mott insulators.

Further evidence favoring the relevance of a sharp dopon band at mean-field level comes from the comparison between the hole and electron doped regimes. Indeed, within the doped-carrier mean-field framework, the only distinction between both regimes arises from the different dopon dispersions. Specifically, in the hole doped case dopons are present in the nodal region whereas in the electron doped regime dopons appear around $(0, \pi)$ and $(\pi, 0)$. This difference leads to a distinct doping evolution of the spectral weight transfer to low energy: (i) in the hole doped case, nodal arcs are formed [Figs. 3.3(d) - 3.3(f)], while (ii) for electron doped parameters low energy spectral weight appears both in the antinodal and nodal regions [Figs. 3.7(d) - 3.7(f)]. This distinction is consistent with experiments.[18,24–26,46,80,81]

Finally, we remark that both AF and dSC correlations are intrinsic to the proximity to the Mott insulating state. Also, as far as our mean-field theory and the comparison to ARPES data are concerned, these are *local* correlations. The fact that either AF or dSC (or maybe both) correlations can acquire long-range order is not the cause, and does not invalidate, the local analysis of the interplay between hole hopping and staggered local moments encoded in the mean-field theory. An interesting insight from the doped-carrier formulation of the $tt't''J$ model is that the physics of spin liquids and staggered spin correlations are not mutually exclusive in doped Mott insulators. Indeed, we find that local AF correlations may affect the phenomenology of doped spin liquid states in consistency with experiments.

3.5.4. *Doping dependent pseudogap energy*

As expressed in Eq. (3.11), and as Ref. 38 extensively discusses, the effective hopping parameters t_2 and t_3 are determined phenomenologically upon fitting the high energy dispersion $\epsilon_{2,\boldsymbol{k}}$ to both numerical results and cuprate ARPES data. The resulting doping dependence of t_2 and t_3 (which is put in by hand) captures the doping evolution of the high energy pseudogap. It should be emphasized that, based on the above input, the doped-carrier mean-field theory correctly predicts a variety of spectral properties observed at lower energy. This agreement with experiments discloses that the doping dependence of the high energy pseudogap scale underlies, for instance, the increase of the low energy spectral arcs' length with doping. Furthermore, it provides extra support to the doped-carrier mean-field approach.

The simple physical picture that underlies the doped-carrier framework, and which we discuss above, provides the basis to understand the doping dependence of the pseudogap energy. Specifically, this doping dependence stems from the coexistence of both staggered moment and singlet valence bond correlations at short-length scales. To understand this fact, recall that t' and t'' are *not* strongly frustrated by AF correlations, whereas d-wave singlet correlations strongly renormalize these hopping parameters. Since the predominance of the latter correlations grows upon doping away from half-filling, effectively, these gradually decrease t' and t'' with increasing doping level. The effective reduction of t' and t'' is captured by the renormalization factor $r(x)$ which multiplies these hopping parameters in Eq. (3.11), which dictates the doping dependence of t_2 and t_3 in such a way that the high energy pseudogap scale reduces with doping.

Hence, we propose that the above energy scale's doping dependence follows the growing importance of local d-wave singlet bond correlations over local staggered moment correlations as we move away from half-filling. In other words, we propose that this doping evolution follows from the decrease with doping of the time scale over which staggered moment spin configurations surround the vacancies. We further remark that the experimentally observed lowering of the the high energy pseudogap scale offers additional support for spin correlations that strongly renormalize t' and t'', as is the case with resonating singlet bonds.

3.6. Summary

Following the introduction in Refs. 37 and 38 of a new slave-particle mean-field approach that captures the crucial interplay between the hole dynamics and the background spin correlations in the low doping and low temperature paramagnetic regime of the $tt't''J$ model, in this paper we use it to address the single hole dynamics in the dSC state near the Mott insulator transition, as observed in ARPES experiments on the high-T_c cuprate materials.

As we explain above, in addition to the model parameters borrowed from band calculations, in this paper we make use of a specific input from experiments, namely, the change with doping of the high energy $\epsilon^{\pm}_{2,\boldsymbol{k}}$ dispersion along $(0,\pi) - (\pi,0)$ in order to reproduce the lowering of the high energy pseudogap energy scale with the change in hole concentration. We remark that there is no such experimental input on the low energy d-wave dispersion $\epsilon^{\pm}_{1,\boldsymbol{k}}$, whose spectral properties follow from theory alone. Notably, these properties agree with experiments, as we briefly summarize below.

In Sec. 3.4 we show that the doped-carrier mean-field theory obtains spectral peaks, which disperse linearly across the Fermi energy and whose weight vanishes at half-filling, that develop above the insulating valence band (Fig. 3.2).[18,24,25,42] The corresponding low energy spectral weight forms arcs around the nodal direction whose length enlarges with doping in accordance with experimental data [Fig. 3.5(b)].[18,24,25] This amounts to the nodal-antinodal dichotomy,[17] that expects enhanced nodal quasiparticle features as obtained by the current mean-field theory, which is consistent with variational Monte Carlo results [Fig. 3.5(c)].[95] Following the two-band nature of the mean-field approximation, a peak-dip-hump spectral structure is obtained in different momentum space regions.[21] Interestingly, the above calculation captures the doping evolution of the spectral intensity of the peak feature at $(0,\pi)$ [Fig. 3.5(d)].[94,96] Along the nodal direction a dispersion kink, which smoothens out with doping, appears at the dip energy $\omega \approx \frac{J}{2}$ [Figs. 3.2(b) and 3.2(c)] in close agreement with experiments.[17–21,27,103] The mean-field theory also predicts the evolution of the minimum gap locus topology with doping [Figs. 3.3(a) - 3.3(c)] in consistency with ARPES data[22,104–107] and variational Monte Carlo computations,[108] and reproduces the strong renormalization of the dispersion which is surprisingly flat around $(0,\pi)$ [Fig. 3.1(d)].[21–23] In addition, the renormalization of the superconducting gap is also consistent with experiments – it deviates from the pure d-wave form in that it flattens around the nodal direction in the hole doped regime [Figs. 3.4(j) - 3.4(l)] and in the antinodal region in the electron doped case [Figs. 3.7(j) and 3.7(k)].[82,110,116] Finally, in Sec. 3.4.4 we show that the differences between hole and electron doped spectra can be rationalized in terms of the fact that AF correlations induce a gap at $(0, \pi)$ in the former case and at $(\frac{\pi}{2}, \frac{\pi}{2})$ in the electron doped regime.

The central point of this paper is to propose a theoretical description of ARPES data of superconducting doped Mott insulators, which deviates in many ways from conventional BCS mean-field behavior. The doped-carrier mean-field approach, which accounts for the role of both short-range staggered moment and d-wave singlet pair correlations, is found to reproduce a broad set of non-trivial experimental observations. Therefore, this paper offers strong support to the fact that local correlations play a fundamental role in the dynamics of electrons in strongly interacting systems. In addition, it illustrates how ARPES data can provide crucial insights to understand the short-range correlations that develop in the cuprate underdoped regime. Specifically, our results support that ARPES data is consistent with the coexistence of short-range AF and dSC correlations throughout a vast range of the high-T_c phase diagram.

Acknowledgments

The authors acknowledge several conversations with P.A. Lee. They also thank C. Nave and P.A. Lee for providing some of their variational Monte Carlo data. This work was supported by the Fundação Calouste Gulbenkian Grant No. 58119 (Portugal), by the NSF Grant No. DMR–04–33632, NSF-MRSEC Grant No. DMR–02–13282 and NFSC Grant No. 10228408. TCR was also supported by the LDRD program of LBNL under DOE #DE-AC02-05CH11231.

References

1. P. W. Anderson, The resonating valence bond state in La_2CuO_4 and superconductivity, *Science*. **235**, 1196, (1987).
2. E. Dagotto, Correlated electrons in high-temperature superconductors, *Rev. Mod. Phys.* **66**, 763, (1994).
3. P. A. Lee, N. Nagaosa, and X.-G. Wen, Doping a Mott insulator: Physics of high-temperature superconductivity, *Rev. Mod. Phys.* **78**, 17, (2006).
4. N. F. Mott, The basis of the electron theory of metals, with special reference to the transition metals, *Proc. Phys. Soc. (London)*. **A62**, 416, (1949).

5. P. W. Anderson, New approach to the theory of superexchange interactions, *Phys. Rev.* **115**, 2, (1959).
6. J. Hubbard, Electron correlations in narrow energy bands, *Proc. Roy. Soc. (London)*. **A276**, 238, (1963).
7. T. Timusk and B. Statt, The pseudogap in high-temperature superconductors: an experimental survey, *Rep. Prog. Phys.* **62**, 61, (1999).
8. C. E. Gough, M. S. Colclough, E. M. Forgan, R. G. Jordan, M. Keene, C. M. Muirhead, A. I. M. Rae, N. Thomas, J. S. Abell, and S. Sutton, Flux-quantization in a high-T_c superconductor, *Nature*. **326**, 855, (1987).
9. J. C. Wynn, D. A. Bonn, B. W. Gardner, Y.-J. Lin, R. Liang, W. N. Hardy, J. R. Kirtley, and K. A. Moler, Limits on spin-charge separation from h/2e fluxoids in very underdoped $YBa_2Cu_3O_{6+x}$, *Phys. Rev. Lett.* **87**, 197002, (2001).
10. D. A. Bonn, J. C. Wynn, B. W. Gardner, Y.-J. Lin, R. Liang, W. N. Hardy, J. R. Kirtley, and K. A. Moler, A limit on spin-charge separation in high-T-c superconductors from the absence of a vortex-memory effect, *Nature*. **414**, 887, (2001).
11. A. Hosseini, R. Harris, S. Kamal, P. Dosanjh, J. Preston, R. Liang, W. N. Hardy, and D. A. Bonn, Microwave spectroscopy of thermally excited quasiparticles in $YBa_2Cu_3O_{6.99}$, *Phys. Rev. B*. **60**, 1349, (1999).
12. Y. Zhang, N. P. Ong, P. W. Anderson, D. A. Bonn, R. Liang, and W. N. Hardy, Giant enhancement of the thermal hall conductivity k_{xy} n the superconductor $YBa_2Cu_3O_7$, *Phys. Rev. Lett.* **86**, 890, (2001).
13. H. Matsui, T. Sato, T. Takahashi, S.-C. Wang, H.-B. Yang, H. Ding, T. Fujii, T. Watanabe, and A. Matsuda, BCS-like Bogoliubov quasiparticles in high-T_c superconductors observed by angle-resolved photoemission spectroscopy, *Phys. Rev. Lett.* **90**, 217002, (2003).
14. M. S. D. G. Hawthorn, R. W. Hill, F. Ronning, S. Wakimoto, H. Zhang, C. Proust, E. Boaknin, C. Lupien, L. Taillefer, R. Liang, D. A. Bonn, W. N. Hardy, R. Gagnon, N. E. Hussey, T. Kimura, M. Nohara, and H. Takagi, Thermal conductivity across the phase diagram of cuprates: Low-energy quasiparticles and doping dependence of the superconducting gap, *Phys. Rev. B*. **67**, 174520, (2003).
15. J. Bardeen, L. N. Cooper, and J. R. Schrieffer, Theory of superconductivity, *Phys. Rev.* **108**, 1175, (1957).
16. J. R. Schrieffer, *Theory of Superconductivity*. (Addison-Wesley, New York, 1964).
17. X. J. Zhou, T. Yoshida, D.-H. Lee, W. L. Yang, V. Brouet, F. Zhou, W. X. Ti, J. W. Xiong, Z. X. Zhao, T. Sasagawa, T. Kakeshita, H. Eisaki, S. Uchida, A. Fujimori, Z. Hussain, and Z.-X. Shen, Dichotomy between nodal and antinodal quasiparticles in underdoped $(La_{2-x}Sr_x)CuO_4$ superconductors, *Phys. Rev. Lett.* **92**, 187001, (2004).
18. F. Ronning, T. Sasagawa, Y. Kohsaka, K. M. Shen, A. Damascelli, C. Kim, T. Yoshida, N. P. Armitage, D. H. Lu, D. L. Feng, L. L. Miller, H. Takagi, and Z.-X. Shen, Evolution of a metal to insulator transition in $Ca_{2-x}Na_xCuO_2$ Cl_2 as seen by angle-resolved photoemission, *Phys. Rev. B*. **67**, 165101, (2003).
19. A. Lanzara, P. V. Bogdanov, X. J. Zhou, S. A. Kellar, D. L. Feng, E. D. Lu, T. Yoshida, H. Eisaki, A. Fujimori, K. Kishio, J.-I. Shimoyama, T. Noda, S. Uchida, Z. Hussain, and Z.-X. Shen, Evidence for ubiquitous strong electron-phonon coupling in high-temperature superconductors, *Nature*. **412**, 510, (2001).

20. A. Lanzara, P. V. Bogdanov, X. J. Zhou, N. Kaneko, H. Eisaki, M. Greven, Z. Hussain, and Z.-X. Shen, Normal state spectral lineshapes of nodal quasiparticles in single layer Bi2201 superconductor, *cond-mat/0412178.* (2004).
21. A. Kaminski, M. Randeria, J. C. Campuzano, M. R. Norman, H. Fretwell, J. Mesot, T. Sato, T. Takahashi, and K. Kadowaki, Renormalization of spectral line shape and dispersion below T_c in $Bi_2Sr_2CaCu_2O_{8+\delta}$, *Phys. Rev. Lett.* **86**, 1070, (2001).
22. A. Ino, C. Kim, M. Nakamura, T. Yoshida, T. Mizokawa, A. Fujimori, Z.-X. Shen, T. Kakeshita, H. Eisaki, and S. Uchida, Doping-dependent evolution of the electronic structure of $La_{2-x}Sr_xCuO_4$ in the superconducting and metallic phases, *Phys. Rev. B.* **65**, 094504, (2002).
23. D. M. King, Z.-X. Shen, D. S. Dessau, D. S. Marshall, C. H. Park, W. E. Spicer, J. L. Peng, Z. Y. Li, and R. L. Greene, Observation of a saddle-point singularity in $Bi_2(Sr_{0.97}Pr_{0.03})_2CuO_{6+\delta}$ and its implications for normal and superconducting state properties, *Phys. Rev. Lett.* **73**, 3298, (1994).
24. T. Yoshida, X. J. Zhou, T. Sasagawa, W. L. Yang, P. V. Bogdanov, A. Lanzara, Z. Hussain, T. Mizokawa, A. Fujimori, H. Eisaki, Z.-X. Shen, T. Kakeshita, and S. Uchida, Metallic behavior of lightly doped $La_{2-x}Sr_xCuO_4$ with a Fermi surface forming an arc, *Phys. Rev. Lett.* **91**, 027001, (2003).
25. Y. Kohsaka, T. Sasagawa, F. Ronning, T. Yoshida, C. Kim, T. Hanaguri, M. Azuma, M. Takano, Z.-X. Shen, and H. Takagi, Angle-resolved photoemission spectroscopy of $(Ca,Na)_2CuiO_2Cl_2$ crystals: Fingerprints of a magnetic insulator in a heavily underdoped superconductor, *J. Phys. Soc. Jpn.* **72**, 1018, (2003).
26. N. P. Armitage, F. Ronning, D. H. Lu, C. Kim, A. Damascelli, K. M. Shen, D. L. Feng, H. Eisaki, Z.-X. Shen, P. K. Mang, N. Kaneko, M. Greven, Y. Onose, Y. Taguchi, and Y. Tokura, Doping dependence of an n-type cuprate superconductor investigated by angle-resolved photoemission spectroscopy, *Phys. Rev. Lett.* **88**, 257001, (2002).
27. A. Damascelli, Z.-X. Shen, and Z. Hussain, Angle-resolved photoemission studies of the cuprate superconductors, *Rev. Mod. Phys.* **75**, 473, (2003).
28. E. Manousakis, The spin-$1/2$ Heisenberg antiferromagnet on a square lattice and its application to the cuprous oxides, *Rev. Mod. Phys.* **63**, 1, (1991).
29. G.Baskaran, Z. Zou, and P. W. Anderson, The resonating valence bond state and high-tc superconductivity - a mean field-theory, *Solid State Commun.* **63**, 973, (1987).
30. G. Baskaran and P. W. Anderson, Gauge theory of high-temperature superconductors and strongly correlated Fermi systems, *Phys. Rev. B.* **37**, R580, (1988).
31. I. Affleck, Z. Zou, T. Hsu, and P. W. Anderson, SU(2) gauge symmetry of the large-U limit of the Hubbard model, *Phys. Rev. B.* **38**, 745, (1988).
32. G. Kotliar and J. Liu, Superexchange mechanism and d-wave superconductivity, *Phys. Rev. B.* **38**, R5142, (1988).
33. P. A. Lee and N. Nagaosa, Gauge theory of the normal state of high-T_c superconductors, *Phys. Rev. B.* **46**, 5621, (1992).
34. X.-G. Wen and P. A. Lee, Theory of underdoped cuprates, *Phys. Rev. Lett.* **76**, 503, (1996).
35. P. A. Lee, N. Nagaosa, T.-K. Ng, and X.-G. Wen, SU(2) formulation of the t-J model: Application to underdoped cuprates, *Phys. Rev. B.* **57**, 6003, (1998).

36. S. Schmitt-Rink, C. M. Varma, and A. E. Ruckenstein, Spectral function of holes in a quantum antiferromagnet, *Phys. Rev. Lett.* **60**, 2793, (1988).
37. T. C. Ribeiro and X.-G. Wen, New mean field theory of the tt't"J model applied to the high-T_c superconductors, *Phys. Rev. Lett.* **95**, 057001, (2005).
38. T. C. Ribeiro and X.-G. Wen, Doped carrier formulation and mean-field theory of the tt't"J model, *Phys. Rev. B.* **74**, 155113, (2006).
39. J. D. Koralek, J. F. Douglas, N. C. Plumb, Z. Sun, A. V. Fedorov, M. M. Murnane, H. C. Kapteyn, S. T. Cundiff, Y. Aiura, K. Oka, H. Eisaki, and D. S. Dessau, Laser based angle-resolved photoemission, the sudden approximation, and quasiparticle-like spectral peaks in $Bi_2Sr_2CaCu_2O_{8+\delta}$, *Phys. Rev. Lett.* **96**, 017005, (2006).
40. C. Kim, P. J. White, Z.-X. Shen, T. Tohyama, Y. Shibata, S. Maekawa, B. O. Wells, Y. J. Kim, R. J. Birgeneau, and M. A. Kastner, Systematics of the photoemission spectral function of cuprates: Insulators and hole- and electron-doped superconductors, *Phys. Rev. Lett.* **80**, 4245, (1998).
41. Z.-X. Shen, D. S. Dessau, B. O. Wells, D. M. King, W. E. Spicer, A. J. Arko, D. Marshall, L. W. Lombardo, A. Kapitulnik, P. Dickinson, S. Doniach, J. DiCarlo, T. Loeser, and C. H. Park, Anomalously large gap anisotropy in the a-b plane of $Bi_2Sr_2CaCu_2O_{8+\delta}$, *Phys. Rev. Lett.* **70**, 1553, (1993).
42. K. M. Shen, F. Ronning, D. H. Lu, W. S. Lee, N. J. C. Ingle, W. Meevasana, F. Baumberger, A. Damascelli, N. P. Armitage, L. L. Miller, Y. Kohsaka, M. Azuma, M. Takano, H. Takagi, and Z.-X. Shen, Missing quasiparticles and the chemical potential puzzle in the doping evolution of the cuprate superconductors, *Phys. Rev. Lett.* **93**, 267002, (2004).
43. A. Ino, C. Kim, M. Nakamura, T. Yoshida, T. Mizokawa, Z.-X. Shen, A. Fujimori, T. Kakeshita, H. Eisaki, and S. Uchida, Electronic structure of $La_{2-x}Sr_xCuO_4$ in the vicinity of the superconductor-insulator transition, *Phys. Rev. B.* **62**, 4137, (2000).
44. B. O. Wells, Z.-X. Shen, A. Matsuura, D. M. King, M. A. Kastner, M. Greven, and R. J. Birgeneau, E versus k relations and many body effects in the model insulating copper oxide $Sr_2CuO_2Cl_2$, *Phys. Rev. Lett.* **74**, 964, (1995).
45. F. Ronning, C. Kim, D. L. Feng, D. S. Marshall, A. G. Loeser, L. L. Miller, J. N. Eckstein, I. Bozovic, and Z.-X. Shen, Photoemission evidence for a remnant fermi surface and a d-wave-like dispersion in insulating $Ca_2CuO_2Cl_2$, *Science.* **282**, 2067, (1998).
46. H. S. Jin, D. H. Lu, N. P. Armitage, W. H. Choi, B. J. Kim, S.-J. Oh, S. H. Moon, H. Eisaki, and C. Kim, Angle resolved photoemission studies on $Sm_{(2-x)}Ce_{(x)}CuO_4$: remnant Fermi surfaces and coupling to (pi,pi) scattering, *J. Phys. Chem. Solids.* **65**, 1403, (2004).
47. V. J. Emery, Theory of high-T_c superconductivity in oxides, *Phys. Rev. Lett.* **58**, 2794, (1987).
48. C. M. Varma, S. Schmitt-Rink, and E. Abrahams, Charge-transfer excitations and superconductivity in ionic metals, *Solid State Commun.* **62**, 681, (1987).
49. F. C. Zhang and T. M. Rice, Effective hamiltonian for the superconducting Cu oxides, *Phys. Rev. B.* **37**, 3759, (1988).
50. N. B. Brookes, G. Ghiringhelli, O. Tjernberg, L. H. Tjeng, T. Mizokawa, T. W. Li, and A. A. Menovsky, Detection of Zhang-Rice singlets using spin-polarized photoemission, *Phys. Rev. Lett.* **87**, 237003, (2001).

51. G. Ghiringhelli, N. B. Brookes, L. H. Tjeng, T. Mizokawa, O. Tjernberg, P. G. Steeneken, and A. A. Menovsky, Probing the singlet character of the two-hole states in cuprate superconductors, *Physica B.* **312-313**, 34, (2002).
52. T. Tohyama and S. Maekawa, Angle-resolved photoemission in high T-c cuprates from theoretical viewpoints, *Supercond. Sci. Technol.* **13**, R17, (2000).
53. T. Tohyama, Asymmetry of the electronic states in hole- and electron-doped cuprates: Exact diagonalization study of the t-t[prime]-t[double-prime]-J model, *Phys. Rev. B.* **70**, 174517, (2004).
54. T. Xiang and J. M. Wheatley, Quasiparticle energy dispersion in doped two-dimensional quantum antiferromagnets, *Phys. Rev. B.* **54**, R12653, (1996).
55. V. I. Belinicher, A. L. Chernyshev, and V. A. Shubin, Single-hole dispersion relation for the real CuO_2 plane, *Phys. Rev. B.* **54**, 14914, (1996).
56. M. S. Hybertsen, E. B. Stechel, M. Schluter, and D. R. Jennison, Renormalization from density-functional theory to strong-coupling models for electronic states in Cu-O materials, *Phys. Rev. B.* **41**, 11068, (1990).
57. E. Pavarini, I. Dasgupta, T. Saha-Dasgupta, O. Jepsen, and O. K. Andersen, Band-structure trend in hole-doped cuprates and correlation with $T_{c,max}$, *Phys. Rev. Lett.* **87**, 047003, (2001).
58. P. E. Sulewski, P. A. Fleury, K. B. Lyons, S.-W. Cheong, and Z. Fisk, Light scattering from quantum spin fluctuations in R_2CuO_4 (R=La, Nd, Sm), *Phys. Rev. B.* **41**, 225, (1990).
59. R. Coldea, S. M. Hayden, G. Aeppli, T. G. Perring, C. D. Frost, T. E. Mason, S.-W. Cheong, and Z. Fisk, Spin waves and electronic interactions in La_2CuO_4, *Phys. Rev. Lett.* **86**, 5377, (2001).
60. K. Tanaka, T. Yoshida, A. Fujimori, D. H. Lu, Z.-X. Shen, X.-J. Zhou, H. Eisaki, Z. Hussain, S. Uchida, Y. Aiura, K. Ono, T. Sugaya, T. Mizuno, and I. Terasaki, Effects of next-nearest-neighbor hopping t[prime] on the electronic structure of cuprate superconductors, *Phys. Rev. B.* **70**, 092503, (2004).
61. A. G. Loeser, Z.-X. Shen, D. S. Dessau, D. S. Marshall, C. H. Park, P. Fournier, and A. Kapitulnik, Excitation gap in the normal state of underdoped $Bi_2Sr_2CaCu_2O_{8+\delta}$, *Science.* **273**, 325, (1996).
62. H. Ding, T. Yokoya, J. C. Campuzano, T. Takahashi, M. Randeria, M. R. Norman, T. Mochiku, K. Kadowaki, and J. Giapintzakis, Spectroscopic evidence for a pseudogap in the normal state of underdoped high-T-c superconductors, *Nature.* **382**, 51, (1996).
63. M. Randeria and J. C. Campuzano, High Tc superconductors: New insights from angle-resolved photoemission, *cond-mat/9709107.* (1997).
64. J. C. Campuzano, H. Ding, M. R. Norman, H. M. Fretwell, M. Randeria, A. Kaminski, J. Mesot, T. Takeuchi, T. Sato, T. Yokoya, T. Takahashi, T. Mochiku, K. Kadowaki, P. Guptasarma, D. G. Hinks, Z. Konstantinovic, Z. Z. Li, and H. Raffy, Electronic spectra and their relation to the (π, π) collective mode in high- tc superconductors, *Phys. Rev. Lett.* **83**, 3709, (1999).
65. M. Eschrig and M. R. Norman, Neutron resonance: Modeling photoemission and tunneling data in the superconducting state of $Bi_2Sr_2CaCu_2O_{8+\delta}$, *Phys. Rev. Lett.* **85**, 3261, (2000).

66. A. Kaminski, H. M. Fretwell, M. R. Norman, M. Randeria, S. Rosenkranz, U. Chatterjee, J. C. Campuzano, J. Mesot, T. Sato, T. Takahashi, T. Terashima, M. Takano, K. Kadowaki, Z. Z. Li, and H. Raffy, Doping-dependent nonlinear Meissner effect and spontaneous currents in high-T_c superconductors, *Phys. Rev. B.* **71**, 014517, (2005).
67. J. E. Hoffman, K. McElroy, D.-H. Lee, K. M. Lang, H. Eisaki, S. Uchida, and J. C. Davis, Imaging quasiparticle interference in $Bi_2Sr_2CaCu_2O_{8+\delta}$, *Science.* **297**, 1148, (2002).
68. A. Hosseini, D. M. Broun, D. E. Sheehy, T. P. Davis, M. Franz, W. N. Hardy, R. Liang, and D. A. Bonn, Survival of the d-wave superconducting state near the edge of antiferromagnetism in the cuprate phase diagram, *Phys. Rev. Lett.* **93**, 107003, (2004).
69. H. Pan, J. P. O'Neal, R. L. Badzey, C. Chamon, H. Ding, J. R. Engelbrecht, Z. Wang, H. Eisaki, S. Uchida, A. K. Gupta, K.-W. Ng, E. W. Hudson, K. M. Lang, and J. C. Davis, Microscopic electronic inhomogeneity in the high-T_c superconductor $Bi_2Sr_2CaCu_2O_{8+x}$, *Nature.* **413**, 282, (2001).
70. K. McElroy, D.-H. Lee, J. E. Hoffman, K. M. Lang, J. Lee, E. W. Hudson, H. Eisaki, S. Uchida, and J. C. Davis, Coincidence of checkerboard charge order and antinodal state decoherence in strongly underdoped superconducting $Bi_2Sr_2CaCu_2O_{8+\delta}$, *Phys. Rev. Lett.* **94**, 197005, (2005).
71. Y. Kohsaka, K. Iwaya, S. Satow, T. Hanaguri, M. Azuma, M. Takano, and H. Takagi, Imaging nanoscale electronic inhomogeneity in the lightly doped Mott insulator $Ca_{2-x}Na_xCuO_2Cl_2$, *Phys. Rev. Lett.* **93**, 097004, (2004).
72. T. Hanaguri, C. Lupien, Y. Kohsaka, D.-H. Lee, M. Azuma, M. Takano, H. Takagi, and J. C. Davis, A 'checkerboard' electronic crystal state in lightly hole-doped $Ca_{2-x}Na_xCuO_2Cl_2$, *Nature.* **430**, 1001, (2004).
73. Y. Gallais, A. Sacuto, T. P. Devereaux, and D. Colson, Interplay between the pseudogap and superconductivity in underdoped $HgBa_2CuO_{4+\delta}$ single crystals, *Phys. Rev. B.* **71**, 012506, (2005).
74. N. Gedik, J. Orenstein, R. Liang, D. A. Bonn, and W. N. Hardy, Diffusion of nonequilibrium quasi-particles in a cuprate superconductor, *Science.* **300**, 1410, (2003).
75. C. Panagopoulos, T. Xiang, W. Anukool, J. R. Cooper, Y. S. Wang, and C. W. Chu, Superfluid response in monolayer high-T_c cuprates, *Phys. Rev. B.* **67**, 220502(R), (2003).
76. D. van der Marel, Anisotropy of the optical conductivity of high-T_c cuprates, *Phys. Rev. B.* **60**, R765, (1999).
77. S. Ono and Y. Ando, Evolution of the resistivity anisotropy in $Bi_2Sr_{2-x}La_xCuO_{6+\delta}$ single crystals for a wide range of hole doping, *Phys. Rev. B.* **67**, 104512, (2003).
78. L. Krusin-Elbaum, T. Shibauchi, and C. H. Mielke, Null orbital frustration at the pseudogap boundary in a layered cuprate superconductor, *Phys. Rev. Lett.* **92**, 097005, (2004).
79. L. Krusin-Elbaum, G. Blatter, and T. Shibauchi, Zeeman and orbital limiting fields: Separated spin and charge degrees of freedom in cuprate superconductors, *Phys. Rev. B.* **69**, 220506(R), (2004).
80. T. Claesson, M. Månsson, C. Dallera, F. Venturini, C. D. Nadaï, N. B. Brookes, and O. Tjernberg, Angle resolved photoemission from $Nd_{1.85}Ce_{0.15}CuO_4$ using high energy photons: A Fermi surface investigation, *Phys. Rev. Lett.* **93**, 136402, (2004).

81. H. Matsui, K. Terashima, T. Sato, T. Takahashi, S.-C. Wang, H.-B. Yang, H. Ding, T. Uefuji, and K. Yamada, Angle-resolved photoemission spectroscopy of the antiferromagnetic superconductor $Nd_{1.87}Ce_{0.13}CuO_4$: Anisotropic spin-correlation gap, pseudogap, and the induced quasiparticle mass enhancement, *Phys. Rev. Lett.* **94**, 047005, (2005).
82. H. Matsui, K. Terashima, T. Sato, T. Takahashi, M. Fujita, and K. Yamada, Direct observation of a nonmonotonic $d_{x^2-y^2}$ wave superconducting gap in the electron-doped high-T_c superconductor, *Phys. Rev. Lett.* **95**, 017003, (2005).
83. T. Tohyama and S. Maekawa, Electronic states in the antiferromagnetic phase of electron-doped high-T_c cuprates, *Phys. Rev. B.* **64**, 212505, (2001).
84. P. W. Leung, Charge carrier correlation in the electron-doped t-J model, *Phys. Rev. B.* **73**, 075104, (2006).
85. T. C. Ribeiro and X.-G. Wen, Tunneling spectra of layered strongly correlated d-wave superconductors, *Phys. Rev. Lett.* **97**, 057003, (2006).
86. T. C. Ribeiro, A. Seidel, J. H. Han, and D.-H. Lee, The electronic states of two oppositely doped Mott insulators bilayers, *Europhys. Lett.* **76**, 891, (2006).
87. T. C. Ribeiro, X.-G. Wen, and A. Vishwanath. Electromagnetic response of high-T_c superconductors – the slave-boson and doped-carrier theories. to be published.
88. C. L. Kane, P. A. Lee, and N. Read, Motion of a single hole in a quantum antiferromagnet, *Phys. Rev. B.* **39**, 6880, (1989).
89. E. Dagotto, A. Nazarenko, and M. Boninsegni, Flat quasiparticle dispersion in the 2d t-J model, *Phys. Rev. Lett.* **73**, 728, (1994).
90. T. Tohyama and S. Maekawa, Role of next-nearest-neighbor hopping in the t-t'-J model, *Phys. Rev. B.* **49**, 3596, (1994).
91. M. Randeria, R. Sensarma, N. Trivedi, and F.-C. Zhang, Particle-hole asymmetry in doped Mott insulators: Implications for tunneling and photoemission spectroscopies, *Phys. Rev. Lett.* **95**, 137001, (2005).
92. M. R. Norman, M. Randeria, H. Ding, and J. C. Campuzano, Phenomenological models for the gap anisotropy of $Bi_2Sr_2CaCu_2O_8$ as measured by angle-resolved photoemission spectroscopy, *Phys. Rev. B.* **52**, 615, (1995).
93. H. Ding, M. R. Norman, T. Yokoya, T. Takeuchi, M. Randeria, J. C. Campuzano, T. Takahashi, T. Mochiku, and K. Kadowaki, Evolution of the Fermi surface with carrier concentration in $Bi_2Sr_2CaCu_2O_{8+\delta}$, *Phys. Rev. Lett.* **78**, 2628, (1997).
94. R. H. He, D. L. Feng, H. Eisaki, J.-I. Shimoyama, K. Kishio, and G. D. Gu, Superconducting order parameter in heavily overdoped $Bi_2Sr_2CaCu_2O_{8+\delta}$: A global quantitative analysis, *Phys. Rev. B.* **69**, 220502(R), (2004).
95. C. P. Nave, D. A. Ivanov, and P. A. Lee, Variational Monte Carlo study of the current carried by a quasiparticle, *Phys. Rev. B.* **73**, 104502, (2006).
96. D. L. Feng, D. H. Lu, K. M. Shen, C. Kim, H. Eisaki, A. Damascelli, R. Yoshizaki, J.-I. Shimoyama, K. Kishio, G. D. Gu, S. Oh, A. Andrus, J. O'Donnell, J. N. Eckstein, and Z.-X. Shen, Signature of superfluid density in the single-particle excitation spectrum of $Bi_2Sr_2CaCu_2O_{8+\delta}$, *Science.* **289**, 277, (2000).
97. H. Ding, J. R. Engelbrecht, Z. Wang, J. C. Campuzano, S.-C. Wang, H.-B. Yang, R. Rogan, T. Takahashi, K. Kadowaki, and D. G. Hinks, Coherent quasiparticle weight and its connection to high- T_c superconductivity from angle-resolved photoemission, *Phys. Rev. Lett.* **87**, 227001, (2001).

98. A. Carrington, A. P. Mackenzie, C. T. Lin, and J. R. Cooper, Temperature dependence of the Hall angle in single-crystal $YBa_2(Cu_{1-x}Co_x)_3O_{7-\delta}$, *Phys. Rev. Lett.* **69**, 2855, (1992).
99. P. C. Pattnaik, C. L. Kane, D. M. Newns, and C. C. Tsuei, Evidence for the van Hove scenario in high-temperature superconductivity from quasiparticle-lifetime broadening, *Phys. Rev. B.* **45**, 5714, (1992).
100. P. Monthoux and D. Pines, Spin-fluctuation-induced superconductivity and normal-state properties of $YBa_2Cu_3O_7$, *Phys. Rev. B.* **49**, 4261, (1994).
101. R. Hlubina and T. M. Rice, Resistivity as a function of temperature for models with hot spots on the Fermi surface, *Phys. Rev. B.* **51**, 9253, (1995).
102. M. R. Norman, Relation of neutron incommensurability to electronic structure in high-temperature superconductors, *Phys. Rev. B.* **61**, 14751, (2000).
103. M. R. Norman, H. Ding, J. C. Campuzano, T. Takeuchi, M. Randeria, T. Yokoya, T. Takahashi, T. Mochiku, and K. Kadowaki, Unusual dispersion and line shape of the superconducting state spectra of $Bi_2Sr_2CaCu_2O_{8+\delta}$, *Phys. Rev. Lett.* **79**, 3506, (1997).
104. A. Fujimori, A. Ino, T. Mizokawa, C. Kim, Z.-X. Shen, T. Sasagawa, T. Kimura, K. Kishio, M. Takaba, K. Tamasaku, H. Eisaki, and S. Uchidau, Chemical potential shift, density of states and Fermi surfaces in overdoped and underdoped $La_{2-x}Sr_xCuO_4$, *J. Phys. Chem. Solids.* **59**, 1892, (1998).
105. T. Kondo, T. Takeuchi, T. Yokoya, S. Tsuda, S. Shin, and U. Mizutani, Hole-concentration dependence of band structure in $(Bi,Pb)_2(Sr,La)_2CuO_{6+\delta}$ determined by the angle-resolved photoemission spectroscopy, *J. Electron Spectr. Relat. Phenom.* **137-140**, 663, (2004).
106. P. V. Bogdanov, A. Lanzara, X. J. Zhou, S. A. Kellar, D. L. Feng, E. D. Lu, H. Eisaki, J.-I. Shimoyama, K. Kishio, Z. Hussain, and Z.-X. Shen, Photoemission study of Pb doped $Bi_2Sr_2CaCu_2O_8$: A Fermi surface picture, *Phys. Rev. B.* **64**, 180505(R), (2001).
107. A. Kaminski, S. Rosenkranz, H. M. Fretwell, M. R. Norman, M. Randeria, J. C. Campuzano, J.-M. Park, Z. Z. Li, and H. Raffy, Change of Fermi-surface topology in $Bi_2Sr_2CaCu_2O_{8+\delta}$ with doping, *Phys. Rev. B.* **73**, 174511, (2006).
108. C. T. Shih, T. K. Lee, R. Eder, C.-Y. Mou, and Y. C. Chen, Enhancement of pairing correlation by t[prime] in the two-dimensional extended t-J model, *Phys. Rev. Lett.* **92**, 227002, (2004).
109. H. C. Fu, J. C. Davis, and D.-H. Lee, On the charge ordering observed by recent STM experiments, *cond-mat/0403001*. (2004).
110. J. Mesot, M. R. Norman, H. Ding, M. Randeria, J. C. Campuzano, A. Paramekanti, H. M. Fretwell, A. Kaminski, T. Takeuchi, T. Yokoya, T. Sato, T. Takahashi, T. Mochiku, and K. Kadowaki, Superconducting gap anisotropy and quasiparticle interactions: A doping dependent photoemission study, *Phys. Rev. Lett.* **83**, 840, (1999).
111. T. Matsuzaki, N. Momono, M. Oda, and M. Ido, Electronic specific heat of $La_{2-x}Sr_xCuO_4$: Pseudogap formation and reduction of the superconducting condensation energy, *J. Phys. Soc. Jpn.* **73**, 2232, (2004).
112. D. Sénéchal and A. M. S. Tremblay, Hot spots and pseudogaps for hole- and electron-doped high-temperature superconductors, *Phys. Rev. Lett.* **92**, 126401, (2004).

113. D. Sénéchal, P. L. Lavertu, M. A. Marois, and A. M. S. Tremblay, Competition between antiferromagnetism and superconductivity in high-T_c cuprates, *cond-mat/0410162*. (2004).
114. C. Dahnken, M. Potthoff, E. Arrigoni, and W. Hanke, Correlated band structure of electron-doped cuprate materials, *cond-mat/0504618*. (2005).
115. T. C. Ribeiro, The short-range correlations of a doped Mott insulator, *cond-mat/0605437*. (2006).
116. G. Blumberg, A. Koitzsch, A. Gozar, B. S. Dennis, C. A. Kendziora, P. Fournier, and R. L. Greene, Nonmonotonic d_{x2-y2} superconducting order parameter in $Nd_{2-x}Ce_xCuO_4$, *Phys. Rev. Lett.* **88**, 107002, (2002).
117. T. Watanabe, T. Miyata, H. Yokoyama, Y. Tanaka, and J.-I. Inoue, Nonmonotonic $d_{x^2-y^2}$-wave superconductivity in electron-doped cuprates viewed from the strong-coupling side, *J. Phys. Soc. Jpn.* **74**, 1942, (2005).
118. T. C. Ribeiro and X.-G. Wen, Possible z2 phase and spin-charge separation in electron doped cuprate superconductors, *Phys. Rev. B*. **68**, 024501, (2003).
119. A. Moreo, S. Haas, A. W. Sandvik, and E. Dagotto, Quasiparticle dispersion of the t-J and Hubbard models, *Phys. Rev. B*. **51**, R12045, (1995).
120. R. Preuss, W. Hanke, and W. von der Linden, Quasiparticle dispersion of the 2d Hubbard model: From an insulator to a metal, *Phys. Rev. Lett.* **75**, 1344, (1995).
121. A. Dorneich, M. G. Zacher, C. Gröber, and R. Eder, Strong-coupling theory for the Hubbard model, *Phys. Rev. B*. **61**, 12816, (2000).
122. C. Gröber, R. Eder, and W. Hanke, Anomalous low-doping phase of the Hubbard model, *Phys. Rev. B*. **62**, 4336, (2000).
123. A. S. Mishchenko and N. Nagaosa, Electron-phonon coupling and a polaron in the t-J model: From the weak to the strong coupling regime, *Phys. Rev. Lett.* **93**, 036402, (2004).

Chapter 4

An introduction to the physics of graphene layers

Eduardo V. Castro[1] *, N. M. R. Peres[2], J. M. B. Lopes dos Santos[1], F. Guinea[3], A. H. Castro Neto[4]

[1] *CFP and Departamento de Física, Faculdade de Ciências Universidade do Porto, P-4169-007 Porto, Portugal*

[2] *Center of Physics and Departament of Physics, Universidade do Minho, P-4710-057, Braga, Portugal*

[3] *Instituto de Ciencia de Materiales de Madrid. CSIC. Cantoblanco. E-28049 Madrid, Spain*

[4] *Department of Physics, Boston University, 590 Commonwealth Avenue, Boston, MA 02215, USA*

In this chapter we revise the basic physics of a single layer and a double layer of graphene. In both cases, and starting from a tight-binding description, we show how to construct the effective continuum models. Also for the single layer and for the bilayer, both with zigzag edges, we discuss the existence of a zero energy band made of surface states localized near the edges of the sample. The spectrum of the single layer in the presence of a magnetic field is studied and its relation with the half-odd integer quantum Hall effect is discussed. For the bilayer the electronic bulk properties are studied and the unconventional quantum Hall effect is addressed. The concept of a biased bilayer is introduced together with its energy spectrum.

Contents

* evcastro@fc.up.pt

4.1. Introduction

A major breakthrough in condensed matter physics took place when K. S. Novoselov *et al.*[1] discovered an electric field effect in atomically thin carbon films. A single layer of these thin carbon films is called graphene and its electric and magneto-electric properties triggered a new research field in condensed matter physics. The manufacture of graphene was followed by the production of other two-dimensional (2D) crystals,[2] which however do not show the same exciting properties of graphene. Applying high magnetic fields to a graphene sample, the Manchester group discovered that in graphene the quantization rule for the Hall conductivity is not the same observed in the 2D electron gas, being given instead by:[3]

$$\sigma_{\text{Hall}} = \pm 4 \left(n + \frac{1}{2} \right) \frac{e^2}{h} , \qquad n = 0, 1, 2, \ldots \tag{4.1}$$

A confirmation of this result was independently obtained by other group.[4] This new quantum Hall effect was predicted by two groups working independently and using different methods.[5,6] As explained by the two groups this new quantization rule for the Hall conductivity is a consequence of the dispersion relation of electrons in the honeycomb lattice. Low energy quasi-particles in the honeycomb lattice have the same energy-momentum relation of particles in the ultra-relativistic regime, i.e., massless particles moving at an effective velocity of light. Particles with this energy dispersion – known as Dirac fermions – have in two dimensions a zero energy mode in perpendicular magnetic field, which is responsible for the unusual quantization rule of the Hall conductivity. The presence of Dirac fermions in graphite received recently experimental confirmation.[7]

More surprises were to be experimentally obtained in double layer graphene, with a quantization rule different from that seen in the single layer.[8] The quantization observed in graphene bilayer was explained by McCann and Fal'ko using a low energy effective model.[9] The experimental energy spectrum of a the graphene bilayer was recently probed by angle resolved photoemission spectroscopy, both for the unbiased and for the biased regimes.[10] In the biased regime the system has a Fermi ring occurring at finite momentum.[11]

From the above it is clear that the electronic properties of atomically thin carbon films depend very much on the number of layers. For example, films with an odd number of layers show the presence of Dirac fermions, which does not occur for even layered films.

4.1.1. *Basic definitions*

The physics of graphene is tightly related to its chemical bonding. The carbon atoms have sp^2 orbitals which lie in the two-dimensional plane forming the crystal. These orbitals do not contribute to the electronic transport however. In addition, the p_z orbitals, which are perpendicular to the plane of carbon atoms, give rise to delocalized $\pi-$orbitals which are responsible for the conducting properties of graphene. From the theoretical point of view, the $\pi-$orbitals can be modeled by a simple tight-binding Hamiltonian with a single orbital per carbon atom; it is this approach we follow in this chapter.

Let us start with a bipartite 2D layer without specifying the geometry. We assume that $\mathbf{a}_1$ and $\mathbf{a}_2$ are the basis vectors, so that any lattice vector $\mathbf{R}$ can be represented as

$$\mathbf{R} = m\mathbf{a}_1 + n\mathbf{a}_2, \tag{4.2}$$

with n, m integers. Moreover, we assume that vectors $\boldsymbol{\delta}_1 ... \boldsymbol{\delta}_z$ connect nearest-neighbors (NN) sites, and are not necessarily equal to the basis vectors (the unit cell may have more than one atom). The system is modeled through a single-orbital tight-binding Hamiltonian,

$$H_0 = -\sum_{\mathbf{R},\boldsymbol{\delta},\sigma} t_{\mathbf{R},\mathbf{R}+\boldsymbol{\delta}} c^\dagger_{\mathbf{R},\sigma} c_{\mathbf{R}+\boldsymbol{\delta},\sigma}, \tag{4.3}$$

where the hopping is nonzero only for NN sites and is assumed to be homogeneous,

$$t_{\mathbf{R},\mathbf{R}+\boldsymbol{\delta}} = \langle \mathbf{R} | H_0 | \mathbf{R} + \boldsymbol{\delta} \rangle \equiv t. \tag{4.4}$$

Many of the relevant properties of the system occur in the presence of a static magnetic field, $\mathbf{B} = \nabla \times \mathbf{A}$. Let us call $H_{\mathbf{B}}$ the Hamiltonian of the system when $\mathbf{B} \neq \mathbf{0}$, and neglect the Zeeman term. The introduction of a magnetic field in the tight-binding approximation reduces to the so called peierls substitution,[12–15] where the matrix elements of $H_{\mathbf{B}}$ in the localized-orbital basis are expressed in terms of H_0 as

$$\langle \mathbf{R} | H_{\mathbf{B}} | \mathbf{R} + \boldsymbol{\delta} \rangle = e^{i\frac{e}{\hbar}\int_{\mathbf{R}}^{\mathbf{R}+\boldsymbol{\delta}} \mathbf{A}\cdot d\mathbf{r}} \langle \mathbf{R} | H_0 | \mathbf{R} + \boldsymbol{\delta} \rangle \ . \tag{4.5}$$

Therefore, in the presence of a magnetic field, the tight-binding Hamiltonian in Eq. 4.3 is given by,

$$H_{\mathbf{B}} = -\sum_{\mathbf{R},\boldsymbol{\delta},\sigma} e^{i\frac{e}{\hbar}\int_{\mathbf{R}}^{\mathbf{R}+\boldsymbol{\delta}} \mathbf{A}\cdot d\mathbf{r}} t_{\mathbf{R},\mathbf{R}+\boldsymbol{\delta}} c^{\dagger}_{\mathbf{R},\sigma} c_{\mathbf{R}+\boldsymbol{\delta},\sigma} \,. \tag{4.6}$$

4.2. The lattice structure of graphene

Let us now focus on a specific lattice: the honeycomb lattice. This lattice is shown in Fig. 4.1 and is made of two interpenetrating triangular sublattices, labeled A and B, where the triangles have side length $a = 2.46\,\text{Å}$. The unit cell has two atoms, each belonging to a different sublattice, and the basis vectors are,

$$\mathbf{a}_1 = a\,\hat{\mathbf{e}}_x\,, \qquad \mathbf{a}_2 = \frac{a}{2}(\hat{\mathbf{e}}_x - \sqrt{3}\,\hat{\mathbf{e}}_y)\,. \tag{4.7}$$

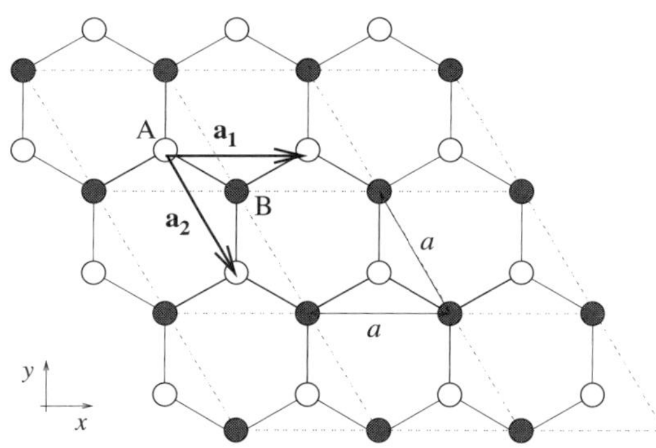

Fig. 4.1. Lattice structure of graphene.

In the honeycomb lattice the Hamiltonian defined by Eq. (4.3) can be written as,

$$H = -t\sum_{\mathbf{R},\sigma} \left[a^{\dagger}_{\sigma}(\mathbf{R})b_{\sigma}(\mathbf{R}) + a^{\dagger}_{\sigma}(\mathbf{R})b_{\sigma}(\mathbf{R}-\mathbf{a}_1) + a^{\dagger}_{\sigma}(\mathbf{R})b_{\sigma}(\mathbf{R}-\mathbf{a}_2) + \text{h.c.}\right], \tag{4.8}$$

where $c^{\dagger}_{\sigma}(\mathbf{R})$ $[c_{\sigma}(\mathbf{R})]$, with $c = a, b$, creates (annihilates) an electron of spin σ on the respective atom (A or B) of cell $\mathbf{R}$.

4.2.1. *Bulk electronic properties*

The bulk properties of the model defined by Eq. (4.8) can be determined imposing periodic boundary conditions (PBC's) to the underlying bravais lattice and applying Fourier transformation. The reciprocal lattice of a triangular lattice is steal a triangular lattice. This is easily seen by explicit construction. The basis vectors of the reciprocal lattice $\mathbf{b}_1$ and $\mathbf{b}_2$ are given through the relation $\mathbf{b}_i \cdot \mathbf{a}_j = 2\pi\,\delta_{ij}$ as usual. Recalling definition of $\mathbf{a}_j$ in Eq. (4.7) we obtain,

$$\mathbf{b}_1 = \frac{2\pi}{a}\left(\hat{\mathbf{e}}_x + \frac{1}{\sqrt{3}}\,\hat{\mathbf{e}}_y\right), \qquad \mathbf{b}_2 = -\frac{2\pi}{a}\frac{2}{\sqrt{3}}\,\hat{\mathbf{e}}_y\,. \tag{4.9}$$

In Fig. 4.2 we show the first Brillouin zone defined by the basis vectors $\mathbf{b}_1$ and $\mathbf{b}_2$, which can be viewed both as a parallelogram or as an hexagon.

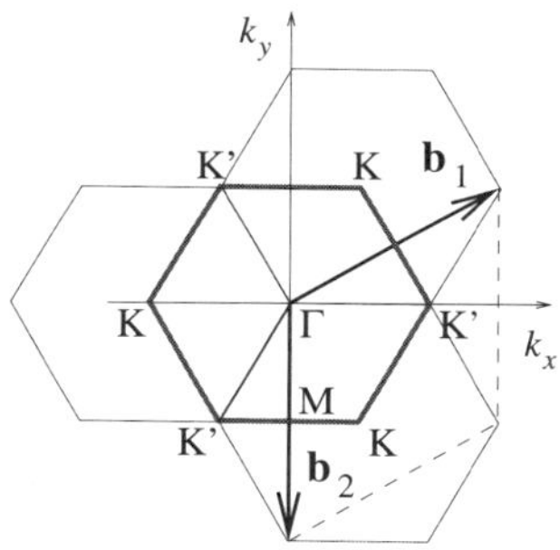

Fig. 4.2. First Brillouin zone of a triangular lattice (sublattice of the honeycomb lattice).

Introducing the Fourier components $a_{\sigma,\mathbf{k}}$ and $b_{\sigma,\mathbf{k}}$ of operators $a_\sigma(\mathbf{R})$ and $b_\sigma(\mathbf{R})$, respectively, we can rewrite Eq. (4.8) as,

$$H = \sum_{\mathbf{k},\sigma} \psi^\dagger_{\sigma,\mathbf{k}} H_\mathbf{k} \psi_{\sigma,\mathbf{k}}, \tag{4.10}$$

where $\psi^\dagger_{\sigma,\mathbf{k}} = [a^\dagger_{\sigma,\mathbf{k}}, b^\dagger_{\sigma,\mathbf{k}}]$ is a two component spinor and $H_\mathbf{k}$ is a 2×2 matrix,

$$H_\mathbf{k} = -t\begin{pmatrix} 0 & s_\mathbf{k} \\ s^*_\mathbf{k} & 0 \end{pmatrix}, \tag{4.11}$$

with $s_\mathbf{k}$ given by,

$$s_\mathbf{k} = 1 + e^{i\mathbf{k}\cdot\mathbf{a}_1} + e^{i\mathbf{k}\cdot\mathbf{a}_2}\,. \tag{4.12}$$

The spinorial form of Eq. (4.11) is a direct consequence of the presence of two atoms per unit cell, or in other words, two sublattices. Taking into account that $\mathbf{k}\cdot\mathbf{a}_1 = ak_x$ and $\mathbf{k}\cdot\mathbf{a}_2 = ak_x/2 - ak_y\sqrt{3}/2$, the secular equation applied to $H_\mathbf{k}$ gives the dispersion relation:

$$\epsilon_\mathbf{k} = \pm t\Big\{3 + 2\big[\cos(ak_x) + 2\cos(ak_x/2)\cos(\sqrt{3}ak_y/2)\big]\Big\}^{1/2}. \tag{4.13}$$

Noting that $\epsilon_{\mathbf{k}} = \pm t|s_{\mathbf{k}}|$ we easily find for the band operators,

$$e_{\mathbf{k}} = \frac{1}{\sqrt{2}}(e^{i\delta_{\mathbf{k}}/2}a_{\sigma,\mathbf{k}} - e^{-i\delta_{\mathbf{k}}/2}b_{\sigma,\mathbf{k}}) \,, \tag{4.14}$$

$$h_{\mathbf{k}} = \frac{1}{\sqrt{2}}(e^{i\delta_{\mathbf{k}}/2}a_{\sigma,\mathbf{k}} + e^{-i\delta_{\mathbf{k}}/2}b_{\sigma,\mathbf{k}}) \,, \tag{4.15}$$

where e and h stand for electron (+) and hole (-) bands, and $\delta_{\mathbf{k}}$ is defined by:

$$\delta_{\mathbf{k}} = s_{\mathbf{k}}/|s_{\mathbf{k}}| \,. \tag{4.16}$$

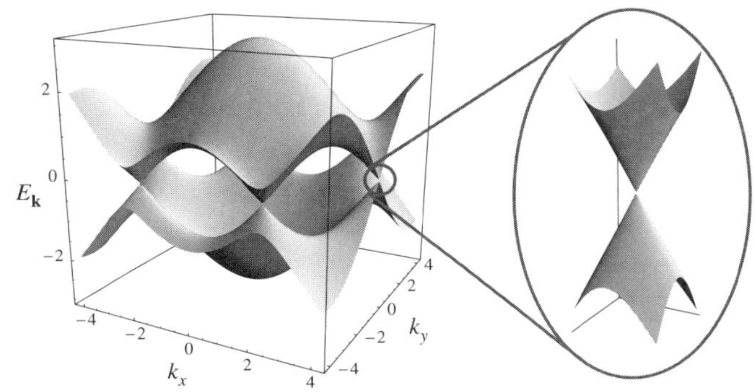

Fig. 4.3. Dispersion relation surface as defined by Eq. (4.13). Energy is given in units of t, k_x and k_y are given in units of $1/a$.

The surface defined by Eq. (4.13) can be seen in Fig. 4.3. A cone like behavior is clearly seen at the corners of first Brillouin zone. Noting that only two of such cones are non-equivalent, for example,

$$\mathbf{K}' = \frac{2\pi}{a}\frac{2}{3}\,\hat{\mathrm{e}}_x \,, \qquad \mathbf{K} = -\frac{2\pi}{a}\frac{2}{3}\,\hat{\mathrm{e}}_x \,, \tag{4.17}$$

we can expand Eq. (4.13) around $\mathbf{K}$ and $\mathbf{K}'$ to get an effective low energy description. In doing so the Dirac-like linear dispersion follows,

$$E(q) \simeq \pm v_F \hbar q, \tag{4.18}$$

where $v_F = ta\hbar^{-1}\sqrt{3}/2$ is the Fermi velocity which substitutes the speed of light, and $q = |\mathbf{q}| = |\mathbf{q}'|$, with

$$\mathbf{q} = \mathbf{k} - \mathbf{K} \,, \qquad \mathbf{q}' = \mathbf{k} - \mathbf{K}' \,. \tag{4.19}$$

In Eq. (4.17) the two Brillouin zone points $\mathbf{K}$ and $\mathbf{K}'$ are usually referred to as Dirac points.

An immediate consequence of this massless Dirac-like dispersion is a cyclotron mass that depends on the electronic density as its square root.[3] The proof is quite simple. The cyclotron mass is defined, within the semiclassical approximation,[16] as

$$m^* = \frac{\hbar^2}{2\pi}\frac{\partial A(E)}{\partial E}\,, \tag{4.20}$$

with $A(E)$ the area in $k-$space enclosed by the orbit and given by

$$A(E) = \pi q^2 = \pi \frac{E^2}{v_F^2 \hbar^2}\,. \tag{4.21}$$

Using Eq. (4.21) in Eq. (4.20) one obtains

$$m^* = \frac{E}{v_F^2} = \frac{q\hbar}{v_F}\,. \tag{4.22}$$

The electronic density, n_e, is related the Fermi momentum, q_F, as $q_F^2/\pi = n_e$ which leads to

$$m^* = \frac{\sqrt{\pi}\hbar}{v_F}\sqrt{n_e}\,. \tag{4.23}$$

Fitting Eq. (4.23) to the experimental data provides an estimation to the Fermi velocity and the hopping parameter, respectively $v_F \approx 10^{-6}\,\mathrm{ms}^{-1}$ and $t \approx 3\,\mathrm{eV}$. Note that v_F is actually 300 times smaller than c, the velocity of light. Nevertheless, the experimental observation of the $\sqrt{n_e}$ dependence of the cyclotron mass provides evidence for the existence of massless Dirac quasi-particles in graphene – note that the usual parabolic (Schrödinger) dispersion implies constant cyclotron mass.

4.2.1.1. *The continuum approximation*

At long wave lengths and low energies we can linearize Eq. (4.11) around $\mathbf{K}$ and $\mathbf{K}'$, and arrive at a continuum approximation for Hamiltonian (4.10). Within this approximation it turns out that π-electrons in the honeycomb lattice are effectively described by the Dirac Hamiltonian.

Let us start by redefining k_x and k_y as,

$$k_x = \frac{1}{2}q_x + \frac{\sqrt{3}}{2}q_y\,, \qquad k_y = -\frac{\sqrt{3}}{2}q_x + \frac{1}{2}q_y\,, \tag{4.24}$$

with $\mathbf{q}$ as in Eq. (4.19). The transformation defined by Eq. (4.24) can be seen as a $\pi/3$ clockwise rotation of the original reference frame. Expanding Eq. (4.11) around $\mathbf{K}$ and taking into account the definitions of k_x and k_y we can write:

$$H_{\mathbf{K}} = v_F \hbar \begin{pmatrix} 0 & k_x - ik_y \\ k_x + ik_y & 0 \end{pmatrix} . \tag{4.25}$$

As regards the second Dirac point $\mathbf{K}'$ we redefine k_x and k_y as,

$$k_x = -\frac{1}{2}q_x - \frac{\sqrt{3}}{2}q_y \,, \qquad k_y = \frac{\sqrt{3}}{2}q_x - \frac{1}{2}q_y \,, \tag{4.26}$$

which is just a $4\pi/3$ clockwise rotation of the original reference frame. Expanding Eq. (4.11) around $\mathbf{K}'$ we obtain,

$$H_{\mathbf{K}'} = v_F \hbar \begin{pmatrix} 0 & k_x + ik_y \\ k_x - ik_y & 0 \end{pmatrix} , \tag{4.27}$$

where k_x and k_y are given by Eq. (4.26), respectively. Introducing the vector of Pauli matrices $\boldsymbol{\sigma} = (\sigma_x, \sigma_y)$ the Dirac Hamiltonian follows easily for Eqs. (4.25) and (4.27),

$$H_{\mathbf{K}}(\mathbf{p}) = v_F \boldsymbol{\sigma} \cdot \mathbf{p} \,, \tag{4.28}$$

$$H_{\mathbf{K}'}(\mathbf{p}) = v_F \boldsymbol{\sigma}^{\dagger} \cdot \mathbf{p} \,, \tag{4.29}$$

where we have used the equality $\mathbf{p} = \hbar\mathbf{k}$. Therefore, in the continuum approximation graphene quasi-particles are described by:

$$H = \sum_{\sigma,\alpha=K,K'} \int d\mathbf{p}\, \psi^{\dagger}_{\sigma,\alpha}(\mathbf{p}) H_{\alpha}(\mathbf{p}) \psi_{\sigma,\alpha}(\mathbf{p}) \,, \tag{4.30}$$

where α is a cone index referring to $\mathbf{K}$ or $\mathbf{K}'$, and $\psi^{\dagger}_{\sigma,\alpha}(\mathbf{p}) = [a^{\dagger}_{\sigma,\alpha}(\mathbf{p}), b^{\dagger}_{\sigma,\alpha}(\mathbf{p})]$ is the spinor of continuum momentum fields, one field per sublattice.

4.2.2. *Ribbons of finite width: Surface states*

As shown in Sec. 4.2.1 low energy quasi-particles in graphene behave as massless Dirac fermions, a consequence of graphene's honeycomb structure. The honeycomb lattice is responsible for another surprising property of graphene revealed in the nineties.[17,18] When the lattice is cut such that a zigzag edge shows up, an extra band of zero energy states localized at the surface appear. These states are usually referred as *surface states*. To see why this is so we consider the ribbon geometry with zigzag edges shown in Fig. 4.4. The ribbon width is such that it has N unit cells in the transverse cross section (y direction). We assume that it has infinite length in the longitudinal direction (x direction).

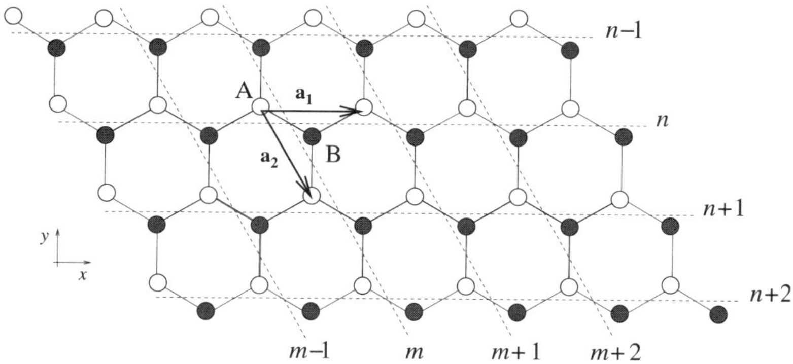

Fig. 4.4. Ribbon geometry with zigzag edges.

We start by rewriting Eq. (4.8) in terms of integer indices m and n introduced in Eq. (4.2) and shown in Fig. 4.4:

$$H = -t \sum_{m,n,\sigma} [a^\dagger_\sigma(m,n) b_\sigma(m,n) + a^\dagger_\sigma(m,n) b_\sigma(m-1,n) + a^\dagger_\sigma(m,n) b_\sigma(m,n-1) + \text{h.c.}] . \quad (4.31)$$

Given that the ribbon is infinite in the $\mathbf{a}_1$ direction we can Fourier transform along Ox introducing the quantum number $k \in [0, 2\pi[$. After this transformation Eq. (4.31) reads,

$$H = -t \int \mathrm{d}k \sum_{n,\sigma} [(1 + e^{ika}) a^\dagger_\sigma(k,n) b_\sigma(k,n) + a^\dagger_\sigma(k,n) b_\sigma(k,n-1) + \text{h.c.}] . \quad (4.32)$$

With the definition $c^\dagger_\sigma(k,n)\,|0\rangle = |c,\sigma,k,n\rangle$, where $c = a, b$, the one-particle eigenstates which are solution of the Schrödinger equation, $H\,|\mu,k,\sigma\rangle = E_{\mu,k}\,|\mu,k,\sigma\rangle$, can be generally expressed as,

$$|\mu,k,\sigma\rangle = \sum_n [\alpha(k,n)\,|a,k,n,\sigma\rangle + \beta(k,n)\,|b,k,n,\sigma\rangle] , \quad (4.33)$$

where $\alpha(k,n)$ stands for the wavefunction amplitude at sites of sublattice A and $\beta(k,n)$ at sites of sublattice B. Applying Hamiltonian (4.32) to $|\mu,k,\sigma\rangle$ we obtain for coefficients α and β the following equations:

$$-t[(1 + e^{-ika})\alpha(k,n) + \alpha(k,n+1)] = E_{\mu,k}\beta(k,n) , \quad (4.34)$$

$$-t[(1 + e^{ika})\beta(k,n) + \beta(k,n-1)] = E_{\mu,k}\alpha(k,n) . \quad (4.35)$$

We have to be careful with the boundary conditions, as the ribbon only exists between $n = 0$ and $n = N - 1$. Thus at the boundary Eqs. (4.34) and (4.35) read,

$$-t(1 + e^{-ika})\alpha(k, N-1) = E_{\mu,k}\beta(k, N-1)\,, \tag{4.36}$$

$$-t(1 + e^{ika})\beta(k, 0) = E_{\mu,k}\alpha(k, 0)\,. \tag{4.37}$$

As surface states show up as a zero-energy band they must be solution of Eqs. (4.34-4.37) with $E_{\mu,k} = 0$:

$$(1 + e^{-ika})\alpha(k, n) + \alpha(k, n+1) = 0\,, \tag{4.38}$$

$$(1 + e^{ika})\beta(k, n) + \beta(k, n-1) = 0\,, \tag{4.39}$$

$$\alpha(k, N-1) = 0\,, \tag{4.40}$$

$$\beta(k, 0) = 0\,. \tag{4.41}$$

The recursive structure of Eq. (4.38) and (4.39) is easily solved in terms of initial amplitudes $\alpha(k, 0)$ and $\beta(k, N-1)$, respectively:

$$\alpha(k, n) = [-2\cos(ka/2)]^n e^{i\frac{ka}{2}n}\alpha(k, 0)\,, \tag{4.42}$$

$$\beta(k, n) = [-2\cos(ka/2)]^{(N-1-n)} e^{-i\frac{ka}{2}(N-1-n)}\beta(k, N-1)\,. \tag{4.43}$$

Note that the initial amplitudes are actually the amplitudes at the edges. This is easily inferred from Fig. 4.4, where the top edge at lattice index $n = 0$ is composed by sites of sublattice A only, whereas the bottom edge at lattice index $n = N - 1$ has only sites of sublattice B. Therefore, requiring the convergence condition $|-2\cos(ka/2)| < 1$, we find that the semi infinite system has a surface state exponentially localized at the zigzag edge. Moreover, this surface state has finite amplitude only at sites of the same sublattice as edge sites. As required by the convergence condition, the surface state only exists for ka in the region $2\pi/3 < ka < 4\pi/3$, which corresponds to $1/3$ of the possible k's. The localization length is given by $\lambda(k) = -1/\log|2\cos(ka/2)|$, and diverges when ka approaches the limits of the convergence region $]2\pi/3, 4\pi/3[$. The amplitudes of the surface state may be written as,

$$|\alpha(k, n)| = \sqrt{\frac{2}{\lambda(k)}}e^{-n/\lambda(k)}\,, \tag{4.44}$$

$$|\beta(k, n)| = \sqrt{\frac{2}{\lambda(k)}}e^{-(N-1-n)/\lambda(k)}\,, \tag{4.45}$$

for an edge whose sites belong to sublattice A and B, respectively.

Although the boundary conditions defined by Eqs. (4.40) and (4.41) are satisfied for solutions (4.42) and (4.43) in the semi infinite system, they are not in the ribbon geometry. In the graphene ribbon the two surface states, which come from both edges, overlap with each other. The bonding and anti-bonding states formed by the two surface states are then the ribbon eigenstates[19] (note that at zero energy there are no other states with which the surface states can hybridize). As bonding and anti-bonding states have a gap in energy the zero energy flat bands of surface states become slightly dispersive, depending on the ribbon width N. However, as surface states are exponentially localized at opposite edges, overlap between the them is only appreciable for ka near $2\pi/3$ and $4\pi/3$. This means that deviations from zero energy flatness can only be seen near these points.

The presence in real samples of the surface states just discussed was experimental observed after scanning tunneling microscopy studies on both zigzag and armchair edges, with the former showing a clear blueprint in the local density of states, as the microscope tip approaches the zigzag edge of the sample.[20,21]

4.2.3. *Graphene in a perpendicular magnetic field*

As we have seen in Sec. 4.1.1 when a perpendicular magnetic field $\mathbf{B} = B\,\hat{\mathbf{e}}_z$ is applied to the system the hopping t acquires a phase φ_{ij} given by,

$$t_{ij}^{(0)} \to t_{ij}^{(B)} = t_{ij}^{(0)} e^{i\varphi_{ij}} = t_{ij}^{(0)} e^{i\frac{e}{\hbar}\int_{\mathbf{R}_i}^{\mathbf{R}_j}\mathbf{A}\cdot d\mathbf{r}}, \tag{4.46}$$

where e is the electron charge, $\hbar$ is the Plank constant divided by 2π, and $\mathbf{A}$ is the vector potential. In order to set the phase φ_{ij} we may choose a gauge and then compute the integral $\int_{\mathbf{R}_i}^{\mathbf{R}_j}\mathbf{A}\cdot d\mathbf{r}$. Or we can do the other way around, choosing the phase such that,

$$\sum_{\text{closed path}} \varphi_{ij} = 2\pi\phi/\phi_0\,, \tag{4.47}$$

where ϕ is the magnetic field flux through the area enclosed in the "path" and $\phi_0 = h/e$ is the flux quantum. Only ϕ is physical, so any phase φ_{ij} will do provided that Eq. (4.47) is satisfied. For example, we can choose the phases as shown in Fig. 4.5 (a), or, equivalently, we can use the phases shown in Fig. 4.5 (b), which are the result of the line integral $\int_{\mathbf{R}_i}^{\mathbf{R}_j}\mathbf{A}\cdot d\mathbf{r}$ for the Landau gauge $\mathbf{A} = (-y, 0)B$.

The approaches discussed above are perfectly equivalent. However, setting the phases through direct computation of the line integral is more systematic, because it guarantees a correct phase even when the hopping structure becomes more complex (as in the second-nearest-neighbor hopping model or in the bilayer system).

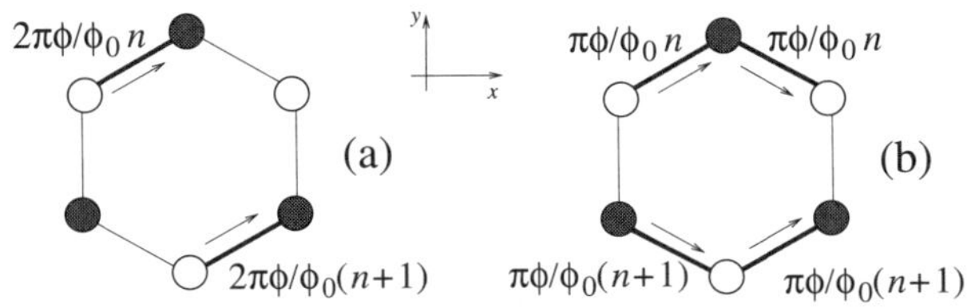

Fig. 4.5. Two possible hopping phases of the tight-binding Hamiltonian in the presence of perpendicular magnetic field.

4.2.3.1. *Landau levels for the zigzag edge sample*

Here we consider a graphene ribbon with zigzag edges. The tight-binding Hamiltonian for a single graphene layer with hopping phases as given in Fig. 4.5 (b) reads,

$$H = -t \sum_{m,n,\sigma} [e^{i\pi \frac{\phi}{\phi_0} n} a_\sigma^\dagger(m,n) b_\sigma(m,n) + e^{-i\pi \frac{\phi}{\phi_0} n} a_\sigma^\dagger(m,n) b_\sigma(m-1,n) + a_\sigma^\dagger(m,n) b_\sigma(m,n-1) + \text{h.c.}] , \quad (4.48)$$

where the gauge $\mathbf{A} = B(-y, 0)$ have been chosen. Fourier transformation along Ox direction gives,

$$H = -t \sum_{k,n,\sigma} [e^{\frac{ika}{2}} 2\cos(\pi \frac{\phi}{\phi_0} n - \frac{ka}{2}) a_\sigma^\dagger(k,n) b_\sigma(k,n) + a_\sigma^\dagger(k,n) b_\sigma(k,n-1) + \text{h.c.}] . \quad (4.49)$$

The eigenproblem $H |\mu, k, \sigma\rangle = E_{\mu,k} |\mu, k, \sigma\rangle$, with $|\mu, k, \sigma\rangle$ as in Eq. 4.33, can be cast in terms of a set of equations called Harper equations,[22] which read:

$$-t[e^{-\frac{ika}{2}} 2\cos(\pi \frac{\phi}{\phi_0} n - \frac{ka}{2}) \alpha(k,n) + \alpha(k,n+1)] = E_{\mu,k} \beta(k,n), \quad (4.50)$$

$$-t[e^{\frac{ika}{2}} 2\cos(\pi \frac{\phi}{\phi_0} n - \frac{ka}{2}) \beta(k,n) + \beta(k,n-1)] = E_{\mu,k} \alpha(k,n). \quad (4.51)$$

As the ribbon only exists between $n = 0$ and $n = N - 1$ at the boundary Eqs. (4.50) and (4.51) are given by,

$$-te^{-\frac{ika}{2}} 2\cos(\pi \frac{\phi}{\phi_0} n - \frac{ka}{2}) \alpha(k, N-1) = E_{\mu,k} \beta(k, N-1) , \quad (4.52)$$

$$-te^{\frac{ika}{2}} 2\cos(\pi \frac{\phi}{\phi_0} n - \frac{ka}{2}) \beta(k,0) = E_{\mu,k} \alpha(k,0) . \quad (4.53)$$

Equations (4.50-4.53) are equivalent to a one-dimensional problem with open boundary conditions, where the dimension of the effective Hilbert space is $2N$. Setting $N = 400$ and $B = 30\,\mathrm{T}$, we show in Fig. 4.6 the obtained spectrum around $E \approx 0$ (top left panel) and the module square of the wavefunctions for zero-energy bands. The wavefunctions are shown for the set of k's indicated by vertical lines in the spectrum, and each panel shows the probability amplitude across the ribbon for the two sublattices.

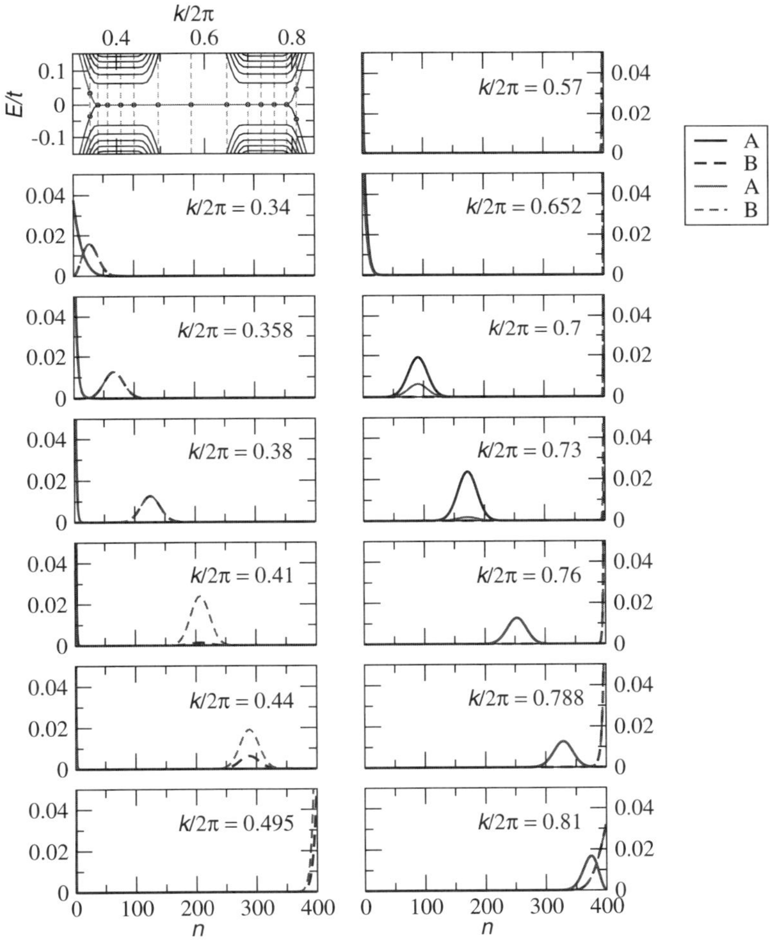

Fig. 4.6. Landau levels (top left) and Landau states (zero energy bands) for the set of k's shown with vertical lines in the top left panel.

As regards the spectrum, the two cones at $ka = 2\pi/3$ and $k'a = 4\pi/3$ present for $B = 0$ give rise to a set of bands with a flat segment (see top left panel in Fig. 4.6). Each non-zero energy flat segment corresponds to a Landau level, as in the usual 2D electron gas.[24] For each k there is a Landau state localized along the ribbon's cross-section, and as k is changed along the segment the localization center moves along the ribbon width. The confining potential, or in other words the lack of lattice outside the ribbon, is responsible for the energy increase (decrease) at both ends of each $E > 0$ ($E < 0$) segment, where edge states show up. The major difference between the 2D electron gas in perpendicular magnetic field and non-zero Landau levels in graphene is the double degeneracy of the last, consequence of the honeycomb lattice.

Now we turn to the zero-energy bands shown in Fig. 4.6 (top left panel). We have seen in Sec. 4.2.2 that zero energy bands are consequence of localized surface states at the zigzag edges of graphene ribbons. In the presence of a perpendicular magnetic field, however, there is also a double-degenerate Landau level at zero energy which coexists with surface states, as perceived from the wavefunctions shown in Fig. 4.6. Zero-energy Landau levels are completely absent in the 2D electron gas, and there presence in graphene is responsible for the unconventional quantum Hall effect. Indeed, using Laughlin-Halperin's argument,[23,24] we can compute the Hall conductivity in the ribbon geometry from the number of edge states crossing the Fermi level. If the chemical potential is above the nth Landau level we have $2n + 1$ edge states crossing the Fermi level. The Hall conductivity, including spin degeneracy, is then given by Eq. (4.1).[5] The same conclusions can be drawn from the armchair ribbon,[25] or even from the continuum limit, as we discuss in the next section.

4.2.3.2. *Landau levels in the continuum approximation*

Here we start directly from the Dirac Eq. (4.30). The magnetic field is introduced through minimal coupling, $\mathbf{p} \to \mathbf{p} - e\mathbf{A}$, where $\mathbf{A}$ is the vector potential. Using the landau gauge $\mathbf{A} = (-y, 0)B$ and changing to position basis Eqs. (4.28) and (4.29) read,

$$H_{\mathbf{K}}(\mathbf{r}) = v_F \begin{pmatrix} 0 & \frac{\hbar}{i}\partial_x + eyB - \hbar\partial_y \\ \frac{\hbar}{i}\partial_x + eyB + \hbar\partial_y & 0 \end{pmatrix}, \tag{4.54}$$

$$H_{\mathbf{K}'}(\mathbf{r}) = v_F \begin{pmatrix} 0 & \frac{\hbar}{i}\partial_x + eyB + \hbar\partial_y \\ \frac{\hbar}{i}\partial_x + eyB - \hbar\partial_y & 0 \end{pmatrix}. \tag{4.55}$$

In order to find the Landau levels and respective Landau states we have to diagonalize Eqs. (4.54) and (4.55). Translational invariance along the x direction suggests that we write the spinor eigenfunction as,

$$\psi_k(x,y) = e^{ikx}\begin{pmatrix} c_1\phi_1^{\mathrm{A}}(y) \\ c_2\phi_2^{\mathrm{B}}(y) \end{pmatrix}, \tag{4.56}$$

both for $\mathbf{K}$ and $\mathbf{K}'$. Moreover, we can parametrize the magnetic field in terms of a new length scale known as *magnetic length*, $l_B^2 = \hbar/|e|B$, and reduce the eigenproblem to,

$$v_F\frac{\hbar}{l_B}\begin{pmatrix} 0 & l_Bk - \frac{y}{l_B} \mp l_B\partial_y \\ l_Bk - \frac{y}{l_B} \pm l_B\partial_y & 0 \end{pmatrix}\begin{pmatrix} c_1\phi_1^{\mathrm{A}} \\ c_2\phi_2^{\mathrm{B}} \end{pmatrix} = E\begin{pmatrix} c_1\phi_1^{\mathrm{A}} \\ c_2\phi_2^{\mathrm{B}} \end{pmatrix}, \tag{4.57}$$

where upper and lower signs stand for $\mathbf{K}$ and $\mathbf{K}'$ points, respectively. Analogously to the 2D electron gas, and despite the matrix form of Eq. (4.57), the functions ϕ_1 and ϕ_2 are solutions of the harmonic oscillator Hamiltonian. To see why this is so, we define $\tilde{y} = y/l_B - l_Bk$ and $\partial_{\tilde{y}} = l_B\partial_y$, and the respective staircase operators,

$$a = \frac{1}{\sqrt{2}}(\tilde{y} + \partial_{\tilde{y}}), \qquad a^\dagger = \frac{1}{\sqrt{2}}(\tilde{y} - \partial_{\tilde{y}}). \tag{4.58}$$

Inserting (4.58) in (4.57), and defining $\phi_n(\tilde{y})$ as the solutions of the harmonic oscillator for which $a^\dagger\phi_n = \sqrt{n+1}\phi_{n+1}$ and $a\phi_n = \sqrt{n}\phi_{n-1}$ hold, we arrive at the following solution of the eigenproblem,

$$\begin{pmatrix} c_1\phi_1^{\mathrm{A}}(y) \\ c_2\phi_2^{\mathrm{B}}(y) \end{pmatrix} = \begin{pmatrix} \phi_n^{\mathrm{A}}(\tilde{y}) \\ \pm\phi_{n+1}^{\mathrm{B}}(\tilde{y}) \end{pmatrix}, \quad \begin{pmatrix} c_1\phi_1^{\mathrm{A}}(y) \\ c_2\phi_2^{\mathrm{B}}(y) \end{pmatrix} = \begin{pmatrix} \phi_{n+1}^{\mathrm{A}}(\tilde{y}) \\ \pm\phi_n^{\mathrm{B}}(\tilde{y}) \end{pmatrix}, \tag{4.59}$$

respectively for $\mathbf{K}$ and $\mathbf{K}'$, with double-degenerate eigenenergies (spin degeneracy apart),

$$E_{\mathbf{K}} = E_{\mathbf{K}'} = \pm v_F\sqrt{2}\frac{\hbar}{l_B}\sqrt{n+1}. \tag{4.60}$$

There is in addition a zero energy solution given by,

$$\begin{pmatrix} c_1\phi_1(y) \\ c_2\phi_2(y) \end{pmatrix} = \begin{pmatrix} 0 \\ \phi_0(\tilde{y}) \end{pmatrix}, \quad \begin{pmatrix} c_1\phi_1(y) \\ c_2\phi_2(y) \end{pmatrix} = \begin{pmatrix} \phi_0(\tilde{y}) \\ 0 \end{pmatrix}, \tag{4.61}$$

at $\mathbf{K}$ and $\mathbf{K}'$ respectively. This last solution is in complete agreement with numerical results from diagonalization of finite ribbons. As can be seen in Fig. 4.6, zero-energy bulk states for $k \gtrsim 2\pi/3$ have finite amplitude at B sublattice only, whereas for $k \gtrsim 4\pi/3$ they have finite amplitude at A sublattice only, which correspond to solutions $\mathbf{K}$ and $\mathbf{K}'$, respectively, in Eq. (4.61). Furthermore, as $\tilde{y}$ is given by $\tilde{y} = y/l_B - l_Bk$ the Landau level center moves across the sample as k changes, as in Fig. 4.6.

As referred in Sec. 4.2.3.1, the unconventional quantum Hall effect observed in graphene is deeply related with the presence of zero-energy Landau levels. Let us call the positive energy levels *electron levels* and the negative ones *hole levels*. As we have seen above, an electron level (or hole level) is twice-degenerated due to the two non-equivalent Dirac points in the Brillouin zone. The zero-energy Landau levels are such that one is an electron level and the other is a hole level. Let us now suppose that concentration of electrons is such that that the chemical potential lies in the gap between the electron level n and the electron level $n+1$. This implies that one has $(2n+1)$ fully occupied electron levels. By taking into account the spin degeneracy we obtain for the Hall conductivity the rule given in Eq. (4.1).

To end this section we note that Eq. (4.60) is completely different from the 2D electron gas result where equally spaced Landau levels occur. The scaling $E \propto \pm\sqrt{n}$ is another signature of the presence of massless relativistic quasi-particles in graphene.

4.3. The graphene bilayer

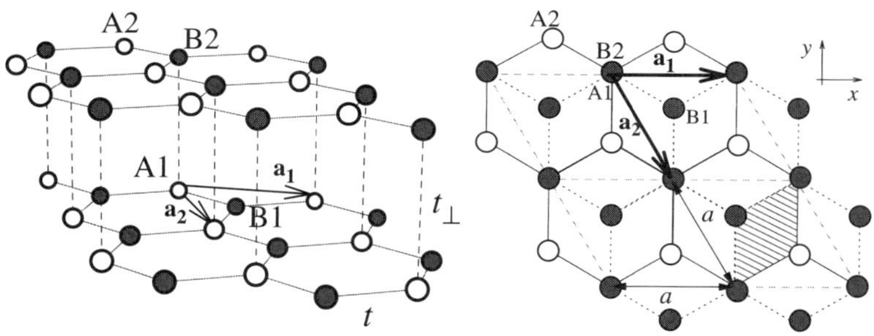

Fig. 4.7. Lattice structure of the graphene bilayer.

The lattice structure of bilayer graphene is shown in Fig. 4.7. It is made up of two graphene layers where the upper layer has its B sublattice on top of sublattice A of the underlying layer (Bernal stacking). As can be seen in Fig. 4.7, bilayer graphene has four atoms per unit cell. The unit cell is the same as in graphene, with $\mathbf{a}_1$ and $\mathbf{a}_2$ given by Eq. (4.7). The simplest Hamiltonian describing noninteracting electrons on the bilayer reads,

$$\begin{aligned} H = &-t\sum_{\mathbf{R},\sigma}[a^\dagger_{1,\sigma}(\mathbf{R})b_{1,\sigma}(\mathbf{R}) + a^\dagger_{1,\sigma}(\mathbf{R})b_{1,\sigma}(\mathbf{R}-\mathbf{a}_1) + a^\dagger_{1,\sigma}(\mathbf{R})b_{1,\sigma}(\mathbf{R}-\mathbf{a}_2))] \\ &-t\sum_{\mathbf{R},\sigma}[b^\dagger_{2,\sigma}(\mathbf{R})a_{2,\sigma}(\mathbf{R}) + b^\dagger_{2,\sigma}(\mathbf{R})a_{2,\sigma}(\mathbf{R}+\mathbf{a}_1) + b^\dagger_{2,\sigma}(\mathbf{R})a_{2,\sigma}(\mathbf{R}+\mathbf{a}_2))] \\ &-t_\perp\sum_{\mathbf{R},\sigma}[a^\dagger_{1,\sigma}(\mathbf{R})b_{2,\sigma}(\mathbf{R})] + \text{h.c.}, \end{aligned} \tag{4.62}$$

where $a^\dagger_{i,\sigma}(\mathbf{R})$ $\left[a_{i,\sigma}(\mathbf{R})\right]$, $b^\dagger_{i,\sigma}(\mathbf{R})$ $\left[b_{i,\sigma}(\mathbf{R})\right]$, creates (annihilates) an electron on position $\mathbf{R}$, with spin $\sigma = \uparrow, \downarrow$ on plane $i = 1, 2$ on sublattice A, and B, respectively. The new energy scale $t_\perp$ parametrizes the hopping of π-electrons between layers. First-principles calculations in graphite suggest $t_\perp \sim t/10$.[26]

4.3.1. *Unbiased bilayer (bulk)*

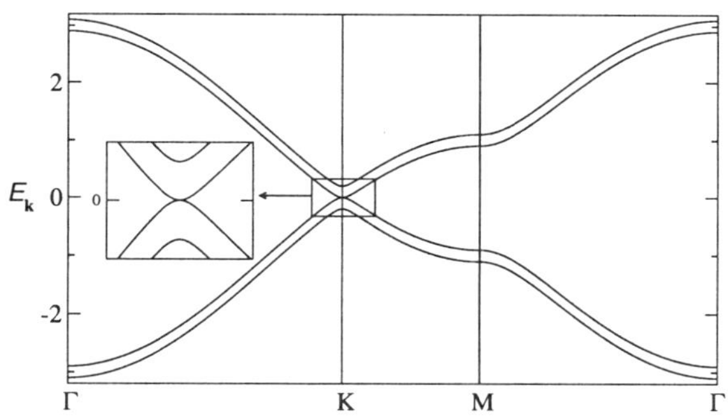

Fig. 4.8. Bilayer band structure for $t_\perp/t = 0.2$. Energy is given in units of t.

The bulk properties of the model defined by Eq. (4.62) can be determined imposing PBC's and applying Fourier transformation, as in Sec. 4.2.1. Introducing the Fourier components $a_{i,\sigma,\mathbf{k}}$ and $b_{i,\sigma,\mathbf{k}}$ of operators $a_{i,\sigma}(\mathbf{R})$ and $b_{i,\sigma}(\mathbf{R})$, respectively, with $i = 1, 2$, we can rewrite Eq. (4.62) as,

$$H = \sum_{\mathbf{k},\sigma} \psi^\dagger_{\sigma,\mathbf{k}} H_{\mathbf{k}} \psi_{\sigma,\mathbf{k}}, \tag{4.63}$$

where $\psi_{\sigma,\mathbf{k}} = [a_{1,\sigma,\mathbf{k}}, b_{1,\sigma,\mathbf{k}}, a_{2,\sigma,\mathbf{k}}, b_{2,\sigma,\mathbf{k}}]$ is a four component spinor, and $H_{\mathbf{k}}$

reads,

$$H_{\mathbf{k}} = -t \begin{pmatrix} 0 & s_{\mathbf{k}} & 0 & t_\perp/t \\ s^*_{\mathbf{k}} & 0 & 0 & 0 \\ 0 & 0 & 0 & s_{\mathbf{k}} \\ t_\perp/t & 0 & s^*_{\mathbf{k}} & 0 \end{pmatrix}, \tag{4.64}$$

with $s_{\mathbf{k}}$ as in Eq. (4.12). From H_k as given in Eq. (4.64) follows the dispersion relation,

$$E_{\mathbf{k}} = \pm\sqrt{\epsilon^2_{\mathbf{k}} + \left(\frac{t_\perp}{2}\right)^2} \pm \frac{t_\perp}{2}, \tag{4.65}$$

where $\epsilon_{\mathbf{k}}$ is the single layer dispersion given in Eq. (4.13). The band structure defined by Eq. (4.65) can be seen in Fig. 4.8 along three directions in the first Brillouin zone (see Fig. 4.2). The spectrum is now composed by four bands, consequence of the number of atoms per unit cell. At the corners of the first Brillouin zone the dispersion is now parabolic, as shown in the inset of Fig. 4.8. Expanding Eq. (4.65) around two non-equivalent corners [Eq. (4.17)] we get,

$$E(q) \approx \begin{cases} t_\perp + t\frac{3a^2}{4t_\perp/t}q^2 \\ \pm t\frac{3a^2}{4t_\perp/t}q^2 \\ -t_\perp - t\frac{3a^2}{4t_\perp/t}q^2 \end{cases}, \tag{4.66}$$

with $q = |\mathbf{q}| = |\mathbf{q}'|$ as in Eq. (4.19).

Bilayer graphene is thus characterized by two carrier types, one gapless and the other with a gap of value $t_\perp$. Both carriers are massive $m^* = 2t_\perp\hbar^2/3a^2t^2$.

The spectrum of a graphene bilayer can be obtained and interpreted in a very graceful way starting with two uncoupled graphene layers. The operators that diagonalize the uncoupled layers can be written as,

$$e_{1,\sigma,\mathbf{k}} = \frac{1}{\sqrt{2}}(e^{i\delta_{\mathbf{k}}/2}a_{1,\sigma,\mathbf{k}} - e^{-i\delta_{\mathbf{k}}/2}b_{1,\sigma,\mathbf{k}}), \tag{4.67}$$

$$h_{1,\sigma,\mathbf{k}} = \frac{1}{\sqrt{2}}(e^{i\delta_{\mathbf{k}}/2}a_{1,\sigma,\mathbf{k}} + e^{-i\delta_{\mathbf{k}}/2}b_{1,\sigma,\mathbf{k}}), \tag{4.68}$$

$$e_{2,\sigma,\mathbf{k}} = \frac{1}{\sqrt{2}}(e^{i\delta_{\mathbf{k}}/2}a_{2,\sigma,\mathbf{k}} - e^{-i\delta_{\mathbf{k}}/2}b_{2,\sigma,\mathbf{k}}), \tag{4.69}$$

$$h_{2,\sigma,\mathbf{k}} = -\frac{1}{\sqrt{2}}(e^{i\delta_{\mathbf{k}}/2}a_{2,\sigma,\mathbf{k}} + e^{-i\delta_{\mathbf{k}}/2}b_{2,\sigma,\mathbf{k}}), \tag{4.70}$$

where e and h stand for electron and hole bands in Eq. (4.13). The phase used in Eqs. (4.67-4.70) is given by $1 + e^{i\mathbf{k}\cdot\mathbf{a}_1} + e^{i\mathbf{k}\cdot\mathbf{a}_2} = |\epsilon_{\mathbf{k}}|e^{i\delta_{\mathbf{k}}}$. The only non-diagonal term in Eq. (4.63), when written in terms of the band operators in Eqs. (4.67-4.70), is the interlayer coupling term,

$$H_{12} = \frac{t_\perp}{2} \sum_{\mathbf{k},\sigma} \left(e^\dagger_{1,\sigma,\mathbf{k}} e_{2,\sigma,\mathbf{k}} + e^\dagger_{1,\sigma,\mathbf{k}} h_{2,\sigma,\mathbf{k}} + h^\dagger_{1,\sigma,\mathbf{k}} e_{2,\sigma,\mathbf{k}} + h^\dagger_{1,\sigma,\mathbf{k}} h_{2,\sigma,\mathbf{k}} + \text{h.c.} \right). \tag{4.71}$$

Equation (4.71) is nothing but a 4-site tight-binding problem in a square, where, as it is well known, there are two degenerate states ($\pm\pi/2$ Bloch states) of zero energy, one of energy $-t_\perp$ (0 Bloch state), and a fourth one at $+t_\perp$ (π Bloch state). This accounts for the four states at the corners of the Brillouin zone [$q = 0$ in Eq. (4.66)]. For other $\mathbf{k}$-values the electron and hole states are no longer degenerate, showing an energy separation of $2\epsilon_{\mathbf{k}}$. However, the electron (hole) states of the two uncoupled layers remain degenerate. For $2\epsilon_{\mathbf{k}} > t_\perp$ we obtain a coupling $t_\perp/2$ between e_1 and e_2 degenerate bands, as well as between h_1 and h_2. The interlayer term H_{12} splits these two pairs of bands by $t_\perp$, whereas the coupling between e and h bands is suppressed by a $t_\perp/2\epsilon_{\mathbf{k}}$ factor.

4.3.1.1. *The continuum approximation for the bilayer Hamiltonian*

At long wave lengths and low energies we can use a continuum approximation for Hamiltonian (4.64) near $\mathbf{K}$ and $\mathbf{K}'$, as we did for the single layer in Sec. 4.2.1.1.

Expanding Eq. (4.12) around $\mathbf{K}$, and redefining k_x and k_y as in Eq. (4.24), we can write Eq. (4.64) as,

$$H_{\mathbf{K}} = v_F \begin{pmatrix} 0 & p_x - ip_y & 0 & t_\perp/v_F \\ p_x + ip_y & 0 & 0 & 0 \\ 0 & 0 & 0 & p_x - ip_y \\ t_\perp/v_F & 0 & p_x + ip_y & 0 \end{pmatrix}. \tag{4.72}$$

If, instead, we expand Eq. (4.12) around $\mathbf{K}'$ and redefine k_x and k_y as in Eq. (4.26), we obtain for Eq. (4.64) the following approximation,

$$H_{\mathbf{K}'} = v_F \begin{pmatrix} 0 & p_x + ip_y & 0 & t_\perp/v_F \\ p_x - ip_y & 0 & 0 & 0 \\ 0 & 0 & 0 & p_x + ip_y \\ t_\perp/v_F & 0 & p_x - ip_y & 0 \end{pmatrix}. \tag{4.73}$$

In Eqs. (4.72) and (4.73) we use $v_F = ta\hbar^{-1}\sqrt{3}/2$ and $p_\xi = \hbar k_\xi$, as in the single layer case. The diagonalization of these two Hamiltonians gives exactly Eq. (4.65) with the graphene dispersion approximated by $\epsilon_{\mathbf{k}} \approx v_F p$, as we would expect.

4.3.1.2. *Landau levels in the continuum approximation for the bilayer*

With Hamiltonians (4.72) and (4.73) in hand it is possible to study the effect of a perpendicular magnetic field applied to the bilayer system in the continuum approximation. The steps followed in Sec. 4.2.3.2, where we studied the case of a single graphene layer, may be easily reproduced in the bilayer case. Within the Landau gauge we end up with the following Hamiltonians describing bilayer graphene in perpendicular magnetic field,

$$H_{\mathbf{K}} = \gamma \begin{pmatrix} 0 & a & 0 & t_\perp/\gamma \\ a^\dagger & 0 & 0 & 0 \\ 0 & 0 & 0 & a \\ t_\perp/\gamma & 0 & a^\dagger & 0 \end{pmatrix}, \qquad H_{\mathbf{K}'} = \gamma \begin{pmatrix} 0 & a^\dagger & 0 & t_\perp/\gamma \\ a & 0 & 0 & 0 \\ 0 & 0 & 0 & a^\dagger \\ t_\perp/\gamma & 0 & a & 0 \end{pmatrix}, \tag{4.74}$$

respectively for $\mathbf{K}$ and $\mathbf{K}'$, where $\gamma = v_F\sqrt{2}\frac{\hbar}{l_B}$, and l_B is the magnetic length as introduced in Sec. 4.2.3.2. The presence in Eq. (4.74) of the staircase operators [Eq. (4.58)] makes the appearance of harmonic oscillator wavefunctions ϕ_n quite natural. Indeed, after simple algebra we verify that, in the present case, the spinor wavefunctions have the form,

$$\psi_k(x,y) = e^{ikx} \begin{pmatrix} c_1\phi_{n+1}^{\mathrm{A1}}(\tilde{y}) \\ c_2\phi_{n+2}^{\mathrm{B1}}(\tilde{y}) \\ c_3\phi_{n}^{\mathrm{A2}}(\tilde{y}) \\ c_4\phi_{n+1}^{\mathrm{B2}}(\tilde{y}) \end{pmatrix}, \qquad \psi_k(x,y) = e^{ikx} \begin{pmatrix} c_1\phi_{n+1}^{\mathrm{A1}}(\tilde{y}) \\ c_2\phi_{n}^{\mathrm{B1}}(\tilde{y}) \\ c_3\phi_{n+2}^{\mathrm{A2}}(\tilde{y}) \\ c_4\phi_{n+1}^{\mathrm{B2}}(\tilde{y}) \end{pmatrix}, \tag{4.75}$$

for $\mathbf{K}$ and $\mathbf{K}'$, respectively, with $\tilde{y} = y/l_B - l_B k$, where the c coefficients are easily found by solving the eigenproblem. The respective Landau levels are double-degenerate, as in the single layer case, and are given by,

$$E_{\mathbf{K}} = E_{\mathbf{K}'} = \pm\frac{\gamma}{\sqrt{2}}\sqrt{t_\perp^2/\gamma^2 + 1 + 2n \pm \sqrt{\left(t_\perp^2/\gamma^2+1\right)^2 + 4nt_\perp^2/\gamma^2}}\,. \tag{4.76}$$

Zero-energy Landau levels also exist in bilayer graphene. However, the bilayer has four zero-energy Landau levels, two per Dirac point, at odds with single layer graphene. The respective spinor wavefunctions read,

$$\psi_k(x,y) = \frac{e^{ikx}}{\sqrt{\gamma^2+t_\perp^2}} \begin{pmatrix} 0 \\ t_\perp\phi_1^{\mathrm{B1}}(\tilde{y}) \\ 0 \\ \gamma\phi_0^{\mathrm{B2}}(\tilde{y}) \end{pmatrix} \qquad \psi_k(x,y) = e^{ikx} \begin{pmatrix} 0 \\ \phi_0^{\mathrm{B1}}(\tilde{y}) \\ 0 \\ 0 \end{pmatrix}, \tag{4.77}$$

at **K** point, and

$$\psi_k(x,y) = \frac{e^{ikx}}{\sqrt{\gamma^2 + t_\perp^2}} \begin{pmatrix} \gamma\phi_0^{\text{A1}}(\tilde{y}) \\ 0 \\ t_\perp \phi_1^{\text{A2}}(\tilde{y}) \\ 0 \end{pmatrix} \qquad \psi_k(x,y) = e^{ikx} \begin{pmatrix} 0 \\ 0 \\ \phi_0^{\text{A2}}(\tilde{y}) \\ 0 \end{pmatrix}, \qquad (4.78)$$

at point $\mathbf{K}'$.

As we have seen in Sec. 4.2.3, the presence of zero-energy Landau levels lead to an unconventional quantum Hall effect, different from that seen in 2D electron gases. In the bilayer, the fourfold degeneracy of zero-energy Landau levels, as opposed to the twofold degeneracy in single layer graphene, is responsible for a new type of unconventional quantum Hall effect, different from that observed in graphene. Following the discussion in the end of Sec. 4.2.3.2, if the electron concentration is such that the chemical potential lies between the n and the $n+1$ levels, the number of occupied electron levels is $(2n+2)$. Taking into account spin degeneracy we obtain for the Hall conductivity in bilayer graphene,

$$\sigma_{\text{Hall}} = \pm 4\,(n+1)\,\frac{e^2}{h}, \qquad n = 0, 1, 2, \ldots \qquad (4.79)$$

in agreement with experiments,[8] and different from Eq. (4.1).

4.3.2. *Biased bilayer (bulk)*

Because of the potential technological applications of double-layer graphene, the study of the energy spectrum of the bilayer with both planes at unequal electric potential is a problem of considerable interest. The possibility of having two planes of a graphene bilayer at different potentials introduces the concept a *biased bilayer*. This situation was already considered in the continuum limit by Guinea *et al.*[11] Here we study its subtleties using the full tight-binding description.

Let us introduce a bias V such that the electrostatic potential is $+V/2$ and $-V/2$ in layer 1 and 2, respectively. The simplest tight-binding Hamiltonian for these system then reads,

$$H_V = H + \frac{V}{2}\sum_{\mathbf{R},\sigma}[n^\dagger_{\text{A1},\sigma}(\mathbf{R}) + n^\dagger_{\text{B1},\sigma}(\mathbf{R}) - n^\dagger_{\text{A2},\sigma}(\mathbf{R}) - n^\dagger_{\text{B2},\sigma}(\mathbf{R})], \qquad (4.80)$$

where H is the unbiased Hamiltonian given by Eq. (4.62), and $n^\dagger_{\text{A}i,\sigma}(\mathbf{R})$ and $n^\dagger_{\text{B}i,\sigma}(\mathbf{R})$, with $i = 1, 2$, are number operators. The spectrum of Eq. (4.80) can be obtained through Fourier transformation, analogously to the unbiased case, and

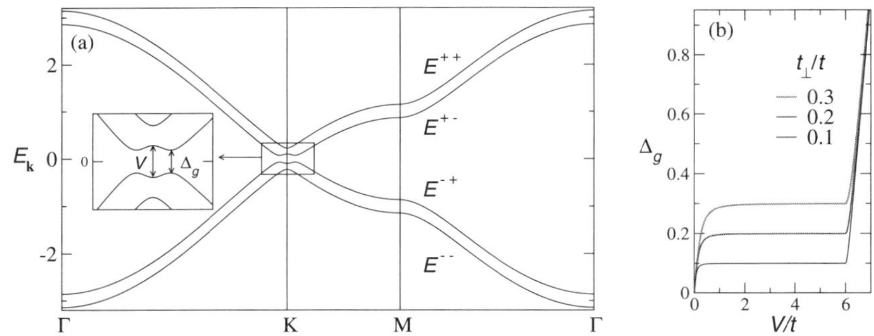

Fig. 4.9. (a) - Band structure of the biased bilayer for $t_\perp/t = 0.2$ and $V = t_\perp$. Energy is given in units of t. (b) - Variation of the gap Δ_g with V for various $t_\perp$ values. The gap is given in units of t.

reads,

$$E_{\mathbf{k}}^{\pm\pm}(V) = \pm\sqrt{\epsilon_{\mathbf{k}}^2 + \frac{t_\perp^2}{2} + \frac{V^2}{4} \pm \sqrt{\frac{t_\perp^4}{4} + (t_\perp^2 + V^2)\epsilon_{\mathbf{k}}^2}}, \tag{4.81}$$

where $\epsilon_{\mathbf{k}}$ is the single layer dispersion given in Eq. (4.13). The band structure defined by Eq. (4.81) is shown in Fig. 4.9 (a). When compared with Fig. 4.8, we see that the $V = 0$ gapless semiconductor has turned into an insulator whose gap (Δ_g) is completely controlled by V. Moreover, as the two bands closer to zero energy $E_{\mathbf{k}}^{\pm-}(V)$ are strongly deformed near the corners of the Brillouin zone, the minimum of $|E_{\mathbf{k}}^{\pm-}(V)|$ no longer occurs at these corners as in the $V = 0$ case. As a consequence the low doping Fermi surface is completely different from the $V = 0$ circle, with a shape controlled by V.

Let us first compute how the gap varies with V. The minimum of band $E_{\mathbf{k}}^{+-}(V)$ [or equivalently, the maximum of $E_{\mathbf{k}}^{--}(V)$] occurs for all $\mathbf{k}$'s satisfying,

$$\epsilon_{\mathbf{k}}^2 = \alpha(V, t_\perp), \tag{4.82}$$

where

$$\alpha(V, t_\perp) = \frac{V^2}{2t^2}\frac{1/2 + (t_\perp/V)^2}{1 + (t_\perp/V)^2}. \tag{4.83}$$

Equation (4.82) can only be solved as long as $\sqrt{\alpha} \leq 3$, where $3t$ is half of the single layer bandwidth. When $\sqrt{\alpha} > 3$ the minimum of $E_{\mathbf{k}}^{+-}(V)$ occurs at the Γ

point. The gap, which is twice the minimum value of $E_{\mathbf{k}}^{+-}(V)$, is thus given by,

$$\Delta_g = \begin{cases} V\sqrt{\frac{(t_\perp/V)^2}{(t_\perp/V)^2+1}} & \sqrt{\alpha} \leq 3 \\ 2t\sqrt{9 + \frac{t_\perp^2}{2t^2} + \frac{V^2}{4t^2} - \sqrt{\frac{t_\perp^4}{4t^4} + 9\frac{t_\perp^2+V^2}{t^2}}} & \sqrt{\alpha} > 3 \end{cases} . \quad (4.84)$$

From Eq. (4.84) it can be seen that both $V \ll t_\perp$ and $V \gg t$ give the same gap behavior $\Delta_g \sim V$. However, there is a region for $t_\perp \lesssim V \lesssim 6t$ where the gap shows a plateau $\Delta_g \sim t_\perp$, as depicted in Fig. 4.9 (b). The plateau ends when $\sqrt{\alpha} = 3$. Again this behavior is easily understood from the point of view of uncoupled layers plus a 4-site tight-binding problem [Eq. (4.71)]. For the unbiased bilayer the two degenerate eigenstates ψ_+ and ψ_- of Eq. (4.71), the so-called $\pm\pi/2$ Bloch states, are given in terms of band the operators defined in Eqs. (4.67-4.70) as,

$$\begin{aligned} \psi_\pm &= \frac{1}{2}(e_1 + e^{\pm i\pi/2}e_2 - h_1 + e^{\mp i\pi/2}h_2) \\ &= \frac{1}{\sqrt{2}}(-e^{-i\delta/2}b_1 + e^{\pm i\pi/2}e^{i\delta/2}a_2) , \end{aligned} \quad (4.85)$$

where momentum and spin indices have been omitted. For $V \ll t_\perp$, near the Dirac points, the bias acts as a perturbation, coupling the two zero energy states $\psi_\pm$. Only the non-diagonal matrix element is non zero and equal to $V/2$. So the splitting, and the gap, is in this limit,

$$\Delta_g = V + O\Big(\frac{V}{t_\perp}\Big) \qquad V \ll t_\perp . \quad (4.86)$$

The more interesting case $V > t_\perp$ is also completely unveiled with this framework. For the uncoupled layers in the presence of a bias we can readily see that there are values of $\mathbf{k}$ at which the hole band of layer 1 crosses the electron band of layer 2. At values of $\mathbf{k}$ such that,

$$2\epsilon_{\mathbf{k}} = V , \quad (4.87)$$

which is just Eq. (4.82) for $t_\perp = 0$, the coupling term given by Eq. (4.71) couples four states of energies $-|\epsilon_{\mathbf{k}}| - V$, $-|\epsilon_{\mathbf{k}}| + V = |\epsilon_{\mathbf{k}}| - V$, and $|\epsilon_{\mathbf{k}}| + V$. For $t_\perp \ll \epsilon_{\mathbf{k}}$ the effect of H_{12} is to lift the degeneracy of the two middle states. Since the coupling between them is $t_\perp/2$, the gap between these states is $t_\perp$. This explains the origin of the plateau in the gap. As V changes the $\mathbf{k}$-values where the bands h_1 and e_2 are degenerate move [Eq. (4.87)], but the gap remains,

$$\Delta_g = t_\perp + O\Big(\frac{t_\perp^2}{V}, \frac{t_\perp^2}{\epsilon_{\mathbf{k}}}\Big) \qquad t_\perp \ll V \leq 6t . \quad (4.88)$$

Eventually, at $V = 6t$ (which is $\sqrt{\alpha} = 3$ for $t_\perp \ll V$), the bands no longer cross and the gap occurs at the Γ point between the h_1 and e_2 bands, which are now very weakly coupled, so that,

$$\Delta_g = V - 6t + O\Big(\frac{t_\perp^2}{V - 6t}\Big) \qquad V > 6t\,. \tag{4.89}$$

It is worth mentioning that experimentally $V \lesssim t_\perp$ always hold, as larger bias values would imply huge unsustainable electric fields due to the small separation between layers.

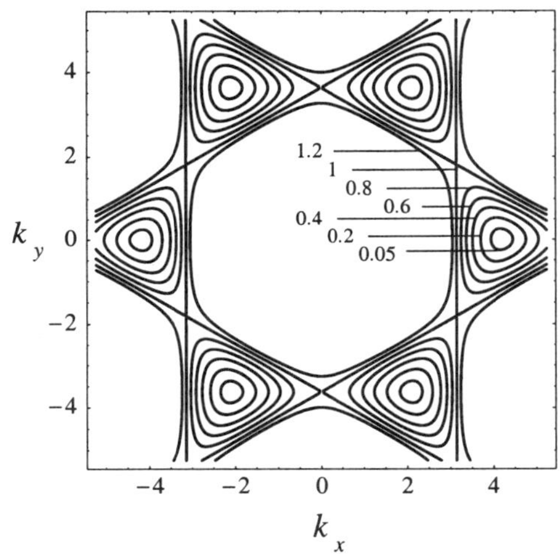

Fig. 4.10. Solution of Eq. (4.82) in k-space. Each line corresponds to a different α value, as indicated in the figure.

Figure 4.10 shows the solution of Eq. (4.82) in k-space for several α values.In a low doping situation, and as long as $\alpha \leq 9$, the Fermi sea acquires a line shape given by the solution of Eq. (4.82), the line width being determined by the doping level. As can be seen in Fig. 4.10, when $\alpha \ll 1$ the Fermi sea approaches a ring, the Fermi ring, centered at the Brillouin zone corners. In particular, the Fermi ring is certainly present for $\alpha \ll t_\perp/t$ [or equivalently $V \ll t_\perp$ from Eq. (4.83)], as shown by Guinea *et al.*[11] The ring acquires a triangular shape as α increases, and for $\alpha = 1$ the Fermi sea is a perfect triangle. This particular α value occurs for $V \sim t$ and marks a change of behavior of the solution of Eq. (4.82). It happens that for $\alpha > 1$ the Fermi see is again a ring, now centered at the Γ point. This Γ-centered Fermi ring is present during the gap plateau shown in Fig. 4.9 (b), its radius gets smaller as α (or equivalently V) increases, and it turns into a small circle centered at Γ when the plateau ends ($\alpha = 9$).

4.3.3. *Surface states for the bilayer with zigzag edges*

In single layer graphene, whenever a zigzag edge shows up a zero-energy band of surface states localized at the edge appears, as we showed in Sec. 4.2.2. Weather zero-energy surface states exist in bilayer graphene in the presence of such extended defects is the subject of this section.

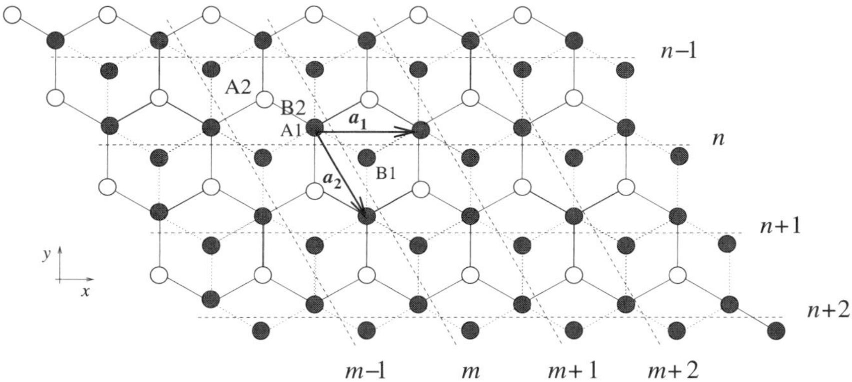

Fig. 4.11. Ribbon geometry with zigzag edges for bilayer graphene.

In order to study the presence of surface states in the bilayer we consider the ribbon geometry with zigzag edges shown in Fig. 4.11. The ribbon width is such that it has N unit cells in the transverse cross section (y direction), and we assume that it has infinite length in the longitudinal direction (x direction).

Following Sec. 4.2.2, we may rewrite Eq. (4.62) in terms of the integer indices m and n introduced in Eq. (4.2) and then Fourier transform along Ox, introducing the quantum number $k \in [0, 2\pi[$. The resultant Hamiltonian reads,

$$\begin{aligned} H = &-t \int \mathrm{d}k \sum_{n,\sigma} [(1 + e^{ika}) a^\dagger_{1,\sigma}(k,n) b_{1,\sigma}(k,n) + a^\dagger_{1,\sigma}(k,n) b_{1,\sigma}(k,n-1)] \\ &-t \int \mathrm{d}k \sum_{n,\sigma} [(1 + e^{-ika}) b^\dagger_{2,\sigma}(k,n) a_{2,\sigma}(k,n) + b^\dagger_{2,\sigma}(k,n) a_{2,\sigma}(k,n+1)] \\ &-t_\perp \int \mathrm{d}k \sum_{n,\sigma} [a^\dagger_{1,\sigma}(k,n) b_{2,\sigma}(k,n)] + \mathrm{h.c.}. \end{aligned} \tag{4.90}$$

With the definition $c^\dagger_{i,\sigma}(k,n)\,|0\rangle = |c_i,\sigma,k,n\rangle$, where $c_i = a_1, b_1, a_2, b_2$, the one-particle eigenstates which are solution of the Schrödinger equation, $H\,|\mu,k,\sigma\rangle = E_{\mu,k}\,|\mu,k,\sigma\rangle$, can be generally expressed as,

$$|\mu,k,\sigma\rangle = \sum_n [\alpha_1(k,n)\,|a_1,k,n,\sigma\rangle + \beta_1(k,n)\,|b_1,k,n,\sigma\rangle + \\ + \alpha_2(k,n)\,|a_2,k,n,\sigma\rangle + \beta_2(k,n)\,|b_2,k,n,\sigma\rangle], \quad (4.91)$$

where $\alpha_i(k,n)$ stands for the wavefunction amplitude at sites of sublattice A_i and $\beta_i(k,n)$ at sites of sublattice B_i, with $i = 1, 2$. As in Sec. 4.2.2, the equations for coefficients α_i and β_i are obtained by applying Hamiltonian (4.90) to $|\mu,k,\sigma\rangle$, and read,

$$E_{\mu,k}\alpha_1(k,n) = -t[(1+e^{ika})\beta_1(k,n) + \beta_1(k,n-1)] - t_\perp\beta_2(k,n), \quad (4.92)$$
$$E_{\mu,k}\beta_1(k,n) = -t[(1+e^{-ika})\alpha_1(k,n) + \alpha_1(k,n+1)], \quad (4.93)$$
$$E_{\mu,k}\alpha_2(k,n) = -t[(1+e^{ika})\beta_2(k,n) + \beta_2(k,n-1)], \quad (4.94)$$
$$E_{\mu,k}\beta_2(k,n) = -t[(1+e^{-ika})\alpha_2(k,n) + \alpha_2(k,n+1)] - t_\perp\alpha_1(k,n). \quad (4.95)$$

We must remember the finite width of the ribbon, and write Eqs. (4.92-4.95) at the boundaries as,

$$E_{\mu,k}\alpha_1(k,0) = -t(1+e^{ika})\beta_1(k,0) - t_\perp\beta_2(k,0), \quad (4.96)$$
$$E_{\mu,k}\beta_1(k,N-1) = -t(1+e^{-ika})\alpha_1(k,N-1), \quad (4.97)$$
$$E_{\mu,k}\alpha_2(k,0) = t(1+e^{ika})\beta_2(k,0), \quad (4.98)$$
$$E_{\mu,k}\beta_2(k,N-1) = -t(1+e^{-ika})\alpha_2(k,N-1) - t_\perp\alpha_1(k,N-1). \quad (4.99)$$

Surface states are solutions of Eqs. (4.92-4.99) with $E_{\mu,k} = 0$. With left hand members equal to zero we can group Eqs. (4.92-4.95) in pairs. Equation (4.93) and (4.95) may be written as,

$$\begin{bmatrix}\alpha_1(k,n+1)\\ \alpha_2(k,n+1)\end{bmatrix} = e^{-i\frac{ka}{2}}\begin{bmatrix} D_k & 0\\ -\frac{t_\perp}{t}e^{i\frac{ka}{2}} & D_k\end{bmatrix}\begin{bmatrix}\alpha_1(k,n)\\ \alpha_2(k,n)\end{bmatrix}, \quad (4.100)$$

and Eqs. (4.92) and (4.94) can be arranged to read,

$$\begin{bmatrix}\beta_2(k,n-1)\\ \beta_1(k,n-1)\end{bmatrix} = e^{i\frac{ka}{2}}\begin{bmatrix} D_k & 0\\ -\frac{t_\perp}{t}e^{-i\frac{ka}{2}} & D_k\end{bmatrix}\begin{bmatrix}\beta_2(k,n)\\ \beta_1(k,n)\end{bmatrix}, \quad (4.101)$$

where $D_k = -2\cos(ka/2)$. The 2×2 matrix in Eqs. (4.100) and (4.101) has the following property:

$$\begin{bmatrix} D_k & 0 \\ p_k & D_k \end{bmatrix}^n = \begin{bmatrix} D_k^n & 0 \\ nD_k^{n-1}p_k & D_k^n \end{bmatrix}. \quad (4.102)$$

Thus, the general solution of Eqs. (4.100) and (4.101) is,

$$\begin{bmatrix} \alpha_1(k,n) \\ \alpha_2(k,n) \end{bmatrix} = e^{-i\frac{ka}{2}n} \begin{bmatrix} D_k^n & 0 \\ -nD_k^{n-1}\frac{t_\perp}{t}e^{i\frac{ka}{2}} & D_k^n \end{bmatrix} \begin{bmatrix} \alpha_1(k,0) \\ \alpha_2(k,0) \end{bmatrix}, \quad (4.103)$$

and

$$\begin{bmatrix} \beta_2(k,N-n-1) \\ \beta_1(k,N-n-1) \end{bmatrix} = e^{i\frac{ka}{2}n} \begin{bmatrix} D_k^n & 0 \\ -nD_k^{n-1}\frac{t_\perp}{t}e^{-i\frac{ka}{2}} & D_k^n \end{bmatrix} \begin{bmatrix} \beta_2(k,N-1) \\ \beta_1(k,N-1) \end{bmatrix}, \quad (4.104)$$

respectively, with $n \geq 1$. Now we require the convergence condition $|D_k| < 1$, which guarantees that Eqs. (4.96-4.99) are satisfied for semi infinite systems with $E_{\mu,k} = 0$. It is then easily seen that the semi infinite bilayer sheet has surface states for ka in the region $2\pi/3 < ka < 4\pi/3$, which corresponds to $1/3$ of the possible k's, as in the graphene sheet. The next question concerns the number of surface states. As any initialization vector is a linear combination of only two linearly independent vectors,

$$\begin{bmatrix} \alpha_1(k,0) \\ \alpha_2(k,0) \end{bmatrix} = \alpha_1(k,0) \begin{bmatrix} 1 \\ 0 \end{bmatrix} + \alpha_2(k,0) \begin{bmatrix} 0 \\ 1 \end{bmatrix}, \quad (4.105)$$

$$\begin{bmatrix} \beta_2(k,N-1) \\ \beta_1(k,N-1) \end{bmatrix} = \beta_2(k,N-1) \begin{bmatrix} 1 \\ 0 \end{bmatrix} + \beta_1(k,N-1) \begin{bmatrix} 0 \\ 1 \end{bmatrix}, \quad (4.106)$$

there are two surface states per edge. Moreover, Eqs (4.103) and (4.104) are surface-state solutions on different sides of the ribbon. When the semi infinite system is considered only one of them survives. In particular, taking the limit $N \to \infty$ the two possible surface states are,

$$\begin{cases} \alpha_1(k,n) = \alpha_1(k,0)D_k^n e^{-i\frac{ka}{2}n} \\ \alpha_2(k,n) = -\alpha_1(k,0)nD_k^{n-1}\frac{t_\perp}{t}e^{-i\frac{ka}{2}(n-1)} \end{cases}, \quad (4.107)$$

and

$$\begin{cases} \alpha_1(k,n) = 0 \\ \alpha_2(k,n) = \alpha_2(k,0)D_k^n e^{-i\frac{ka}{2}n} \end{cases}, \quad (4.108)$$

for $n \geq 1$ and initialization given by Eq. (4.105). Though linearly independent, it is clear that surface states (4.107) and (4.108) are not orthogonal, except at

$ka = \pi$. It is convenient to orthogonalize solution (4.107) with respect to solution (4.108). After simple algebra we find for the orthonormalized surface states the following real space dependence,

$$\begin{cases} \alpha_1(k,n) = \alpha_1(k,0) D_k^n e^{-i\frac{ka}{2}n} \\ \alpha_2(k,n) = -\alpha_1(k,0) D_k^{n-1} \frac{t_\perp}{t} e^{-i\frac{ka}{2}(n-1)} \left(n - \frac{D_k^2}{1-D_k^2}\right) \end{cases}, \tag{4.109}$$

and

$$\begin{cases} \alpha_1(k,n) = 0 \\ \alpha_2(k,n) = \alpha_2(k,0) D_k^n e^{-i\frac{ka}{2}n} \end{cases}, \tag{4.110}$$

where the normalization constants are given by,

$$|\alpha_1(k,0)|^2 = \frac{(1-D_k^2)^3}{(1-D_k^2)^2 + t_\perp^2/t^2}, \tag{4.111}$$

$$|\alpha_2(k,0)|^2 = 1 - D_k^2. \tag{4.112}$$

Equations (4.109) and (4.110) are valid for $n \geq 0$, except at $ka = \pi$ where the only nonzero amplitudes are $\alpha_1(k,0)$ and $\alpha_2(k,1) = -\alpha_1(k,0)t_\perp/t$ for a surface state of the first type [Eq. (4.109)], and $\alpha_2(k,0)$ for a surface state of the second type [Eq. (4.110)], with normalization constants given by $|\alpha_1(k,0)|^2 = 1/(1+t_\perp^2/t^2)$ and $|\alpha_2(k,0)|^2 = 1$, respectively.

The solution given by Eq. (4.110) is exactly the same as that found for a single graphene layer,[18] where the only sites with nonzero amplitude belong to the A sublattice of layer 2, the one disconnected from the other layer. Solution (4.109) is a surface state that can only be found in bilayer graphene, as it has finite amplitudes in both layers. The sites of non-vanishing amplitude for this surface state occur at sublattice A of layer 2, as in the other solution, and at sublattice A of layer 1, which is connected to the other layer through $t_\perp$. Had we grown the semi infinite sheet from the other side of the ribbon, and two similar surface states would have appeared in the opposite edge with nonzero amplitudes at sites of the B sublattices. As concerns the localization length (λ), it is easily seen from Eqs. (4.109) and (4.110) that both solutions have the same $\lambda = -1/\ln|D_k|$. Nevertheless, the solution given by Eq. (4.109) has a linear dependence in n which enhances its penetration into the bulk, as seen in Fig. 4.12 (a) and (b), where we show the charge density of the surface state given by Eq. (4.109) in each layer at two different k values [the charge density $|\alpha_1(k,n)|^2$ shown in Fig. 4.12 can also be seen as the solution given by Eq. (4.110) for $|\alpha_2(k,n)|^2$, apart from a normalization factor].

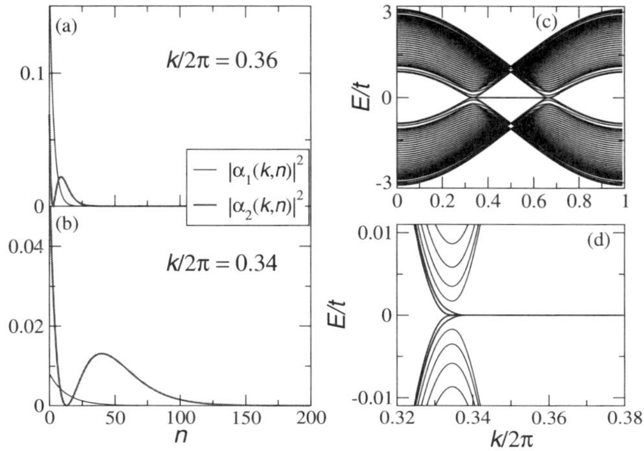

Fig. 4.12. (a) - Charge density of the surface states at $k/2\pi = 0.36$. (b) - The same as in (a) at $k/2\pi = 0.34$. (c) - Energy spectrum for a graphene bilayer ribbon with zigzag edges for $N = 400$. (d) - Zoom in of panel (c). The interlayer coupling was set to $t_\perp/t = 0.2$ in all panels.

The band structure of a bilayer ribbon with zigzag edges is shown in Fig. 4.12 (c) for $N = 400$. We can see the partly flat bands at $E = 0$ for k in the range $2\pi/3 \leq k \leq 4\pi/3$, corresponding to four surface states, two localized states per edge. The zoom in shown in Fig. (4.12) (d) for $k \approx 2\pi/3$ clearly shows that the flat bands are four. Strictly speaking, the surface states given by Eqs. (4.109) and (4.110) [and the other two resulting from Eq. (4.104)] are eigenstates of the semi infinite system only. In the ribbon the overlapping of these four surface states leads to a slight dispersion and non-degeneracy. However, as long as the ribbon width is sufficiently large, this effect is only important at $k \simeq 2\pi/3$ and $k \simeq 4\pi/3$ where λ is large enough for the overlapping to be appreciable.[19] As Eq. (4.109) has a deeper penetration into the bulk its degeneracy is lifted first, as can be seen in Fig. 4.12 (d).

When the two layers are made inequivalent by the bias the surface states are strongly affected. The semi infinite biased system has only one surface state given by Eq. (4.110), as the surface state having finite amplitudes at both layers [Eq. (4.109)] ceases to be an eigenstate of Hamiltonian (4.90) in the presence of the bias term. In Fig. 4.13 we show the band structure of a bilayer ribbon for different values of the bias, $V = t_\perp/10$ [(a) and (d)], $V = t_\perp/2$ [(b) and (e)], and $V = t_\perp$ [(c) and (f)].Two partly flat bands for k in the range $2\pi/3 \leq k \leq 4\pi/3$ are clearly seen at $E = \pm V/2$. These are bands of surface states localized at opposite ribbon edges, with finite amplitudes on a single layer [Eq. (4.110) and

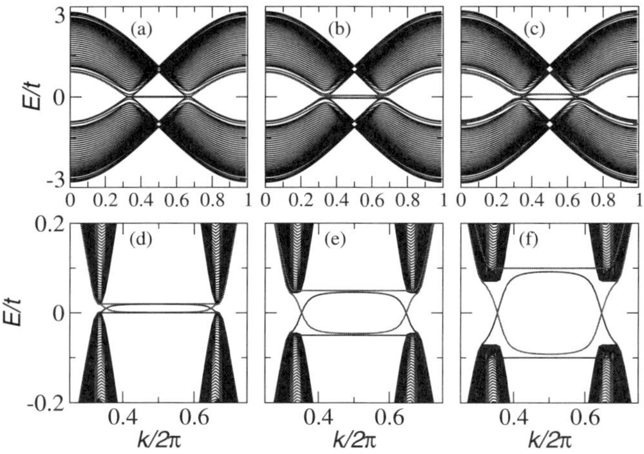

Fig. 4.13. Energy spectrum for a bilayer ribbon with zigzag edges for different values of the applied bias: $V = t_\perp/10$ (a), $V = t_\perp/2$ (b), $V = t_\perp$ (c). (d)-(f) – Zoom in of panels (a) to (c), respectively. The interlayer coupling was set to $t_\perp/t = 0.2$ and $N = 400$ in all panels.

its counterpart for the other edge]. Also evident is the presence of two dispersive bands crossing the gap, showing that the bilayer with zigzag edges is actually gapless, even for $V \neq 0$. These bands result from hybridization of the surface states given by Eq. (4.109) with delocalized bulk states. In particular, near $k \approx \pi$ they are reminiscent of those surface states as this k-value has the largest gap to bulk states, strongly reducing hybridization. In fact, we can understand both the closeness of the dispersive bands to $E \approx \pm V/2$ for $k \approx \pi$ and their crossing at $E = 0$ near the Dirac points by using perturbation theory on V/t. As surface states living at opposite edges have an exponentially small overlapping, and those belonging to the same edge were chosen to be orthogonal, we can treat the solution given by Eq. (4.109) and its counterpart for the other edge separately. Starting with Eq. (4.109), the first order energy shift induced by the applied bias is $E_k = V/2(\langle n_1^k \rangle - \langle n_2^k \rangle)$, where $\langle n_1^k \rangle$ and $\langle n_2^k \rangle$ give the probability of finding the localized electron in layer 1 and 2, respectively. The value of this quantities is easily obtained from Eq. (4.109) through real space summation, which gives,

$$\langle n_1^k \rangle = \frac{(1 - D_k^2)^2}{(1 - D_k^2)^2 + t_\perp^2/t^2}, \tag{4.113}$$

$$\langle n_2^k \rangle = \frac{t_\perp^2/t^2}{(1 - D_k^2)^2 + t_\perp^2/t^2}. \tag{4.114}$$

The band dispersion may then be written as,

$$E_k = \pm \frac{V}{2} \frac{(1 - D_k^2)^2 - t_\perp^2/t^2}{(1 - D_k^2)^2 + t_\perp^2/t^2}, \tag{4.115}$$

where the minus sign stands for the band of states localized at the other edge. Note that for $k \approx \pi$ we have $D_k \to 0$, and thus $E_k \approx \pm V/2$, as can be seen in Fig. 4.13. This means that for $k \approx \pi$ the surface state given by Eq. (4.109) is essentially localized at layer 1, which is clearly seen from Eqs. (4.113) and (4.114) as long as $t_\perp/t \ll 1$. However, for $1 - D_k^2 = t_\perp/t$ the energy shift [Eq. (4.115)] is zero, which leads to band crossing. For $t_\perp \ll t$ we can expand around the Dirac points, $k_0^\pm = 2\pi/3, 4\pi/4$. If $k = k_0 + \delta k$, the crossing takes place for,

$$\delta k = \pm \frac{t_\perp}{\sqrt{3}t}, \tag{4.116}$$

each sign being assigned to each Dirac point. As can be seen from Eq. (4.116), the value of δk is independent of the bias voltage. Moreover, its quantitative estimation compares fairly well with the numerical results shown in Fig. 4.13, where the various approximations are valid ($V \leq t_\perp \ll t$). The gap finally opens for $V \gtrsim t$.

4.4. Summary

In this chapter we gave a basic introduction to the physics of single and double graphene layers. For a single graphene layer we have discussed the Dirac like nature of the elementary excitations and computed the cyclotron mass of the charge carriers, showing that its dependence on the electronic density is much different from other two-dimensional systems. We also considered finite graphene ribbons with zigzag edges and showed that this system has a band of zero energy states. We have also discussed the nature of the Landau levels when a magnetic field perpendicular to the graphene plane is present. For the graphene bilayer we have studied its electronic spectrum both with and without an applied bias, with special attention to the formation of a the Fermi ring ground state. We have discussed the quantum Hall effect in the double layer, and obtained its unusual quantization rule. For the graphene bilayer with zigzag edges we have shown the existence of a new kind of zero energy states.

Acknowledgments

E.V.C. and J.M.B.L.S. were supported by FCT grant No. SFRH/BD/13182/2003 and EU through POCTI (QCAIII). N.M.R.P. acknowledges ESF Science

Programme No. INSTANS 2005-2010 and FCT and EU Grant No. POCTI/FIS/58133/2004. F.G. was supported by MEC (Spain) grant No. FIS2005-05478-C02-01 and EU contract 12881 (NEST). A. H. C. N was supported through NSF grant DMR-0343790.

References

1. K. S. Novoselov, A. K. Geim, S. V. Morozov, D. Jiang, Y. Zhang, S. V. Dubonos, I. V. Grigorieva, and A. A. Firsov, Electric field effect in atomically thin carbon films, *Science*. **306**, 666, (2004).
2. K. S. Novoselov, D. Jiang, F. Schedin, T. J. Booth, V. V. Khotkevich, S. V. Morozov, and A. K. Geim, Two-dimensional atomic crystals, *Proc. Nat. Acad. Sc.* **102**, 10451, (2005).
3. K. S. Novoselov, A. K. Geim, S. V. Morozov, D. Jiang, M. I. Katsnelson, I. V. Grigorieva, S. V. Dubonos, and A. A. Firsov, Two-dimensional gas of massless dirac fermions in graphene, *Nature*. **438**, 197, (2005).
4. Y. Zhang, Y.-W. Tan, H. L. Stormer, and P. Kim, Experimental observation of the quantum hall effect and berry's phase in graphene, *Nature*. **438**, 201–204, (2005).
5. N. M. R. Peres, F. Guinea, and A. H. Castro Neto, Electronic properties of disordered two-dimensional carbon, *Phys. Rev. B*. **73**, 125411, (2006).
6. V. P. Gusynin and S. G. Sharapov, Unconventional integer quantum hall effect in graphene, *Phys. Rev. Lett.* **95**, 146801, (2005).
7. S. Y. Zhou, G. H. Gweon, J. Graf, A. V. Fedorov, C. D. Spataru, R. D. Diehl, Y. Kopelevich, D. H. Lee, S. G. Louie, and A. Lanzara, First direct observation of dirac fermions in graphite, *Nature Physics*. **2**, 595, (2006).
8. K. S. Novoselov, E. McCann, S. V. Morozov, V. I. Falko, M. I. Katsnelson, U. Zeitler, D. Jiang, F. Schedin, and A. K. Geim, Unconventional quantum hall effect and berry's phase of 2pi in bilayer graphene, *Nature Physics*. **2**, 177–180, (2006).
9. E. McCann and V. I. Fal'ko, Landau-level degeneracy and quantum hall effect in a graphite bilayer, *Phys. Rev. Lett.* **96**, 086805, (2006).
10. T. Ohta, A. Bostwick, T. Seyller, K. Horn, and E. Rotenberg, Controlling the electronic structure of bilayer graphene, *Science*. **313**, 951, (2006).
11. F. Guinea, A. H. Castro Neto, and N. M. R. Peres, Electronic states and landau levels in graphene stacks, *Phys. Rev. B*. **73**, 245426, (2006).
12. R. Peierls, Zur theorie des diamagnetismus von leitungselektronen, *Z. Phys.* **80**, 763, (1933).
13. J. M. Luttinger, The effect of a magnetic field on electrons in a periodic potential, *Phys. Rev.* **84**, 814–817, (1951).
14. M. Graf and P. Vogl, Electromagnetic fields and dielectric response in empirical tight-binding theory, *Phys. Rev. B*. **51**, 4940–4949, (1995).
15. T. B. Boykin, Electromagnetic coupling and gauge invariance in the empirical tight-binding method, *Phys. Rev. B*. **63**, 245314, (2001).
16. N. W. Ashcroft and N. D. Mermin, *Solid State Physics*, (Saunders College, Philadelphia, PA, 1976).

17. M. Fujita, K. Wakabayashi, K. Nakada, and K. Kusakabe, Peculiar localized state at zigzag graphite edge, *J. Phys. Soc. Jpn.* **65**, 1920–1923, (1996).
18. K. Nakada, M. Fujita, G. Dresselhaus, and M. S. Dresselhaus, Edge state in graphene ribbons: Nanometer size effect and edge shape dependence, *Phys. Rev. B.* **54**, 17954–17961, (1996).
19. K. Wakabayashi, M. Fujita, H. Ajiki, and M. Sigrist, Electronic and magnetic properties of nanographite ribbons, *Phys. Rev. B.* **59**, 8271–8282, (1999).
20. Y. Niimi, T. Matsui, H. Kambara, K. Tagami, M. Tsukada, and H. Fukuyama, Scanning tunneling microscopy and spectroscopy studies of graphite edges, *Appl. Surf. Sci.* **241**, 43, (2005).
21. Y. Kobayashi, K.-I. Fukui, T. Enoki, K. Kusakabe, and Y. Kaburagi, Observation of zigzag and armchair edges of graphite using scanning tunneling microscopy and spectroscopy, *Phys. Rev. B.* **71**, 193406, (2005).
22. P. G. Harper, Single band motion of conduction electrons in a uniform magnetic field, *Proc. Phys. Soc. Lond. A.* **68**, 874, (1955).
23. R. B. Laughlin, Quantized hall conductivity in two dimensions, *Phys. Rev. B.* **23**, 5632–5633, (1981).
24. B. I. Halperin, Quantized hall conductance, current-carrying edge states, and the existence of extended states in a two-dimensional disordered potential, *Phys. Rev. B.* **25**, 2185–2190, (1982).
25. D. A. Abanin, P. A. Lee, and L. S. Levitov, Spin-filtered edge states and quantum hall effect in graphene, *Phys. Rev. Lett.* **96**, 176803, (2006).
26. J. C. Charlier, X. Gonze, and J. P. Michenaud, First-principles study of the electronic properties of graphite, *Phys. Rev. B.* **43**, 4579–4589, (1991).

Chapter 5

Anomalous Hall effect

V. K. Dugaev[1], M. Taillefumier[2,3], B. Canals[2], C. Lacroix[2], P. Bruno[4]

[1]*Department of Mathematics and Applied Physics, Rzeszów University of Technology, 35-959 Rzeszów, Poland and Departamento de Física and CFIF, Instituto Superior Técnico, Av. Rovisco Pais, 1049-001 Lisboa, Portugal and Frantsevich Institute for Problems of Materials Science, Vilde 5, 58001 Chernovtsy, Ukraine*

[2]*Laboratoire Louis Néel, CNRS, 25 Av. des Martyrs, 38042 Grenoble, Cedex 9, France*

[3]*Department of Physics, Norwegian University of Science and Technology, N-7491 Trondheim, Norway*

[4]*Max-Planck-Institut für Mikrostrukturphysik, Weinberg 2, 06120 Halle, Germany*

We review the theory and main physical mechanisms of the anomalous Hall effect in magnetic metals and semiconductors. Recently proposed mechanisms of the chirality-induced Hall effect, topological Hall effect, and the intrinsic mechanism of the anomalous Hall effect are reviewed. In the latter case, we discuss in more details the problem of defects and scattering from impurities. Our consideration is mostly based on the original articles of the authors.

Contents

5.1. Introduction

The Anomalous Hall effect (AHE) has been known for a very long time,[1] and since then it still fascinates the physicists by the variety of proposed ideas to understand the physics of this phenomenon, numerous hot discussions and some controversial results. At the first glance, the AHE is like the ordinary Hall effect in metals or semiconductors: if we place a sample with an electric current into an external magnetic field H perpendicular to the current, then there occurs a voltage in the direction perpendicular to both the direction of current and the magnetic field. It is well known that the ordinary Hall effect is related to the Lorentz force acting on electrons moving in the field H.

In the magnetic materials, there exist an analogue of this effect without the external field, i.e., the role of the magnetic field plays the internal magnetization. It is usually called "anomalous" or "extraordianary" Hall effect. Even though the magnetization in ferromagnets is accompanied by the magnetic induction acting on electrons like a magnetic field in vacuum, the magnitude of AHE cannot be explained by the Lorentz force mechanism.

Like the ordinary Hall effect, the AHE is an important tool to characterize the magnetic state of metals and semiconductors. The magnitude of the AHE is determined by the internal magnetization. Therefore, it enables to find the Curie temperature by measuring the temperature dependence of the AHE. Besides, the AHE can be used to determine the orientation of the magnetization vector.

A possible mechanism of the AHE has been proposed about fifty years ago by Luttinger and Karplus[2] but this pioneer work induced a lot of discussions and criticisms, and gradually the general opinion has inclined to a different explanation. Luttinger and Karplus suggested that the AHE can be related to a complex energy structure of metals and semiconductors, which includes the effect of spin-orbit (SO) interaction on the energy spectrum of electrons. They assumed that the impurities are not so important. On the contrary, the mechanisms proposed by Smit[3] and Berger[4] were associated with the scattering from impurities including a spin-flip scattering due to the SO interaction.

Recently, some new ideas occurred in the theory of AHE. First, it was proposed that the AHE can be induced by the chirality of magnetic ordering in inhomogeneous ferromagnets.[5–7] In the case of a smooth variation of the magnetization, the corresponding theory can be formulated in terms of the Berry phase of electrons moving in the adiabatically varying magnetization.

The other intriguing idea was expressed by Jungwirth et al.[8] and Onoda et al.[9]

They revised the approach of Ref.[2] presenting it in an elegant form of the Berry phase of electron moving in the momentum space. The necessary condition for the Berry-phase induced AHE in this case is a nontrivial topology of the energy bands, which in turn results from the SO interaction relating the crystal potential to an effective field acting on the spin.

5.2. Scattering from impurities

The mechanisms proposed by Smit and Berger are usually called the "skew scattering" and "side-jump", respectively. A simple physical interpretation of the skew scattering and side-jump mechanisms is based on a picture of the scattering event, which takes into account the anisotropy of the scattering amplitude (in case of skew scattering) and a lateral displacement of the electron wavepacket during the scattering (side-jump).[10] In both cases the SO interaction plays the key role.

Let us consider a model of the ferromagnet with a homogenious magnetization field $\mathbf{M}$ oriented along the axis z and a SO relativistic term (we take the units with $\hbar = 1$)

$$H = \int d^3r\, \psi^\dagger(\mathbf{r}) \left[-\frac{\nabla^2}{2m} - M\sigma_z - \frac{i\,\lambda_0^2}{4} \left(\boldsymbol{\sigma} \times \nabla V(\mathbf{r}) \right) \cdot \nabla + V(\mathbf{r}) \right] \psi(\mathbf{r}) \,, \tag{5.1}$$

where m is the electron effective mass, λ_0 is a constant, which measures the strength of the SO interaction, $V(\mathbf{r})$ is a random potential created by impurities or defects, $\boldsymbol{\sigma} = (\sigma_x, \sigma_y, \sigma_z)$ are the Pauli matrices, and $\psi^\dagger \equiv \left(\psi_\uparrow^\dagger,\, \psi_\downarrow^\dagger \right)$ is the spinor field, corresponding to electrons with spin up and down orientations.

We assume that the potential $V(\mathbf{r})$ is short-ranged with zero mean value, $\langle V(\mathbf{r}) \rangle = 0$, where the angle brackets mean the configurational averaging over all realizations of $V(\mathbf{r})$. We can characterize this potential by its second, γ_2, and third, γ_3, momenta, denoting $\langle V(\mathbf{r}_1)\, V(\mathbf{r}_2) \rangle = \gamma_2\, \delta(\mathbf{r}_1 - \mathbf{r}_2)$ and $\langle V(\mathbf{r}_1)\, V(\mathbf{r}_2)\, V(\mathbf{r}_3) \rangle = \gamma_3\, \delta(\mathbf{r}_1 - \mathbf{r}_3)\, \delta(\mathbf{r}_2 - \mathbf{r}_3)$. It should be emphasized that the constants γ_2 and γ_3 are the parameters, characterizing not only the strength of the disorder potential, but also the statistical properties of the random field. When the potential $V(\mathbf{r})$ is created by impurities, distributed randomly at some points $\mathbf{R}_i$, we have $V(\mathbf{r}) = \sum_i v(\mathbf{r} - \mathbf{R}_i)$. It results in $\gamma_2 = N_i v_0^2$ and $\gamma_3 = N_i v_0^3$, where N_i is the impurity concentration, and v_0 is the matrix element of the short-ranged potential of one isolated impurity, $v(\mathbf{r} - \mathbf{R}_i) = v_0\, \delta(\mathbf{r} - \mathbf{R}_i)$. In the case of purely Gaussian potential, we should take $\gamma_3 = 0$.

Calculating the matrix elements of the Hamiltonian (5.1) in momentum representation, one obtains

$$H=\sum_{\mathbf{k}}\psi^{\dagger}_{\mathbf{k}}\left(\frac{k^2}{2m}-M\sigma_z\right)\psi_{\mathbf{k}}+\sum_{\mathbf{kk'}}\psi^{\dagger}_{\mathbf{k}}V_{\mathbf{k}-\mathbf{k'}}\left[1+\frac{i\,\lambda_0^2}{4}\,(\mathbf{k}\times\mathbf{k'})\cdot\boldsymbol{\sigma}\right]\psi_{\mathbf{k'}}\,, \tag{5.2}$$

where $V_{\mathbf{k}}$ is the Fourier transform of the potential $V(\mathbf{r})$. The second term in Eq. (5.2) describes the SO scattering from impurities.

To find the expression for current density operator $\mathbf{j}(t)$, we switch on an electromagnetic field $\mathbf{A}(t)$ in a gauge-invariant way, $\mathbf{k}\to(\mathbf{k}-e\mathbf{A}/c)$, and calculate the derivative

$$j_\alpha=-c\,\frac{\delta H}{\delta A_\alpha}\,, \tag{5.3}$$

which gives us

$$j_\alpha=\sum_{\mathbf{kk'}}\psi^{\dagger}_{\mathbf{k}}\left[\frac{e}{m^*}\left(k_\alpha-\frac{eA_\alpha}{c}\right)\delta_{\mathbf{kk'}}+\frac{ie\,\lambda_0^2}{4}\,V_{\mathbf{k}-\mathbf{k'}}\,\epsilon_{\alpha\beta\gamma}\,(k'_\beta-k_\beta)\,\sigma_\gamma\right]\psi_{\mathbf{k'}}\,. \tag{5.4}$$

where $\epsilon_{\alpha\beta\gamma}$ is the unit antisymmetric tensor. According to Eq. (5.4), the SO interaction contributes to the current vertex in the Feynman diagrams of the conductivity tensor.[11]

5.2.1. *Side-jump*

Calculating the loop diagrams with the current vertex modified by the SO interaction (5.4) we find for the off-diagonal (Hall) conductivity[12]

$$\sigma^{(sj)}_{xy}=-\frac{ie^2\lambda_0^2\,\gamma_2}{4\pi m}\,\mathrm{Tr}\sum_{\mathbf{kk'}}\sigma_z\,k_y^2\,G^R_{\mathbf{k}}\,G^A_{\mathbf{k}}\,\left(G^R_{\mathbf{k'}}-G^A_{\mathbf{k'}}\right)\,, \tag{5.5}$$

where the retarded (R) and advanced (A) Green functions at the Fermi surface are diagonal matrices

$$G^{R,A}_{\mathbf{k}}=\mathrm{diag}\left(\frac{1}{\mu-\varepsilon_\uparrow(k)\pm i/(2\tau_\uparrow)}\,,\;\frac{1}{\mu-\varepsilon_\downarrow(k)\pm i/(2\tau_\downarrow)}\right). \tag{5.6}$$

Here $\varepsilon_{\uparrow,\downarrow}(k)=k^2/(2m^*)\mp M$ are the energy spectra of spin-up and spin-down electrons, respectively, μ is the chemical potential, and $\tau_{\uparrow,\downarrow}$ are the corresponding relaxation times. The relaxation times are determined by the scattering from the random potential, and they are equal to $\tau_{\uparrow,\downarrow}=(2\pi\nu_{\uparrow,\downarrow}\gamma_2)^{-1}$, where $\nu_\uparrow$ and $\nu_\downarrow$ are the densities of states for spin-up (majority) and spin-down (minority) electrons at the Fermi level.

After calculating the integrals in Eq. (5.5), we find the side-jump anomalous Hall conductivity[10]

$$\sigma_{xy}^{(sj)} = \frac{e^2\lambda_0^2}{6} \left(\nu_\downarrow\, k_{F\downarrow}\, v_{F\downarrow} - \nu_\uparrow\, k_{F\uparrow}\, v_{F\uparrow}\right), \tag{5.7}$$

where $k_{F\uparrow,\downarrow}$ and $v_{F\uparrow,\downarrow}$ are the momenta and velocities of majority and minority electrons at the Fermi surfaces, respectively. The AHE conductivity (5.7) does not depend on the impurity relaxation time and, correspondingly on the impurity density.

5.2.2. *Skew scattering*

In frame of the skew scattering mechanism, we take into account the third-order corrections due to scattering from impurities, keeping the first order of SO-depending matrix elements. Without quantum corrections, the relevant Feynman diagrams for the skew scattering mechanism are presented in Refs. [10,12] Calculating these diagrams, we find

$$\sigma_{xy}^{(ss)} = \frac{\pi e^2\lambda_0^2\gamma_3}{18\gamma_2} \left(\nu_\downarrow^2 k_{F\downarrow}^2 v_{F\downarrow}^2 \tau_\downarrow - \nu_\uparrow^2 k_{F\uparrow}^2 v_{F\uparrow}^2 \tau_\uparrow\right). \tag{5.8}$$

In this formula, the factor γ_3/γ_2 contains the information about both the strength of the random potential and its statistical properties. The Hall conductivity (5.8) is proportional to the impurity relaxation time. Correspondingly, the contribution of the skew scattering to the Hall conductivity is growing in the limit of low impurity density. However, this mechanism of AHE is vanishingly small in the case of a Gaussian disorder like in alloy compounds.

5.3. Spin chirality and the topological Hall effect

In this section we consider a model of 2D electron gas in a smoothly varying magnetization field $\mathbf{M}(\mathbf{r})$ and demonstrate that the nonvanishing spin chirality can be another source of the AHE in disordered magnetic materials The Hamiltonian, acting on the spinor wave function of electrons, has the following form

$$H = -\frac{\nabla^2}{2m} - \boldsymbol{\sigma}\cdot\mathbf{M}(\mathbf{r}). \tag{5.9}$$

We assume the magnitude of M to be constant, $\mathbf{M}(\mathbf{r}) = M\,\mathbf{n}(\mathbf{r})$, and the three-dimensional unit vector $\mathbf{n}(\mathbf{r})$ to be a slowly varying function of coordinates.

It is convenient to use the local gauge transformation $T(\mathbf{r})$, which makes the quantization axis oriented along vector $\mathbf{n}(\mathbf{r})$ at each point.[14] It obeys the unitary condition $T^\dagger(\mathbf{r})\,T(\mathbf{r}) = 1$ and transforms the last term in Eq. (5.9) as

$$T^\dagger(\mathbf{r})\;[\boldsymbol{\sigma}\cdot\mathbf{n}(\mathbf{r})]\;T(\mathbf{r}) = \sigma_z. \tag{5.10}$$

The transformation (5.10) corresponds to the rotation of the quantization axis from z-axis to the position along vector $\mathbf{n}(\mathbf{r})$.

The transformed Hamiltonian has the form of Hamiltonian of electrons moving in a gauge potential $\mathbf{A}(\mathbf{r})$, and in a constant magnetization field along the axis z

$$H = -\frac{1}{2m}\,(\nabla - i\mathbf{A}(\mathbf{r}))^2 - M\,\sigma_z\,, \tag{5.11}$$

where

$$\mathbf{A}(\mathbf{r}) = i\,T^\dagger(\mathbf{r})\,\frac{\partial}{\partial\mathbf{r}}\,T(\mathbf{r}). \tag{5.12}$$

The components of $\mathbf{A}(\mathbf{r})$ can be found using Eqs. (5.10) and (5.12). Hamiltonian (5.11) with matrix field $\mathbf{A}(\mathbf{r})$ determined by Eq. (5.12) contains the terms proportional to σ_x and σ_y, which induce transitions between the spin-polarized states. In the following we can assume that the probability of spin-flip processes is small. It corresponds to the adiabaticity condition,[13] which can be formulated as a small variation of the effective spin-flip potential $V_{sf}\sim k_F/m\xi$ (where ξ is the characteristic length of the variation of magnetization) with respect to the spin splitting M

$$\frac{\epsilon_F}{M\,k_F\xi} \ll 1. \tag{5.13}$$

Assuming adibaticity (5.13), in the case of Fermi level in the lower spin-up electron band, we can neglect the spin-down energy band and restrict ourselves by considering the following truncated Hamiltonian of electrons without off-diagonal in spin terms

$$\tilde{H} = -\frac{1}{2m}\left(\frac{\partial}{\partial\mathbf{r}} - i\,\boldsymbol{a}(\mathbf{r})\right)^2 + V(\mathbf{r}), \tag{5.14}$$

where the new vector potential reads

$$a_i(\mathbf{r}) = -\frac{n_x\,\partial_i n_y - n_y\,\partial_i n_x}{2(1+n_z)}, \tag{5.15}$$

and the potential relief $V(\mathbf{r})$ results from second order in $\mathbf{A}(\mathbf{r})$ terms in Eq. (5.12)

$$V(\mathbf{r}) = \frac{1}{8m}\,(\partial_i n_\mu)^2\,, \tag{5.16}$$

and $\mu = x, y, z$. The vector potential $\boldsymbol{a}(\mathbf{r})$ acts only on the electrons within the spin-up band. Hamiltonian (5.14) describes spinless electrons in the gauge field $\boldsymbol{a}(\mathbf{r})$ and potential relief $V(\mathbf{r})$. The Hamiltonian of electrons belonging to the spin-down subband differs from (5.14) only by the sign of gauge field $\boldsymbol{a}(\mathbf{r})$. Assuming the adiabaticity of the electron motion, we can neglect the transitions between the spin subbands.

The topological field can be defined now as

$$b_t(\mathbf{r}) = \frac{\phi_0}{2\pi}\,\mathrm{rot}\,\boldsymbol{a} = \frac{\phi_0}{2\pi}\left(\partial_x a_y - \partial_y a_x\right), \tag{5.17}$$

where $\phi_0 = hc/e$ is the flux quantum. The topological field $b_t(\mathbf{r})$ acts on the spin-polarized electrons like the ordinary magnetic field. In particular, just like an external magnetic field H leading to the ordinary Hall effect in a system with electric current, the topological field b_t induces the anomalous Hall current in the direction perpendicular to the electric current flow. In the 2D system under consideration, the topological field is directed perpendicular to the 2D plane. It should be noted that the Hall current induced by the topological field $b_t(\mathbf{r})$ is the *local* current. The necessary condition of a nonzero macroscopic Hall effect is $\overline{b_t(\mathbf{r})} \neq 0$, where the average is over the whole 2D plane.

Using Eqs. (5.15) and (5.17) we find

$$b_t(\mathbf{r}) = -\frac{\phi_0}{4\pi}\,\epsilon_{\mu\nu\lambda}\, n_\mu\, (\partial_x n_\nu)\, (\partial_y n_\lambda), \tag{5.18}$$

where $\epsilon_{\mu\nu\lambda}$ is the unit antisymmetic tensor. The integral

$$\Omega(L) = \frac{1}{2}\int_S d^2\mathbf{r}\ \epsilon_{\mu\nu\lambda}\, n_\mu\, (\partial_x n_\nu)\, (\partial_y n_\lambda) \tag{5.19}$$

is the Berry phase calculated as the spherical angle spanned by an area S inside the contour L in $\mathbf{n}$-space. This results from the mapping of a closed contour L_0 in the plane to the contour L in the mapping space S_2. In the 2D case with a constant magnetization at infinity, one can compactify 2D plane to a sphere S_2. Then the vector

$$Q_\mu(\mathbf{r}) = \frac{1}{4\pi}\,\epsilon_{\mu\nu\lambda}(\partial_x n_\nu)(\partial_y n_\lambda) \tag{5.20}$$

can be identified as the topological current, and the quantity

$$I = \int d^2\mathbf{r}\ n_\mu(\mathbf{r})\ Q_\mu(\mathbf{r}) \tag{5.21}$$

is the topological invariant, i.e., and integer corresponding to the number of covering the mapping space.[15]

An example of the 2D electronic structure with a periodic distribution of magnetization $\mathbf{M}(\mathbf{r})$ such that $\overline{\mathbf{M}(\mathbf{r})} = 0$ and $\overline{b_t(\mathbf{r})} \neq 0$ was proposed in Ref.[16] It presents a structure with a 2D layer of the diluted magnetic semiconductor on top of a periodic lattice of magnetic nanocylinders.[17] Due to the purely topological nature of the AHE in this case, it was called *topological Hall effect*. Note that the occurrence of the topological Hall effect is not necessarily related to any SO interaction.

5.4. Complex energy spectrum and Berry phase in momentum space

The mechanism of AHE proposed by Karplus and Luttinger[2] differs from any of the mechanisms considered before. Its modern interpretation is related to the nontrivial topology of wavefunctions in the momentum space. This possibility can be realized in different models like, for example, the Kane model of the valence band in III-V semiconductors or a model with SO interaction related to the electric field near the structure interface. As an example, in this section we consider a model of 2DEG in a homogeneous magnetization field $\mathbf{M}_0$ directed along the axis z, with the Rashba SO interaction[18] and with a number of randomly distributed "weak" impurities. As we demonstrate below, the inclusion of impurities may be crucial for this mechanism, even in the limit of vanishingly small impurity density. This is a natural source and an explanation of many controversies in different theoretical approaches.

The Hamiltonian without the random field related to impurities reads

$$H_0 = \varepsilon_{\mathbf{k}} + \alpha\,(\sigma_x\, k_y - \sigma_y\, k_x) - M\sigma_z\ , \tag{5.22}$$

where $\varepsilon_{\mathbf{k}} = k^2/2m$, α is the coupling constant of the Rashba SO interaction. Due to the SO coupling, the components of velocity operators, $v_i = \partial H/\partial k_i$, are matrices in the spin space

$$v_x = \frac{k_x}{m} - \alpha\,\sigma_y, \qquad v_y = \frac{k_y}{m} + \alpha\,\sigma_x. \tag{5.23}$$

We include into consideration the scattering from impurities, described by the disorder potential $V(\mathbf{r})$. We assume that the disorder potential is short-range and weak, so that we can treat it in the Born approximation of the impurity scattering. Thus, the Hamiltonian of our model is $H = H_0 + V(\mathbf{r})$, and any physical variables including the Hall conductivity should be calculated with a corresponding averaging over the random potential $V(\mathbf{r})$. Note that such a model does not take into account the SO interaction in the impurity potential $V(\mathbf{r})$. In this section we concentrate on the mechanism induced only by the SO interaction associated with the crystal-lattice potential.

The conductivity tensor within the usual Kubo formalism is

$$\sigma_{ij}(\omega) = \frac{e^2}{\omega}\,\mathrm{Tr}\int \frac{d\varepsilon}{2\pi}\,\left\langle \hat{v}_i\,\hat{G}(\varepsilon+\omega)\,\hat{v}_j\,\hat{G}(\varepsilon)\right\rangle , \tag{5.24}$$

where $\hat{G}(\varepsilon) = (\varepsilon - H)^{-1}$ is the operator Green function of electrons described by the Hamiltonian H, which includes the disorder, $\hat{v}_i$ is the corresponding velocity operator, the trace goes over any eigenstates in the space of operator H (e.g., eigenstates of the Hamiltonian H itself), and $\langle ... \rangle$ means the disorder average. After averaging on disorder, the off-diagonal conductivity can be presented in the form

$$\sigma_{xy}(\omega) = \frac{e^2}{\omega}\,\mathrm{Tr}\int \frac{d\varepsilon}{2\pi}\,\frac{d^2\mathbf{k}}{(2\pi)^2}\,\mathcal{T}_x(\varepsilon,\omega)\,G_{\mathbf{k}}(\varepsilon+\omega)\,v_y\,G_{\mathbf{k}}(\varepsilon), \tag{5.25}$$

where the trace runs over the spin states, and the Green functions $G_{\mathbf{k}}(\varepsilon)$ are the disorder-averaged functions in momentum representation.

In the following calculation of the static off-diagonal conductivity $\sigma_{xy}(\omega = 0)$, we will be interested in the case when the density of impurities N_i is small but finite. It also includes the "clean limit" of $N_i \to 0$, which can be physically realized in samples with vanishingly small concentration of impurities and defects. Correspondingly, when we calculate the static conductivity tensor of a clean sample using Eq. (5.25), we take the limit of $\omega \to 0$ before the limit of $N_i \to 0$.

In accordance with the well-known result of Středa,[10,19] Eq. (5.25) leads to two terms in the off-diagonal conductivity, one of which, σ_{xy}^{I}, is due to the electron states near the Fermi energy (it corresponds to the non-geometric contribution), and the other one, σ_{xy}^{II}, is related to the contribution of all occupied electron states below the Fermi energy (identified as the Berry-phase-induced intrinsic mechanism). Note that in some early considerations assuming a "pure limit", the σ_{xy}^{I} contribution related to impurities, was completely neglected. It should be emphasized that this was essentially wrong. Performing the calculations, which include the scattering from impurities, we find an additional contribution, which is also intrinsic, i.e., it does not vanish in the limit of $N_i \to 0$. This is quite similar to the side-jump mechanism of AHE.[10] As we see, the additional contribution to AHE comes from the states at the Fermi surface, for which the finite relaxation time due to impurity scattering is important. If we totally neglect the scattering from impurities, we miss this term. The real structures always have some impurities or defects. Therefore, we can justify taking the limit of $\omega \to 0$ before $N_i \to 0$ as a physical realization of the clean system as a system with very small but still nonzero density of impurities.

In the absence of impurity scattering, the electron Green function can be found using Hamiltonian (5.22)

$$G_{\mathbf{k}}^{0}(\varepsilon) = \frac{\varepsilon - \varepsilon_{\mathbf{k}} + \mu + \alpha\,(k_y\sigma_x - k_x\sigma_y) - M\sigma_z}{(\varepsilon - E_{\mathbf{k},+} + \mu + i\delta\,\mathrm{sign}\,\varepsilon)\,(\varepsilon - E_{\mathbf{k},-} + \mu + i\delta\,\mathrm{sign}\,\varepsilon)}\,, \tag{5.26}$$

where μ is the chemical potential and the electron energy spectrum $E_{\mathbf{k}}$ consists of two branches, which we label as "+" and "−" corresponding mostly to the spin up and down electrons, respectively (even though they contain an admixture of the opposite spin due to the SO interaction),

$$E_{\mathbf{k},\pm} = \varepsilon_{\mathbf{k}} \mp \lambda(k), \tag{5.27}$$

and we denote $\lambda(k) = \left(M^2 + \alpha^2 k^2\right)^{1/2}$. We consider a general case when the chemical potential μ can be situated in both spin up and down subbands, corresponding to $\mu > M$. When only the spin up subband is filled with electrons, $-M < \mu < M$, only the contribution of the filled subband, $E_{\mathbf{k},+}$, should be kept.

5.4.1. *Self energy*

Due to the scattering from impurities and defects, the Green function (5.26) is modified. It is important to include the effect of scattering for the correct evaluation of the contribution to the off-diagonal conductivity from the Fermi surface. We consider the model of disorder created by weak short-range scatterers, which can be treated in the Born approximation. The corresponding self energy of electrons is calculated as

$$\Sigma_i\left(\varepsilon\right) = N_i\, V_0^2 \int \frac{d^2\mathbf{k}}{(2\pi)^2}\, G^0_{\mathbf{k}}(\varepsilon), \tag{5.28}$$

where V_0 is the Fourier transform of the impurity potential, and N_i is the impurity density. The result of calculation using Eqs. (5.26) and (5.28) can be presented as

$$\Sigma_i\left(\varepsilon\right) = -i\left(\Gamma + \tilde{\Gamma}\,\sigma_z\right), \tag{5.29}$$

where

$$\Gamma = \frac{\pi N_i V_0^2 \operatorname{sign}\varepsilon}{2}\left(\nu_+ + \nu_-\right), \qquad \tilde{\Gamma} = \frac{\pi N_i V_0^2 M \operatorname{sign}\varepsilon}{2}\left(\frac{\nu_+}{\lambda_+} - \frac{\nu_-}{\lambda_-}\right), \tag{5.30}$$

$\lambda_\pm \equiv \lambda(k_\pm)$, $\nu_\pm$ are the densities of states at the energy surfaces $\varepsilon = E_{\mathbf{k},\pm}$ of two different subbands for $\mu > -M$, and $k_\pm$ are the momenta of the majority and minority electrons, respectively.

Taking into account the self-energy correction (5.280), we find the Green function of the electron system with impurities

$$G_{\mathbf{k}}(\varepsilon) = \frac{\varepsilon + i\,\Gamma - \varepsilon_{\mathbf{k}} + \mu + \alpha(k_y\sigma_x - k_x\sigma_y) - \sigma_z(M + i\,\tilde{\Gamma})}{\left(\varepsilon - E_{\mathbf{k},+} + \mu + i\Gamma_+\right)\left(\varepsilon - E_{\mathbf{k},-} + \mu + i\Gamma_-\right)} \tag{5.31}$$

where

$$\Gamma_\pm \equiv \frac{\operatorname{sign}\varepsilon}{2\tau_\pm} = \Gamma \pm \tilde{\Gamma}\,\frac{M}{\lambda_\pm}. \tag{5.32}$$

The values of τ_+ and τ_- determined by Eq. (5.32) are the relaxation times of electrons in different energy subbands.

It should be noted that the relaxation times τ_+ and τ_- are equal for $\varepsilon > M$. This results from the assumed model of electron-impurity interaction, which does not depend on spin. If the energy ε is located within the gap, $-M < \varepsilon < M$, then $\nu_- = 0$, and the relaxation times τ_+, τ_- differ strongly.

5.4.2. *Vertex correction*

The equation for the vertex $\mathcal{T}_i(\varepsilon, \omega)$ in Eq. (5.25) can be presented using the Feynman diagrams with the impurity ladder.[11] For a short-range impurity potential, this equation has the following form

$$\mathcal{T}_i(\varepsilon, \omega) = v_i + N_i V_0^2 \int \frac{d^2\mathbf{k}}{(2\pi)^2} G_{\mathbf{k}}(\varepsilon)\, \mathcal{T}_i(\varepsilon, \omega)\, G_{\mathbf{k}}(\varepsilon + \omega). \tag{5.33}$$

In the limit of $\omega \to 0$, the integral is not zero only at the Fermi surface, i.e., for energies $\varepsilon \ll \tau_{\uparrow,\downarrow}^{-1}$. We assume that the density of impurities is low, which corresponds to the large relaxation times $\tau_{\uparrow,\downarrow}$. Thus, we calculate the vertex part for $\varepsilon = 0$, otherwise we take $\mathcal{T}_i = v_i$. Denoting $\mathcal{T}_x = \mathcal{T}_x(\varepsilon = 0,\ \omega \to 0)$, we find the solution of Eq. (5.33) in the form

$$\mathcal{T}_x = a k_x + b\sigma_x + c\sigma_y \tag{5.34}$$

with some coefficients a, b, and c. In the limit of $N_i \to 0$, $a = 1/m$ and $b = 0$, and the constant c is defined by

$$c = -\alpha \left[1 + 2N_i V_0^2 \int \frac{d^2\mathbf{k}}{(2\pi)^2} \frac{\varepsilon_{\mathbf{k}}\,(\mu - \varepsilon_{\mathbf{k}})}{D_+ D_-} \right] \times \left[1 - N_i V_0^2 \int \frac{d^2\mathbf{k}}{(2\pi)^2} \frac{(\mu - \varepsilon_{\mathbf{k}})^2 - M^2}{D_+ D_-} \right]^{-1}, \tag{5.35}$$

where $D_\pm = (\mu - E_{\mathbf{k},+} \pm i/2\tau_+)\,(\mu - E_{\mathbf{k},-} \pm i/2\tau_-)$.

If we take the low-impurity-density limit $b = 0$, then the role of the vertex correction reduces to a renormalization of the SO constant α due to impurities. The finite constant b in (5.34) is an impurity-induced SO interaction term. In the limit of $N_i \to 0$, this interaction term is proportional to the impurity density, and it leads to an additional contribution to the off-diagonal conductivity which *does not depend* on N_i.[20]

5.4.3. *Contribution of states below the Fermi energy*

The integration over energy ε in Eq. (5.25) leads to the separation of σ_{xy} into two parts, $\sigma_{xy} = \sigma_{xy}^{I} + \sigma_{xy}^{II}$, corresponding to the contribution of energy states near the Fermi surface and the states below the Fermi energy, respectively.[10,19]

First we calculate the contribution from the states below the Fermi energy, σ_{xy}^{II}. Using (5.23), (5.25) and (5.31), we find σ_{xy}^{II} as an integral over momentum with the Fermi-Dirac functions $f(E_{\mathbf{k},\pm})$. Thus, it accounts for the contribution of all states with $E_{\mathbf{k},+} < \mu$ and $E_{\mathbf{k},-} < \mu$. In the static limit of $\omega \to 0$ we obtain

$$\sigma_{xy}^{II} = -4e^2 M\alpha^2 \int \frac{d^2\mathbf{k}}{(2\pi)^2}\, \frac{f(E_{\mathbf{k},+}) - f(E_{\mathbf{k},-})}{(E_{\mathbf{k},+} - E_{\mathbf{k},-})^3} \,. \tag{5.36}$$

Using Eq. (5.27) we calculate the integral over momentum and obtain finally

$$\sigma_{xy}^{II} = \frac{e^2}{4\pi} \left[1 - \frac{M}{\lambda_+} - \theta(\mu - M) \left(1 - \frac{M}{\lambda_-} \right) \right]. \tag{5.37}$$

In the limit of weak SO interaction, $\alpha k_{F,\pm} \ll M$, we get from (5.37)

$$\sigma_{xy}^{II} \simeq \frac{e^2}{2\pi}\, \frac{m\alpha^2}{M} \left[\theta(M - \mu)\, \frac{\mu + M}{2M} + \theta(\mu - M) \right]. \tag{5.38}$$

The expression for σ_{xy}^{II} can be presented in a different form demonstrating the topological character of this contribution. Let us introduce a 3D unit vector $\mathbf{n}(\mathbf{k})$ at each point of two-dimensional momentum plane

$$\mathbf{n}(\mathbf{k}) = \left(\frac{\alpha k_y}{\lambda(k)}\,,\, -\frac{\alpha k_x}{\lambda(k)}\,,\, -\frac{M}{\lambda(k)} \right). \tag{5.39}$$

By using the $\mathbf{n}$-field (5.39), we parameterize the manifold of 2×2 Hermitian matrices corresponding to Hamiltonian (5.22), since $H_0 = \varepsilon_{\mathbf{k}} + \lambda(k)\, \boldsymbol{\sigma} \cdot \mathbf{n}(\mathbf{k})$. At $\mathbf{k} = 0$, the vector $\mathbf{n}$ is perpendicular to the $\mathbf{k}$-plane, whereas for large $|\mathbf{k}| \gg M/\alpha$, it lies in the $\mathbf{k}$-plane and is oriented perpendicular to the momentum $\mathbf{k}$. The dependence $\mathbf{n}(\mathbf{k})$ is a mapping of the $\mathbf{k}$-plane to the unit sphere S_2, for which the total $\mathbf{k}$-plane maps onto the lower hemisphere of S_2.

Using (5.39) we find that in terms of the $\mathbf{n}$-field, the contribution σ_{xy}^{II} can be written as

$$\sigma_{xy}^{II} = -\frac{e^2}{2} \int \frac{d^2\mathbf{k}}{(2\pi)^2}\, [f(E_{\mathbf{k},+}) - f(E_{\mathbf{k},-})]\, \epsilon_{\alpha\beta\gamma}\, n_\alpha\, \frac{\partial n_\beta}{\partial k_x}\, \frac{\partial n_\gamma}{\partial k_y} \,. \tag{5.40}$$

The integral

$$\Omega = \frac{1}{2} \int \frac{d^2\mathbf{k}}{(2\pi)^2}\, [f(E_{\mathbf{k},+}) - f(E_{\mathbf{k},-})]\, \epsilon_{\alpha\beta\gamma}\, n_\alpha\, \frac{\partial n_\beta}{\partial k_x}\, \frac{\partial n_\gamma}{\partial k_y} \tag{5.41}$$

is the spherical angle on S_2 enclosed by two contours L_+ and L_-, where $L_\pm$ are the mappings of the Fermi surfaces (circles) $E_{\mathbf{k}_F,+} = \mu$ and $E_{\mathbf{k}_F,-} = \mu$ to the sphere S_2, respectively.

The contribution σ_{xy}^{II} can be also presented in a form of the Berry phase in momentum space.[8,9,21] For this purpose we use the chiral basis $|\mathbf{k}n\rangle$, in which the Green functions are diagonal

$$\sigma_{xy}^{II}(\omega) = \frac{e^2}{\omega} \int \frac{d\varepsilon}{2\pi} \sum_{\mathbf{k}} \sum_{n \neq m} (v_x)_{nm} G_{\mathbf{k}\,mm}(\varepsilon + \omega)\, (v_y)_{mn}\, G_{\mathbf{k}\,nn}(\varepsilon)\,. \tag{5.42}$$

After integrating over energy, this result can be presented as

$$\sigma_{xy}^{II} = e^2 \sum_n \sum_{\mathbf{k}} f(E_{\mathbf{k}n}) \left(\frac{\partial A_y(\mathbf{k}n)}{\partial k_x} - \frac{\partial A_x(\mathbf{k}n)}{\partial k_y} \right), \tag{5.43}$$

where

$$A_\alpha(\mathbf{k}n) = -i \left\langle \mathbf{k}n \left| \frac{\partial}{\partial k_\alpha} \right| \mathbf{k}n \right\rangle \tag{5.44}$$

is the gauge potential in the momentum space related to the transformation of the Hamiltonian H_0 to the diagonal form. In a general case, this transformation is local in the $\mathbf{k}$-space, leading to the nonvanishing gauge potential $A_\alpha(\mathbf{k}n)$.

In the model of 2DEG with Rashba Hamiltonian we can calculate explicitly the eigenfunctions

$$\langle \mathbf{k}, \pm| = \sqrt{\frac{\lambda(k) \pm M}{2\lambda(k)}} \left(1\,,\ -\frac{i\alpha(k_x - ik_y)}{M \pm \lambda(k)} \right). \tag{5.45}$$

Then using (5.44) we find the gauge potential

$$\mathbf{A}(\mathbf{k}, \pm) = \left(-\frac{\alpha^2 k_y}{2\lambda(k)\,[M \pm \lambda(k)]}\,,\ \frac{\alpha^2 k_x}{2\lambda(k)\,[M \pm \lambda(k)]} \right), \tag{5.46}$$

and from Eq. (5.43) we come again to the same result of Eq. (5.36). Note that (5.46) can be also found as a gauge potential corresponding to the local transformation of vector field (5.39) to the homogenous field oriented along axis z in the momentum space (like in the case of local transformations in the real space[16]).

In the 2D case, the flux of curl of the gauge potential $\mathbf{A}(\mathbf{k}, \pm)$ in Eq. (5.43) through the surface $E_{\mathbf{k},\pm} = \mu$ can be presented as the circulation of vector $\mathbf{A}(\mathbf{k}, \pm)$ along the circle (Fermi surface in 2D). In other words, the contribution of the filled states below the Fermi surface can be also presented by the integral of the gauge field over the Fermi surface. As shown by Haldane[24] such a reduction of the integral in momentum space takes place in any dimensionality, in accordance with the Landau concept of the Fermi liquid stating that the transport properties are fully determined by the properties of electrons near the Fermi surface.

5.4.4. *Role of impurities*

When calculating the integral over ε in Eq. (5.25), we also find a contribution to the Hall conductivity from the vicinity of Fermi surface. In the case of strong SO interaction, such calculation should include the impurity-induced correction to the vertex part ($b \neq 0$).[25] Here we present the results for a weak SO coupling, when $\alpha k_F \ll M$. Then in the second order of the SO coupling constant we obtain

$$\sigma^I_{xy} = -e^2 M\alpha\alpha^* \int \frac{d^2\mathbf{k}}{(2\pi)^2} \frac{1}{(E_{\mathbf{k},+} - E_{\mathbf{k},-})^2} \left[\left(1 - \frac{\tau_+}{\tau_-}\right) \times \left(-\frac{\partial f(E_{\mathbf{k},+})}{\partial \varepsilon}\right) + \left(1 - \frac{\tau_-}{\tau_+}\right) \left(-\frac{\partial f(E_{\mathbf{k},-})}{\partial \varepsilon}\right) \right]. \quad (5.47)$$

where we denote $\alpha^* \equiv c$ the renormalized constant of SO interaction (5.35). In Eq. (5.47) the presence of factor $(-\partial f/\partial \varepsilon)$ restricts (5.47) to the integral over Fermi surface (we consider the low-temperature limit of $T/(\mu + M) \ll 1$). Calculating the integral over momentum (5.47), we find

$$\sigma^I_{xy} = -\frac{e^2\alpha\alpha^*}{4M} \left[\nu_+ \left(1 - \frac{\tau_+}{\tau_-}\right) + \nu_- \left(1 - \frac{\tau_-}{\tau_+}\right) \right], \quad (5.48)$$

where the second term is zero if the chemical potential is in the gap and $\nu_- = 0$. The striking property of this expression is that σ^I_{xy} neither depend on the impurity density N_i nor the magnitude of electron-impurity interaction V_0 because it includes only a ratio of the relaxation times in the subbands, τ_+/τ_-.

If the chemical potential is not located in the "gap", $\mu > M$, then using Eqs. (5.32) and (5.48) we find $\sigma^I_{xy} = 0$.[22,23] It turns out that this result is exact for any magnitude of the SO interaction.[20] It should be emphasized that the total Hall conductivity $\sigma_{xy} = \sigma^I_{xy} + \sigma^{II}_{xy}$ is not vanishing at $\mu > M$ due to the nonvanishing topological contribution σ^{II}_{xy}.

5.5. Conclusions

We discussed main mechanisms and related physical ideas in the theory of anomalous Hall effect. It is well known that the history of research in this field had several dramatic misunderstandings (see, for example, Ref.[26]), and some basic problems are not completely resolved and understood by now. To our knowledge, the main point of confusion is related to the role of impurities. Our consideration shows that the observed Hall conductivity essentially depends on the ratio between the frequency and impurity relaxation rate. The real static limit ($\omega \to 0$) implies that the relaxation rate is large even in the case of low impurity density. Thus the "pure limit" can be reached by different ways, so that we may expect different properties of a mesoscopic structure without any impurities at all. However, the common understanding of the pure limit does not imply the absence of any single impurity or defect but a reasonably low concentration of impurities.

This work was supported by the FCT Grant No. POCI/FIS/58746/2004 in Portugal and by the STCU Grant No. 3098 in Ukraine.

References

1. C. M. Hurd, *The Hall Effect in Metals and Alloys* (Plenum Press, New York, 1972); *The Hall Effect and Its Applications*, edited by C. L. Chien and C. R. Westgate (Plenum, New York, 1979).
2. R. Karplus and J. M. Luttinger, Hall effect in ferromagnets, *Phys. Rev.* **95**, 1154 (1954);
   J. M. Luttinger, Theory of the Hall effect in ferromagnetic substances, *Phys. Rev.* **112**, 739 (1958).
3. J. Smit, The spontaneous hall effect in ferromagnetics II, *Physica* **24**, 39 (1958).
4. L. Berger, Side-jump mechanism for the Hall effect in ferromagnets, *Phys. Rev. B* **2**, 4559 (1970);
   Application of the side-jump model to the Hall effect and Nernst effect in ferromagnets, *Phys. Rev. B* **5**, 1862 (1972).
5. P. Matl, N. P. Ong, Y. F. Yan, Y. Q. Li, D. Studebaker, T. Baum, and G. Doubinina, Hall effect of the colossal magnetoresistance manganite $La_{1-x}Ca_xMnO_3$, *Phys. Rev. B* **57**, 10248 (1998);
6. J. Ye, Y. B. Kim, A. J. Millis, B. I. Shraiman, P. Majumdar, and Z. Tesanovic, Berry phase theory of the anomalous Hall effect: application to colossal magnetoresistance manganites, *Phys. Rev. Lett.* **83**, 3737 (1999).
7. S. Onoda and N. Nagaosa, Spin chirality fluctuations and anomalous Hall effect in itinerant ferromagnets, *Phys. Rev. Lett.* **90**, 196602 (2003).
8. T. Jungwirth, Q. Niu, and A. H. MacDonald, Anomalous Hall effect in ferromagnetic semiconductors, *Phys. Rev. Lett.* **88**, 207208 (2002);
   D. Culcer, A. MacDonald, and Q. Niu, Anomalous Hall effect in paramagnetic two-dimensional systems, *Phys. Rev. B* **68**, 045327 (2003);
   T. Jungwirth, J. Sinova, K. Y. Wang, K. W. Edmonds, R. P. Campion, B. L. Gallagher, C. T. Foxon, Q. Niu, and A. H. MacDonald, Dc-transport properties of ferromagnetic (Ga,Mn)As semiconductors, *Appl. Phys. Lett.* **83**, 320 (2003);
   D. Culcer, J. Sinova, N. A. Sinitsyn, T. Jungwirth, A. H. MacDonald, and Q. Niu, Semiclassical spin transport in spin-orbit-coupled bands, *Phys. Rev. Lett.* **93**, 046602 (2004);
9. M. Onoda and N. Nagaosa, Topological nature of anomalous Hall effect in ferromagnets, *J. Phys. Soc. Jpn.* **71**, 19 (2002); Quantized anomalous Hall effect in two-dimensional ferromagnets: Quantum Hall effect in metals, *Phys. Rev. Lett.* **90**, 206601 (2003);
   Z. Fang, N. Nagaosa, K. S. Takahashi, A. Asamitsu, R. Mathieu, T. Ogasawara, H. Yamada, M. Kawasaki, Y. Tokura, and K. Terakura, The anomalous Hall effect and magnetic monopoles in momentum space, *Science* **302**, 92 (2003).

10. A. Crépieux and P. Bruno, Theory of the anomalous Hall effect from the Kubo formula and the Dirac equation, *Phys. Rev. B* **64**, 014416 (2001).
11. A. A. Abrikosov, L. P. Gorkov, and I. E. Dzyaloshinski, *Methods of Quantum Field Theory in Statistical Physics* (Dover, New York, 1963).
12. V. K. Dugaev, A. Crépieux, and P. Bruno, Localization corrections to the anomalous Hall effect in a ferromagnet, *Phys. Rev. B* **64**, 104411 (2001);
A. Crépieux, J. Wunderlich, V. K. Dugaev, and P. Bruno, Anomalous Hall effect and localization corrections in a ferromagnet, *J. Magn. Magn. Mater.* **242-245**, 464 (2002).
13. M. V. Berry, Quantal phase factors accompanying adiabatic changes, *Proc. R. Soc. Lond. A* **392**, 45 (1984).
14. G. Tatara and H. Fukuyama, Resistivity due to a domain wall in ferromagnetic metal, *Phys. Rev. Lett.* **78**, 3773 (1997).
15. E. Fradkin, *Field Theories of Condensed Matter Systems* (Addison-Wesley, Reading, 1991).
16. P. Bruno, V. K. Dugaev, and M. Taillefumier, Topological Hall effect and Berry phase in magnetic nanostructures, *Phys. Rev. Lett.* **93**, 096806 (2004).
17. K. Nielsch, R.B. Wehrspohn, J. Barthel, J. Kirschner, U. Gösele, S. F. Fischer, and H. Kronmüller, Hexagonally ordered 100 nm period nickel nanowire arrays, *Appl. Phys. Lett.* **79**, 1360 (2001);
K. Nielsch, R. B. Wehrspohn, J. Barthel, J. Kirschner, S. F. Fischer, H. Kronmüller, T. Schweinböck, D. Weiss, and U. Gösele, High density hexagonal nickel nanowire array, *J. Magn. Magn. Mater.* **249**, 234 (2002).
18. Yu. A. Bychkov and E. I. Rashba, Properties of two-dimensional electron gas with broken degeneracy of spectrum, *Pis'ma v Zh. Eksp. Teor. Fiz.* **39**, 64 (1984) [JETP Lett. **39**, 78 (1984);
Oscillatory effects and the magnetic susceptibility of carriers in inversion layers, *J. Phys. C* **17**, 6093 (1984).
19. P. Středa, Theory of quantised Hall conductivity in two dimensions, *J. Phys. C* **15**, L717 (1982).
20. N. A. Sinitsyn, J. E. Hill, H. Min, J. Sinova, and A. H. MacDonald, Charge and spin Hall conductivity in metallic graphene, *Phys. Rev. Lett.* **97**, 106804 (2006) and J. Sinova, private communication (unpublished).
21. Y. Yao, L. Kleinman, A. H. MacDonald, J. Sinova, T. Jungwirth, D.-S. Wang, E. Wang, and Q. Niu, First principles calculation of anomalous Hall conductivity in ferromagnetic bcc Fe, *Phys. Rev. Lett.* **92**, 037204 (2004).
22. V. K. Dugaev, P. Bruno, M. Taillefumier, A. Canals, C. Lacroix, Anomalous Hall effect in a two-dimensional electron gas with spin-orbit interaction, *Phys. Rev. B* **71** 224423 (2005).
23. J. Inoue, T. Kato, Y. Ishikawa, H. Itoh, G. E. W. Bauer, and L. W. Molenkamp, Vertex corrections to the anomalous Hall effect in spin-polarized two-dimensional electron gases with a Rashba spin-orbit interaction, *Phys. Rev. Lett.* **97**, 046604 (2006).
24. F. D. M. Haldane, Berry Curvature on the Fermi Surface: Anomalous Hall Effect as a Topological Fermi-Liquid Property, *Phys. Rev. Lett.* **93**, 206602 (2004).

25. N. A. Sinitsyn, A. H. MacDonald, T. Jungwirth, V. K. Dugaev, J. Sinova, Anomalous Hall effect in 2D Dirac band: link between Kubo-Streda formula and semiclassical Boltzmann equation approach, *Phys. Rev. B* **75**, 045315 (2007).
26. J. Sinova, T. Jungwirth, and J. Černe, Magnetotransport and magnetooptical properties of ferromagnetic (III,Mn)V semiconductors: a review, *Int. J. Mod. Phys. B* **18**, 1083 (2004).

Chapter 6

Dynamics and domain growth in quantum spin systems

V. Turkowski

Department of Physics and Astronomy, University of Missouri-Columbia, Columbia, MO 65211, USA
turkowskiv@missouri.edu

V. Rocha Vieira, P.D. Sacramento

CFIF and Departamento de Física, Instituto Superior Técnico, Av. Rovisco Pais, 1049-001 Lisboa, Portugal

In this contribution, we review some of our recent results on the nonequilibrium properties of the spin $S = 1/2$ Heisenberg ferromagnet. We consider the situation when the system is coupled to a phonon heat bath and/or in the presence of an external time-dependent magnetic field. The problem is studied by means of a path integral approach using the Majorana fermion representation for the spin operators. In particular, we consider the relaxation of the magnetization in the case when the magnetic field suddenly changes its direction. Another important case considered in this paper is the process of spinodal decomposition, or magnetic domain growth after the temperature is lowered below the critical value. We compare some of our results with the corresponding results in the case of classical spin models and some other models, and discuss possible applications of the results.

Contents

6.1. Introduction

The nonequilibrium behavior of a quantum spin system coupled to a heat bath or in the presence of an external field is a hot topic in theoretical condensed matter physics. One of the most important reasons for this is a huge number of applications of the results, especially in the field of magnetic recording technologies. One of the possible ways for a further advance in the ultra-high density information storage is nanomagnetism (see, for example Ref. 1). It is extremely important to understand the condition of the stability of small magnetic systems with respect to fluctuations or changes of temperature and an external magnetic field. Recently, an enormous progress was made in the field of the manipulation of magnetic matter by heating with short (femtosecond) laser pulses. For example, by using this method, one can create ferromagnetic order from initially antiferromagnetic FeRh thin films.[2] This technology has a great potential to be used in the field of the thermomagnetic writing, and it is very important to understand theoretically fast nonequilibrium processes in such systems. Another important problem is the problem of the magnetic bubble growth. These bubbles are domains of reversed magnetization created in thin ferromagnetic films due to demagnetization effects, when the external magnetic field is weak.[3–5] They have been observed in films with an uniaxial anisotropy. It is important to understand conditions of the stability of these magnetic domains with respect to a temperature and external magnetic field change, and also the time-dependence of their size. There are also some other experiments which require a quantitative description. For example, the problem of a large relaxation time in recently discovered big magnetic molecules.[6–9] There are several theoretical works where an attempt to give an explanation of this phenomenon was performed,[10–12] but it is necessary do develop a more quantitative description. It is important to understand the behavior of magnets in time-dependent, for example oscillating, magnetic fields. Such processes often take place in real systems in technological applications. However, there is a lack of complete understanding of the processes which take place in such systems, since this problem is very complicated. The nonequilibrium behavior of spin systems could show many surprises. For example, recently it was shown that some ferromagnetic models in the presence of a driving field show a dynamical phase transition with an hysteresis, with a nonzero average value of the physical variable, like magnetization, that undergoes such an hysteresis.[13,14]

Besides many technological applications, the problem of the theoretical description of the nonequilibrium properties of a quantum spin system in presence of a time-dependent field and coupled to a heat reservoir degrees of freedom is an important problem of theoretical physics by itself, since it is formally related to many similar problems in other branches of physics from cosmology to superconductors (see, for example Refs. 15–19).

Theoretical studies of the nonequilibrium properties of spin systems were mostly performed in the framework of the Ising model, because of its simplicity. The equilibrium properties of this model are explored well enough, and there has been a tremendous progress in understanding the out-of-equilibrium behavior, like the effect of external fields, change of temperature in the clean case and in the case of the presence of disorder.[20,21] The case of many component quantum spin models is not so well explored, because of the complexity of the problem. Some progress has been made recently due to the development of Monte Carlo methods (see, for example Ref. 14).

In this paper, we analyze some of the nonequilibrium properties of the Heisenberg ferromagnet by using the powerful analytical method of the Majorana fermion representation for the spin $S = \frac{1}{2}$ operators.[22,23] There are many other theoretical methods to study quantum vector spin $S = 1/2$ models, like the Holstein-Primakoff[24] and the Schwinger[25] boson representation for the spin operators and the Green function decoupling method (see, for example Ref. 26), along with numerical approaches like quantum Monte Carlo method (see Ref. 27 and references therein). The Majorana fermion representation formalism has a big advantage in comparison with many other approaches by the fact that the problem is reduced to a fermionic problem, which allows one to use the nonequilibrium closed-time-path Green function formalism[28,29] by applying the Wick theorem in the diagrammatic technique. This technique can't be applied directly for spin operators. This makes the results obtained by other approaches less controllable. In the Holstein-Primakoff and Schwinger boson representations one can also use the bosonic diagrammatic technique, however, in the Holstein-Primakoff case it is necessary to make an $1/S$-expansion, which is less suitable in the small spin $S = 1/2$ case, and in the case of the Schwinger boson representation one needs to use an additional constraint for the two kinds of boson operators. One can avoid this problem by using the Majorana fermion representation for the spin operators. In the formalism we use, the Green functions are defined on a three-branch complex time contour (Fig. 6.1). There are also other methods to study nonequilibrium phenomena, like the two-branch time representation formalism (see, for example Ref. 30), the four-branch time contour thermofield dynamics method[31] and the quantum Boltzmann equation approach.[32]

We study the effect of an external magnetic field by considering the case when the magnetic field suddenly changes its direction. Especially interesting case is the case when the magnetic field flips $\vec{H} \to -\vec{H}$. It is known from the relativistic scalar ϕ^4-theory in the presence of such a field, that there is a critical value of the field, below which the order parameter does not change the sign after the field flipping.[33,34] We consider this situation in the case of the quantum Heisenberg model and show that a similar effect also takes place in this case.[35]

The effects of the temperature change are studied in the case when the temperature of the Heisenberg ferromagnet is quenched below the critical value. It is well-know that in this case the system is in an unstable state, and magnetic domains, which spontaneously appear in the system as a result of a phase separation process, start to grow in order to approach the equilibrium homogeneously magnetized state.[20,21] This problem is well explored in the case of the scalar ϕ^4-model with non-conserved order parameter (see, for example Refs. 20,21,36–38). This case corresponds to the case of the Ising model. It is well established that the domain radius grows in this case as $\sqrt{t}$. A similar kind of problems was studied in relativistic physics where, in particular, the relativistic three-dimensional weak-coupled ϕ^4-theory[15,39,40] and a $(1+1)$ dimensional model with an additional ϕ^6-term[41] were analyzed. It was found that the $\sqrt{t}$-dependence also takes place in these cases at short times.

We analyze the case of quantum spins, which can be described by an effective boson vector theory.[42] We show that the time-dependence of the domain radius changes from $\sqrt{t}$ to t as time increases. This is different from the classical vector theory case, where the time-dependence is $\sqrt{t}$.[43,44]

This paper is organized as follows. In Section 6.2, we present the general formalism to study the spin $S = 1/2$ Heisenberg ferromagnet coupled to a phonon heat bath in the presence of an external spatially inhomogeneous and time-dependent magnetic field by using the Majorana fermion representation for the spin operators. The magnetization time-dependence in the case when the magnetic field changes its direction is reviewed in Section 6.3. In Section 6.4, we consider the process of the magnetic domain growth in an effective vector model of the Heisenberg ferromagnet in the case when the temperature is lowered below the critical value. The conclusions and a brief discussion of applications of the results are presented in Section 6.5.

6.2. Formalism

The Heisenberg Hamiltonian coupled to a phonon heat bath and in the presence of an external magnetic field can be written in the following form:

$$\hat{H} = -\frac{1}{2}\sum_{ij} J_{ij}\vec{S}_i\vec{S}_j - \sum_i \vec{H}_i\vec{S}_i + \sum_{qi} c_{qi}\vec{S}_i\vec{X}_{qi} + \frac{1}{2}\sum_{qi}(\frac{\vec{p}^2_{qi}}{m_{qi}} + m_q\omega^2_{qi}\vec{X}^2_{qi}), \tag{6.1}$$

where $J_{ij} > 0$ is the nearest neighbor ferromagnetic coupling with the amplitude J, i, j are site coordinates, and H is an external, in general time- and space-dependent, magnetic field. The spins are coupled with the coupling c_{qi} to the phonon bath modes q with coordinate $\vec{X}_{qi}$, momentum $\vec{p}_{qi}$ and frequency ω_{qi}.

To study the properties of the system, it is convenient to calculate the normalized generating functional:

$$\mathcal{Z} = \frac{Tr\left[\hat{T}e^{-\int_0^{\beta_f}\hat{H}_b d\tau}e^{-\int_0^{\beta_i}\hat{H}_s d\tau}e^{-i\int_{t_f}^{t_i}\hat{H}dt}e^{-i\int_{t_i}^{t_f}\hat{H}dt}\right]}{Tr\left[\hat{T}e^{-\int_0^{\beta_f}\hat{H}_b d\tau}\right]Tr\left[\hat{T}e^{-\int_0^{\beta_i}\hat{H}_s d\tau}\right]}, \tag{6.2}$$

where $\hat{H}_s$ and $\hat{H}_b$ are the spin and bath parts of the Hamiltonian, respectively. We assume that initially the spin and the phonon subsystems are decoupled. They have different temperatures $T_i = \frac{1}{\beta_i}$ and $T_f = \frac{1}{\beta_f}$, respectively. Indices "i" and "f" stand for the initial temperature and the final temperature of the spin subsystem. Obviously, the final equilibrium temperature coincides with the heat bath temperature. We consider the case when the spin-phonon coupling is turned on at time $t = t_i$. It is difficult to calculate the generating functional Eq. (6.2), since the spin operators don't satisfy standard boson or fermion (anti-)commutation relations. In particular, in the case of fermions, one can use the standard equilibrium quantum many-body theory rules to calculate different operator averages, like in the generating functional (6.2). The difference from the equilibrium case is in the time contour, which is complex in this case (Fig.6.1). In this case, all time orderings of operators must be performed along this contour.

Thus, it is convenient to express spin operators in terms of the fermion operators. In fact, it can be done in the case of the spin $\frac{1}{2}$:[22,45]

$$\vec{S}_i = -\frac{i}{2}\vec{\zeta}_i \times \vec{\zeta}_i. \tag{6.3}$$

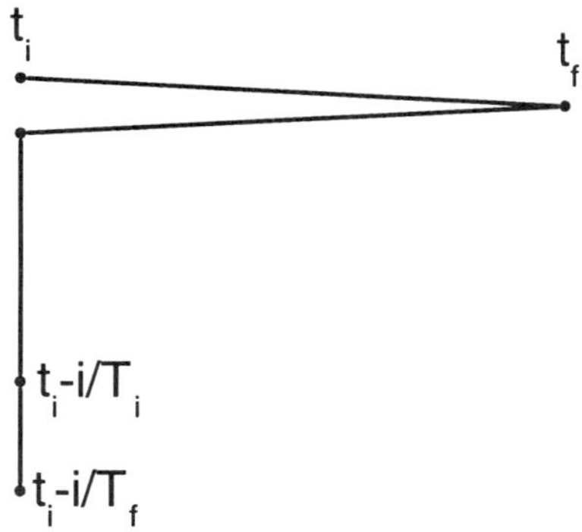

Fig. 6.1. Integration contour for the time variable. The direction is $t_i \to t_f \to t_i \to t_i - i/T_i, t_i - i/T_f$.

In this case, the numerator of the partition function (6.2) can be calculated by using the path integral approach:

$$\int D\vec{\zeta}D\vec{X}\exp\left(-i\int_C dt\left[-\frac{i}{2}\sum_i \vec{\zeta}_i\frac{d}{dt}\vec{\zeta}_i - \frac{1}{2}\sum_{ij}J_{ij}\vec{S}_i\vec{S}_j - \sum_i \vec{S}_i\vec{H}_i\right.\right.$$
$$\left.\left.+\sum_{qi}c_{qi}\vec{S}_i\vec{X}_{qi} + \sum_{qi}\vec{X}_{qi}\tilde{D}^{-1}_{qij}\vec{X}_{q,j}\right]\right) \tag{6.4}$$

in terms of the fermion and phonon field variables. $\tilde{D}$ is the phonon propagator and the time integration contour C is presented in Fig. 6.1. In order to evaluate this integral, it is convenient to transform the expression under the integral to a quadratic form in the Majorana fermions. One can do this by making the Hubbard-Stratonovich transformation for the spin operators introducing the vector field $\vec{\Phi}_i(t)$. The integration over the phonon fields in Eq. (6.2) is also possible yielding:

$$\mathcal{Z} = \int \frac{D\vec{\zeta}D\vec{\Phi}}{\sqrt{\det(2\pi\tilde{J})}}\exp\left(-i\int_C dt\left[-\frac{i}{2}\sum_i \vec{\zeta}_i\frac{d}{dt}\vec{\zeta}_i\right.\right.$$
$$\left.\left.+\frac{1}{2}\sum_{ij}(\vec{\Phi}_i - \vec{H}_i)\tilde{J}^{-1}_{ij}(\vec{\Phi}_j - \vec{H}_i) - \sum_i \vec{\Phi}_i\vec{S}_i\right]\right). \tag{6.5}$$

obtained by making the shift $\vec{\Phi}_i \to \vec{\Phi}_i - \vec{H}_i$, in the vector field (for details, see Ref. 42).

In Eq. (6.5), the effective interaction $\tilde{J}_{ij}$ is a sum of the ferromagnetic nearest neighbor coupling J_{ij} and the local phonon influence functional $i\Delta_{ij}(t-t')$ (self-interaction):

$$\tilde{J}_{ij}(t-t') \rightarrow J_{ij}\delta(t-t') + i\Delta_{ij}(t-t'), \tag{6.6}$$

where

$$\Delta_{ij}(t-t') = \alpha_{ij}(t-t')\theta(t-t') + \alpha^*_{ij}(t-t')\theta(t'-t), \tag{6.7}$$

and $\alpha_{ij}(t-t')$ is the phonon-phonon correlation function, which can be easily found from Eq. (6.1). It can be approximated in the following way:[16]

$$\alpha_{ij}(t-t') \simeq \delta_{ij} A \int_0^\infty d\omega \omega^s e^{-\omega/\omega_c} \left[\frac{e^{i\omega(t-t')}}{e^{\beta_f \omega} - 1} + \frac{e^{-i\omega(t-t')}}{1 - e^{-\beta_f \omega}} \right]. \tag{6.8}$$

In this expression, A, s and ω_c are effective phonon bath parameters. In particular, A is an effective spin-phonon coupling, ω_c is the phonon energy cut-off. The exponential parameter s corresponds to the "subohmic", "ohmic" and "superohmic" cases, when $0 < s < 1$, $s = 1$ and $s > 1$. These effective parameters can be obtained from the exact phonon part of the Hamiltonian Eq. (6.1).

With the substitution of (6.3) into Eq. (6.5), the formal integration over the fermion degrees of freedom can be performed, and one can get the following expression for the free energy functional:

$$\beta F = \frac{i}{2} \int_C dt \sum_{i,j} (\vec{\Phi}_i - \vec{H}_i) \tilde{J}^{-1}_{ij} (\vec{\Phi}_j - \vec{H}_j) - \frac{1}{2} \sum_i \operatorname{Tr} \ln i G^{-1}_{ii}(\vec{\Phi}_i), \tag{6.9}$$

where

$$G^{ab}_{ij}(t,t') = -i \langle \hat{T}_c \zeta^a_i(t) \zeta^b_j(t') \rangle \tag{6.10}$$

is the contour time-ordered Majorana fermion Green function. This function satisfies the following equation of motion:

$$\left(i \frac{d}{dt_1} \delta^{pm} - i\varepsilon^{pmn} \Phi^l_i(t_1) \right) G^{mn}_{ij}(t_1, t_2) = \delta^{pn} \delta(t_1 - t_2). \tag{6.11}$$

In order to find the magnetic properties of the system, one needs to find this function. Since the spin operators can be expressed in terms of the Majorana operators, any spin operator average can be found from the Majorana fermion operator averages, by using the Green function Eq. (6.11). To find this Green function, one can minimize the expression for the functional Eq. (6.9). This gives an additional equation which connects the Green function with the effective field $\vec{\Phi}_j$:

$$\vec{\Phi}_i(t) = \vec{H}_i(t) + \int_C dt' \tilde{J}_{ij}(t,t') \vec{M}_j(t'), \tag{6.12}$$

where the mean-field magnetization $M_j^l(t)$ is

$$M_j^l(t) = \frac{\delta}{i\delta\Phi_j^l(t)}\frac{1}{2}Tr\ln iG^{-1}(\vec{\Phi}_j) = \frac{1}{4}\varepsilon^{lsr}(G_{jj}^{sr}(t^+,t) + G_{jj}^{sr}(t^-,t)). \quad (6.13)$$

The system (6.11)-(6.13) can't be solved exactly in general, when an inhomogeneous time-dependent magnetic field is applied in the presence of a heat bath. In some particular cases it can be analyzed analytically. In particular, when the external field is homogeneous and has a very simple time-dependence, like the case when the field changes its direction at some moment of time (see the following Section). In this case, one can solve the mean-field equation (6.11) for the Green function $G_{ij}^{mn}(t_1,t_2)$ and substitute the solution into Eq. (6.12). This will give a simple equation for the mean-field $\vec{\Phi}_i^0$, which can be solved exactly. The fluctuation corrections can be taken into account in this case (see below). In general, however, the situation is much more complicated and one needs to solve the system (6.11)-(6.13) numerically. In fact, one can discretize the time contour Fig. 6.1. In this case, the Green function becomes a complex square matrix. Formal solution of Eq. (6.11) can be obtained in the case of the symmetric phase space representation for the Majorana fermions:[45]

$$G^{-1} = -2i\begin{pmatrix} \hat{C} & -\hat{1} & \hat{1} & \ldots & -\hat{1} \\ \hat{1} & \hat{C} & -\hat{1} & \ldots & \hat{1} \\ -\hat{1} & \hat{1} & \hat{C} & \ldots & -\hat{1} \\ \ldots & \ldots & \ldots & \ldots & \ldots \\ \hat{1} & -\hat{1} & \hat{1} & \ldots & \hat{C} \end{pmatrix}, \quad \text{where} \quad \hat{C} = \begin{pmatrix} 0 & \Phi_i^z & -\Phi_i^y \\ -\Phi_i^z & 0 & \Phi_i^x \\ \Phi_i^y & -\Phi_i^x & 0 \end{pmatrix} \times \frac{\Delta t}{2}. \quad (6.14)$$

In this case, the problem is reduced to the solution of the matrix equation (6.12) for the effective field $\vec{\Phi}_i$. This equation can be solved by iteration, for example. One can start from some initial set $\vec{\Phi}_i(t)$, substitute it into Eq. (6.14), invert the Green function, and then substitute this function into the right hand side of Eq. (6.12), and obtain a new solution for the field $\vec{\Phi}_i(t)$. The procedure can be repeated until the field $\vec{\Phi}_i(t)$ converges to the exact numerical solution with a given precision. In practice, however, it is difficult to solve the problem in this way, especially in the nonhomogeneous case and at long times, when the matrix size is very large, and the convergence is very slow, if possible at all. Therefore, semianalytical approaches are preferable in this case. We use some of them in this paper.

6.3. Spin system coupled to a time-dependent magnetic field

In this Section, we review some of our recent results on the response of the system to an external homogeneous time-dependent magnetic filed.[35] Namely, we consider the case when a constant magnetic field $\vec{H} = (0, 0, H^z)$ suddenly changes its direction. Unfortunately, it is difficult to solve this problem exactly even in the case of one spin, since the discretization (6.14) leads to very large matrices in the case when one needs to study the long-time response of the system. Therefore, it is necessary to use some analytical approximations.

The mean-field solution of the system (6.11)-(6.13) can be obtained analytically in the spherical basis with unit vectors $\vec{a}_\pm = \frac{1}{\sqrt{2}}(\vec{e}_x \pm i\vec{e}_y)$, $\vec{a}_0 = \vec{e}_z$.[46] Namely,

$$G_\alpha(t-t') = i\left[-\frac{\theta_C(t-t')}{e^{-\beta_i \alpha \Phi^{(0)}}+1} + \frac{\theta_C(t'-t)}{1+e^{\beta_i \alpha \Phi^{(0)}}}\right] e^{-i\alpha\Phi^{(0)}(t-t')}, \quad (6.15)$$

where $\theta_C(t'-t)$ is the theta-function on the contour C (Fig. 6.1).

The mean-field solution (6.13) can't describe properly the response to a time-dependent field, since it gives a time-independent magnetization defined by[35]

$$M^{(0)} = \frac{1}{2}\tanh\left(\frac{\beta_i(H + 2dJM^{(0)})}{2}\right). \quad (6.16)$$

Therefore, it is necessary to go beyond this approximation. We calculate the magnetization by taking into account second order fluctuations of the field $\phi_i^l(t) = \Phi_i^l(t) - \Phi_i^{(0)l}$. In this case, integration over $\phi_i^l(t)$ gives the one-loop correction $F^{(1)}$ to the functional Eq. (6.9). In this case, the magnetization correction is:[47]

$$M_i^{k(1)}(t) = \frac{\delta(-\beta F^{(1)})}{\delta H_i^k(t)} = [\delta_{is}\delta^{nr}\delta(t_2-t_3)$$
$$+ i\bar{\chi}_{ib}^{na}(t_2,t_4)D_{bs}^{ar}(t_4,t_3)]B_{sij}^{rnl}(t_3,t_2,t_1)D_{ji}^{lk}(t_1,t), \quad (6.17)$$

where

$$\bar{\chi}_{sj}^{lm}(t,t') = -\frac{1}{2}\varepsilon^{ll_1l_2}\varepsilon^{mm_1m_2}G_{ij}^{l_2m_1}(t,t')G_{ij}^{m_2l_1}(t',t), \quad (6.18)$$

$$D_{sj}^{pm-1}(t,t') = \tilde{J}_{sj}^{pm-1}\delta(t,t') - i\bar{\chi}_{sj}^{pm}(t,t') \quad (6.19)$$

$$B_{sij}^{rnl}(t_3,t_2,t_1) = \frac{i}{2}\frac{\delta\bar{\chi}_{si}^{rn}(t_3,t2)}{\delta\phi_j^l(t_1)}$$
$$= \frac{1}{4}\varepsilon^{rr_1r_2}\varepsilon^{nn_1n_2}\varepsilon^{ll_1l_2}(G_{ji}^{l_2n_1}(t_1,t_2)G_{is}^{n_2r_1}(t_2,t_3)G_{sj}^{r_2l_1}(t_3,t_2)$$
$$+ G_{js}^{l_2r_1}(t_1,t_3)G_{si}^{r_2n_1}(t_3,t_2)G_{ij}^{n_2l_1}(t_2,t_1)), \quad (6.20)$$

and the total magnetization becomes

$$M_i^k(t) = M_i^{k(0)}(t) + M_i^{k(1)}(t) \quad (6.21)$$

Let us note that the functions $G, \bar{\chi}$ and B are diagonal in the site indices. The Green's functions G in these expressions are calculated self-consistently by means of the equations (6.12) and (6.13).

6.3.1. *The case of one spin*

In this Subsection, we consider the case of one spin coupled to a magnetic field $\vec{H}(t) = (0, 0, H_z(t))$, which changes it direction by the angle π or $\pi/2$ at time $t = t_i$. At this moment of time we also turn on the spin-phonon coupling, so the temperature begins to change from T_i to the temperature of the heat bath T_f. We explore the spin relaxation in the case of different heat bath parameters. For simplicity, we assume in Eqs. (6.7) and (6.8) that the spin-phonon coupling is weak and that only the x-component of the spin is coupled to the heat bath ($\Delta^{nn} \sim \delta^{nx}$), and make the approximation $e^{-\omega/\omega_c} \simeq \theta(\omega_c - \omega)$, introducing a cut-off in the frequency space.

In the case of the flipped magnetic field $\vec{H} \to -\vec{H}$ and lowering the temperature, the magnetization relaxes from the initial value $M^{(0)} = \frac{1}{2}\tanh(\frac{\beta_i H}{2})$ to the final value $M_f^{(0)} = -\frac{1}{2}\tanh(\frac{\beta_f H}{2})$ with increasing time. The time-dependence of the magnetization calculated in the one-loop approximation in the case of the heat bath parameter $s = 0.5$ is presented in Fig. 6.2. It decreases as $(t - t_i)^2$ at short times and is a linear function of $t - t_i$, when $(t - t_i) \gg 1/|H|$. Similar behavior was found at different values of s,[35] however the magnetization changes with time slower with increasing s. This is caused by the lowering of the effective phonon attraction with larger s. In fact, in this case the effective phonon density of states, which is proportional to ω^s, is lowered at small energies, which are the most important in the spin-phonon coupling. It can be shown that at long times the magnetization depends on time as

$$M^{z(1)}(t) = M_f^{z(1)} + (M^{(0)z} - M_f^{z(1)})e^{-\lambda(t-t_i)}, \tag{6.22}$$

where λ is defined by the magnetic field and by the heat bath parameters.[35] It is important that the magnetization flips at any value of the magnetic field. It will be shown in the following Subsection that the situation is different in the case of many interacting spins.

A similar behavior of the system takes place in the case when the magnetic field is rotated on a $\pi/2$-angle: $\vec{H}_i = (0, 0, H) \to \vec{H}_f = (0, H, 0)$ (Fig.6.3). The magnetization oscillates and changes its direction towards the field direction.

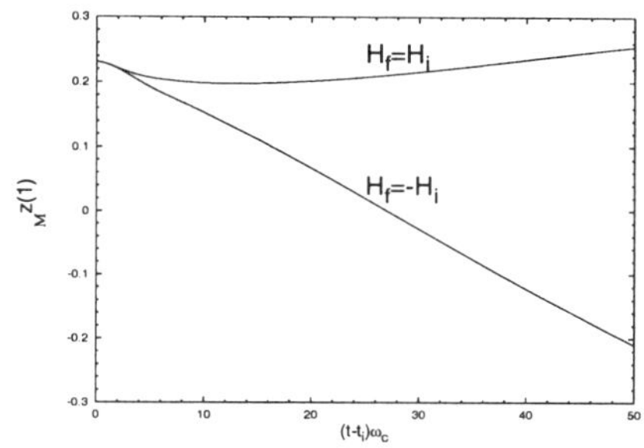

Fig. 6.2. Time-dependence of one-spin magnetization after lowering temperature for the cases of magnetic field flipping and constant magnetic field at $s = 0.5, A = 0.025, H/\omega_c = 0.1, T_i/\omega_c = 0.1$ and $T_f/\omega_c = 0.02$ (Reproduced with permission of Elsevier from Ref. 35).

6.3.2. *The case of many spins*

The problem of the time-dependence of the magnetization in a many-spin system is complicated by the spin-spin interaction. The role of this interaction results in many nontrivial effects. Here we show that even on the mean-field level the system demonstrates a very interesting behavior. In the mean-field approximation, the many-spin Hamiltonian can be substituted by a one spin Hamiltonian

$$H_{spin}(t) = -[2dJM^{(0)} + H^z(t)]S^z(t)$$

coupled to an effective field $2dJM^{(0)} + H^z(t)$. The initial magnetization $M^{(0)}$ is defined by

$$M^{(0)} = \frac{1}{2}\tanh\left(\frac{\beta_i(H + 2dJM^{(0)})}{2}\right). \tag{6.23}$$

We consider again the case of the magnetic field flipping $\vec{H} \to -\vec{H}$. Numerical results for the one-loop magnetization show that the time-dependence of the magnetization and its final value strongly depend on the relation $|H|/dJ$.[35] In particular, as it follows from the numerical calculations, the magnetization doesn't flip when the value of the magnetic field is lower than some critical value H_{cr}. A naïve estimation shows that the value of the field amplitude H must be stronger than the ferromagnetic coupling energy $2dJM^{(0)}$. Putting $\frac{1}{2}\tanh(\frac{\beta_i H}{2})$ instead of $M^{(0)}$ and equalizing it with H gives the expression, which connects H_{cr} with J and T_i:

$$\frac{H_{cr}}{dJ} = \tanh(\frac{H_{cr}}{T_i}). \tag{6.24}$$

A more rigorous derivation of this equation can be found in Ref. 35. The solution of Eq. (6.24) is presented in Fig. 6.4. At low temperatures, the critical value

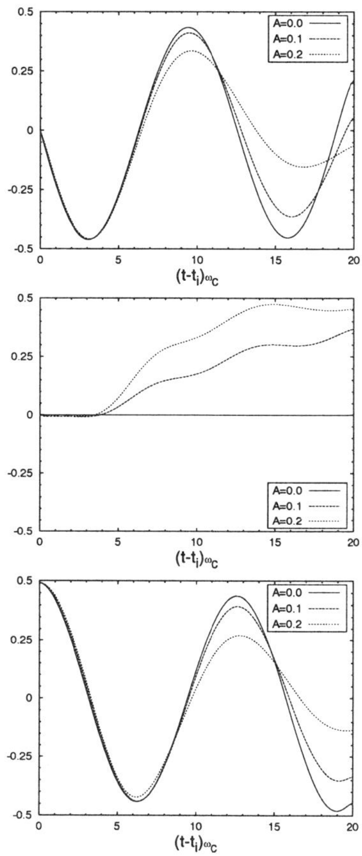

Fig. 6.3. Time-dependence of one-spin magnetization after the $\pi/2$-rotation of the magnetic field $\vec{H}_i = (0, 0, H) \rightarrow \vec{H}_f = (0, H, 0)$ at different values of the spin-phonon coupling A. Other parameters are fixed at $s = 0.5, H/\omega_c = 0.5, T_i/\omega_c = T_f/\omega_c = 0.1$ (Reproduced with permission of Elsevier from Ref. 35).

almost doesn't depend on the temperature, since in this case the ferromagnetic coupling is almost temperature-independent. As the temperature increases to the critical value, it requires a weaker field to destroy the ferromagnetic order, which becomes weaker in this case. When the field is lower than the critical field, the magnetization does not flip. It precesses with a large amplitude, similar to the case of the relativistic ϕ^4-theory.[33,34]

An even more interesting behavior of the Heisenberg model was observed in the case of an oscillating field.[13,14] In this case, the system shows a dynamical phase transition with an hysteresis loop for the order parameter.

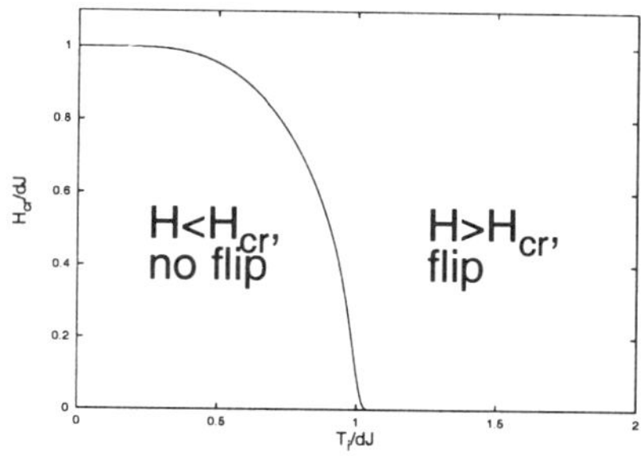

Fig. 6.4. Temperature-dependence of H_{cr}. Mean-field critical temperature of ferromagnetic transition is $T_c = 0.5dJ$ (Reproduced with permission of Elsevier from Ref. 35).

6.4. Domain growth

In this Section, we consider the behavior of the system in the case when the temperature is lowered below the critical value. We don't study the effect of the external field here and assume that it is constant in space and time and that its value is small. We apply the field in order to stabilize the magnetization as time increases. For simplicity, we also don't include a spin-phonon coupling, and assume that the temperature suddenly changes to a fixed value T.

After lowering the temperature, the system is in an unstable phase. As a result, small magnetic domains (bubbles), created by fluctuations, will grow in time in order to approach the final ferromagnetic state. The time dependence of the radius of such domains was studied in detail in the case of the Ising model. It was found that the bubble radius grows with time as $\sqrt{t}$.[20,21,36–38] Here we consider the case of a vector spin model, and show that the time dependence of the bubble radius is different from the Ising model case at longer times.[42] This problem can't be solved exactly from Eqs. (6.11)-(6.13), since in the nonhomogeneous case the dimension of the discretized Green function (6.14) is very large, and the solution of the problem can take an enormously long time. One can obtain an approximate analytical solution by expanding the functional Eq. (6.9) in powers of the deviations of the field $\vec{\phi} = \vec{\Phi} - \vec{\Phi}^{(0)}$ from the initial value $\vec{\Phi}^{(0)}$. This approximation is valid at short and intermediate times, but the case of late times must be studied using more complicated methods. We consider the fourth order approximation in the deviation field, which is usually enough to take into account the main nonlinear effects, when the field is not very large. The minimization of the functional Eq. (6.9) with respect to the deviation field will give a system of equations for the field components, which can be analyzed analytically in different limits.

The expression for the generalized free energy (6.9) has the following form in the fourth order approximation, in the long-wave limit and in the cylindrical coordinates $\vec{\phi} = (\phi_\perp \cos(\varphi), \phi_\perp \sin(\varphi), \phi^z)$:[42]

$$
\begin{aligned}
F = iT \int_c dt \int d\mathbf{x} \Big\{ & -\tilde{j}^z \phi^z(\mathbf{x},t) + \frac{1}{2}\phi_\perp(\mathbf{x},t)\left[\gamma_1 \frac{d^2}{dt^2} + (\tilde{m}^{xx})^2\right]\phi_\perp(\mathbf{x},t) \\
& + \frac{1}{2}(\tilde{m}^{zz})^2 \phi^{z2}(\mathbf{x},t) + \frac{1}{2}\gamma_2 \phi_\perp^2(\mathbf{x},t)\frac{d\varphi(\mathbf{x},t)}{dt} \\
& + \frac{a^2}{16dT_c}\left[(\nabla\phi_\perp(\mathbf{x},t))^2 + \phi_\perp^2(\mathbf{x},t)(\nabla\varphi(\mathbf{x},t))^2 + (\nabla\phi^z(\mathbf{x},t))^2\right] \\
& + \frac{1}{3!}\tilde{C}^{xxz}\phi_\perp^2(\mathbf{x},t)\phi^z(\mathbf{x},t) + \frac{1}{3!}\tilde{C}^{zzz}\phi^{z3}(\mathbf{x},t) \\
& + \frac{1}{4!}\tilde{\lambda}^{xxxx}\phi_\perp^4(\mathbf{x},t) + \frac{1}{4!}\tilde{\lambda}^{xxzz}\phi_\perp^2(\mathbf{x},t)\phi^{z2}(\mathbf{x},t) + \frac{1}{4!}\tilde{\lambda}^{zzzz}\phi^{z4}(\mathbf{x},t)\Big\},
\end{aligned}
\tag{6.25}
$$

where the coefficients can be found in Ref. 42. It is important that the mass coefficients m^2 become negative at $T < T_c$ and all other coefficients are positive, except λ^{zzzz}, which is negative at $H > 1.316T$. We assume that the field is low enough so λ^{zzzz} is positive. Positivity of the fourth order coefficients is necessary in order to have a stable ϕ^4-theory with a finite minimum at equilibrium.

Minimization of Eq. (6.25) with respect to different field components leads to three nonlinear equations. For simplicity, we assume that the angle order parameter, φ, time dependence can be approximated by a linear function of time, so the order parameter oscillates around the z-direction.[42] Therefore, one basically remains with two independent variables $\phi_\perp$ and ϕ^z. The equations for these two variables can be solved within some approximation in order to study the dynamics of the system. In particular, it is possible to find the structure factor $S^{ll}(\mathbf{r},t) = \langle \phi^l(\mathbf{r},t)\phi^l(0,t)\rangle$, or its Fourier transform $S^{ll}(\mathbf{k},t) = \int d^d\mathbf{r} \exp(-i\mathbf{kr}) S^{ll}(\mathbf{r},t) = \langle \phi^l_{\mathbf{k}}(t)\phi^l_{-\mathbf{k}}(t)\rangle$, which describes the time-dependence of the bubble size.[15]

Below we consider separately the free case and the interacting case and show that the behavior of the system is very different in the two cases.

6.4.0.1. *Free case*

The system of equations for the field components has a simple form in the free case:

$$
\left[\gamma_1 \frac{d^2}{dt^2} + (\tilde{m}^{xx})^2 - \frac{a^2}{8dT_c}\nabla^2\right]\phi_\perp(\mathbf{x},t) = 0, \tag{6.26}
$$

$$\left[(\tilde{m}^{zz})^2 - \frac{a^2}{8dT_c}\nabla^2\right]\phi^z(\mathbf{x},t) = \tilde{j}^z. \tag{6.27}$$

The equations are decoupled and only the plane component of the field $\phi_\perp$ is time-dependent. It is necessary to note that the masses for the transverse and longitudinal components of the field are different in general. Moreover, the square of the masses can have different signs for these components. However, in the limit of small H/T they coincide. This rotational symmetry breaking is caused by the magnetic field.

The system of equations (6.26), (6.27) can be easily solved in the momentum representation,[42] and it allows one to calculate an approximate expression for the scaled correlation function for the transverse field components, in the real space representation:

$$\bar{S}(\mathbf{r},t) = \int \frac{d\mathbf{k}}{(2\pi)^d} e^{i\mathbf{kr}} \left[S(\mathbf{k},t) - S(\mathbf{k},0)\right]. \tag{6.28}$$

The scaling properties of this function allow one to find the domain size time-dependence.

In the dimensionless variables

$$\kappa = k/m_f, \quad \tau = m_f t, \quad x = r/m_f, \quad L^2 = m_i^2/m_f^2, \tag{6.29}$$

an approximate expression for the scaled correlation function (6.28) can be obtained:

$$\bar{S}(x,\tau) = g\left(\frac{x^2}{\tau}\right)\bar{S}(0,\tau), \tag{6.30}$$

where the local correlation function is

$$\bar{S}(0,\tau) = \frac{m_f^{d-2}T_i(L^2+1)}{2^d\pi^{d/3}L^2}\frac{e^{2\tau}}{\tau^{d/2}}, \tag{6.31}$$

and the scaled space dependent part of the correlation function is:

$$g\left(\frac{x^2}{\tau}\right) \simeq e^{-x^2/4\tau}, \quad d=1$$
$$g\left(\frac{x^2}{\tau}\right) \simeq e^{-x^2/2\tau}, \quad d=2$$
$$g\left(\frac{x^2}{\tau}\right) \simeq \frac{\sin(x/\sqrt{\tau})}{x/\sqrt{\tau}}e^{-x^2/4\tau}, \quad d=3. \tag{6.32}$$

It follows from (6.31) that, at long times, the fluctuations inside the domain $\bar{S}(0,\tau) = <\phi^2(0,\tau)>$, are the strongest in the 1D case, and they decrease when the dimensionality of the system is increased, as it should be.

The scaling of the correlation function arguments x^2/τ shows that the domain size grows as $R \sim \sqrt{t}$ in all dimensions, similarly to the 3D case.[15]

6.4.0.2. *The interacting case*

In the interacting case, the system of equations for the order parameter is much more complicated. We shall use the Hartree approximation, similarly to the relativistic scalar theory case studied in Ref. 15. We also consider the limit of small H/T. In this case, all λ coefficients are equal to each other and H-independent and the coefficients $\tilde{j}$ and $\tilde{C}$ are small and proportional to H/T. Therefore, in this case, their role is not very important. In general, they play an important role, especially in the case of the phase transitions. In the classical theory, they do not change the rate of the domain growth, where only the domain boundary curvature defines the bubble radius velocity (for a review, see for example Ref. 43).

The coupled system of equations for the local in space correlation functions $\langle\phi_\perp^2(t)\rangle \equiv \langle\phi_\perp(\mathbf{0},t)\phi_\perp(\mathbf{0},t)\rangle$ and $\langle\phi^{z2}(t)\rangle \equiv \langle\phi^z(\mathbf{0},t)\phi^z(\mathbf{0},t)\rangle$ can be found in this case. To do this, one has to solve the system of equations for the field components in the Hartree approximation:

$$\left[\frac{d^2}{dt^2}+\mathbf{k}^2+m^2+\frac{\lambda}{2}\left(\langle\phi_\perp^2(t)\rangle-\langle\phi_\perp^2(0)\rangle+\langle\phi^{z2}(t)\rangle-\langle\phi^{z2}(0)\rangle\right)\right]\phi_{\perp\mathbf{k}}^{\pm}(t)=0, \tag{6.33}$$

$$\begin{aligned}\left[m^2+\mathbf{k}^2+\frac{\lambda}{2}\left(\langle\phi_\perp^2(t)\rangle-\langle\phi_\perp^2(0)\rangle+\langle\phi^{z2}(t)\rangle-\langle\phi^{z2}(0)\rangle\right)\right]\phi_{z\mathbf{k}}^{\pm}(t)\\ =j-C\left(\langle\phi_\perp^2(t)\rangle-\langle\phi_\perp^2(0)\rangle+\langle\phi^{z2}(t)\rangle-\langle\phi^{z2}(0)\rangle\right),\end{aligned} \tag{6.34}$$

As it follows from these equations, the system displays time-dependent behavior only when both, parallel and perpendicular, components of the order parameter are nonzero. In order to find the local correlation functions, it is necessary to solve the system of equations (6.33) and (6.34) numerically. This system can be simplified by using an approximate relation, which connects the local correlators for the parallel and perpendicular components of the order parameter:

$$\begin{aligned}\langle\phi^{z2}(t)\rangle-\langle\phi^{z2}(0)\rangle= &\left(\frac{H}{4T}\right)^2\left(1-\frac{\langle\phi_\perp^2(t)\rangle-\langle\phi_\perp^2(0)\rangle}{2T^2}\right)^2A(m_i^2,m^2,T_i)\\ &-B(m_i^2,T_i),\end{aligned} \tag{6.35}$$

where $A(m_i^2,m^2,T_i)$ and $B(m_i^2,T_i)$ are coefficients, which depend on the model parameters. Therefore, it is necessary to solve only one equation. The relation (6.35) already suggests the answer of how the vector field behaves in the unstable phase. $\langle\phi_\perp^2(t)\rangle$ must be an oscillating function of time with a large amplitude of the oscillations (see below), the amplitude of the oscillations decreases with time and $\langle\phi_\perp^2(t)\rangle$ approaches a small positive value. As it follows from (6.35), $\langle\phi^{z2}(t)\rangle$ is also an oscillating function of time with the same period as $\langle\phi_\perp^2(t)\rangle$. Then, $\langle\phi^{z2}(t)\rangle - \langle\phi^{z2}(0)\rangle$ approaches the equilibrium value $\sim (H/(4T))^2 \simeq ((1/2)\tanh(H/2T))^2$. This can be shown by solving the equation for $\langle\phi_\perp^2(t)\rangle - \langle\phi_\perp^2(0)\rangle$ at $\langle\phi^{z2}(t)\rangle - \langle\phi^{z2}(0)\rangle = 0$ in Eq. (6.33), since $\langle\phi^{z2}(t)\rangle - \langle\phi^{z2}(0)\rangle$ is small at short times. This case was considered in Ref. 15 in three dimensions. It was shown that the solution for $\langle\phi_\perp^2(t)\rangle - \langle\phi_\perp^2(0)\rangle$ in the 3D case is an oscillating function of time with the oscillation amplitude decaying exponentially with time. The domain size increases with time as $\xi_D(t) = (8\sqrt{2})^{1/2}\sqrt{t\xi(0)}$, where in our notations $\xi(0) \sim \sqrt{J/T}$. We present results of calculations for the correlation function for both components of the order parameter $\phi_\perp$ and ϕ^z in the 3D case in Fig. 6.5. In the calculations, we used the relation (6.35), between the field components.

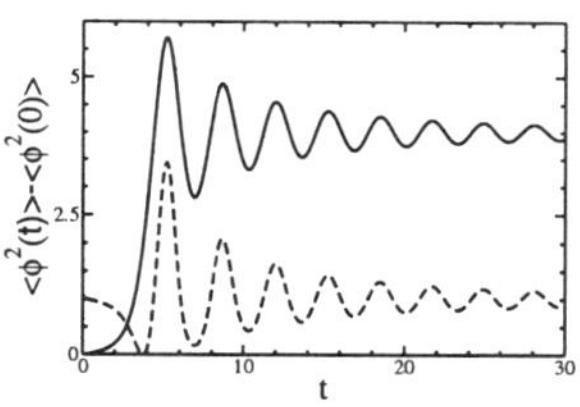

Fig. 6.5. Time-dependencies of the correlation functions for perpendicular field component (solid line) and renormalized parallel component (dashed line) in the 3D case. The model parameters are $m_i^2 = 1$, $m_f^2 = -1$, $T_i = 1$, $\lambda = 0.5$ (Reproduced with permission of the American Physical Society from Ref. 42).

As it follows from Fig. 6.5, the fluctuations initially grow until $m^2 + \frac{\lambda}{2}(\langle\phi^2(t)\rangle - \langle\phi^2(0)\rangle)$ becomes positive. The condition

$$m^2 + \frac{\lambda}{2}\left(\langle\phi_\perp^2(t)\rangle - \langle\phi_\perp^2(0)\rangle\right) = 0 \tag{6.36}$$

can be used to define the spinodal time, or time when the instabilities start to disappear. Our calculations show that a similar behavior takes place in the 2D and 1D cases, but the spinodal time is decreasing when the dimensionality of the system becomes lower.[42] As it follows from Eq. (6.31), the correlations are stronger in low dimensions, since $\langle\phi^2(t)\rangle \sim t^{-d/2}$.

After the spinodal time, the correlation function starts to approach the equilibrium value

$$\langle \phi_\perp^2(t) \rangle - \langle \phi_\perp^2(0) \rangle = \frac{2|m^2|}{\lambda}. \tag{6.37}$$

The parallel and perpendicular correlation functions have the same time dependence, except at very short time. Therefore, the correlation functions, and the bubble radius, for both components must have the same time-dependence. Finally, the system relaxes to the equilibrium state, and the z-component of the order parameter approaches its equilibrium value defined by the final temperature and the magnetic field. It is important that the ϕ^z has time-dependence only when j and the C-parameters are different from zero. Otherwise, Eq. (6.34) has only the trivial solution $U^{z\pm}(\mathbf{k}, t) = 0$. The parameters j and C are finite only in the case of finite external field. Therefore, the magnetization evolves in time to its equilibrium value directed along z-axis only when an external field H is applied. It is enough to have an extremely small magnetic field to get this symmetry broken equilibrium state.

To find the explicit time dependence of the bubble radius in the interacting case, one has to solve the system (6.33), (6.34). Then it is necessary to find if the correlation function arguments satisfy some scaling condition, like x^2/t in the free case. This scaled variable will define the domain size time dependence ($x \sim \sqrt{t}$ in the free case). As it follows from our previous analysis, the effective mass remains negative until the spinodal time. Therefore, the correlation function in the interacting case has the same time- and space-dependence as in the free case, if the coupling λ is weak. Our case corresponds to a weakly coupled theory, since we make an expansion of the free energy in powers of the field, and the coupling should be small in this case. Thus, at early times, i.e. at times smaller than the spinodal time, the domains grow as $\sqrt{t}$ in our effective model. The presence of the cubic terms does not change this result, since the cubic coupling parameter C is also assumed to be small. It would be extremely interesting to generalize these results to the case of stronger fields, where the role of the cubic terms can be nontrivial.

6.4.0.3. *Scaling analysis at long times*

It is actually possible to show, using a scaling analysis, that the long-time bubble radius time dependence in the Hartree approximation is $R \sim t$. This analysis in the classical scalar case is presented in Ref. 43, for example.

In fact, at long times, the order parameter equation can be written as

$$\left[\frac{d^2}{dt^2} + \mathbf{k}^2 - a(t)\right] \phi_{\perp\mathbf{k}}(t) = 0, \tag{6.38}$$

where

$$a(t) = -m^2 - \frac{\lambda}{2}\int \frac{d^d k}{(2\pi)^d}\left(\langle \phi_{\perp\mathbf{k}}(t)\phi_{\perp-\mathbf{k}}(t)\rangle + \langle \phi^z_{\mathbf{k}}(t)\phi^z_{-\mathbf{k}}(t)\rangle\right). \tag{6.39}$$

This function approaches zero when $t \to \infty$. We can show analytically that at long times

$$S(\mathbf{r}, t) \sim \int_0^\infty dx x^{d-3}\mu_d\left(\frac{r}{t}x\right)\sin^2 x, \tag{6.40}$$

where $\mu_d(\kappa x)$ is the d-dimensional measure of integration:

$$\begin{aligned} \mu_1(\kappa x) &= \frac{1}{\pi}\cos(\kappa x), \quad d = 1 \\ \mu_2(\kappa x) &= \frac{1}{\pi^2}\int_0^{\pi/2} d\varphi \cos(\kappa x \cos(\varphi)), \quad d = 2 \\ \mu_3(\kappa x) &= \frac{1}{2\pi^2}\frac{\sin(\kappa x)}{\kappa x}, \quad d = 3. \end{aligned} \tag{6.41}$$

(for details, see Ref. 42). Therefore, the solution in the Hartree approximation shows that the magnetic domains in the Heisenberg model should grow with time as $R \sim t$ at longer times.

6.5. Conclusions

In this paper, we reviewed some of the nonequilibrium properties of the ferromagnetic Heisenberg model in the presence of an external time-dependent magnetic field and in the case when the system temperature is lowered below the critical value. We considered a microscopic description for the vector spin system directly instead of using an effective bosonic theory from the start. Representing the spin operators by Majorana fermions we constructed the path integral representation for the spin system. Also, in some cases, we used a physical way to introduce the relaxation to equilibrium coupling the system to a heat bath. For simplicity we used a phonon heat bath. The spin-phonon coupling can be integrated exactly introducing an effective time-dependent interaction between the spins.

In the first part of the paper, we considered the case when the direction of the magnetic field is suddenly changed. As expected, the behavior of the system strongly depends on the phonon bath properties. It was shown that there is a critical value of the magnetic field below which the magnetization does not relax to the equilibrium value after the magnetic field flipping, similarly to the relativistic scalar ϕ^4-theory case, where there is also a critical value of the external source.[33,34]

In the second part of the paper, we have studied the process of the quantum spinodal decomposition in an effective vector boson theory of the Heisenberg ferromagnet in a weak external magnetic field. This theory is similar to the relativistic theory with additional linear and cubic terms. It was shown that the magnetic domains grow only in the case when the field contains both parallel and perpendicular field components. The perpendicular component correlations grow faster at early times and at late times the equilibrium state is established with the magnetization parallel to the external magnetic field. Both parallel and perpendicular component correlations grow with time as $\sqrt{t}$ at short times in different space dimensionalities. This result is similar to the well-known results for the classical ϕ^4-theory, which correspond to the Ising model case, to the classical vector model and to the relativistic scalar model at short times. Contrarily to the classical cases, the domain grows as t at longer times, as we have shown by using the Hartree approximation to solve the equation for the order parameter.

The results presented here may have some practical applications, especially in the field of the storage and reading of the magnetically recorded information. It is important there to understand the stability of the magnetic domains with respect to temperature and external field fluctuations.

There are some open problems in this direction which still remain unsolved. A specially important question is the study of non-homogeneous problems, with magnetic domains similar to those in the recording media. In would be very important to solve the problem beyond the Hartree approximation, and for the case of strong magnetic fields. As it was shown in this paper, these problems can be resolved in principle in a framework of the presented Majorana formalism. We believe that it will be possible in the nearest future, when more powerful computers will become available.

To conclude, we would like to mention that the nonequilibrium Majorana fermion method for spin systems and the results reviewed in this paper can be used in modeling different magnetic devices, including nanomagnetic devices,[1] for example, nanoscale heterostructures of ferromagnetic and normal materials.[48] The Majorana fermion formalism can also be used to study many related nonequilibrium problems, like for example the Kondo lattice problem for heavy fermions (see, for example Ref. 49, and 50, where some of the equilibrium and nonequilibrium Kondo lattice problems were studied, correspondingly) and the insulating and weakly doped phases of high-temperature superconductors, which can be described by resonating valence bond models for the two-dimensional spin $S = 1/2$ antiferromagnet.[51]

References

1. S.D. Bader, Colloquium: Opportunities in nanomagnetism, *Rev. Mod. Phys.* **78**, 1-15 (2006).
2. B. Bergman, G. Ju, J. Hohfeld et al, Identifying growth mechanisms for laser-induced magnetization in FeRh, *Phys. Rev. B* **73**, 060407 (2006).
3. A.H. Bobeck, E. Della Torre, *Magnetic Bubbles*, Selected Topics in Solid State Physics (Ed.E.P. Wahlfarth), North Holland, Amsterdam (1975).
4. A.H. Eschenfelder, *Magnetic Bubble Technology*, Springer Series in Solid-State Sciences (Eds.M. Cardona, P. Fulde, H-J. Queisser), Springer-Verlag, Berlin (1981).
5. A.O. Caldeira, K. Furuya, Quantum Nucleation of Magnetic Bubbles in a 2-D Anisotropic Heisenberg Model, *J. Phys. C* **21**, 1227-1241 (1988).
6. O. Kahn, *Molecular Magnetism*, VCH, New York, 1993.
7. D. Gatteschi, A. Caneschi, L. Pardi, and R. Sessoli, Large Clusters of Metal Ions: The Transition from Molecular to Bulk Magnets, *Science* **265**, 1054-1058 (1994).
8. A. Caneschi, A. Gatteschi, J. Laugier, R. Ray et al, Preparation, crystal structure, and magnetic properties of an oligonuclear complex with 12 coupled spins and an S = 12 ground state, *J. Am. Chem. Soc.* **110**, 2795-2799 (1988).
9. R. Sessoli, D. Gatteschi, A. Caneschi, and M. Novak, Magnetic bistability in a metal-ion cluster, *Nature* (London) **365**, 141-143 (1993).
10. J. Villain, F. Hartmann-Bourton, R. Sessoli, and A. Rettori, Magnetic relaxation in big magnetic molecules, *Europhys. Lett.* **27**, 159-164 (1994).
11. P. Politi, A. Rettori, F. Hartmann-Bourton, J. Villain, Tunneling in Mesoscopic Magnetic Molecules, *Phys. Rev. Lett.* **75**, 537-540 (1995).
12. N.V. Prokof'ev and P.C.E. Stamp, Low-Temperature Quantum Relaxation in a System of Magnetic Nanomolecules, *Phys. Rev. Lett.* **80**, 5794-5797 (1998).
13. B.K. Chakrabarti, M. Acharyya, Dynamic transitions and hysteresis, *Rev. Mod. Phys.* **71**, 847-859 (1999).
14. M. Acharyya, Nonequilibrium phase transitions in model ferromagnets: a review, *Int. Journ. of Mod. Phys. C* **16** 1631-1670 (2005).
15. D. Boyanovsky, Quantum spinodal decomposition, *Phys. Rev. E* **48**, 767-771 (1993).
16. A.O. Caldeira and A.J. Leggett, Path integral approach to quantum Brownian motion, *Physica* **121A**, 587-616 (1983).
17. L. Dolan and R. Jackiw, Symmetry behavior at finite temperature,*Phys. Rev. D* **9**, 3320-3341 (1974); A.J. Niemi and G.F. Semenoff, Finite-temperature quantum field theory in Minkowski space, *Ann. Phys.* (N.Y.) **152**, 105-129 (1984).
18. A.J. Leggett, S. Chakravarty, A.T. Dorsey et al, Dynamics of the dissipative two-state system, *Rev. Mod. Phys.* **59**, 1-85 (1987).
19. E. Abrahams and T. Tsuneto, Time Variation of the Ginzburg-Landau Order Parameter, *Phys. Rev.* **152**, 416-432 (1966).
20. J.D. Gunton, M. Droz, *Introduction to the Theory of Metastable and Unstable States*, Springer-Verlag, Berlin (1983).
21. M. Grant, J.D. Gunton, Temperature dependence of the dynamics of random interfaces, *Phys. Rev. B* **28**, 5496-5506 (1983).
22. F.A. Berezin and M.S. Marinov, Particle spin dynamics as the Grassmann variant of classical mechanics, *Ann. of Phys.* **104**, 336-362 (1977).

23. V.R. Vieira, Kondo lattice: Renormalization study using a new pseudofermion representation, *Phys. Rev. B* **23**, 6043-6054 (1981).
24. T. Holstein and H. Primakoff, Field Dependence of the Intrinsic Domain Magnetization of a Ferromagnet, *Phys. Rev.* **58**, 1098-1113 (1940).
25. J. Schwinger, in *"Quantum Theory of Angular Momentum"*, Eds. L.C. Biedenharn H. van Dan (Academic, New York, 1965).
26. H. Nakano and M. Takahashi, Magnetic properties of quantum Heisenberg ferromagnets with long-range interactions, *Phys. Rev. B* **52**, 6606-6610 (1995).
27. O.N. Vassiliev, M.G. Cattam, and I.V. Rojdestvenski, Quantum Monte Carlo study of a Heisenberg spin system in two dimensions with long-range interactions *Journ. of Appl. Phys.* **89**, 7329-7331 (2001).
28. J. Schwinger, Brownian Motion of a Quantum Oscillator, *J. Math. Phys.* **2**, 407-432 (1961).
29. K.V. Keldysh, Diagram technique for nonequilibrium processes, *Sov. Phys. JETP* **20**, 1018-1026 (1965).
30. J. Rammer and H. Smith, Quantum field-theoretical methods in transport theory of metals, *Rev. Mod. Phys.* **58**, 323-359 (1986).
31. H. Umezawa, H. Matsumoto, and M. Tachiki, *Thermo Field Dynamics and Condensed States* (North-Holland, Amsterdam, 1982); H. Matsumoto, Y. Nakano, H. Umezawa, F. Mancini, and M. Marinaro, Thermo Field Dynamics in Interaction Representation, *Prog. Theor. Phys.* **70**, 599-602 (1983).
32. G.D. Mahan, Quantum transport equation for electric and magnetic fields, *Phys. Rep.* **145**, 251-318 (1987).
33. F.J. Cao and H.J. de Vega, Nonequilibrium dynamics in quantum field theory at high density: The tsunami, *Phys. Rev. D* **63**, 045021 (2001).
34. F.J. Cao and H.J. de Vega, Out of equilibrium nonperturbative quantum field dynamics in homogeneous external fields, *Phys. Rev. D* **65**, 045012 (2002).
35. V. Turkowski, V.R. Vieira, and P.D. Sacramento, Non-equilibrium properties of the Heisenberg model in a time-dependent magnetic field, *Physica A* **327**, 461-476 (2003).
36. K. Kawasaki, M.C. Yalabik, and D.J. Gunton, Growth of fluctuations in quenched time-dependent Ginzburg-Landau model systems, *Phys. Rev. A* **17**, 455-470 (1978).
37. P.S. Sahni, G.S. Grest, and S.A. Safran, Temperature Dependence of Domain Kinetics in Two Dimensions, *Phys. Rev. Lett.* **50**, 60-63 (1983).
38. T. Ohta, D. Jasnow, and K. Kawasaki, Universal Scaling in the Motion of Random Interfaces, *Phys. Rev. Lett.* **49**, 1223-1226 (1982).
39. D. Boyanovsky, H.J. de Vega, R. Holman, Non-Equilibrium Dynamics of Phase Transitions: From the Early Universe to Chiral Condensates, *preprint hep-th/9412052*.
40. D. Boyanovsky, H.J. de Vega, R. Holman, Erice Lectures on Inflationary Reheating, in *Proceedings of the 5th Erice Chalonge School on Astrofundamental Physics*, edited by N. Sánchez and A. Zichichi (World Scientific, Singapore, 1997); *preprint hep-ph/9701304*.
41. Y. Bergner, L.M.A. Bettencourt, A step beyond the bounce: Bubble dynamics in quantum phase transitions, *Phys. Rev. D* **68**, 025014 (2003).
42. V.M. Turkowski, P.D. Sacramento, and V.R. Vieira, Domain growth in the Heisenberg ferromagnet: Effective vector theory of the S=1/2 model, *Phys. Rev. B* **73**, 214437 (2006).

43. A.J. Bray, Theory of phase-ordering kinetics, *Adv. Phys.* **43**, 357-459 (1994).
44. S. Dattagupta, S. Puri, *Dissipative phenomena in condensed matter: some applications*, Springer, New York (2004).
45. V.R. Vieira, Finite-temperature real-time field theories for spin 1/2, *Phys. Rev. B* **39**, 7174-7195 (1989).
46. V.R. Vieira and I.R. Pimentel, Relevance of the imaginary-time branch in real-time formalisms for thermodynamic equilibrium: Study of the Heisenberg model, *Phys. Rev. B* **39**, 7196-7204 (1989).
47. P.D.S. Sacramento and V.R. Vieira, The Helmholtz free-energy functional for quantum spin-1/2 systems, *J. Phys. C* **21**, 3099-3131 (1988).
48. Ya. Tserkovnyak, A. Brataas, G.E.W. Bauer, and B.I. Halperin, Nonlocal magnetization dynamics in ferromagnetic heterostructures, *Rev. Mod. Phys.* **77**, 1375 (2005).
49. P. Coleman, E. Miranda, and A. Tsvelik, Possible realization of odd-frequency pairing in heavy fermion compounds, *Phys. Rev. Lett.* **70**, 2960 - 2963 (1993).
50. W. Mao, P. Coleman, C. Hooley, and D. Langreth, Spin Dynamics from Majorana Fermions, *Phys. Rev. Lett.* **91**, 207203 (2003).
51. P.A. Lee, N. Nagaosa, and X.-G. Wen, Doping a Mott insulator: Physics of high-temperature superconductivity, *Rev. Mod. Phys.* **78**, 17 (2006).

Chapter 7

Nonequilibrium dynamical mean-field theory of strongly correlated electrons

V. Turkowski

Department of Physics and Astronomy, University of Missouri-Columbia, Columbia, MO 65202, USA
turkowskiv@missouri.edu

J.K. Freericks

Department of Physics, Georgetown University, Washington, D.C. 20057, USA
freericks@physics.georgetown.edu

We present a review of our recent work in extending the successful dynamical mean-field theory from the equilibrium case to nonequilibrium cases. In particular, we focus on the problem of turning on a spatially uniform, but possibly time varying, electric field (neglecting all magnetic field effects). We show how to work with a manifestly gauge-invariant formalism, and compare numerical calculations from a transient-response formalism to different types of approximate treatments, including the semiclassical Boltzmann equation and perturbation theory in the interaction. In this review, we solve the nonequilibrium problem for the Falicov-Kimball model, which is the simplest many-body model and the easiest problem to illustrate the nonequilibrium behavior in both diffusive metals and Mott insulators. Due to space restrictions, we assume the reader already has some familiarity both with the Kadanoff-Baym-Keldysh nonequilibrium formalism and with equilibrium dynamical mean-field theory; we provide a guide to the literature where additional details can be found.

Contents

7.1. Introduction

Dynamical mean-field theory (DMFT) was introduced in 1989[1] shortly after Metzner and Vollhardt[2] proposed scaling the hopping matrix element as the inverse square root of the spatial dimension to achieve a nontrivial limit where the many-body dynamics are local. Since then, the field has blossomed to the point where nearly all model many-body problems have now been solved,[3] and much recent work has focused on applying DMFT principles to real materials calculations.[4] Little work has emphasized nonequilibrium aspects of the many-body problem, where the strongly correlated system is driven by an external field that can possibly sustain a nonequilibrium steady state. In this contribution, we will review recent work that has been completed on expanding DMFT approaches into the nonequilibrium realm. We will show how to work with so-called gauge-invariant Green functions[5] to illustrate that one can carry out calculations in a form that manifestly is independent of the gauge chosen to describe the driving fields. This approach is different from our previously published work, where we worked solely with Green functions in the Hamiltonian gauge (where the scalar potential vanishes).

We examine the problem of strongly correlated electrons driven by a spatially uniform electric field in the limit of infinite dimensions,[6–8] where DMFT can be applied to solve the problem exactly. In infinite dimensions, the self-energy of the electrons is local, and the lattice problem can be mapped onto the problem of an impurity coupled to an effective time-dependent field (which is adjusted so that the impurity Green function and the local Green function on the lattice are identical). The impurity problem in the dynamical mean field can be solved exactly for many different cases. In equilibrium, a large number of strongly correlated models have been solved in infinite dimensions, like the Falicov-Kimball model,[1,10,11] the Hubbard model,[12–14] the periodic Anderson model,[15,16] and the Holstein model[17,18] (for a reviews, see Refs. 3 and 19). Recently, there has been a significant effort in combining DMFT with density functional theory (DFT) to describe properties of real materials when DFT is insufficient to properly describe the electron-electron interactions (see Ref. 4 for a review). It is now generally believed that DMFT is a good approximation to the many-body problem in three dimensions, and it can accurately describe strong electron-electron correlation effects in bulk systems.

The first attempt to employ DMFT to describe nonequilibrium properties of a strongly correlated model was made by Schmidt and Monien in Ref. 20, where they studied the spectral properties of the Hubbard model in the presence of a time-dependent chemical potential by using iterated perturbation theory (PT). Recently, we have developed a generalized nonequilibrium DMFT formalism to study the response of correlated electrons to a spatially uniform time-dependent electric field and applied that formalism to the Falicov-Kimball model.[6–8] The Falicov-Kimball model,[21] is the simplest model for strongly correlated electrons that demonstrates long range order and undergoes a metal-to-Mott-insulator transition. It consists of two kinds of electrons: conducting c-electrons and localized f-electrons, which interact through an on-site Coulomb repulsion. The model was introduced to describe valence-change and metal-insulator transitions[21] in rare-earth and transition-metal compounds. It was reinvented as a model to describe crystal formation[22] resulting from the Pauli exclusion principle. DMFT was actually developed with the original solution of the Falicov-Kimball model in infinite dimensions[1,10,11] and now there is an almost complete understanding of its general properties (for a review, see Ref. 19). We extended the equilibrium formalism to the nonequilibrium case, where we numerically solved a system of the equations for the Green function and self-energy defined on a complex time contour (see Fig. 7.1) by using the Kadanoff-Baym-Keldysh nonequilibrium Green function formalism.[23,24]

In this review, we summarize the successes of recent work to generalize DMFT to nonequilibrium problems with a focus on solutions of the spinless Falicov-Kimball model on an infinite-dimensional hypercubic lattice in the presence of an external time-dependent electric field. There are many interesting and surprising results which differ from semiclassical predictions (such as those made from the Boltzmann equation solution). In addition to the exact solutions, we also present results for the noninteracting case and for the case of second-order perturbation theory in the interaction. In particular, we analyze the limitations of the perturbation theory approximation, especially in studying (long-time) steady-state behavior.

7.2. General nonequilibrium formalism

The nonequilibrium properties of a quantum many-particle system can be studied by calculating the contour-ordered Green function in momentum space:

$$\begin{aligned} G^c_{\mathbf{k}}(t_1,t_2) &= -i\langle \hat{\mathrm{T}}_c c_{\mathbf{k}H}(t_1) c^\dagger_{\mathbf{k}H}(t_2)\rangle \\ &= \frac{-i\mathrm{Tr}\left\{e^{-\beta H(-t_{\max})}\hat{\mathrm{T}}_c \exp[-i\int_c dt H_I(t)] c_{\mathbf{k}I}(t_1) c^\dagger_{\mathbf{k}I}(t_2)\right\}}{\mathrm{Tr} e^{-\beta H(-t_{\max})}}, \end{aligned} \tag{7.1}$$

defined on the complex time-contour presented in Fig. 7.1 (see, for example, Ref. 25). Since the system is initially in equilibrium, the ther-

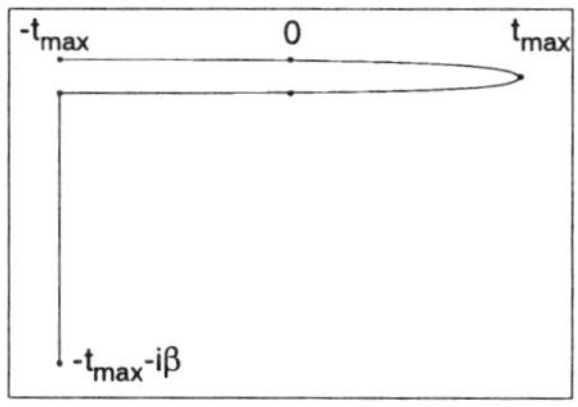

Fig. 7.1. The complex Kadanoff-Baym-Keldysh time contour for the two-time Green functions in the nonequilibrium case. The time increases from $-t_{max}$ (left point on the top branch) along the contour to t_{max} then decreases back to $-t_{max}$ and then runs parallel to the imaginary axis to $-t_{max} - i\beta$. We consider the situation when the electric field is turned on at $t = 0$, so the vector potential is nonzero for $t > 0$. We assume that both t_1 and t_2 lie somewhere on the contour.

mal average in Eq. (7.1) is performed with the equilibrium density matrix $\exp[-\beta H(-t_{\max})]/\mathrm{Tr}\exp[-\beta H(-t_{\max})]$ with respect to the initial Hamiltonian $H(-t_{\max})$ with vanishing electric field (the symbol $\beta = 1/T$ is the inverse temperature). The operator indices H and I in Eq. (7.1) stand for the Heisenberg and Interaction representations, respectively. In this formalism, familiar quantum many-body techniques derived in equilibrium can also be used in the nonequilibrium case, except that now the time ordering $\hat{\mathrm{T}}_c$ of the operators is along the complex contour. In particular, the Schwinger-Dyson equation, which connects the contour-ordered Green function with the electron self-energy $\Sigma^c_{\mathbf{k}}(t_1, t_2)$, remains valid:

$$G^c_{\mathbf{k}}(t_1, t_2) = G^{0c}_{\mathbf{k}}(t_1, t_2) + \int_c dt \int_c d\bar{t} G^{0c}_{\mathbf{k}}(t_1, t)\Sigma^c_{\mathbf{k}}(t, \bar{t})G^c_{\mathbf{k}}(\bar{t}, t_2), \quad (7.2)$$

where the matrix product of the continuous matrix operators is accomplished by line integrals over the contour.

In DMFT, we work with the local Green function, which is found by summing the momentum-dependent Green function over all momenta. We then map the many-body problem on the lattice to an impurity problem, but in a dynamical mean field that mimics the hopping of electrons onto and off of the given site. It turns out that one needs the full freedom available with the three-branch contour to find the proper dynamical mean field to map the impurity onto the lattice. Hence, our approach will work with the less common Green functions on the three-branch contour, as opposed to a simpler two-branch contour, which we work with when we discuss the perturbative approach on the lattice. One can find the time-ordered, anti-time-ordered, lesser, greater, retarded, advanced and thermal Green functions on this contour.[26] For example, the retarded Green function, which is related to the density of quantum states, is

$$G^R_{\mathbf{k}}(t_1, t_2) = -i\theta(t_1 - t_2)\langle\{c_{\mathbf{k}H}(t_1), c^\dagger_{\mathbf{k}H}(t_2)\}_+\rangle, \quad (7.3)$$

where the braces indicate the anticommutator of the two operators, and the lesser Green function, which is related to how the electrons are distributed amongst the quantum states, satisfies

$$G^{<}_{\mathbf{k}}(t_1,t_2) = i\langle c^{\dagger}_{\mathbf{k}H}(t_2)c_{\mathbf{k}H}(t_1)\rangle. \tag{7.4}$$

Both of these functions can be extracted from $G^{c}_{\mathbf{k}}$.

7.3. Nonequilibrium dynamical mean-field theory for the Falicov-Kimball model

The spinless Falicov-Kimball model[21] consists of two kinds of electrons: conduction c-electrons and localized f-electrons. They interact with each other through an on-site Coulomb repulsion U. The model Hamiltonian has the following form in the absence of any external fields:

$$\mathcal{H} = -\sum_{\langle ij\rangle} t_{ij}c^{\dagger}_i c_j - \mu\sum_i c^{\dagger}_i c_i - \mu_f\sum_i f^{\dagger}_i f_i + U\sum_i f^{\dagger}_i f_i c^{\dagger}_i c_i, \tag{7.5}$$

where $t_{ij} = t$ is the nearest-neighbor hopping matrix element for the c-electrons, μ and μ_f are chemical potentials of the c- and f-electrons, correspondingly. Due to the Pauli principle, there is no local cc- and ff-electron interaction in the spinless case. The Hamiltonian in Eq. (7.5) can also be regarded as an approximation to the spin $s = 1/2$ Hubbard model, where spin-up (c-) electrons move in a frozen background of the localized spin-down (f-electrons). We consider the problem on the infinite-dimensional ($d \to \infty$) hypercubic lattice at half-filling, when the particle densities of the c- and f-electrons are equal to 0.5. In this limit, the hopping parameter is renormalized in the following way:[2] $t = t^*/2\sqrt{d}$. In the limit of infinite dimensions, one can solve the equilibrium problem for the conduction electrons exactly at any temperature, particle concentration and Coulomb repulsion. The key simplification, which allows one to obtain the exact solution as $d \to \infty$, comes from the fact that the electron self-energy is momentum-independent.[1,27] Although that original work was performed in equilibrium, Langreth's rules[28] guarantee that it also holds for the nonequilibrium case.

Nowadays, most of the equilibrium properties of the model, including the phase diagram, are well known (see Ref. 19). In particular, the model demonstrates a Mott transition when $n_c + n_f = 1$ at some critical value of the Coulomb repulsion,[29] which depends on the particular value of n_c (n_f is equal to $1 - n_c$ in this case). In the insulating phase, the density of states $A(\omega)$ is not equal to zero for frequencies inside the "gap region", but is exponentially suppressed, except for $\omega = 0$. Therefore, the density of states actually demonstrates a pseudogap in the insulating phase, which is an artifact of the fact that the infinite-dimensional hypercubic lattice has a Gaussian density of states for the noninteracting problem, which does not have a finite bandwidth. Another important feature is the behavior of the imaginary part of the self-energy for frequencies close to zero in the "metallic" phase: $\mathrm{Im}\Sigma(\omega) \sim -c + c'\omega^2$ (c and $c' > 0$ and independent of temperature), which differs from the standard Fermi liquid behavior $\mathrm{Im}\Sigma(\omega) \sim -a(T) - b\omega^2$ (a and $b > 0$ and $a(T) \to 0$ as $T \to 0$). This means that there are no long-lived Fermi liquid quasiparticles in the model.

We are interested in the case when the system is coupled to an external electric field $\mathbf{E}(\mathbf{r}, t)$. This field can be expressed by a scalar potential $\varphi(\mathbf{r}, t)$ and by a vector potential $\mathbf{A}(\mathbf{r}, t)$ in the following way:

$$\mathbf{E}(\mathbf{r}, t) = -\nabla\varphi(\mathbf{r}, t) - \frac{1}{c}\frac{\partial \mathbf{A}(\mathbf{r}, t)}{\partial t}. \tag{7.6}$$

We assume that the electric field is spatially uniform and choose the temporal or Hamiltonian gauge for the electric field: $\varphi(\mathbf{r}, t) = 0$. In this case, the electric field is introduced into the Hamiltonian by means of the Peierls substitution for the hopping matrix:[30]

$$t_{ij} \to t_{ij} \exp\left[-\frac{ie}{\hbar c}\int_{\mathbf{R}_i}^{\mathbf{R}_j} \mathbf{A}(\mathbf{r}, t) d\mathbf{r}\right] = t_{ij} \exp\left[\frac{ie}{\hbar c}\mathbf{A}(t) \cdot (\mathbf{R}_i - \mathbf{R}_j).\right], \tag{7.7}$$

where the last formula holds for a spatially uniform field where we take $\mathbf{A}(t) = -\mathbf{E}ct\theta(t)$ for a uniform field turned on at $t = 0$.

We assume that it is safe to neglect magnetic field effects, because the electric field varies slow enough in time (recall Maxwell's equations say that a time-varying electric field creates a time varying magnetic field). This approximation is valid when the electric field is smooth enough in time that the magnetic fields can be ignored. Another way of describing this is that we assume our electric field is always spatially uniform, even though it has a time dependence, which is not precisely a solution of Maxwell's equations, but is approximately so.

The electric field introduced into the Hamiltonian Eq. (7.5) results in a time-dependent shift of the momentum in the free electron dispersion relation:

$$\epsilon\left(\mathbf{k} - \frac{e\mathbf{A}(t)}{\hbar c}\right) = -2t\sum_{l=1}^{d} \cos\left[a\left(k^l - \frac{eA^l(t)}{\hbar c}\right)\right]. \tag{7.8}$$

It is convenient to consider the case, when the electric field lies along the elementary cell diagonal:[31]

$$\mathbf{A}(t) = A(t)(1, 1, ..., 1). \tag{7.9}$$

In this case, the free electron spectrum

$$\epsilon\left(\mathbf{k}-\frac{e\mathbf{A}(t)}{\hbar c}\right)=\cos\left(\frac{eaA(t)}{\hbar c}\right)\epsilon(\mathbf{k})+\sin\left(\frac{eaA(t)}{\hbar c}\right)\bar{\varepsilon}(\mathbf{k}), \tag{7.10}$$

depends on only two energy functions:

$$\epsilon(\mathbf{k})=-2t\sum_l\cos(ak^l) \tag{7.11}$$

and

$$\bar{\varepsilon}(\mathbf{k})=-2t\sum_l\sin(ak^l). \tag{7.12}$$

Of course, when the field vanishes, the energy spectra in Eq. (7.10) reduces to the standard spectra in Eq. (7.11) for free electrons on the hypercubic lattice. In the limit of an infinite dimensional hypercubic lattice, one can calculate the joint density of states for the two energy functions in Eqs. (7.11) and (7.12),[32]

$$\rho_2(\epsilon,\bar{\varepsilon})=\frac{1}{\pi t^{*2}a^d}\exp\left[-\frac{\epsilon^2}{t^{*2}}-\frac{\bar{\varepsilon}^2}{t^{*2}}\right]. \tag{7.13}$$

Below we use atomic units, putting all fundamental constants, except the electron charge e, to be equal to one: $a=\hbar=c=t^*=1$.

To solve the problem of the response of the conduction electrons to an external electric field, we use a generalized nonequilibrium DMFT formalism.[8] The electron Green functions and self-energies are functions of two time arguments defined on the complex time-contour in Fig. 7.1. Since the action for the Falicov-Kimball model is quadratic in the conduction electrons, the Feynman path integral over the Kadanoff-Baym-Keldysh contour can be expressed by the determinant of a continuous matrix operator with arguments defined on the contour. Because the concentration of localized particles on each site is conserved, one can calculate the trace over the fermionic variables. It is possible to show that the self-energy remains local in the limit of infinite dimensions in the presence of a field; start with the equilibrium perturbation theory expansion for the self-energy[27] and then apply Langreth's rules[28] to the self-energy diagrams, which say that every nonequilibrium diagram is obtained from a corresponding equilibrium diagram, with the time variables now defined on the Kadanoff-Baym-Keldysh contour.

The generalized system of nonequilibrium DMFT equations for the contour ordered Green function $G(t_1,t_2)$, self-energy $\Sigma(t_1,t_2)$ and an effective dynamical mean-field $\lambda(t_1,t_2)$ can be written in analogy with the equilibrium case[12] as follows:

$$G(t_1,t_2) = \sum_{\mathbf{k}}[G_{\mathbf{k}}^{(0)-1} - \Sigma]^{-1}(t_1,t_2), \tag{7.14}$$

$$G_0(t_1,t_2) = [G^{-1} + \Sigma]^{-1}(t_1,t_2), \tag{7.15}$$

$$\lambda(t_1,t_2) = G_{0imp}^{-1}(t_1,t_2;\mu) - G_0^{-1}(t_1,t_2), \tag{7.16}$$

$$G(t_1,t_2) = (1-w_1)G_0(t_1,t_2) + w_1[G_{0imp}^{-1}(\mu - U) - \lambda]^{-1}(t_1,t_2), \tag{7.17}$$

where $G_{\mathbf{k}}^{(0)}(t_1,t_2)$ is the noninteracting electron Green function in the presence of an external time-dependent electric field, which can be calculated analytically (see below) and $G_{0imp}(t_1,t_2;\mu)$ is the free impurity Green function in a chemical potential μ. The symbol w_1 is the average number of the f-electrons per site. In our case, $w_1 = 1/2$.

The momentum summation in Eq. (7.14) can be performed by introducing the two energy functions Eqs. (7.11) and (7.12) and using the joint density of states in Eq. (7.13): $\sum_{\mathbf{k}} F_{\mathbf{k}} = \int d\epsilon \int d\bar{\epsilon}\rho_2(\epsilon,\bar{\epsilon})F_{\epsilon,\bar{\epsilon}}$ whenever the summand $F_{\mathbf{k}}$ depends on momentum only through the two energy functions. The system of equations (7.14)-(7.17) formally resembles the corresponding system in the equilibrium case, except now we have to work with a two-time formalism on the contour, rather than being able to Fourier transform the relative time to a frequency. And, because we are working with the contour-ordered Green functions, which depend on the distribution of electrons, we need to be careful to treat how the chemical potential is shifted by U when we perform the trace over the f-electrons.

The system of equations (7.14)-(7.17) can be solved by iteration as follows. One starts with an initial self-energy matrix, for example the equilibrium self-energy. Substitution of this function into Eq. (7.14) gives the Green function. Then, from Eq. (7.16) one can find the effective dynamical mean-field $\lambda(t_1,t_2)$, which allows one to find the new value for the Green function $G(t_1,t_2)$ from Eq. (7.17). After that, one finds the new self-energy $\Sigma(t_1,t_2)$ from the impurity Dyson equation and the dynamical mean field. The calculations are repeated until the difference between the old and new values for the self-energy $\Sigma(t_1,t_2)$ are smaller than some desired precision (usually 10^{-6} in relative error).

In practice, to solve this system numerically, one needs to discretize the complex time contour Fig. (7.1) with some step Δt along the real axis and $\Delta\tau$ along the imaginary axis. In this case, the functions in Eqs. (7.14)-(7.17) become general complex square matrices. In order to study the long time behavior, one needs to choose the value of t_{max} large enough. The precision of the solution strongly depends on the value of the discretization step Δt, which must be small enough. Therefore, in order to get a precise long time solution it is necessary to use large complex square matrices in Eqs. (7.14)-(7.17). This causes some constraints connected with the machine memory and the computational time. In our calculations, we used the time step Δt ranging from 0.1 to 0.0167 and matrices up to order 4900×4900. The two-energy integration in Eq. (7.14) was performed by using a Gaussian integration scheme (for details, see Ref. 6). Since each energy is independent of each other, the algorithm parallelizes naturally.

It is important to find ways to benchmark this nonequilibrium DMFT algorithm, to ensure that it is accurate. The simplest way to do this is to calculate the equilibrium results within the nonequilibrium formalism and compare those results with the results obtained by the equilibrium DMFT approach. One of the most important elements is a proper choice of the discretization step Δt of the contour. These equilibrium calculations help us choose the step size Δt to be small enough to obtain accurate results (see Ref. 6). Another useful way to check the accuracy of the solution is to calculate the moments of the retarded electron spectral functions $A(t_{ave}, \omega) = \sum_{\mathbf{k}}(-1/\pi)\mathrm{Im}G_{\mathbf{k}}(t_{ave}, \omega)$, where t_{ave} is the average time and ω is the electron frequency arising from a Fourier transform of the relative time (see below). We have found[7] that the lowest spectral moments in the Falicov-Kimball model can be calculated exactly, and they are time-independent even in the presence of a time-dependent electric field. In particular, when a spatially homogeneous time-dependent electric field is applied, one can find for the zeroth and first two retarded spectral moments:

$$\int_{-\infty}^{\infty} d\omega A^R(t_{ave}, \omega) = 1, \tag{7.18}$$

$$\int_{-\infty}^{\infty} d\omega \omega A^R(t_{ave}, \omega) = -\mu + U n_f = 0, \tag{7.19}$$

$$\int_{-\infty}^{\infty} d\omega \omega^2 A^R(t_{ave}, \omega) = \frac{1}{2} + \mu^2 - 2U\mu n_f + U^2 n_f = \frac{1}{2} + \frac{U^2}{4}, \tag{7.20}$$

where the second equality holds in the half-filled case. We estimate the accuracy of the discretization of the contour by calculating the spectral moments and comparing them with the exact analytical results in Eqs. (7.18)-(7.20).[7] In general, one needs to reduce the discretization size as the interaction strength increases. This is clearly seen in the equilibrium case, where the numerics can be well controlled because there is no dependence on the energy $\bar{\varepsilon}$. Surprisingly, in the presence of a field, one can use a somewhat larger discretization size, especially for moderate to large fields.

7.4. Gauge invariance and physical observables

In nonequilibrium problems, we work with two-time Green functions because the system no longer has time-translation invariance. Wigner[33] first realized that it is more physical to express results in terms of average and relative coordinates, where the dependence on the average coordinates drops out in equilibrium. In our case, the relative and average times satisfy

$$t = t_1 - t_2, \quad t_{ave} = \frac{t_1 + t_2}{2} \tag{7.21}$$

while for the spatial coordinates we have

$$\mathbf{r} = \mathbf{r}_1 - \mathbf{r}_2, \quad \mathbf{r}_{ave} = \frac{\mathbf{r}_1 + \mathbf{r}_2}{2}; \tag{7.22}$$

note that at this point we are restricting the time coordinates to lie on the real axis piece of the contour since the imaginary axis piece is not important for determining physical properties on the lattice (the full structure is only needed for the self-consistent DMFT loop, not for calculating any physical properties once the self-energy has been determined). We want to be able to convert the relative time and space coordinates into frequency and momentum via a Fourier transformation. Since we are working with a uniform electric field, we expect that the system will have no average spatial coordinate dependence, because it is spatially homogeneous. The easiest way to construct the right transformation is to create a Fourier transformation that makes the gauge-invariance of the problem manifest; the result is then called the gauge-invariant Green function, which depends only on the fields, not on the scalar or vector potentials.[5] The procedure is somewhat technical, but completely straightforward. The starting point is a generalized Fourier transformation

$$G(\mathbf{k}, \omega, \mathbf{r}_{ave}, t_{ave}) = \int d^d r \int dt \exp[iW(\mathbf{k}, \omega, \mathbf{r}, t, \mathbf{r}_{ave}, t_{ave})] \times G(\mathbf{r}, t, \mathbf{r}_{ave}, t_{ave}), \tag{7.23}$$

with W being a complicated function of its variables, in general. In equilibrium, when there is no external space- and time-dependent electric field, the Green function doesn't depend on the average coordinates t_{ave} and $\mathbf{r}_{ave}$, and the transform (7.23) is the well-known Fourier transformation with $W(\mathbf{k}, \omega, \mathbf{r}, t, \mathbf{r}_{ave}, t_{ave}) = t\omega - \mathbf{r} \cdot \mathbf{k}$.

The situation is more complicated when an external field is present. In this case, the field is introduced by using a specific gauge for the scalar and vector potential (we work with the Hamiltonian gauge). It is important to have a Green function on the left hand side of Eq. (7.23), which doesn't depend on the choice of gauge so all results are manifestly independent of the scalar and vector potentials. Therefore, we need to construct a function $W(\mathbf{k}, \omega, \mathbf{r}, t, \mathbf{r}_{ave}, t_{ave})$ in Eq. (7.23), which makes $G(\mathbf{k}, \omega, \mathbf{r}_{ave}, t_{ave})$ invariant under the gauge transformation:

$$\varphi(\mathbf{r}_1, t_1) \rightarrow \varphi(\mathbf{r}_1, t_1) - \frac{\partial \chi(\mathbf{r}_1, t_1)}{\partial t_1}, \tag{7.24}$$

$$\mathbf{A}(\mathbf{r}_1, t_1) \rightarrow \mathbf{A}(\mathbf{r}_1, t_1) + \nabla \chi(\mathbf{r}_1, t_1), \tag{7.25}$$

where $\chi(\mathbf{r}_1, t_1)$ is an arbitrary function. The χ function must also be used in the local unitary gauge transformation of the fermion operators:

$$c(\mathbf{r}_1, t_1) \rightarrow \exp[ie\chi(\mathbf{r}_1, t_1)]c(\mathbf{r}_1, t_1), \tag{7.26}$$
$$c^\dagger(\mathbf{r}_2, t_2) \rightarrow \exp[-ie\chi(\mathbf{r}_2, t_2)]c^\dagger(\mathbf{r}_2, t_2), \tag{7.27}$$

since it corresponds to the phase picked up by the fermions as a result of the local gauge transformation. Obviously, the Green function on the right hand side of Eq. (7.23) is not generically invariant in this case:

$$G(\mathbf{r}_1, t_1; \mathbf{r}_2, t_2) \rightarrow \exp[ie(\chi(\mathbf{r}_1, t_1) - \chi(\mathbf{r}_2, t_2))]G(\mathbf{r}_1, t_1; \mathbf{r}_2, t_2). \tag{7.28}$$

However, it is possible to show that its transform in Eq. (7.23) is invariant, when one chooses[5]

$$\begin{aligned} W(\mathbf{k}, \omega, \mathbf{r}, t, \mathbf{r}_{ave}, t_{ave}) = \int_{-1/2}^{1/2} & d\lambda \{t[\omega + e\varphi(\mathbf{r}_{ave} + \lambda \mathbf{r}, t_{ave} + \lambda t)] \\ & -\mathbf{r} \cdot [\mathbf{k} + e\mathbf{A}(\mathbf{r}_{ave} + \lambda \mathbf{r}, t_{ave} + \lambda t)]\} \end{aligned} \tag{7.29}$$

(for details, see Ref. 34).

In the case of a spatially homogeneous electric field in the Hamiltonian gauge with $\varphi(\mathbf{r}, t) = 0$, which we study in this chapter, this transformation is

$$\tilde{G}(\mathbf{k}, t, \mathbf{r}_{ave}, t_{ave}) \rightarrow G\left(\mathbf{k} - \frac{1}{t}\int_{-t/2}^{t/2} e\mathbf{A}(t_{ave} + \bar{t})d\bar{t}, t, \mathbf{r}_{ave}, t_{ave}\right), \tag{7.30}$$

because the function W just involves a shift of the momentum; note that the Green function is actually independent of $\mathbf{r}_{ave}$ in this case. Hence, the gauge invariant Green function in the momentum representation contains a shift of the momentum, which depends on both the relative and average time coordinates. We consider the case when a constant electric field is turned on at time $t = 0$: $\mathbf{A}(t) = -\mathbf{E}t\theta(t)$.

Then the momentum shift is

$$\mathbf{k} \rightarrow \mathbf{k} - e\mathbf{E}\Big[t_{ave}\theta(t_{ave} - |t/2|) + \left(-\frac{t_{ave}^2}{2t} + \frac{t_{ave}}{2} - \frac{t}{8}\right)\theta(-t/2 - |t_{ave}|) + \left(\frac{t_{ave}^2}{2t} + \frac{t_{ave}}{2} + \frac{t}{8}\right)\theta(t/2 - |t_{ave}|)\Big]. \tag{7.31}$$

Note that this shift does not depend on the relative time coordinate t for long times, $t_{ave} > |t/2|$. However, in general, one has to first shift the momentum, and then Fourier transform the relative time to a frequency. It is important that the time-dependent momentum shift takes place for some negative average times (if the absolute value of the relative time is large enough, then either t_1 or t_2 is larger than 0 and hence "sees" the field).

The shift of the momentum becomes particularly simple for equal time Green functions, such as those needed to calculate the current flowing or to determine the distribution of the electrons amongst the quantum states. In this case, $t = 0$, and the momentum is shifted by $-e\mathbf{E}t_{ave}$ if $t_{ave} > 0$. Therefore, gauge invariant Green functions can be obtained from the Hamiltonian gauge Green functions by simply shifting the momentum by $-e\mathbf{E}t_{ave}$. Note that local quantities, like the local density of states or the local distribution function are always gauge invariant, because they are summed over momentum, and if the shift is the same for each momentum value, then we still sum over all the momentum points in the Brillouin zone. In cases where the relative time is nonzero, the transformation from the Green function in a particular gauge to the gauge-invariant Green function must be handled with care. Finally, one should note that in the steady state, where $t_{ave} \rightarrow \infty$, the momentum shift is also simple ($-e\mathbf{E}t_{ave}$); it turns out that the retarded and advanced Green functions depend only on the relative time, but the lesser, greater, and Keldysh Green functions depend on both the average and relative time because there is an average-time-dependent shift of the momentum in Fermi-Dirac distribution functions. Caution must be used in trying to directly find the steady-state Green functions, because the Dyson equation is modified, since the momentum shift does not remove all average time dependence in internal variables that are integrated over in the $G_0\Sigma G$ term.

7.5. Bloch electrons in infinite dimensions

The work presented in this section is based on Ref. 31 where the original solution for Bloch electrons in a field was given. There the work focused on the Hamiltonian gauge, here we discuss the gauge-invariant formalism.

Bloch[35] and Zener[36] originally showed that when electrons are placed on a perfect lattice, with no scattering, the current oscillates due to Bragg reflection of the wavevector as it evolves to the Brillouin-zone boundary. Here we show how to analyze this problem on the infinite-dimensional hypercubic lattice. The noninteracting problem can be solved exactly in the case of an arbitrary time-dependent electric field. In particular, the noninteracting contour-ordered Green function is (in the Hamiltonian gauge[31]):

$$\begin{aligned} G^{c0}_{\mathbf{k}}(t_1,t_2) = \quad & i[f(\epsilon(\mathbf{k})-\mu)-\theta_c(t_1,t_2)]\exp[i\mu(t_1-t_2)] \\ & \times \exp\left[-i\int_{t_2}^{t_1} d\bar{t}\epsilon\left(\mathbf{k}-e\mathbf{A}(\bar{t})\right)\right], \end{aligned} \tag{7.32}$$

where $f[\epsilon(\mathbf{k})-\mu)] = 1/\{1+\exp[\beta(\epsilon(\mathbf{k})-\mu)]\}$ is the Fermi-Dirac distribution (half-filling corresponds to $\mu = 0$). The symbol $\theta_c(t_1,t_2)$ is equal to one if t_1 lies after t_2 on the contour, and is zero otherwise. Note that the Green function in Eq. (7.32) is also used in the system of equations (7.14)-(7.17) to solve the interacting problem.

When we have a constant electric field directed along the diagonal and turned on at $t = 0$, each component of the vector potential satisfies $A(t) = -Et\theta(t)$. Then the integral that appears in the exponent of Eq. (7.32) is

$$\begin{aligned} & \theta(-|t/2|-t_{ave})\epsilon(\mathbf{k})t \\ & +\theta(-t/2-|t_{ave}|) \\ & \times\left[\frac{\epsilon(\mathbf{k})(\sin eE(t_{ave}-t/2)+t_{ave}+t/2)-\bar{\varepsilon}(\mathbf{k})(\cos eE(t_{ave}-t/2)-1)}{eE}\right] \\ & +\theta(t/2-|t_{ave}|) \\ & \times\left[\frac{\epsilon(\mathbf{k})(\sin eE(t_{ave}+t/2)-t_{ave}+t/2)+\bar{\varepsilon}(\mathbf{k})(\cos eE(t_{ave}+t/2)-1)}{eE}\right] \\ & +\theta(t_{ave}-|t/2|)\Big[\epsilon(\mathbf{k})(\sin eE(t_{ave}+t/2) \\ & \quad -\sin eE(t_{ave}-t/2))+\bar{\varepsilon}(\mathbf{k})(\cos eE(t_{ave}+t/2)-\cos eE(t_{ave}-t/2))\Big]\Big/eE \end{aligned} \tag{7.33}$$

when expressed in terms of the Wigner coordinates.

To get the gauge-invariant Green function, we now shift the momentum as shown in Eq. (7.31); note that the shift is done both for the momentum in the exponent, and for the momentum in the Fermi-Dirac distribution. Two of the four cases for the exponent in Eq. (7.33) are easy to work out for the gauge-invariant Green functions. The first is the $\theta(-|t/2|-t_{ave})$ term which remains unchanged and the second is the $\theta(-t/2-|t_{ave}|)$ term, which becomes $2\epsilon(\mathbf{k})\sin(eEt/2)$.

Note that both these exponents are independent of the average time. The average time enters for the other two terms, and in the argument of the Fermi-Dirac distribution.

The retarded $G_{\mathbf{k}}^{R0}(t_1, t_2)$ and lesser $G_{\mathbf{k}}^{<0}(t_1, t_2)$ Green functions can be obtained from Eq. (7.32), by replacing the prefactor $[f(\epsilon(\mathbf{k}) - \mu) - \theta_c(t_1, t_2)]$ by $-\theta(t_1 - t_2)$ and $f(\epsilon(\mathbf{k}) - \mu)$, correspondingly. One needs to shift the momentum accordingly to get the gauge-invariant retarded and lesser Green functions. Note that at long times the gauge-invariant retarded Green function depends only on relative time.

Since the electrical current is found from the time derivative of the polarization operator, the current operator is determined by taking the commutator of the Hamiltonian (in a particular gauge) with the polarization operator. The result, for the αth component of the current-density operator is

$$\mathbf{j}_\alpha(t_{ave}) = e \sum_{\mathbf{k}} \frac{\partial \epsilon(\mathbf{k} - e\mathbf{A}(t_{ave}))}{\partial k_\alpha} c_{\mathbf{k}}^\dagger(t_{ave}) c_{\mathbf{k}}(t_{ave}), \tag{7.34}$$

where we have emphasized that the operator is evaluated with a vanishing relative time ($t = 0$). We want the expectation value of the current operator, which is found by taking expectation value of the expression in Eq. (7.34) and noting that each component gives the same result for a field pointing along the diagonal. The expectation value of the product $c_{\mathbf{k}}^\dagger(t_{ave}) c_{\mathbf{k}}(t_{ave})$ can be replaced by the lesser Green function $G_{\mathbf{k}}^<(t_{ave}, 0)$. So we have

$$j(t_{ave}) = ie \sum_{\mathbf{k}} \bar{\varepsilon}\left(\mathbf{k} - e\mathbf{A}(t_{ave})\right) G_{\mathbf{k}}^<(t_{ave}, 0) = ie \sum_{\mathbf{k}} \bar{\varepsilon}\left(\mathbf{k}\right) \tilde{G}_{\mathbf{k}}^<(t_{ave}, 0), \tag{7.35}$$

where the second equality comes from the transformation to the gauge-invariant Green function. The summation over momentum can be converted to a double integral over the joint density of states in Eq. (7.13). Substituting in the expression for the lesser Green function, yields

$$j(t_{ave}) = \frac{et^*}{4\sqrt{d}\pi} \sin(eEt_{ave}) \int d\epsilon \frac{df(\epsilon - \mu)}{d\epsilon} \rho(\epsilon), \tag{7.36}$$

where the single-particle density of states is

$$\rho(\epsilon) = \int d\bar{\varepsilon} \rho_2(\epsilon, \bar{\varepsilon}) = \frac{1}{\sqrt{\pi}} \exp\left[-\epsilon^2\right]. \tag{7.37}$$

The current is a periodic function of time, even though the field is time-independent; this effect is called a Bloch oscillation.[37] The period of the oscillation is equal to $2\pi/eE$. In order to see this oscillation in real solids, one needs to prepare a system where the scattering time is longer than the period of the

Bloch oscillations. In solids, the scattering time is much shorter than the oscillation period, so this effect is not observed. However, Bloch oscillations are seen in semiconductor superlattices, where the period of oscillations is much shorter due to the larger lattice spacing. As we show in the following Sections, the effects of strong electron-electron correlations modify the Bloch oscillations significantly, but the driven oscillations in large fields survive for a surprisingly long time.

Now we discuss the time-dependence of the density of states (DOS) for non-interacting electrons in a constant electric field. The DOS is found by using the Wigner time coordinates in Eq. (7.21), and making a Fourier transformation of the corresponding Green functions with respect to the relative time coordinate. In particular, the local DOS is

$$A(t_{ave},\omega) = -\frac{1}{\pi}\mathrm{Im}\int_0^\infty dt e^{i\omega t} G^R_{loc}(t_{ave},t). \tag{7.38}$$

Since this is a local quantity, summed over all momenta, it is automatically gauge-invariant. The local retarded Green function can be obtained from Eq. (7.32). It is possible to show[31] that the steady state ($t_{ave} \to \infty$), for the case of a constant electric field turned on at $t = 0$, has a retarded Green function which satisfies

$$G^R_{loc}(t_{ave} \to \infty, t) = -i\theta(t)\exp\left[\frac{1}{2e^2E^2}\left\{\cos(eEt) - 1\right\}\right]. \tag{7.39}$$

Substitution of this expression into Eq. (7.38), and evaluating the Fourier transform with respect to the relative time, yields the steady state DOS, which consists of a set of delta-functions with different amplitudes (called the Wannier-Stark ladder[38]). The distance between the delta-function peaks is equal to eE. The weight of these peaks is:[31]

$$w_N = \frac{2}{e^2E^2}\int_0^{2\pi} du \cos(Nu)\exp\left(\frac{t^{*2}}{2e^2E^2}[\cos u - 1]\right), \tag{7.40}$$

for the Nth Bloch frequency, $\omega_N = eEN$. It takes an infinite amount of time for the delta functions to develop.

In Fig. 7.2, we show how the DOS evolves from the time the field is turned on, at $t_{ave} = 0$, to a large time. The DOS remains Gaussian for $t_{ave} < -2$ and then develops large oscillations as t_{ave} increases. Though the DOS oscillates and acquires negative values in the transient regime, it is possible to show (numerically) that its first three moments always satisfy the relevant sum rules.

7.6. Exact solution

In this Section we present the results for the interacting case,[6,8,9] where we vary the Coulomb repulsion through the metal-insulator transition that occurs at $U_c = \sqrt{2}$.

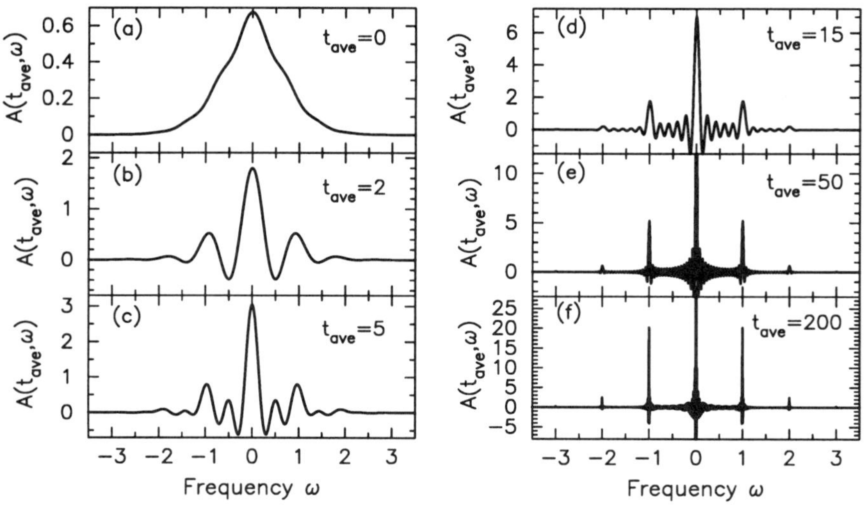

Fig. 7.2. Density of states for noninteracting electrons with $eE = 1$ at different values of time t_{ave} (be aware that the vertical scale changes from plot to plot). Note how the build up of the delta function at the Bloch frequencies is slow.

The problem is solved by numerically solving the DMFT loop in Eqs. (7.14)-(7.17).

Once the Green functions and self-energy have been found by self-consistently solving the DMFT equations, we can extract the momentum-dependent lesser Green function and use Eq. (7.35) to find the current. In these calculations, we used the Green functions in a particular gauge, but one could easily shift to the gauge-invariant Green functions if desired. When the field is small, and the correlations are small, we see a damping of the Bloch oscillations, as expected. This is shown in the left panel of Fig. 7.3. One can see that the Bloch oscillations maintain their periodicity, but are damped as the scattering increases. As we start to approach the metal-insulator transition at $U \approx 1.414$, one can see the character of the oscillations start to change. As we move into the insulating phase, as shown in the right panel, the character of the oscillations changes completely, and we no longer see the regular Bloch structure. The oscillations seem to survive to much longer times than would be expected from a Boltzmann equation type of analysis. It remains unclear whether the steady state has some residual oscillations, or it goes to a constant value as predicted by semiclassical ideas.

Even more surprising is the fact that when the field is large, the current displays two anomalous features: (i) first, its decay is much slower than expected

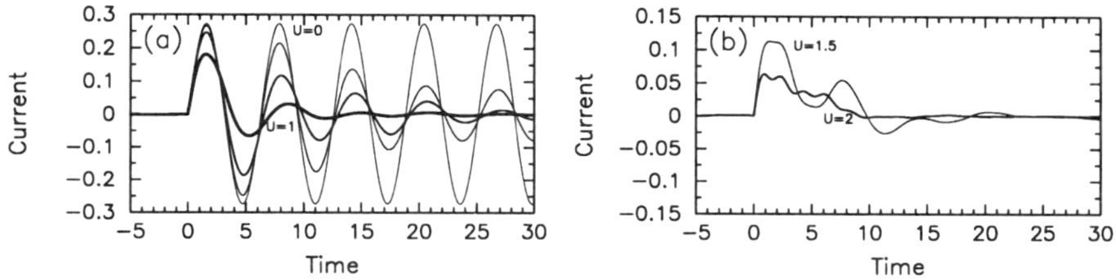

Fig. 7.3. Electric current for different values of U with $\beta = 10$: (a) metals ($U = 0$, $U = 0.25$, $U = 0.5$, and $U = 1$) and (b) insulators ($U = 1.5$ and $U = 2$).

from a semiclassical approach, where the relaxation time is inversely proportional to the imaginary part of the self-energy at the chemical potential, which is proportional to $1/U^2$ and (ii) the current develops beats with a beat frequency proportional to $1/U$. An example of this behavior is shown in Fig. 7.4. These beats are always present in the metallic phases (for large U), but disappear once one moves into the insulator.

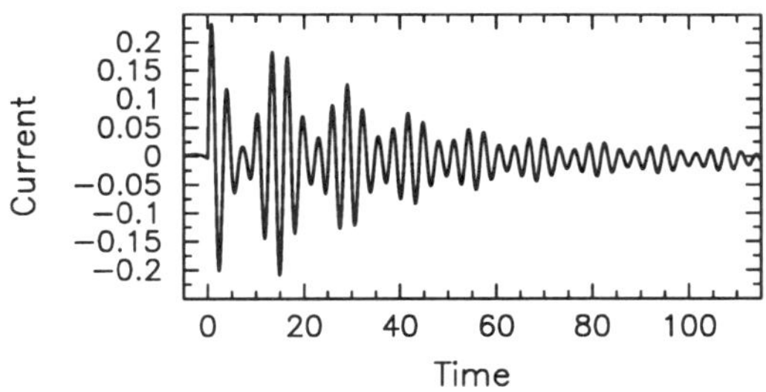

Fig. 7.4. Time-dependence of the current for $U = 0.5$, $E = 2.0$, and $\beta = 10$. Note how the current has beats in its time dependence and that the decay of the current is rather slow.

The time-dependence of the density of states can be calculated from Eq. (7.38). Here, we present some results for the case when the system is initially in the metallic phase, (Fig. 7.5). What we find is that for small fields, the delta function peaks of the Wannier-Stark ladder get broadened, but the structure is still readily apparent. But as we increase the field strength, the behavior qualitatively changes, and in the long-time limit, the system evolves into a peaked structure, where the peaks are maximal near the edges of minibands, which are spaced apart in size by U, and the DOS has a local minimum in the center, where the Wannier-Stark peak used to appear. This is also behavior that is quite surprising.

As the scattering increases, the DOS approaches the steady state value rela-

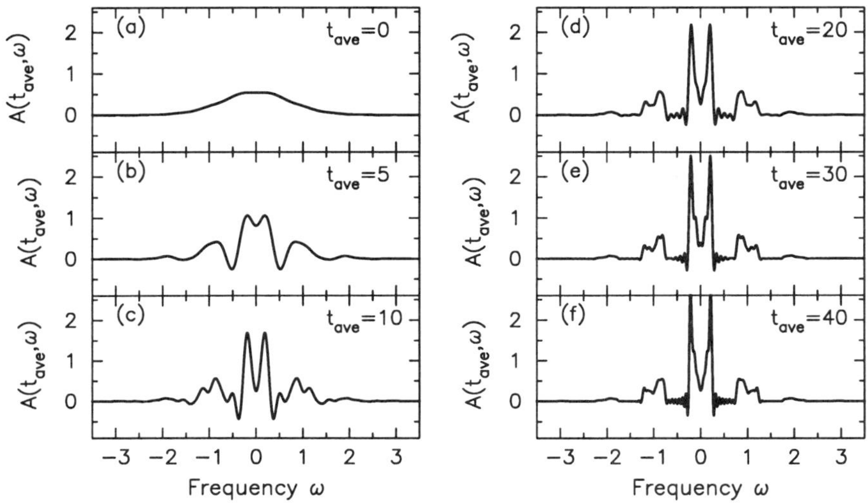

Fig. 7.5. Density of states for different average times from $t_{ave} = 0$ to $t_{ave} = 40$ for $U = 0.5$, $E = 1$ and $\beta = 10$. Note how the DOS develops split peaks, separated by $U = 0.5$ around the Bloch frequencies (integers here).

tively quickly. This illustrates the dichotomy between the average time, which is important for determining the current, and the relative time, which is important for determining the DOS. The decay is rapid as a function of relative time, but is much slower as a function of average time.

7.7. Perturbation theory

A perturbative analysis can be performed directly on the lattice.[40] In this case, we do not need any DMFT loop, and we can restrict the contour to be solely on the real axis. As described above, the perturbation theory is similar for the equilibrium and nonequilibrium cases, with the only significant changes being that one needs to calculate with time-ordered objects along the contour and one needs to use noninteracting Green functions in the field. A strictly truncated expansion for the self-energy to second order in U is equal to the usual Hartree-Fock term (which vanishes at half filling) plus a second-order term which satisfies

$$\Sigma^{c(2)}(t_1, t_2) = U^2 w_1(1 - w_1) G^{c0}_{loc}(t_1, t_2); \tag{7.41}$$

one can determine the retarded and lesser self-energies from this in a straightforward fashion. It turns out that this truncated perturbation theory is most accurate at short times—in essence, the perturbation series expansion is an expansion in a power series in time away from the time the field was turned on. In more conventional perturbation series in terms of frequency-dependent Green functions, the perturbation series is most accurate at high frequencies, and least accurate at low frequencies. After performing a Fourier transform, one can immediately see that this is equivalent to the perturbation theory being most accurate at short times and breaking down at long times. Indeed, we find that this perturbative treatment cannot reproduce the steady-state behavior at long times.

As a benchmark of our calculation, we compare the equilibrium self-energy found from a numerical solution of the DMFT equations to the perturbative result for small U. We find quite good agreement in the small-U range, and the equilibrium case appears to be fairly accurate up to $U \approx 0.5$. For larger U values the perturbation theory breaks down—it is not capable of properly describing the Mott-insulating phase.

Next we analyze the time-dependence of the electric current calculated by second order perturbation theory in the case when a constant electric field is turned on at time $t = 0$. Before presenting the second-order perturbative solution for the current, we briefly review the corresponding results from a semiclassical Boltzmann equation approximation. As was mentioned above, these results are qualitatively different from the exact solution.

In the Boltzmann equation approach, one introduces a nonequilibrium quasiparticle distribution function $f^{\mathrm{non}}(\mathbf{k}, t) = -iG^{<}_{\mathbf{k}}(t, t)$, which satisfies the following phenomenological equation:

$$\frac{\partial f^{\mathrm{non}}(\mathbf{k}, t)}{\partial t} + e\mathbf{E}(t) \cdot \nabla_{\mathbf{k}} f^{\mathrm{non}}(\mathbf{k}, t) = -\frac{1}{\tau}[f^{\mathrm{non}}(\mathbf{k}, t) - f(\mathbf{k})], \tag{7.42}$$

with the boundary condition:

$$f^{\mathrm{non}}(\mathbf{k}, t = 0) = f(\mathbf{k}) = \frac{1}{\exp[\beta(\epsilon(\mathbf{k}) - \mu)] + 1}. \tag{7.43}$$

This equation can be solved exactly (see, for example Ref. 40). Substitution of this expression instead of $-iG^{<}$ into Eq. (7.35) allows one to calculate the semiclassical current. This semiclassical current approaches a steady state as time goes to infinity. In particular, in the infinite-dimensional limit, one obtains:

$$\begin{aligned} j(t) = &-\frac{e}{\sqrt{d}} \frac{eE\tau}{1 + e^2E^2\tau^2} \int d\epsilon \rho(\epsilon) \epsilon f(\epsilon) \\ &\times \left[1 - (\cos(eEt) - eE\tau \sin(eEt))\, e^{-t/\tau}\right]. \end{aligned} \tag{7.44}$$

Therefore, the current is a strongly oscillating function of time for $t \ll \tau$, and it approaches the steady-state value

$$j^{\mathrm{steady}} = \frac{eE\tau}{1 + e^2E^2\tau^2} j_0, \tag{7.45}$$

where

$$j_0 = -\frac{e}{\sqrt{d}} \int d\epsilon \rho(\epsilon) \epsilon f(\epsilon), \tag{7.46}$$

as $t/\tau \to \infty$. The steady-state current amplitude is proportional to E in the case of a weak field (the linear-response regime), and then becomes proportional to $1/E$ at $eE\tau \to \infty$. The amplitude of the current goes to zero in this nonlinear regime with the field amplitude increasing. One would naively expect that the second-order perturbation theory would give similar results in the case of a weak Coulomb repulsion, since one can extract an effective scattering time for the equilibrium limit of the Falicov-Kimball model with small U: $\tau = 1/(\pi^2 U^2)$.[39] However, it will be shown below, that the behavior of the current calculated in second-order perturbation theory is rather different from the Boltzmann equation case, and closer to the exact numerical result at short times.

The electric current in the second-order perturbation theory can be calculated by substituting the expression for the second-order lesser Green function into Eq. (7.35). In this case

$$j(t) = \frac{ie}{\sqrt{d}} \int d\epsilon \int d\bar{\varepsilon} \rho_2(\epsilon, \bar{\varepsilon}) \left[\bar{\varepsilon} \cos\left(eA_\alpha(t)\right) - \epsilon \sin\left(eA_\alpha(t)\right)\right] G^<_{\epsilon,\bar{\varepsilon}}(t,t). \tag{7.47}$$

It is difficult to find exact analytical expressions for the current, except for some limiting cases. Of course, in the limit $U = 0$ we recover the free electron case result:

$$j(t) = j_0 \sin\left(eEt\right), \tag{7.48}$$

where the amplitude j_0 of the Bloch oscillations is given by Eq. (7.46). Therefore, the general expression for the time-dependence of the electric current in a strictly truncated second-order perturbation theory expansion can be written as:

$$j(t) = j_0 \sin\left(eEt\right) + U^2 j_2(t). \tag{7.49}$$

The electric current is a superposition of an oscillating part and some other piece proportional to U^2. Obviously this cannot produce a constant steady-state current for all small U's, because the function j_2 is independent of U. This is a clear indication that the perturbation theory will hold only for short times.

Numerical results for the time-dependence of the electric current calculated from Eq. (7.47) at $eE = 1$ and different values of U are presented in Fig. 7.6 (dashed lines). Note how the current oscillates for all times within our finite time window. We also show the corresponding Boltzmann equation solution and the exact solution. We use two different values for the Boltzmann equation—one

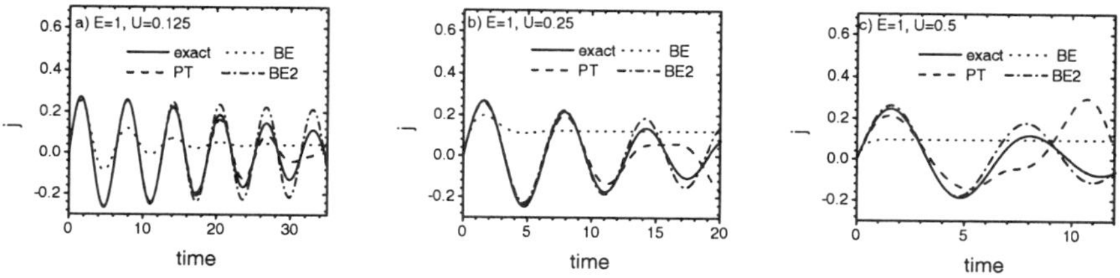

Fig. 7.6. Perturbative expansion for the electric current as a function of time for $E = 1.0, \beta = 10$ and different values of U (dashed lines). The solid and dotted lines correspond to the exact DMFT solution and the Boltzmann equation (BE) solution, respectively. The dash-dotted lines (BE2) are the Boltzmann equation result with a phenomenological relaxation time $\tau = \alpha/(\pi^2 U^2)$; $\alpha = 20$ in Figs. a) and b), and $\alpha = 36.5$ in Fig. c).

fixes the relaxation time to the prediction from the equilibrium solution, while the other adjusts the relaxation time to obtain the best fit. Comparison of the perturbation theory and the Boltzmann equation solution shows that they are close at short times, but at longer times the PT current remains oscillating, while the Boltzmann equation solution approaches a steady state. Moreover, at times longer than $\sim 2/U$ the perturbation theory breaks down showing an oscillating current with increasing amplitude. At times shorter than $2/U$ the perturbation theory solution is close to the exact result, displaying an oscillating current with decreasing amplitude. It is also possible to fit the Boltzmann equation results to the exact and PT solution at short times if one chooses the relaxation time $\tau = \alpha/(\pi^2 U^2)$, where $\alpha \sim 20 - 30$, which is much larger than $\alpha = 1$ in the case of the second order perturbation theory.[39] These results clearly show that the semiclassical approach, with one effective time variable that is damped on the timescale of the relaxation time is not sufficient to describe the behavior in the quantum case.

The perturbation theory calculations at different values of the electric field give results similar to the results presented in Fig. 7.6. Numerical analysis shows that the agreement between the perturbation theory results and the exact results is better when eE is larger than U. In fact, it is possible to find an analytical expression for the current in the case of large electric fields:

$$j(t) \simeq -\frac{e}{\sqrt{d}} \int d\epsilon \rho(\epsilon) \epsilon f(\epsilon) \left[1 - U^2 B(\beta) - \frac{U^2}{4} t^2 \right] \sin(eEt), \qquad (7.50)$$

where $B(\beta)$ is a positive decreasing function of temperature:[40] $0.25 < B < 0.5$.

7.8. Conclusions

To conclude, we have presented some results on the nonequilibrium properties of the Falicov-Kimball model of strongly correlated electrons in the limit of infinite dimensions. Despite the simplicity of the model, the solutions show that strong electron-electron correlations result in nontrivial behavior. The dynamical mean-field theory approximation is believed to be a precise method to solve strongly correlated problems in three dimensions. Therefore, we believe that some of the results, like the long-time Bloch oscillations of the current, beats in the current for strong fields and the splitting of the Wanier-Stark peaks in the DOS could be observed in bulk systems with dominant electron-electron scattering in the presence of a strong electric field. Such a field can be present, in particular, in nanostructures, where a moderate external electric potential can produce strong (uniform) electric fields due to the small size of the systems. One might also be able to observe this behavior in mixtures of heavy and light atoms trapped in pulled optical lattices. In addition, we demonstrated that the perturbation theory solution cannot be used to study the long-time behavior of the system. It would be interesting to generalize these results to more complicated models and to lower dimensions, where we expect qualitatively similar behavior.

Acknowledgments

We thank Antti-Pekka Jauho, Alexander Joura, Joseph Serene and Veljko Zlatić for valuable discussions. We would like to acknowledge support by the National Science Foundation under grant number DMR-0210717 and by the Office of Naval Research under grant number N00014-05-1-0078. Supercomputer time was provided by the DOD HPCMO at the ASC and ERDC centers (including a 2006 CAP project) and by a National Leadership Computing System grant from NASA.

References

1. U. Brandt and C. Mielsch, Thermodynamics and correlation functions of the Falicov-Kimball model in large dimensions, *Z. Phys. B: Condens. Matter* **75**, 365–370 (1989).
2. W. Metzner and D. Vollhardt, Correlated lattice Fermions in $d = \infty$ dimensions, *Phys. Rev. Lett.* **62**, 324–327 (1989).
3. A. Georges, G. Kotliar, W. Krauth, and M. J. Rozenberg, Dynamical mean-field theory of strongly correlated fermion systems and the limit of infinite dimensions, *Rev. Mod. Phys.* **68**, 13–125 (1996).

4. G. Kotliar, S. Y. Savrasov, K. Haule, V. S. Oudovenko, O. Parcollet, and C. A. Marianetti, Electronic structure calculations with dynamical mean-field theory, *Rev. Mod. Phys.* **78**, 865–951 (2006).
5. R. Bertoncini and A. P. Jauho, Gauge-invariant formulation of the intracollisional field effect including collisional broadening, *Phys. Rev. B* **44**, 3655-3664 (1991).
6. J. K. Freericks, V. M. Turkowski, and V. Zlatić, Real-time formalism for studying the nonlinear response of "smart" materials to an electric field, *Proceedings of the HPCMP Users Group Conference 2005, Nashville, TN, June 28–30, 2005*, edited by D. E. Post (IEEE Computer Society, Los Alamitos, CA, 2005), pp. 25–34.
7. V. M. Turkowski and J. K. Freericks, Spectral moment sum rules for strongly correlated electrons in time-dependent electric fields, *Phys. Rev. B* **73**, 075108 (2006); Erratum: *Phys. Rev. B* **73**, 209902(E) (2006).
8. J. K. Freericks, V. M. Turkowski, and V. Zlatić, Nonequilibrium dynamical mean-field theory, *Phys. Rev. Lett.* **97**, 266408 (2006).
9. J. K. Freericks, V. M. Turkowski, and V. Zlatić, Nonlinear response of strongly correlated materials to large electric fields, in *Proceedings of the HPCMP Users Group Conference 2006, Denver, CO, June 26–29, 2006*, edited by D. E. Post (IEEE Computer Society, Los Alamitos, CA, 2006), *to appear*.
10. U. Brandt and C. Mielsch, Thermodynamics of the Falicov-Kimball model in large dimensions II, *Z. Phys. B: Condens. Matter* **79**, 295–299 (1990).
11. U. Brandt and C. Mielsch, Free energy of the Falicov-Kimball model in large dimensions, *Z. Phys. B: Condens. Matter* **82**, 37–41 (1991).
12. M. Jarrell, Hubbard model in infinite dimensions: A quantum Monte Carlo study, *Phys. Rev. Lett.* **69**, 168–171 (1992).
13. M. J. Rozenberg, X. Y. Zhang, and G. Kotliar, Mott-Hubbard transition in infinite dimensions, *Phys. Rev. Lett.* **69**, 1236–1239 (1992).
14. R. Bulla, Zero temperature metal-insulator transition in the infinite-dimensional Hubbard model, *Phys. Rev. Lett.* **83**, 136–139 (1999).
15. M. Jarrell, Symmetric periodic Anderson model in infinite dimensions, *Phys. Rev. B* **51**, 7429-7440 (1995) .
16. Th. Pruschke, R. Bulla, and M. Jarrell, Low-energy scale of the periodic Anderson model, *Phys. Rev. B* **61**, 12799–12809 (2000).
17. J. K. Freericks, M. Jarrell, and D. J. Scalapino, Holstein model in infinite dimensions, *Phys. Rev. B* **48**, 6302–6314 (1993).
18. D. Meyer, A. C. Hewson, and R. Bulla, Gap formation and soft phonon mode in the Holstein model, *Phys. Rev. Lett.* **89**, 196401 (2002).
19. J. K. Freericks and V. Zlatić, Exact dynamical mean field theory of the Falicov-Kimball model, *Rev. Mod. Phys.* **75**, 1333–1382 (2003).
20. P. Schmidt and H. Monien, Nonequilibrium dynamical mean-field theory of a strongly correlated system, *preprint cond-mat/0202046*.
21. L. M. Falicov and J. C. Kimball, Simple model for semiconductor-metal transitions: SmB_6 and Transition-Metal Oxides, *Phys. Rev. Lett.* **22**, 997–999 (1969).
22. T. Kennedy and E. H. Lieb, An itinerant electron model with crystalline or magnetic long range order, *Physica A* **138**, 320–358 (1986).
23. L. P. Kadanoff and G. Baym, *Quantum Statistical Mechanics* (Benjamin, New York, 1962).

24. L. V. Keldysh, Diagram technique for nonequilibrium processes, *Zh. Eksp. Teor. Fiz.* **47**, 1515–1527 (1964) [*Sov. Phys.-JETP* **20**, 1018–1026 (1965)].
25. J. Rammer and H. Smith, Quantum field-theoretical methods in transport theory of metals, *Rev. Mod. Phys.* **58**, 323–359 (1986).
26. M. Wagner, Expansions of nonequilibrium Green's functions, *Phys. Rev. B* **44**, 6104–6117 (1991).
27. W. Metzner, Linked-cluster expansion around the atomic limit of the Hubbard model, *Phys. Rev. B* **43**, 8549–8563 (1991).
28. D. C. Langreth, Linear and non-linear response theory with applications, in *Linear and Nonlinear Electron Transport in Solids*, NATO Advanced Study Institute Series B, Vol. **17**, edited by J. T. Devreese and E. van Doren (Plenum, New York/London, 1976), p. 3–32.
29. D. O. Demchencko, A. V. Joura, and J. K. Freericks, Effect of particle-hole asymmetry on the Mott-Hubbard metal-insulator transition, *Phys. Rev. Lett.* **92**, 216401 (2004).
30. R. E. Peierls, On the theory of the diamagnetism of conduction electrons, *Z. Phys.* **80**, 763–791 (1933); A. P. Jauho and J. W. Wilkins, Theory of high-electric-field quantum transport for electron-resonant impurity systems, *Phys. Rev. B* **29**, 1919–1938 (1984).
31. V. Turkowski and J. K. Freericks, Nonlinear response of Bloch electrons in infinite dimensions, *Phys. Rev. B* **71**, 085104 (2005).
32. P. Schmidt, Time-dependent dynamical mean-field theory, *Diplome thesis*, University of Bonn (1999).
33. E. Wigner, On the quantum correction for thermodynamic equilibrium, *Phys. Rev.* **40**, 749–759 (1932).
34. H. Haug and A.-P. Jauho, *Quantum Kinetics in Transport and Optics of Semiconductors* (Springer-Verlag, Berlin/Heidelberg, 1996).
35. F. Bloch, On the quantum mechanics of electrons in crystal lattices, *Z. Phys.* **52**, 555–600 (1928).
36. C. Zener, A theory of the electrical breakdown of solid dielectrics, *Proc. R. Soc. (London) Ser. A* **145**, 523–529 (1934).
37. N. W. Ashcroft and N. D. Mermin, *Solid State Physics* (Holt, Rinehart and Winston, New York, 1976).
38. G. H. Wannier, Dynamics of band electrons in electric and magnetic fields, *Rev. Mod. Phys.* **34**, 645–655 (1962).
39. J. K. Freericks and V. M. Turkowski, Steady state nonequilibrium dynamical mean-field theory and the quantum Boltzmann equation, *J. Phys.: Confer. Ser.* **35**, 39–52 (2006).
40. V. Turkowski and J.K. Freericks, Nonequilibrium perturbation theory of the spinless Falicov-Kimball model, *preprint, cond-mat/0610640*.

PART 2
Strongly Correlated Magnetic Systems

Chapter 8

Introduction of effective interactions in Real Space Renormalization Group techniques

Mohamad Al Hajj, Nathalie Guihéry and Jean Paul Malrieu

Laboratoire de Chimie et Physique Quantiques, IRSAMC/UMR5626, Université Paul Sabatier, 118 route de Narbonne, F-31062 Toulouse cedex 4, FRANCE

It is tempting to apply the elegant proposal of the Real Space Renormalization Group (RSRG) method to the treatment of heavy fermion lattices. The method proceeds through the definition of blocks and a truncation of the Hilbert space to products of the lowest eigenstates of the blocks. As such the method is not accurate enough. A great improvement is obtained when one defines effective interactions (instead of the bare ones) between the products of the selected block eigenstates. These effective interactions are obtained from the exact treatment of pairs or trimers of blocks through the effective Hamiltonian theory of Bloch. If the blocks and the effective inter-block Hamiltonian maintain an isomorphism between the block lattice and the starting one, the exponential scale change may be iterated to convergence. The qualitative and quantitative potentialities of the method are illustrated on a series of one-dimensional (1D) and two-dimensional (2D) spin lattices (non frustrated and frustrated), concerning the cohesive energies, the localization of quantum phase transitions and the spin gaps.

Contents

8.1. Introduction

The research of efficient methods for the study of the low energy solutions of model Hamiltonians for extended systems of strongly correlated fermions remains a subject of intensive efforts in the community of Solid State theoreticians. When periodic lattices are concerned many methods go through the treatment of finite clusters and the extrapolation on the size of the clusters. While this last step is reliable for 1D systems it becomes rather hazardous for 2D and 3D systems. The present chapter intends to present a method which starts with the exact diagonalization of finite clusters but tends to approach the energy of the periodic system through an iterative change of scale, in the spirit of the Real Space Renormalization Group. It differs from the original suggestion of Wilson[1] by a systematic use of the Theory of Effective Hamiltonians, in order to renormalize the interactions appearing during the successive steps of the treatment. The original proposal by K. G. Wilson consists in an iterative truncation of the Hilbert space, concentrating on the lowest eigenstates of the problem, at each step. It proceeds through a succession of group-renormalization transformations and the elimination of the degrees of freedom which are not important for the low-energy physics of the lattice. The original version of the method is extremely elegant, since the iterative process converges on physically relevant accumulation points which may indicate the localization of a quantum phase transition when the parameters of interactions vary, or the closure of an excitation gap at the thermodynamic limit. The RSRG furnishes a satisfactory treatment of the single-impurity Kondo problem but fails to reproduce the properties of numerous spin or electron lattices.[2,3] The reason of this failure has been attributed[4] to the use of finite clusters with fixed bondary conditions. Although this attribution is questionable, it has led to the proposition of several solutions, among which the Density Matrix Renormalization Group (DMRG)[4–6] appeared as the most efficient one. Its efficiency for the treatment of 1D systems is remarkable but it can hardly be applied to 2D lattices (and even to thick quasi-1D systems).

The method[7] which we shall present does not reach the same accuracy as DMRG for 1D systems but it is applicable to 2D and 3D lattices. It keeps the elegance of the original idea of Wilson but greatly improves its efficiency. It is closely related with the so-called Contractor Renormalization (CORE) method,[8–11] which also defines effective interactions between selected eigenstates of finite clusters. The CORE method has received numerous applications during the last years, but it deviates from the RSRG spirit when it is stopped at the first iteration as it is the case in most applications. Let us present the RSRG method and the introduction of effective interactions in its scheme.

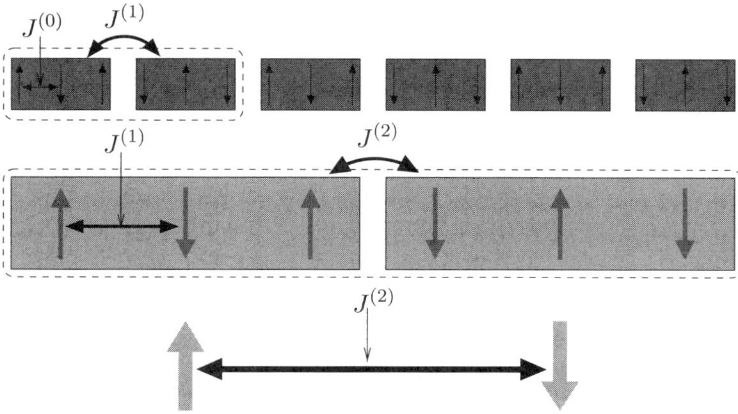

Fig. 8.1. Schematic presentation of the renormalisation with effective interactions.

8.2. The method

8.2.1. *General presentation*

The Fig. 8.1 presents a simple illustration of the first iterations of the RSRG method with effective interactions (RSRG-EI) in the problem of an anti-ferromagnetic (AF) chain with a spin coupling J between nearest neighbor sites. The process begins with the subdivision of a finite site cluster, hereafter called superblock (SB), into a small number of smaller clusters, the blocks, involving an odd number N_s of sites. For a sake of simplicity the blocks will be taken identical, but this is not complulsory. In the Fig. 8.1 $N_s = 3$ and the superblock contains 2 blocks. In the next step we shall only consider the ground state eigen-functions of each block. This will lead to a drastic truncation of the Hilbert space. The ground state of the block is a non-degenerate doublet ($M_s = \pm 1/2$). The method consists in considering the block as a supersite possessing a superspin of $M_s = \pm 1/2$. It then becomes possible to treat, for the same computational effort as at the first step, a superblock of N_s supersites, i.e., $2 \times N_s^2$ (i.e., 18 in our example) elementary sites. When iterating the procedure the system size increases exponentially to infinity. This is the specification of the original idea of Wilson. But instead of calculating directly the interactions between the functions of the truncated Hilbert space through the true Hamiltonian, i.e., the integrals $J^{(0)}$ between the sites connecting two blocks, we shall evaluate an effective Hamiltonian ruling the interactions between the supersites, according to a procedure which may be applied at each iteration. This procedure exploits the spectrum of the superblock and implicitely takes into account the effect of the excited states of the blocks which have been kept out of the truncated Hilbert space.

8.2.2. *Choice of the target and model spaces*

Before starting the procedure one must identify the states of the superblocks from which one hopes to describe the desired physics. These states must belong to the so-called target space of the effective Hamiltonian extraction procedure. The Hamiltonian matrix for the superblock is diagonalized through the Lanczos[12] or the Davidson[13] methods. N_m eigenvectors $\Psi_m^{(0)}$ of the superblock, usually those of lowest energies,

$$H_{SB} \mid \Psi_m^{(0)}\rangle = E_m^{(0)} \mid \Psi_m^{(0)}\rangle, \tag{8.1}$$

where (0) concerns the iteration number, will define the target space, hereafter labelled S. The superblock is then divided in a small number N_B of blocks, which are treated exactly

$$H_B|\phi_i^{(0)}\rangle = e_i^{(0)}|\phi_i^{(0)}\rangle. \tag{8.2}$$

The N eigenstates $\phi_i^{(0)}$ of lowest energies will be kept to build the the truncated Hilbert space of the superblock. This space, called model space and labelled S_0, is spanned by the direct products of the N eigunfunctions $\phi_i^{(0)}$ of each block.

$$S_0 = \{\Phi_m^{(0)}\} = \{\bigotimes_i \phi_i^{(0)}\}, \qquad dim(S_0) = N^{N_B} = N_m. \tag{8.3}$$

The isodimensionality of S and S_0 is required to establish a bi-univocal correspondance between the target and the model spaces. The values of N_m, N_i and N_B must be chosen accordingly. If P_0 is the projector relative to the model space S_0 :

$$P_0 = \sum_{m=1}^{N_m} |\Phi_m^{(0)}\rangle\langle\Phi_m^{(0)}|. \tag{8.4}$$

The extraction of the effective Hamiltonian is only possible if the N_m projections $|\tilde{\Psi}_m^{(0)}\rangle$ of the vectors of the target space onto the model space

$$|\tilde{\Psi}_m^{(0)}\rangle = P_0|\Psi_m^{(0)}\rangle, \tag{8.5}$$

are non-zero and linearly independant. One may alternatively write P_0 :

$$P_0 = \sum_{m=1}^{N_m} |\tilde{\Psi}_m^{(0)}\rangle\langle(\tilde{\Psi}_m^{(0)})^{\perp}|, \tag{8.6}$$

where $(\tilde{\Psi}_m^{(0)})^{\perp}$ is the biorthogonal vector of $\tilde{\Psi}_m^{(0)}$. One must actually notice that the projections of the orthogonal eigenvectors of the target space onto the model space may be non-orthogonal. Their overlap matrix is

$$s_{mn} = \langle \tilde{\Psi}_m^{(0)} | \tilde{\Psi}_n^{(0)} \rangle, \tag{8.7}$$

and the bi-orthogonal vector is :

$$|(\tilde{\Psi}_m^{(0)})^{\perp}\rangle = s^{-1} | \tilde{\Psi}_m^{(0)} \rangle. \tag{8.8}$$

The quality of the extracted effective Hamiltonian depends on the norm of these projections, which should be as large as possible to have a good correspondance between the model space and the target space.

8.2.3. *Extraction of the effective Hamiltonian and iteration of the procedure*

We make use of the formalism proposed by Bloch,[14] which entirely determines the effective Hamiltonian matrix from the N_m eigenvalues and eigenvectors of the exact Hamiltonian involved in the target space. The effective Hamiltonian is such that its eigenvalues are the eigenvalues of the exact Hamiltonian, and its eigenvectors are the projections of the corresponding exact eigenvectors onto the model space :

$$(H^{eff})^{(1)} | \tilde{\Psi}_m^{(0)} \rangle = E_m^{(0)} | \tilde{\Psi}_m^{(0)} \rangle. \tag{8.9}$$

The spectral decomposition of the effective Hamiltonian is given by :

$$(H^{eff})^{(1)} = \sum_{m=1}^{N_m} E_m^{(0)} | \tilde{\Psi}_m^{(0)} \rangle \langle (\tilde{\Psi}_m^{(0)})^{\perp} |. \tag{8.10}$$

The effective interactions are the matrix elements of $(H^{eff})^{(1)}$, defined by the preceding equation, between the model space basis vectors Φ_m^0. Notice that the effective Hamiltonian is not necessarily Hermitian. The slight modification suggested by des Cloizeaux[15] enables one to produce a Hermitian effective Hamiltonian.

It is possible to decompose the effective Hamiltonian (which will be used as the true Hamiltonian at the next iteration) in terms of intra-block and inter-block operators. If the superblock is composed of two blocks only one may write :

$$(H^{eff})^{(1)} = \sum_{i=1}^{N_B} (H_i^{eff})^{(1)} + \sum_{<ij>} (H_{ij}^{eff})^{(1)}, \tag{8.11}$$

where i and j label the blocks. Since the space is spanned by products of block eigenvectors, the first term is nothing but the sum of the eigenvalues of the blocks. The second term introduces two-body corrections to the diagonal elements and effective interactions between the blocks i and j. Of course if the superblock concerns three blocks three-block operators will be introduced in the effective Hamiltonian.

The lattice is now constituted of the blocks of the starting iteration. One may consider a superblock consisting in N_{SB} blocks. The Hamiltonian $H_{SB}^{(1)}$ is the previously obtained effective Hamiltonian and the indices run on the blocks or supersites. Returning to the logics of the first iteration one may define effective interactions between blocks of blocks, i.e., a new effective Hamiltonian ($H_{SB}^{(2)}$), and the procedure may be iterated to the convergence, where some of the interactions between the super/super/... blocks vanish.

Some important remarks have to be done here. If the blocks were built from an even number of sites and had a ground state singlet and if one only retains that state to define the model space, the first iteration would only produce scalar interactions and would result in an additive energy systematics in terms of block energies and block interactions.[8] In general the nature of the effective Hamiltonian may be different from that of the starting Hamiltonian, being either simpler or more complex, depending on the number of eigenstates retained for the blocks to define the model spaces . Similarly the connectivity between the blocks may be different from that between the sites of the initial lattice. In order to obtain a higher accuracy most of the applications of the CORE method usually retain several eigenstates per block and the effective Hamiltonian at the first iteration is rather complex and can only be solved once for a few-block cluster. On the contrary one may start from a rather complex Hamiltonian or a complex lattice and produce at the first step a simpler Hamiltonian and/or a simpler lattice. One may for instance start from an exact or a Hubbard Hamiltonian and produce a spin Hamiltonian, or start from a ladder and map it on a simple chain. This simplification is an important issue of the renormalization concept, but the most fascinating applications are those where the formal structure of the Hamiltonian and structural features of the lattice are kept along the iterations. In such a case the information obtained at the first iteration is sufficient to predict the evolution of the interactions along the iterations, the ratio of which will go to fixed values, defining fixed points or accumulation points. In some cases the iterations lead to summable analytic series. The next section will illustrate these potentialities on a series of spin lattices.

8.3. Illustrative applications

8.3.1. *Cohesive energies of the regular 1D spin chain and of the 2D square spin lattice*

The Heisenberg Hamiltonian is written in this subsection as :

$$H = \sum_{<ij>} 2J(\vec{S}_i \vec{S}_j - 1/4), \tag{8.12}$$

where $< ij >$ is a couple of nearest-neigbor sites. The magnetic coupling J is positive for AF systems. Compared to the usual expression $H = \sum_{<ij>} J\vec{S}_i\vec{S}_j$ our expression differs by a factor 2 and by a constant 1/4 which gives an energy zero to the ferromagnetic state.

For the 1D chain the blocks are linear fragments, have an odd number of sites and a doublet ground state. The superblock may be constituted of two adjacent blocks A and B. Its lowest eigenstates are the singlet ground state Ψ_S^{AB} and the first triplet excited state Ψ_T^{AB} of respective energies E_S^{AB} et E_T^{AB}. They constitute the target space. The model space is constituted of the products of the doublet ground states D^A and D^B of the blocks A and B, of energy $E_D^A = E_D^B$. Let us label a and b, and $\bar{a}$ and $\bar{b}$ the $M_s = 1/2$ and $M_s = -1/2$ respectively components of these doublets. The projector on S_0 is :

$$P_0 = |a\bar{b}\rangle\langle a\bar{b}| + |\bar{a}b\rangle\langle\bar{a}b|, \tag{8.13}$$

for the $S_z = 0$ subspace. Reducing the model space to the products of the ground state doublets of the blocks one assimilates these blocks to effective spins *up* and *down* at the next iteration. Hence the effective Hamiltonian at this step remains an Heisenberg Hamiltonian for a 1D chain,

$$(H^{eff})^{(1)} = \sum_{i=1}^{N_B}(H_i^{eff})^{(1)} + \sum_{<ij>}[B^{(1)} + 2J^{(1)}(\vec{S}_i\vec{S}_j - 1/4)], \tag{8.14}$$

where $J^{(1)}$ is the effective spin coupling : $J^{(1)} = \langle\bar{a}b|(H^{eff})^{(1)}|a\bar{b}\rangle$ and $B^{(1)} = \langle\bar{a}b|(H^{eff})^{(1)}|\bar{a}b\rangle - 2E_D^A + J^{(1)}$ is obtained from the energy of the ferromagnetic solution. The Hamiltonian matrix in the $a\bar{b}$, $\bar{a}b$ basis is :

$$(H^{eff})^{(1)} = \begin{pmatrix} 2E_D^A + B^{(1)} - J^{(1)} & J^{(1)} \\ J^{(1)} & 2E_D^A + B^{(1)} - J^{(1)} \end{pmatrix}. \tag{8.15}$$

From the eigenvalues E_S^{AB} and E_T^{AB} one may express the effective interactions $J^{(1)}$ and $B^{(1)}$:

$$E_S^{AB} = 2E_D^A + B^{(1)} - 2J^{(1)}, \tag{8.16}$$

$$E_T^{AB} = 2E_D^A + B^{(1)}. \tag{8.17}$$

Repeating the same process generates the same equations, leading to effective interactions $J^{(2)}$ and $B^{(2)}$. The isomorphism is kept along the iterations. The eigenvalues of the blocks are invariant in terms of $J^{(n)}$ units n and the energy of the block $A^{(n)}$ becomes

$$\begin{cases} A^{(n)} = N_s A^{(n-1)} + n_B B^{(n-1)} + \alpha J^{(n-1)}, \\ B^{(n)} = n_C B^{(n-1)} + \beta J^{(n-1)}, \\ J^{(n)} = \gamma J^{(n-1)}, \end{cases} \tag{8.18}$$

where n_B is the number of bonds inside the block, n_C is the number of bonds between two adjacent blocks, and α, β and γ are iteration independant parameters. Since these equations are true at first iteration one may express $A^{(n)}$, $B^{(n)}$ and $J^{(n)}$ as functions of the initial interactions. Their expressions can by generalized to any dimension :

$$\begin{cases} A^{(n)} \rightarrow N_s^n [A^{(0)} + \frac{n_B}{N_s - n_C} B^{(0)} + \frac{\alpha}{N_s - \gamma} J^{(0)} + \frac{\beta n_B}{(N_s - n_C)(N_s - \gamma)} J^{(0)}], \\ B^{(n)} \rightarrow n_C^n [B^{(0)} + \frac{\beta}{n_C - \gamma} J^{(0)}], \\ J^{(n)} = \gamma^n J^{(0)}. \end{cases} \tag{8.19}$$

For 1D chain $n_C = 1$ and $n_B = N_s - 1$, the above equations lead to the following value of the cohesive energy in terms of $J^{(0)}$:

$$\epsilon_{1D} = \lim_{n \rightarrow \infty} N_s^{-n} A^{(n)} = \frac{\alpha + \beta}{N_s - \gamma} = \frac{2[E_T^{AB} - E_D^A]}{2N_s - E_T^{AB} + E_S^{AB}}. \tag{8.20}$$

Table 8.1 gives the values obtained from 10, 14, 18 and 22 site superblocks and illustrates the dramatic benefit of the RSRG-EI iterations. The cohesive energy value has been extrapolated using the BST algorithm[16] giving $-1.386205J$, in excellent agreement with the exact value[17] -2Jln2=-1.386294J.

The square 2D lattice can be treated similarly by defining square (3×3) 9-site blocks, which have a non-degenerate doublet ground state, and may be considered as superspins. The resulting lattice of blocks is isomorphous to the starting one. The logics is the same as for 1D chain. The cohesive energy is given by Eq. 8.19 with $N_s = 9$, $n_B = 12$, and $n_C = 3$,

$$\epsilon_{2D} = \lim_{n \rightarrow \infty} N_s^{-n} A^{(n)} = \frac{\alpha + 2\beta}{9 - \gamma} = \frac{4E_T^{AB} - 6E_D^A}{18 - E_T^{AB} + E_S^{AB}}. \tag{8.21}$$

Table 8.1. Cohesive energies (in J units) obtained from blocks of various sizes (N_s) and RSRG-EI values the 1D chain. The extrapolated value is -1.386205.

N_s	5	7	9	11
block on-site energy	-1.171154	-1.238925	-1.274738	-1.296744
superblock on-site energy	-1.301607	-1.325246	-1.338556	-1.347097
RSRG cohesive energy	-1.392270	-1.390113	-1.388950	-1.388253

The so-obtained value is $-2.33231J$,[7,18] which compares favorably with the most precise value of Quantum Monte Carlo (QMC) calculation[19] $-2.33868J$ and Dressed Cluster Method (DCM)[20,21] $-2.33856J$. Since the amplitude of the inter-block spin-couplings $J^{(n)}$ decreases exponentially along the iterative procedure according to Eq. 8.19, the lowest excitation energy tends to zero and both the 1D and 2D systems are gapless.

8.3.2. *Accumulation points, illustration on the dimerized and the frustrated 1D chains*

Let us consider first the *dimerized 1D spin chain*, i.e., an infinite system with alternating spin couplings $J_1^{(0)} = \bar{J}^{(0)}(1+\delta^{(0)})$ and $J_2^{(0)} = \bar{J}^{(0)}(1-\delta^{(0)})$, where $\bar{J}^{(0)}$ is a mean coupling (which here is fixed to 1) and $\delta^{(0)}$ is a dimerization parameter varying from 0 to 1. In order to study that problem within our RSRG approach, we have to consider a block with an odd number of sites N_s, presenting a strong interaction $J_1^{(0)}$ on one border and a weak interaction $J_2^{(0)}$ on the other border. Actually when the $J_2^{(0)\prime}/J_1^{(0)}$ ratio increases, the unpaired electron (effective spin) tends to localize on the external site connected to the system by a weak bond. There are, therefore, two possible types of dimers (see Fig. 8.2). The first one concerns those terminating with strong interactions for which the splitting into two blocks goes through a strong interaction, while in the other type of dimers, terminating with two weak interactions, the two blocks are connected by a weak interaction and the interaction between the two remote effective spins will

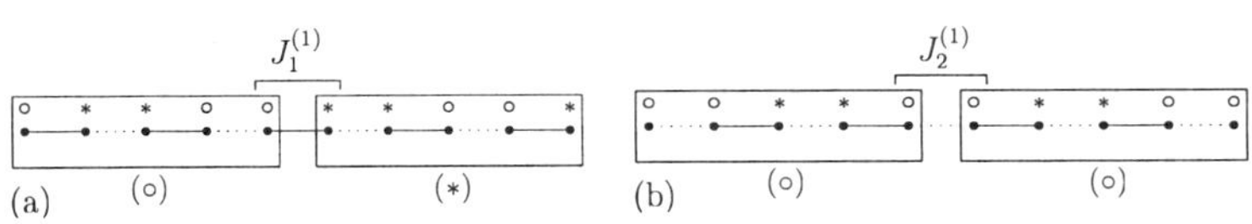

Fig. 8.2. Definition of the two-block superblocks for the extraction of effective interactions in the dimerized 1D chain. Reprinted figure with permission from reference J. P. Malrieu and N. Guihéry, Phys. Rev. B, 63, 085110 (2001). Copyright (2006) by the American Physical Society.

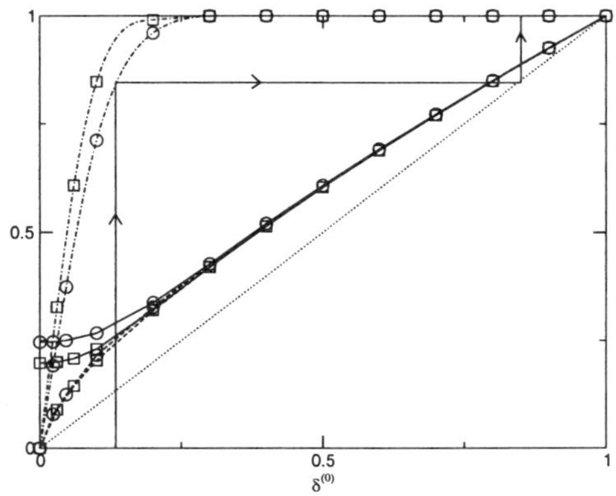

Fig. 8.3. Dimerized 1D AF chain : evolution of the effective inter-block coupling $\bar{J}^{(1)}$ (full line), of the renormalized distortion $\delta^{(1)}$ (dash-dot), and of the converged spin-gap (dot) as a function of the initial dimerization parameter $\delta^{(0)}$, for 9-site (○) and 11-site (□) blocks. Reprinted figure with permission from reference J.-P. Malrieu, and N. Guihéry, Phys. Rev. B, 63, 085110 (2001). Copyright (2006) by the American Physical Society.

be weak. Actually the values for $J_1^{(n)}$ remain finite along the iterations while the values of $J_2^{(n)}$ decrease rapidly to zero.

The logical structure of the problem is pictured in Fig. 8.3 which gives the dependance of $\bar{J}^{(1)}$ and of $\delta^{(1)}$ on $\delta^{(0)}$ for $N_s = 7$ and $N_s = 9$. The iterative process appears as a stair progression towards the accumulation point characterized by $\delta^{(0)} = 1$, $\delta^{(1)} = 1$, and $\bar{J}^{(1)} = 1$. Since the curve $\delta^{(1)} = f(\delta^{(0)})$ reaches the accumulation point with a zero slope, the convergence to the accumulation point is obtained in a finite number of steps. The value of $\bar{J}^{(n)}$ at convergence remains finite and the system presents a finite gap. Numerical calculation have been performed for N_s =3, 5, 7, 9, and 11 and extrapolated.[7] The energy gap has been fitted and follows a power law $\Delta E_{ST} = 4\bar{J}^{(0)}\delta^{0.71}$. A very close power law $\Delta E_{ST} = \alpha\delta^{2/3}$ has been established[22] in the case of the dimerized and frustrated 1D chain at the critical ratio $J_{NNN}/J_{NN} = 0.2411$ (J_{NNN} and J_{NN} being, respectively, the second- and first-nearest-neighbor couplings, (see below)). When the ratio is lower than that critical value, logarithmic corrections appear. One may therefore consider the result of the RSRG-EI method as satisfactory.

An other type of accumulation point appears in the case of the *frustrated 1D chain*. Let us call J_{NN} and J_{NNN} the AF interactions between the nearest-neighbor and next-nearest-neighbor sites respectively

$$H = 2J_{NN}\sum_i \vec{S}_i \vec{S}_{i+1} + 2J_{NNN}\sum_i \vec{S}_i \vec{S}_{i+2}. \qquad (8.22)$$

In order to maintain an isomorphism with the starting lattice along iterations one must define superblocks consisting of three identical blocks ABC, for instance $N_{SB} = 21$ and $N_s = 7$. The ground states of the blocks are again non-degenerate doublets which we may label a, b, and c respectively. For $S_z = 1/2$ the model space is spanned by the products $\bar{a}bc$, $a\bar{b}c$ and $ab\bar{c}$, which generate one quartet and two doublet states. The target space is composed of the two lowest doublet eigenstates D_1 and D_2 and of the lowest quartet eigenstate Q. Due the symmetry of the superblock the knowledge of the three lowest eigenvalues, respectively E_{D_1}, E_{D_2} and E_Q is sufficient to define the effective interactions between the blocks :

$$J_{NN}^{(1)} = (E_Q - E_{D_1})/3, \quad J_{NNN}^{(1)} = (2E_Q - 3E_{D_2} + E_{D_1})/6.$$

It is interesting to look at the $j^{(1)} = J_{NNN}^{(1)}/J_{NN}^{(1)}$ ratio as a function of the $j^{(0)} = J_{NNN}^{(0)}/J_{NN}^{(0)}$ ratio, as plotted in Fig. 8.4. It appears that for small values of the $j^{(0)}$, $j^{(1)}$ is larger than $j^{(0)}$, but the ratio $j^{(1)}/j^{(0)}$ tends to 1 when $j^{(0)}$ approaches 0.24. The minimal difference between $j^{(1)}$ and $j^{(0)}$ decreases when the size of the blocks increases. An extrapolation on the size of the blocks leads to a value of $j_c^{(0)} = 0.2419$ for which $j^{(1)} = j^{(0)}$. This value defines a fixed point of the problem. It compares quite well with the most accurate value of this critical point ($j^{(0)} = 0.24117$).[23] Starting the iteration procedure from a value $j^{(0)} < j_c^{(0)}$, an infinite number of iterations is required to reach the accumulation point. At each step the new value of $J_{NN}^{(n+1)}$ is obtained by multiplying $J_{NN}^{(n)}$ by the value of $J_{NN}^{(1)}$

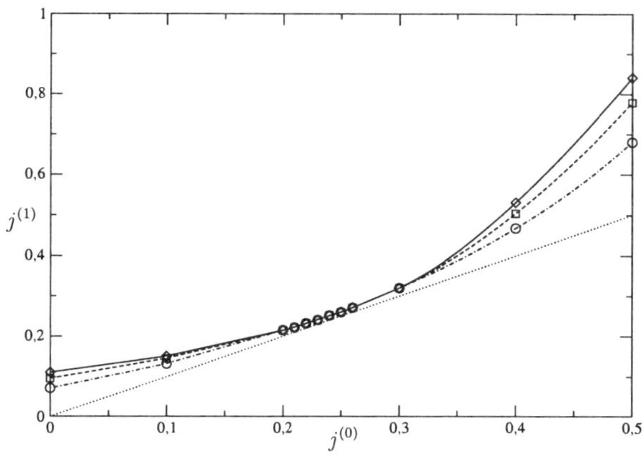

Fig. 8.4. 1D frustrated chain: evolution of the ratio $j^{(1)} = J_{NNN}^{(1)}/J_{NN}^{(1)}$ as a function of $j^{(0)} = J_{NNN}^{(0)}/J_{NN}^{(0)}$. Reprinted figure with permission from reference J.-P. Malrieu, and N. Guihéry, Phys. Rev. B, 63, 085110 (2001). Copyright (2006) by the American Physical Society.

(which is always smaller than 1) corresponding to a value of $j^{(0)}$ equal to $j^{(n)}$,

$$J_{NN}^{(n+1)} = J_{NN}^{(1)}(j^{(n)}) \times J_{NN}^{(n)}.$$

The effective interactions decrease at each iteration leading to a zero value of J_{NN}^{∞}. Consequently the gap defined as the energy difference between the quartet and the lowest doublet $\Delta E_{DQ} = 3J_{NN}^{\infty}$ is zero. This example illustrates the case of an attractive critical point for the domain $j^{(0)} < j_c^{(0)}$.

8.3.3. *The anisotropic square 2D spin lattice*

The spin 1/2 anisotropic Heisenberg Hamiltonian on an infinite 2D square lattice may be used as a model problem to test the ability of RSRG-EI to treat a phase-transition phenomenon. This Hamiltonian is given by

$$H = J \sum_{\langle i,j \rangle} (S_i^x S_j^x + S_i^y S_j^y + \lambda S_i^z S_j^z), \tag{8.23}$$

where $\langle i, j \rangle$ runs over all pairs of nearest-neighbor sites. This 2D square lattice model has no exact solution and has therefore been the subject of numerous calculations in the recent past, the most accurate of which are certainly recent QMC calculations.[19] At $\lambda = -1$ a first-order transition takes place between the ferromagnetic phase and a planarlike phase, in which the spins in the ground state wave function lie in the XY plane. This so-called XY polarized function Φ_0^{XY} is such that the sites of one sublattice bear

$$X = (\alpha + \beta)/\sqrt{2}, \tag{8.24}$$

where α and β are the usual spin up and spin $down$ functions, and those of the other sublattice bear

$$X = (\alpha - \beta)/\sqrt{2}. \tag{8.25}$$

If one works in the basis of (X, Y) functions instead of (α, β) ones, this XY polarized function will appear as the leading configuration for the $-1 < \lambda < 1$ domain. At $\lambda = 1$ (isotropic Hamiltonian) a transition to an Ising like phase occurs. Actually, for $\lambda \to \infty$ the Hamiltonian becomes an Ising Hamiltonian and the ground state becomes the Ising fully spin alternate function $\Phi_0 = \alpha\beta\alpha\beta...$, which is also the leading configuration for $\lambda > 1$. The best QMC calculations (see for instance[19]) confirm the first-order character of the phase transition.

The RSRG-EI treatment again considers square 9-site blocks and superblocks of two blocks (i.e., $18 = 3 \times 6$ sites). One wants to establish the effective energies of, and the interactions between, the four products of ground state wave functions which define a model space ab, $a\bar{b}$, $\bar{a}b$, $\bar{a}\bar{b}$. Diagonalizing the exact Hamiltonian for the AB superblock one may identify the eigenstates $\Psi_{T^+}(S_z = 1)$, $\Psi_{T^-}(S_z = -1)$, $\Psi_{T^0}(S_z = 0)$, $\Psi_{S^0}(S_z = 0)$, which have the largest projections on the model space, and their energies $E_{T^+} = E_{T^-}$, E_{T^0}, and E_{S^0}. The three energies can be seen as the eigenvalues of a new anisotropic Hamiltonian

$$H_{AB}^{(1)} = J_{AB}^{(1)}\lambda^{(1)}\left[S_{z_A}S_{z_B} - 1/4\right] + \frac{J_{AB}^{(1)}}{2}\left[S_A^+S_B^- + S_A^-S_B^+\right] + E_A + E_B + \Delta E_{AB}. \tag{8.26}$$

Hence

$$\begin{cases} E_{T^+} = E_A + E_B + \Delta E_{AB}, \\ E_{T^0} = -J^{(1)}\lambda^{(1)}/2 + J^{(1)}/2 + E_A + E_B + \Delta E_{AB}, \\ E_{S^0} = -J^{(1)}\lambda^{(1)}/2 - J^{(1)}/2 + E_A + E_B + \Delta E_{AB}, \end{cases} \tag{8.27}$$

from which one obtains

$$\begin{cases} J^{(1)} = E_{S^0} - E_{T^0}, \\ J^{(1)}\lambda^{(1)} = 2E_{T^+} - E_{T^0} - E_{S^0}. \end{cases} \tag{8.28}$$

These equations define a new anisotropic Heisenberg Hamiltonian between blocks. The process may be repeated, treating a block of 9 blocks and a superblock of 18 blocks, until convergence is achieved.

The qualitative key points in that problem are the facts that

- for $\lambda = 1$, $\lambda^{(1)} = 1$, the problem remains isotropic; (it is fixed point)
- for $\lambda > 1$, $\lambda^{(1)} > \lambda$, the anisotropy is increased in the direction of an Ising problem; and
- for $\lambda < 1$, $\lambda^{(1)} < \lambda$, the anisotropy increases in the opposite direction towards a pure XY problem.

For graphical purposes, using $\lambda = \tan\Phi$, the anisotropic Hamiltonian may been written as

$$H = J\big[(S_zS_z)\sin\Phi + (S_xS_x + S_yS_y)\cos\Phi\big]. \tag{8.29}$$

The isotropic case corresponds to $\Phi = -\pi/4$, the XY problem to $\Phi = \pi/2$, and the Ising situation to $\Phi = -\pi/2$. Figure 8.5 reports the evolution of $\Phi^{(1)}$ as a function of Φ. The iterative process, starting from as new value Φ, leads to a new anisotropy angle $\Phi_1 = \Phi^{(1)}(\Phi)$. The qualitative nature of the phase transition appears dramatically. Starting from $\Phi > \pi/4$, $\Phi^{(1)}$ increases rapidly. As seen from Fig. 8.5 the process converges in a few steps to the $\Phi_n = \pi/2$ fixed point, i.e., to an Ising problem. Oppositely, starting from $\Phi < \pi/4$, Φ_1 decreases. The

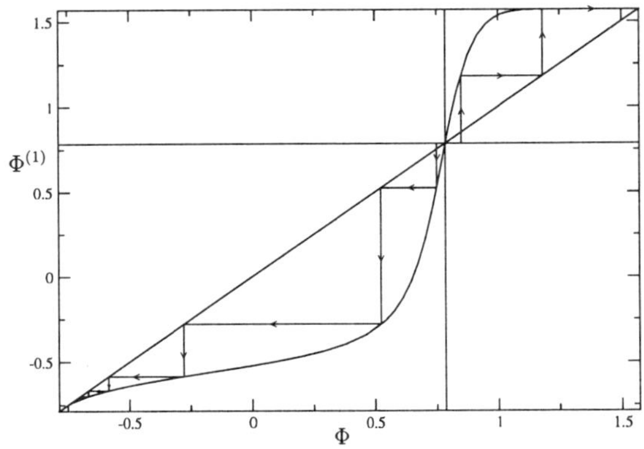

Fig. 8.5. RSRG study: anisotropic parameter $\Phi^{(1)}$ after the first iteration, as a function of the initial anisotropic parameter Φ Eq. 8.29. The stairs illustrate the convergence of the iterative procedure to the fixed points, Ising on the right side, XY on the left side. Reprinted figure with permission from reference M. Al Hajj, N. Guihéry, J.-P. Malrieu and P.Wind, Phys. Rev. B, 70, 094415 (2004). Copyright (2006) by the American Physical Society.

fixed point on that $\lambda < 1$ side is the pure XY problem ($\lambda = -1, \Phi = -\pi/4$), but the curve $\Phi^{(1)} = f(\Phi)$ is tangent to the line of slope one $\Phi^{(1)} = \Phi$ for $\Phi = -\pi/2$. Hence, the fixed point is in principle reached in an infinite number of steps. This problem illustrates the existence of fixed point characterizing the phase transition and two types of accumulation points, one concerning a gapped phase, the other one concerning a gapless phase.

The phase transition may be analyzed from the dependance of the cohesive energy on the λ value. Fig. 8.6 reports the RSRG-EI calculated cohesive energy. The underestimation of the cohesive energy by the RSRG-EI method with respect to the QMC values is the same as for the isotropic case (see subsection 3.1) and never exceeds 2% (see Fig. 8.6).

A discontinuity clearly appears in the curve $E_{coh} = f(\lambda)$ from the insert of Fig. 8.6. The existence of a discontinuity was not *a priori* evident, since the quantities $J^{(1)}$ and $\lambda^{(1)}$ are continuous functions of λ. The discontinuity comes from the fact that the iterations tend to different fixed points for $\lambda > 1$ and $\lambda < 1$.

We have checked the behavior of the gap as a function of λ. Spin-wave theory[24] predicts that it should follow the law,

$$\Delta E = 2(\lambda^2 - 1)^{1/2}. \tag{8.30}$$

Previous numerical works[25] have shown that the excitation energies are significantly lower, by a factor close to 0.5. An interpolation $0.86634\left(\lambda^2 - 1\right)^{1/2}$ fits well our calculated values. Hence the RSRG-EI method appears to be able to

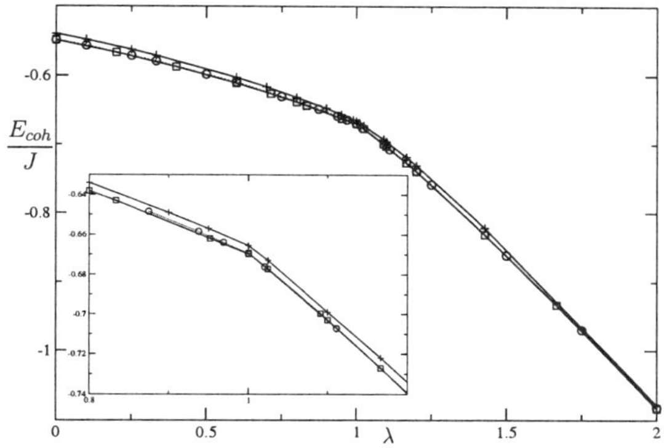

Fig. 8.6. Cohesive energy around the isotropic point. (○) QMC,[19] (□) DCM,[21] (+) RSRG-EI. Reprinted figure with permission from reference M. Al Hajj, N. Guihéry, J.-P. Malrieu and P.Wind, Phys. Rev. B, 70, 094415 (2004). Copyright (2006) by the American Physical Society.

evaluate excitation energies as well as cohesive energies.

8.3.4. *The frustrated 2D Shastry-Sutherland lattice*

It is interesting to apply the RSRG-EI method to frustrated 2D lattices since the QMC method, which may be considered as the most reliable treatment for non-frustrated 2D or 3D lattices, cannot be applied in these cases. The $SrCu_2(BO_3)_2$ lattice[28] is a famous 2D AF system presenting a spin gap and free from long range order. The Copper atoms are of d^9 character and can be seen as $S = 1/2$ spins. This lattice may be considered as a realization of the Shastry-Sutherland model[29] which can be schematized as a square lattice, with J' AF coupling between nearest neighbors, and diagonal AF interactions J in one plaquette over two, as pictured in Fig. 8.7. This system is supposed to obey the corresponding Heisenberg Hamiltonian

$$H = 2J' \sum_{[i,j]} \left(\overrightarrow{S_i}\overrightarrow{S_j} - \frac{1}{4} \right) + 2J \sum_{\langle i,j \rangle} \left(\overrightarrow{S_i}\overrightarrow{S_j} - \frac{1}{4} \right) \qquad (8.31)$$

where the couples $[i, j]$ concern the bonds of the square lattice and $\langle i, j \rangle$ the connected pairs of next nearest neighbor atoms (oblique bonds). The Shastry-Sutherland Hamiltonian has been widely studied by theoreticians (for a review see[30]), varying the J'/J ratio. For small values of J'/J the ground state is a product of singlets strictly localized on the oblique bonds.

The real material would correspond to a ratio $J'/J = 0.635$, close to the critical value where this phase disappears. Oppositely when J is small, the perturbed 2D square lattice is gapless with long range order. Other works (for instance[31])

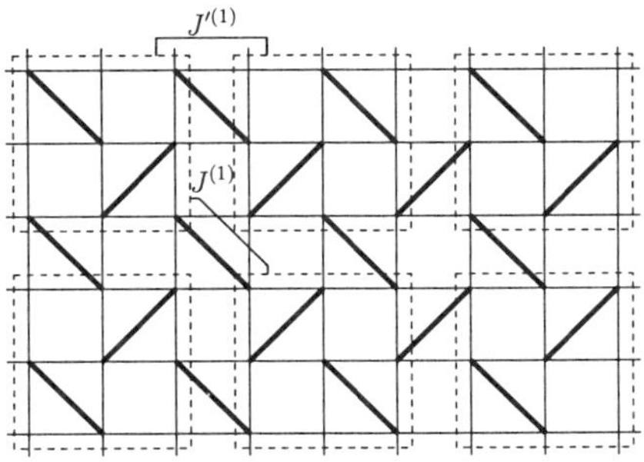

Fig. 8.7. Definition of the blocks and of their effective interactions for the Shastry-Sutherland lattice.

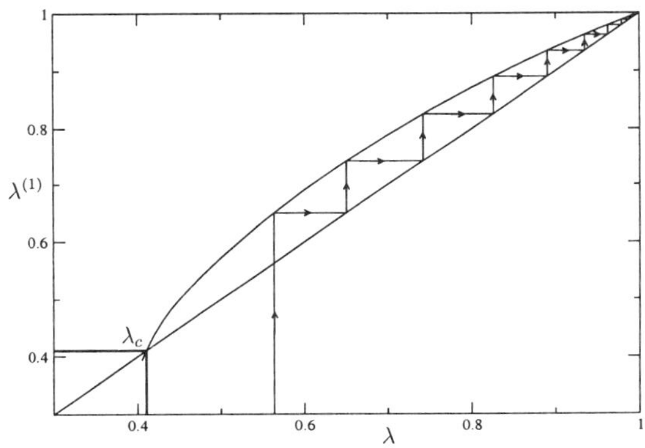

Fig. 8.8. Evolution of the first iteration ratio $\lambda^{(1)} = J'^{(1)}/(J'^{(1)} + J^{(1)})$ as a function of λ. The phase transition appears for $\lambda^{(1)} = \lambda$ at λ_c. The stairs illustrate the steps of the RSRG-EI iterations toward the ($J = 0$) non-frustrated 2D accumulation point, and the gapless character of this phase. Reprinted with permission from reference M. Al Hajj and J.P. Malrieu, Phys. Rev. B 72, 094436 (2000). Copyright (2006) by the American Physical Society.

have suggested the existence of an intermediate phase between $J'/J = 0.69$ and 0.83.

The RSRG-EI method has been employed to study the phase transition in this system, starting from 9 sites (3×3) square blocks pictured in Fig. 8.7, which seems relevant when J' is larger than J. They define an isomorphic lattice at each step. The first iteration of λ defines a ratio $\lambda^{(1)} = J'^{(1)}/(J'^{(1)} + J^{(1)}) = f(\lambda)$. Figure 8.8 gives the evolution in the $\lambda > 0.4$ ($J'/J > 0.666$) regime.

One sees that, for $\lambda_c = 0.41032$ ($(J'/J)_c = 0.69583$), $\lambda^{(1)} = \lambda$. This represents a fixed point of the problem. For $\lambda > \lambda_c$ the ratio $\lambda^{(n)}$ increases at each iteration, going to the $\lambda = 1$ accumulation point in an infinite number of steps since $(\partial\lambda^{(1)}/\partial\lambda) \neq 0$ (actually close to 0.5). Hence the system is gapless for $J'/J > (J'/J)_c$. For $J'/J < (J'/J)_c$, $J'^{(1)}/J^{(1)}$ becomes larger than $(J'/J)_c$, the procedure diverges since it becomes impossible to identify the target eigenvectors. This value of the critical ratio $(J'/J)_c$ compares well with other estimates but its relevance is questioned by the possible existence of an intermediate phase (see below).

The evolution of the cohesive energy for $(J'/J) > (J'/J)_c$ is consistent with the preceding analysis. The cohesive energy in the gapless phase is almost a linear function of λ, which goes to -2.33231 for $\lambda = 1$. With this definition of the Hamiltonian the energy of the product of bond singlets is equal to -1 whatever J'/J. The energy obtained from RSRG-EI iterations is plotted in Fig. 8.9. The consistence of the two independent criteria, i) $\lambda^{(1)} = f(\lambda)$ (Fig. 8.8), and ii) crossing of the cohesive energies $E_{coh} = f(\lambda)$ (Fig. 8.9), is very good. The first one gives $(J'/J)_c = 0.69583$, while for the second $(J'/J)_c = 0.67224$.

Regarding the existence of intermediate phase, which is supposed to be of columnar or plaquette nature, one may consider rectangular blocks. If one uses $2N_s$ sites blocks, with a non degenerate singlet ground state ψ_A^0 of energy E_A^0, the ground state of the lattice is studied from the product of the ground state of the blocks $\Psi^0 = \prod_A \psi_A^0$ according to the simplest version of the CORE method.[8] The knowledge of the ground state exact energy E_{AB}^0 of the AB dimers enables one to define an effective interaction v_{AB} between the blocks[8] as

$$v_{AB} = E_{AB}^0 - E_A^0 - E_B^0 \tag{8.32}$$

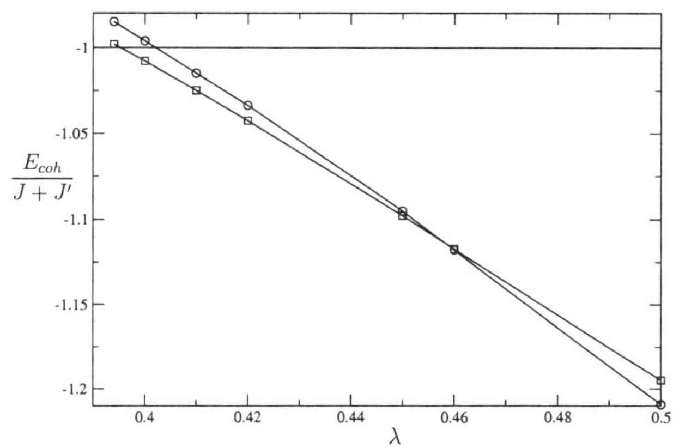

Fig. 8.9. Evolution of the cohesive energy as a fonction of λ. The horizontal line concerns the dimer phase, ○ RSRG-EI from (3×3) site square blocks, □ CORE with (2×6) sites columnar blocks.

and the energy per block is

$$E_A = E_A^0 + \frac{1}{2}\sum_B v_{AB}. \tag{8.33}$$

We have considered 12 (2×6) sites columnar blocks built from three aligned plaquettes. Interestingly enough the calculated cohesive energy is the lowest one in the interval $0.656 < J'/J < 0.845$. This result supports the existence of an intermediate plaquette phase and agrees with previous works[30,31] on its domain.

8.4. Conclusion

The Real Space Renormalization Group method, proposed by Wilson, is an extremely seducing idea for the treatment of periodic lattices of electrons or spins. The basic idea consists in a scale change (from sites to blocks of sites) and a truncation of the Hilbert space to products of the lowest eigenstates of the blocks. In this simple version the method is not quantitative. It becomes efficient when the interaction beteween the products of selected eigenstates are not taken through the bare Hamiltonian, but are reevaluated as effective interactions from the exact treatment of pairs (or trimers) of blocks, using the theory of effective Hamiltonians proposed by Bloch and des Cloizeaux. The procedure circumvents the fixed boundary conditions defects of the original RSRG. In many applications (especially most of those of the CORE method) the blocks are small but a number of eigenvectors of the blocks are kept and the effective Hamiltonian is more complex than the original one and the study stops at the treatment of a cluster of blocks. However one may consider larger blocks and keep a very low number (for instance one) of eigenvectors per block. In such a case the effective Hamiltonian keeps a very simple form. A proper design of the blocks may produce a lattice of blocks isomorphous to the starting one and the exponential scale-change may be iterated, producing a very elegant picture of the problem.

The first iteration already contains the crucial informations and enables one to study the quantum phase transitions occuring when the ratio of the various intersite interactions vary. Fixed points and accumulation points clearly emerge, as well as the gapped or gapless character of the different phases. One must insist on the very moderate computational cost of the method, which is limited by the research of the lowest roots of the superblock problem. Due to this bottleneck the method can only handle blocks of moderate sizes.

A few examples have been given in this chapter concerning spin lattices, namely the cohesive energy of the 1D AF chain and of the 2D square lattice, the phase transition occuring in the anisotropic 2D square lattice, in the 1D frustrated lattice and in the frustrated 2D Shastry-Sutherland lattice, the behavior of the gap in the dimerized 1D chain and the anisotropic 2D lattice. Other works have concerned the phase transition in the 2D 1/5-depleted and the chekerboard lattices, as well as the spin gaps in coupled ladders.[26] The method has been employed to settle a parity law concerning the spin gap of graphitic ribbons.[32] In many problems the scale change may be employed qualitatively to map a lattice on a simpler (already known) lattice. The method may be employed to treat Hubbard or more complex Hamiltonians. One may also mention that the scale change has been used in a so-called Renormalized Excitonic Method[33] which handles blocks with an even number of sites and a non-degenerate singlet ground state and works in a space where all blocks except one are in the ground state, the excitation moving on the lattice. Again the hopping integral of the excitation is extracted from the exact treatment of dimers of blocks. The method brings informations which are consistent with and complementary to those obtained by RSRG. The combination of the scale change from sites to blocks with effective Hamiltonian theory definitely appears as a very promizing tool.

Acknowledgments : Nathalie Guihéry thanks V. Rocha Vieira and P. Sacramento for their kind hospitality in the CFIF group and for stimulating exchanges.

References

1. K. G. Wilson, The renormalization group: Critical phenomena and the Kondo problem, *Rev. Mod. Phys.* **47**, 773 (1975).
2. J. W. Bray and S. T. Chui, Computer renormalization-group calculations of $2k_F$ and $4k_F$ correlation functions of the one-dimensional Hubbard model, *Phys. Rev. B* **19**, 4876 (1979) ; S. T. Chui and J. W. Bray, Computer renormalization-group technique applied to the Hubbard model, *Phys. Rev. B* **18**, 2426 (1978) ; J. E. Hirsh, Renormalization-group study of the Hubbard model, *Phys. Rev. B* **22**, 5259 (1980).
3. P. A. Lee, Real-Space Scaling Studies of Localization, *Phys. Rev. Lett.* **42**, 1492 (1979).
4. S. R. White and R. M. Noack, Real-space quantum renormalization groups, *Phys. Rev. Lett.* **68**, 3487 (1992).
5. S. R. White, Density-matrix algorithms for quantum renormalization groups, *Phys. Rev. B* **48**, 10345 (1993).
6. S. R. White, Density matrix formulation for quantum renormalization groups, *Phys. Rev. Lett.* **69**, 2863 (1992).
7. J.-P. Malrieu and N. Guihéry, Real-space renormalization group with effective interactions, *Phys. Rev. B* **63**, 85110 (2001).
8. C. J. Morningstar and M. Weinstein, Contractor renormalization group technology and exact Hamiltonian real-space renormalization group transformations, *Phys. Rev. D* **54**, 4131 (1996).

9. M. Weinstein, Contractor renormalization group and the Haldane conjecture, *Phys. Rev. B* **63**, 174421 (2001).
10. E. Altman and A. Auerbach, Plaquette boson-fermion model of cuprates, *Phys. Rev. B* **65**, 104508 (2002).
11. S. Capponi and D. Poilblanc, Charge density correlations in $t-J$ ladders investigated by the contractor-renormalization method, *Phys. Rev. B* **66**, 180505 (2002).
12. C. Lanczos, An iterative method for the solution of the eigenvalue problem of linear differential and integral operators, *J. Res. Natl. Bur. Stand.* **45**, 255 (1950).
13. E. R. Davidson, The Iterative Calculation of a Few of the Lowest Eigenvalues and Corresponding Eigenvectors of Large Real-Symmetric Matrices, *J. Comp. Phys.* **17**, 87 (1975) ; E. R. Davidson, Monster Matrices: Their Eigenvalues and Eigenvectors, *Computers in Physics* **7**, No. 5, 519 (1993).
14. C. Bloch, Sur la Théorie des Perturbations des États Liés, *Nucl. Phys.* **6**, 329 (1958).
15. J. Des Cloizeaux, Extension d'une formule de Lagrange à des problèmes de valeurs propres, *Nucl. Phys.* **20**, 321 (1960).
16. M. Henkel and G. Schütz, Finite-lattice extrapolation algorithms, *J. Phys. A; Math. Gen.* **21**, 2617 (1988).
17. H. A. Bethe, On the Theory of Metals, I. Eigenvalues and Eignefunctions of a Linear Chain of Atoms, *Z. Phys.* **74**, 205 (1931); L. Hulthén, Over the exchange problem of a crystal, *Ark. Mat. Astron. Fys. A* **26**, 106 (1938); E.H. Lieb and F.Y. Wu, Absence of Mott Transition in an Exact Solution of the Short-Range, One-Band Model in One Dimension, *Phys. Rev. Lett.* **20** 1445 (1968).
18. T. P. Živković, B. L. Sandleback, T. G. Schmalz, and D. J. Klein, Heisenberg model for the square-planar lattice and fragments, *Phys. Rev. B* **41**, 2249 (1990).
19. H.-Q. Lin, J, S. Flynn, and D. D. Betts, Exact diagonalization and quantum Monte Carlo study of the spin-1/2 XXZ model on the square lattice, *Phys. Rev. B* **64**, 214411 (2001).
20. P. Wind, N. Guihéry, and J.-P. Malrieu, Approximation of an infinite periodic system by a self-consistent embedding of a finite cluster: The dressed-cluster method, *Phys. Rev. B* **59**, 2556 (1999).
21. M. Al Hajj, N. Guihéry, J.-P. Malrieu, and P. Wind, Theoretical studies of the phase transition in the anisotropic two-dimensional square spin lattice, *Phys. Rev. B* **70**, 094415 (2004).
22. M. C. Cross and D. S. Fisher, A new theory of the spin-Peierls transition with special relevance to the experiments on TTFCuBDT, *Phys. Rev. B* **19**, 402 (1979).
23. S. Eggert, Numerical evidence for multiplicative logarithmic corrections from marginal operators, *Phys. Rev. B* **54**, 9612 (1996) ; K. Okamoto and K. Nomura, Fluid-dimer critical point in $S = \frac{1}{2}$ antiferromagnetic Heisenberg chain with next nearest neighbor interactions, *Phys. Lett. A* **169**, 433 (1992) ; R. Chitra, S. Pati, H. R. Krishnamurthy, D. Sen, and S. Ramasesha, Density-matrix renormalization-group studies of the spin-1/2 Heisenberg system with dimerization and frustration, *Phys. Rev. B* **52**, 6581 (1995).
24. P. W. Anderson, An Approximate Quantum Theory of the Antiferromagnetic Ground State, *Phys. Rev.* **86**, 694 (1952).
25. T. Barnes, K. J. Cappon, E. Dagotto, D. Kotchan, and E. S. Swanson, Critical behavior of the two-dimensional anisotropic Heisenberg antiferromagnet: A numerical test of spin-wave theory, *Phys. Rev. B* **40**, 8945 (1989).

26. M. Al Hajj, N. Guihéry, J.-P. Malrieu, and B. Bocquillon, Real space renormalization group with effective interactions: applications to 2-D spin lattices, *Eur. J. Phys. B* **41**, 11 (2004).
27. M. Al Hajj and J.-P. Malrieu, Phase transitions in the Shastry-Sutherland lattice, *Phys. Rev. B*, **72** 094436 (2005).
28. R. W. Smith and D. A. Keszler, Synthesis, structure, and properties of the orthoborate $SrCu_2(BO_3)_2$, *J. Solid State Chem.* **93**, 430 (1991).
29. B. S. Shastry and B. Sutherland, Exact ground state of a quantum mechanical antiferromagnet, *Physica (Amsterdam)* **108B**, 1069 (1981).
30. S. Miyahara and K. Ueda, Theory of the orthogonal dimer Heisenberg spin model for $SrCu_2(BO_3)_2$, *J. Phys.: Condens. Matter.* **15**, R327 (2003).
31. A. Koga and N. Kawakami, Quantum Phase Transitions in the Shastry-Sutherland Model for $SrCu_2(BO_3)_2$, *Phys. Rev. Lett.* **84**, 4461 (2000).
32. M. Al Hajj, F. Alet, S. Capponi, M. B. Lepetit, J.-P. Malrieu, and S. Todo, Parity law of the singlet-triplet gap in graphitic ribbons, *Eur. J. Phys. B* **51**, 517 (2006).
33. M. Al Hajj, J.-P. Malrieu, and N. Guihéry, Renormalized excitonic method in terms of block excitations: Application to spin lattices, *Phys. Rev. B* **72**, 224412 (2005).

Chapter 9

Spin glasses

I. R. Pimentel

Department of Physics and CFTC, University of Lisbon,
Av. Prof. Gama Pinto 2, 1649-003 Lisboa, Portugal,
E-mail: iveta@cii.fc.ul.pt

A review is given on the theory of Ising spin glasses in a magnetic field. We consider the nature of the low-temperature phase, below the *freezing* transition, and present the "replica symmetry breaking" and the "droplet" pictures, which have been proposed to describe spin glass behavior. The spin glass transitions in zero and nonzero magnetic field are analyzed within the renormalization group, which indicates that there is no Almeida-Thouless transition, i.e., no spin glass transition occurs in a finite magnetic field.

Contents

9.1. Introduction

Spin glasses[1–5] are especially interesting disordered magnetic systems, with competing ferromagnetic and antiferromagnetic interactions, which generate frustration. These systems exhibit a *freezing* transition, at a temperature T_f, to a low-temperature phase where the magnetic moments, or spins, are aligned in random directions, presenting a broad spectrum of relaxation times, that extends from microscopic to macroscopic time scales.

Quenched disorder and frustration are the two key features of spin glasses, which reveal one of the most complex forms of behavior in condensed matter physics. The attempt to understand their properties has led to the development of new concepts, and experimental and computational techniques, which have had an impact on a variety of other subjects, such as combinatorial optimization, neural networks, protein folding and prebiotic evolution.

Examples of spin glasses are found in metallic alloys, e.g., *CuMn*, with RKKY interactions between the spins, and in insulators, e.g., $Eu_xSr_{1-x}S$, with ferromagnetic and antiferromagnetic exchange interactions between first and second neighbors, respectively. The *freezing* of the spins is characterized by the observation of a cusp in the low-frequency susceptibility in low field, absence of periodic long-range magnetic order, and onset of slow relaxation and remanence. The static susceptibility obtained by cooling the system in an applied field has a higher value than that obtained by first cooling in zero field and then applying the field. Also, the thermoremanent magnetization measured by cooling in a field and then removing it, is greater than the isothermal remanent magnetization, obtained by first cooling, then applying and finally removing the field. These observations indicate that the spin glass phase has many metastable states, with a distribution of energy barriers separating them. In fact, because of frustration which arises from the spins receiving contradictory ordering instructions, there is no unique spin configuration that simultaneously satisfies all the interactions. The spin glass free energy is then expected to exhibit a landscape with many "valleys", as illustrated in Fig. 9.1.

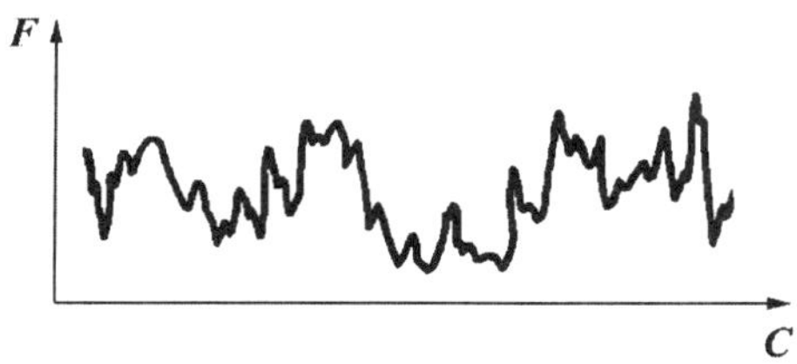

Fig. 9.1. Schematic view of the free energy of a spin glass as a function of the configuration space.

Despite the enormous amount of work dedicated over the past three decades to the study of spin glasses, no consensus has yet been reached on the most fundamental properties of these systems, namely, the nature and complexity of the glassy phase and the existence of a transition in a nonzero magnetic field. Edwards and Anderson[6] introduced a lattice model for the spin glass, with short-range interactions, and used a replica method to perform the average over quenched disorder. Two different pictures have since been proposed for the spin glass. One is the mean field theory for the Ising spin glass, provided by the Parisi[7] solution for the infinite-range, Sherrington-Kirkpatrick[8] model, which predicts a glassy phase described by an infinite number of pure states organized in an ultrametric structure, and a phase transition occurring in a magnetic field. The alternative is the "droplet" model,[9–11] which claims that the real, short-range systems behave quite differently, the glassy phase being described by only two pure states, related by a global inversion of the spins, and no phase transition occurring in a magnetic field. The first picture results from replica symmetry breaking, while in the second picture there is no replica symmetry breaking. The spin glass transition in a field observed in the Serrington-Kirkpatrick model, was found by de Almeida and Thouless.[12] A fundamental step towards the understanding of the spin glass, lies in the investigation of how the fluctuations, associated into the finite-range interactions, modify the mean-field picture.

In this review we concentrate on the equilibrium properties of Ising spin glasses in a magnetic field. We consider the nature of the low-temperature phase, and present the two main theories, the "replica symmetry breaking" solution and the "droplet" model, that have been proposed to describe spin glass behavior. The spin glass transitions in zero and nonzero magnetic field are analyzed within the renormalization group, discussing the effects of fluctuations. For a review on the dynamics of spin glasses, not considered here, see, e.g., Ref. 13.

9.2. Edwards-Anderson Model and Spin Glass Order Parameters

The Edwards-Anderson[6] (EA) model, which contains the essential features of spin glasses, is described by the Hamiltonian

$$H = -\sum_{\langle i,j\rangle} J_{ij}\mathbf{S}_i \cdot \mathbf{S}_j \tag{9.1}$$

where N spins $\mathbf{S}_i$, are located on a regular d-dimensional lattice, with bonds J_{ij} between nearest-neighbor spins only, that are independent random variables. Most experimental spin glasses are Heisenberg spin systems with some anisotropy, which may lead to Ising-like behavior. It is then common to consider the simplest case of an Ising model, with a single spin component S_i. For the random bonds, it is usual to take a Gaussian distribution with mean zero and variance J^2, or a bimodal distribution $\pm J$.

One wishes to characterize the "ordering" in the spin glass phase. As the spins are frozen in random directions, the system exhibits a nonzero thermally averaged local magnetization $\langle S_i \rangle$, but a vanishing global magnetization, $m = (1/N)\sum_i \langle S_i \rangle$. The characterization of the order in a spin glass is therefore not simple, and furthermore one has to address the possibility of nontrivial broken ergodicity. Let us consider that below the *freezing* temperature, the space of states is divided into many "valleys", separated by infinite barriers of the free energy *F*, so that each of those "valleys" corresponds to a stable state. Each "valley", which we label by an index *a*, is characterized by a free energy F_a, a statistical weight $P_a \propto \exp(-\beta F_a)$, and a local magnetization $\langle S_i \rangle^a$. The spin glass order parameter is defined as,

$$q = \left[\langle S_i \rangle^2 \right]_{av} = \left[\left(\sum_a P_a \langle S_i \rangle^a \right)^2 \right]_{av} \tag{9.2}$$

where $\langle\ \rangle$ denotes thermal average, and $[]_{av}$ represents average over the bond disorder. The quantity *q* measures the mean square local equilibrium magnetization. Previously, Edwards and Anderson had proposed the order parameter $q_{\mathrm{EA}} = [\sum_a P_a (\langle S_i \rangle^a)^2]_{av}$, which measures the mean square "single-valley" local magnetization, averaged over all possible "valleys". The order parameter q differs from q_{EA} in having "intervalley" contributions, and hence provides a more complete description of the spin glass phase.

We define the "overlap" of states by,

$$q_{ab} = \frac{1}{N} \sum_i \langle S_i \rangle^a \langle S_i \rangle^b \tag{9.3}$$

which measures the correlation between states. The statistical properties of the overlaps are described by the probability distribution

$$P(q) = \left[\sum_{ab} P_a P_b \delta(q - q_{ab}) \right]_{av} . \tag{9.4}$$

It is the distribution function *P*(*q*) that characterizes the structure of the space of states in spin glasses. For a system with just two states, *P*(*q*) is just the sum of a pair of delta functions, whereas if there is nontrivial broken ergodicity, *P*(*q*) may have a continuous part that indicates the possibility of overlaps between various states.

At the critical *freezing* temperature T_f, the spin glass susceptibility, which is defined as,

$$\chi_{SG} = \frac{\beta^2}{N} \sum_{i,j} \left[(\langle S_i S_j \rangle - \langle S_i \rangle \langle S_j \rangle)^2 \right]_{av} \tag{9.5}$$

diverges. It turns out that χ_{SG} is measurable through the "nonlinear susceptibility" χ_{nl}, defined as the coefficient of $-h^3$ in the expansion of the magnetization in powers of the external field.

In order to obtain the physical quantities one needs to average over the quenched disorder. The replica method allows to calculate the average of the free energy, in terms of the average of the partition function of n replicas of the original system Z^n, with n a positive integer, as

$$F_{av} = -\frac{1}{\beta}\left[\ln Z\right]_{av} = -\lim_{n\to 0}\frac{1}{\beta n}\left(\left[Z^n\right]_{av} - 1\right) \tag{9.6}$$

where

$$Z^n = \mathrm{Tr}_{S^\alpha}\exp\left(-\beta\sum_{\alpha=1}^{n} H(J, S^\alpha)\right) \tag{9.7}$$

$\beta = 1/T$, and we have introduced the replica label $\alpha = 1, \ldots, n$. The average is now easy to perform, leading to

$$\left[Z^n\right]_{av} = \mathrm{Tr}_{S^\alpha}\exp\left(-\beta H_{eff}(S^\alpha)\right). \tag{9.8}$$

Thus, the original disordered system is converted into a uniform one, described by an effective Hamiltonian, which in addition to the spin symmetry of the original model, exhibits permutation symmetry in the replica space.

One can express the spin glass order parameter in terms of replicas, with

$$q^{\alpha\beta} = \left\langle S_i^\alpha S_i^\beta \right\rangle \tag{9.9}$$

for any pair of replicas $\alpha \neq \beta$, where $\langle\ \rangle$ refers to a thermal average in an ensemble with effective Hamiltonian H_{eff},

$$q = \lim_{n\to 0}\frac{1}{n(n-1)}\sum_{\alpha\neq\beta} q^{\alpha\beta}, \tag{9.10}$$

q_{EA} is identified with the largest $q^{\alpha\beta}$, and

$$P(q) = \lim_{n\to 0}\frac{1}{n(n-1)}\sum_{\alpha\neq\beta}\delta(q - q^{\alpha\beta}). \tag{9.11}$$

Comparing (9.11) with (9.4), one can see that there is an intimate connection between broken ergodicity and broken replica symmetry. The replica method has in fact led to much of the progress in the theory of spin glasses. The technique of replica symmetry breaking allows to describe systems that possess many equilibrium states.

9.3. Mean Field Theory

A first natural approach to the study of the short-range EA spin glass model is to consider the mean field solution. This is equivalent to the solution of the Sherrington-Kirkpatrick[8] (SK) model, which corresponds to the infinite-range version of the EA model, with an appropriate scaling of the interactions with the number of spins.

9.3.1. *The SK Model*

For the SK model with Ising spins, $S_i = \pm 1$, in a uniform magnetic field h, given by the Hamiltonian

$$H = -\sum_{(i,j)} J_{ij} S_i S_j - h \sum_i S_i \tag{9.12}$$

where the sum (i,j) runs over all pairs of sites, and a Gaussian distribution of random bonds J_{ij}, $P(J_{ij}) = (2\pi\Delta^2)^{-1/2} \exp(-J_{ij}/2\Delta^2)$, with variance $\Delta^2 = J^2/N$, Eq. (9.8) becomes

$$[Z^n]_{av} = \mathrm{Tr}_{S^\alpha} \exp\left(\frac{(\beta J)^2}{4N} \sum_{i,j} \sum_{\alpha,\beta} S_i^\alpha S_i^\beta S_j^\alpha S_j^\beta + \beta h \sum_i \sum_\alpha S_i^\alpha \right) . \tag{9.13}$$

One can simplify the four-spin interaction via a Hubbard-Stratonovich transformation, which leads to a field-theoretical representation of the spin-glass model in terms of replica fields $q^{\alpha\beta}$, with

$$[Z^n]_{av} = \int D\hat{q} \exp\left(-\beta n F(\hat{q})\right) \tag{9.14}$$

where

$$\beta F(\hat{q}) = N \left[-\frac{(\beta J)^2}{4} + \frac{(\beta J)^2}{2n} \sum_{(\alpha\beta)} (q^{\alpha\beta})^2 - \frac{1}{n} \ln \mathrm{Tr}_{S^\alpha} \exp V(\hat{q}) \right] \tag{9.15}$$

and

$$V(\hat{q}) = (\beta J)^2 \sum_{(\alpha\beta)} q^{\alpha\beta} S^\alpha S^\beta + \beta h \sum_\alpha S^\alpha . \tag{9.16}$$

The fields $q^{\alpha\beta}$ are defined on an $n(n-1)/2$ – dimensional replica space (since $q^{\alpha\alpha} = 0$, $q^{\alpha\beta} = q^{\beta\alpha}$), and the sums are over pairs $(\alpha\beta)$ of distinct replicas. The notation $D\hat{q} \propto \prod_{(\alpha<\beta)} dq^{\alpha\beta}$ is used.

In the thermodynamic limit, $N \to \infty$, the partition function (9.14) with (9.15), can be evaluated by steepest descents. The saddle-point is determined by the condition $\delta F/\delta q^{\alpha\beta} = 0$, or, equivalently by,

$$q^{\alpha\beta} = \langle S^\alpha S^\beta \rangle = \frac{\mathrm{Tr}_{S^\alpha}\left[S^\alpha S^\beta \exp V(\hat{q})\right]}{\mathrm{Tr}_{S^\alpha} \exp V(\hat{q})} \quad . \tag{9.17}$$

The quantity $F(\hat{q})$ with $\hat{q}$ corresponding to the saddle-point, gives in the limit $n \to 0$ the free energy of the system.

9.3.2. *Replica-Symmetric Ansatz*

The replica symmetric (RS) ansatz corresponds to consider that all the replica fields are equal:

$$q^{\alpha\beta} = q, \quad \text{all } \alpha \neq \beta. \tag{9.18}$$

The order parameter q is then given by

$$q = \int_{-\infty}^{+\infty} \frac{dx}{\sqrt{2\pi}} \exp(-x^2/2) \tanh^2\left(\beta J x\sqrt{q} + \beta h\right). \tag{9.19}$$

In zero magnetic field, $h = 0$, the solution of (9.19) gives a phase transition at a critical temperature $T_f = J$: $q = 0$ for $T \geq T_f$, while $q \neq 0$ for $T < T_f$, with $q = \tau + \mathrm{O}(\tau^2)$, for $\tau = (T_f - T)/T_f \ll 1$, i.e., in the vicinity of T_f. However, the RS solution turns out to be unstable in the low-temperature phase, and replica symmetry breaking is then required.

In a nonzero magnetic field, $h \neq 0$, q is finite in the high-temperature phase. The spin glass transition in a field, found by de Almeida and Thouless[12] (AT), occurs along a line in the field-temperature plane, $T_f(h)$, the AT line, which marks the instability of the RS solution against replica symmetry breaking. One naturally finds that $T_f(0) = T_f$.

The stability of the RS solution, can be analyzed by looking at the normal modes of fluctuations of $q^{\alpha\beta}$ about the RS value, $q^{\alpha\beta} = q + \phi^{\alpha\beta}$. One expands (9.15) to second order in the fluctuations $\phi^{\alpha\beta}$, and determine the eigenvalues of the Hessian matrix, $M_{\alpha\beta,\gamma\delta} = (n\beta/N)\delta^2 F/\delta q^{\alpha\beta}\delta q^{\gamma\delta}$, that is,

$$M_{\alpha\beta,\gamma\delta} = (\beta J)^2 \left[\delta^{Kr}_{\alpha\beta,\gamma\delta} - (\beta J)^2 \left(\langle S^\alpha S^\beta S^\gamma S^\delta \rangle - \langle S^\alpha S^\beta \rangle \langle S^\gamma S^\delta \rangle\right)\right] \tag{9.20}$$

evaluated at the RS saddle-point. There are $n(n-1)/2$ independent modes, divided into three sets characterized by different symmetries: the longitudinal (L), with 1 mode, symmetric under the interchange of all replica indices; the anomalous (A), with $(n-1)$ modes, symmetric under the interchange of all but one of the replica indices; and the replicon (R), with $n(n-3)/2$ modes, symmetric under the interchange of all but two of the replica indices; respectively, with eigenvalues m_L, m_A and m_R.[12,14] In the limit $n \to 0$, one finds

$$m_L = m_A = (\beta J)^2 \left[1 - (\beta J)^2(1 - 4q + 3r)\right] \tag{9.21}$$

$$m_R = (\beta J)^2 \left[1 - (\beta J)^2(1 - 2q + r)\right] \tag{9.22}$$

with q given by (9.19), and

$$r = \int_{-\infty}^{+\infty} \frac{dx}{\sqrt{2\pi}} \exp(-x^2/2) \tanh^4\left(\beta J x \sqrt{q} + \beta h\right). \tag{9.23}$$

The longitudinal and anomalous eigenvalues are positive. The instability is determined by the replicon eigenvalue, which vanishes at the AT line and becomes negative bellow this line. The AT line is then characterized by the condition $m_R = 0$. In the region of small fields h, and near the zero-field critical temperature T_f, the AT line takes the form,

$$1 - T_f(h)/T_f = (3/4)^{1/3}(h/J)^{2/3}. \tag{9.24}$$

The AT line represents second-order phase transitions. In zero field all the modes become critical at the transition temperature ($m_L = m_A = m_R = 0$), while in a finite field only the replicon mode becomes critical at the AT line ($m_R = 0, m_L = m_A \simeq h^{2/3}$).

9.3.3. *Replica Symmetry Breaking: Parisi's Ansatz*

The search of a solution with replica symmetry breaking (RSB) was motivated by the possibility of nontrivial broken ergodicity. As the theory is quite involved at general temperatures one restricts as usual to the region near T_f. Equation (9.15) is expanded in powers of $q^{\alpha\beta}$, keeping only the first relevant terms, so

$$\beta F(\hat{q}) = \lim_{n \to 0} \frac{N}{n} \left[-\frac{1}{2}\tau \sum_{(\alpha,\beta)} (q^{\alpha\beta})^2 - \frac{1}{6} \sum_{(\alpha,\beta,\gamma)} q^{\alpha\beta} q^{\beta\gamma} q^{\gamma\alpha} \right.$$
$$\left. - \frac{1}{12} \sum_{(\alpha,\beta)} (q^{\alpha\beta})^4 - \frac{1}{2} h^2 \sum_{(\alpha,\beta)} q^{\alpha\beta} \right] \tag{9.25}$$

where $\tau = (T_f - T)/T_f$, and the sums are over distinct replicas, (the $q^{\alpha\beta}$ incorporate a factor $(\beta J)^2$ and h a factor β). One notes that, it is the quartic term that makes the RS solution unstable below T_f, and it is this term which is responsible for RSB. There are other quartic terms, which can however be neglected in lowest order in τ.

Parisi[7] proposed an ansatz for the RSB pattern, which can be described as follows. Consider $q^{\alpha\beta}$ as a symmetric $n \times n$ matrix with zeros on its diagonals. One starts with the RS form, in which all the off-diagonal elements have the same value, q_0. We then divide the $n \times n$ matrix into blocks of size $m_1 \times m_1$, and in the diagonal blocks replace q_0 by q_1, leaving q_0 in the off-diagonal blocks. Each of the $m_1 \times m_1$ blocks on the diagonal is subdivided further into $m_2 \times m_2$ sub-blocks, and in the diagonal sub-blocks q_1 is replaced by q_2. This procedure of subdivision of the diagonal blocks is repeated infinitely many times, with $n > m_1 > m_2 > \ldots > 1$, which in the $n \to 0$ limit, must be turned around into, $0 < m_1 < m_2 < \ldots < 1$. The m_i become continuous $m_i \to x$, and

$$q^{\alpha\beta} = q_k \to q(x), \quad 0 \leq x \leq 1. \tag{9.26}$$

For small reduced temperature τ and field h, the order parameter is given by

$$q(x) = \begin{cases} q(0), & 0 \leq x \leq x_0 \\ x/2, & x_0 \leq x \leq x_1 \\ q(1), & x_1 \leq x \leq 1 \end{cases} \tag{9.27}$$

where $x_1 = 2q(1)$, $x_0 = 2q(0)$, and $q(1) \simeq \tau$, $q(0) \simeq (h/J)^{2/3}$. As the field increases, the low-x plateau rises, and once it reaches the height of the second plateau, the only solution is x-independent, that is, the RSB disappears. The point at which this happens corresponds to the AT transition line.

The study of the stability of the Parisi solution is a rather difficult task that was carried out by De Dominicis and Kondor.[15] They found that the Hessian matrix evaluated at the RSB saddle-point has no negative eigenvalues and hence the solution is stable. There are however some zero eigenvalues, which tells that the system is apparently very "soft".

With the Parisi ansatz for the $q^{\alpha\beta}$, the distribution of overlaps is given by

$$P(q) = \int_0^1 \delta(q - q(x))dx = \frac{dx(q)}{dq} \tag{9.28}$$

where $x(q)$ is the inverse function of $q(x)$. It contains a delta-function spike at $q = q(1)$ and a continuous part with a finite weight down to $q(0)$. The largest overlap $q_{\mathrm{M}} = q(1)$ must be the single-state order parameter q_{EA}, and in zero magnetic field $q(0) = 0$. In Fig. 9.2, we show a schematic representation of the distribution of overlaps $P(q)$, according to the RSB picture provided by the Parisi solution. Hence, there exist many states resembling each other in many degrees.

The phase diagram expected for a spin glass in a magnetic field is represented in Fig. 9.3, where according to the Parisi solution, the AT line marks the transition from a paramagnetic to a spin glass phase with RSB.

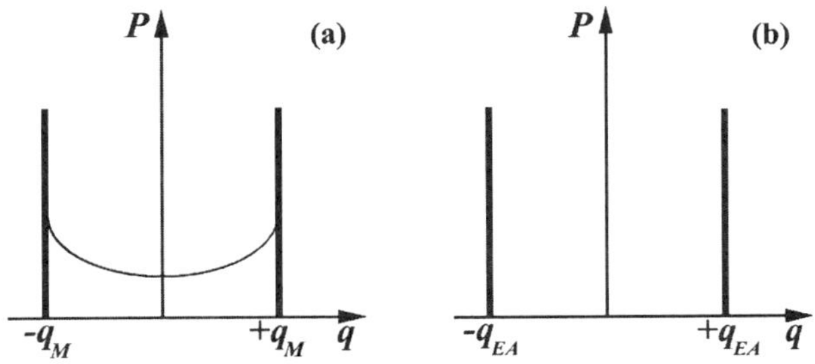

Fig. 9.2. Behavior of the probability distribution $P(q)$, according to: (a) the RSB picture; (b) the "droplet" picture. In zero magnetic field.

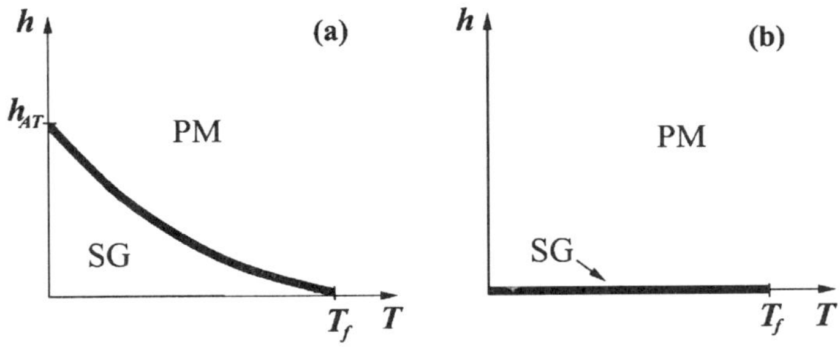

Fig. 9.3. h-T phase diagram for the transition from a paramagnetic (PM) to a spin glass (SG) phase, according to: (a) the RSB picture, $T_f(h)$ represents the AT line; (b) the "droplet" picture, a spin-glass phase exists only for $h = 0$.

A particular feature of the Parisi solution is ultrametricity. It is defined by the condition: given three arbitrary states α, β, γ, the mutual overlaps between these states, verify

$$q^{\alpha\beta} = q^{\alpha\gamma} \leq q^{\beta\gamma}, \quad \forall \alpha, \beta, \gamma. \tag{9.29}$$

Ultrametricity expresses a kind of hierarchical structure in the organization of the states, which can be represented in terms of a tree, as illustrated in Fig. 9.4. The tree describes the hierarchical fragmentation of the space of states in the spin glass into the "valleys".

9.4. Short-Range Models

There has been much debate on whether the behavior of realistic short-range spin glasses, in physical dimensions, can be described by the results of mean field

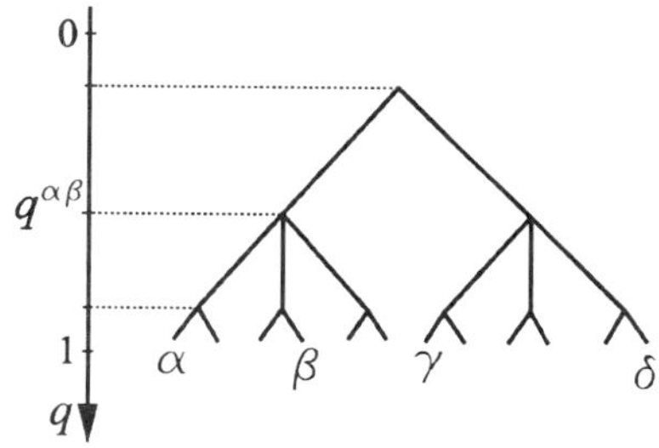

Fig. 9.4. Tree representation of Parisi's RSB scheme. The end points are the replica states.

theory, which corresponds to the infinite-range or infinite-dimensional model. The key point is to consider the effects of fluctuations on the phase transition and the low-temperature phase of spin glasses. Renormalization group approaches are fundamental to understand the phase transition and critical behavior of short-range models. Computer simulations have also played a major role in the investigation of the short-range models.

9.4.1. *The "Droplet" Model*

Fisher and Huse[9] proposed an alternative theory, the "droplet" model, to describe short-range Ising spin glasses. It consists of a phenomenological scaling theory, that was motivated by a series of numerical "domain-wall" renormalization-group studies.[10,11] The model assumes a phase transition in which the global spin-reversal symmetry is broken and there are only two pure-states related by this global symmetry, so that the broken ergodicity is trivial. It also assumes that the dominant low-lying excitations are droplets of coherently flipped spins, which at length scale L have a typical free energy that scales as

$$F_L \propto Y L^{\theta}. \tag{9.30}$$

Y is a "generalized stiffness", and θ is a fundamental exponent that characterizes the spin glass phase, with $\theta \leq (d-1)/2$.

Numerical results[10,11,16,17] show that $\theta < 0$ for $d = 2$ ($\theta \simeq -0.29$) while $\theta > 0$ for $d = 3$ ($\theta \simeq 0.19$), which implies that there is no spin glass order at finite temperatures, i.e., $T_f = 0$, in two-dimensional systems, while in three-dimensional systems there should be spin glass order at finite temperatures, i.e., $T_f > 0$. The lower critical dimension d_l, below which the transition temperature T_f is zero, is identified with the highest value of d where $\theta \leq 0$, and one believes that $d_l = 2.5$ for short-range Ising spin glasses.[18,19]

The randomness induces free-energy barriers, the typical barriers scaling as $B \propto bL^{\psi}$, with $\theta \leq \psi \leq d-1$.[20] A picture with many "valleys" is still expected for the free energy, the distinction with respect to the case of nontrivial broken ergodicity being that two "valleys", corresponding to fully ordered configurations, are somewhat deeper than all the other "valleys", which correspond to metastable configurations, with large clusters of spins overturned. The droplets are assumed to have a fractal surface, with fractal dimension $d_s < d$, where d is the space dimension. Numerical studies have indicated that $d_s \simeq 1.26$ in two-dimensional systems,[21] and $d_s \simeq 2.6$ in three-dimensional systems.[22]

In the droplet model the order parameter distribution $P(q)$ is just the sum of a pair of delta functions at $q = \pm q_{\mathrm{EA}}$, which differs from the RSB form predicted by the Parisi solution for the SK model, as illustrated in Fig. 9.2. The hierarchy of states and ultrametricity found in this model does not exist in the droplet model. Also, in the Parisi solution for the SK model, excitations that flip a finite fraction of spins have a surface that is space-filling, i.e. $d_s = d$,[23] in contrast to the droplet model.

The effect of a magnetic field on the spin glass phase can be viewed as follows. In a magnetic field the droplets formed have a domain wall energy $\propto YL^{\theta}$ and a field energy $\propto h\sqrt{q}L^{d/2}$, where q is the EA order parameter. The size of the droplets is set by the crossover length where the two energies become similar, giving

$$L_h \propto (Y/\sqrt{q}h)^{1/(d/2-\theta)} . \tag{9.31}$$

If $\theta < d/2$, there will be large reorientations of the spins, and the spin glass state will be unstable against an infinitesimal field. The condition $\theta < d/2$ is indeed verified in $d = 3$, and as a result the magnetic field removes the spin-glass transition. Hence, the AT line, which in the SK model marks the onset of spin glass order in a magnetic field, is absent here. The corresponding phase diagram is illustrated in Fig. 9.3.

Fisher and Huse argued that there should, however, be a "dynamic AT line", observable in experiments with fixed timescales or frequencies. In the droplet model the dynamics of the spin glass phase is described by activation over barriers. The relaxation time associated to a barrier, that has to be surmounted to flip a region of size L, is of Arrhenius form: $\tau(L) \propto \tau_0 \exp(bL^{\psi}/T)$. In a magnetic field there is a maximum relaxation time, $\tau_{\max}$, associated with scales of the order of the domain size L_h given by (9.31). The system will fall out of equilibrium on experimental time scales when $\tau_{\max}$ is comparable to the observation time $\tau_{\exp}$, giving

$$h^2 \propto (Y^2/q)\,[b/T \ln(\tau_{\exp}/\tau_0)]^{(d-2\theta)/\psi} \tag{9.32}$$

for the dynamic AT line. Near T_f, the spin glass correlation length diverges as $\xi_{SG} \propto (T_f - T)^{-\nu}$, the order parameter behaves as $q \propto (T_f - T)^{\beta}$, and scaling implies that $Y \propto (T_f - T)^{\theta\nu}$ and $b \propto (T_f - T)^{\psi\nu}$, which yields $h^2 \propto (T_f - T)^{\varphi}$, where $\varphi = d\nu - \beta$ is the usual crossover exponent.[24]

The scaling theory of spin glasses can be formulated in a real space renormalization group approach. A block of spins of linear dimensions L is considered, and the difference in the free energy between periodic and anti-periodic boundary conditions in one of the directions, $\Delta F(L)$, is regarded as an effective bond strength on length scale L. One then analyses, either numerically or by an approximate renormalization such as the Migdal-Kadanoff scheme, the way the distribution of bond strengths $P(\Delta F(L))$ changes with scale. The behavior of the system is determined by a zero-temperature fixed point.[11]

The spin glass phase exhibits a special characteristic that is chaos. Chaos means that the spin configuration of an equilibrium state is drastically changed by an arbitrarily small shift in the couplings or the temperature. The existence of chaos in the spin glass phase was first pointed out in the mean field theory[25] and later analyzed in the scaling theories.[9,26,27] At zero temperature, consider a low-energy excitation of the system, a droplet of reversed spins of size L, with a typical energy $\propto JL^{\theta}$, and introduce a small random change in the couplings of typical size δJ. This will change the surface energy by an amount $\propto \delta J L^{d_s/2}$, since the surface of the droplet has a fractal dimensionality d_s. The ground state will then become unstable on scales longer than,

$$L_J^* \propto (J/\delta J)^{1/\varsigma}, \quad \varsigma = d_s/2 - \theta, \tag{9.33}$$

obtained by equating the two energies. At a finite temperature, the free energy of the droplet scales as $F_L \propto YL^{\theta}$, and the surface entropy as $S_L \propto \sigma L^{d_s/2}$, which implies that a small temperature change δT will lead to instability of the ground state on length scales larger than,

$$L_T^* \propto (Y/\sigma\delta T)^{1/\varsigma}, \quad \varsigma = d_s/2 - \theta. \tag{9.34}$$

For Ising spin glasses the condition $\zeta > 0$, is ensured by the inequalities $d_s > (d-1)$ and $\theta \leq (d-1)/2$, and the instability will always occur. ζ is the Lyapunov exponent characterizing the chaotic behavior.

Numerical studies, by Monte Carlo simulations, of chaos in short-range Ising spin glasses, show that it occurs with respect to both coupling and temperature changes, in two-dimensional systems, where $T_f = 0$, with a unique exponent $\zeta \simeq 1.0$,[28] and also in three-dimensional systems, where $T_f > 0$, with similar scaling functions for the coupling and temperature chaos with an exponent $\zeta \simeq 1.16$.[29] It is however hard to observe temperature chaos, which has been explained by a renormalization group calculation.[30] A recent analytical computation proved the existence of temperature chaos in the SK model, but the effect is exceedingly small, namely of the ninth order in perturbation theory.[31]

9.4.2. *Beyond Mean Field Theory*

A natural way to investigate the behavior of short-range Ising spin glasses, is to start by deriving a field theory that describes the short-range model, and expand about the mean-field solution, which corresponds to the infinite-range model, in order to consider the effects of the fluctuations associated into the short-range interactions.

The averaged replicated partition function for a short-range Ising spin glass, the generalization to short-range interactions of (9.14) with (9.15), can be built by standard techniques, introducing local replica fields $Q_i^{\alpha\beta}$.[14] To construct a perturbation expansion around the mean-field solution, one separates the field $Q_i^{\alpha\beta}$ into

$$Q_i^{\alpha\beta} = Q^{\alpha\beta} + \phi_i^{\alpha\beta} \tag{9.35}$$

where $Q^{\alpha\beta}$ represents the uniform, mean-field value of the order parameter and $\phi_i^{\alpha\beta}$ are the fluctuations around it. One obtains, for the partition function of an Ising spin glass with nearest-neighbor interactions, and a Gaussian distribution of bonds with variance J^2, the form[32,33]

$$[Z^n]_{av} \simeq \int D\phi \exp(-L^{(0)} - L^{(1)} - L^{(2)} - L^{(3)} - \ldots) \tag{9.36}$$

where, after Fourier transform into momenta space, one has, for contributions up to cubic order,

$$L^{(0)} = N\left[\frac{1}{2}\Theta^{-1}\sum_{(\alpha\beta)}(Q^{\alpha\beta})^2 - \ln\mathrm{Tr}_{S^\alpha}\exp V_s(\hat{Q})\right] \tag{9.37}$$

with

$$V_s(\hat{Q}) = \sum_{(\alpha\beta)} Q^{\alpha\beta}S^\alpha S^\beta + \beta h\sum_\alpha S^\alpha \tag{9.38}$$

and

$$L^{(1)} = \sqrt{N}\sum_{(\alpha\beta)}\left[\Theta^{-1}Q^{\alpha\beta} - \langle S^\alpha S^\beta\rangle\right]\phi^{\alpha\beta}_{\mathbf{p}=\mathbf{0}} \tag{9.39}$$

$$L^{(2)} = \frac{1}{2}\sum_{(\alpha\beta),(\gamma\delta)}\sum_{\mathbf{p}}\phi^{\alpha\beta}_{\mathbf{p}}M_{\alpha\beta,\gamma\delta}(\mathbf{p})\phi^{\gamma\delta}_{-\mathbf{p}} \tag{9.40}$$

$$L^{(3)} = -\frac{1}{6}\frac{1}{\sqrt{N}} \sum_{(\alpha\beta),(\gamma\delta),(\mu\nu)} \sum_{\mathbf{p}_1,\mathbf{p}_2,\mathbf{p}_3} W_{\alpha\beta,\gamma\delta,\mu\nu}\phi^{\alpha\beta}_{\mathbf{p}_1}\phi^{\gamma\delta}_{\mathbf{p}_2}\phi^{\mu\nu}_{\mathbf{p}_3} \tag{9.41}$$

the sum over momentum in (9.41) being constrained to momentum conservation: $\mathbf{p}_1 + \mathbf{p}_2 + \mathbf{p}_3 = \mathbf{0}$. One has $\Theta = z(\beta J)^2$, where $z = 2d$ is the spin coordination number, and the average $\langle\ \rangle$ is defined with the Boltzmann weight $\exp(V_s)$. The term $L^{(0)}$ corresponds to the mean field theory, with the order parameter $Q^{\alpha\beta}$ being given by the stationary condition $L^{(1)} = 0$. In $L^{(2)}$ the masses $M_{\alpha\beta,\gamma\delta}(\mathbf{p})$ include momentum dependence, and the long-wavelength ($\mathbf{p} \ll \mathbf{1}$) behavior of the system is described by keeping as usual only the terms up to second order in the expansion in momentum. The masses $M_{\alpha\beta,\gamma\delta}$ and the couplings $W_{\alpha\beta,\gamma\delta,\mu\nu}$ are expressed in terms of spin correlations. An appropriate rescaling of the fields, with a corresponding rescaling of the masses and the couplings, allows to write the mass operator in standard form, i.e., with the coefficient of the momentum equal to unity,

$$M_{\alpha\beta,\gamma\delta}(\mathbf{p}) = \mathbf{p}^2\delta^{Kr}_{\alpha\beta,\gamma\delta} + z\left[\delta^{Kr}_{\alpha\beta,\gamma\delta} - \Theta\left(\left\langle S^\alpha S^\beta S^\gamma S^\delta\right\rangle - \left\langle S^\alpha S^\beta\right\rangle\left\langle S^\gamma S^\delta\right\rangle\right)\right] \tag{9.42}$$

and

$$\begin{aligned} W_{\alpha\beta,\gamma\delta,\mu\nu} = (z\Theta)^{3/2}\Big[&\left\langle S^\alpha S^\beta S^\gamma S^\delta S^\mu S^\nu\right\rangle - \left\langle S^\alpha S^\beta\right\rangle\left\langle S^\gamma S^\delta S^\mu S^\nu\right\rangle \\ &- \left\langle S^\gamma S^\delta\right\rangle\left\langle S^\alpha S^\beta S^\mu S^\nu\right\rangle - \left\langle S^\mu S^\nu\right\rangle\left\langle S^\alpha S^\beta S^\gamma S^\delta\right\rangle \\ &+ 2\left\langle S^\alpha S^\beta\right\rangle\left\langle S^\gamma S^\delta\right\rangle\left\langle S^\mu S^\nu\right\rangle\Big]. \end{aligned} \tag{9.43}$$

The expansion in the fluctuations corresponds to an expansion in $1/z$.

9.4.3. *Critical Behavior*

In order to study critical phenomena, either in zero field or in a finite magnetic field, one approaches the transition from the high-temperature phase, which is replica-symmetric, so to avoid the complexity of the glassy phase. Replica symmetry, $Q^{\alpha\beta} = Q$, allows three distinct components for the mass: $M_{\alpha\beta,\alpha\beta}(\mathbf{p}) = \mathbf{p}^2 + M_1$, $M_{\alpha\beta,\alpha\gamma}(\mathbf{p}) = M_2$, $M_{\alpha\beta,\gamma\delta}(\mathbf{p}) = M_3$, and eight distinct components for the cubic interaction: $W_{\alpha\beta,\beta\gamma,\gamma\alpha} = W_1$, $W_{\alpha\beta,\alpha\beta,\alpha\beta} = W_2$, $W_{\alpha\beta,\alpha\beta,\beta\gamma} = W_3$, $W_{\alpha\beta,\alpha\beta,\gamma\delta} = W_4$, $W_{\alpha\beta,\beta\gamma,\gamma\delta} = W_5$, $W_{\alpha\beta,\alpha\gamma,\alpha\delta} = W_6$, $W_{\alpha\beta,\alpha\gamma,\delta\mu} = W_7$, $W_{\alpha\beta,\gamma\delta,\mu\nu} = W_8$. It is, however, more convenient to work directly with a field theory defined in terms of the eigenfields, which corresponds to a block diagonalization of the mass operator into the replicon (*R*), anomalous (*A*), and longitudinal (*L*) modes, with an appropriate cubic interaction. This can be obtained by introducing the following representation.[32–34] Any replica field $\phi^{\alpha\beta}$ can be decomposed in its projections onto the *L*, *A*, and *R* subspaces,

$$\phi^{\alpha\beta} = \phi_L^{\alpha\beta} + \phi_A^{\alpha\beta} + \phi_R^{\alpha\beta} \tag{9.44}$$

which, according to the symmetry properties in replica space, can be represented as: $\phi_L^{\alpha\beta} = \phi_L$, a longitudinal field replica-independent, $\phi_A^{\alpha\beta} = (\phi_A^\alpha + \phi_A^\beta)/2$, an anomalous field given by a one-replica field ϕ_A^α, with the condition $\sum_\alpha \phi_A^\alpha = 0$, and a replicon field $\phi_R^{\alpha\beta}$ that depends on two replica indices, with the conditions $\sum_{\beta(\neq\alpha)} \phi_R^{\alpha\beta} = 0,\ \alpha = 1, \ldots, n$. One obtains

$$L^{(2)} = \frac{1}{2}\sum_{\mathbf{p}} \left[(\mathbf{p}^2 + m_L)\frac{n(n-1)}{2}\phi_L(\mathbf{p})\phi_L(-\mathbf{p}) \right. \tag{9.45}$$

$$\left. +(\mathbf{p}^2 + m_A)\frac{(n-2)}{4}\sum_\alpha \phi_A^\alpha(\mathbf{p})\phi_A^\alpha(-\mathbf{p}) + (\mathbf{p}^2 + m_R)\sum_{(\alpha\beta)} \phi_R^{\alpha\beta}(\mathbf{p})\phi_R^{\alpha\beta}(-\mathbf{p}) \right]$$

and

$$L^{(3)} = -\frac{1}{\sqrt{N}}\sum_{\mathbf{p}_1,\mathbf{p}_2,\mathbf{p}_3} \left[\frac{1}{6}g_1 \sum_{\alpha,\beta,\gamma} \phi_R^{\alpha\beta}(\mathbf{p}_1)\phi_R^{\beta\gamma}(\mathbf{p}_2)\phi_R^{\gamma\alpha}(\mathbf{p}_3) \right. \tag{9.46}$$

$$\left. + \frac{1}{12}g_2 \sum_{\alpha,\beta} \phi_R^{\alpha\beta}(\mathbf{p}_1)\phi_R^{\alpha\beta}(\mathbf{p}_2)\phi_R^{\alpha\beta}(\mathbf{p}_3) + \ldots + \frac{1}{6}g_8\phi_L(\mathbf{p}_1)\phi_L(\mathbf{p}_2)\phi_L(\mathbf{p}_3) \right].$$

The masses, m_L, m_A and m_R, respectively for the longitudinal, anomalous and replicon modes, and the couplings, $g_1 = g_{RRR}$, $g_2 = g'_{RRR}$, $g_3 = g_{RRA}$, $g_4 = g_{RRL}$, $g_5 = g_{RAA}$, $g_6 = g_{AAA}$, $g_7 = g_{AAL}$, $g_8 = g_{LLL}$, between the different modes, can be explicitly calculated in terms of the masses M_i, $i = 1$, 2, 3, and the couplings W_i, $i = 1, \ldots, 8$.[32,33] The structure of the masses and the couplings depends on the mean-field value of the order parameter. Because the anomalous and replicon fields are not independent, the free propagators are nondiagonal and involve projector operators. We have then obtained a general replica-symmetric field theory for the spin glass, which works directly with the eigenfields, and allows to perform a standard perturbation expansion.[33]

At the mean-field level, where the order parameter is given by,

$$\Theta^{-1}Q = \langle S^\alpha S^\beta \rangle = \frac{\mathrm{Tr}_{S^\alpha}\left[S^\alpha S^\beta \exp V_s(Q)\right]}{\mathrm{Tr}_{S^\alpha} \exp V_s(Q)} \tag{9.47}$$

one finds the usual results, that is a transition in zero field, which occurs at $\Theta = 1$, i.e., at a temperature $T_f = \sqrt{z}J$, where all the modes become critical, and a transition in nonzero field, along the AT line, with only the replicon mode critical.

For spin glasses, the upper critical dimension, above which mean field theory applies, is $d_u = 6$, since the lowest-order interaction is cubic in the fields. The momentum-space renormalization group (RG) approach allows to analyze the critical behavior and to calculate the critical exponents as an expansion in powers of $\epsilon = 6 - d$.

The spin-glass transition in zero field was first studied within the RG by Harris *et al.*[35] Bray and Roberts[34] considered the case of a nonzero field, and carried out a RG study, in which they retained only the replicon modes to calculate the critical behavior at the AT line. Pimentel *et al.*[33] studied the spin-glass transition in a field, within the complete set of RG equations, containing the replicon, anomalous, and longitudinal modes. This allowed to discuss, in a common framework, the transitions in zero and nonzero fields, and the crossover region around the zero-field critical point, thus investigating the role that a small magnetic field plays in the transition. The renormalization-group equations were obtained by standard methods of integration of degrees of freedom over an infinitesimal momentum shell, $e^{-dl}\Lambda < |\mathbf{p}| < \Lambda$, at a cutoff $\Lambda \simeq 1$. The leading, one-loop, order approximation for the masses and the couplings corresponds to, respectively, "bubble" and "triangle" diagrams. The expression for the complete set of renormalization equations, for the three masses m_i, $i = L, A, R$, and the eight couplings g_i, $i = 1, \ldots, 8$, can be seen in Ref. 33.

In zero magnetic field, $h = 0$, and above the transition, $T > T_f$, the order parameter is zero, $Q = 0$, and there is indeed only one mass and one coupling, i.e., there is a symmetry in which, $m_L = m_A = m_R = m$, $g_2 = 0$ and the couplings g_i, with $i = 3, \ldots, 8$, are linearly related to $g_1 = g$. The renormalization group equations in the generalized parameter space, reduce then to, in the limit $n \to 0$,

$$\frac{dm}{dl} = (2 - \eta)m + 2g^2 \frac{1}{(1+m)^2} \tag{9.48}$$

$$\frac{dg}{dl} = \frac{1}{2}(\varepsilon - 3\eta)g - 2g^3 \frac{1}{(1+m)^3} \tag{9.49}$$

with,

$$\eta = -g^2 \frac{2(1+3m)}{3(1+m)^4}. \tag{9.50}$$

These equations have a nontrivial fixed-point: $m^* = -\epsilon/2$, $(g^*)^2 = \epsilon/2$, which corresponds to the spin-glass transition in zero-field, as found by Harris *et al.*[35]

In a finite magnetic field, $h \neq 0$, the order parameter above the transition, $T > T_f$, is nonzero, $Q \neq 0$, which generates a splitting of the masses and the couplings, and it is then necessary to work with all the three masses and the eight cubic couplings. One finds that in the generalized parameter space, there is a fixed-point associated to the spin-glass transition in zero field, which is stable along a direction that has the zero-field symmetry, but it is unstable in other directions in the coupling space. This observation explains the existence of a stable zero-field fixed point, which however becomes unstable in the presence of a small magnetic field. On the AT line, the order parameter takes the value $Q \simeq h^{2/3}$, and only the replicon modes are critical $m_R = 0$, the longitudinal and the anomalous modes having $m_L = m_A \simeq h^{2/3}$. Under the renormalization group transformation, the longitudinal and the anomalous modes scale out of the problem, and only the replicon modes remain, with the reduced set of equations,

$$\frac{dm_R}{dl} = (2 - \eta_R)m_R - \left[4g_1^2 - 8g_1g_2 + \frac{11}{4}g_2^2\right]\frac{1}{(1+m_R)^2} \tag{9.51}$$

$$\frac{dg_1}{dl} = \frac{1}{2}(\varepsilon - 3\eta_R)g_1 + \left[14g_1^3 - 18g_1^2g_2 + \frac{9}{2}g_1g_2^2 + \frac{1}{8}g_2^3\right]\frac{1}{(1+m_R)^3} \tag{9.52}$$

$$\frac{dg_2}{dl} = \frac{1}{2}(\varepsilon - 3\eta_R)g_2 + \left[24g_1^2g_2 - 30g_1g_2^2 + \frac{17}{2}g_2^3\right]\frac{1}{(1+m_R)^3} \tag{9.53}$$

with,

$$\eta_R = \left[4g_1^2 - 8g_1g_2 + \frac{11}{4}g_2^2\right]\frac{(1+3m_R)}{3(1+m_R)^4}. \tag{9.54}$$

These equations are equivalent to the ones studied by Bray and Roberts.[34] The analysis of these equations shows the existence of some fixed points, which are however unphysical. As a result, no physical stable fixed point is found to describe the AT transition.[33,34] This implies that below six dimensions, there is no second-order spin-glass transition in a finite magnetic field, possibly because the fluctuations may drive the transition to become first order, or the fluctuations may destroy the transition.

The critical behavior associated to the spin-glass transition in zero-field is described as follows, respectively, for the order parameter, the susceptibility, the specific heat, and the correlation length, as $T \to T_f$,

$$q \propto (T_f - T)^{\beta}\,, \quad \chi_{SG} \propto |T - T_f|^{-\gamma}\,, \tag{9.55}$$

$$C \propto |T - T_f|^{-\alpha}\,, \quad \xi_{SG} \propto (T - T_f)^{-\nu}\,, \tag{9.56}$$

and the correlation function, and the q dependence on the conjugate field, at $T = T_f$, as

$$G_{SG}(p) \propto p^{-(2-\eta)}, \qquad q \propto \left(h^2\right)^{1/\delta}. \tag{9.57}$$

The critical exponents were calculated by the renormalization group, and to lowest order in ϵ, it was found:[33,35]

$$\nu = \frac{1}{2} + \frac{5}{12}\varepsilon, \qquad \eta = -\frac{\varepsilon}{3}, \tag{9.58}$$

the other exponents being obtained via the scaling relations, $\alpha + 2\beta + \gamma = 2$, $\beta = (d - 2 + \eta)\nu/2$, $\gamma = (2 - \eta)\nu$, $\delta = (d + 2 - \eta)/(d - 2 + \eta)$.

As mentioned, the zero-field fixed-point becomes unstable in the presence of a small magnetic field, and crossover exponents were calculated, which were related to zero-field critical exponents.[33] It is however not clear where the crossover leads, since no physical stable fixed point was found that would describe the spin glass transition in a magnetic field.

Computer simulations give convincing evidence that there exists a finite-temperature phase transition in the three-dimensional short-range Ising spin glass, in zero magnetic field.[36] The critical exponents were calculated numerically by Monte Carlo simulations and high-temperature series expansions. Although there is some spread in the values for the critical exponents, an analysis of the results suggest that three-dimensional Ising spin glasses obey universality, in the sense that the critical exponents depend only on the dimensionality of space and the symmetry of the order parameter, but not on the bond distribution, i.e. , they are essentially the same for, e.g., the Gaussian or the bimodal distribution ($\nu \approx 2.4$, $\eta \approx -0.38$).[37] In the two-dimensional Ising spin glasses, the transition occurs at zero temperature, and it has been found that the systems fall into different universality classes: for the Gaussian distribution ($\theta < 0$), $\eta = 0$ and $\nu = -1/\theta$ ($\nu \approx 3.5$),[17] while for the bimodal distribution ($\theta = 0$), $\nu = \infty$ (exponential scaling) and η is nonzero ($\eta \approx 0.14$).[38] To compare the critical exponents calculated with experimental values measured in real spin glasses see, e.g., Refs. 1, 3 and 4.

Extensive numerical simulations have also been carried out on the three-dimensional short-range Ising spin glass, in order to investigate whether there is a spin glass transition in a magnetic field. While some authors[39,40] claim that there is some evidence of a phase transition occurring in a field, others[41–43] state that the spin glass phase does not survive in any finite field, and hence there is no phase transition in a field, i.e., no AT line. From the experimental point of view, there have been reports of the observation of irreversibility lines in the h-T plane, which were interpreted as "AT lines". However, the analysis of recent experiments led to the conclusion that no phase transition exists in a magnetic field.[44–46] The apparent "AT lines" seen in experiments, probably correspond to a dynamical effect due to failure of the system to reach equilibrium.

9.4.4. *Glassy Phase*

In order to study the low-temperature phase of short-range Ising spin glasses, with broken replica symmetry, one needs to consider the expansion (9.36) with (9.37)-(9.41), around the Parisi solution of the SK model, and now one should also include the contribution of the quartic term $(Q_i^{\alpha\beta})$,[4] which is responsible for replica symmetry breaking. The development of the theory has proved to be quite difficult, even at the level of the quadratic fluctuations, and the results obtained so far are reviewed in Ref. 47.

The structure of the states in the low-temperature phase of short-range Ising spin glasses has been investigated by different methods. Most numerical work, by Monte Carlo simulations or other techniques, has concentrated on the calculation of the distribution $P(q)$ of the overlaps q. Nontrivial overlap distributions were indeed found for both the SK model and the three-dimensional short-range Ising spin glass. Some authors[23,39,48,49] claim that many features of the short-range Ising spin glass are common to the mean field theory, and hence are well described in terms of Parisi's replica symmetry breaking. However, others[50] state that the low-temperature phases of the SK model and the three-dimensional short-range Ising spin glass are different: while in the former there is ultrametricity, in the latter there is strong evidence for lack of ultrametricity, and therefore the structure determined by the mean field theory is not valid for the short-range spin glass. Analytical work[51] has been critic of the replica symmetry breaking description for the low-temperature phase of short-range Ising spin glasses, and favor the droplet model. The droplet model has also been supported by numerical calculations.[52,53] The search for a satisfactory description of the low-temperature phase of short-range Ising spin glasses continues.

9.5. Conclusion

There is still controversy about the nature of ordering in short-range Ising spin glasses below the *freezing* temperature. Two scenarios have been proposed for the spin-glass phase. The replica symmetry breaking theory assumes that in the short-range spin glasses there are infinitely many equilibrium states, organized in an ultrametric structure, as found in the Parisi solution for the infinite-range spin glass model. It is however not clear whether the qualitative features of the mean field theory apply for the experimentally relevant short-range Ising spin glasses. In turn, the droplet model considers that in the spin-glass phase there are only two states, which are simply related by a global spin inversion, and hence there is no replica symmetry breaking. While the results predicted by the droplet model are intuitively appealing, it consists of a phenomenological scaling theory, supported by some real space renormalization formulation, but still requiring a microscopic justification. It would be desirable to have a replica field theory of interacting fluctuations to describe the low-temperature phase of short-range Ising spin glasses.

In the absence of a magnetic field both the mean field theory and the droplet model predict the existence of a phase transition, but in the presence of a magnetic field the situation is different: mean field predicts a phase transition occurring in a field along the de Almeida-Thouless line, whereas the droplet model shows that the magnetic field destroys the spin-glass phase. However, going beyond mean field theory and considering the effects of the fluctuations associated into the short-range interactions within a renormalization group treatment, it is found that there is no physical fixed-point to describe the AT line, i.e., there is no second-order spin-glass transition in a magnetic field. This result implies agreement with the droplet model, according to which there is no AT line.

At present, there is intensive research work dedicated to the study of the out-of-equilibrium dynamics in spin glasses, which have exhibited remarkable aging effects. The explanation of these phenomena is essential for a deeper understanding of spin glasses.

Acknowledgments

The author would like to specially thank C. De Dominicis for the fruitful interaction during the research work done in collaboration and also acknowledges the discussions with T. Temesvari.

References

1. K. Binder and A. P. Young, Spin glasses: experimental facts, theoretical concepts, and open questions, *Rev. Mod. Phys.* **58**, 801 (1986).
2. M. Mézard, G. Parisi, and M. A. Virasoro, *Spin Glass Theory and Beyond* (World Scientific, Singapore, 1987).
3. K. H. Fischer and J. H. Hertz, *Spin Glasses* (Cambridge University Press, Cambridge, 1991).
4. J. A. Mydosh, *Spin Glasses, An Experimental Introduction* (Taylor & Francis, London, 1993).

5. A. P. Young, Ed., *Spin Glasses and Random Fields* (World Scientific, Singapore, 1998).
6. S. F. Edwards and P. W. Anderson, Theory of spin glasses, *J. Phys. F: Met. Phys.* **5**, 965 (1975).
7. G. Parisi, Infinite number of order parameters for spin-glasses, *Phys. Rev. Lett.* **43**, 1754 (1979) ; A sequence of approximated solutions to the S-K model for spin glasses, *J. Phys. A* **13**, L115 (1980) ; Order parameter for spin glasses: a function of the interval $0-1$, **13**, 1101 (1980) ; Magnetic properties of spin glasses in a new mean field theory, **13**, 1887 (1980) ; Order parameter for spin glasses, *Phys. Rev. Lett.* **50**, 1946 (1983).
8. D. Sherrington and S. Kirkpatrick, Solvable model of a spin-glass, *Phys. Rev. Lett.* **35**, 1792 (1975).
9. D. S. Fisher and D. A. Huse, Ordered phase of short-range Ising spin-glasses, *Phys. Rev. Lett.* **56**, 1601 (1986); Equilibrium behavior of the spin-glass ordered phase, *Phys. Rev. B* **38**, 386 (1988).
10. W. L. McMillan, Scaling theory of Ising spin glasses, *J. Phys. C* **17**, 3179 (1984).
11. A. J. Bray and M. A. Moore, Scaling theory of the ordered phase of spin glasses, in *Heidelberg Colloquium on Glassy Dynamics and Optimization*, Eds. J. L. van Hemmen and I. Morgenstern (Springer, Heidelberg, 1986), p. 121; Lower critical dimension of Ising spin glasses: a numerical study, *J. Phys. C* **17**, L463 (1984) ; Nonanalytic magnetic field dependence of the magnetisation in spin glasses, **17**, L613 (1984).
12. J. R. L. de Almeida and D. J. Thouless, Stability of the Sherrington-Kirkpatrick solution of a spin glass model, *J. Phys. A* **11**, 983 (1977).
13. J.-P. Bouchaud, L. Cugliandolo, J. Kurchan and M. Mézard, Out of equilibrium dynamics in spin-glasses and other glassy systems, in Ref. 5, p. 161.
14. A. J. Bray and M. A. Moore, Replica symmetry and massless modes in the Ising spin glass, *J. Phys. C* **12**, 79 (1979).
15. C. De Dominicis and I. Kondor, Eigenvalues of the stability matrix for Parisi solution of the long-range spin-glass, *Phys. Rev. B* **27**, 606 (1983) ; On spin-glass fluctuations, *J. Phys. (Paris) Lett.* **45**, L205 (1984) ; Gaussian propagators for the Ising spin-glass below T_c, *J. Phys. (Paris) Lett.* **46**, L1037 (1985).
16. A. K. Hartmann, Scaling of stiffness energy for three-dimensional $\pm J$ Ising spin glasses, *Phys. Rev. E* **59**, 84 (1999).
17. H. G. Katzgraber, L. W. Lee and A. P. Young, Correlation length of the two-dimensional Ising spin glass with Gaussian interactions, *Phys. Rev. B* **70**, 014417 (2004).
18. C. Amoruso, E. Marinari, O. C. Martin and A. Pagnani, Scalings of domain wall energies in two dimensional Ising spin glasses, *Phys. Rev. Lett.* **91**, 087201 (2003).
19. S. Boettcher, Stiffness of the Edwards-Anderson model in all dimensions, *Phys. Rev. Lett.* **95**, 197205 (2005).
20. C. Amoruso, A. K. Hartmann and M. Moore, Determining energy barriers by iterated optimization: the two-dimensional Ising spin glass, *Phys. Rev. B* **73**, 184405 (2006).
21. A. K. Hartmann and A. P. Young, Large-scale low-energy excitations in the two-dimensional Ising spin glass, *Phys. Rev. B* **66**, 094419 (2002).
22. M. Palassini and A. P. Young, Nature of the Spin Glass State, *Phys. Rev. Lett.* **85**, 3017 (2000).

23. E. Marinari and G. Parisi, Effects of changing the boundary conditions on the ground state of Ising spin glasses, *Phys. Rev. B* **62**, 11677 (2000).
24. A. J. Bray, The ordered phase of a spin glass, *Comments Cond. Mat. Phys.* **14**, 21 (1988).
25. G. Parisi, Spin glasses and replicas, *Physica A* **124**, 523 (1984).
26. A. J. Bray and M. A. Moore, Chaotic nature of the spin-glass phase, *Phys. Rev. Lett.* **58**, 57 (1987).
27. J. R. Banavar and A. J. Bray, Chaos in spin glasses: a renormalization-group study, *Phys. Rev. B* **35**, 8888 (1987).
28. M. Ney-Nifle and A. P. Young, Chaos in a two-dimensional Ising spin glass, *J. Phys. A* **30**, 5311 (1987).
29. H. G. Katzgraber and F. Krzakala , Temperature and disorder chaos in three-dimensional Ising spin glasses, cond-mat/0606180 (2006).
30. T. Aspelmeier, A. J. Bray and M. A. Moore, Why temperature chaos in spin glasses is hard to observe, *Phys. Rev. Lett.* **89**, 197202 (2002).
31. T. Rizzo and A. Crisanti, Chaos in temperature in the Sherrington-Kirkpatrick model, *Phys. Rev. Lett.* **90**, 137201 (2003).
32. T. Temesvari, C. De Dominicis and I. R. Pimentel, Generic replica symmetric field-theory for short range Ising spin glasses, *Eur. Phys. J. B* **25**, 361 (2002).
33. I. R. Pimentel, T. Temesvari and C. De Dominicis, Spin-glass transition in a magnetic field: a renormalization group study, *Phys. Rev. B* **65**, 224420 (2002).
34. A. J. Bray and S. A. Roberts, Renormalisation-group approach to the spin glass transition in finite magnetic fields, *J. Phys. C* **13**, 5405 (1980).
35. A. B. Harris, T. C. Lubensky and J. H. Chen, Critical properties of spin-glasses, *Phys. Rev. Lett.* **36**, 415 (1976).
36. H. G. Ballesteros, A. Cruz, L. A. Fernández, V. Martín-Mayor, J. Pech, J. J. Ruiz-Lorenzo, A. Tarancón, P. Téllez, C. L. Ullod and C. Ungil, Critical behavior of the three-dimensional Ising spin glass, *Phys. Rev. B* **62**, 14237 (2000).
37. H. G. Katzgraber, M. Körner and A. P. Young, Universality in three-dimensional Ising spin glasses: a Monte Carlo study, *Phys. Rev. B* **73**, 224432 (2006).
38. H. G. Katzgraber and L. W. Lee, Correlation length of the two-dimensional Ising spin glass with bimodal interactions, *Phys. Rev. B* **71**, 134404 (2005).
39. E. Marinari, G. Parisi, and J. J. Ruiz-Lourenzo, Numerical Simulations of Spin Glass Systems, in Ref. 5, p. 59.
40. F. Krzakala, J. Houdayer, E. Marinari, O. C. Martin and G. Parisi, Zero-temperature responses of a 3d spin glass in a magnetic field, *Phys. Rev. Lett.* **87**, 197204 (2001).
41. J. Houdayer and O. C. Martin, Ising spin glasses in a magnetic field, *Phys. Rev. Lett.* **82**, 4934 (1999).
42. J. Lamarcq, J.-P. Bouchaud and O. C. Martin, Local excitations of a spin glass in a magnetic field, *Phys. Rev. B* **68**, 012404 (2003).
43. A. P. Young and H. G. Katzgraber, Absence of an Almeida-Thouless line in three-dimensional spin glasses, *Phys. Rev. Lett.* **93**, 207203 (2004).
44. P. Nordblad and P. Svedlindh, Experiments in spin glasses, in Ref. 5, p. 1.
45. J. Mattsson, T. Jonsson, P. Nordblad, H. Aruga Katori and A. Ito, No phase transition in a magnetic field in the Ising spin glass $Fe_{0.5}Mn_{0.5}TiO_3$, *Phys. Rev. Lett.* **74**, 4305 (1995).

46. P. E. Jönsson, H. Takayama, H. Aruga Katori and A. Ito, Dynamical breakdown of the Ising spin-glass order under a magnetic field, *Phys. Rev. B* **71**, 180412(R) (2005).
47. C. De Dominicis, I. Kondor and T. Temesvari, Beyond the Sherrington- Kirkpatrick model, in Ref. 5, p. 119.
48. E. Marinari, G. Parisi, F. Ricci-Tersenghi, J. J. Ruiz-Lourenzo, and F. Zuliani, Replica symmetry breaking in short-range spin glasses: theoretical foundations and numerical evidences, *J. Stat. Phys.* **98**, 973 (2000).
49. E. Marinari, O. C. Martin and F. Zuliani, Equilibrium valleys in spin glasses at low temperature, *Phys. Rev. B* **64**, 184413 (2001).
50. G. Hed, A. Young and E. Domany, Lack of ultrametricity in the low-temperature phase of three-dimensional Ising spin glasses, *Phys. Rev. Lett.* **92**, 157201 (2004).
51. C. M. Newman and D. L. Stein, Non-mean-field behavior of realistic spin glasses, *Phys. Rev. Lett.* **76**, 515 (1996); Ordering and broken symmetry in short-ranged spin glasses, *J. Phys.: Condens. Matter* **15**, R1319 (2003).
52. M. A. Moore, H. Bokil and B. Drossel, Evidence for the droplet picture of spin glasses, *Phys. Rev. Lett.* **81**, 4252 (1998).
53. A. K. Hartmann and M. A. Moore, Corrections to scaling are large for droplets in two-dimensional spin glasses, *Phys. Rev. Lett.* **90**, 127201 (2003).

Chapter 10

Competition between several model Hamiltonians in half-doped manganites

Roland Bastardis and Nathalie Guihéry

Laboratoire de Chimie et Physique Quantiques, IRSAMC, Université Paul Sabatier

118, route de Narbonne, 31062 Toulouse cedex

*nathalie@irsamc.ups-tlse.fr**

Ab initio calculations combined with the effective Hamiltonian theory of Bloch provide a rational way to determine model Hamiltonians. The embedded cluster approach is the most reliable method of extraction of effective interactions for the study of highly correlated material. In the specific case of half-doped manganites, several model Hamiltonians can be considered to reproduce the local physics generated by the interactions between the magnetic sites according to the position of the doping holes. While a double exchange mechanism takes place between the Mn sites if the holes are localized on the metals, a purely magnetic Heisenberg Hamiltonian should be considered for a localization of the holes on the bridging oxygens. For intermediate situations in which both elements share the doping holes, a truncated Hubbard model which treats variationaly double exchange and Heisenberg configurations seems to be the most appropriate. This model can be mapped on both simpler double exchange and Heisenberg Hamiltonians. The analytical spectrum of the Heisenberg model in the case of two metals bridged by a magnetic oxygen is identical (except for one state) to the double exchange one, for a peculiar relation between the electronic interactions of the two models. Finally, the most appropriate hamiltonians is a refined double exchange model which combines the Anderson-Hazegawa and the Girerd-Papaefthymiou antiferromagnetic contributions.

Contents

*The laboratoire de Chimie et Physique Quantiques is UMR 5626 of the CNRS.

10.1. Introduction

The complexity of the description of highly correlated materials comes from both the infinite size of the system and the multireference character of their correlated wave-functions. Most of the methods available for the study of collective effects in such periodic lattices cannot handle the exact electronic Hamiltonian. Their reliability rest on the accuracy of the used effective Hamiltonians. *Ab initio* calculations performed using explicitely correlated methods combined with the effective Hamiltonian theory of Bloch[1] provides a rational way to extract these simpler model Hamiltonians. The embedded cluster approach[2–5] is at the moment the most reliable available method of extraction. It consists in a correlated study of the local interactions involved in a fragment of the material embedded in the adequate crystal-like environment. The method proceeds through an identification of the exact Hamiltonian spectrum with the effective Hamiltonian spectrum. The extraction can be performed i) from the single spectrum by expressing the effective interactions in terms of energy spacings ii) using both the spectrum and wave-functions projections onto the model space when additional equations are required. In all cases, wave-functions are used to control the procedure since only those states which have a large projection onto the model space can be correctly reproduced by the effective Hamiltonian. When the extraction is equivocal, this criterium can also be used in order to discriminate between different modelizations.

The present chapter is devoted to the comparison of the adequacy of different model Hamiltonians, in particular double exchange, purely magnetic Heisenberg and truncated Hubbard models, for the description of manganites.[8] These compounds have been extensively studied for their remarkable property of colossal magnetoresistance[9,10] which can be used for industrial applications like information storage for instance. Nevertheless, the richness[11] of their phase diagrams originating from strong interplay between electronic, magnetic and structural factors makes them attractive from a fundamental point of view too.[12] For instance, $La_{0.5}Ca_{0.5}MnO_3$ presents two phase transitions when increasing temperature: It is antiferromagnetic below the Néel temperature T_N and paramagnetic for $T > T_{CO}$. For intermediate temperature $T_N < T < T_{CO}$ and according to the formal oxidation degrees of Mn sites (equal proportion of Mn(IV) and Mn(III)), the electronic ground state was seen as a charge ordered (CO) phase exhibiting a CE-type ordering of the magnetic moments[13,14] for a long time. *Ab initio* studies[4] performed on these crystal structures confirm the CE magnetic order. This description has however been questioned by several authors.[15–20] Daoud-Aladine and co-workers[15] reported a crystal structure determination of the closely related $Pr_{0.6}Ca_{0.4}MnO_3$ material in which all Mn ions are identical (intermediate valence state $Mn^{3.5+}$, corresponding to a resonance between $Mn^{3+}O^{2-}Mn^{4+}$ and $Mn^{4+}O^{2-}Mn^{3+}$). The crystal structure suggests the trapping of electrons within pairs of Mn sites involving a local double exchange process. The proposed electronic structure is interpreted as a Zener polaron ordering, in reference to the Zener double exchange model.[21] Here again *ab initio* calculations[4] performed on the corresponding experimentally determined crystal structure confirm the existence of a polaronic order. These calculations finally show that charge, orbital and magnetic orders are strongly depending on the crystal structure. Both orders (CE phase or Zener polarons) are actually compatible with theoretical results obtained from the two kinds of crystal structures (CO or not). Such a controversy cannot therefore be assessed from explicitely correlated *ab initio* calculations as long as more refined crystal structures are not available. As a counter part, all *ab initio* calculations exhibit strong O to Mn charge transfer, resulting in a partial localization of the holes on the bridging oxygens. UHF and DFT periodic calculations[17–19] on $La_{0.5}Ca_{0.5}MnO_3$ even lead to a dominantly $Mn^{3+}O^{-}Mn^{3+}$ ground state wave-function. The present chapter focuses on the impact of the participation of the oxygen on the double exchange mechanism. For this purpose we have considered the crystal determination of A. Daoud Aladine *et al.* where symmetric dimers ruled by a simple double exchange mechanism are invoked. According to the position of the holes, different electronic structures are expected for the Zener polarons :

- If the holes are localized on the Mn sites, the double exchange mechanism induces a resonance between $Mn^{3+}O^{2-}Mn^{4+}$ and $Mn^{4+}O^{2-}Mn^{3+}$. The t_{2g}-like electrons are unpaired and essentially localized on each ion, while one e_g-like electron is delocalized between two Mn ions. The double exchange mechanism is expected to rule the nature and energy spacings of the eight states in the low-energy spectrum of the Mn pairs that arise from this seven unpaired electron system.

- If the holes are localized on the bridging oxygen, the corresponding charge distribution suggests a dominant purely magnetic local order ($Mn^{3+}O^{-}Mn^{3+}$) in which the dimers would be ferrimagnetic entities involving a magnetic oxygen and therefore nine unpaired electrons. The model Hamiltonian which provides a relevant description of such a local electronic order is a Heisenberg Hamiltonian.
- For an intermediate situation in which the Mn and O sites share the holes, both double exchange and Heisenberg electronic configurations should be considered and a Hubbard-type Hamiltonian in which they are treated variationally is expected to be the most appropriate model Hamiltonian.

In order to choose between these different model Hamiltonians, let us confront both their spectrum and ground state wave-function to those of the exact electronic Hamiltonian in the case of a dimer of Mn sites (the Zener polaron of the $Pr_{0.6}Ca_{0.4}MnO_3$ material) embedded in the adequate crystal-like environnement.

10.2. The double exchange model

The double exchange mechanism takes place in mixed valence complexes. In the considered case, only configurations compatible with a $Mn^{3.5+}O^{2-}Mn^{3.5+}$ electronic structure can *a priori* be described within a double exchange model (see Fig. 10.1(a)). Let us call $a_{2,3,4}$ and $b_{2,3,4}$ the t_{2g}-like orbitals and a_1 and b_1 the e_g-like ones respectively localized on the left (a_i) and right (b_i) Mn sites. Due to Jahn-Teller distortions only one e_g orbital is occupied. The simpler double exchange model[21,22] is based on the idea that the spectrum of the dimer can be reproduced by considering the metallic ions in their atomic ground states (infinite Hund's coupling). Local ground state on each Mn site can be either a quintet Q_2^i or a quartet $Q_{3/2}$ state according to the position of the extra electron (e_g-like in the present case).

The dominant electronic interaction is the hopping integral t of the e_g-like electron between two essentially metallic orbitals. In the considered bimetallic complex, the delocalization of this electron will generate two octet states O_+ and O_-, in the $S_z = 7/2$ subspace. When considering lower S_z components of the local ground states, the coupling between the corresponding double exchange determinants through the hopping integral t generates states of lower spin multiplicity, namely two sextet S_- and S_+, two quartet Q_- and Q_+ and two doublet D_- and D_+ states. As shown by Girerd and Papaefthymiou,[23,24] the electronic circulation in the t_{2g}-like orbitals introduces an antiferromagnetic contribution of Heisenberg type that stabilizes the lowest and intermediate spin states compared to the

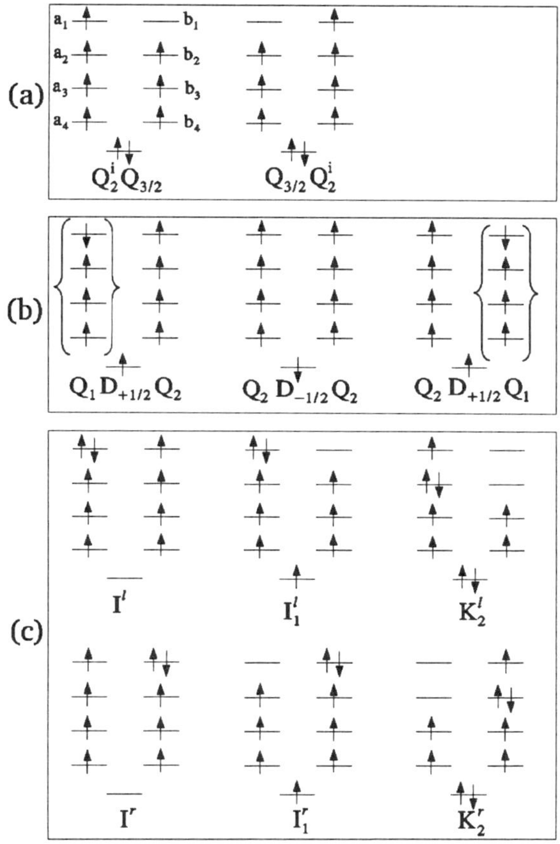

Fig. 10.1. Picture of the determinants considered in the (S_z=7/2 subspace) : (a) Neutral double exchange determinants, (b) Neutral charge transfer (Heisenberg) determinants; The function $(Q_1)_l = \frac{1}{2}(\bar{a_1}a_2a_3a_4 + a_1\bar{a_2}a_3a_4 + a_1a_2\bar{a_3}a_4 + a_1a_2a_3\bar{a_4})$ is a linear combination of determinants differing by the position of the down spin on the left Mn site, the brackets join the orbitals which can bear the down spin, (c) ionic or neutral determinants having a doubly occupied metallic orbital indicated by the index. Index 1 means a_1 (b_1) in I_1^l (I_1^r respectively). Reused with permission from Roland Bastardis, Nathalie Guihéry, Nicolas Suaud, and Coen de Graaf, Journal of Chemical Physics, 125, 194708 (2006). Copyright 2006, American Institute of Physics.

octet ones. Let us call J the overall antiferromagnetic exchange integral between the Mn ions. In the particular case of a symmetric homonuclear bimetallic complex, the eigenenergies of the usual double exchange model (here noted the GP model in reference to Girerd and Papaefthymiou) are analytically known. Energies $E^{GP}(S, \pm)$ of the different states of total spin S are given by the expression :

$$E^{GP}(S,\pm) = \pm\frac{t}{S_m^{DE}+\frac{1}{2}}\left(S+\frac{1}{2}\right) - \frac{J}{2}\left[S(S+1) - S_m^{DE}(S_m^{DE}+1)\right], \tag{10.1}$$

where S_m^{DE} is the highest total spin (m stands for maximal) of the double exchange model and the zero of energy is taken as the mean value of the highest-spin states energies. The first term of Eq. (10.1) is generally dominant and accounts for the appearance of a high spin ground state, i.e. a ferromagnetic order. Indeed, the delocalization of the extra electron occurs between the orbitals having the largest overlap, and therefore leading to the largest hopping integral t. Due to the presence of a bridging ligand, this integral is expected to be dominated by the through-ligand contribution. The optimal a_1 and b_1 orbitals for the double exchange model will be slightly twisted d_{z^2} orbitals presenting large delocalization tails on the $2p_z$ bridging oxygen orbital (the Z axis being the intermetallic axis).

10.3. The Heisenberg model

The Heisenberg Hamiltonian works in the space spanned by the nine-unpaired-electron configurations corresponding to a $Mn^{3+}O^{-}Mn^{3+}$ electronic structure (see Fig. 10.1(b)). Here again the magnetic sites are taken in their atomic ground state. Each Mn ion has therefore four unpaired electrons arranged in a Quintet state, the five S_z components of which will be noted Q^i_{-2}, Q^i_{-1}, Q^i_0, Q^i_1 and Q^i_2. The bridging oxygen has one unpaired electron in a local and orthogonal $2p_z$-like orbital leading to a doublet ground state having the two $D_{-1/2}$ and $D_{1/2}$ S_z components. In the considered case, the Heisenberg Hamiltonian can be written as follows :

$$\mathbf{H}^H = -J_1\mathbf{S}_l.\mathbf{S}_O - J_1\mathbf{S}_r.\mathbf{S}_O - J_2\mathbf{S}_l.\mathbf{S}_r, \tag{10.2}$$

where l and r stand for the left and right identical Mn ions, the magnetic interaction between each Mn ion and the oxygen one is J_1 while J_2 parametrizes the overall spin exchange interaction between the two Mn ions. The operator $\mathbf{S}^2$ can be written as

$$\mathbf{S}^2 = (\mathbf{S}_l + \mathbf{S}_r + \mathbf{S}_O)^2 = \mathbf{S}_r^2 + \mathbf{S}_l^2 + \mathbf{S}_O^2 + 2(\mathbf{S}_l.\mathbf{S}_O + \mathbf{S}_r.\mathbf{S}_O + \mathbf{S}_r.\mathbf{S}_l). \tag{10.3}$$

Replacing $\mathbf{S}_r.\mathbf{S}_l$ by its expression as a function of $\mathbf{S}^2$ and using $\mathbf{S}' = \mathbf{S}_r + \mathbf{S}_l$ and $\mathbf{S}'.\mathbf{S}_O = \frac{1}{2}\left(\mathbf{S}^2 - \mathbf{S}'^2 - \mathbf{S}_O^2\right)$ the Hamiltonian can be rewritten

$$\mathbf{H}^H = -\frac{J_1 - J_2}{2}\left[\mathbf{S}^2 - \mathbf{S}'^2 - \mathbf{S}_O\right] - \frac{J_2}{2}\left[\mathbf{S}^2 - \mathbf{S}_l^2 - \mathbf{S}_r^2 - \mathbf{S}_O^2\right]. \quad (10.4)$$

Since the oxygen has only one unpaired electron, i.e. $S_O = 1/2$, the values of S' can only be $S' = S \pm \frac{1}{2}$, and an analytical expression of the eigenenergies can be found

$$\begin{aligned} E^H(S,\pm) = {} & \frac{J_1 - J_2}{2}\left[\frac{1}{2} \pm \left(S + \frac{1}{2}\right)\right] \\ & - \frac{J_2}{2}\left[S(S+1) - \frac{1}{2}\left(S_m^H - \frac{1}{2}\right)^2 - \frac{3}{4}\right], \end{aligned} \quad (10.5)$$

where $S_m^H = S_l + S_r + \frac{1}{2}$ is the highest value that S can take. Let us note that, for the highest spin multiplicity state, only the $E(S_m^H, -)$ root has a physical meaning since S' has the single value $S' = S - 1/2$.

10.4. Extraction of the double exchange and Heisenberg models from the *ab initio* spectrum

10.4.1. *Confrontation of the model spectra with the ab initio spectrum*

Model spectra will be confronted with the spectrum computed using explecitely correlated *ab initio* methods.[6,7]

The effective electronic interactions of the double exchange and Heisenberg Hamiltonians are extracted from the *ab initio* spectrum using the effective Hamiltonian theory.[1] In the two here-considered models, the number of equations is higher than the number of effective model interactions. Different values of these interactions can therefore be extracted from the energy differences between the calculated states. Interactions have therefore been optimized in order to reproduce at best the computed energies of the two octet, two sextet, two quartet and two doublet states and are reported in Table 10.1. The mean error per state is calculated as the difference between the computed energies and the modelized ones (using the optimized interactions) divided by the spectrum width and the number of states. Let us call ϵ the percentage of mean error. The values of the extracted interactions are congruent with what could be expected for such a material. One should note the large value of t which actually rationalises the trapping of one hole inside a dimer of Mn sites, i.e. the polaron.

Using these electronic interactions and the Eq. (10.1) and Eq. (10.5), the spectra of the two model Hamiltonians (including the decuplet ($2S$+1=10) state) have

Table 10.1. Effective electronic interactions (in eV), and percentage of error of the double exchange (GP) and Heisenberg (Heis.) models.

	J_1	J_2	t	J	ϵ
Heis.	0.59	0.07	-	-	3.33
GP	-	-	1.05	0.07	3.33

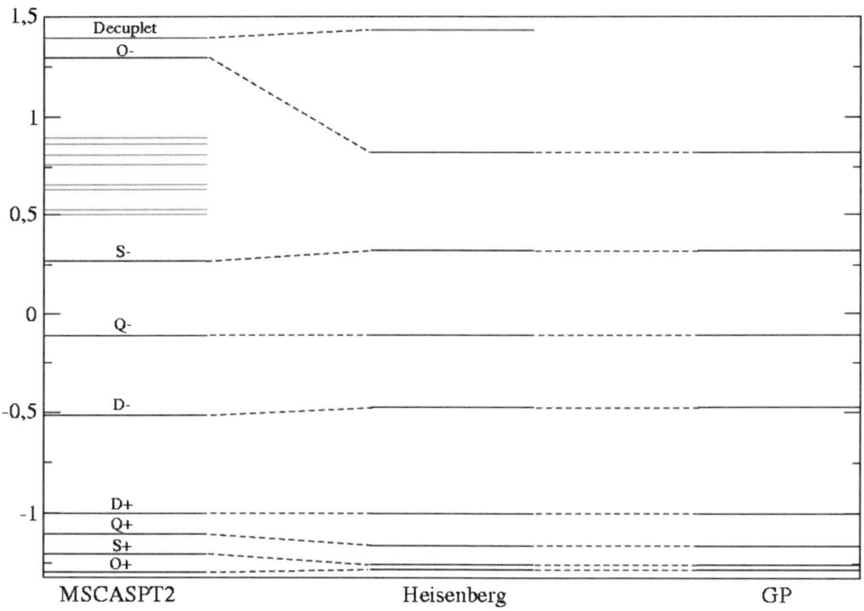

Fig. 10.2. Confrontation of the double exchange and Heisenberg model spectra to the *ab initio* MS-CASPT2 calculated one.

been calculated and are represented in front of the *ab initio* spectrum in Fig. 10.2. A second family of states which is actually ruled by a different physics and will therefore be discarded in the following discussion is intercalated in the upper part of the here-considered spectrum.

Surprisingly enough the two model spectra are identical except for one state, the decuplet state wich does not belong to the double exchange model space. As can be checked in Table 10.1, the error (calculated or the eight lowest states) of both models is also of course strictly identical. The physics of the two models is different and the obtention of a single and identical spectrum is absolutly not expected. In order to understand the origin of the identity of the two spectra, let us come back to the Eq. (10.1) and Eq. (10.5) from which they have been obtained. If one uses the same zero of energy in the two models, for instance the zero of energy of the double exchange model, the expression of the energies of the Heisenberg model becomes :

$$E^H(S,\pm) = \pm\, \frac{J_1 - J_2}{2}\left(S + \frac{1}{2}\right) - \frac{J_2}{2}[S(S+1) - S_m^{DE}(S_m^{DE}+1)].\quad(10.6)$$

By comparing Eq. (10.1) and Eq. (10.6), one reaches the following important conclusion : The spectra of the Heisenberg and double exchange models are strictly identical (except for the highest spin state) if the electronic interactions of the two models fulfill the following conditions :

$$t = \frac{(J_1 - J_2)}{2}\left(S_m^{DE} + \frac{1}{2}\right) \quad \text{and} \quad J = J_2. \quad (10.7)$$

This general property of the two model Hamiltonians makes impossible the choice between the two models from the comparison of their spectrum (experimentally or theoretically determined). Since the spectroscopic study does not provide any decisive argument to choose between these two models, let us confront their ground state wave-functions with the *ab initio* one.

10.4.2. *Confrontation of the model ground state wave-functions with the ab initio one*

According to the effective Hamiltonian theory of Bloch,[1] the adequacy of the extracted effective Hamiltonian can directly be appreciated by the weight of the model wave-functions in the *ab initio* ones. In order to compare the weights of the two ground-state model wave-functions, appropriate orbitals (i.e. which maximizes the projections of the *ab initio* wave-functions onto both model spaces) must be determined first. Indeed, the orbital set in which is defined any model Hamiltonian is always implicit. While the here-calculated *ab initio* wave-function is invariant under the active orbital rotations, the values of the wave-function projections onto both model spaces strongly depend on the choosen orbital set. For

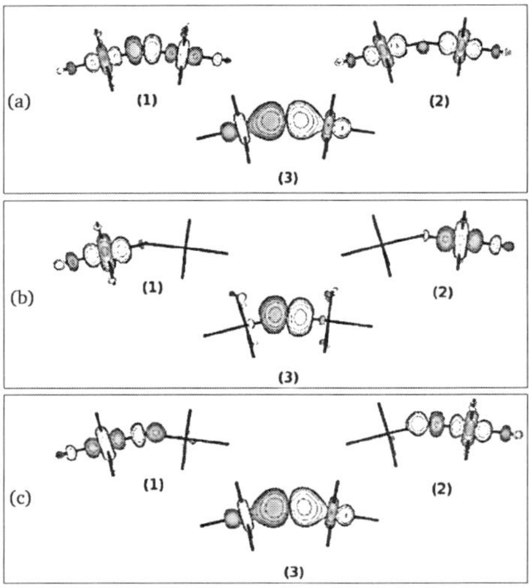

Fig. 10.3. (a) Symmetry adapted natural orbitals, (b) strongly localized orbitals, (c) semi-localized orbitals. Reused with permission from Roland Bastardis, Nathalie Guihéry, Nicolas Suaud, and Coen de Graaf, Journal of Chemical Physics, 125, 194708 (2006). Copyright 2006, American Institute of Physics.

the Heisenberg Hamiltonian, strongly localized orbitals on the three centers (Mn and O) constitute the best orbital set (in which the occupation of the bridging oxygen site is close to 1). On the contrary, semi-localized orbitals having large tails on the neighbouring sites are more adapted to the double exchange model since this set maximizes the occupation of the oxygen (which is assumed to be O^{2-} in the double exchange mechanism). The three orbitals which differ in the two models are represented in Fig. 10.3. They have been obtained from the canonical ones by unitary transformations.

The weights of the two model wave-functions in the *ab initio* ground state wave-function are 0.72 for the double exchange model and 0.74 for the Heisenberg one. Once more we are facing an indecidable situation. While these values are important enough to ensure a correct representability of the physics of the polaron by one or the other model, the similarity of the projections prevents us from deciding which model is the most appropriate. Let us note however that while these models are not identical, a non-negligeable and common part of the physics is described by both models, since the projections are both higher than 50 %. For a deeper insight into the physics of the system, let us consider a variational treatment of both Heisenberg and double exchange configurations, i.e. a Hubbard-like model.

10.5. A truncated Hubbard model

10.5.1. *Theory*

The Hubbard Hamiltonian works on a space spanned by the electronic configurations obtained by distributing all magnetic electrons in all magnetic orbitals. When the number of open-shells per center is large, as it is the case in manganites, most of the so-obtained distributions are di-ionic, tri-ionic and tetra-ionic configurations corresponding for instance to electronic structures compatible with $Mn^+O^-Mn^{5+}$, MnO^-Mn^{6+} or even $Mn^-O^-Mn^{7+}$. The energy of these functions being very high, their weight in the lowest Hubbard states wave-functions is negligible and they will not be considered in the here-developped truncated Hubbard model. The following partition of the configurational space is performed (see Fig. 10.1(a,b,c)) :

- Heisenberg and double exchange configurations belong to the model space and are treated variationaly.
- Neutral and ionic configurations which are coupled through a hopping integral to the configurations belonging to the model space are treated at the second order of perturbation.
- All other configurations are neglected.

Let us now introduce the interactions of the model. The energy of the Heisenberg configurations is set to zero. The double exchange configurations are at the energy Δ and are coupled to the Heisenberg configurations by the hopping integral $t_{pd} = t_{2p_z a_1} = -t_{2p_z b_1}$ between the $2p_z$ orbital of the bridging oxygen and the a_1 (b_1) orbitals of the Mn ions. The through-space hopping integral $t_{a_1 b_1} = t'$ between the a_1 and b_1 orbitals is expected to be small since the essential contribution to the effective hopping between the two Mn sites goes through the oxygen ligand. Three types of configurations are treated at the second order of perturbation :

- the ionic configurations I in which the two electrons of the oxygen $2p_z$ orbitals are excited in one e_g-like orbital (a_1 or b_1) are set to the energy U and coupled through the t_{pd} hopping integral to the Heisenberg (or single charge transfer) configurations. They are partly responsible for the J_1 interaction in the Heisenberg model.
- the ionic determinants I_1^l and I_1^r in which the couple (a_1, b_1) of orbitals bears a double occupation are at the energy U_1 and are coupled to the double exchange configurations by the hopping t_{pd}. Since the perturbative treatment of these configurations results in a global shift of the diagonal matrix elements of the double exchange configurations, their effect can be incorporated in Δ.

On the contrary, all ionic configurations $I^{l,r}_{2,3,4}$ (corresponding to a double occupation in the respective $(a_{2,3,4}, b_{2,3,4})$ couple of orbitals) have a differential effect on the Heisenberg configurations. They are responsible for the J_2 exchange integral appearing in the Heisenberg configurations energy.

- as well, the $K^{l,r}_{2,3,4}$ neutral or ionic (according to the position of the extra electron) determinants having one t_{2g}-like doubly occupied orbital are incorporated through the J exchange integral appearing in the double exchange configurations energy. Since $S_z \leq 5/2$ in these determinants, they only contribute to the sextet, quartet and doublet states.

The resolution of this model leads to the following energy expression valid for the nine states spanned by the model space :

$$
\begin{aligned}
E^v(S,\pm) = \frac{1}{2}\Bigg\{ & \Delta \pm \frac{t'}{S_m^H + \frac{(-1)^{S_m^H - S_m^{DE}}}{2}} + \left(J_2 - \frac{t_{pd}^2}{nU}\right)\left[S_m^H + \frac{1}{2} \mp \left(S + \frac{1}{2}\right)\right] \\
& + (J_2 - \frac{J}{2})\left[S(S+1) - S_m^{H^2}\right] + \left((-1)^{(S_m^H - S_m^{DE})}\frac{J}{2} - J_2\right) S_m^H \\
& - \Bigg\{ \frac{4t_{pd}^2}{n}\left[S_m^H + \frac{1}{2} \mp \left(S + \frac{1}{2}\right)\right] + \Bigg(\pm \frac{t'}{S_m^H + \frac{(-1)^{S_m^H - S_m^{DE}}}{2}} \\
& - \left(J_2 - \frac{t_{pd}^2}{nU}\right)\left[S_m^H + \frac{1}{2} \mp \left(S + \frac{1}{2}\right)\right] - (J_2 + \frac{J}{2})\left[S(S+1) - S_m^{H^2}\right] \\
& + \left((-1)^{(S_m^H - S_m^{DE})}\frac{J}{2} + J_2\right) S_m^H + \Delta \Bigg)^2 \Bigg\}^{1/2} \Bigg\},
\end{aligned} \tag{10.8}
$$

where n is the number of open-shell orbitals per Mn site.

10.5.2. *Confrontation of the truncated Hubbard model spectrum to the exact Hamiltonian one and discussion*

Here again the electronic interactions of the truncated Hubbard model are optimized in order to fit at best the *ab initio* spectrum. They are reported in Table 10.2 as well as the corresponding percentage of error ϵ. In order to reduce the number of variables in the optimisation procedure, we have neglected t' (which is expected to be negligible) and assumed $J_2 = J$. These two exchange integrals are of the same order of magnitude and as it will be shown in the next section, their extracted values are not precise when the Non-Hund states are not explicitely introduced in the model space.

It is interesting to note that Δ is large and positive, i.e., the Heisenberg configurations are lower in energy than the double exchange ones in the strongly localized orbitals set. The values of the exchange integrals while being optimized under constraint are close to those of the Heisenberg and double exchange models. The value of the t_{pd} hopping integral is larger than the commonly accepted values in such materials[25–27] confirming the polaronic order in this crystal structure. The large value of J_1 reveals the partially covalent interaction between the Mn atoms and the bridging oxygen which is also confirmed by the large ratio t_{pd}/Δ.
Figure 10.4 reproduces the truncated Hubbard model spectrum and the *ab initio* computed one. The common double exchange and Heisenberg spectrum is also reported for comparison.
The comparison of the truncated-Hubbard model spectrum with the *ab initio* one and with the previously calculated simpler model spectrum shows that the improvement brought by the variational truncated-Hubbard model is not noticeable, despite of the large number of parameters (or electronic interactions) of this complex model. Looking both at the obtained errors (3.17 % to be compared to 3.33 %) and at the qualitative features of the calculated spectra, it seems that the truncated Hubbard could be mapped on the simpler models.
Let us see how the delocalization of the orbitals between the metal and the ligand affects the physical content of the double exchange and Heisenberg models.

10.5.3. *Possible mapping of the truncated Hubbard model on the simpler double exchange and Heisenberg models*

The configurational expansion of the double exchange and Heisenberg ground-state wave-functions depends on the mixing between the metal and the ligand orbitals (see Fig. 10.3). In order to get this expansions in the semi-localized orbital set, let us first determine the expressions of the semi-localized orbitals as functions of the strongly localized ones. One should note that the t_{2g}-like orbitals are strongly localized in both orbital sets. Due to symmetry requirements, only

Table 10.2. Effective electronic interactions (in eV), and mean error of the model Hamiltonians. Heis. stands for Heisenberg, GP for the usual double exchange model and TH for the truncated Hubbard model.

	J_1	J_2	t	J	t_{pd}	U	Δ	ϵ
Heis.	0.59	0.07	-	-	-	-	-	3.33
GP	-	-	-1.05	0.07	-	-	-	3.33
TH	-	0.054	-	0.054	2.39	14.7	4.09	3.17

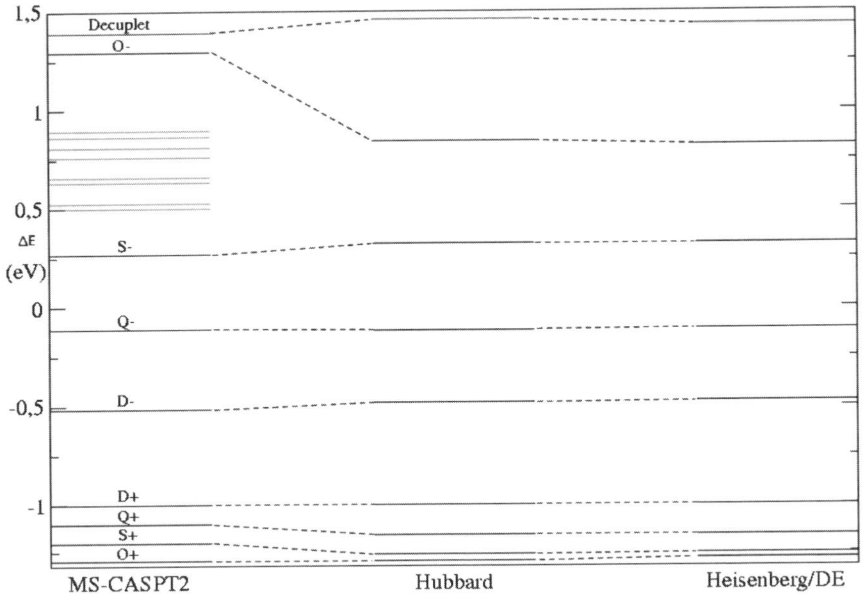

Fig. 10.4. Confrontation of the MS-CASPT2 spectrum to the model spectra. Reused with permission from Roland Bastardis, Nathalie Guihéry, Nicolas Suaud, and Coen de Graaf, Journal of Chemical Physics, 125, 194708 (2006). Copyright 2006, American Institute of Physics.

mixings between the antibonding linear combination of the e_g-like orbital and the $2p_z$ orbital of the bridging oxygen are possible. Let us call p' the mostly ligand orbital that would result from this mixing and u the mostly metallic one. In order to get left (l') and right (r') centered semi-localized orbitals of dominantly metallic character, symmetric and antisymmetric combinations of the bonding denoted g and antibonding denoted u of the e_g-like orbitals can be performed. In the general case the so-obtained three orbitals can be written :

$$p' = \cos\phi\ 2p_z + \sin\phi \frac{a_1 - b_1}{\sqrt{2}} = \cos\phi\ 2p_z + \sin\phi\ u, \tag{10.9}$$

$$r' = \frac{1}{\sqrt{2}}[g + u] = \frac{1}{2}[(1 + \cos\phi)b_1 + (1 - \cos\phi)a_1 - \sqrt{2}\sin\phi\ 2p_z], \tag{10.10}$$

$$l' = \frac{1}{\sqrt{2}}[g - u] = \frac{1}{2}[(1 - \cos\phi)b_1 + (1 + \cos\phi)a_1 + \sqrt{2}\sin\phi\ 2p_z], \tag{10.11}$$

where ϕ is the variable angle describing the mixing between the ligand and the metallic orbitals. The expression of the double exchange octet ground state wave-function using these semi-localized orbitals is :

$$\begin{aligned} |\Psi^{DE}_{rot}\rangle &= \cos^2\phi\,|\Psi^{DE}\rangle - \frac{\cos\phi\,\sin\phi}{\sqrt{2}}|\mathrm{I}^-\rangle + \sin^2\phi|\mathrm{I}_1^+\rangle \\ &\quad - \frac{3}{2\sqrt{2}}\cos\phi\,\sin\phi\left[|\Psi^{\mathrm{H}}\rangle - \frac{1}{\sqrt{3}}|\Psi^*\rangle\right], \end{aligned} \tag{10.12}$$

where $|\Psi^*\rangle$ is a linear combination of functions involving a local ground state on one Mn site and an excited local Non-Hund state on the other Mn site.

$$|\Psi^*\rangle = \frac{1}{\sqrt{2}}|\mathrm{T}_1\mathrm{D}_+\mathrm{Q}_2^{\mathrm{i}} + \mathrm{Q}_2^{\mathrm{i}}\mathrm{D}_+\mathrm{T}_1\rangle, \tag{10.13}$$

where T_1 is a local triplet state. The expression of the Heisenberg ground state wave-function in the semi-localized orbitals is :

$$\begin{aligned} |\Psi^{H}_{rot}\rangle &= \frac{3}{2\sqrt{2}}\cos\phi\,\sin\phi|\Psi^{DE}\rangle - \frac{3}{4}\sin^2\phi\,|\mathrm{I}^-\rangle - \frac{3}{2\sqrt{2}}\cos\phi\,\sin\phi|\mathrm{I}_1^+\rangle \\ &\quad + \left(\cos^2\phi - \frac{1}{8}\sin^2\phi\right)|\Psi^{H}\rangle + \frac{3\sqrt{3}}{8}\sin^2\phi\,|\Psi^*\rangle, \end{aligned} \tag{10.14}$$

where I^- and I_1^+ are the ionic states (see Fig. 10.1). It is interesting to note that both Ψ^{H}_{rot} and Ψ^{DE}_{rot} are expanded on the leading configurations of the problem owing to the mixing of the $2p_z$ and e_g-like orbitals of the metals. In particular, for $\phi \neq 0$ they describe the mixing of the double exchange and Heisenberg configurations, although they are orthogonal for the same value of ϕ. One may conclude that the hybridization of the orbitals allows the description of the physics of the system in a simpler model i.e. the truncated Hubbard model can be mapped onto either the Heisenberg or the double exchange model. Using the same arguments, it has already been shown by Zhang and Rice[28] in the case of doped cuprates that a two-band model can be mapped on the simpler t-J model. Let us notice here that, in the case of a system having several open shells per center, the hybrization

does not only ensure the validity of the Heisenberg and/or the double exchange model. It also permits one to introduce non-Hund configurations that are coupled to the model space. These configurations are responsible for the prevalence of a strong Hund's coupling (parallel spins) in the t_{2g}-like orbitals due to the poor delocalization of these strongly correlated electrons while a doublet arrangement of the electrons prevails in the p', l' and r' orbitals.
The Equations (10.12) and (10.14) show that the hybridization introduces neutral excited Non-Hund states of Heisenberg type. These states are not directly coupled with the considered model spaces and only contribute to the energy at the fourth order of perturbations. On the contrary, the excited Non-Hund state of the double exchange model which is coupled with the double exchange configurations is not taken into account in the usual (GP) double exchange model.

10.6. Role of the excited Non-Hund states : A refined double exchange model

The model space of the Anderson-Hasegawa (AH) model is not only constituted of the products of local ground-states. It is extended to products of a single excited atomic state on one Mn site by an atomic ground-state on the other Mn site. The so-obtained functions interact with the configurations of the usual double exchange model space through a term proportional to the hopping integral t. Since these new functions are also built from neutral determinants having only singly occupied metallic orbitals, their energy is expected to be lower than those of the neutral and ionic determinants with double occupancy accounted for by the exchange integral J of the usual (GP) double exchange model. In one Mn ion, the only non-Hund state (expressed in the local S_z=1 subspace) that produces an interaction with the usual double exchange model space is the following: $(3a_1a_2a_3\bar{a_4} - a_1a_2\bar{a_3}a_4 - a_1\bar{a_2}a_3a_4 - \bar{a_1}a_2a_3a_4)/\sqrt{12}$. Let us call ΔE its relative energy. A variational treatment of the non-Hund states of the two Mn ions combined with the Girerd-Papaefthymiou antiferromagnetic contribution leads to the following refined energy expression of the DE states:

$$E(S,\pm) = \frac{1}{2}\left[\Delta E - \sqrt{(\Delta E)^2 + 4t\left(t \mp \frac{S+1/2}{S_m^{DE}+1/2}\Delta E\right)}\right] - \frac{J}{2}\left[S(S+1) - S_m^{DE}\left(S_m^{DE}+1\right)\right]. \tag{10.15}$$

The intermediate and low spin states are stabilized by this interaction while the energies of the octet states are not affected by the introduction of the non-Hund states in the model space. This contribution is therefore antiferromagnetic.

Table 10.3. Electronic interactions (in eV) and percentage of error of the different double exchange models. The Zener model only includes the hopping integral interaction; GP stands for the usual double exchange model, AH for the Anderson-Hazegawa model and GP/AH for the combined model.

	t	J	ΔE	ϵ
Zener	1.08	-	-	19.76
GP	1.05	0.07	-	3.33
AH	1.37	-	2.34	2.52
GP/AH	1.29	0.03	2.56	0.70

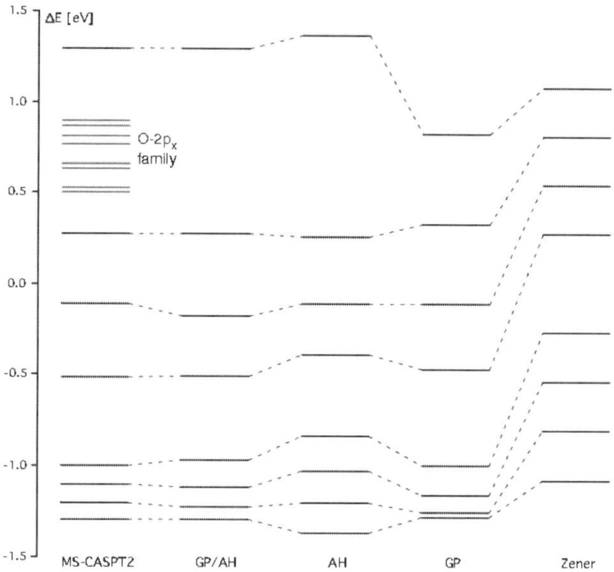

Fig. 10.5. Comparison of *ab initio* spectrum (column 1) with the outcomes of the combined (GP/AH) model (column 2), the Anderson/Hasegawa (AH) model (column 3), the usual (GP) model (column 4), and the Zener model (column 5). Reprinted figure with permission from reference Roland Bastardis, Nathalie Guihéry, and Coen de Graaf, Phys. Rev. B, 74, 014432 (2006). Copyright (2006) by the American Physical Society.

The electronic interactions optimized introducing separately the different antiferromagnetic contributions are reported in Table 10.3 while the corresponding spectra are represented in Fig. 10.5. The comparison of the different double exchange models brings the following conclusions :

- The introduction of antiferromagnetic contributions (either by the GP exchange integral or by the non-Hund states) considerably improves the modelization. The error of a pure Zener model in which only the hopping integral is considered is almost 20 %.
- While the exchange integral J of the GP and combined GP/AH models are physically identical, the GP exchange integral J value is artificially enhanced by the optimization of the extracted interactions. The GP spin exchange J should be proportional to the hopping integrals in the t_{2g}-like orbitals, which are actually very small due to the weak overlap between the corresponding orbitals. However, the larger J better fits the calculated spectra and phenomenologically mimicks the antiferromagnetic contribution of the non-Hund states (explicitely treated in the AH and the AH/GP models).
- The Anderson-Hazegawa model overestimates the role of the non-Hund states resulting in a larger t and smaller ΔE because of the absence of the GP antiferromagnetic contribution.
- The combined treatment of both antiferromagnetic effects in the GP/AH model provides accurate effective interactions and reproduces with a high accuracy the *ab initio* spectrum. The mean error is only 0.7 % for this combined model.

10.7. Conclusion

An analytical expression of the Heisenberg energies in the case of two metallic centers (having several open-shells) and bridged by a magnetic oxygen has been derived. Surprisingly enough the resulting spectrum is analytically identical to the usual double exchange one, except for one state. Since the physics supported by each model is different, the analysis of the exact Hamiltonian ground state wavefunction should enable one to determine the most appropriate model. Actually, in the case of half-doped manganites neither the spectrum nor the wavefunction analysis bring any decisive arguments to settle the question. It is likely that such undecidability would be encountered in experimental information.
Nevertheless the legitimity of the use of both double exchange and purely magnetic models for the description of oxygen-bridged mixed valence systems has been shown. Actually a variational Hubbard-like model does not qualitatively improves the reproduction of the *ab initio* spectrum. It has been demonstrated that this complex truncated Hubbard model could be mapped on both simpler double exchange and Heisenberg models owing to hybridization of the metal and ligand orbitals. The consistency of the effective interaction values optimized for the three different models has also been shown.

The main missing effect of the considered Hamiltonians is due to the interactions of atomic excited states i.e. the Non-Hund states. Their explicit consideration in the double exchange modelization results in a dramatic improvment of the reproduction of the calculated spectrum.
The Zener Hamiltonian proposes an oversimplified description of the Zener polaron physics. A more accurate description is obtained after the introduction of antiferromagnetic contributions. The antiferromagnetic contribution arising from the local non-Hund state through the strong interaction with the bridging oxygen accounts for covalency effects between the double exchange configurations, enhancing the hopping integral value of the extra electron. The combined GP/AH model, includes both the antiferromagnetic contributions due to the electronic circulation in the t_{2g}-like orbitals and the local non-Hund states. The quality of the resulting spectrum finally shows that a double exchange mechanism dominates the electronic structure of the Zener polaron.

References

1. C. Bloch and J. Horowitz, Sur la détermination des premiers états d'un système de fermions dans le cas dégénéré, *Nucl. Phys.* **8**, 91 (1958).
2. C.J. Calzado and J. P. Malrieu, Proposal of an extended t-J Hamiltonian for high-Tc cuprates from ab initio calculations on embedded clusters, *Phys. Rev. B* **63**, 214520 (2001).
3. N. Suaud, A. Gaita-Arinño, J.M. Clemente-Juan, J. Sanchez Marin and E. Coronado, Electron Delocalization in Mixed-Valence Keggin Polyoxometalates. Ab Initio Calculation of the Local Effective Transfer Integrals and Its Consequences on the Spin Coupling, *J. Am. Chem. Soc.* **124**, 15134 (2002).
4. C. de Graaf, C. Sousa and R. Broer, Ab initio study of the charge order and Zener polaron formation in half-doped manganites, *Phys. Rev. B* **70**, 235104 (2004).
5. F. Illas, I. de P. R. Moreira, C. de Graaf, O. Castell and J. Casanovas, Absence of collective effects in Heisenberg systems with localized magnetic moments, *Phys. Rev. B* **56**, 5069 (1997).
6. R. Bastardis, N. Guihéry and C. de Graaf, Ab initio study of the Zener polaron spectrum of half-doped manganites: Comparison of several model Hamiltonians, *Phys. Rev. B* **74**, 014432 (2006).
7. R. Bastardis, N. Guihéry, N. Suaud and C. de Graaf, Competition between double exchange and purely magnetic Heisenberg models in mixed valence systems: Application to half-doped manganites, *J. Chem. Phys.* **125**, 194708 (2006).
8. G. Jonker and J. van Santen, Ferromagnetic compounds of manganese with perovskite structure, *Physica (Amsterdam)* **16**, 337 (1950).
9. S. Jin, M. Mc Cormack, T. Tiefel and R. Ramesh, Colossal magnetoresistance in La-Ca-Mn-O ferromagnetic thin films (invited), *J. Appl. Phys.* **76**, 6929 (1994).
10. R. M. Kusters, J. Singleton, D. A. Keen, R. Mc Greevy and W. Hayes, Magnetoresistance measurements on the magnetic semiconductor Nd0.5Pb0.5MnO3, *Physica*

(Amsterdam) **155B**, 362 (1989); A. Urushibara, Y. Moritomo, T. Arima, A. Asamitsu, G. Kido and Y. Tokura, Insulator-metal transition and giant magnetoresistance in La1-xSrxMnO3, *Phys. Rev. B* **51**, 14103 (1995).
11. E. Dagotto, Open questions in CMR manganites, relevance of clustered states and analogies with other compounds including the cuprates, *New J. Phys.* **7**, 67 (2005).
12. J. B. Goodenough, Theory of the Role of Covalence in the Perovskite-Type Manganites [La, M(II)]MnO3, *Phys. Rev.* **100**, 564 (1955).
13. E.O. Wollan and W.C. Koehler, Neutron Diffraction Study of the Magnetic Properties of the Series of Perovskite-Type Compounds [(1-x)La, xCa]MnO3, *Phys. Rev.* **100**, 545 (1955).
14. P. G. Radaelli, D. E. Cox, M. Marezio and S.-W. Cheong, Charge, orbital, and magnetic ordering in La0.5 Ca0.5 MnO3s, *Phys Rev. B* **55**, 3015 (1997).
15. A. Daoud-Aladine, J. Rodríguez-Carvajal, L. Pinsard-Gaudart, M. T. Fernandez-Díaz and A. Revcolevschi, Zener Polaron Ordering in Half-Doped Manganites, *Phys. Rev. Lett.* **89**, 97205 (2002).
16. J. García, M. C. Sánchez, J. Blasco, G. Subías and M. G. Proietti, Analysis of the x-ray resonant scattering at the Mn K edge in half-doped mixed valence manganites, *J. Phys. Condens. Matter* **13**, 3243 (2001).
17. G. Zheng and C. H. Patterson, Ferromagnetic polarons in La0.5Ca0.5MnO3 and La0.33Ca0.67MnO3, *Phys. Rev. B* **67**, 220404 (R) (2003).
18. V. Ferrari, M. Towler and P. B. Littlewood, Oxygen Stripes in La0.5Ca0.5MnO3 from Ab Initio Calculations, *Phys. Rev. Lett.* **91**, 227202 (2003).
19. C. H. Patterson, Competing crystal structures in La0.5Ca0.5MnO3: Conventional charge order versus Zener polarons, *Phys. Rev. B* **72**, 085125 (2005).
20. D. V. Efremov, J. van den Brink and D. I. Khomskii, Bond- versus site-centred ordering and possible ferroelectricity in manganites, *Nature Materials* **3**, 853 (2004).
21. C. Zener, Interaction between the d-Shells in the Transition Metals. II. Ferromagnetic Compounds of Manganese with Perovskite Structure, *Phys. Rev.* **82**, 403 (1951).
22. P. W. Anderson and H. Hasegawa, Considerations on Double Exchange, *Phys. Rev.* **100**, 675 (1955).
23. J.-J. Girerd, V. Papaefthymiou, K.K. Surerus and E. Münck, Double exchange in iron-sulfur clusters and a proposed spin-dependent transfer mechanism, *Pure and Appl. Chem.* **61**, 805 (1989).
24. V. Papaefthymiou, J.-J. Girerd, I. Moura, J.J. G. Moura and E. Münck, Moessbauer study of D. gigas ferredoxin II and spin-coupling model for Fe3S4 cluster with valence delocalization, *J. Am. Chem. Soc.* **109**, 4703 (1987).
25. A. Yu. Ignatov and N. ali, Mn K-edge XANES study of the La1-xCaxMnO3 colossal magnetoresistive manganites, *Phys. Rev. B* **64**, 014413 (2001).
26. H. Meskine, H. Hönig and S. Satpathy, Orbital ordering and exchange interaction in the manganites, *Phys. Rev. B* **64**, 094433 (2001).
27. K. Yonemitsu, A. R. Bishop and J. Lorenzana, Magnetism and covalency in the two-dimensional three-band Peierls-Hubbard model, *Phys. Rev. B* **47**, 8065 (1993).
28. F. C. Zhang and T. M. Rice, Effective Hamiltonian for the superconducting Cu oxides, *Phys. Rev. B* **37**, 3759 (1988).

Chapter 11

Disorder in the double exchange model

V. M. Pereira, E. V. Castro and J. M. B. Lopes dos Santos

Centro de Física do Porto e Departamento de Física
Faculdade de Ciências, Universidade do Porto, 4169-007 Porto, Portugal

Some materials, most noticeably the *manganites*, show, in certain composition ranges, concomitant para-ferromagnetic and metal-insulator transitions. It was precisely in the context of the experimental discovery of this remarkable correlation between transport and magnetism in the case of the manganites, that Zener proposed a magnetic exchange mechanism, *double exchange*, in which charge transport and magnetic correlations are closely inter-dependent.

In this article we review studies which addressed the conditions under which a simple double exchange model can present a Anderson-like metal-insulator transition when it orders ferromagnetically. We present arguments that show that intrinsic disorder in some calcium doped manganites is much higher than has generally been admitted in the literature. Nevertheless, such a model is a dramatic over simplification in the case of the manganites, in which orbital degeneracy, antiferromagnetic interactions, and strong electron-lattice coupling play an important part in the physics.

We also review some recent work on another class of compounds showing very large magnetoresistance at the Curie temperature, the europium hexaborides, in which is is argued that a variety of optical, transport and magnetic properties can be well understood as a manifestation of Anderson localization in the context of a double exchange model, without the complicating factors that are present in the manganites.

Contents

11.1. Introduction

It appears that the designation "double exchange" was first introduced in the physics literature in Zener's 1951 article.[1] Zener's proposal of a new magnetic exchange mechanism was motivated by Jonker and Van Santen's discovery of a remarkable correlation between the transport and the magnetic properties of a series of doped manganese oxides of chemical formula $La_{1-x}D_xMnO_3$, where $D = Ca$, Sr, Ba is a divalent alkaline–earth ion.[2]

The crystal structures of these manganese oxides, *manganites*, are based on the ideal perovskite structure of $CaTiO_3$: the magnetic manganese ions occupy the corners of a cube, the oxygen ions the corresponding edges and the trivalent rare-earth, La, or its divalent replacement, the center of the cube (see Fig. 11.1). In the end compound, $x = 0$, one expects the manganese ions to be in a

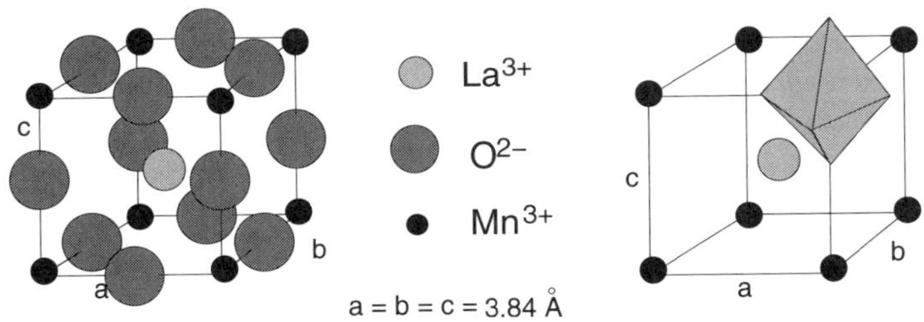

Fig. 11.1. (a) Ideal perovskite structure; (b) the octahedral environment of a manganese ion.

trivalent, d^4 configuration ($La^{3+}Mn^{3+}O_3^{2-}$); the replacement of La^{3+} by the divalent ion, D^{2+}, leads to the appearance of Mn^{4+}, ions in d^3 configuration, $La^{3+}_{1-x}D^{2+}_x\left(Mn^{3+}_{1-x}Mn^{4+}_x\right)O_3^{2-}$. Zener was thus lead to consider the physics of a $Mn - O - Mn$ bond. The configuration $Mn^{3+} - O^{2-} - Mn^{4+}$ can resonate with the degenerate one, $Mn^{4+} - O^{2-} - Mn^{3+}$, through a virtual state $Mn^{3+} - O^{-} - Mn^{3+}$. The electron is transferred between Mn ions with no change in its spin state and the strong intra-atomic Hund coupling in the manganese d–shell requires complete alignment of all electronic spins. Therefore, if the spins of the two manganese ions are not aligned, the hopping amplitude between the lowest energy configurations is correspondingly reduced. The mobility of the extra hole in Mn^{4+} is clearly related to the magnetic arrangement of the manganese spins. The band energy of these holes is lower for a ferromagnetic configuration and this constitutes an interaction which Zener proposed as the origin of the ferromagnetism observed by Jonker and Van Santen for compositions $x = 0.2 \sim 0.4$. Because the full process involves an electron transfer from the oxygen p orbital to a d orbital in Mn^{4+}, and from a d orbital in Mn^{3+} to the oxygen, Zener termed it double exchange.

In fact, the magnetism of these compounds proved, very early, to be extremely complex. Still in the fifties, Wollan and Koehler[3] did a series of beautiful neutron scattering experiments and identified several different anti-ferromagnetic phases in $La_{1-x}Ca_xMnO_3$ in addition to the ferromagnetism studied by Jonker and Van Santen. Very important theoretical work followed shortly. Goodenough[4] clarified several aspects of the magnetic interactions in the manganites, Anderson and Hasegawa[5] placed the double exchange mechanism in a firmer, more quantitative formulation and, later, De Gennes[6] explored the interplay between the double exchange mechanism and direct Mn-Mn antiferromagnetic interactions, in shaping the phase diagram of the manganites series. Still, and somewhat surprisingly, the study of these interesting compounds was overshadowed by more pressing concerns, only to be revived in the nineties, with a veritable explosion of interest, following the discovery of very large ("colossal") magnetoresistance (CMR) at room temperature, first for Barium compounds ($La_{1-x}Ba_xMnO_3$)[7] and later in the Calcium series.[8,9] These studies showed that the para-ferromagnetic transition could be accompanied by a spectacular decrease in resistance, and a change from a insulator–like behavior ($d\rho/dT < 0$, ρ resistivity, T temperature) above the Curie temperature, T_c, to metal–like behavior below ($d\rho/dT > 0$). Near the transition a large negative magnetoresistance is observed; Jin *et al.*[8] reported $-\left(\rho(H) - \rho(0)\right)/\rho(0) \sim 1000$, for field values of a few Tesla.

The double exchange mechanism immediately comes to mind as a possible explanation for this observation of concomitant Metal-Insulator (MI) and para-ferromagnetic transitions; we have already seen that ferromagnetic alignment of manganese spins enhances charge mobility.

In this article we present a brief review of studies MI transition in the context of the double exchange model. We will see that the MI transition in the manganites involves considerably more ingredients than just the double exchange mechanism. We will also discuss another series compounds, the europium hexaborides, in which a rather remarkable set of magnetic, optical, and transport properties can be explained rather simply in terms of the intrinsic connection of transport and magnetism implied by the double exchange mechanism.

11.2. The Double Exchange Hamiltonian

Let us look more carefully into the possible states of a manganese ion in the manganite series. In an octahedral environment, the five d-orbital split into a threefold degenerate representation, t_{2g}, with d_{xy}, d_{zx} and d_{yz} orbitals, and twofold one, e_g, with $d_{3z^2-r^2}$ and $d_{x^2-y^2}$ orbitals. The latter, being directed towards the negative oxygen ions, have higher energy. The corresponding energy separation is of order of an electron-Volt, $\varepsilon(e_g) - \varepsilon(t_{2g}) \sim 1$ eV.[10] In the Mn^{4+} ion, with a d^3 configuration, the three electrons occupy t_{2g} orbitals with parallel spins due to the large intra-atomic Hund Coupling. The Mn^{3+} ion has an extra electron in one e_g orbital. For a composition $La_{1-x}D_xMnO_3$, we have $1-x$ electrons per manganese site in e_g orbitals.

The simplest Hamiltonian that we can use to describe the motion of this extra electron must include the following ingredients:

(i) each manganese ion has a localized spin $S = 3/2$ corresponding to the three occupied t_{2g} orbitals with electronic spins aligned by the Hund coupling;
(ii) the e_g electron can hop between nearest neighbor manganese ions with a given amplitude $-t$;
(iii) there is ferromagnetic Hund coupling, J_H , between the spin of this band electron and and the local t_{2g} spin.

This description leads directly into what is called the Ferromagnetic Kondo Model (FKM)

$$H_{FKM} = -t \sum_{\langle ij \rangle \sigma} a^\dagger_{i\sigma} a_{j\sigma} - J_H \sum_{i\sigma\sigma'} (\mathbf{S}_i \cdot \boldsymbol{\tau}_{\sigma\sigma'}) \, a^\dagger_{i\sigma} a_{i\sigma'}, \tag{11.1}$$

where $\mathbf{S}_j$ is the local t_{2g} spin, $a^\dagger_{i\sigma}$, $a_{i\sigma'}$ are the creation and destruction operators for an e_g orbital on site i and $\boldsymbol{\tau} = (\tau_x, \tau_y, \tau_z)$ are Pauli matrices. Two further simplifications are required to reach the Hamiltonian which will be the focus of this review.

(1) When the local spin quantum number is large, $S \gg 1$, we can treat the local spins as classical vectors; in the manganites $S = 3/2$. We will see that in the Europium Hexaborides $S = 7/2$, corresponding to the a $4f^7$configuration of the Eu^{2+} ion.
(2) By itself the first term in the Hamiltonian (11.1) describes a band of width $W = 2zt$ where z is the coordination number; the second term gives two energy levels in each site, with energies $\pm J_H S$, corresponding to electronic spin parallel and anti parallel to the local t_{2g} spin direction. In the limit where

$2J_HS \gg W$ we can assume the spectrum splits into two bands, the lower one being formed out of the local states of energy, $-J_HS$, *i.e.* , the states in which the conduction electron spin is parallel to the local spin $\mathbf{S}_i$. It is therefore advantageous, in this limit, to change into a spin basis of eigenstates of $\boldsymbol{\tau} \cdot \hat{\mathbf{n}}_i$, where $\hat{\mathbf{n}}_i$ denotes the direction of spin on site i , $\mathbf{S}_i = S\hat{\mathbf{n}}_i$.

In fact, this change of basis can be done quite generally, with no assumptions on the parameters of the Hamiltonian.

Let θ_i, ϕ_i denote the polar and azimuthal angles of the spin at site i, $\mathbf{S}_i = S(\sin\theta_i\cos\phi_i, \sin\theta_i\sin\phi_i, \cos\theta_i)$. $\mathcal{R}(\theta_i,\phi_i) = \mathcal{R}_z(\phi_i)\mathcal{R}_y(\theta_i)$ denotes the rotation operator that maps a global coordinate system to a local one with a z' axis coinciding with the direction of $\mathbf{S}_i$. In the eigenbasis of σ_z (global z direction), $\langle\sigma|\,\mathcal{R}(\theta_i,\phi_i)\,|\sigma'\rangle \equiv \mathbf{R}_{\sigma\sigma'}(\theta_i,\phi_i)$, with

$$\mathbf{R}(\theta,\phi) = \begin{bmatrix} e^{i\phi/2}\cos\frac{\theta}{2} & e^{i\phi/2}\sin\frac{\theta}{2} \\ -e^{-i\phi/2}\sin\frac{\theta}{2} & e^{-i\phi/2}\cos\frac{\theta}{2} \end{bmatrix}. \tag{11.2}$$

The following relations define the operators, $c_{i\Uparrow}$ and $c_{i\Downarrow}$, for the spin states with $\pm 1/2$ projection along the direction of the spin $\mathbf{S}_i$, $|\Uparrow\rangle = \mathcal{R}(\theta_i,\phi_i)\,|\uparrow\rangle\,, |\Downarrow\rangle = \mathcal{R}(\theta_i,\phi_i)\,|\downarrow\rangle$:

$$a_{i\sigma} = \sum_{\sigma'} \mathbf{R}_{\sigma\sigma'}(\theta_i,\phi_i)c_{i\sigma'}. \tag{11.3}$$

We may now rewrite the FKM using this local spin basis, instead of the global one, by inserting the representation of Eq. (11.3) in (11.1). The J_H term becomes diagonal in this representation:

$$-J_H\sum_{i\sigma\sigma'}\left(\mathbf{S}_i\cdot\sigma_{\sigma\sigma'}\right)a^\dagger_{i\sigma}a_{i\sigma'} = -J_HS\sum_i\left(c^\dagger_{i\Uparrow}c_{i\Uparrow} - c^\dagger_{i\Downarrow}c_{i\Downarrow}\right). \tag{11.4}$$

The hopping term is

$$-t\sum_{\langle ij\rangle\sigma}a^\dagger_{i\sigma}a_{j\sigma} = -t\sum_{\langle ij\rangle\sigma\sigma'}\left[\mathbf{R}^\dagger(\theta_i,\phi_i)\cdot\mathbf{R}(\theta_j,\phi_j)\right]_{\sigma\sigma'}c^\dagger_{i\sigma}c_{j\sigma'} + h.c. \tag{11.5}$$

and so,

$$\begin{aligned}\mathcal{H}_{FKM} &= -J_HS\sum_i\left(c^\dagger_{i\Uparrow}c_{i\Uparrow} - c^\dagger_{i\Downarrow}c_{i\Downarrow}\right)\\ &\quad - t\sum_{\langle ij\rangle\sigma\sigma'}\left[\mathbf{R}^\dagger(\theta_i,\phi_i)\cdot\mathbf{R}(\theta_j,\phi_j)\right]_{\sigma\sigma'}c^\dagger_{i\sigma}c_{j\sigma'} + h.c.\end{aligned} \tag{11.6}$$

The double exchange model is obtained in the limit $2J_HS \gg W$ by retaining only the low energy states of spin parallel to $\mathbf{S}_i$. The J_H term becomes an unimportant energy shift, $-J_H S N_e$, and

$$\mathcal{H}_{DE} = -\sum_{\langle ij \rangle} t(\mathbf{S}_i, \mathbf{S}_j)\, c_{i\Uparrow}^\dagger c_{j\Uparrow} \tag{11.7}$$

with

$$t(\mathbf{S}_i, \mathbf{S}_j) = t\left(e^{-i(\phi_i - \phi_j)/2} \cos\frac{\theta_i}{2} \cos\frac{\theta_j}{2} + e^{i(\phi_i - \phi_j)/2} \sin\frac{\theta_i}{2} \sin\frac{\theta_j}{2} \right). \tag{11.8}$$

Since there is only one possible spin state in the low energy sector, the electrons can be treated as *spinless fermions* and the spin label safely omitted. For this Hamiltonian a filled band corresponds to one electron per site, since there is only one spin state per site. In a manganite $La_{1-x}D_xMnO_3$ there is a Mn^{4+} ion for each divalent D^{2+}, so x is the concentration of *holes* in a filled band. Apart from a unimportant constant, the double-exchange Hamiltonian has the same form for holes and electrons.

The Hamiltonian of Eq. (11.7) is deceptively simple. It appears to describe a simple non-interacting tight binding system. But the hopping depends on the local spin degrees of freedom; in a paramagnetic phase the electronic system is intrinsically disordered. The local spins, on the other hand, are coupled only by the fact that their global configuration determines the kinetic energy of the electronic system: this Hamiltonian displays in the simplest possible fashion the relation between transport and magnetism, that Zener suggested as the essence of his double exchange mechanism.

This Hamiltonian and its parent, the FKM, grossly misrepresent the complexity of the physics present in the manganites. However, it is not the intention of this review to cover in detail the physics of manganites, which has been addressed in various excellent works,[10–14] but to answer the simpler question of whether the double exchange Hamiltonian, and slight variants of it, contains the possibility of concomitant para-ferromagnetic and MI transitions. But before we proceed let us refer some of the complications arising in the manganites.

Other magnetic interactions besides double exchange Consider, for instance, the $x = 1$ compound $CaMnO_3$. In this case all manganese ions are in Mn^{4+} state with three t_{2g} electrons: there is no double exchange. Since processes of charge excitation like $Mn^{4+} - O^{2-} - Mn^{4+} \to Mn^{3+} - O^{2-} - Mn^{5+}$ are very costly in energy due to the intra-atomic Coulomb terms, this system is a Mott-Hubbard insulator. As is common in such cases, it orders antiferromagnetically (G type structure) with a Néel temperature ~ 100 K. We expect Anderson's super-exchange mechanism to give rise an antiferromagnetic coupling between the t_{2g}

spins and to be present for any composition.

The e_g level is doubly degenerate. We do not expect double occupancy in the e_g orbitals because of intra-atomic Coulomb correlation. In fact the $x = 0$ compound of the series, $LaMnO_3$, which has one e_g electron per manganese ion, is a Mott Hubbard insulator. With this restriction, the orbital degree of freedom can be represented by a isospin variable in which we associate $I_z = \pm 1/2$ with the two e_g orbitals $|3z^2 - r^2\rangle$, $|x^2 - y^2\rangle$.[15] The hopping term has to be rewritten as

$$H_{cin} = -\sum_{\langle ij \rangle} \sum_{\gamma\gamma'\sigma} (\mathbf{t}_{ij})_{\gamma\gamma'} \, a^{\dagger}_{i\gamma\sigma} a_{j\gamma'\sigma} \,, \tag{11.9}$$

in which $\mathbf{t}_{ij}$ is a 2×2 matrix in orbital spin space. This orbital degeneracy is essential in understanding the magnetic structure, as already pointed out by Goodenough,[4] because the magnetic ordering is accompanied by orbital (isospin) ordering. Maezono *et al.*[15] discuss this issue in great detail in their study of the phase diagram of manganese oxides.

Jahn-Teller effect and e_g degeneracy It is well known that orbital degeneracy in a high symmetry environment is an unstable situation (Jahn-Teller effect). The degeneracy is lifted by a lattice distortion with a lowering of symmetry, since the gain in electronic energy is linear in the distortion and the cost in elastic energy is quadratic. This well known Jahn-Teller effect shows up in $LaMnO_3$, most noticeably, as a stretching of the oxygen octahedra along one axis in the basal plane, in a direction alternating by 90° from one site to the neighboring one. In a

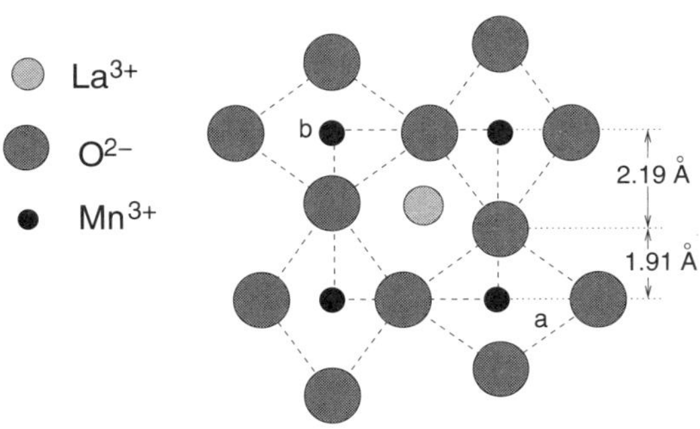

Fig. 11.2. The dominant Jahn-Teller distortion in $LaMnO_3$ is a stretching of the oxygen octahedron along an axis in the basal plane alternating in direction. There is also a much smaller stretching along the c-axis.[12]

Mn^{4+} configuration there is no reason for this distortion. In order to predict the observed sequence of low energy phases, as x is varied, the Jahn-Teller effect has to be considered.[15] This shows that the coupling to the lattice cannot be ignored in the manganites and some authors have vehemently argued that it plays an essential role in the MI transition.[16,17]

11.3. Double Exchange and Disorder in the Manganites

11.3.1. *Localization and Off-diagonal Disorder*

In a paramagnetic phase the double exchange Hamiltonian (Eq. (11.7)) has off-diagonal disorder, i.e., random hopping amplitudes. States at the edges of the band should be localized. If the Fermi level lies in a region of localized states in the paramagnetic phase, one expects insulating behavior; as one enters the ferromagnetic phase, the spins align, the hopping becomes more uniform and the mobility edge should move towards the band edge. Near the Curie temperature spins align easily with a magnetic field and a large negative magnetoresistance is expected.

Varma[18] first proposed that in low doping manganites the Fermi level lies in a region of localized states in the paramagnetic phase (PM), crossing over to extended states as the doping is increased, thereby implying the above mentioned mechanism as explanation for the large magnetoresistance near the Curie temperature for $0.1 < x < 0.3$. The hopping between sites i and j can be expressed in the angle, θ_{ij}, between spins $\mathbf{S}_i$ and $\mathbf{S}_j$:

$$t\left(\mathbf{S}_i, \mathbf{S}_j\right) = t_{ij} = te^{i\phi_{ij}} \cos\frac{\theta_{ij}}{2}\,. \tag{11.10}$$

In a paramagnetic, uncorrelated, phase one can easily compute the distribution of $|t_{ij}|$,

$$\begin{aligned} P\left(|t_{ij}|\right) &= \frac{1}{4\pi}\int d\phi d\theta sin\theta\delta\left(|t_{ij}| - t\cos\frac{\theta}{2}\right) \\ &= 2\frac{|t_{ij}|}{t}, \qquad 0 \le |t_{ij}| \le t, \end{aligned} \tag{11.11}$$

and calculate the mean and variance of $|t_{ij}|$:

$$\bar{t} = \langle |t_{ij}| \rangle = \frac{2}{3}t \tag{11.12}$$

$$\sigma_t^2 = \left\langle |t_{ij}|^2 \right\rangle - \langle |t_{ij}| \rangle^2 = \frac{1}{18}t^2 \tag{11.13}$$

Based on previous studies of random hopping models with a semicircular distribution, of Economou and Antoniou,[19] Varma estimated that these fluctuations in the hopping amplitude were sufficient to localize a significant percentage of the states in the band (certainly more than 10% if the Fermi level is to cross the mobility edge in the range $0.1 < x < 0.3$), though no precise values of this percentage were provided.

More precise estimates of the fraction of localized states were given by Li *et al.*[20] and Sheng *et al.*[21] These authors used the transfer matrix method[22,23] to estimate the location of the mobility edge, and numerical diagonalization of clusters of $10 \times 10 \times 10$ sites (with the direction of each spin chosen randomly from a uniform distribution on a sphere) to calculate the density of states (DOS) and locate the Fermi energy as a function of doping, in the PM phase. The mobility edge occurs very close to the band edge and less than 0.5% of the states are localized.[20] This result would seem to imply that we can discard Anderson localization as a relevant factor in the physics of manganites, for dopings $0.1 < x < 0.3$ (recall that x is the concentration of itinerant holes).

However, for each carrier introduced in the system, there is a $La^{3+} \longrightarrow D^{2+}$ substitution. The corresponding change in the Coulomb field shifts the site energy of a hole in a manganese site at a distance R by $\Delta\epsilon_d = -e^2/4\pi\varepsilon R$, where ε is the dielectric constant of the material. We can take this effect into account by including a random site energy term (Anderson disorder) in the double exchange Hamiltonian:

$$\mathcal{H}_{DE} = -\sum_{ij} t\left(\mathbf{S}_i, \mathbf{S}_j\right) c_i^\dagger c_j + \sum_i \epsilon_i c_i^\dagger c_i \tag{11.14}$$

Sheng *et al.*[21,24] modeled the site disorder term with a uniform probability distribution for $-W/2 \leq \epsilon_i \leq W/2$ and plotted the mobility edge in a (E, W) plane in the paramagnetic (PM) and ferromagnetic (FM) states (Fig. 11.3). They concluded that a MI transition occurs when the system orders ferromagnetically, for $0.2 < x < 0.5$ provided the diagonal disorder is strong enough, $12t < W < 16.5t$.

The plausibility of such large values of the disorder parameter was questioned by Pickett and Singh.[25] In particular, they looked at the $x = 1/3$ concentration, and performed LDA calculations of band structure for a periodic structure of $La_2CaMn_3O_9$ with a tetragonal unit cell containing an La–Ca–La set of planes. There are two nonequivalent Mn sites in this structure, one with eight La^{3+} and the other with four Ca^{2+} and four La^{3+} nearest neighbors. The local density of states at the manganese sites showed a difference of $\Delta\epsilon_{\mathrm{Mn}} = 0.5\,\mathrm{eV}$ between the band edges for these two types of sites, which was interpreted as arising from the

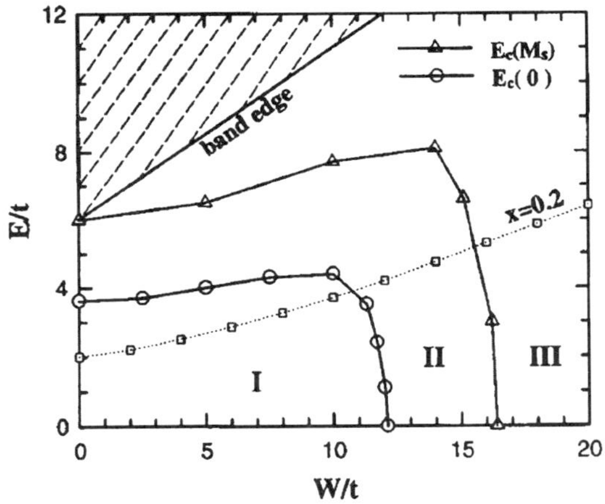

Fig. 11.3. Phase diagram in the E/t *vs* W/t plane: open triangles represent the mobility edge in the ferromagnetic phase (FM) and the circles in the paramagnetic one (PM). Only the upper half of the band is shown and holes fill down from the top. The dotted line denotes the position of the Fermi energy for a hole concentration of $x = 0.2$. A system with (W, E_F) in region **I** is metallic in the PM and FM phase; in region **II** the mobility edge crosses the Fermi level in the transition PM $\rightarrow$ FM. For concentrations $0.2 < x < 0.5$ this requires $12t < W < 16.5t$. Reproduced from.[24]

different charges $Ca^{2+}$and La^{3+},

$$\Delta\epsilon_{\mathrm{Mn}} = 4V = 4 \times \frac{e^2}{4\pi\varepsilon_0\varepsilon R}, \tag{11.15}$$

where $R = \sqrt{3}a/2$ is the La $-$ Mn distance, and V the energy shift from a single Ca^{2+} ion. From the calculated value of $\Delta\epsilon_{\mathrm{Mn}}$ a dielectric constant $\varepsilon \approx 33$ is obtained, which gives $V \approx 0.13\,\mathrm{eV} \approx 0.6t$ with the hopping parameter $t \approx 0.2\,\mathrm{eV}$.[10]

Such a dielectric constant, however, is quite unlikely. Firstly, we should note that Eq. (11.15) is actually a microscopic description, where $R \approx 2.2\,\text{Å}$. Neglecting metallic screening we should get a relative permittivity, ε, closer to unity. Furthermore, infrared reflectivity measurements on $La_{0.67}Ca_{0.33}MnO_3$ give a high-frequency dielectric constant $\varepsilon_\infty \approx 7.5$ at 78 K.[26] Finally, notice that Eq. (11.15) is a special case where only first nearest-neighbors (La/Ca sites) contribute to the local potential. A more realistic situation should account for next nearest-neighbors contributions.

The value $\Delta\epsilon_{\mathrm{Mn}} = 0.5\,\mathrm{eV}$ found by Pickett and Singh[25] is reproduced if second and third nearest-neighbors are taken into account with $\varepsilon_1 \approx 10$ and $\varepsilon_3 \approx 17$, ε_i being the dielectric constant for the ith shell. Note that the two nonequivalent Mn sites have the same second nearest-neighbors' environment, and therefore ε_2

does not enter the calculation of $\Delta\epsilon_{\mathrm{Mn}}$. Therefore, the results of the LDA calculation are not inconsistent with a more plausible value of ε. In what follows we will keep only first and second nearest-neighbors contributions, with $\varepsilon_1 = \varepsilon_2 \approx 10$, in the calculation of the probability distribution for the site energies, as Pickett and Singh[25] did in their work. Such a value of the dielectric constant yields, from Eq. (11.15), $V \approx 0.43\,\mathrm{eV} \approx 2.1t$. Knowing this parameter, one can calculate the actual distribution of site energies due to a random placement of Ca^{2+} ions.[25,27] The resulting coarse grained distribution is shown in Fig. 11.4 (full line), and the site energy's relative probability is shown in the inset. The distribution is approximately Gaussian with a root mean square (RMS) deviation

$$\sqrt{\langle\epsilon_i^2\rangle - \langle\epsilon_i\rangle^2} \simeq 4.6t, \tag{11.16}$$

as obtained by fitting with a Gaussian distribution (dashed line in Fig. 11.4). A rectangular distribution with the same RMS deviation has $W \approx 15.9t$; well in the range required for a MI transition at the Curie temperature, $12t < W < 16.5t$ as found by Sheng *et al.*.[24]

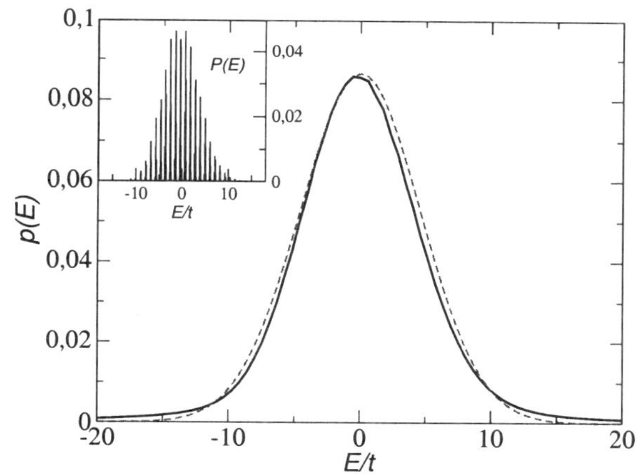

Fig. 11.4. The full line shows the probability distribution of Mn-site energies due to random placement of La^{3+} (2/3 probability) and Ca^{2+} (1/3 probability) on first and second neighbor sites, obtained by substituting $\delta-$functions by Lorentzians with half width of t at half maximum. A dielectric constant of $\varepsilon \approx 10$ was used for both shells. The dashed line is the fit to the full line with a Gaussian distribution. In the inset is shown the true discrete Mn-site energy probability.

Some remarks regarding the conclusions of Picket and Singh's work[25] are in place. The authors did the same analysis which lead us to Fig. 11.4 and Eq. (11.16), but arrived at a different conclusion about the strength of disorder in colossal magnetoresistive manganites. The main difference between their and our analysis is the value of the dielectric constant. Even though they say that the effect of first and second shells are taken into account with $\varepsilon \approx 10$ their distribution probability was clearly obtained for $\varepsilon \approx 33$. As a consequence they found for the associated Gaussian distribution a RMS of $1.3t$. This means that a rectangular distribution with the same RMS deviation has $W \approx 4.5t$; well below the values required for a MI transition at the Curie temperature, $12t < W < 16.5t$ as found by Sheng *et al.*[24]

One of us[27] performed a transfer matrix calculation in a model in which the site energies are calculated from a random distribution of the dopant ions Ca^{2+}. In this case the site disorder is parametrized by x and by the parameter V defined above. First and second shells of La/Ca-sites were taken into account assuming equal dielectric constant, with V given by Eq. (11.15). The Density of States (DOS) was calculated for clusters of $60 \times 60 \times 60$ sites using the Recursion Method.[28] The main result of this work is presented in Fig. 11.5. For each concentration, critical values of V, at which the mobility edge and the Fermi level coincide, were calculated in the PM and FM phases. A value of V between these two implies a crossing of the Fermi level and the mobility edge when the system orders. A value of $V > 2.5t$ is sufficient to give rise to an Anderson MI transition for some concentrations. While it is still higher than the estimate based on Pickett and Singh's work,[25] $V \approx 2.1t$, it is sufficiently close to cast some doubt on a straightforward dismissal of a role of Anderson localization in the magnetoresistance of the manganites. Note that, in this model the critical value of disorder does not vary monotonically with x and shows a maximum at around $x \sim 0.1$. One should bear in mind that, in this model, changing x also changes the distribution of site energies, and so V does not, by itself, characterize the disorder.

11.3.2. *Summary*

Colossal magnetoresistive manganites are intrinsically disordered materials: spin disorder at high temperatures in the PM phase, substitutional disorder due to the presence of both divalent a trivalent ions randomly distributed, and structural disorder arising from *rare-earth/alkaline-earth* ionic size mismatch. Its role in the MI transition and associated CMR effect in manganites, however, has been disregarded due to the assumed small amounts of disorder present in this compounds.[13,25] Here we showed that a careful analysis of the Mn-site energies arising from random distribution of divalent and trivalent ions produces a probability distribution with a RMS deviation of $4.6t \approx 0.9\,\text{eV}$. Such a RMS is a consequence of a parametrization of screened Coulomb energies for which an energy shift $V \approx 2.1t \approx 0.43\,\text{eV}$ is assumed to show up in a Mn-site whenever a nearest-

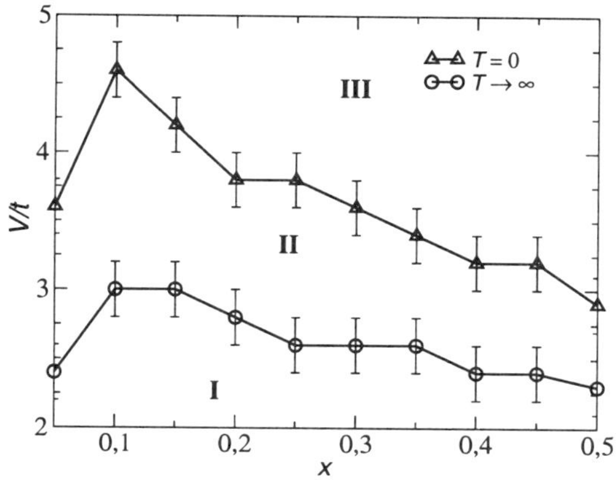

Fig. 11.5. Critical value of V, the energy shift in a manganese ion due to a nearest neighbor replacement $La^{3+} \to D^{2+}$, in the PM (circles) and FM phases (triangles), as a function of x, the concentration of divalent ions. The site energies are obtained by accounting for both first and second shells of La/Ca-sites with equal dielectric constant. A MI transition with ferromagnetic ordering is possible only for a system in region **II**. The estimated value of V from the work of Pickett and Singh[25] is around $V \approx 2.1t$.[27]

neighbor replacement $La^{3+} \to D^{2+}$ occurs. That RMS value already places the system in the disorder window for which a MI transition occurs when the magnetic transition FM $\to$ PM takes place. Furthermore, a double exchange model with a realistic parametrization of on-site disorder was shown to undergo a MI transition at the FM $\to$ PM transition for $V \approx 3t$.[27] This V value is slightly larger than the expected $V \approx 2.1t$, but it is sufficiently close to show that disorder must be considered at least on the same foot as the coupling to the lattice.[29] Moreover, the value $V \approx 2.1t$ only takes into account the random distribution of divalent and trivalent ions. The presence of *rare-earth/alkaline-earth* ionic size mismatch is expected to produce $V \sim 3t$.[30]

Even though this shows that manganites are strongly disordered systems, Anderson localization can never be the full story. We have mentioned the strong coupling to the lattice, and the presence of direct antiferromagnetic interactions. There is ample evidence that the DEM model with antiferromagnetic interactions is unstable toward phase separation of hole-rich metallic regions and hole-poor antiferromagnetic, insulating regions,[10] and this phenomenon certainly plays a role in the transport properties of the manganites. However, this Anderson localization scenario in a DEM model has recently been proposed for a completely different system, the europium hexaborides, in which some of the complications of the manganites are not present. We review this work in the remaining part of this article.

11.4. Eu-based Hexaborides

11.4.1. *Introduction*

Stable rare-earth hexaborides are inclusion compounds of the clathrate type, in which a rare earth ion lives in the midst of an enclosing framework of boron octahedra.[31] Hexaborides have the full cubic symmetry (P_{m3m}) and a crystal structure as detailed in Fig. 11.6. The elementary unit cell is constituted by the rare earth plus eight B atoms. The boron cage is held through both inter and intra-octahedral

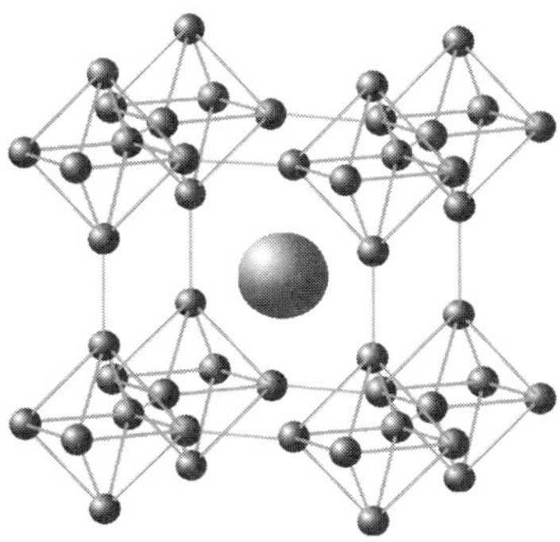

(a)

Parameter	Value
a	4.1849 Å
d_{inter}	1.6964 Å
d_{intra}	1.7596 Å
$d_{\text{Eu-B}}$	3.0783 Å
$\nu_{\text{u.cell}}$	7.330×10^{-23} cm^3

(b)

Fig. 11.6. Crystal structure of EuB_6. (a) Representation of the real-space disposition of the rare-earth (red/center) and the B_6^{2-} framework (blue) in the hexaborides. (b) Relevant lattice parameters for EuB_6, after Blomberg *et al.*[32]

covalent B–B bonds, which account for the rigidity of hexaborides, their low coefficients of thermal expansion and high melting points. The interstitial atom, on the other hand, binds weekly to the boron enclosure, as can be indirectly observed through the phonon modes of the compound.[33] According to the earliest electronic structure investigations, the atomic orbitals of the B atoms hybridize into 10 bonding and 14 anti-bonding orbitals, creating a deficit of 2 electrons per unit cell, insofar as the 6 borons only contribute with their 18 valence electrons.[34,35] The saturation of the bonding orbitals is hence afforded by the transfer of 2 electrons from the metallic atom and, on account of that, it was advanced that divalent hexaborides (like EuB_6 or CaB_6) would be insulators, whereas their trivalent counterparts (e.g. YB_6, LaB_6) should exhibit clear metallic conductivity,[36] a view that seemed to tally with the scanty measurements available at the time.[37] But even though trivalent hexaborides are generally good metals, and the electron counting argument works, the situation is not as clear for the divalent hexaborides.

We will take particular note of the divalent EuB_6, the only ferromagnetic hexaboride and an intriguing metal, and the derived series $Eu_{1-x}Ca_xB_6$. There is no doubt that Eu adopts the divalent configuration in the hexaborides,[32,38–41] and its half-filled $4f$ shell leaves a localized $S = 7/2$ spin at the center of each cube, in the depiction of Fig. 11.6. EuB_6 shares many aspects with the manganites, not only at a physical level, but also in their late comeback to popularity: although experiments on EuB_6 have been available for a long time (see Geballe[39] and references therein), they became notorious only very recently after the discovery of their CMR properties and other intriguing features to which we now turn.

11.4.2. *Overview of Phenomenology*

Electronic Structure It is known from as early as the time of Pauling's landmark works on the nature of chemical bonds[42] that Boron is a chemically notorious element. Its valence in many metallic species is not describable by usual oxidation states, and the ideas of covalent bonding are not straightforwardly applicable.[34] Some divalent hexaborides — EuB_6 included — started to call attentions with the appearance of the first contradicting first principles calculations. As alluded above, the earliest calculations which pointed to a polar semiconductor behavior in the hexaborides,[36] were later challenged by the self consistent APW bandstructure calculations of Hasegawa and Yanase.[43] These showed that, depending on the details of the calculation (muffin-tin, or non-muffin-tin approximation) the outcome could be either semimetallic, with an overlap of conduction and valence bands at the X point in the Brillouin Zone (BZ), or a direct gap semiconductor, also at X. From this point on, a debate emerged around the nature of the bandstructure near this X point. In 1997 LDA+FPLAPW calculations[44] revealed a tiny band overlap, and a period of time followed during which the semimetallic scenario seemed to be accepted, also on account of some apparently consistent experimental surveys of the Fermi surface.[45] However, very recently, a quantity of first principles calculations definitely opened the debate, clearly demonstrating that different approximations to DFT, as tiny as such differences might be, can yield disparate and even contradictory results for the band structure near the X point. So, for example, using the GW approximation, on can find sizeable gaps of ~ 0.8 eV[46] or ~ 0.3 eV,[47] but also an increased band overlap,[47,48] depending on the many-body formulation employed. The latest calculations yield a gap of ~ 0.8 eV, within a so-called weighted density approximation.[49] Figure 11.7(a) gives an example of these conflicting results.

As far as EuB_6 is concerned, things have been even more involved since its

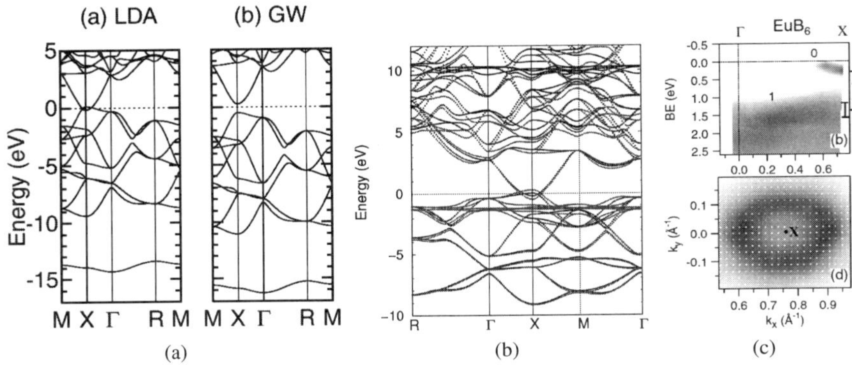

Fig. 11.7. Bandstructure of some hexaborides. (a) LDA and LDA+GW calculations in CaB_6, featuring a small overlap and a sizeable gap, reproduced from Tromp *et al.*[46] (b) LDA+U forEuB_6, featuring a band overlap and slight spin-splitting, after Kunes and Pickett.[50] (c) Bandstructure near the X point and a snapshot of a Fermi surface section obtained from ARPES in EuB_6, reproduced from Denlinger *et al.*[51]

electronic structure has been many times extrapolated from those calculations, most of which done in the context of CaB_6, and not accounting for the f orbitals of Eu.[50] As a consequence, the theoretical electronic structure of EuB_6 remains an unsettled issue to this day.

Fortunately, progress on the experimental front has provided clearer insights. In particular, ARPES and X-ray emission spectroscopy measurements[51,52] clearly show now the presence of a gap of $\sim$ 1 eV at the X point for CaB_6, SrB_6 and EuB_6, and that the latter has an ellipsoidal, pocket-like, Fermi surface with the Fermi level lying near the bottom of the conduction band (Fig. 11.7(c)). Electronic tunneling[53] and low-temperature thermoelectric power measurements[54] corroborate these findings. EuB_6 emerges therefore as a metal with a Fermi level lying at the very bottom of the conduction band,[51] a band with a strong $5d$(Eu) character.[44,47]

Transport and Magnetism EuB_6 outstands among the divalent hexaborides due to its robust ferromagnetism stemming from the $^8S_{7/2}$ state adopted by the cation, which produces an effective Curie-Weiss magnetic moment $\mu_{eff} = 7.94\,\mu_B$. This figure is confirmed by several susceptibility, magnetization and specific heat measurements,[38,39,55–57] which also reveal that long range magnetic order is established below $T_C \simeq 15\,K$. EuB_6 is a remarkably soft ferromagnet with no detectable hysteresis, and negligible remanent magnetization or coercive fields.[56,58] The magnetic response is generally isotropic but, under applied fields, a tiny but interesting anisotropic behavior emerges below T_C, and the system seemingly develops a T and H dependent easy axis.[56] In addition, T_C shows a pressure dependence consistent with the enhancement of bulk ferromagnetism.[59]

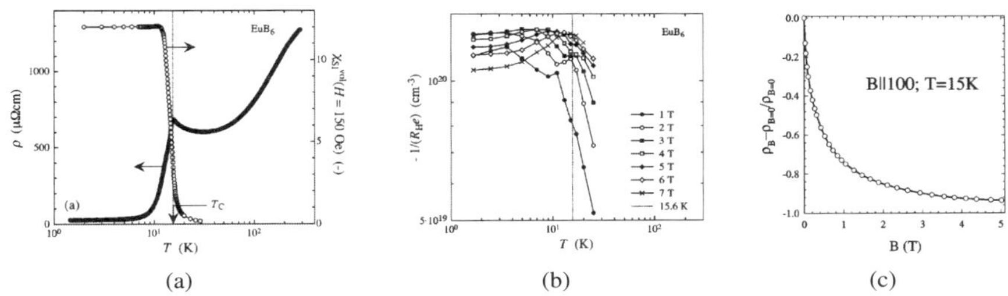

Fig. 11.8. Magnetotransport in EuB_6. (a) DC electrical resistivity and magnetic susceptibility, according to Paschen *et al.*[60] (b) Temperature and field dependence of the carrier concentration, *ibid.* (c) Magnetoresistance at T_C, reproduced from Sullow *et al.*[56]

The unconventional nature of EuB_6 is best perceived when the intricate correlations between electronic and magnetic degrees of freedom are analyzed. Already during the seminal transport measurements,[38,61] EuB_6 revealed itself as an extremely good metal, with residual resistivities of $\sim 10\mu\Omega$.cm, or less,[45,56] in a clear challenge against the early theoretical proposals of the 60's and 70's. The measured carrier densities in EuB_6 are typically of the order of $n_e \sim 10^{-3}$ electrons per unit cell,[38,60,62] tallying with the small electron pockets seen in ARPES (Fig. 11.7(c)). The conduction electrons are attributed to the presence of defects in the boron framework,[32,63] which produce an excess of electrons in the system that occupy the states in the conduction band. Notwithstanding such small densities, residual resistivity ratios of $\rho(T \simeq 300\text{K})/\rho(T \simeq 0)) \sim 50$–100, are commonly obtained.[56]

The hallmark of transport in EuB_6 is the clear feature seen in the resistivity, $\rho(T)$, as T is lowered below T_C. As shown in Fig. 11.8(a), the onset of FM order (signaled by the sharp increase in the magnetic susceptibility) is accompanied by a precipitous drop of the electrical resistivity.[38,56,60] Such steep plunging is preceded by a cusp-shaped upturn in $\rho(T)$ that develops slightly above T_C, and can be very accurately used to determine the Curie temperature by pure electrical means.[56,60] Such cusp features are typical evidence for spin fluctuations interfering with transport.[64] This hint is confirmed by the CMR response of EuB_6 under external magnetic fields,[61] which can be as high as 100% in the vicinity of T_C [56,60,65] (Fig. 11.8). Such conspicuous MR, caught the eye of many condensed matter physicists [a] in a time where the CMR manganites were being heavily studied, which, in turn, fostered still more intriguing discoveries.

One such puzzling aspect is the fact that, the carrier density in EuB_6 is not only extremely small, but displays a reproducible temperature and field dependence. As soon as the FM sets in, n_e is significantly enhanced with decreasing temperatures, and can vary by a factor of 3 between $T > T_C$ and $T \simeq 0$ K, as show in Fig. 11.8(b). This behavior is intriguing insofar as it is not related with the anomalous Hall effect characteristic of some FM metals,[60,66] and suggests that the localized spins influence the transport at a deeper level — much beyond scattering effects — implying some sort of influence in the electronic structure.

Magneto-optical Behavior More intriguing behavior appears when looking at reflectivity signals: EuB_6 shows a giant blue-shift of the unscreened plasma edge, ω_p, with temperature, never seen before.[67,68] As displayed in Fig. 11.9(a), the reflectivity spectrum displays a typical metallic behavior, with a clearly defined plasma threshold in the far infra-red (FIR), at about 2200 cm^{-1}. As a consequence, the optical conductivity exhibits a consistent Drude-like shape (Fig. 11.9(b)). But as soon as the FM sets in, ω_p increases markedly in such a way that it varies by a factor of almost 3 between T_C and $T \ll T_C$.[68,69] This is accompanied by a considerable transfer of spectral weight from high energy to the FIR region. The effect occurs irrespective of whether the temperature is lowered below T_C, or an external magnetic field is applied to the sample. But it is of the highest significance that ω_p scales solely with the magnetization, M, as demonstrated in 2002 by Broderick *et al.*,[69] and shown in Fig. 11.9(c).

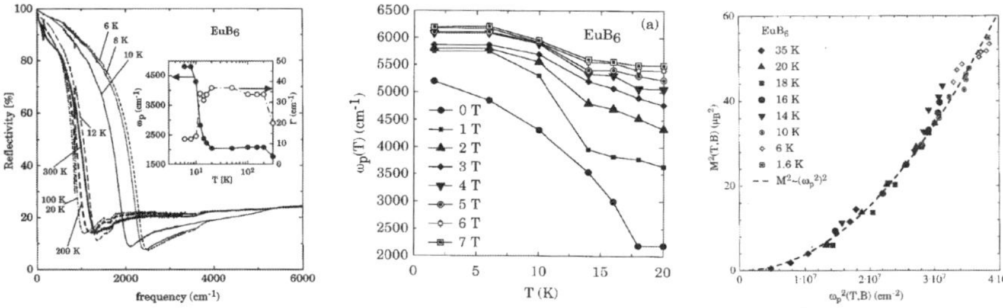

(a) (b) (c)

Fig. 11.9. Magneto-optical behavior of EuB_6. (a) FIR reflectivity spectra at different temperatures, reproduced from Degiorgi *et al.*[68] (b) T dependence of ω_p, at different magnetic fields, as reported by Broderick *et al.*[69] (c) Scaling of ω_p with the magnetization, *ibid.*

[a] Especially because they are structurally simpler than the manganites, and don't display the Jahn-Teller effect.

The magneto-optical Kerr effect in EuB_6 is also exceptional, attaining maximum Kerr rotations among the very largest values ever observed.[70] A resonance in the polarization rotation occurs near 0.3 eV, which is also blue-shifted with H, thus following the plasma edge. This is strong evidence for an interplay between the free electron components (the conduction band electrons) and the localized electron components (the f electrons), in the sense that such interplay is known to cause a strong resonance in the Kerr angle at frequencies coinciding with the plasma edge,[71] and has been observed in other f-electron systems.[72] Finally, Raman scattering experiments in the FIR unveil a diffusive response characteristic of a collision-dominated electronic scattering, except for a narrow temperature range ($T_C < T < T_m$), where a broad Gaussian peak develops around 50–100 cm, and carries all the signatures characteristic of being induced by the presence of magnetic polarons.[73–75] One is lead to the conclusion that the PM–FM transition in EuB_6 is mediated by a polaronic phase, and that these polarons are involved in precipitating the transition into the long-range, ordered, magnetic phase.

Influence of Doping When Ca is substituted for Eu in the family $Eu_{1-x}Ca_xB_6$, ferromagnetism weakens, with smaller values of T_C consistently obtained the higher the doping strength, x, and CaB_6 exhibiting no FM at all.[62,76] Concurrently, the massive drop in $\rho(T)$ below T_C, characteristic of $x = 0$, is smaller and the distinctive metallic character evolves into a bad metal behavior at low/intermediate dopings,[60,77] ending up in a typical semiconducting behavior at the extreme limit of CaB_6.[62] Therefore, the system clearly undergoes a Metal-Insulator (MI) transition induced by the doping level. It is observed that, for $x = 0.4$, both the resistivity[77] and ω_p[78] scale exponentially with the magnetization, which is clearly distinct from the linear scaling observed in pure EuB_6. In addition, this correlates with the fact that the behavior in $\rho(T)$ suggests that the $x = 0.4$ compound is already on (or quite near) the insulating side of the MI transition.

11.4.3. *Double Exchange and $Eu_{1-x}Ca_xB_6$*

11.4.3.1. *Theoretical approaches to EuB_6*

Such a wealth of intriguing behavior has encouraged several theoretical investigations of their underlying mechanisms, and many of them were almost inevitably influenced by the lingering LDA suggestion of the semimetallic character of EuB_6.

In the wake of their investigations of pressure dependent transport and magnetism, Cooley *et al.*[59] suggested that magnetism is driven by a RKKY-type interaction between carriers and local moments, ruling out earlier proposals like superexchange or the Bloemberger-Rowland mechanism.

Kunes and Pickett[50] formulated a model with overlapping electron and hole bands, plus a local magnetic coupling to the Eu spin. Their results from LDA+U serve to extract the model parameters and show that holes and electrons couple with opposite signs to the local magnetic moments. The consequences of this and of the selective splitting of majority and minority bands are studied within mean field, and a semimetallic scenario is proposed.

Later, Lin and Millis[79] improved on such approach by solving the same problem within DMFT obtaining the magnetic phase diagram as a function of band overlap, which includes the possibility of metamagnetic transitions. The spin splitting of the bands plays an important role in this model, and the authors address the variation in specific heat and ω_p at T_C, although a clear difficulty in reproducing all T_C, ω_p and C_V with the same set of parameters remains.

In Calderón *et al.*,[80] in order to reconcile the magnitude of the polaron-induced spin-flip peak in Raman with the assumed values for the exchange coupling of the carriers to the local spins, a semimetallic scenario was suggested, wherein the itinerant electrons are responsible for the magnetic (RKKY-like) coupling, and holes, coupling much weakly to the local moments, lie at the origin of the Stokes signal associated with polarons.

Finally, Wigger *et al.*,[57] presented a detailed analysis of the possible scattering mechanisms contributing to the magneto-transport. Within their interpretation, EuB_6 is a heavily self-doped, strongly compensated n-type semiconductor, with carriers arising from Eu and B defects. The interplay between contributions from orbital and magnetic scattering is explored to explain resistivity and MR measurements, although their fit entails a temperature dependent carrier density and the mechanism for the Hall resistivity appears off by orders of magnitude. A mechanism of charge transfer between spin-splitted valence, conduction, donor and acceptor bands is then employed to address the variation in carrier density and plasma frequency with temperature and magnetic field. More recently, the same authors extended this treatment to address the doped family $Eu_{1-x}Ca_xB_6$, with emphasis on the contributions from defect scattering as the doping is increased.[81] In all these investigations, the presence of the an electron and a hole band crossing the Fermi level is an important ingredient, even in the cases where the starting ground state at $T > T_C$, $H = 0$ is not semimetallic (as in Wigger *et al.*). A different approach is provided by the DE model, and is discussed below in more detail.

11.4.3.2. *The Double Exchange Model for EuB_6*

The development of a microscopic model for EuB_6 hinging upon the DE mechanism,[82] draws support from the phenomenology described above. In particular, EuB_6 is a good metal characterized by a very small carrier density which, at T_C is of the order of 10^{-3} electrons per unit cell. For definiteness, we can take the reference values $n_e(T \gg T_C) \simeq 0.003$ and $n_e(T \ll T_C) \simeq 0.009$, as reported by Paschen *et al.*[60] Its magnetism is attributed to local spins of high magnitude ($S = 7/2$), and the conduction band arises from the $5d$ orbitals of the cation. Hence, the conduction band electrons itinerate among the Eu sites, interacting magnetically with the local $7/2$ spin through Hund's coupling. In the DE approach, the valence band is physically inert and separated from the conduction one by a sizeable gap as reported in ARPES.[51]

FKM, Low Densities and the DE Limit The Hamiltonian describing conduction electrons hopping in a tridimensional cubic lattice, and coupled to local spins at each lattice site is the FKM Hamiltonian, introduced before in Eq. (11.1). The high local spin allows us to take its classical limit, as in the manganites, but the consideration of the other parameters needs more care. In the manganites, the DE limit of Eq. (11.7) is justified by the large exchange to hopping ratio. The relevant parameters are still those two, but now, according to the literature,[44,45,50,57,79] $t \sim 0.5-1$ eV and $J_H S \sim 0.5-0.7$, which is clearly more delicate. The key point lies in the extremely reduced carrier density characteristic of these hexaborides, which places the Fermi level very close to the bottom of the band. The relevant comparison is the Fermi energy which will be consequently small in comparison with J_H. To appreciate this reasoning one can take the FKM Hamiltonian (11.6) written in the local spin basis and assume a mean-field decoupling of the $\Uparrow$ and $\Downarrow$ bands. If $t \sim J_H$ then the most part of the two sub-bands will overlap, but for such low values of n_e as the ones found in EuB_6, E_F still lies below the overlapping region (Fig. 11.10). Since we are working with local spin bands, this means that only states from the lower sub-band are relevant, exactly as in the DE limit. Of course, this pictorial argument is still crude, especially because the hybridization between the two sub-bands is non-perturbative, and begs a more quantitative investigation. What characterizes the DE limit is that the electron always keeps its spin parallel to the local moment's at every visited site. This means that the local spin polarization per electron (in the direction of the local $4f$ spin), defined as

$$m = \frac{1}{N_e} \sum_{i=1}^{N} \left\langle c_{i\Uparrow}^{\dagger} c_{i\Uparrow} - c_{i\Downarrow}^{\dagger} c_{i\Downarrow} \right\rangle_{GS}, \tag{11.17}$$

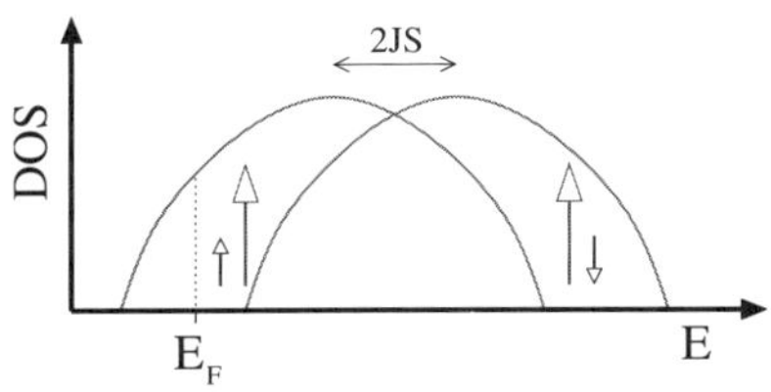

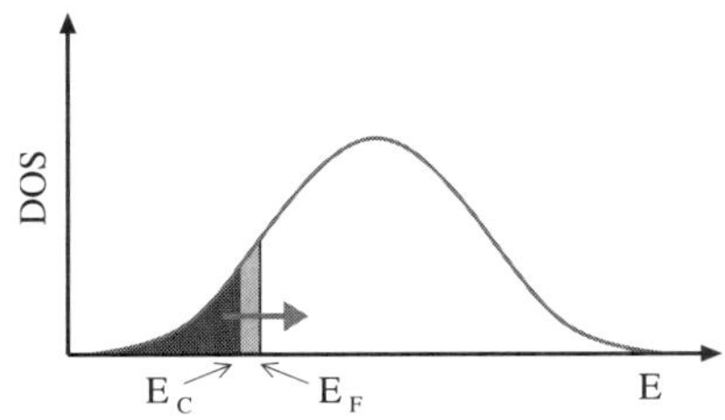

Fig. 11.10. (a) Schematic representation of the mean-field decoupling of the sub-bands in the local basis mentioned in the text. (b) Depiction of the close proximity between E_F and E_C in EuB_6 in the PM phase. The arrow signals the doping induced variation in E_C.

is equal to unity for electron densities satisfying $n_e \leq 1$[b]. Therefore, the local spin polarization is a good quantity to ascertain how close we are to the DE limit for arbitrary t/J_H. The parameter m has been numerically calculated by the authors and A. Castro-Neto for the Hamiltonian (11.6), as a function of electron density and magnetization, and without approximations. It was found that, at the lowest densities and in the PM phase ($M = 0$), $m \simeq 0.5$ for the worst case scenario of $J_H S = t$, and $m \simeq 0.7$ for $J_H S = 2t$. The value $m \simeq 0.5$ means that only 25% of the electron states have an *"upper band character"*, and m goes rapidly to unity as the magnetization is increased from zero.[83] Therefore even for $J_H \sim t$, the DE limit is an acceptable approximation to the FKM, provided that the carrier density is small enough. This justifies and motivates the application of Eq. (11.7) as the working Hamiltonian in the context of Eu-based hexaborides.

Anderson Localization The situation is now similar to the simplest models of manganites, except for the very small carrier density, which turns out to be of paramount relevance. At any nonzero temperature, there will be thermal fluctuations of the local spins, generating a disordered background for the itinerant electrons. The maximally disordered state is attained at T_C, when $M \to 0$, and the so-called mobility edge (E_C), is farther from the bottom of the band. As already discussed, this non-diagonal disorder is rather weak in the sense that the percentage of localized states[c] is notoriously small, being $\sim 0.3\%$ at $M = 0$[20,83] (see also § 11.3.1).

But this figure, though tiny, corresponds to a density of localized states ~ 0.003, which is of utmost pertinence to the hexaborides! The experimental values of n_e between 0.003 and 0.009 inferred from the Hall effect, correspond

[b]In the strict DE limit ($J_H \gg t$), m is unity by construction.
[c]That is, the number of electronic band states below the mobility edge.

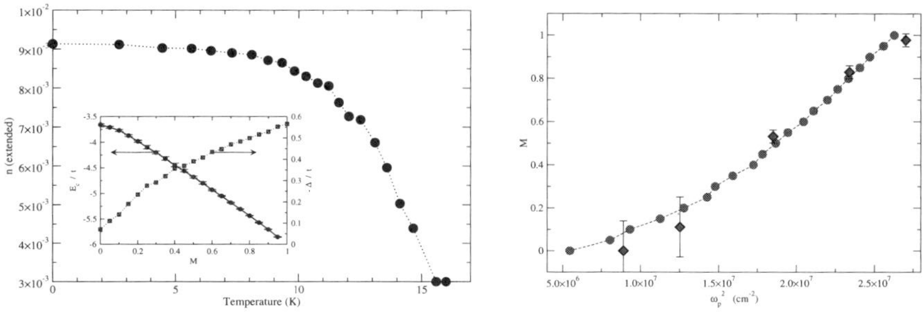

(a) (b)

Fig. 11.11. Theoretical n_e and ω_p for EuB_6 within DE. (a) The plot of $n_e(T)$ obtained for the DE model as discussed in the text; the inset shows the evolution of E_C and the mobility gap as M is varied. (b) The theoretical curve of ω_p vs M (circles) and its comparison with the experimental results (diamonds). Reproduced from Pereira *et al.*[82]

to the density of carriers lying between the mobility edge and E_F, for only they contribute to transport[d]. These are the *extended* carriers, whose density is of the same order of the *localized* carriers for $T > T_C$. By investigating the relative position of E_C and E_F, and its dependence on the magnetization/temperature, Pereira *et al.*[82] were able to reproduce the variation of electron density in the FM phase of EuB_6, as seen in the experiments.[60] The mechanism is essentially due to the fact that when T descends below T_C, the electronic disorder is progressively attenuated, causing a drift of the mobility edge towards the band edge, and the concomitant release of the formerly localized states. The interplay between the M dependencies of both E_C and E_F results in a net increase in the number of extended states with M, as shown in Fig. 11.11(a).

This was also the first appearance in the literature of the trajectory of $E_C(M)$ for the DEM. At the same time, the plasma frequency for a single-particle Hamiltonian as (11.7) can be formally obtained with resort to the optical sum rules and the Kubo formula for the conductivity.[84,85] The contribution from the extended (metallic) states to ω_p so obtained is shown in Fig. 11.11(b), where it has been plotted as a function of M, and superimposed with the experimental data. The agreement in the variation of both n_e or ω_p with the values seen experimentally is quite remarkable, even more so when the only model parameters are the carrier density (fixed by the experimental values) and the hopping t (which is obtained by fitting ω_p to the experimental data, resulting in $t \simeq 0.55$ eV[82]). Other experimental findings regarding EuB_6 that this DE approach is able to reproduce include the

[d]Since $T_C \simeq 15$K and $t \sim 1$eV, the electronic subsystem can be considered at zero temperature for practical purposes.

magnitude of the precipitous drop of $\rho(T)$ below T_C, the large MR at and around T_C, and the scaling of ω_p with M.[83]

Magnetic Polarons Since the evidence towards a polaronic phase mediating the PM–FM transition is compelling, the possibility of the stabilization of magnetic polarons solely via the DE mechanism was subsequently investigated.[86] Under such conditions the magnetic polaron (an electron self trapped in a local FM cloud, embedded in a global PM background) can only emerge from a favorable competition between entropy and trapping energy. One of the results is that the DE alone does support a polaronic phase in the regime of densities and interval of temperatures ($T_C < T < T_m$) compatible with EuB_6. In addition, the polaronic phase induces a significant decrease in the theoretical T_C. This is also interesting for, without consideration of this phase, the theoretical T_C overestimates the real value,[82] and the polaronic phase brings it to the correct range around 15 K.

The known problem of an instability towards phase separation in the DE model at low densities[87–92] has also been addressed in this same context. In the presence of Coulomb interactions, the phase separation is frustrated by the large electrostatic energy price of confining charge to a given region of the system. It is found that, for reasonable values of the dielectric constant, the tendency for charge neutrality (which favors small radius of electron rich regions) and the kinetic energy of localization, lead to a strong suppression of the phase separation region, in temperature and electron concentration, and its replacement by a polaronic phase.[83]

11.4.3.3. *The Double Exchange Model and $Eu_{1-x}Ca_xB_6$*

The Eu→Ca substitution, has two simultaneous implications. It dilutes the local spin sub-system (Ca^{2+} is non-magnetic) and, at the same time, reduces the number of lattice sites available for the electronic hopping. Thus it is not surprising that magnetism weakens and percolation signatures transpire from the transport measurements in the doped series.[81] From the above is seems clear that disorder and Anderson localization plays a determinant role in the magneto-optics and transport of EuB_6. Doping, being effectively a site dilution for the electrons, configures a much strong form of electronic disorder, and the question of Anderson localization then becomes a matter of quantum percolation. Since Ca and Eu are isovalent, it has been proposed that the primary effect of doping would be the displacement of E_C towards the band center (Fig. 11.10(b)). Now, following the DE picture, E_C and E_F are already quite close to each other in EuB_6, and it is clear that at some critical doping, x_{MI}, these two energy scales will meet, and a metal–insulator transition ensues. For higher dopings the system should be an Anderson insulator, although magnetism is still possible on account of the finite localization lengths that still permit some itinerancy, and thus indirect magnetic coupling between the remaining local spins. Magnetism should then cease near the percolation threshold, as follows from the phase diagram proposed by the authors and A. Castro-Neto in 2004 (Fig. 11.12(a)).

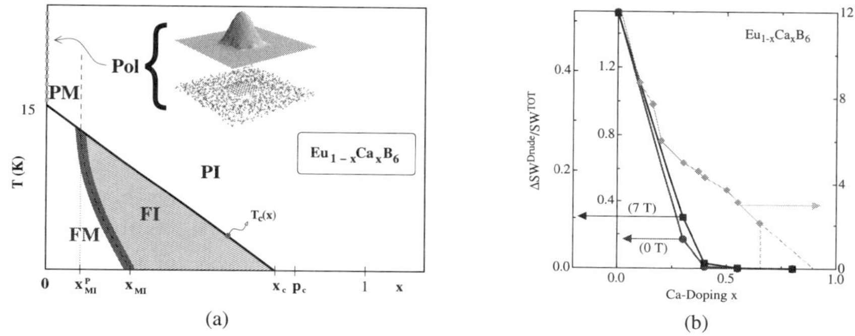

Fig. 11.12. (a) The phase diagram proposed by Pereira *et al.*[82] for $Eu_{1-x}Ca_xB_6$. The inset depicts a magnetic polaron, with the electronic wavefunction and the underlying local magnetization profile, applicable at $x = 0$. PM: paramagnetic metal, FM: Ferromagnetic metal, PI: paramagnetic insulator, FI: ferromagnetic insulator, Pol: Polaronic phase. (b) Variation of the Drude spectral weight in $\sigma_1(\omega)$, ΔSW^{Drude}, compared with the variation in T_C across the Ca-doped series. Reproduced from Caimi *et al.*[93]

In 2006 Caimi *et al.*[93] undertook an excellent set of magneto-optical experiments in the $Eu_{1-x}Ca_xB_6$ series. Their measurements provided a map of the free carrier contribution to the total spectral weight in the optical conductivity, $\sigma(\omega)$, and studied its variation between the PM and FM phase (ΔSW^{Drude}), revealing that it behaves as shown in Fig. 11.12(b). The experimental curves feature a Drude spectral weight that varies between PM and FM phases only up to $x \simeq 0.4$, although magnetism persists up to percolation. Such behavior of ΔSW^{Drude} confirms that, at low doping, $E_C(M = 0)$ remains below $E_F(M = 0)$ and that the magnetization–induced change in E_C is still significant to be seen in the optical response. $x \simeq 0.4$ is presumably the point at which the two energies meet, something also following from the behavior of the resistivity.[62,77] Past this doping level, E_C lies above E_F, in such a way that the onset of FM is not sufficient to bring it back again across the Fermi level, and so there is no significant change in the Drude spectral weight upon spin polarization: the system is an insulator. The results of Fig. 11.12(b) constitute a magneto-optical counterpart of the phase diagram in Fig. 11.12(a), and, in view of the above, provide a strong support as to the DE interpretation of the relevant mechanisms at play in these Eu-based hexaborides.

Thus, whereas the simple DE model with a magnetization/doping induced change in the mobility edge is just one of the several ingredients presumably at the origin of the notable properties of manganites, in the hexaborides it seems to be one unavoidable factor.

11.5. Conclusion

In the double exchange model the ferromagnetic transition is always accompanied by a decrease in electronic disorder. We have reviewed studies that addressed the possibility that this decrease in electronic disorder induces a insulator-metal transition at T_C, both in the context of the manganites and the hexaborides. We concluded that some manganites are strongly disordered, close to an Anderson transition, in contrast with what appears to be the dominant opinion in the literature. Nevertheless, it is in the europium hexaborides that we find the clearest evidence that the DE mechanism is a key factor in a consistent interpretation of a variety of magnetic, optical and transport properties.

Acknowledgements

The authors would like to acknowledge the co-authorship of A. Castro-Neto in some of the work reviewed here and many enlightening discussions and important insights from N. M. R. Peres, L. Degiorgi, Paco Guinea and Y. G. Pogorelov.

This work has been supported by FCT (Portugal) through the grants with references SFRH/BD/4655/2001 (VMP) and SFRH/BD/13182/2003 (EVC), and through the program POCI2010.

References

1. C. Zener, Interaction between the d-shells in the transition metals. II. Ferromagnetic compounds of manganese with perovskite structure, *Phys. Rev.* **82**, 403 – 405, (1951).
2. G. H. Jonker and J. H. Van Santen, Ferromagnetic compounds of manganese with perovskite structure, *Physica.* **16**, 337 – 349, (1950).
3. E. O. Wollan and W. C. Koehler, Neutron diffraction study of the magnetic properties of the series of perovskite-type compounds $[(1-x)$La, xCa]MnO_3, *Phys. Rev.* **100**, 545 – 563, (1955).
4. J. B. Goodenough, Theory of the role of covalence in the perovskite-type manganites [La, M(II)]MnO_3, *Phys. Rev.* **100**, 564 – 573, (1955).
5. P. W. Anderson and H. Hasegawa, Considerations on double exchange, *Phys. Rev.* **100**, 675 – 681, (1955).

6. P. G. de Gennes, Effects of double exchange in magnetic crystals, *Phys. Rev.* **1**, 141 – 154, (1960).
7. R. von Helmolt, J. Wecker, B. Holzapfel, L. Schultz, and K. Samwer, Giant negative magnetoresistance in perovskitelike $La_{2/3}Ba_{1/3}MnO_x$ ferromagnetic films, *Phys. Rev. Lett.* **71**, 2331 – 2333, (1993).
8. S. Jin, T. H. Tiefel, M. McCormack, R. A. Fastnacht, R. Ramesh, and L. H. Chen, Thousandfold change in resistivity in magnetoresistive La-Ca-Mn-O films, *Science.* **264**, 413 – 415, (1994).
9. M. McCormack, S. Jin, T. H. Tiefel, R. M. Fleming, J. M. Phillips, and R. Ramesh, Very large magnetoresistance in perovskite-like La-Ca-Mn-O thin films, *Appl. Phys. Lett.* **64**, 3045 – 3047, (1994).
10. E. Dagotto, T. Hotta, and A. Moreo, Colossal magnetoresistant materials: The key role of phase separation, *Phys. Rep.* **344**, 1 – 153, (2001).
11. Y. Tokura, Ed., *Colossal Magnetoresistive Oxides.* (Gordon & Breach, London, 2000).
12. V. M. Loktev and Y. G. Pogorelov, Peculiar physical properties and the colossal magnetoresistance of manganites (review), *Low Temp. Phys.* **26**, 171 – 193, (2000).
13. D. M. Edwards, Ferromagnetism and electron-phonon coupling in the manganites, *Adv. Phys.* **51**, 1259 – 1318, (2002).
14. M. B. Salamon and M. Jaime, The physics of manganites: Structure and transport, *Rev. Mod. Phys.* **73**, 583 – 628, (2001).
15. R. Maezono, S. Ishihara, and N. Nagaosa, Phase diagram of manganese oxides, *Phys. Rev. B.* **58**, 11583 – 11596, (1998).
16. A. J. Millis, B. I. Shraiman, and R. Mueller, Dynamic jahn-teller effect and colossal magnetoresistance in La1-xSrxMnO3, *Phys. Rev. Lett.* **77**, 175 – 178, (1996).
17. H. Röder, J. Zang, and A. R. Bishop, Lattice effects in the colossal-magnetoresistance manganites, *Phys. Rev. Lett.* **76**, 1356 – 1359, (1996).
18. C. M. Varma, Electronic and magnetic states in the giant magnetoresistive compounds, *Phys. Rev. B.* **54**, 7328 – 7333, (1996).
19. E. N. Economou and P. D. Antoniou, Localization and off-diagonal disorder, *Solid State Commun.* **21**, 285–288, (1977).
20. Q. M. Li, J. Zang, A. R. Bishop, and C. M. Soukoulis, Charge localization in disordered colossal-magnetoresistance manganites, *Phys. Rev. B.* **56**, 4541 – 4544, (1997).
21. L. Sheng, D. Y. Xing, D. N. Sheng, and C. S. Ting, Theory of colossal magnetoresistance in $R_{1-x}A_x\text{MnO}_3$, *Phys. Rev. Lett.* **79**, 1710 – 1713, (1997).
22. A. Mackinnon and B. Kramer, One-parameter scaling of localization length and conductance in disordered systems, *Phys. Rev. Lett.* **47**, 1546 – 1549, (1981).
23. A. Mackinnon and B. Kramer, The scaling theory of electrons in disordered solids - additional numerical results, *Z. Phys. B - Condens. Matter.* **53**, 1 – 13, (1983).
24. L. Sheng, D. Xing, D. Sheng, and C. Ting, Metal-insulator transition in the mixed-valence manganites, *Phys. Rev. B.* **56**, R7053 – R7056, (1997).
25. W. E. Pickett and D. J. Singh, Chemical disorder and charge transport in ferromagnetic manganites, *Phys. Rev. B.* **55**, R8642 – R8645, (1997).
26. A. V. Boris, N. N. Kovaleva, A. V. Bazhenov, P. J. M. van Bentum, T. Rasing, S.-W. Cheong, A. V. Samoilov, and N.-C. Yeh, Infrared studies of a $La_{0.67}Ca_{0.33}MnO_3$ single crystal: Optical magnetoconductivity in a half-metallic ferromagnet, *Phys. Rev. B.* **59**, R697 – R700, (1999).

27. E. V. Castro. Efeitos de desordem em manganites (in portuguese). Master's thesis, Universidade de Aveiro, Portugal, (2003).
28. R. Haydock. The recursive solution of the schrödinger equation. In eds. H. Ehrenreich, F. Seitz, and D. Turnbull, *Solid State Physics*, vol. 35, p. 215. Academic Press, New York, (1980).
29. E. E. Narimanov and C. M. Varma, Transition temperature and magnetoresistance in double-exchange compounds with moderate disorder, *Phys. Rev. B*. **65**, 024429, (2001).
30. J. Alonso, L. Fernandez, F. Guinea, V. Laliena, and V. Martin-Mayor, Interplay between double-exchange, superexchange, and lifshitz localization in doped manganites, *Phys. Rev. B*. **66**, 104430, (2002).
31. J. Etourneau and P. Hagenmuller, Structure and physical features of the rare-earth borides, *Philos. Mag. B*. **52**, 589, (1985).
32. M. K. Blomberg, M. J. Merisalo, M. M. Korsukova, and V. N. Gurin, Single-crystal X-ray diffraction study of NdB6, EuB6 and YbB6, *J. Alloys and Compounds*. **217**, 123, (1995).
33. D. Mandrus, B. C. Sales, and R. Jin, Localized vibrational mode analysis of the resistivity and specific heat of LaB_6, *Phys. Rev. B*. **64**, 12302, (2001).
34. H. C. Longuet-Higgins and M. D. V. Roberts, The Electronic Structure of the Borides MB[6], *Proc. R. Soc. A*. **224**, 336, (1954).
35. W. N. Lipscomb and D. Britton, Valence Structure of the Higher Borides, *J. Chem. Phys*. **33**, 275, (1960).
36. M. Yamazaki, Group-Theoretical Treatment of the Energy Bands in Metal Borides MeB6, *J. Phys. Soc. Jpn*. **12**, 1, (1956).
37. R. W. Johnson and A. H. Daane, Electron Requirements of Bonds in Metal Borides, *J. Chem. Phys*. **38**, 425, (1963).
38. Z. Fisk, D. C. Johnston, B. Cornut, S. von Molnar, S. Oseroff, and R. Calvo, Magnetic, transport, and thermal properties of ferromagnetic EuB_6, *J. Appl. Phys*. **50**, 1911, (1979).
39. T. H. Geballe, B. T. Matthias, K. Andres, J. P. Maita, A. S. Cooper, and E. Corenzwit, Magnetic Ordering in the Rare-Earth Hexaborides, *Science*. **160**, 1443, (1968).
40. J. A. Clack, J. D. Denlinger, J. W. Allen, D. M. Poirier, C. G. Olson, Z. Fisk, D. Young, and P. Canfield, Resonant, Core Level and Angle Resolved Photoemission Studies of Rare Earth Hexaborides, *Unpublished*.
41. L.-P. Li, G.-S. Li, W.-H. Su, X.-D. Zhao, and X. Liu, A Mössbauer study of $La_{1-x}Eu_xB_6$ compounds synthesized at high pressure and temperature, *Hyperfine Interactions*. **128**, 409, (2000).
42. L. Pauling, The Nature of the Chemical Bond. II. The One-Electron Bond and the Three-Electron Bond., *J. Am. Chem. Soc*. **53**, 3225, (1931).
43. A. Hasegawa and A. Yanase, Electronic structure of CaB_6, *J. Phys. C: Solid State Physics*. **12**, 5431, (1979).
44. S. Massidda, A. Continenza, T. M. de Pascale, and R. Monnier, Electronic structure of divalent hexaborides, *Z. Phys. B*. **102**, 83, (1996).
45. M. C. Aronson, J. L. Sarrao, Z. Fisk, M. Whitton, and B. L. Brandt, Fermi surface of the ferromagnetic semimetal, EuB_6, *Phys. Rev. B*. **59**, 4720, (1999).
46. H. J. Tromp, P. van Gelderen, P. J. Kelly, G. Brocks, and P. A. Bobbert, CaB_6: A New Semiconducting Material for Spin Electronics, *Phys. Rev. Lett*. **87**, 16401, (2001).

47. H. Kino, F. Aryasetiawan, K. Terakura, and T. Miyake, Abnormal quasiparticle shifts in CaB6, *Phys. Rev. Lett.* **66**, 121103R, (2002).
48. C. O. Rodriguez, R. Weht, and W. E. Pickett, Electronic Fine Structure in the Electron-Hole Plasma in SrB_6, *Phys. Rev. Lett.* **84**, 3903, (2000).
49. Z. Wu, D. Singh, and R. Cohen, Electronic Structure of Calcium Hexaboride within the Weighted Density Approximation, *Phys. Rev. B.* **69**, 193105, (2004).
50. J. Kunes and W. E. Pickett, Kondo and anti-Kondo coupling to local moments in EuB_6, *Phys. Rev. B.* **69**, 165111, (2004).
51. J. D. Denlinger, J. A. Clack, J. W. Allen, G.-H. Gweon, D. M. Poirier, and C. G. Olson, Bulk Band Gaps in Divalent Hexaborides, *Phys. Rev. Lett.* **89**, 157601, (2002).
52. S. Souma, H. Komatsu, T. Takahashi, R. Kaji, T. Sasaki, Y. Yokoo, and J. Akimitsu, Electronic Band Structure and Fermi Surface of CaB_6 Studied by Angle-Resolved Photoemission Spectroscopy, *Phys. Rev. Lett.* **90**, 27202, (2003).
53. J.-S. Rhyee, B. H. Oh, B. K. Cho, M. H. Jung, H. C. Kim, Y. K. Yoon, J. H. Kim, and T. Ekino, Formation of mid-gap states and ferromagnetism in semiconducting CaB_6, *cond-mat/0310068*. (2003).
54. K. Giannò, A. V. Sologubenko, H. R. Ott, A. D. Bianchi, and Z. Fisk, Low-temperature thermoelectric power of cab_6, *Journal of Physics: Condensed Matter*. **14**, 1035, (2002).
55. W. Henggeler, H.-R. Ott, D. P. Young, and Z. Fisk, Magnetic ordering in EuB_6, investigated by neutron diffraction, *Solid State Commun.* **108**, 929, (1998).
56. S. Süllow, I. Prasad, M. C. Aronson, J. L. Sarrao, Z. Fisk, D. H. A. H. Lacerda, M. F. Hundley, A. Vigliante, and D. Gibbs, Structure and magnetic order of EuB_6, *Phys. Rev. B.* **57**, 5860, (1998).
57. G. Wigger, R. Monnier, H. R. Ott, D. Young, and Z. Fisk, Electronic transport in EuB_6, *Phys. Rev. B.* **69**, 125118, (2004).
58. R. G. Goodrich, N. Harrison, J. J. Vuillemin, A. Teklu, D. W. Hall, Z. Fisk, D. Young, and J. Sarrao, Fermi surface of ferromagnetic EuB, *Phys. Rev. B.* **58**, 14896, (1998).
59. J. C. Cooley, M. C. Aronson, J. L. Sarrao, and Z. Fisk, High pressures and ferromagnetic order in EuB_6, *Phys. Rev. B.* **56**, 14541, (1997).
60. S. Paschen, D. Pushin, M. Schlatter, P. Vonlanthen, H. R. Ott, D. P. Young, and Z. Fisk, Electronic transport in $Eu_{1-x}Ca_xB_6$, *Phys. Rev. B.* **61**, 4174, (2000).
61. C. N. Guy, S. von Molnar, J. Etourneau, and Z. Fisk, Charge transport and pressure dependence of T_C of Single Crystal Ferromagnetic EuB_6, *Solid State Commun.* **33**, 1055, (1980).
62. J.-S. Rhyee, B. K. Cho, and H.-C. Ri, Electrical transport properties and small polarons in $Eu_{1-x}Ca_xB_6$, *Phys. Rev. B.* **67**, 125102, (2003).
63. R. Monnier and B. Delley, Point Defects, Ferromagnetism, and Transport in Calcium Hexaboride, *Phys. Rev. Lett.* **87**, 157204, (2001).
64. M. E. Fisher and J. S. Langer, Resistive Anomalies at Magnetic Critical Points, *Phys. Rev. Lett.* **20**, 665, (1968).
65. R. Urbano, P. Pagliuso, C. Rettori, P. Schlottmann, J. Sarrao, A. Bianchi, S. Nakatsuji, Z. Fisk, E. Velazquez, and S. B. Oseroff, Gradual transition from insulator to semimetal of $Ca_{1-x}Eu_xB_6$ with increasing Eu concentration, *Phys. Rev. B.* **71**, 184422, (2005).
66. E. M. Pugh and N. Rostoker, Hall Effect in Ferromagnetic Materials, *Rev. Mod. Phys.* **25**, 151, (1953).

67. S. Broderick, L. Degiorgi, H. Ott, J. Sarrao, and Z. Fisk, Polar Kerr rotation of the ferromagnet EuB_6, *Eur. Phys. J. B.* **33**, 47, (2003).
68. L. Degiorgi, E. Felder, H. R. Ott, J. L. Sarrao, and Z. Fisk, Low-Temperature Anomalies and Ferromagnetism of EuB_6, *Phys. Rev. Lett.* **79**, 5134, (1997).
69. S. Broderick, B. Ruzicka, L. Degiorgi, H. R. Ott, J. L. Sarrao, and Z. Fisk, Scaling between magnetization and Drude weight in EuB_6, *Phys. Rev. B.* **65**, 121102(R), (2002).
70. S. Broderick, L. Degiorgi, H. R. Ott, J. L. Sarrao, and Z. Fisk, Giant magneto-optical response of ferromagnetic EuB_6, *Eur. Phys. J. B.* **27**, 3, (2002).
71. H. Feil and C. Haas, Magneto-Optical Kerr Effect, Enhanced by the Plasma Resonance of Charge Carriers, *Phys. Rev. Lett.* **58**, 65, (1986).
72. F. Salghetti-Drioli. *Magneto-optical Kerr effect in the plasma-edge region : evidence for interplay between Drude term and interband transitions*. PhD thesis, ETH, Zürich, (1999).
73. P. Nyhus, S. Yoon, M. Kauffman, S. L. Cooper, Z. Fisk, and J. Sarrao, Spectroscopic study of bound magnetic polaron formation and the metal-semiconductor transition in EuB_6, *Phys. Rev. B.* **56**, 2717, (1997).
74. D. Heiman, P. A. Wolff, and J. Warnock, Spin-flip Raman scattering, bound magnetic polaron, and fluctuations in (Cd,Mn)Se, *Phys. Rev. B.* **27**, 4848, (1983).
75. D. L. Peterson, D. U. Bartholomew, U. Debska, A. K. Ramdas, and S. Rodriguez, Spin-flip Raman scattering in n-type diluted magnetic semiconductors, *Phys. Rev. B.* **32**, 323, (1985).
76. J.-S. Rhyee, B. H. Oh, B. K. Cho, H. C. Kim, and M. H. Jung, Magnetic properties in Ca-doped Eu hexaborides, *Phys. Rev. B.* **67**, 212407, (2003).
77. G. A. Wigger, C. Wälti, , H. R. Ott, A. D. Bianchi, and Z. Fisk, Magnetization-dependent electronic transport in Eu-based hexaborides, *Phys. Rev. B.* **66**, 212410, (2002).
78. A. Perucchi, G. Caimi, H. Ott, L. Degiorgi, A. Bianchi, and Z. Fisk, Optical evidence for a spin-filter effect in the charge transport of $Eu_{0.6}Ca_{0.4}B_6$, *Phys. Rev. Lett.* **92**, 67401, (2003).
79. C. Lin and Andrew.J.Millis, Dynamical Mean Field Theory of Temperature and Field Dependent Band Shifts in Magnetically Coupled Semimetals: Applocation to EuB6, *cond-mat/0407706*. **71**, 75111, (2004).
80. M. Calderon, L. Wegener, and P. Littlewood, Evaluation of evidence for magnetic polarons in EuB_6, *Phys. Rev. B.* **70**, 92408, (2004).
81. G. Wigger, C. Beeli, E. Felder, H. Ott, A. Bianchi, and Z. Fisk, Percolation and Colossal Magnetoresistance in Eu-based Hexaborides, *Phys. Rev. Lett.* **93**, 147203, (2004).
82. V. M. Pereira, J. M. B. Lopes dos Santos, E. V. Castro, and A. H. Castro Neto, Double Exchange Model for Magnetic Hexaborides , *Phys. Rev. Lett.* **93**, 147202, (2004).
83. V. M. Pereira. *Disorder and Localization Effects in Correlated Electronic Systems*. PhD thesis, University of Porto, Porto, Portugal, (2006).
84. P. F. Maldague, Optical spectrum of a Hubbard chain, *Phys. Rev. B.* **16**, 2437, (1977).
85. F. Wooten, *Optical Properties of Solids*. (Academic Press, New York, 1972).
86. V. M. Pereira, J. M. B. L. dos Santos, and A. H. C. Neto, The Double Exchange Model at Low Densities, *cond-mat/0505741*. (2005).
87. E. L. Nagaev, Instability of the double-exchange-induced canted antiferromagnetic ordering, *Phys. Rev. B.* **58**, 2415, (1998).

88. S. Yunoki, J. Hu, A. L. Malavezzi, A. Moreo, N. Furukawa, and E. Dagotto, Phase Separation in Electronic Models for Manganites, *Phys. Rev. Lett.* **80**(4), 845, (1998).
89. D. P. Arovas, G. Gómez-Santos, and F. Guinea, Phase separation in double-exchange systems, *Phys. Rev. B.* **59**, 13569, (1999).
90. M. Y. Kagan, D. I. Khomskii, and M. V. Mostovoy, Double-exchange model: phase separation versus canted spins, *Eur. Phys. J. B.* **12**, (1999).
91. J. L. Alonso, L. A. Fernández, F. Guinea, V. Laliena, and V. MartíÂn-Mayor, Variational Mean-Field Approach to the Double-Exchange Model, *Phys. Rev. B.* **63**, 54411, (2001).
92. J. L. Alonso, L. A. Fernández, F. Guinea, V. Laliena, and V. Martín-Mayor, Discontinuous Transitions in Double-Exchange Materials, *Phys. Rev. B.* **63**(64416), (2001).
93. G. Caimi, A. Perucchi, L. Degiorgi, H. Ott, V. Pereira, A. Castro-Neto, A. Bianchi, and Z. Fisk, Magneto-optical evidence of double exchange in a percolating lattice, *Phys. Rev. Lett.* **96**, 016403, (2006).

Chapter 12

Spin transport in magnetic nanowires with domain walls

V. K. Dugaev[a,b], M. A. N. Araújo[a,c], V. Rocha Vieira[a], P. D. Sacramento[a], J. Barnaś[d] and J. Berakdar[e]

a) CFIF and Departamento de Física, Instituto Superior Técnico, Av. Rovisco Pais, 1049-001 Lisbon, Portugal

b) Department of Mathematics and Applied Physics, Rzeszów University of Technology, Al. Powstańców Warszawy 6, 35-959 Rzeszów, Poland and Frantsevich Institute for Problems of Materials Science, National Academy of Sciences of Ukraine, Vilde 5, 58001 Chernovtsy, Ukraine

c) Departamento de Física, Universidade de Évora, P-7000-671, Évora, Portugal

d) Department of Physics, Adam Mickiewicz University, Umultowska 85, 61-614 Poznań and Institute of Molecular Physics, Polish Academy of Sciences, Smołuchowskiego 17, 60-179 Poznań, Poland

e) Institut für Physik, Martin-Luther Universität Halle-Wittenberg, Heinrich-Damerow-Straße 4 - Nanotechnikum Weinberg, 06120 Halle, Germany

We review briefly the problem of electron transport in magnetic nanowires with thin domain walls. Transmission of electrons in such structures is associated with charge and spin currents leading to the occurrence of a spin torque that acts on the domain wall. Experimentally, the properties of such structures are manifested as a large magnetoresistance, current-induced motion of the domain wall, generation of spin currents, etc. The effect of electron interactions on the scattering from a sharp domain wall is also considered in more details. Using a renormalization group approach for the interactions, we obtain scaling equations for the scattering amplitudes. The RG equations obtained are independent of the single-particle model for the domain wall. We describe the nature of the zero temperature fixed points. For repulsive interactions, the wall reflects all incident electrons at the fixed points. However, the interactions determine whether this reflection is accompanied by spin reversal or not. In one of the fixed points the wall flips the spin of all incident electrons, generating a finite spin current without an associated charge current. It is also shown that the RG flow affects short walls more quickly than long walls, implying that correlations have a more important effect on short walls.

Contents

12.1. Introduction

It is commonly recognized by now that magnetic micro- and nanowires with domain walls (DWs) have excellent perspectives for spintronic applications.[1–4] This is mostly related to the easy control of the DWs by means of an external magnetic field or an electric current. For example, the DW can be put into motion by a magnetic field of 1 kOe or by current density of 10^7 A/cm^2. On the other hand, the transport properties of magnetic wires are strongly affected by the presence of DWs. In particular, it was recently demonstrated that the magnetoresistance of a microwire, associated with DW, can be very large – up to 2000% and more.[5,6]

In this review article we concentrate on the theory of spin and charge transport in the presence of a DW. The key point of our description is that we consider rather thin domain walls having a width L which is comparable or smaller than the electron wavelength λ_F. This can be completely unrealistic assumption in the case of classical 3D ferromagnets with DWs. However, the DW width in nanowires can be much smaller than in the bulk – of the order of atomic size.[7] Besides, for semiconductor magnetic nanowires the typical carrier wavelength λ_F can be much larger than in metallic ferromagnets.

The main reason why we focus on thin DWs is that in the opposite case of thick DWs, $L \gg \lambda_F$, the effect of the DW on the charge and spin conductivity is very weak. This is because the motion of electrons in a smoothly varying magnetic profile is purely adiabatic. Correspondingly, the probability of reflection or any spin-flip processes is exponentially small. This was first demonstrated by Cabrera and Falicov[8] and later confirmed by many others.[9–11] The interesting point about the effect of a smooth DW is that the absolute DW resistance is not necessarily positive but can be also negative[12] due to an interplay of contributions to the conductivity from different spin channels.

The problem of the spin and the charge transport in the case of a thin DW was considered in Refs.[13–16] The simplest case of $L \ll \lambda_F$ can be treated analytically because the problem reduces to the calculation of reflection and transmission coefficients for scattering from a spin-dependent δ-potential. The effective potential located at the DW affects the charge and spin conductivity and also creates current-induced spin currents and spin torque acting on the DW.

Up to now the role of electron-electron (e-e) interactions in the transport properties of magnetic wires was not sufficiently studied. In the case of thick DW, it was found that due to the interactions there arise a charge profile near the DW.[11] It can result in an electromagnetic interaction of DWs. In the case of DWs in magnetic nanowires, one can anticipate that the role of e-e interactions is crucial as it suppresses strongly the transmission through any localized potential. This problem has been studied thoroughly for the nonmagnetic 1D systems,[17] and also considered recently for the magnetic wires.[16,18] Here we discuss the results of our consideration for the one-channel magnetic nanowires.

12.2. Model and scattering states

Let us consider at first non-interacting electrons described by a parabolic energy band, propagating in a spatially nonuniform magnetization field $\mathbf{M}(\mathbf{r})$. The system is described by the Hamiltonian

$$H = -\frac{\hbar^2 \nabla^2}{2m} - J\,\boldsymbol{\sigma} \cdot \mathbf{M}(\mathbf{r}) \tag{12.1}$$

where J is the exchange integral and $\boldsymbol{\sigma} = (\sigma_x, \sigma_y, \sigma_z)$ are the Pauli matrices. For a domain wall with its center localized at $z = 0$, we assume $\mathbf{M}(z) = [M_0 \sin\varphi(z),\, 0,\, M_0 \cos\varphi(z)]$, where $\varphi(z)$ varies from zero to π for z changing from $z = -\infty$ to $z = +\infty$. Let the characteristic length scale of this change be L (DW width).

When DW is laterally constrained, the number of quantum transport channels can be reduced to a small number. In the extreme case only a single conduction channel is active. In such a case, one can restrict considerations to the corresponding one-dimensional model, and rewrite the Hamiltonian (12.1) as

$$H = -\frac{\hbar^2}{2m}\frac{d^2}{dz^2} - JM_z(z)\,\sigma_z - JM_x(z)\,\sigma_x\,. \tag{12.2}$$

Although this model describes only a one-channel quantum wire, it is sufficient to account qualitatively for some of the recent observations. Apart from this, it can be rather easily generalized to the case of a wire with a few conduction channels.

In the following description we use the basis of scattering states. The asymptotic form of a state corresponding to the wave incoming to the DW from the left (taken sufficiently far from DW) can be written as

$$\psi_\uparrow(z) = \begin{cases} \left(\mathrm{e}^{ik_\uparrow z} + r_\uparrow\, \mathrm{e}^{-ik_\uparrow z}\right)|\uparrow\rangle + r_{\uparrow f}\, \mathrm{e}^{-ik_\downarrow z}\,|\downarrow\rangle\,, & z \ll -L, \\ t_\uparrow\, \mathrm{e}^{ik_\downarrow z}\,|\uparrow\rangle + t_{\uparrow f}\, \mathrm{e}^{ik_\uparrow z}\,|\downarrow\rangle\,, & z \gg L, \end{cases} \tag{12.3}$$

where $k_{\uparrow(\downarrow)} = \sqrt{2m(E \pm M)}/\hbar$, with $M = JM_0$, and E denoting the electron energy. The scattering state (12.3) describes the electron wave in the spin majority channel incident from $z = -\infty$, which is partially reflected into the spin-majority and spin-minority channels, and also partially transmitted into these two channels. The coefficients $t_\uparrow$ and $t_{\uparrow f}$ are the transmission amplitudes without and with spin reversal, respectively, whereas $r_\uparrow$ and $r_{\uparrow f}$ are the relevant reflection amplitudes. The scattering states corresponding to the electron wave incident from $z = -\infty$ in the spin-minority channel have a similar form. Also similar form have the scattering states describing electron waves incident from the right to left.

In a general case, the transmission and reflection coefficients are calculated numerically. When $k_{F\uparrow(\downarrow)}L \ll 1$, they can be calculated analytically. Upon integrating the Schrödinger equation $H\psi_\uparrow = E\psi_\uparrow$ from $z = -\delta$ to $z = +\delta$, and assuming $L \ll \delta \ll k_{\uparrow(\downarrow)}^{-1}$, one obtains

$$-\frac{\hbar}{2m}\left(\left.\frac{d\psi_\uparrow}{dz}\right|_{z=+\delta} - \left.\frac{d\psi_\uparrow}{dz}\right|_{z=-\delta}\right) - \lambda\,\sigma_x\,\psi_\uparrow(z=0) = 0 \tag{12.4}$$

for each of the scattering states, where

$$\lambda \simeq \frac{J}{\hbar}\int_{-\infty}^{\infty} dz\; M_x(z). \tag{12.5}$$

Equation (12.4) has the form of a spin-dependent condition for electron transmission through a δ-like potential barrier located at $z = 0$. The magnitude of the parameter λ in Eq. (12.5) can be estimated as $\lambda \simeq JM_0L/\hbar = ML/\hbar$.

Using the full set of scattering states and the condition (12.4), together with the wave function continuity condition, one finds the transmission amplitudes

$$t_{\uparrow,\downarrow} = \frac{2v_{\uparrow,\downarrow}(v_\uparrow + v_\downarrow)}{(v_\uparrow + v_\downarrow)^2 + 4\lambda^2}\,, \qquad t_{\uparrow,\downarrow f} = \frac{4i\lambda\, v_{\uparrow,\downarrow}}{(v_\uparrow + v_\downarrow)^2 + 4\lambda^2}\,, \tag{12.6}$$

where $v_{\uparrow,\downarrow} = \hbar k_{\uparrow(\downarrow)}/m$ denotes the electron velocity in the spin-majority and spin-minority channels.

According to (12.6), the magnitude of spin-flip transmission coefficient can be estimated and, taking $\varepsilon_F \sim M$, one obtains $|t_f|^2 \sim (Mk_F L\varepsilon_F)^2 \ll 1$. Thus, a sharp DW can be considered as an effective barrier for the spin-flip transmission. On the other hand, the probability of spin conserving transmission is much larger, $|t/t_f|^2 \sim \varepsilon_F\varepsilon_0/M^2 \gg 1$. This means that the electron spin does not follow adiabatically the magnetization direction when it propagates through the wall, but its orientation is rather fixed.

12.3. Resistance of a thin domain wall

To calculate conductance of the system under consideration, one can start from the current operator

$$\hat{\jmath}(z) = e\,\psi^\dagger(z)\,\hat{v}\,\psi(z), \tag{12.7}$$

where $\hat{v} = \hat{p}/m$ is the velocity operator, whereas $\psi^\dagger(z)$ and $\psi(z)$ are the electron field operators in the spinor form. In the linear response regime and using the scattering states we come to the Landauer-Büttiker formula for the conductance

$$G = \frac{e^2}{2\pi\hbar}\left(\frac{v_\downarrow}{v_\uparrow}|t_\uparrow|^2 + |t_{\uparrow f}|^2 + \frac{v_\uparrow}{v_\downarrow}|t_\downarrow|^2 + |t_{\downarrow f}|^2\right), \tag{12.8}$$

where all the velocities and transmission coefficients are taken at the Fermi level. When $k_{F\uparrow(\downarrow)}L \ll 1$, using Eq. (12.6), one can write the conductance in the form

$$G = \frac{4e^2}{\pi\hbar}\,\frac{v_\uparrow v_\downarrow (v_\uparrow + v_\downarrow)^2 + 2\lambda^2\left(v_\uparrow^2 + v_\downarrow^2\right)}{\left[(v_\uparrow + v_\downarrow)^2 + 4\lambda^2\right]^2}. \tag{12.9}$$

In the limit of $v_\uparrow = v_\downarrow$ and $\lambda \to 0$, we obtain the conductance of a one-channel spin-degenerate wire, $G_0 = e^2/\pi\hbar$. In the regime of ballistic transport G_0 is also the conductance of the wire without DW.

The variation of the conductance G with the wall width L (Fig. 12.1) was calculated from Eq. (12.8), with the transmission coefficients determined numerically. Thus, the results shown in Fig. 12.1 are valid for arbitrary value of $k_F L$. In the limit of $k_F L \ll 1$, the results shown in Fig. 12.1 coincide with those obtained from Eq. (12.9). The conductance in the presence of a domain wall is substantially smaller than in the absence of the wall. Accordingly, the associated magnetoresistance (defined as a difference between resistances with and without DW) can be large. For example, for $p = M/\epsilon_F = 0.9$ in Fig. 12.1 the magnetoresistance is equal to about 70% (which corresponds to $G/G_0 = 0.6$). It should be noted that in a real magnetoresistance experiment on magnetic semiconductor nanowires,

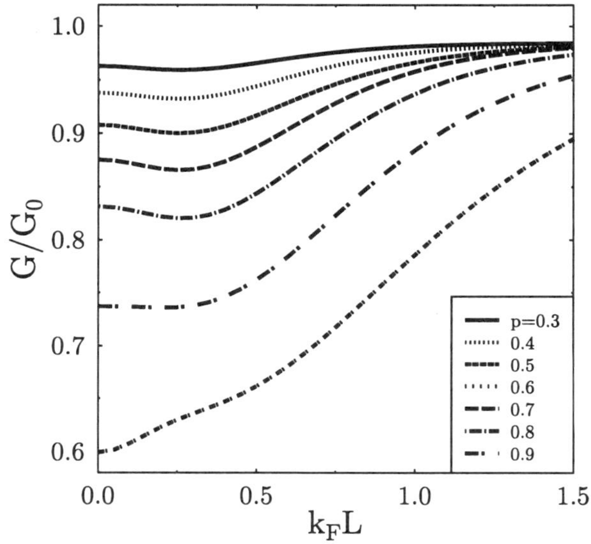

Fig. 12.1. Conductivity of the one-channel magnetic wire as a function of the DW width for different values of $p = M/\varepsilon_F$. Reproduced from Ref.[13]

for which the inequality $k_F L \gg 1$ can be easily fulfilled, one can have more than one domain wall. Accordingly, the magnetoresistance effect can be significantly enhanced.

12.4. Spin current

When the electric current is spin polarized and when there is some asymmetry between the two spin channels, the flow of charge is accompanied by a flow of spin (angular momentum). The z-component of the spin current can be calculated from the following definition of the corresponding spin-current operator

$$\hat{j}^s_z(z) = \psi^\dagger(z)\, \sigma_z\, \hat{v}\, \psi(z). \tag{12.10}$$

Using the scattering states one arrives in the linear response regime (limit of small bias voltage U) at the following formulas for the spin current j^s_z:

$$j^s_z(z < -L,\, z > L) = \frac{eU}{2\pi\hbar} \left(\frac{v_\downarrow}{v_\uparrow} \left|t_\uparrow\right|^2 \pm \left|t_{\uparrow f}\right|^2 - \frac{v_\uparrow}{v_\downarrow} \left|t_\downarrow\right|^2 \mp \left|t_{\downarrow f}\right|^2 \right). \tag{12.11}$$

Using Eqs. (6) we find for $k_F L \ll 1$

$$j_z^s(z > L) = -\frac{8eU}{\pi\hbar} \frac{\lambda^2 \left(v_\uparrow^2 - v_\downarrow^2\right)}{\left[(v_\uparrow + v_\downarrow)^2 + 4\lambda^2\right]^2} \tag{12.12}$$

and $j_z^s(z < -L) = -j_z^s(z > L)$.

It should be noted that spin-flip scattering due to DW does not allow to separate spin channels like it was in the case for homogeneous ferromagnets. Defining the spin conductance as $G_s = j_z^s/U$, one can write for $z > 0$

$$G_s = -\frac{8e}{\pi\hbar} \frac{\lambda^2 \left(v_\uparrow^2 - v_\downarrow^2\right)}{\left[(v_\uparrow + v_\downarrow)^2 + 4\lambda^2\right]^2} . \tag{12.13}$$

Thus, G_s is negative for $z > 0$ and positive for $z < 0$.

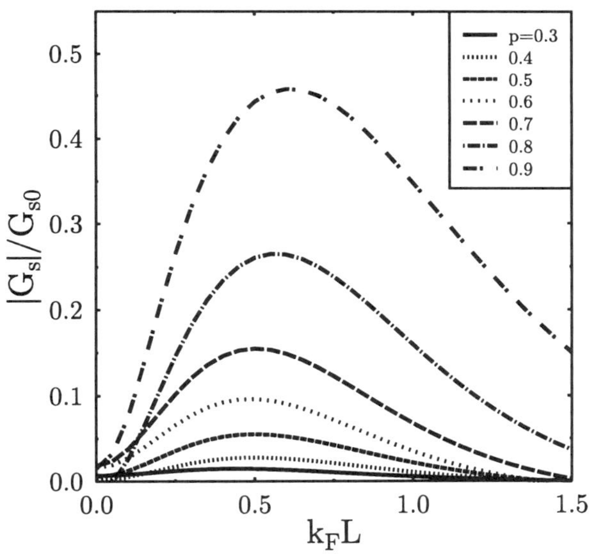

Fig. 12.2. Spin conductivity of the wire as a function of the DW width for different values of p. Reproduced from Ref.[13]

In a nonmagnetic case we have $v_\uparrow = v_\downarrow$ and therefore $G_s = 0$. In the case considered here, $G_s = 0$ when there is no DW. Let us introduce the spin conductance for one (spin-up) channel only, $G_{s0} = e/2\pi\hbar$. The relative spin conductance in the presence of DW, G_s/G_{s0}, calculated using Eq. (12.11) and with numerically found transmission coefficients, is shown in Fig. 12.2 as a function of the DW width L and for the indicated values of the parameter p. It corresponds to the spin current outside the region of the domain wall. The spin current inside the wall is not conserved because of the spin-flip transitions.

In accordance with (12.11) and (12.6), the nonzero spin current in a one-channel wire with DW is due to a difference in spin-flip transmissions for spin-up and spin-down channels: the corresponding transmission coefficient turns out to be larger for faster (majority) electrons.

12.5. Spin polarization due to the domain wall

Spin dependent reflections from the wall lead to additional spin polarization of electrons near the wall. The distribution of spin density created by the wall can be calculated using the basis of scattering states. The z-component of the spin density operator is

$$\hat{S}_z(z) = \psi^\dagger(z)\, \sigma_z\, \psi(z). \tag{12.14}$$

Calculating the average value we find that the spin density contains a constant part corresponding to the spin density in the absence of DW, as well as the z-dependent part $\delta S_z(z)$ created by the reflection from the wall,

$$\begin{aligned} \delta S_z(z) &= \frac{1}{\pi}\int_0^{k_{F\uparrow}} dk\; r_{R\uparrow}\, \cos(2k_\uparrow z) - \frac{1}{\pi}\int_0^{k_{F\downarrow}} dk\; r_{R\downarrow}\, \cos(2k_\downarrow z), \;\; (z < -L), \\ &= \frac{1}{\pi}\int_0^{k_{F\uparrow}} dk\; r_{L\uparrow}\, \cos(2k_\uparrow z) - \frac{1}{\pi}\int_0^{k_{F\downarrow}} dk\; r_{L\downarrow}\, \cos(2k_\downarrow z), \;\; (z > L), \end{aligned} \tag{12.15}$$

where indices R and L refer to the waves incoming from the left and right, respectively. In accordance with (12.15), the spin dependent reflections from the wall create spatial oscillations of the electron spin density. These oscillations are similar to the Friedel oscillations of charge in a nonmagnetic metal. However, one should point out here that in addition to the above calculated spin polarization, there is also a nonequilibrium spin polarization due to flowing current.[19]

12.6. Current-induced spin torque

Now we consider the spin torque transferred to the DW in the presence of a steady current of spin polarized charge carriers. It is more convenient to change slightly notation in this section and use the 1D model with the wire along the axis x (instead of along the axis z). When considering scattering of electrons from a magnetic moment $\mathbf{M}(x)$ we assume that the moment is frozen on the scale of the characteristic times of electron motion. This assumption renders possible the calculation of the torque as in the case of a static DW.

12.6.1. *Scattering from a single magnetic moment in a magnetic wire*

First we calculate the torque in the case of a magnetic wire with magnetization $\mathbf{M}$ oriented along the axis x for $x < 0$ (left of the wall) and in the opposite direction for $x > 0$ (right of the wall). We also introduce an additional frozen magnetic moment $\mathbf{M}_0 = M_0\,(n_x,\, n_y,\, 0)$ located at the point $x = 0$. For definiteness, let the vector $\mathbf{M}_0$ lie in the $x - y$ plane. The corresponding Hamiltonian is

$$H = -\frac{\hbar^2}{2m}\frac{d^2}{dx^2} + JM\,\sigma_x\,\mathrm{sgn}\,(x) + JM_0\,\mathbf{n}\cdot\boldsymbol{\sigma}\,\delta(x). \tag{12.16}$$

Let us consider the torque created by spin-polarized electron waves coming from the left. The scattering functions for the Hamiltonian (12.16) and the transmission amplitudes can be calculated like in Sec. 2. Using Eq. (12.10) we calculate all the components of spin current associated with the incoming spin-up and down waves

$$j^s_{\uparrow,\downarrow x}(x) = \begin{cases} \pm v_{\uparrow,\downarrow}\,(1 - |r_{\uparrow,\downarrow}|^2) \pm v_{\downarrow,\uparrow}\,|r_{\uparrow,\downarrow f}|^2, & x < 0, \\ \pm v_{\downarrow,\uparrow}\,|t_{\uparrow,\downarrow}|^2 \mp v_{\uparrow,\downarrow}\,|t_{\uparrow,\downarrow f}|^2, & x > 0, \end{cases} \tag{12.17}$$

$$j^s_{\uparrow,\downarrow y}(x) = \begin{cases} \pm t_{\uparrow,\downarrow f}\,\mathrm{Im}\left[v_{\uparrow,\downarrow}(e^{ik_+x} - r_{\uparrow,\downarrow}e^{-ik_-x}) + v_{\downarrow,\uparrow}(e^{-ik_+x} + r^*_{\uparrow,\downarrow}e^{ik_-x})\right], \\ \qquad\qquad x < 0, \\ t_{\uparrow,\downarrow f}\,\mathrm{Im}\left[\mp v_{\uparrow,\downarrow}\,t^*_{\uparrow,\downarrow}\,e^{ik_-x} \pm v_{\downarrow,\uparrow}\,t_{\uparrow,\downarrow}\,e^{-ik_-x}\right], \qquad x > 0, \end{cases} \tag{12.18}$$

$$j^s_{\uparrow,\downarrow z}(x) = \begin{cases} t_{\uparrow,\downarrow f}\,\mathrm{Re}\left[v_{\uparrow,\downarrow}(e^{ik_+x} - r_{\uparrow,\downarrow}e^{-ik_-x}) - v_{\downarrow,\uparrow}(e^{-ik_+x} + r^*_{\uparrow,\downarrow}e^{ik_-x})\right], \\ \qquad\qquad x < 0, \\ t_{\uparrow,\downarrow f}\,\mathrm{Re}\left[v_{\uparrow,\downarrow}t^*_{\uparrow,\downarrow}e^{ik_-x} + v_{\downarrow,\uparrow}t_{\uparrow,\downarrow}e^{-ik_-x}\right], \qquad x > 0, \end{cases} \tag{12.19}$$

where $k_\pm = k_\uparrow \pm k_\downarrow$ and $v_{\uparrow,\downarrow} = \hbar k_{\uparrow,\downarrow}/m$. Note that the transverse components of the spin currents, $j^s_{\uparrow y}(x)$ and $j^s_{\uparrow z}(x)$, are nonzero for $x < 0$ and for $x > 0$. As we see from (12.18) and (12.19), the transverse components of the spin current are oscillating functions of x. The nonconservation of spin current in the magnetic wire is related to indirect magnetic interactions accompanying the inhomogeneous distribution of the spin density. In the nonmagnetic case, corresponding to the limit of $k_- \to 0$, it reduces to the conservation of spin current at $x < 0$ and $x > 0$.

The spin torque acting on the moment $\mathbf{M}_0$ can be calculated as the transferred spin current at the point $x = 0$,

$$T_\mu = j^s_\mu(-\delta) - j^s_\mu(+\delta). \tag{12.20}$$

Using (17)-(19) we find

$$T_{\uparrow x} = 2v_\uparrow \,\mathrm{Re}\, t_\uparrow + (v_\uparrow + v_\downarrow)\left(|t_{\uparrow f}|^2 - |t_\uparrow|^2\right), \tag{12.21}$$

$$T_{\uparrow y} = -2t_{\uparrow f}\,(v_\uparrow + v_\downarrow)\;\mathrm{Im}\, t_\uparrow, \tag{12.22}$$

$$T_{\uparrow z} = 2t_{\uparrow f}\left[v_\uparrow - (v_\uparrow + v_\downarrow)\;\mathrm{Re}\, t_\uparrow\right]. \tag{12.23}$$

These results for the torque can also be obtained from the equation of motion of the magnetic moment $\mathbf{M}_0$,

$$T_\mu = -\frac{JM_0}{\hbar}\,\epsilon_{\mu\nu\lambda}\, n_\nu\, S_\lambda(0), \tag{12.24}$$

where $\mathbf{n}$ is the unit vector along $\mathbf{M}_0$, and $\epsilon_{\mu\nu\lambda}$ is the unit antisymmetric tensor. Here the net spin at $x = 0$ is

$$S_{\uparrow,\downarrow\, x}(0) = \frac{4k^2_{\uparrow,\downarrow}\left[(k_\uparrow + k_\downarrow)^2 + J_0^2\,(n_x^2 - n_y^2)\right]}{[(k_\uparrow + k_\downarrow)^2 + J_0^2]^2}\,, \tag{12.25}$$

$$S_{\uparrow,\downarrow\, y}(0) = \frac{8J_0^2\, k^2_{\uparrow,\downarrow}\, n_x n_y}{[(k_\uparrow + k_\downarrow)^2 + J_0^2]^2}\,, \tag{12.26}$$

$$S_{\uparrow,\downarrow\, z}(0) = \mp\frac{8J_0\, k^2_{\uparrow,\downarrow}\,(k_\uparrow + k_\downarrow)\, n_y}{[(k_\uparrow + k_\downarrow)^2 + J_0^2]^2}\,, \tag{12.27}$$

with $J_0 = 2JmM_0/\hbar^2$. In the case of a fully spin polarized electron gas, only the spin-up spin current components Eqs. (12.17)-(12.19) are relevant (corresponding to the majority electrons). Accordingly, in these equations we should substitute $k_\uparrow \to i\kappa_\uparrow$, where $\kappa_\uparrow$ is real.

12.6.2. *Local torque in the magnetic wire with a thin domain wall*

Let us consider again the magnetic wire with a single DW corresponding to the magnetization $\mathbf{M}$ along the axis x for $x < L$ and opposite to the axis x for $x > L$. Now we are going to calculate the torque acting locally on the moments within the DW.

Upon applying a small voltage, an electric current can flow in the wire. We assume the current in the negative x axis direction. If the only imperfection in the wire is the DW, one can assume a jump $\Delta\phi$ in the electrostatic potential at the wall, and both the charge and the spin currents can be calculated as integrals over the energies in the interval between ε_{FR} and $\varepsilon_{FL} = \varepsilon_{FR} + e\,\Delta\phi$, where ε_{FL} and ε_{FR} are the Fermi levels on the left and right sides. In the limit of small voltage,

$|e\,\Delta\phi| \ll \varepsilon_F$, the transport is linear and is associated with electrons at the Fermi level.

We assume the electrons approaching the DW from the left are spin-polarized according to the magnetization direction in the left part of the wire. The incoming electrons are scattered from a large number of magnetic moments in the wall. To calculate the transmission of electrons through the DW, we take the perturbation created by the total magnetic moment $\widetilde{\mathbf{M}}(x) = \sum_i \mathbf{M}_i\,\delta(x - x_i)$, where $\mathbf{M}_i$ is the localized moment at the point $x = x_i$, and all of the moments $\mathbf{M}_i$ are located within a region of the wall width, $|x_i| < L$, which is assumed to be small as compared to the wavelength of electrons, $k_{F\uparrow,\downarrow}L \ll 1$. Electron scattering from the total moment $\widetilde{\mathbf{M}}(x)$ located within a region much smaller than λ can be described using the spin-dependent δ-potential. Then, in the limit of small voltage, the current takes the form $j_0 = G\,\Delta\phi$ with G from Eq. (12.11), where we should use the transmission coefficients $\tilde{t}_{\uparrow,\downarrow}$ $\tilde{t}_{\uparrow,\downarrow f}$ for scattering of electrons from an effective moment $\mathbf{M}_{eff} \simeq \int_{-L}^{+L} \widetilde{\mathbf{M}}(x)\,dx$. In the DW with the assumed magnetization profile, the effective moment, $\mathbf{M}_{eff}$, is oriented along the y axis. The transmission coefficients $\tilde{t}_\uparrow$, $\tilde{t}_{\uparrow f}$ and $\tilde{t}_\downarrow$, $\tilde{t}_{\downarrow f}$ for the Hamiltonian (6) can be found like in Sec. 2 taking $n_x = 0$, $n_y = 1$ and substituting $J_0 \to \tilde{J}_0 \equiv 2mgM_{eff}/\hbar^2$. The magnitude of M_{eff} is $M_{eff} \simeq \int_{-L}^{L} M_y(x)\,dx$.

The spin current calculated in the linear response approximation includes the sum of partial spin currents

$$\mathbf{j}^s(x) = \frac{e\,\Delta\phi}{2\pi\hbar}\left(\frac{\tilde{\mathbf{j}}^s_\uparrow(x)}{v_\uparrow} + \frac{\tilde{\mathbf{j}}^s_\downarrow(x)}{v_\downarrow}\right), \tag{12.28}$$

where the components of $\tilde{\mathbf{j}}^s_{\uparrow,\downarrow}$ can be found using (12.17)-(12.19) with the substitution $t_{\uparrow,\downarrow}$, $t_{\uparrow,\downarrow f} \to \tilde{t}_{\uparrow,\downarrow}$, $\tilde{t}_{\uparrow,\downarrow f}$. The appearance of $v_\uparrow$ and $v_\downarrow$ in the denominators of (12.28) is related to the 1D density of states for spin-up and spin-down electrons. The spin current components perpendicular to the axis x are oscillating functions, and the wavelength of the oscillations is determined by the inverse momentum at the Fermi level. Hence, the oscillation wavelength of the transverse component of the spin current is much larger than the DW width.

It is worth noting that in 3D systems, the transverse component of the spin current decays due to the integration over momentum in the DW plane. In metallic ferromagnets, the decay is very fast due to the large electron Fermi momentum. However, there is an additional nonvanishing spin transfer for the transverse component in the 3D case.

We can also calculate the net spin density induced by the external current j_0. It can be found as the expectation value of the spin σ_μ in the scattering state of the incoming electrons, integrated over all energies between ε_F and $\varepsilon_F + e\,\Delta\phi$, as in the calculation of the charge and spin currents. Then we obtain

$$\mathbf{S}(0) = \frac{e\,\Delta\phi}{2\pi\hbar}\left(\frac{\tilde{\mathbf{S}}_\uparrow(0)}{v_\uparrow} + \frac{\tilde{\mathbf{S}}_\downarrow(0)}{v_\downarrow}\right), \tag{12.29}$$

where $\tilde{\mathbf{S}}_{\uparrow,\downarrow}(0)$ can be found using (25)-(27) with $n_x = 0$ and the substitution $J_0 \to \tilde{J}_0$ corresponding to the scattering from the effective moment $\mathbf{M}_{eff}$.

Finally, we find the torque acting on a single localized moment in the domain wall. For this purpose we use Eq. (12.24) with $\mathbf{S}(0)$ from (12.2), describing the spin accumulation created by scattering from the domain wall as a whole. The result can be presented in a general form

$$\mathbf{T}(x) = \frac{j_0}{e}\left[\eta\,\mathbf{n}\times(\mathbf{n}\times\mathbf{s}) + \zeta\,\mathbf{n}\times\mathbf{s}\right]. \tag{12.30}$$

where

$$\eta = \frac{J_0\,\tilde{J}_0\,(k_\downarrow^2 - k_\uparrow^2)}{2k_\uparrow k_\downarrow(k_\uparrow + k_\downarrow)^2 + \tilde{J}_0^2(k_\uparrow^2 + k_\downarrow^2)}\,, \tag{12.31}$$

$$\zeta = -\frac{J_0\,(k_\uparrow + k_\downarrow)^2\,[(k_\uparrow + k_\downarrow)^2 - \tilde{J}_0^2]}{2\left[2k_\uparrow k_\downarrow(k_\uparrow + k_\downarrow)^2 + \tilde{J}_0^2(k_\uparrow^2 + k_\downarrow^2)\right]}\,, \tag{12.32}$$

and s is the unit vector along the spin polarization corresponding to magnetization $\mathbf{M}$ at $x < -L$. As we see from (12.31) and (12.32), both coefficients strongly depend on the parameters describing the ferromagnet and on the parameters of the wall.

12.7. Effect of interaction on the transmission of electrons through a thin domain wall

We now study the effect of electron interactions in the system introduced in section 12.2, where we will introduce some changes in notation for convenience. We write the single-particle part of the Hamiltonian as ($J > 0$):

$$\hat{H}_0 = -\frac{\hbar^2}{2m}\frac{d^2}{dz^2} + \hbar V\delta(z) + JM_z(z)\hat{\sigma}_z + \hbar\lambda\delta(z)\hat{\sigma}_x\,, \tag{12.33}$$

with $M_z(z \to \pm\infty) = \pm M_0$. The term $\hbar\lambda\delta(z)\hat{\sigma}_x$ describes spin dependent scattering due to the $M_x(z)$ component (as given in equation 12.5) and V is a potential (spin independent) scattering term which may be present because sharp DW's are usually achieved by making constrictions in wires. The specific DW shape and length will determine the strength of the spin-flip scattering $\hbar\lambda$.

An incident electron from the left with momentum k and spin $\uparrow$ (or $\downarrow$) can be transmitted to the $z > 0$ region preserving its spin, but changing momentum from k to k^- (or k^+), given by $k^\pm = \sqrt{k^2 \pm 4JM_0 m/\hbar^2}$ (spin-$\uparrow$ electrons are spin majority for $z < 0$ and spin-minority for $z > 0$). Therefore, $k^\pm(k)$ is actually function of the incident momentum k and, obviously, $k^+(k^-(k)) = k$. If the transmission occurs with spin reversal, the momentum k remains unchanged. The reflection amplitudes for spin-σ electrons with or without spin reversal shall be denoted by r'_σ and r_σ, respectively. The same convention applies to the transmission amplitudes t'_σ and t_σ. In this notation, the scattering amplitudes found in section 12.2 read:

$$t_{\uparrow(\downarrow)}(k) = \frac{2\,(v + v^\mp + 2iV)\,v}{(v + v^\mp + 2iV)^2 + 4\lambda^2} = r_{\uparrow(\downarrow)}(k) + 1\,, \qquad (12.34)$$

$$t'_{\uparrow(\downarrow)}(k) = \frac{4i\lambda v}{(v + v^\mp + 2iV)^2 + 4\lambda^2} = r'_{\uparrow(\downarrow)}(k)\,, \qquad (12.35)$$

with $v = \hbar k/m$, $v^\pm = \hbar k^\pm/m$, where the upper (lower) sign refers to $\uparrow$ ($\downarrow$). Henceforth $\epsilon(\pm p, \sigma)$ denotes the scattering state energy with momentum $+p$ (or $-p$) and spin σ, incident from the left (or right).

But we are concerned with electron interactions. The latter can be described by the well known g-ology model:

$$\begin{aligned}\hat{H}_{int} = {} & g_{1,\alpha,\beta} \int \frac{dk_1 dq}{(2\pi)^2} \hat{a}^\dagger_{k_1,\alpha} \hat{b}^\dagger_{k_2,\beta} \hat{a}_{k_2+q,\beta} \hat{b}_{k_1-q,\alpha} \\ & + g_{2,\alpha,\beta} \int \frac{dk_1 dq}{(2\pi)^2} \hat{a}^\dagger_{k_1,\alpha} \hat{b}^\dagger_{k_2,\beta} \hat{b}_{k_2+q,\beta} \hat{a}_{k_1-q,\alpha}\,, \end{aligned} \qquad (12.36)$$

where Greek letters denote spin indices, and the summation convention over repeated indices is used. The operators in (12.36) are right- moving ($\hat{a}_{q\sigma}$) and left-moving ($\hat{b}_{q\sigma}$) plane-wave states. The coupling constants g_1 and g_2 describe back and forward scattering processes between electrons moving in opposite directions, respectively.

Because the Fermi momentum is spin dependent, we distinguish between $g_{1(2)\uparrow}$, which describes the interaction between spin-majority particles (that is spin-$\uparrow$ on the left and spin-$\downarrow$ on the right of the barrier) and $g_{1(2)\downarrow}$, which describes the interaction between spin-minority particles (that is spin-$\downarrow$ on the left and spin-$\uparrow$ on the right of the barrier).

We use $g_{1(2)\perp}$ to denote interaction between particles with opposite spin. The forward scattering process between particles which move in the same direction will not affect the transmission amplitudes, although it will renormalize the Fermi velocity.[20] This effect is equivalent to an effective mass renormalization and the electrons with different spin orientations may turn out to have different effective masses.

How will the interactions affect the transmission and reflection amplitudes given in (12.34) and (12.35)? We may think of the interacting problem in terms of a Hartree-Fock (HF) picture:[16] already at the non-interacting level, the DW produces Friedel oscillations in the Fermi sea. These are charge and spin oscillations because there are two Fermi Surfaces (FS's): a small one for spin minority particles and a larger FS for spin majority particles.

This Fermi sea produces a spatially oscillating HF potential which is going to be felt by an incident electron on the DW. One can calculate the HF wavefunction for this electron perturbatively in the interactions $g_{1(2)\sigma}$. Therefore, one can obtain perturbative corrections (denoted by δt_σ, $\delta t'_\sigma$, etc) to equations (12.34) and (12.35).

It turns out that in a one-dimensional system such corrections are divergent near the Fermi level, already in first order. The divergence is logarithmic, proportional to $\log(|\epsilon'|/D)$, where ϵ' denotes the energy of the scattered electron and D is the bandwidth. A similar divergence occurs in the Kondo problem and the usual way to deal with it is the poor man's renormalization procedure: reducing the bandwidth D, step by step, by removing states near the band edge. This state elimination is compensated by renormalization of t_σ at each step.

Applying this procedure and noting that $t_\sigma + \delta t_\sigma$ remains invariant as D is reduced, one can write down the differential equation:

$$dt_\sigma + \frac{\partial\, \delta t_\sigma}{\partial D}\, dD = 0\,,$$

which describes the renormalization group (RG) flow of t_σ. The bandwidth can be reduced until it becomes equal to the temperature.

The term (12.36) forces us to write the scattering states in second quantized form. To this end, we introduce operators $\hat{c}_{k,\sigma}$ and $\hat{d}_{k,\sigma}$ for the eigenstates corresponding to electrons incident from the left and right, respectively. We take $k > 0$. A state $\hat{c}_{k,\sigma}$, for instance, should be a linear combination of right and left moving plane waves in such a way that it contains an incident electron from the left which corresponds to a right-moving plane wave state $\hat{a}_{k,\sigma}$ on the negative z-semi-axis. It also contains a reflected electron occupying a left-moving plane-wave state $\hat{b}_{-k,\pm\sigma}$ on the negative z-semi-axis and a transmitted particle occupying a right-moving plane-wave state $\hat{a}_{k,\pm\sigma}$ on the positive z-semi-axis.

The actual expression for $\hat{c}_{k,\sigma}$ is more complicated than this because restricting a plane-wave state to a semi-axis is going to cause uncertainty in the momentum due to the Heisenberg principle. This means that the transmitted particle, for instance, is actually a linear combination of all $\hat{a}_{p,\pm\sigma}$ with $p > 0$.

Consider, for instance, the Matsubara propagator

$$\mathcal{G}_\sigma(\tau) = -\langle T_\tau e^{-\int \hat{H}_{int}(\tau')d\tau'} \hat{a}_{p,\sigma}(\tau)\hat{c}^\dagger_{p',\sigma} \rightarrow_0 , \tag{12.37}$$

where $\langle \ldots \rightarrow_0$ denotes the average in the Fermi sea *of non-interacting scattering states*. The zero-order propagator for $\sigma = \uparrow$ is:

$$\mathcal{G}^{(0)}_\uparrow(i\omega) = \frac{i}{i\omega - \epsilon(p', \uparrow)} \left[\frac{1}{p - p' + i0} - \frac{t_\uparrow(p')}{p - p'^- - i0} \right] , \tag{12.38}$$

where 0 denotes a positive infinitesimal. The poles in the denominators identify the semi-axis on which the electron behaves as a right moving plane wave. The transmission amplitude appears associated with the denominator $p - p'^- - i0$ which, for the variable p, gives a pole in the upper half plane. The meaning of this pole is that the transmitted particle is right-moving in the $z > 0$ half-axis. We can calculate the first order correction to $\mathcal{G}$, in which a pole in $p - p'^- - i0$ will appear. In analogy with (12.38), the corresponding residue will be identified with $-\delta t_\uparrow(p')$, *i.e.*, minus the transmission amplitude correction.

The diagrammatic representation of $\mathcal{G}^{(1)}_\uparrow$ is shown in Fig. 12.3 and it provides a simple physical description of the virtual processes going on. The horizontal lines represent the electron scattered by the Hartree-Fock potential of the Fermi sea.

The latter is represented by the closed loop. Consider, for instance, the upper left diagram: an electron, initially in state $c_{p',\uparrow}$ close to the Fermi level passes through the DW as a right-moving ($\hat{a}$) particle. Then, it is reflected (from $\hat{a}$ to $\hat{b}$ particle) while exchanging momentum q with the Fermi sea on the $z > 0$ semi-axis. Finally, it is reflected by the DW again, becoming a spin-up right moving particle of momentum p. According to the physical interpretation of the diagrams, we always know on which side of the DW the interaction with the Fermi sea (closed loop in the diagram) is taking place. The diagram is logarithmically divergent if the Fermi sea can provide exactly the momentum that is needed to keep the electron always near the Fermi level during the intermediate virtual steps.

In order to write down the RG equations, we introduce a variable $\xi = \log(D/D_0)$, which is integrated from 0 to $\log(T/D_0)$, corresponding to the fact that the bandwidth is progressively reduced from $D = D_0$ to $D = T$. Fermi level velocities are denoted by $v_\pm$ and Fermi wavevectors by $k_{F\pm}$ for majority or minority spin particles. It is convenient to rewrite the interaction parameters as $g_\uparrow = (g_{2\uparrow} - g_{1\uparrow})/4hv_+$, $g_\downarrow = (g_{2\downarrow} - g_{1\downarrow})/4hv_-$, $g_\perp = g_{2\perp}/2h(v_+ + v_-)$. The scaling differential equations for the transmission amplitudes read:

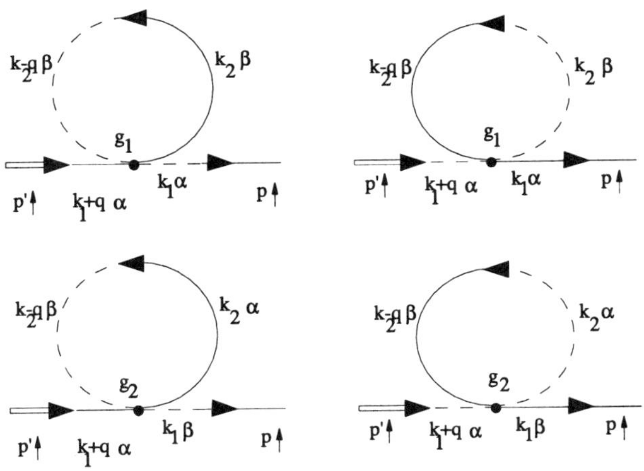

Fig. 12.3. Feynman diagrams for the first order contribution $\mathcal{G}^{(1)}$ to the propagator (12.37). The scattering state is represented by a double line, the $\hat{a}$ ($\hat{b}$) particle is represented by a continuous (dashed) line. The loop represents the Hartree-Fock potential of the Fermi sea. The scattered electron exchanges momentum q with the Fermi sea. Reproduced from Ref. *Phys. Rev. B* **74**, 224429 (2006).

$$\frac{dt_\uparrow}{d\xi} = g_\downarrow \left[r_\downarrow^* r_\downarrow t_\uparrow + r_\downarrow^* r_\uparrow' t_\downarrow' \right] + g_\uparrow \left[r_\uparrow^* r_\uparrow t_\uparrow + r_\uparrow^* r_\uparrow' t_\uparrow' \right]$$
$$+ g_\perp \left[r_\downarrow'^* r_\uparrow' t_\uparrow + r_\uparrow'^* r_\downarrow t_\uparrow' + r_\uparrow'^* r_\uparrow t_\downarrow' + r_\downarrow'^* r_\uparrow' t_\uparrow \right], \qquad (12.39)$$

$$\frac{dt_\uparrow'}{d\xi} = 2g_\downarrow r_\downarrow^* r_\downarrow' t_\uparrow + 2g_\uparrow r_\uparrow^* r_\uparrow t_\uparrow' + 2g_\perp \left[r_\downarrow'^* r_\uparrow t_\uparrow + r_\uparrow'^* r_\downarrow' t_\uparrow' \right]. \qquad (12.40)$$

Equations for the reflection amplitudes $r_\sigma(p')$ and $r_\sigma'(p')$ can be obtained from the propagators $-\langle T_\tau \hat{b}_{p,\pm\sigma}(\tau) \hat{c}_{p',\sigma}^\dagger \rightarrow$. The equation for $r_\uparrow(p')$ is

$$\frac{dr_\uparrow}{d\xi} = g_\uparrow \left[r_\uparrow^* r_\uparrow r_\uparrow + r_\uparrow^* t_\uparrow' t_\uparrow' \right] + g_\downarrow \left[r_\downarrow^* t_\uparrow t_\downarrow + r_\downarrow^* r_\downarrow' r_\uparrow' \right]$$
$$+ g_\perp \left[r_\uparrow'^* r_\downarrow' r_\uparrow + r_\downarrow'^* r_\uparrow' r_\uparrow + r_\uparrow'^* t_\downarrow t_\uparrow' + r_\downarrow'^* t_\uparrow t_\uparrow' \right] - g_\uparrow r_\uparrow \qquad (12.41)$$

and the equation for $r_\uparrow'(p')$ is

$$\frac{dr_\uparrow'}{d\xi} = g_\uparrow \left[r_\uparrow^* r_\uparrow r_\uparrow' + r_\uparrow^* t_\uparrow' t_\uparrow \right] + g_\downarrow \left[r_\downarrow^* t_\uparrow t_\downarrow' + r_\downarrow^* r_\downarrow r_\uparrow' \right]$$
$$+ g_\perp \left[r_\uparrow'^* r_\downarrow r_\uparrow + r_\downarrow'^* r_\uparrow' r_\uparrow' + r_\uparrow'^* t_\downarrow' t_\uparrow' + r_\downarrow'^* t_\uparrow t_\uparrow \right] - g_\perp r_\uparrow'. \qquad (12.42)$$

The scaling equations for spin-$\downarrow$ amplitudes follow from the above by simply inverting the spin and velocity indices. All one has to do now is to insert single-particle values for the scattering amplitudes as initial values for scaling. The initial amplitudes may be obtained from any effective single particle model for the DW. In what follows, we shall use (12.34) and (12.35).

12.7.1. *Fixed points*

We now analyze the nature of the zero temperature fixed points predicted by the RG equations. The parameters of the model which enter the scaling equations are $g_\uparrow$, $g_\downarrow$, $g_\perp$, and the ratio v_-/v_+. For repulsive interactions ($g_\uparrow, g_\downarrow, g_\perp > 0$) the system flows to insulator fixed points, as expected.

We distinguish two regimes: the one with λ/v_+ larger than about 0.1 and the one with λ/v_+ smaller than about 0.1, as shown in Fig. 12.4.

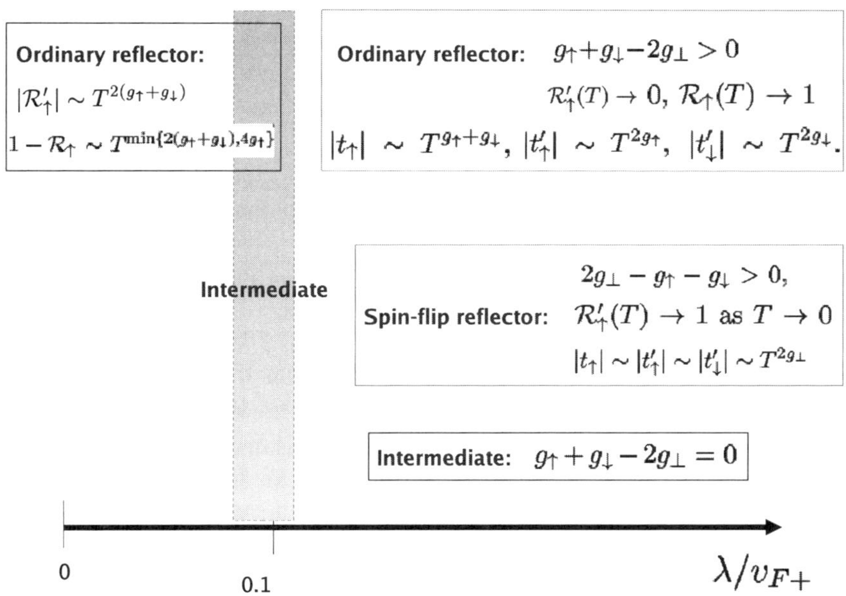

Fig. 12.4. Schematic representation of the nature of the $T = 0$ insulator fixed points for repulsive interactions.

The conservation of the charge current for an incident spin-up electron can be expressed by the relation

$$1 = |r_\uparrow|^2 + \frac{v_-}{v_+}|r'_\uparrow|^2 + |t'_\uparrow|^2 + \frac{v_-}{v_+}|t_\uparrow|^2 .$$

The reflection coefficient for a spin-up incident electron is $\mathcal{R}_\uparrow = |r_\uparrow|^2$ if its spin is preserved upon reflection. The reflection coefficient with a spin flip is $\mathcal{R}'_\uparrow = (v_-/v_+)|r'_\uparrow|^2$.

In the regime where λ/v_+ larger than about 0.1 we obtain:

$$\mathcal{R}'_\uparrow(T) = \frac{\frac{\mathcal{R}'_{\uparrow,0}}{1-\mathcal{R}'_{\uparrow,0}}\left(\frac{T}{D_0}\right)^{2(g_\uparrow+g_\downarrow-2g_\perp)}}{1+\frac{\mathcal{R}'_{\uparrow,0}}{1-\mathcal{R}'_{\uparrow,0}}\left(\frac{T}{D_0}\right)^{2(g_\uparrow+g_\downarrow-2g_\perp)}} . \tag{12.43}$$

where $\mathcal{R}'_{\uparrow,0}$ denotes the initial non-interacting value.

Hence, if $2g_\perp - g_\uparrow - g_\downarrow > 0$ then $\mathcal{R}'_\uparrow(T) \to 1$ as $T \to 0$. The DW reflects all incident electrons while additionally reversing their spin. That means that an incident spin polarized current is reflected back with its polarization reversed. The DW behaves as a 100% "spin-flip reflector" at zero temperature, generating a finite net spin current but no charge current, since the incident and reflected charge currents cancel.

If $g_\uparrow + g_\downarrow - 2g_\perp > 0$ we have $\mathcal{R}'_\uparrow(T) \to 0$, $\mathcal{R}_\uparrow(T) \to 1$. The DW reflects then all incident electrons while preserving their spin. This is the "ordinary" insulator fixed point where no spin-flip occurs.

If $g_\uparrow + g_\downarrow - 2g_\perp = 0$ then both $r'_\sigma(T)$ and $r_\sigma(T)$ tend to finite values. The scaling equations for $t_\uparrow$, t'_σ, with constant reflection amplitudes, become a linear algebraic 3 by 3 system. The eigenvalues of the matrix give the temperature exponents and each transmission amplitude will be a linear combination of the three powers of T. For decreasing temperature there may be crossovers from one exponent to the other and the lowest one dominates as $T \to 0$.

For smaller values of λ/v_+ (*i.e.*, smaller than about 0.1), the system flows to a fixed point where $R'_\uparrow$ vanishes about as fast as the transmissions and $R_\sigma \to 1$. This is again the "ordinary" insulator fixed point. This time, it occurs because the spin-flip produced by the bare DW was already small.

We can estimate λ/v_+ by assuming, for instance, that $\vec{M}(z) = M_0 \cos\theta(z)\hat{\vec{z}} + M_0 \sin\theta(z)\hat{\vec{x}}$ with $\cos\theta(z) = \tanh(z/L)$, where L is the length of the DW. It follows that λ/v_+ is proportional to the DW length:

$$\frac{\lambda}{v_+} = \pi m \frac{JM_0}{\hbar^2 k_{F+}^2}(Lk_{F+}). \tag{12.44}$$

The condition for the DW to be smaller than the Fermi wavelength is $Lk_{F+} < 2\pi$. We can compare how fast the RG flow affects short and longer DW's. Figure

12.5 shows non-renormalized ($\xi = 0$) and renormalized (for $\xi = -1$) reflection coefficients: a thick DW causes weak spin-flip reflection and correlations hardly modify this behavior unless renormalization is taken much farther than $\xi = -1$. A sharp DW (inset), however, is strongly renormalized which results in a large magneto-resistance effect.

One can make estimates of physical parameters for real systems. The smaller Fermi wavelength is that of spin-majority electrons, $2\pi/k_{F+}$. The ratio v_-/v_+ depends on the degree of polarization of the electron system. In a 1D nonmagnetic system there is a single Fermi momentum, k_F, for up and down electrons and a Fermi energy $E_F = \hbar^2 k_F^2/(2m)$. Once the system becomes magnetized, the Zeeman shift of the bands, $\Delta E/2 = JM_0$, and the two new Fermi momenta, $k_{F\pm}$, satisfy

$$\frac{k_{F\pm}}{k_F} = 1 \pm \frac{\Delta E}{4E_F} . \tag{12.45}$$

Inserting this result in Eq.(12.44) above, we obtain

$$\frac{\lambda}{v_+} = \pi \frac{(\Delta E/4E_F)}{[1 + (\Delta E/4E_F)]^2} (Lk_{F+}) . \tag{12.46}$$

In the full polarization limit $k_{F-} = 0$, $k_{F+} = 2k_F$, and Eq. (12.46) gives $\lambda/v_+ \approx 0.79\, Lk_{F+}$. Typical values for a non-fully polarized system are $E_F = 90\,\mathrm{meV}$ and $\Delta E = 30\,\mathrm{meV}$.[6] In this case we have $v_-/v_+ = 0.84$ and Eq. (12.46) gives $\lambda/v_+ \approx 0.22 Lk_{F+}$. For Lk_{F+} not exceeding about 2π, the system can flow to any of the fixed points described above.

12.8. Summary and concluding remarks

We have presented in this paper a theoretical description of the resistance of a magnetic microjunction with a constrained domain wall at the contact. In the limit of $k_{F\uparrow(\downarrow)}L \ll 1$, the electron transport across the wall was treated effectively as electron tunneling through a spin-dependent potential barrier. For such narrow and constrained domain walls the electron spin does not follow adiabatically the magnetization direction, but its orientation is rather fixed. However, the domain wall produces some mixing of the spin channels.

The calculations carried out in the paper were restricted to a limiting case of a single quantum transport channel. Accordingly, the system was described by a one-dimensional model. However, such a simple model turned out to describe qualitatively rather well the basic physics related to electronic transport through constrained domain walls, although the magnetoresistance obtained is still smaller than in some experiments. In realistic situations one should use a more general

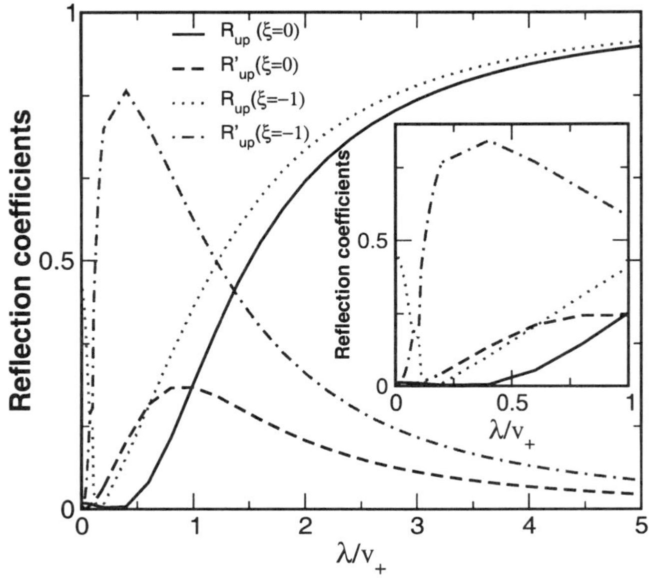

Fig. 12.5. Reflection coefficients vs. DW width ($\lambda/v_+ \propto Lk_{F+}$). The parameters are: $g_\uparrow = 1$, $g_\downarrow = 0.9$, $g_\perp = 1.1$ and $v_+/v_- = 0.8$.

model. When the domain wall does not cause transition between different channels, then the description presented here can be applied directly to the multichannel case by simply adding contributions from different channels.

A domain wall leads to spin dependent scattering of conduction electrons. Therefore, it also leads to a net spin polarization at the wall, which oscillates with the distance from the wall, similarly to Friedel oscillations of charge density near a nonmagnetic defect in a nonmagnetic metal. We have calculated the equilibrium component of this spin polarization.

Electron interactions have a more important effect on the transmission through thin domain walls than in the case of long domain walls. The repulsive interactions make the wall become insulating at T=0 and will also determine whether the reflection of the incident current occurs in the same spin channel or in the opposite spin channel.

This research was supported by Portuguese program POCI under Grant POCI/FIS/58746/2004, by Ministry of Science and Higher Education (Poland) as a research project in years 2006-2009, and by STCU Grant No. 3098 in Ukraine.

References

1. J. Grollier, P. Boulenc, V. Cros, A. Hamzič, A. Vaurès, A. Fert, and G. Faini, Switching a spin valve back and forth by current-induced domain wall motion, *Appl. Phys. Lett.* **83**, 509 (2003).
2. A. Yamaguchi, T. Ono, S. Nasu, K. Miyake, K. Mibu, and T. Shinjo, Real-space observation of current-driven domain wall motion in submicron magnetic wires, *Phys. Rev. Lett.* **92**, 077205 (2004).
3. M. Yamanouchi, D. Chiba, F. Matsukura, and H. Ohno, Current-induced domain-wall switching in a ferromagnetic semiconductor structure, *Nature* **428**, 539 (2004).
4. E. Saitoh, H. Miyajima, T. Yamaoka, and G. Tatara, Current-induced resonance and mass determination of a single magnetic domain wall, *Nature* **432**, 203 (2004).
5. H. D. Chopra and S. Z. Hua, Ballistic magnetoresistance over 3000% in Ni nanocontacts at room temperature, *Phys. Rev. B* **66**, 020403(R) (2002);
   H. D. Hua and H. D. Chopra, 100,000 % ballistic magnetoresistance in stable Ni nanocontacts at room temperature, *Phys. Rev. B* **67**, 060401(R) (2003).
6. C. Rüster, T. Borzenko, C. Gould, G. Schmidt, L. W. Molenkamp, X. Liu, T. J. Wojtowicz, J. K. Furdyna, Z. G. Yu, and M. E. Flatté, Very large magnetoresistance in lateral ferromagnetic (Ga,Mn)As wires with nanoconstrictions, *Phys. Rev. Lett.* **91**, 216602 (2003).
7. P. Bruno, Geometrically Constrained Magnetic Wall, *Phys. Rev. Lett.* **83**, 2425 (1999).
8. G. G. Cabrera and L. M. Falicov, Theory of residual resistivity of Bloch walls 1. Paramagnetic effects, *Phys. Status Solidi B* **61**, 539 (1974);
   Theory of residual resistivity of Bloch walls 2. Inclusion of diamagnetic effects, *ibid* **62**, 217 (1974).
9. G. Tatara and H. Fukuyama, Resistivity due to a domain wall in ferromagnetic metal, *Phys. Rev. Lett.* **78**, 3773 (1997).
10. A. Brataas, G. Tatara, and G. E. W. Bauer, Ballistic and diffuse transport through a ferromagnetic domain wall, *Phys. Rev. B* **60**, 3406 (1999).
11. V. K. Dugaev, J. Barnaś, A. Łusakowski, and Ł. A. Turski, Electrons in a ferromagnetic metal with a domain wall, *Phys. Rev. B* **65**, 224419 (2002)
12. R. P. van Gorkom, A. Brataas, and G. E. W. Bauer, Negative domain wall resistance in ferromagnets, *Phys. Rev. Lett.* **83**, 4401 (1999).
13. V. K. Dugaev, J. Berakdar, and J. Barnaś, Reflection of electrons from a domain wall in magnetic nanojunctions, *Phys. Rev. B* **68**, 104434 (2003);
    V. K. Dugaev, J. Barnaś, J. Berakdar, V. I. Ivanov, W. Dobrowolski, and V. F. Mitin, Magnetoresistance of a semiconducting magnetic wire with a domain wall, *ibid* **71**, 024430 (2005).
14. V. K. Dugaev, J. Berakdar, and J. Barnaś, Tunable conductance of magnetic nanowires with structured domain walls, *Phys. Rev. Lett.* **96**, 047208 (2006).
15. V. K. Dugaev, V. R. Vieira, P. D. Sacramento, J. Barnaś, M. A. N. Araújo, and J. Berakdar, Current-induced motion of a domain wall in a magnetic nanowire, *Phys. Rev. B* **74**, 054403 (2006).
16. M. A. N. Araújo, V. K. Dugaev, V. R. Vieira, J. Berakdar, and J. Barnaś, Role of electron correlations in transport through domain walls in magnetic nanowires, cond-mat/0602399; Transmission of correlated electrons through sharp domain walls in

magnetic nanowires: a renormalization group approach, *Phys. Rev. B* **74**, 224429 (2006).

17. C. L. Kane and M. P. A. Fisher, Transport in a one-channel Luttinger liquid, *Phys. Rev. Lett.* **68**, 1220 (1992); Resonant tunneling in an interacting one-dimensional electron gas, *Phys. Rev. B* **46**, 7268 (1992).
18. R. G. Pereira and E. Miranda, Domain-wall scattering in an interacting one-dimensional electron gas, *Phys. Rev. B* **69**, 140402(R) (2004).
19. U. Ebels, A. Radulescu, Y. Henry, L. Piraux, and K. Ounadjela, Spin acumulation and domain wall magnetoresistance in 35 nm Co Wires, *Phys. Rev. Lett.* **84**, 983 (2000).
20. K. A. Matveev, *et al*, Tunneling in one-dimensional non-Luttinger electron liquid, *Phys. Rev. Lett.* **71**, 3351 (1993);
    D. Yue, *et al*, Conduction of a weakly interacting one-dimensional electron gas through a single barrier, *Phys. Rev. B* **49**, 1966 (1994).

PART 3

Quantum Coherent Systems

Chapter 13

Density correlations of an ultra-cold quantum gas in the vicinity of Bose-Einstein condensation

José Viana-Gomes[1,2] *, Denis Boiron[1], Michael Belsley[2]

[1]*Laboratoire Charles Fabry de l'Institut d'Optique, CNRS, Univ Paris-sud, Campus Polytechnique, RD 128, 91127 Palaiseau Cedex, France*

[2]*Departamento de Física, Universidade do Minho, Campus de Gualtar, 4710-057 Braga, Portugal*

In this Chapter we present a brief overview of the physics of an ideal ultra cold quantum gas formed within a harmonic trapping potential focusing especially on its coherence properties. Recently it has become experimentally feasible to study these particle correlations. We include a short introduction to the relevant experimental techniques and their limitations. Measurements are typically carried out on atomic clouds released from the trap and left to expand under the influence of gravity. We model the appropriate ballistic expansion for a non-interacting atomic cloud and derive the corresponding one-body and two-body correlation functions. We conclude by summarizing recent measurements of the second order particle correlations present within a falling cloud of metastable Helium atoms close to the Bose-Einstein condensation point.

Contents

*zgomes@fisica.uminho.pt

13.1. Introduction

In 1925, Albert Einstein predicted that if a ideal gas of bosonic atoms were cooled below a certain transition temperature it would undergo a phase transition to a new state where a macroscopic fraction of the atoms would occupy the same fundamental state of the system[1]creating a highly coherent atomic ensemble. As Einstein pointed out, this remarkable statement is a consequence of the *statistics* of identical particles with integral spin, which had been recently derived by himself and Satyendra Nath Bose.[2] Despite this early prediction, it only became possible in the 1990's to create a Bose-Einstein condensate (BEC) in dilute atomic samples, as Einstein had originally imagined.

In these experiments the gas is strongly localized in both coordinate and momentum spaces. To avoid the formation of dimers (and also to reduce losses due to inelastic collisions) the sample is very dilute, typically millions of times less dense than an ideal gas at atmospheric pressure and room temperature. This leads to extremely small phase transition *critical temperatures*, typically in the range of microkelvin[a], and has long constituted a considerable challenge for experimental physicists.

The first atomic BECs were obtained in 1995. The impact was so great that only six years later E. A. Cornell, W. Ketterle and C. E. Wieman received the Nobel prize in 2001 "for the achievement of Bose-Einstein condensation in dilute gases of alkali atoms, and for early fundamental studies of the properties of the condensates". One remarkable experiment reported by Wolfgang Ketterle's group at MIT demonstrated that when two independent BECs were superimposed they interfere[4] in much the same way as coherent light. This was the first clear demonstration of first order coherence of the associated atomic quantum field, which could be characterized through the visibility of interference fringes. Other impressive experimental achievements were the realization of pulsed and CW atom

[a]In condensed matter systems critical temperatures are much higher. For example, the superfluidity of liquid helium takes place at 2.18° K.[3]

lasers[5–8] and the observation of the interference of two matter-wave beams emitted from two spatially separated regions of the same BEC.[9]

These pioneering experiments have verified that in many senses, below the BEC condensation threshold, bosonic atoms become coherent in phase and degenerate in energy, much like the stimulated emission of a single mode laser beam. The similarities between coherent atoms and photons[10] allow many of the key ideas of quantum optics to be directly carried over to describe coherent atom optics. However there are several key differences between atoms and photons; atoms have both mass and internal states that have no counterparts in photons. An especially important difference is that atoms can also interact directly with each other, without requiring a nonlinear medium that mediates the interaction between photons. For example, it is possible to carry out an atomic four-wave mixing experiment in which three different coherent atomic beams interact in a vacuum to generate a fourth beam.[11]

One of the landmark experiments in quantum optics was carried out by Robert Hanbury Brown and Richard Twiss in 1956 to measure the second order temporal coherence properties of thermal light.[12] Second order coherence corresponds to the correlation function of the squared modulus of the field. From a particle point of view it quantifies density correlations and is related to the conditional probability of finding one particle at a certain location given that another particle is present at some other location. This first *HBT* experiment showed that photons originating from a thermal source have the tendency to be detected close together, an effect usually referred as *bosonic bunching* (cf. 13.2). This type of particle correlation arises from exchange symmetry effects and exists even in the absence of interactions between the particles. This was clearly demonstrated in a second experiment that detected a second order correlations between photons coming from widely separated points of a star.[13]

The first measurement of density correlations with atoms was carried out by Yasuda et al.[14] on an ultra-cold atomic beam of the bosonic isotope ^{20}Ne at a temperature far from quantum degeneracy. It confirmed for the first time that thermal bosonic atoms, like thermal light, also display bunching. We will describe later in this chapter a similar experiment carried out by the "Optique atomique" group of the laboratoire Charles Fabry de l'Institut d'Optique. This experiment headed by D. Boiron, A. Aspect and C. I. Westbrook used metastable helium atoms and was able to produce atomic clouds at the vicinity of the quantum degeneracy (cf. 13.7).[15] We will from here on refer to this experiment simply by "He* experiment". Bunching was observed for thermal atomic clouds, while for Bose-Einstein condensed clouds, the *coherent* atoms were detected with a perfectly random temporal sequence analogous to what had already been observed for single mode laser light.[16] This behavior has also been investigated by direct inspection of an atom laser´s particle statistics by Esslinger's group[17] and through the study of the cloud's density fluctuations in a handful of other experiments.[18–21] Anti-bunching effects in dilute fermionic atomic systems have also been observed in two very recent experiments.[22,23] One of them[22] used also a metastable Helium sample but of the fermionic isotope ^{3}He, with the microchannel plate (MCP) detection system we describe later in this Chapter.

Analogous correlations measurements of massive particles, with both bosons and fermions, have also been studied in the field of nuclear physics[24–28] and using low energy electrons.[29,30] The unique properties of cold atoms however make ultra-cold quantum gases an especially attractive testing ground for exploring many-body quantum systems. Both ferimonic and bosonic systems can be studied. These gases are tunable to an extent that is unparalleled in condensed matter physics; it is possible to vary over a significant range their density, temperature, as well as the effective dimensionality of confinement. Using Feshbach resonances[31,32] it is even possible to change the strength and the sign of the interparticle interaction potential.

The results obtained in all the referred experiments, although not unexpected, represent an important step in the experimental exploration of a field that might be called quantum atom optics. However, there are also important unanswered questions that one might hope to address with these experiments. For example, the second order correlation function can probe interesting features of the physics of cold gases in the vicinity of the critical transition associated with the BEC formation and associated quantum bosonic effects and critical fluctuations.

This Chapter's plan

The Chapter is organized as follows. We begin in section 13.2 with a short introduction to the classic Hanbury Brown Twiss experiment, using this to physically motivate the study of second order correlations and to introduce some general definitions from Glauber's theory of quantum coherence. This allows us to briefly review the main results of first and second order coherence theory in optics and, in section 13.3, its generalization for a quantum field of massive particles. In section 13.4 we present a succinct description of the techniques used for producing and detecting a BEC. These have become standard methods used in the manipulation of cold quantum gases. We focus in particular on the technique of evaporative cooling and on the single atom detection scheme based on a microchannel plate (MCP) and a delay line position sensitive detector (PSD) used in the He* experiment .

Section 13.5 briefly introduces the theoretical description of an ultra-cold cloud of bosons in thermal equilibrium inside a harmonic trap. Particular care must be taken to describe the cloud when it is close to the Bose-Einstein transition point where the influence of the ground state is particularly important. Usually, the theoretical modeling of the correlation function treats the sample at thermal equilibrium inside the trap. However, most of the experiments (with one exception[21]) measure the correlation signal of the cloud after having been released from the trap and expanded under the influence of gravity. Consequently in Section 13.6 we describe the ballistic expansion of the falling cloud, under the admittedly limiting approximation of a non-interacting gas, by propagating the eigenstates of the harmonic potential using a Green's function formalism. This formalism is then used to derive the appropriate expressions for the intensity correlation function in the atomic flux in section 13.6.2.

Finally we conclude in section 13.7 by presenting a brief summary of the most important results from the He* experiment characterizing the second order correlations of an ultra-cold gas of He*.

13.2. Correlations and the *Hanbury Brown and Twiss* experiment

In this section we briefly review some key concepts of coherence theory by first considering classical optical sources and then generalizing to quantum fields. We begin with a short discussion of the landmark proposal of R. Hanbury Brown and R. Twiss to describe the intensity correlations present in a thermal light field. The subsequent laboratory experiment is often viewed as being the driving force that opened up the field of quantum optics.

13.2.1. *The stellar HBT experiment and the transverse coherence length*

Consider the scheme illustrated in Fig.13.1, where light from an incoherent wave source, such as a star, travels towards two detectors, D_1 and D_2, a distance R away. For simplicity, we start by considering a one-dimensional monochromatic source with a spatial profile given by $A(x)$ and a characteristic size $2s_\perp$. The two independent detectors are separated from each other by a distance l, small compared to the distance to the source, $R \gg s_\perp, l$.

The scalar wave field emitted by the small source segment dx that arrives at the detector placed at the distance $\mathbf{r} \gg s_\perp, 2\pi/k$ can be written as

$$du(\mathbf{r}) = A(x)dx\,\frac{e^{i\mathbf{k}\cdot\mathbf{r}+i\phi(x)}}{|\mathbf{r}|}.$$

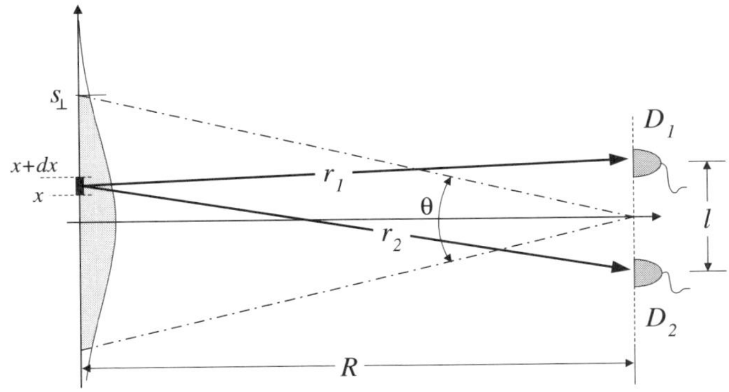

Fig. 13.1. An incoherent extended source emits radiation detected by two detectors, D_1 and D_2, located a distance R from the source. The correlation between the photocurrents generated by the two detectors is related to the angular size θ of the source by the van Cittert-Zernike theorem. This was first used by R. Hanbury Brown and R. Twiss to measure the angular diameter of a star.

Here $\phi(x)$ represents the phase of the wave front emitted at the location x of the source. The total intensity detected by D_1 is then proportional to

$$I_1 \propto \left| \int du(\mathbf{r}_1) \right|^2 \simeq \frac{1}{R^2} \left| \int dx A(x) \exp\left[i\frac{k}{2R}\left(x - \frac{1}{2}l\right)^2 + i\phi(x) \right] \right|^2 ,$$

where we have used the approximation $r_1 \simeq R + (x - \frac{1}{2}l)^2/2R$ (see Fig.13.1). A similar expression holds for the intensity detected by D_2 with the substitution $r_2 \simeq R + (x + \frac{1}{2}l)^2/2R$. If we assume that the phases $\phi(x)$ are random and uncorrelated, averaging the above expression over the statistics of the field (for example by assuming that $du(\mathbf{r})$ is a Gaussian distributed zero mean random variable) effectively converts the above integral into an incoherent sum over the individual contributions from each separate source element. In this case the above expression simplifies to

$$\langle I_1 \rangle = S \int dx\, |A(x)|^2 = \langle I_2 \rangle.$$

where S is a constant that incorporates the detector sensitivity and geometric factors such as $1/R^2$. Since each source element is assumed to be independent and incoherent the detected intensity is independent of the detector position provided $l \ll R$. However, if we compute instead the *correlation* in the intensities registered by the two detectors $\langle I_1\, I_2 \rangle$, again assuming that the phases of the waves emitted at each spatial location on the source are independent so that

$$\langle e^{i\phi(x_1) - i\phi(x_1') + i\phi(x_2) - i\phi(x_2')} \rangle = \delta(x_1 - x_1')\delta(x_2 - x_2') + \delta(x_1 - x_2')\delta(x_1 - x_2')$$

we obtain

$$\langle I_1 I_2\rangle = \langle I_1\rangle\langle I_2\rangle + S^2 \left| \int dx\, |A(x)|^2 \, e^{i\frac{k}{R}xl} \right|^2 , \tag{13.1}$$

a result that depends on both R and l. This result is just the (one-dimensional) *van Cittert-Zernike* theorem,[33] which states that the second order spatial correlation of an incoherent source is related to the Fourier transform of the intensity distribution across the source.[34] To work out an explicit result we continue by assuming that $A(x)$ has a gaussian profile given by

$$|A(x)|^2 = \frac{|A_0|^2}{\sqrt{2\pi}\, s_\perp} \exp\left[-x^2/2s_\perp^2\right].$$

Substituting this expression in Eq.13.1 and normalizing the result by the square of the total detected intensity of the source we obtain

$$g^{(2)}(l) = \frac{\langle I_1 I_2\rangle}{\langle I_1\rangle\langle I_2\rangle} = 1 + \exp\left[-l^2/2{l_\perp}^2\right], \tag{13.2}$$

where $l_\perp = R/(ks_\perp) = 1/(k\theta) = 1/\Delta k$, is the *transverse coherence length* of the source, with $\theta = s_\perp/R$ being the source's *angular size* as seen by the detector and $\hbar\Delta k$ being the effective spread in photon momenta seen by the detector. Thus, by determining the correlation between the intensities registered by the two detectors while varying the distance l between them, it is possible, using Eq.13.2, to measure the angular size θ of the source. This was the rationale behind the original proposal of R. Hanbury Brown and R. Twiss to measure the angular diameter of Sirius, a star at 8.6 light years from the Earth.[13]

Note that had the source been phase coherent with each source element emitting with an identical constant phase then the average over the phase statistics would have been unnecessary and the correlation in the intensities registered by the two detectors would be simply

$$\langle I_1\, I_2\rangle = \langle I_1\rangle\langle I_2\rangle,$$

yielding a normalized second order correlation of $g^{(2)} = 1$. Thus in the limit of superposed detectors ($l \to 0$) the correlation observed when the source consists of incoherent (thermal) emitters is predicted to be twice as large as is observed for the corresponding phase coherent source. This behavior is usually referred to as *photon bunching* and expresses the tendency for thermal photons to be detected in closely spaced pairs in a second order correlation measurement.

13.2.2. *The quantum description*

This result raised a great deal of controversy at the time it was obtained. How is it possible for independent source elements to interfere? How can one obtain phase information by correlating signals that are proportional to the modulus squared of the fields? Does it not violate Dirac's famous statement that photons only interfere with themselves? These questions are best answered by considering the quantum theory of photon detection and correlation that was developed by Roy Glauber and others in a series of articles in the 1960s.[35,36] "For his contribution to the quantum theory of optical coherence" Glauber was awarded half of the 2005 Nobel Prize in Physics.

According to Glauber the quantum mechanical electric field operator can be written as a superposition of positive and negative frequency parts

$$\hat{\mathbf{E}}(\mathbf{r},t) = \hat{\mathbf{E}}^{(+)}(\mathbf{r},t) + \hat{\mathbf{E}}^{(-)}(\mathbf{r},t),$$

where

$$\hat{\mathbf{E}}^{(+)}(\mathbf{r},t) = \sum_k \hat{\epsilon}_{\mathbf{k}} E_{\mathbf{k}} \hat{a}_{\mathbf{k}} e^{-i\omega_k t + i\mathbf{k}\cdot\mathbf{r}} \quad \text{and} \quad \hat{\mathbf{E}}^{(-)}(\mathbf{r},t) = \sum_k \hat{\epsilon}_{\mathbf{k}} E^*_{\mathbf{k}} \hat{a}^\dagger_{\mathbf{k}} e^{i\omega_k t - i\mathbf{k}\cdot\mathbf{r}}$$

Here $\hat{a}^\dagger_{\mathbf{k}}$ and $\hat{a}_{\mathbf{k}}$ are the familiar photon (*bosonic*) creation and annihilation operators while $E_{\mathbf{k}} = i\sqrt{\hbar\omega_k/\epsilon_0 V}$ with V the quantization volume.

Optical detectors usually generate a signal through the photoelectric effect resulting from the absorption of an incident photon. Therefore only the annihilation operator contributes and the transition probability for the detector to absorb one photon from the field at a position $\mathbf{r}$ between times t and $t + dt$ is proportional to

$$p_1(\mathbf{r},t)dt \propto \sum_f |\langle f|\hat{\mathbf{E}}^{(+)}(\mathbf{r},t)|i\rangle|^2 = \langle i|\hat{\mathbf{E}}^{(-)}(\mathbf{r},t)\hat{\mathbf{E}}^{(+)}(\mathbf{r},t)|i\rangle$$

where $|i\rangle$ is the initial state of the field and $|f\rangle$ is the final state in which the field could be found after the process is concluded. The final state being unobserved we summed over a complete set of states. If the initial state of the field is not precisely known one typically resorts to a statistical description by introducing the density operator for the field, ρ so that

$$p_1(\mathbf{r},t)dt \propto Tr[\rho\hat{\mathbf{E}}^{(-)}(\mathbf{r},t)\hat{\mathbf{E}}^{(+)}(\mathbf{r},t)]$$

This leads directly to a definition of the first order correlation function of the field as

$$\begin{aligned} G^{(1)}(\mathbf{r}_1,t_1;\mathbf{r}_2,t_2) &= Tr[\rho\hat{\mathbf{E}}^{(-)}(\mathbf{r}_2,t_2)\hat{\mathbf{E}}^{(+)}(\mathbf{r}_1,t_1)] \\ &= \left\langle \hat{\mathbf{E}}^{(-)}(\mathbf{r}_2,t_2)\hat{\mathbf{E}}^{(+)}(\mathbf{r}_1,t_1) \right\rangle \end{aligned} \tag{13.3}$$

In an analogous manner the joint probability for observing one photoionization at the position $\mathbf{r}_1$ between the times t_1 and $t_1 + dt_1$ and a second at the position $\mathbf{r}_2$ between the times t_2 and $t_2 + dt_2$ will be given by

$$p_2(\mathbf{r}_1, t_1; \mathbf{r}_2, t_2) dt_1 dt_2 \propto Tr[\rho \hat{\mathbf{E}}^{(-)}(\mathbf{r}_1, t_1) \hat{\mathbf{E}}^{(-)}(\mathbf{r}_2, t_2) \hat{\mathbf{E}}^{(+)}(\mathbf{r}_2, t_2) \hat{\mathbf{E}}^{(+)}(\mathbf{r}_1, t_1)]$$

which leads to the definition of the second order correlation function appropriate to the *HBT* experiment

$$G^{(2)}(\mathbf{r}_1, t_1; \mathbf{r}_2, t_2) = \langle \hat{\mathbf{E}}^{(-)}(\mathbf{r}_1, t_1) \hat{\mathbf{E}}^{(-)}(\mathbf{r}_2, t_2) \hat{\mathbf{E}}^{(+)}(\mathbf{r}_2, t_2) \hat{\mathbf{E}}^{(+)}(\mathbf{r}_1, t_1) \rangle . \quad (13.4)$$

Note that the normal operator ordering and time ordering present in this expression follow naturally from the condition that we are interested the correlation of pairs of photons, that is the correlation between one photon detected by D_1 with a second photon detected by D_2. We can now use this expression to explore the physics behind the Hanbury Brown Twiss effect.

We follow the development of M.O. Scully and M.S. Zubairy[37] and consider an ideal "bare-bones" situation in which the incident light field consists of two independent photons, with the same overall momentum but traveling in slightly different directions $\mathbf{k}$ and $\mathbf{k}'$. Writing the incident field as $|\psi\rangle = |1_{\mathbf{k}} 1_{\mathbf{k}'}\rangle$ the second order correlation function at equal times becomes

$$G^{(2)}(\mathbf{r}_1, t; \mathbf{r}_2, t) = \langle 1_{\mathbf{k}} 1_{\mathbf{k}'} | \hat{\mathbf{E}}^{(-)}(\mathbf{r}_1, t) \hat{\mathbf{E}}^{(-)}(\mathbf{r}_2, t) \hat{\mathbf{E}}^{(+)}(\mathbf{r}_2, t) \hat{\mathbf{E}}^{(+)}(\mathbf{r}_1, t) | 1_{\mathbf{k}} 1_{\mathbf{k}'} \rangle .$$

Inserting a complete set of states and noting that $|1_{\mathbf{k}} 1_{\mathbf{k}'}\rangle$ is a two photon state that is reduced to the vacuum by the operator $\hat{\mathbf{E}}^{(+)}(\mathbf{r}_2, t)\hat{\mathbf{E}}^{(+)}(\mathbf{r}_1, t)$ one has

$$\begin{aligned} G^{(2)}&(\mathbf{r}_1, t; \mathbf{r}_2, t) = \\ &= \sum_{\{n\}} \langle 1_{\mathbf{k}} 1_{\mathbf{k}'} | \hat{\mathbf{E}}^{(-)}(\mathbf{r}_1, t) \hat{\mathbf{E}}^{(-)}(\mathbf{r}_2, t) | \{n\} \rangle \langle \{n\} | \hat{\mathbf{E}}^{(+)}(\mathbf{r}_2, t) \hat{\mathbf{E}}^{(+)}(\mathbf{r}_1, t) | 1_{\mathbf{k}} 1_{\mathbf{k}'} \rangle \\ &= |\langle 0 | \hat{\mathbf{E}}^{(+)}(\mathbf{r}_2, t) \hat{\mathbf{E}}^{(+)}(\mathbf{r}_1, t) | 1_{\mathbf{k}} 1_{\mathbf{k}'} \rangle|^2 . \end{aligned}$$

Using the fact that $\hat{\mathbf{E}}^{(+)}(\mathbf{r}_i, t) = E_{\mathbf{k}}(\hat{a}_{\mathbf{k}} e^{-i\omega t + i\mathbf{k}\cdot\mathbf{r}_i} + \hat{a}_{\mathbf{k}'} e^{-i\omega t + i\mathbf{k}'\cdot\mathbf{r}_i})$ we have

$$\begin{aligned} \langle 0 | &\hat{\mathbf{E}}^{(+)}(\mathbf{r}_2, t) \hat{\mathbf{E}}^{(+)}(\mathbf{r}_1, t) | 1_{\mathbf{k}} 1_{\mathbf{k}'} \rangle = \\ &= E_{\mathbf{k}}^2 e^{-i2\omega t} \langle 0 | \hat{a}_{\mathbf{k}} e^{i\mathbf{k}\cdot\mathbf{r}_2} \hat{a}_{\mathbf{k}'} e^{i\mathbf{k}'\cdot\mathbf{r}_1} | 1_{\mathbf{k}} 1_{\mathbf{k}'} \rangle + E_{\mathbf{k}}^2 e^{-i2\omega t} \langle 0 | \hat{a}_{\mathbf{k}'} e^{i\mathbf{k}'\cdot\mathbf{r}_2} \hat{a}_{\mathbf{k}} e^{i\mathbf{k}\cdot\mathbf{r}_1} | 1_{\mathbf{k}} 1_{\mathbf{k}'} \rangle \\ &= E_{\mathbf{k}}^2 e^{-i2\omega t} (e^{i\mathbf{k}\cdot\mathbf{r}_2 + i\mathbf{k}'\cdot\mathbf{r}_1} + e^{i\mathbf{k}'\cdot\mathbf{r}_2 + i\mathbf{k}\cdot\mathbf{r}_1}) \end{aligned}$$

giving

$$G^{(2)}(\mathbf{r}_1, t; \mathbf{r}_2, t) = 2|E|^4 \{1 + \cos[(\mathbf{k}' - \mathbf{k}) \cdot (\mathbf{r}_1 - \mathbf{r}_2)]\} . \quad (13.5)$$

The extra bunching term arises from the fact that the correlation measurement does not specify which detector (D1 or D2) detected which photon (k or k') and the two possibilities must both be considered (see Fig.13.2). Of course the sinusoidal dependence of the interference term is due to the fact that we have detected only two photons. For a finite sized source emitting many photons the interference term would still depend on the spread of momentum of the detected photons,Δk , but would take on a form similar to that derived above in Eq(13.2).

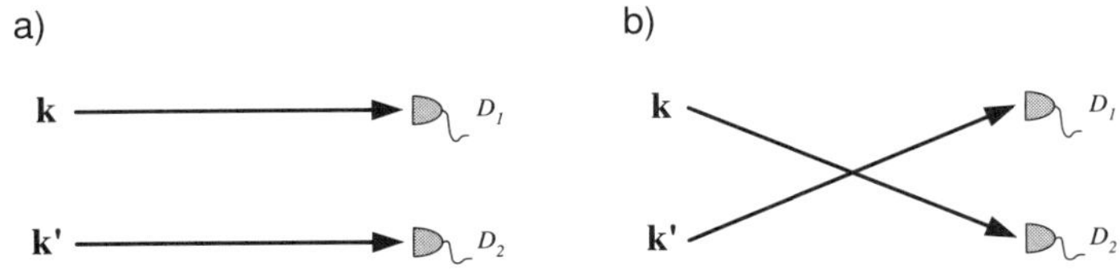

Fig. 13.2. The possibility of both of the above detection sequences produces an interference term in Eq.13.5.

We take this opportunity to note that the sign of the interference term will depend on the quantum symmetry of the particle. For example if the detected particles were fermions the above analysis would carry straight through except that the bosonic annihilation operators would be replaced by their fermionic counterparts, $\hat{a}_{\mathbf{k}} \rightarrow \hat{b}_{\mathbf{k}}$. We would then have to calculate

$$\begin{aligned}\langle 0|\hat{b}_{\mathbf{k}}\hat{b}_{\mathbf{k}'}|1_{\mathbf{k}}1_{\mathbf{k}'}\rangle &= \langle 0|\hat{b}_{\mathbf{k}}\hat{b}_{\mathbf{k}'}\hat{b}^{\dagger}_{\mathbf{k}}\hat{b}^{\dagger}_{\mathbf{k}'}|0\rangle \\ &= -\langle 0|\hat{b}_{\mathbf{k}}\hat{b}^{\dagger}_{\mathbf{k}}|0\rangle\langle 0\hat{b}_{\mathbf{k}'}\hat{b}^{\dagger}_{\mathbf{k}'}|0\rangle \;= -1,\end{aligned}$$

and

$$\begin{aligned}\langle 0|\hat{b}_{\mathbf{k}'}\hat{b}_{\mathbf{k}}|1_{\mathbf{k}}1_{\mathbf{k}'}\rangle &= \langle 0|\hat{b}_{\mathbf{k}'}\hat{b}_{\mathbf{k}}\hat{b}^{\dagger}_{\mathbf{k}}\hat{b}^{\dagger}_{\mathbf{k}'}|0\rangle \\ &= +\langle 0\hat{b}_{\mathbf{k}'}\hat{b}^{\dagger}_{\mathbf{k}'}|0\rangle\langle 0|\hat{b}_{\mathbf{k}}\hat{b}^{\dagger}_{\mathbf{k}}|0\rangle = +1,\end{aligned}$$

where we have used the appropriate fermionic anti-commutation relations. The final result for the detection of two independent fermions is

$$G^{(2)}(\mathbf{r}_1, t; \mathbf{r}_2, t) = 2|E|^4\{1 - \cos[(\mathbf{k}' - \mathbf{k}) \cdot (\mathbf{r}_1 - \mathbf{r}_2)]\},$$

essentially the same expression as for bosons but with a sign change in front of the interference terms. One would thus expect to observe an *anti-bunching* effect when detecting two independent fermions.

The laboratory *HBT* experiment and the coherence time

This treatment can be easily generalized to include multi-mode fields. As a last optics example we treat a situation that corresponds closely to the laboratory experiment that Hanbury Brown and Twiss used to verify their idea. They used inhomogeneously broadened thermal light from the 435.8 nm line of a mercury lamp focused onto a pinhole to produce a transversely coherent source and a beam splitter to allow them to place the two detectors at effectively the same spatial location, corresponding to $l = 0$ in the above analysis. In this case the incident field is well approximated by a plane wave. If the wave is furthermore linearly polarized we can treat it as a unidirectional scalar wave. Then we can write

$$\hat{\mathbf{E}}^{(+)}(\mathbf{r},t) = \sum_k E_k \hat{a}_k e^{-i\omega_k t + i\mathbf{k}\cdot\mathbf{r}},$$

which when substituted into the expressions for the first and second order correlation functions Eqs.13.3 and 13.4 gives

$$G^{(1)}(\mathbf{r},t_1;\mathbf{r},t_2) = \left\langle \sum_{k,l} E_l^* E_k \hat{a}_l^\dagger \hat{a}_k e^{i\omega_l t_2 - i\omega_k t_1} \right\rangle, \tag{13.6}$$

and

$$G^{(2)}(\mathbf{r},t_1;\mathbf{r},t_2) = \\ = \left\langle \sum_{k,l,m,n} E_k^* E_l^* E_m E_m \hat{a}_k^\dagger \hat{a}_l^\dagger \hat{a}_m \hat{a}_n e^{i(\omega_k - \omega_n)t_2 + i(\omega_l - \omega_m)t_1} \right\rangle. \tag{13.7}$$

For a *multi-mode thermal field* only pairwise operator orderings survive the average over the field statistics expressed using the grand canonical ensemble. In this case $\langle \hat{a}_l^\dagger \hat{a}_k \rangle = \langle \hat{a}_k^\dagger \hat{a}_k \rangle \delta_{k,l}$ and

$$\langle \hat{a}_i^\dagger \hat{a}_j^\dagger \hat{a}_k \hat{a}_l \rangle = \langle \hat{a}_i^\dagger \hat{a}_i \rangle \delta_{i,l} \langle \hat{a}_j^\dagger \hat{a}_j \rangle \delta_{j,k} + \langle \hat{a}_i^\dagger \hat{a}_i \rangle \delta_{i,k} \langle \hat{a}_j^\dagger \hat{a}_j \rangle \delta_{j,l}$$

so that

$$G^{(1)}_{mm}(\mathbf{r},t_1;\mathbf{r},t_2) = \sum_k |E_k|^2 \langle \hat{a}_k^\dagger \hat{a}_k \rangle$$

and

$$G^{(2)}_{mm}(\mathbf{r},t_1;\mathbf{r},t_2) = \sum_{k,l} |E_k|^2 |E_l|^2 \langle \hat{a}_k^\dagger \hat{a}_k \rangle \langle \hat{a}_l^\dagger \hat{a}_l \rangle \{1 + e^{i\omega_k (t_1 - t_2) - i\omega_l (t_1 - t_2)}\}.$$

For time stationary fields only the time difference $\delta t = t_1 - t_2$ is important. In this case the normalized second temporal correlation function for multi-mode fields becomes

$$g^{(2)}_{mm}(\delta t) = 1 + \frac{|\sum_k |E_k|^2 \langle \hat{a}_k^\dagger \hat{a}_k \rangle e^{i\omega_k (\delta t)}|^2}{\sum_k |E_k|^2 \langle \hat{a}_k^\dagger \hat{a}_k \rangle}.$$

The second term in the above equation is essentially the normalized Fourier transform of the source's power spectrum. If we take a continuum limit for the modes and assume, as for the spatial correlations above, a gaussian profile for the spectrum with a central frequency ω_0 and a width characterized by $\Delta\omega$,

$$\sum_k |E_k|^2 \rightarrow \int d\omega S(\omega)S(\omega) \text{ with } S(\omega) = \frac{1}{\sqrt{2\pi}\,\Delta\omega}\exp\left(-\frac{(\omega-\omega_0)^2}{2\Delta\omega^2}\right),$$

as is appropriate for an inhomogeneously broadened source, we obtain,

$$g^{(2)}_{mm}(\delta t) = \frac{\langle I_1(t)I_2(t+\delta t)\rangle}{\langle I_1(t)\rangle\langle I_2(t+\delta t)\rangle} = 1 + \exp\left[-\tfrac{1}{2}(\delta t\Delta\omega)^2\right].$$

The decay of observed bunching with increasing delay time, δt is essentially a manifestation of the Wiener-Khintchine theorem and allows one to measure the coherence time of the source. On the other hand if the source emits in a pure single mode (*sm*) then

$$\hat{\mathbf{E}}^{(+)}_{sm}(\mathbf{r},t) = E_k \hat{a}_k e^{-i\omega_k t + i\mathbf{k}\cdot\mathbf{r}}$$

and the normalized second order temporal correlation function becomes

$$g^{(2)}_{sm}(\delta t) = \frac{\langle \hat{a}^\dagger_k \hat{a}^\dagger_k \hat{a}_k \hat{a}_k\rangle}{\langle \hat{a}^\dagger_k \hat{a}_k\rangle} = 1 + \frac{\langle(\Delta n)^2\rangle - \langle n\rangle}{\langle n\rangle}.$$

Here n is the mean number of photons in the field. Single mode laser fields are known to have Poissonian statistics for which $\langle(\Delta n)^2\rangle = \langle n\rangle$. The first experimental demonstration of this was carried out by Arecchi in 1965[16] using a He-Ne laser as light source.

We hasten to note however that the above expression for the second order temporal correlation function in optics does directly not carry over to the case of a falling (finite) thermal cloud of massive particles. Apart from the fact that the above relation has been derived for a continuous wave and not a pulsed wave, the key difference in this latter case is the dispersion in velocities of the massive particles that is not present for photons. In fact, the *longitudinal* correlations observed for a finite falling ideal thermal atomic cloud are well described by the *van Cittert-Zernike* transverse spatial correlations described above in Eq(13.2) as we shall demonstrate later in this chapter. As a first step in this process we now discuss the case of massive particle correlations.

13.3. Massive particle correlations

13.3.1. *Definitions*

The above formalism for optical fields can easily be generalized to describe the correlations that can be measured between massive particles. Within the framework of second quantization, massive particles are described by the field operators,

$$\begin{cases} \hat{\Psi}^{\dagger}(\mathbf{r},t) &= \sum_{\mathbf{m}} \psi^{*}_{\mathbf{m}}(\mathbf{r},t)\, \hat{a}^{\dagger}_{\mathbf{m}} \\ \hat{\Psi}(\mathbf{r},t) &= \sum_{\mathbf{m}} \psi_{\mathbf{m}}(\mathbf{r},t)\, \hat{a}_{\mathbf{m}} \end{cases}. \tag{13.8}$$

In this expression $\{\psi_{\mathbf{m}}\}$ represents an appropriate complete set of wavefunctions that describe the center of mass motion of the particles, for instance the harmonic oscillator wave functions in the case that the particles find themselves confined in an harmonic trap [b]. The operators $\hat{a}^{\dagger}$ and $\hat{a}$ are the usual bosonic creation and annihilation operators which obey the commutation relations $[\hat{a}^{\dagger}_{\mathbf{m}}, \hat{a}_{\mathbf{m}'}] = \delta_{\mathbf{m},\mathbf{m}'}$ and $[\hat{a}^{\dagger}_{\mathbf{m}}, \hat{a}^{\dagger}_{\mathbf{m}'}] = [\hat{a}_{\mathbf{m}}, \hat{a}_{\mathbf{m}'}] = 0$. Since we are dealing with a bosonic field in thermal equilibrium, we have $\langle \hat{a}^{\dagger}_{\mathbf{j}} \hat{a}_{\mathbf{k}} \rangle = \delta_{\mathbf{j}k} \langle n_{\mathbf{j}} \rangle$ with the states's occupation $\langle n_{\mathbf{m}} \rangle$ given by the Bose-Einstein distribution

$$\langle n_{\mathbf{m}} \rangle = \frac{\mathcal{Z}}{\exp\left(\beta \varepsilon_{\mathbf{m}}\right) - \mathcal{Z}}. \tag{13.9}$$

Here $\varepsilon_{\mathbf{m}} = \hbar \sum_{\alpha} \left[\left(m_{\alpha} + \frac{1}{2}\right) \omega_{\alpha}\right]$ and $\beta = 1/k_B T$, with T the temperature and k_B the Boltzmann constant. In this expression the gas fugacity, $\mathcal{Z}$, is related to the chemical potential μ by $\mathcal{Z} = \exp(\mu/k_B T)$.

In direct analogy to the optical expression, the equal time first order particle field correlation function is,

$$G^{(1)}(\mathbf{r}, t; \mathbf{r}', t) = \langle \hat{\Psi}^{\dagger}(\mathbf{r}, t) \hat{\Psi}(\mathbf{r}', t) \rangle. \tag{13.10}$$

Setting $\mathbf{r} = \mathbf{r}'$, this expression is just the cloud´s density at the location $\mathbf{r}$,

$$n(\mathbf{r}, \mathbf{t}) = \langle \hat{\Psi}^{\dagger}(\mathbf{r}, \mathbf{t}) \hat{\Psi}(\mathbf{r}, \mathbf{t}) \rangle = G^{(1)}(\mathbf{r}, t; \mathbf{r}, t). \tag{13.11}$$

Evaluated at different points in space, this function contains an interference term that characterizes the fringe pattern which might be observed in a Michelson's or Young's double slit type of experiments.

Similarly the second order correlation function can be written as,

$$G^{(2)}(\mathbf{r}, t; \mathbf{r}', t') = \langle \hat{\Psi}^{\dagger}(\mathbf{r}, t) \hat{\Psi}^{\dagger}(\mathbf{r}', t') \hat{\Psi}(\mathbf{r}', t') \hat{\Psi}(\mathbf{r}, t) \rangle. \tag{13.12}$$

This expression contains a statistical average over the normal ordered product of creation and annihilation operators, $\langle \hat{a}^{\dagger}_{\mathbf{j}} \hat{a}^{\dagger}_{\mathbf{k}} \hat{a}_{\mathbf{l}} \hat{a}_{\mathbf{n}} \rangle$ For a thermal (non-degenerate) ideal gas described within the framework of the grand canonical ensemble, only pairwise orderings survive, just as in the case of the multi-mode optical field above. As a result

$$\begin{aligned} G^{(2)}_{th}(\mathbf{r}, t; \mathbf{r}', t') &= \langle \hat{\Psi}^{\dagger}(\mathbf{r}, t) \hat{\Psi}(\mathbf{r}, t) \rangle \langle \hat{\Psi}^{\dagger}(\mathbf{r}', t') \hat{\Psi}(\mathbf{r}', t') \rangle + |\langle \hat{\Psi}^{\dagger}(\mathbf{r}', t') \hat{\Psi}(\mathbf{r}, t) \rangle|^2 \\ &= n(\mathbf{r}, t) n(\mathbf{r}', t') + |G^{(1)}_{th}(\mathbf{r}, t; \mathbf{r}', t')|^2. \end{aligned} \tag{13.13}$$

[b] In this case the wavefunctions would of course be time independent.

which when normalized gives

$$g_{th}^{(2)}(\mathbf{r},t;\mathbf{r}',t') = 1 + |g_{th}^{(1)}(\mathbf{r},t;\mathbf{r}',t')|^2. \tag{13.14}$$

13.3.2. *The influence of the ground state population*

For a degenerate atomic cloud, the factorization we used to derive the expression of Eq.13.13 is no longer valid. The calculation of $\langle \hat{a}_{\mathbf{j}}^\dagger \hat{a}_{\mathbf{k}}^\dagger \hat{a}_{\mathbf{l}} \hat{a}_{\mathbf{n}} \rangle$ was done, in the spirit of Wick's theorem, assuming that the system can be adequately described within the grand canonical ensemble. This ensemble assumes the existence of a particle reservoir and is known to lead to unphysically large fluctuations of the condensate at very low temperatures.[38] However, in the thermodynamic limit this pathology disappears in the more realistic case when interatomic interactions are included[c]. Alternatively, the excessive fluctuations in the number of condensed atoms are also eliminated if the system is restricted to a finite number of non-interacting particles by employing the canonical ensemble.[40]

Based on calculations carried out by Politzer,[40] Naraschewski and Glauber[41] have proposed that one can maintain a grand canonical ensemble description if one adds a correction term by subtracting the contribution of the ground state in the *canonical ensemble* $\langle \hat{a}_{\mathbf{0}}^\dagger \hat{a}_{\mathbf{0}} \rangle^2 \delta_{\mathbf{k0}} \delta_{\mathbf{l0}} \delta_{\mathbf{m0}} \delta_{\mathbf{n0}}$. Then denoting the ground-state density by $n_0(\mathbf{r})$, the corrected second order correlation function[d] is,

$$G^{(2)}(\mathbf{r},\mathbf{r}') = n(\mathbf{r})n(\mathbf{r}') + |G^{(1)}(\mathbf{r},\mathbf{r}')|^2 - n_0(\mathbf{r})n_0(\mathbf{r}'). \tag{13.15}$$

For a non-degenerate cloud, the ground state density is negligible and the normalized correlation function $g^{(2)}(\mathbf{r},\mathbf{r}')$ is well described by Eq.13.14, decreasing from 2 to 1 as $|\mathbf{r}-\mathbf{r}'|$ increases to values well beyond the characteristic transverse coherence length. The opposite situation occurs for a pure BEC cloud at $T = 0$ where only the ground-state is occupied. Then, we have

$$|G^{(1)}(\mathbf{r},\mathbf{r}')|^2\Big|_{T=0} = n_0(\mathbf{r})n_0(\mathbf{r}') \text{ and also } g^{(2)}(\mathbf{r},\mathbf{r}')\Big|_{T=0} = 1,$$

independent of the spatial separation between $\mathbf{r}$ and $\mathbf{r}'$.

Systems with a correlation function given by Eq.13.15, are said to exhibit bunching at high temperature for separations smaller than the correlation length and no bunching in the condensed phase.

[c] A similar situation occurs to the amplitude of a single-mode laser, where the fluctuations are damped by gain saturation.[39]

[d] In Ref. 40 it is also shown that the largest deviation between descriptions employing the grand canonical and canonical ensembles is expected to occur near the BEC transition temperature. A yet to be done detailed experimental investigation of correlations in atomic clouds near $T = T_c$ could help establish the degree of accuracy of the prescription given in Eq.13.15.

13.3.3. *Correlation functions in the momentum space*

In almost all experiments, and particularly in the He* experiment, the correlation signal is not measured directly within the trapped atomic cloud. Rather the atoms are detected only after having been released from the trap and allowed to expand during a certain *time of flight*. The initial size of the the trapped cloud is relatively small, and for a sufficiently long time of flight, neglecting interatomic interactions, the atomic flux measured at the detector will reflect the initial momentum distribution of the atoms in the trapped just before release.

The results we have discussed above for the correlation functions in position space, all have analogs in momentum space. In fact the correlation functions in the two reciprocal spaces are closely related. For a trapped cloud at thermal equilibrium, the following relationships can be easily derived:

$$\int d\mathbf{p}\, G^{(1)}(\mathbf{p},\mathbf{p})e^{-i\mathbf{p}.\mathbf{r}/\hbar} = \int d\mathbf{R}\, G^{(1)}(\mathbf{R}-\mathbf{r/2},\mathbf{R}+\mathbf{r/2})$$

$$\int d\mathbf{r}\, G^{(1)}(\mathbf{r},\mathbf{r})e^{i\mathbf{q}.\mathbf{r}/\hbar} = \int d\mathbf{P}\, G^{(1)}(\mathbf{P}-\mathbf{q/2},\mathbf{P}+\mathbf{q/2})$$

These two equations express the fact that $i)$ the spatial correlation length is related to the width of the momentum distribution and, $ii)$ the momentum correlation length is related to the width of the spatial distribution, *i.e.* the size of the cloud.

No simple and equally general relationship holds for the second order correlation functions. This is because, close to the BEC transition temperature, the population density in the ground state is not negligible. Then the special contribution of the ground state, the last term in Eq.13.15, must be included and its contribution depends on the details of the confining potential. On the other hand, for an ideal gas far from the transition temperature one can neglect the ground state density, make the approximation that the correlation length is very short, neglect commutators such as $[\hat{\mathbf{r}},\hat{\mathbf{p}}]$, and then write the thermal density operator as $\hat{\rho} \propto \exp\left(-\beta\frac{\hat{\mathbf{p}}^2}{2M}\right)\, \exp\left[-\beta V(\hat{\mathbf{r}})\right]$, with $V(\mathbf{r})$ the trapping potential. These approximations lead to:

$$G^{(2)}(\mathbf{p},\mathbf{p}') = \rho_{eq}(\mathbf{p})\rho_{eq}(\mathbf{p}') + |G^{(1)}(\mathbf{p},\mathbf{p}')|^2$$

and

$$G^{(1)}(\mathbf{P}-\mathbf{q/2},\mathbf{P}+\mathbf{q/2}) \sim e^{-\beta\frac{\mathbf{P}^2}{2m}} \int d\mathbf{r}\; e^{-\beta V(\mathbf{r})} e^{-i\frac{\mathbf{q}.\mathbf{r}}{\hbar}}$$

One sees that in this limit, the interesting part of $G^{(2)}$ in momentum space is proportional to the square of the Fourier transform of the density distribution and independent of the mean momentum $\mathbf{P}$. This result is entirely equivalent to its optical analog, the *van Cittert-Zernike* theorem, referred to above in the discussion of the stellar *HBT* experiment.

Equivalently, we can write

$$G^{(1)}(\mathbf{R}-\mathbf{r/2},\mathbf{R}+\mathbf{r/2}) \sim e^{-\beta V(\mathbf{R})} \int d\mathbf{p}\, e^{-\beta\frac{\mathbf{p}^2}{2M}}\, e^{i\frac{\mathbf{r}.\mathbf{p}}{\hbar}}$$

$$G^{(1)}(\mathbf{R}-\mathbf{r/2},\mathbf{R}+\mathbf{r/2}) \sim e^{-\beta V(\mathbf{R})}\; e^{-\frac{|\mathbf{r}|^2 M}{2\beta\hbar^2}}$$

which implies that

$$|G^{(1)}(\mathbf{R}-\mathbf{r/2},\mathbf{R}+\mathbf{r/2})|^2 \sim\; e^{-(|\mathbf{r}|/l^{coh})^2},$$

where $l^{coh} = \lambda_T/\sqrt{2\pi}$ is the spatial coherence length with $\lambda_T = \sqrt{2\pi\hbar^2/Mk_BT}$ the *de Broglie* thermal wavelength and M the atomic mass. For temperatures well above the BEC transition temperature, we have $g^{(2)}(\mathbf{r},\mathbf{r}') = 1 + |g^{(1)}(\mathbf{r},\mathbf{r}')|^2$, thus under these conditions, l^{coh} is isotropic and independent of the specific form of the trapping potential.

This concludes our short introduction to the key ideas of coherence theory in quantum optics. We will now turn to briefly discuss some of the experimental issues connecting with the creation of ultra cold quantum gases.

13.4. General introduction to Bose-Einstein condensation. The He*experiment

Traditionally, textbooks present Bose-Einstein condensation by considering a homogeneous ideal gas for which the BEC critical transition takes place when the cloud's atomic density n and temperature T satisfy the relation,[42]

$$n \times \lambda_T^3 = \zeta\,(3/2) \equiv 2.612. \tag{13.16}$$

For low enough temperatures, the atoms' wave functions spread out sufficiently

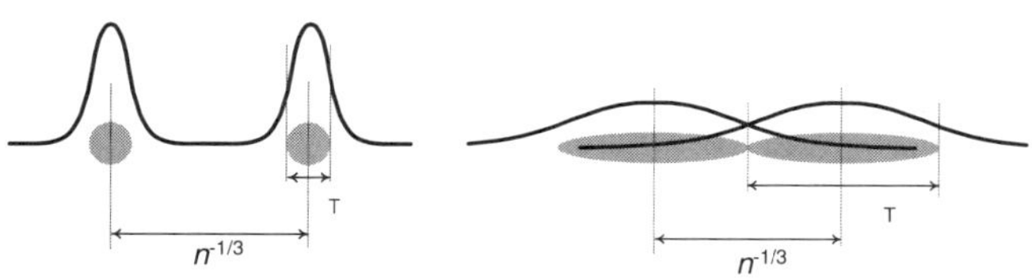

Fig. 13.3. An intuitive interpretation of Eq.13.16 for the relation between the critical density and the temperature (or else the phase space density) in a dilute atomic cloud. The phase transition takes place when the atoms start overlapping, which is for $\lambda_T \sim n^{-1/3}$, with λ_T the *de Broglie* thermal wavelength and n the cloud's density.

to overlap with their neighbors, becoming indiscernible from each other. This begins to happen at the critical transition temperature when the atoms's *de Broglie* thermal wavelength becomes comparable to the spatial separation between the particles, as sketched in Fig.13.3.

In practice, the atomic cloud has to be trapped and its density is, in general, inhomogeneous. In the case of a harmonically trapped cloud, the relation in Eq.13.16 is still valid if one replaces the homogeneous density by the cloud's *peak density*, $n(\mathbf{0})$, the density at the center of the trap. The factor $n(\mathbf{0}) \times \lambda_T^3$ plays the role of a phase space density. According to Eq.13.16, for attaining degeneracy, it must exceed $\zeta(3/2)$.

13.4.1. *Road map to attain Bose-Einstein condensation in dilute atomic gases*

To get to the point where it was possible to achieve BEC in a dilute atomic sample, many important new experimental techniques were developed over the last forty years. These may be divided into two main groups: i) optical trapping and cooling and ii) magnetic trapping and evaporative cooling.

The first laboratory demonstration of optical trapping of macroscopic objects dates from the beginning of the 70's[43e] and of neutral atoms in the early 80's.[44,45] The first realizations of optical cooling followed soon after[46,47] with the development of optical molasses and of the magneto-optical trap.

The basic idea behind cooling atoms by light can be understand by considering a fluorescence cycle.[48] When an atom absorbs or emits a photon both energy and momenta are conserved. All absorbed photons come from the laser beam (with well defined momenta) whereas the florescence photons are emitted spontaneously in random directions. As a result the atom is submitted to a net force in the direction of the laser beam, a *radiation pressure force*. The amplitude of this force depends on the detuning of the laser light relative to the linecenter of the atomic transition. This detuning in turn depends upon the atomic velocity though the Doppler effect. If two counter-propagating laser beams are detuned to energies lower than the atomic rest-frame transition frequency, the atoms will tend to absorb more photons from the laser beam that travels in a direction opposite to the atom's motion, leading to a effective cooling of the atomic cloud. This mechanism is referred as the *Doppler cooling* and is one of the various technique employed to cool atoms by laser.

Simultaneously, Metcalf *et al.* managed to magnetically trap sodium atoms after being optically cooled down[49] and also hydrogen by Hess *et al.*, precooling

[e]This leads to the development of a widespread experimental technique used in many different areas of science for the manipulation of small objects with light, known as optical tweezers.

the gas using cryogenic techniques.[50] The first realization of evaporative cooling was realized in the 80's, within the efforts to attain BEC in a spin polarized sample of hydrogen.[51]

However, BEC was only achieved in 1995 and with alkalis: rubidium (^{87}Rb),[52] sodium (^{23}Na)[53] and lithium (^{7}Li).[54] Still within the alkalis, today there are BEC experiments with potassium (^{41}K),[55] with another isotope of rubidium (^{85}Rb)[56] and also with cesium (^{133}Cs),[57] this one using an *optical dipole trap*.[58] Also using this type of trap, ytterbium (^{74}Yb)[59] and chromium (^{52}Cr)[60] have recently attained condensation. The pioneering atomic specie, hydrogen, was only condensed in 1998.[61]

13.4.2. *The metastable Helium Bose-Einstein condensate*

In 2001, the Helium isotope ^{4}He*also joined the group of condensed atomic species[62,63] [f]. It was the first atomic species to be condensed not in its electronic fundamental state but rather in the metastable electronic excited state 2^3S_1, with a life time of 9000 seconds and internal energy of 20 eV.

There are two important reasons to use metastable Helium (He*). Unlike the ground state atom, it has a closed optical transition to the excited triplet 2^3P_2 state that can be addressed with available laser sources, essential to optically trap and cool the sample[67,68] and also to use standard optical detection schemes as absorption, fluorescence and refractive measurements.[69] Secondly, the state 2^3S_1 has a permanent magnetic dipole moment, needed for the magnetic trapping.

Diagnostic tools unique to the He* experiment

The He* experiment allows the use of unique diagnostic tools. Due to the atoms's internal energy, which is sufficient to extract an electron from a metallic plate, it is possible to detect a falling atomic cloud with a micro-channel plate (MCP).[70] This device works as an electron multiplier and produces a signal proportional to the atomic flux that arrives at its sensitive surface (see Fig.13.4). The extremely good MCP time response and high gain allows single atom detection, which is very difficult to achieve in conventional BEC experiments based upon optical imaging. This made possible the measurement of the atom's correlations within a falling cloud, an experiment that is conceptually similar to the original *HBT* experiment of 1956[12] but realized with *massive particles* (cf. 13.7).

The He* metastability also leads to the possibility of the existence of ionizing collisions among He* atoms. However, the magnetic polarization of the sample

[f] He* condensation was also attained in Amsterdam[64] and, recently, also in Camberra.[65] The Amsterdam's group has also achieved degeneracy in the fermionic isotope ^{3}He using a two-color magneto-optical trap[66] and sympathetic cooling.

and the fact that these collisions don't conserve the total spin of the incoming atoms leads to their strong suppression[71] [g]. These processes are an extra benefit because their low rate is nevertheless easily detectable and proportional to the cloud's density. Ion detection is thus a new, *non-destructive* and *real-time* observation tool for studies of the BEC formation kinetics. We will come back to this issue latter on.

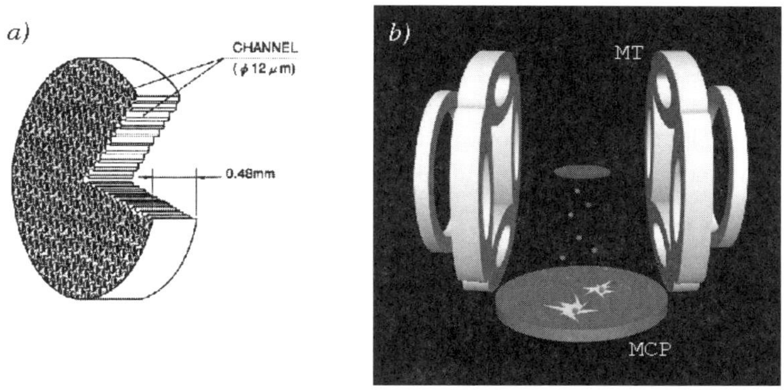

Fig. 13.4. (*a*) a *microchannel plate* (MCP): it is a thin sheet made of millions of very small electron multipliers, the microchannels, oriented parallel to one another in a honey comb structure. (*b*) *magnetic trap* (MT) in the *clover leaf* configuration and the atom's detection scheme used in the He* experiment.

13.4.3. *The magnetic trap and evaporative cooling*

The combination of a magnetic trap (MT) and evaporative cooling techniques constituted the final breakthrough in the achievement of *Bose-Einstein condensation* in dilute atomic samples. However, these techniques are just the final step involved in the production of a BEC.

The magneto-optical trap

To load a magnetic trap with atoms, one needs first to confine them into a cloud cold enough to be held by the shallow magnetic potential (typically, several millikelvin). This is achieved loading first a slow[68] and transversely cooled[72] atomic beam into a so called *magneto-optical trap* (MOT) [h].

The MOT is the first trap used in the experiment and, unlike the MT it can be

[g] This fact is of major importance since it allows one to create stable trapped clouds with long lifetimes and consequently permit one to carry out evaporative cooling.

[h] Some experiments, notably those made on a microchip, use a gas dispensers to charge the MOT instead of an atomic jet.

loaded from a fairly hot atomic source or even from an atomic jet. It is made of three pairs of red-detuned, circularly polarized and counter-propagating laser beams that are made to cross at right angles at the center of the trap, where a pair of anti-Helmoltz coils produces a non-homogeneous magnetic field. This field gradient produces a *Zeeman* shift in the magnetic sub-levels of the atomic transition used for optical manipulation. For atoms traveling outwards the trap's center, this shift is canceled out by the *Doppler* effect. This results in a radiation pressure towards the trap center, confining and cooling down the atoms into a small volume of space around the trap's center. Typically, the resulting trapped cloud is a few millimeters wide with a few hundred million atoms at a temperature of about 1 mK. This corresponds to a phase density still several orders of magnitude smaller than the one predicted by Eq.13.16 to attain the critical transition.

The magnetic trap

Neutral atoms can be magnetically trapped if they have a permanent magnetic dipole moment. This made alkalai atoms the ideal candidates for the initial experiments that used MTs. Immersed in a magnetic field, the magnetic dipole has a interaction energy given by $V(\mathbf{r}) = -\boldsymbol{\mu}\cdot\mathbf{B}(\mathbf{r})$. Classically, $\boldsymbol{\mu}$ processes around the magnetic field $\mathbf{B}$ at the *Larmor precession frequency*, normally much larger than the atom's oscillation frequency in their movement inside the trap. The dipoles follow *adiabatically* the magnetic field $\mathbf{B}$ and preserve, at all times, their initial magnetic spin polarization. Under these conditions, the trapping potential may be well approximated by $V(\mathbf{r}) = g_L\mu_B m|\mathbf{B}(\mathbf{r})|$, where m is the projection of the total angular momentum along the quantization axis parallel to $\mathbf{B}$, g_L is the Landé g-factor and μ_B is the Bohr magneton. The atoms are trapped in a local minimum of this potential, which as mentioned above, must be several times deeper than the cloud's thermal energy $\sim k_BT$.

In free space, *Maxwell*'s equations prevent the existence of a local maxima for magnetostatic fields.[73] Consequently, MTs rely on the creation of a local field minima, that confines the *low field seeking* atoms into the trap's center.

A simple way to produce a non-homogeneous magnetic field is using a *clover leaf* coils configuration[74] (see Fig.13.4−b), an alternative configuration of the *Ioffe-Pritchard* quadrupolar trap[75,76] used in the He* experiment. The trap's axial confinement is assured by a pair of coils in a slightly elongated Helmholtz configuration, while the transverse confinement is due to a quadrupolar magnetic field produced by four additional pairs of coils. An extra pair of Helmholtz coils (not shown in the Figure) compensates the axial field in such a way that the filed minimum is non-zero. This is essential to in avoid Majorana losses. With this compensation the resulting magnetic field in the region of its minimum is harmonic along the all three spatial dimensions.

In the He* experiment, the trap has a cylindrical symmetry (*cigar shaped*) with an axial and transversal oscillation frequencies of $\omega_x/2\pi \sim 50$ Hz and $\omega_y/2\pi = \omega_z/2\pi \sim 1150$ Hz, respectively.

Evaporative cooling

To further cool the atoms and attain the critical phase density, the magnetically confined atoms are subjected to forced evaporation. This process is sometimes pictured as the one by which hot coffee is cooled down by blowing on its surface. By evaporating some of the hottest water molecules, those that remain *trapped* inside the cup thermalize at a colder temperature. In the case of the atoms, the *blowing* is achieved by applying a radio-frequency (`rf`) oscillating magnetic field that couples out of the `MT` some of the atoms, by changing their spin state to a non trapping magnetic sub-level with, for instance, $m = 0$. This is accomplished by tuning the `rf`-field to the transition between this state and the trapping one, which are Zeeman split by the inhomogeneous trapping magnetic field. Since this splitting gets larger at the outer regions of the trap where only the hotter atoms have access, it is possible to eject them from the trap. The general idea of this process is depicted in Fig.13.5 and further explained in its caption.

Fig. 13.5. The evaporative cooling: (a) general scheme of two magnetic sublevels $m = 0$ and $m = +1$ in the presence of an inhomogeneous magnetic field. The trapping state corresponds to the $m = +1$. A rf-field tuned to the $m = +1 \rightarrow m = 0$ transition spin-flips some of the trapped atoms which are ejected from the trap. This transition energy depends on the Zeeman splitting and is higher in the outer regions of the trap potential where only atoms with high kinetic energy have access. Evaporative cooling is accomplished by slowly ramping the rf-field energy from very high energies when the cloud is still very far from degeneracy down to an energy a few times larger than the BEC chemical potential, where the system undergoes the phase transition; (b) Throughout evaporation the cloud's distribution function is continuously truncated at smaller and smaller energies, rethermalizing at lower temperatures. If the process is slow enough, the system evolves through quasi-equilibrium states.

The successive decrease of the cloud's temperature is accomplished by slowly ramping down the `rf`-frequency from a value that corresponds to the initial cloud's temperature to one comparable to the gas critical temperature. The cloud's continuous re-thermalization depends on a high rate of elastic collisions, characterized entirely at the low energy limit by the $s - wave$ scattering length, a. To evolve through almost thermal equilibrium states and loose the fewest number of atoms possible, evaporative cooling must proceed slowly. In spite of this, to decreasing the cloud's temperature three orders of magnitude to the microkelvin regime, only a very small fraction of the initial trapped atoms remain trapped at to the end of the process [i].

The sample's life time. The need of ultrahigh vacuum

For a spin polarized sample, the two and three body recombination processes (in the He* case, ionizing collisions) are unimportant for most of the evaporation cycle because the samples' density is small. However, there are always inelastic collisions of the trapped atoms with background molecules inside the science chamber that ultimately limits the sample's *life time* and thus determines how long the evaporation cycle may last. This explains why in this type of experiments it is important to have a very good vacuum. In the He* experiment, to get a life time of around one minute, the vacuum inside the science chamber must be smaller than 10^{-10} mbar (which corresponds to a free mean path of a few kilometers). Evaporation takes half a minute and we create a condensed cloud once 40 seconds.

The BEC growth and the critical phase transition in the He* experiment

When the cloud's density becomes close to the critical value for BEC formation (10^{12} cm^{-3} for He*), the two- and three-body collisions dominate the ion production. The observed ion signal is thus proportional to square and cube of the cloud's density [j]. Figure13.6 shows the evolution of the ion rate during the two last seconds of evaporation. In addition, it shows several *time of flight* (*TOF*) signals obtained by direct detection of the atoms using the `MCP` after being released from the trap. These *TOF* signals were obtained from different experimental runs; the `MT` was switched-off at the times indicated in the ion signal graph. A comparison of the *TOF* and ion data shows that the appearance of a narrow structure in the *TOF* sinal corresponds to an abrupt change in the slope of the ion signal. This slope change may be considered to be a signature of the phase transition.[80] At $t = 0$, in Fig.13.6, a pure condensate is formed, the ion signal is close to its

[i] As an example, in the He* experiment, the `MT` is loaded from the `MOT` with about 10^9 atoms, while the degenerate cloud has typically less than 10^6 atoms.

[j] For a quantitative analysis of the ion signal, knowledge of the two- and three-body rate constants is essential[78] as well as of the elastic scattering length.[79]

maximum value and the *TOF* takes the shape of the characteristic Thomas-Fermi inverted parabola.[42]

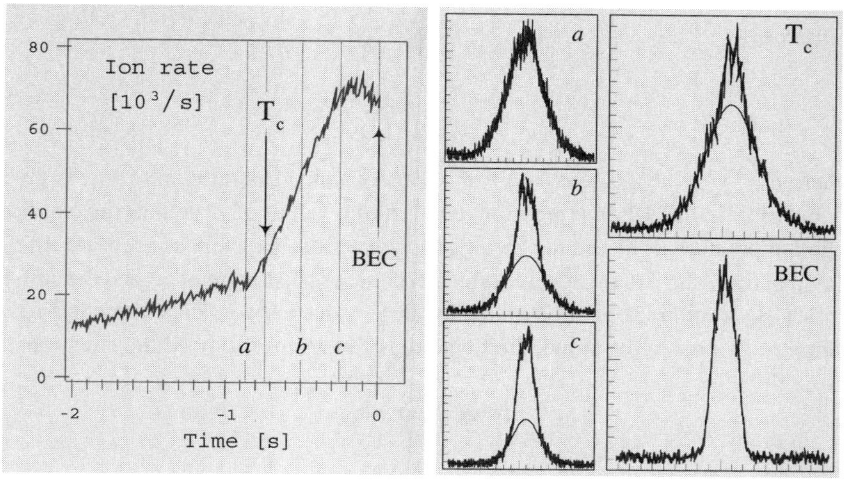

Fig. 13.6. The graph on the left shows a single-shot measurement of the ion flux when the *rf*-frequency ramps down from 1400kHz ($t = -2$ s) to 1000kHz ($t = 0$) at which point a pure BEC is formed. The five graphs on the right hand side are plots of typical *time of flight* signals when the atomic cloud is released from the trap at the instants of time referred in the ion signal curve (all of them come from different experimental runs). Two special cases correspond to a cloud at the critical phase transition and to the pure BEC. The *TOF* signals evolve from a gaussian type curve at higher temperatures, to a inverted parabola shape, which is the signature of the BEC within the Thomas-Fermi approximation.[42] In the curve at T_c as well as in the degenerate cases b and c, the temperature may be derived from the width of the gaussian fitted to the *TOF* tails while the condensate chemical potential may be inferred from a fit to the central structure. Because of the continuous loss of atoms due to the evaporation, the ion signal attains its maximum value before the pure BEC.[81]

13.5. An ideal and non degenerate atomic gas in a harmonic trap

13.5.1. *Atomic density in thermal equilibrium*

For clouds at very low temperatures, the magnetic trap potential may be well approximated by a harmonic oscillator (h.o.) potential [k], $V(\mathbf{r}) = \sum_\alpha \frac{1}{2} M\omega_\alpha^2 r_\alpha^2$, where ω_α are the oscillation frequencies of the trap and M is the atomic mass. In the ideal gas case, the total Hamiltonian is simply $\hat{H} = \frac{\hat{\mathbf{p}}^2}{2M} + V(\hat{\mathbf{r}})$, which has the discrete set of eigenfunctions

$$\psi_{\mathbf{m}}(\mathbf{r}) = \prod_\alpha A_{m_\alpha}\, e^{-r_\alpha^2/2\sigma_\alpha^2}\, H_{m_\alpha}(r_\alpha/\sigma_\alpha). \qquad (13.17)$$

[k]From here one we will use the Greek letter α to denote spatial coordinates.

with $A_{m_\alpha} = (\sqrt{\pi}\sigma_\alpha\ 2^{m_\alpha}\ m_\alpha!)^{-1/2}$ a normalization factor. The trap size is characterized, in each direction of space, by the length scale $\sigma_\alpha = (\hbar/M\omega_\alpha)^{1/2}$.

In thermal equilibrium, the density is given by Eq.13.11, which may also be written as

$$n(\mathbf{r}) = \langle \mathbf{r}|\hat{\rho}|\mathbf{r}\rangle = \sum_{\mathbf{m}=0}^{\infty} \psi_{\mathbf{m}}(\mathbf{r})\psi^*_{\mathbf{m}}(\mathbf{r})\ \langle n_{\mathbf{m}}\rangle \tag{13.18}$$

where $\hat{\rho} = \sum_{\mathbf{m}} |\psi_{\mathbf{m}}\rangle\langle\psi_{\mathbf{m}}|\ \langle n_{\mathbf{m}}\rangle$ is the density matrix operator and $\langle n_{\mathbf{m}}\rangle$ is given by Eq.13.9. In this latter expression, two limiting values of $\mathcal{Z}$ defines the classical high temperature limit and the zero temperature Bose-Einstein condensate. In the classical limit since many levels of the harmonic oscillator are occupied the sum in Eq.13.18 should be continued to include large values of $\mathbf{m}$. Since there is a fixed number of atoms in the cloud, the fugacity is constrained to fulfil the equation

$$N = \int d\mathbf{r}\, n(\mathbf{r}).$$

For a fixed value of N, the fugacity decreases for increasing values of the temperature and, in the limit of very high temperature, tends to zero. In the opposite limit as the temperature decreases to zero, only the lowest energy level is occupied and

$$N = N_0 = \frac{\mathcal{Z}}{\exp(\beta\varepsilon_{\mathbf{0}}) - \mathcal{Z}}$$

where $\varepsilon_{\mathbf{0}} = \frac{3}{2}\hbar\widetilde{\omega}$ is the *zero point energy*, with $\widetilde{\omega} = (\omega_x + \omega_y + \omega_z)/3$ the arithmetic mean of the trap frequencies. As N_0 must be positive, this equation defines a maximum value of the fugacity

$$\mathcal{Z}_{max} = \exp(-\beta\varepsilon_{\mathbf{0}})\,\frac{N_0}{N_0+1} \approx \exp(-\beta\varepsilon_{\mathbf{0}}), \tag{13.19}$$

where we have also assumed $N_0 \gg 1$. For this case, the maximum value of the chemical potential is simply the zero point energy $\mu = \varepsilon_{\mathbf{0}}$.

The temperature Green's function

There are several alternative strategies to evaluate the expression in Eq.13.18.[38,82] Here we will make use of the single particle h.o. Green's function, $\mathcal{G}^{ho}(\mathbf{r},\mathbf{r}',t) = \langle \mathbf{r}'|\exp(-i\hat{H}t/\hbar)|\mathbf{r}\rangle$,[83]

$$\begin{aligned}\mathcal{G}^{ho}(\mathbf{r},\mathbf{r}';t) &= \sum_{\mathbf{m}} \psi^*_{\mathbf{m}}(\mathbf{r}')\exp\left(-i\frac{\epsilon_{\mathbf{m}}}{\hbar}t\right)\psi_{\mathbf{m}}(\mathbf{r}) \\ &= \frac{(2\pi)^{-3/2}}{\overline{\sigma}^3}\prod_\alpha \sqrt{\frac{-i}{\sin(\omega_\alpha t)}} \times \exp\left[i\frac{(r_\alpha^2+r_\alpha'^2)\cos(\omega_\alpha t) - 2r_\alpha r'_\alpha}{2\sigma_\alpha^2\sin(\omega_\alpha t)}\right],\end{aligned} \tag{13.20}$$

where $\overline{\sigma} = \prod_\alpha \sigma_\alpha^{1/3}$ is the geometric mean trap size. Since we are interested in the particle distribution at thermal equilibrium we preform a Wick rotation and replace the (real) time in this expression by $-i\hbar\beta$, a purely imaginary quantity inversely proportional to the temperature. We obtain

$$\mathcal{G}^{ho}(\mathbf{r}, \mathbf{r}'; \beta) = \sum_{\mathbf{m}} \psi^*_{\mathbf{m}}(\mathbf{r}') \exp\left(-\beta\epsilon_{\mathbf{m}}\right)\psi_{\mathbf{m}}(\mathbf{r}), \tag{13.21}$$

which we may refer as a *temperature Green's function*.[82] Expressing Eq.13.9 as $\sum_{l=1}^{\infty} \mathcal{Z}^l \exp\left(l\beta\varepsilon_{\mathbf{m}}\right)$ we may write the first order correlation function of a Bose gas in thermal equilibrium trapped inside a harmonic potential as

$$G^{(1)}(\mathbf{r}, \mathbf{r}') = \langle \mathbf{r}|\hat{\rho}|\mathbf{r}'\rangle = \sum_{l=1}^{\infty} \mathcal{Z}^l \mathcal{G}^{ho}(\mathbf{r}, \mathbf{r}'; l\beta). \tag{13.22}$$

It is convenient to redefine the fugacity so that its maximum value is bounded by one. This can be done by the rescaling $\mathrm{Z} = \mathcal{Z} \exp\left(-\frac{3}{2}\beta\hbar\widetilde{\omega}\right)$. Then Eq.13.22 can be rewritten as

$$G^{(1)}(\mathbf{r}, \mathbf{r}') = \sum_{l=1}^{\infty} Z^l \mathcal{G}(\mathbf{r}, \mathbf{r}'; l\tau), \tag{13.23}$$

where we have used shorthand definition

$$\tau_\alpha = \beta\hbar\omega_\alpha. \tag{13.24}$$

Explicitly, the Green's function is given by

$$\mathcal{G}(\mathbf{r}, \mathbf{r}'; \tau) = \\ = \frac{1}{(\sqrt{\pi}\overline{\sigma})^3} \prod_\alpha \sqrt{\frac{1}{1-e^{-2\tau_\alpha}}} \exp\left[-\left(\frac{r_\alpha + r'_\alpha}{2\sigma_\alpha}\right)^2 \mathrm{th}\left(\frac{\tau_\alpha}{2}\right) - \left(\frac{r_\alpha - r'_\alpha}{2\sigma_\alpha}\right)^2 \mathrm{cth}\left(\frac{\tau_\alpha}{2}\right)\right], \tag{13.25}$$

This results in the cloud's density, being given by

$$n(\mathbf{r}) = G^{(1)}(\mathbf{r}, \mathbf{r}') = \frac{1}{\lambda_T^3} \sum_l Z^l \prod_\alpha \sqrt{\frac{2\tau_\alpha}{1-e^{-2l\tau_\alpha}}} \exp\left[-\left(\frac{r_\alpha}{\sigma_\alpha}\right)^2 \mathrm{th}\left(\frac{\tau_\alpha}{2}\right)\right]. \tag{13.26}$$

This expression reduces in the limit that $T \to 0$ to the ground state density [1]

$$n_0(\mathbf{r}) = \frac{Z}{1-Z}|\psi_0(\mathbf{r})|^2. \tag{13.27}$$

[1]Note that this is the BEC density for the ideal gas case only, where the gaussian density profile corresponds to the h.o. ground state. In a more realistic case, interatomic interactions can not be disregarded and the BEC density profile assumes the shape of an inverted parabola (see the inset graph BEC in Fig.13.6), well described by the Gross-Pitaevskii equation in the Thomas-Fermi approximation.[42]

In the high temperature limit, $\tau_\alpha \ll 1$ and only the first term of the sum will contribute significatively to the expression. In this limit the atomic cloud density simplifies to that given by the Boltzmann approximation

$$n(\mathbf{r}) = \frac{Z}{\lambda_T^3} \prod_\alpha \exp\left[-\frac{r_\alpha^2}{2s_\alpha^2}\right] = \frac{Z}{\lambda_T^3} \exp\left[-\beta V(\mathbf{r})\right], \tag{13.28}$$

where s_α is thermal cloud size, given in each spacial direction α by

$$s_\alpha = \frac{v_T}{\omega_\alpha} \tag{13.29}$$

with $v_T = \sqrt{k_B T/M}$ a convenient measure of the thermal velocity. The cloud's maximum density occurs at the center of the trap for $\mathbf{r} = 0$ and is equal to $Z\lambda_T^{-3}$.

The semi-classical expressions

An intermediate regime between the two limiting cases, corresponding to $T = 0$ of Eq.13.27 and the classical thermal cloud of Eq.13.28, may also be re-caste to a simple expression. If again we take the limit where $\tau_\alpha \ll 1$, we may simplify the square-root on the right hand side of Eq.13.26 to $l^{-1/2}$. Similarly, the argument of the exponential simplifies to $-\frac{1}{2}l\alpha^2/s_\alpha^2 = -l\beta V(r_\alpha)$ and the density may be written as

$$n(\mathbf{r}) = \frac{1}{\lambda_T^3} g_{3/2}\left(Z \exp\left[-\beta V(\mathbf{r})\right]\right) \tag{13.30}$$

where $g_{3/2}(x)$ is the polylogarithmic function [m] of order $3/2$, defined for an arbitrary order u as

$$g_u(x) = \sum_{l=1}^{\infty} \frac{x^l}{l^u}. \tag{13.31}$$

The approximation that leads to the expression in Eq.13.30 for the density is known as the *semi-classical approximation* and is used for computing simple analytical expressions of the cloud's density at temperatures close to the critical temperature. The evaluation of Eq.13.30 at the center of the cloud gives the critical peak density, which agrees with that of Eq.13.16 obtained using simple arguments.

There is no similar approximate semi-classical expression for the first order correlation function (*i.e.* for $\mathbf{r} \neq \mathbf{r}'$). To compute it we normally need to work-out the full expression of Eq.13.23. Moreover, as we will see in the following, even for the cloud's density the semi-classical expression works well only when it is integrated over at least one spatial dimension, as happens in typical time

[m] This function is usually referred to as the *Bose function* among physicists.

of flight detection schemes such as video image recording (one integration) or the `MCP` signal (two integrations). This expression leads to algebraic difficulties when it is used, for example, for defining the cloud's peak density at the critical temperature.

13.5.2. *Definition of the critical temperature of an ideal gas confined in a harmonic trap*

The *critical temperature* and condensed fraction: standard definitions

The phase transition critical temperature, T_c, is usually defined as the temperature at which the saturated excited states population is equal to the total number of atoms,[42] *i.e.*

$$\sum_{\mathbf{m}\neq 0} N_{\mathbf{m}}(Z=1,T_c) = N. \tag{13.32}$$

The total number of atoms, N, is given by the integral of Eq.13.26,

$$N = \sum_{l=1}^{\infty} Z^l \frac{1}{\prod_\alpha \left(1-e^{-l\tau_\alpha}\right)}. \tag{13.33}$$

Within the semi-classical approximation, this expression simplifies to

$$N \simeq \frac{1}{\overline{\tau}^3}\, g_3(Z) \tag{13.34}$$

and, neglecting the ground state population, the critical temperature and the fraction of condensed atoms become

$$T_c^0 = \frac{\hbar\overline{\omega}}{k_B}\left(\frac{N}{\zeta(3)}\right)^{1/3} \quad \text{and} \quad \frac{N_0}{N} = 1-\left(\frac{T}{T_c^0}\right)^3. \tag{13.35}$$

The *finite size effect*

The expressions in Eq.13.35 were obtained in the semi-classical approximation, which considers only the lowest order terms in τ in the series expansion of Eq.13.33. If the term in τ^{-2} is also included, Eq.13.34 is corrected to

$$N \simeq \frac{1}{\overline{\tau}^3}\left[g_3(Z) + \tfrac{3}{2}\tilde{\tau} g_2(Z)\right], \tag{13.36}$$

where the extra term is generally referred as the *finite size effect* correction. This correction[42] modifies the expressions in Eq.13.35, producing terms proportional to $N^{-1/3}$ which are negligible when N is very large, as is usually the case.

Despite being a good approximation for the total number of atoms, as we referred before, the semi-classical approximation fails if used to describe the density at the center of the trap when the fugacity is near one. This can be seen by writing the lowest order correction of Eq.13.30

$$n(\mathbf{r}) = \lambda_T^{-3}\left\{g_{3/2}\Big(Z\exp\left[-\beta V(\mathbf{r})\right]\Big) + \tfrac{3}{2}\tilde{\tau}\, g_{1/2}\Big(Z\exp\left[-\beta V(\mathbf{r})\right]\Big)\right\}. \quad (13.37)$$

The integral of this expression gives Eq.13.36, which is well defined for every value of Z. However, if Eq.13.37 is used for computing the peak density we find a divergence for $Z = 1$ ($g_{1/2}(x)$ diverges logarithmically at $x = 1$). A single spatial integration of this expression avoids the divergence.

The ground state contribution

The divergence contained in Eq.13.37 can obviously be avoided if the fugacity is limited to values strictly smaller than one. We show now that, with the definition for the critical temperature given in Eq.13.32, this is in fact always the case.

The semi-classical approximation simplifies $\mathrm{th}\left(\frac{1}{2}l\tau_\alpha\right)$ to $\frac{1}{2}l\tau_\alpha$, which is only appropriate for small enough values of l. For increasing l's, the hyperbolic tangent is limited by one, whereas $\frac{1}{2}l\tau_\alpha$ grows linearly with l. Examining Eq.13.26, we see that this approximation is equivalent to neglecting the ground state contribution (cf. Eq.13.27).

We may cure this defect by including, *by hand*, the ground state density in Eq.13.37. This changes the total number of atoms to

$$N = \overline{\tau}^{\,-3}\left[g_3(Z) + \tfrac{3}{2}\tilde{\tau}g_2(Z)\right] + \frac{Z}{1-Z}. \quad (13.38)$$

Since N is finite, this expression implies that Z should be strictly smaller than one.

The critical peak density

With the usual criteria for defining the critical temperature (cf. Eq.13.32), the inclusion of the ground state population modifies the critical fugacity from one to a smaller value we will denote by Z_c. If we disregard the finite size effects, this value is equal to[84]

$$Z_c \simeq 1 - \overline{\tau}^{\,3/2}\zeta(2)^{-1/2}, \quad (13.39)$$

which is different from one, under typical experimental conditions, by less than 1%. Nonetheless, this correction does change the degeneracy parameter at the critical phase transition and the habitual criteria of Eq.13.16 is modified to

$$\lambda_T^3 n(0)\Big|_{T=T_c} = \zeta(3/2) + 2\sqrt{2\zeta(2)} \approx 6.24. \quad (13.40)$$

This result shows that the contribution of the ground state density to the cloud's peak density, which is neglected in the *standard* expression in Eq.13.16, is in fact even larger than the density of the excited states. This effect is linked to the pathological behavior of the ground-state density in the thermodynamic limit, i.e. the infinite compressibility of an ideal gas.[3]

No equally simply algebraical result can be obtained if finite size effects are included. A numerical treatment of this may be found in Ref. 84.

13.5.3. *Second order correlation. The different regimes*

We now return to the discussion of the correlation functions, for the particular case of an ideal gas trapped in a harmonic potential.

Substituting the results of Eqs.13.23 and 13.25 in Eq.13.15 results in an explicit expression for the second order correlation function $G^{(2)}(\mathbf{r}, \mathbf{r}')$. As for in the case of the cloud's density, $G^{(2)}(\mathbf{r}, \mathbf{r}')$ has different behaviors depending on the cloud's temperature and whether it is i) far above, ii) in the vicinity but still above or, else, iii) below the phase transition critical temperature. We start by considering the simplest case, $T \gg T_c$.

***i*) The high-temperature limit**

In the high temperature limit, $Z \to 0$ and one recovers the Maxwell-Boltzmann distribution by keeping only the $l = 1$ term in the sum and keeping only leading terms in the expansion of all the factors in τ_α. Eq.13.22 simplifies to

$$G^{(1)}(\mathbf{r}, \mathbf{r}') = \frac{N\overline{\tau}}{\lambda_T^3} e^{-\sum_\alpha \frac{\tau_\alpha}{2}\left(\frac{r_\alpha + r'_\alpha}{2\sigma_\alpha}\right)^2} e^{-\pi\left(\frac{\mathbf{r}-\mathbf{r}'}{\lambda_T}\right)^2}, \tag{13.41}$$

with N given by Eq.13.34.

The first exponential factor on the right hand side of Eq.13.41 depends on $\frac{1}{2}(\mathbf{r} + \mathbf{r}')$ and clearly accounts for the cloud's density at that location (compare with Eq.13.30). This factor makes $G^{(1)}(\mathbf{r}, \mathbf{r}')$ tend to vanish if either $\mathbf{r}_\alpha$ or $\mathbf{r}'_\alpha$ becomes much larger than the cloud's size, $s_\alpha = \sigma_\alpha/\sqrt{\tau_\alpha}$. The second exponential is the *coherence term* as it depends on $|\mathbf{r} - \mathbf{r}'|$. The characteristic length is isotropic and proportional to λ_T, the thermal *de Broglie* wavelength. This results from the fact that, within this approximation the momentum distribution is also isotropic (cf. Eq.13.49). On the other hand, the normalized second order correlation function is, strictly speaking, *not* isotropic. However, if $\tau_\alpha \ll 1$ we obtain the simple formula

$$g^{(2)}(\delta\mathbf{r}) = 1 + \exp\left[-\left(\delta\mathbf{r}/l^{coh}\right)^2\right], \tag{13.42}$$

with $l^{coh} = \lambda_T/\sqrt{2\pi}$ the correlation length. The correlation function in Eq.13.42 presents bunching at $\delta\mathbf{r} = 0$ where $g^{(2)} = 2$, falling down to $g^{(2)} = 1$ when $\delta\mathbf{r} \gg l^{coh}$.

ii) **Quasi-degenerate case**

For temperatures close to but above the Bose-Einstein transition temperature, one has to keep the summation over the index l, in Eq.13.23. In this case, the terms with increasing values of l contribute more as the temperature decreases. It becomes clear from the expressions in Eqs.13.23 and 13.25 that the correlation length near the center of the trap ($r_\alpha, r'_\alpha \ll s_\alpha$) will increase and that the normalized correlation function is no longer Gaussian. Far from the center, only the terms with small values of l are important and the correlation function remains almost Gaussian. Thus, in general, close to degeneracy the correlation length is no longer a constant and becomes position-dependent.

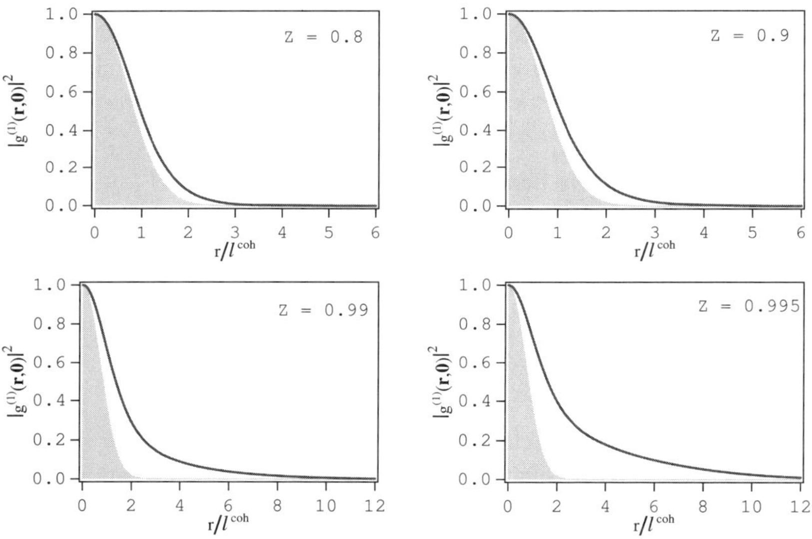

Fig. 13.7. Two-body normalized correlation function $g^{(1)}(\mathbf{r}, \mathbf{0})$ for an atomic cloud of bosons at $T = 1\ \mu$K and four different fugacities. The solid line represents the exact calculation using Eqs.13.23 and 13.25. Shaded regions correspond to the high temperature limit curve (cf. Eq.13.42), with $l^{coh} = \lambda_T/\sqrt{2\pi}$ the high temperature limit correlation length. The number of atoms in the examples shown ranges from $N \approx 125 \times 10^3$ for $Z = 0.8$ to $N \approx 170 \times 10^3$ for $Z = 0.995$.

In Fig.13.7 we trace some examples of the function $|g^{(1)}(\mathbf{r}, \mathbf{0})|$ for cloud fugacities from $Z = 0.8$, which is far from degeneracy, to $Z = 0.995$, where the temperature is close to the critical one. The graphs in this Figure also show, in the shaded region, the corresponding function in the high temperature limit. The departure of $|g^{(1)}(\mathbf{r}, \mathbf{0})|$ from a gaussian shape is already obvious for $Z = 0.9$ and, for fugacities closer to one, it presents a long tail signifying the build up of a long range correlation among the atoms within the cloud.

iii) **Degenerate case**

Increasing further the fugacity to $Z \lesssim 1$ leads to a saturation of the excited states

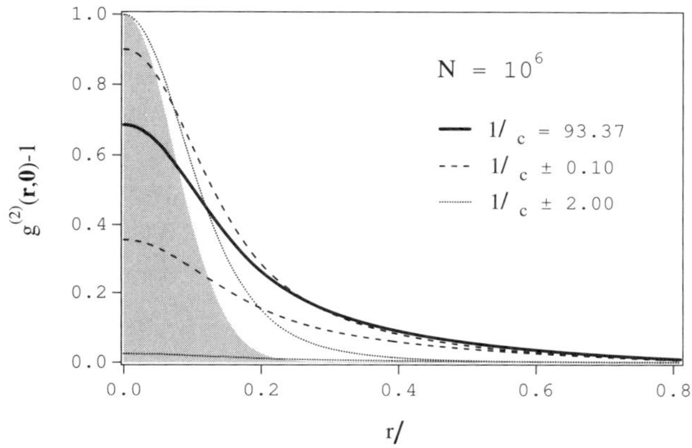

Fig. 13.8. Two-body normalized correlation function at the trap center, $g^{(2)}(\mathbf{r}, 0)$ for 10^6 atoms confined in a isotropic harmonic trap as function of the position $\mathbf{r}$ and for various temperatures around transition temperature. The horizontal axis has as units the size of the harmonic oscillator wave function σ. The thick solid line corresponds to the transition temperature $T_c = 93.37\ \hbar\omega/k_B$, with $N = 10^6$. The top dashed and dotted lines correspond to temperatures higher than T_c. The thermal de Broglie wavelength is $\sim 0.26\ \sigma$. The effect of the ground state population is clearly visible in the reduction of $g^{(2)}(0,0)$, and in the rapid flattening out of the correlation function slightly below T_c.

and a proper calculation of the second order correlation function must take into account the presence of a macroscopic population in the ground state. The second order correlation function is now given by Eq.13.15. Normalized it becomes

$$g^{(2)}(\mathbf{r}, \mathbf{r}') = 1 + \frac{|G^{(1)}(\mathbf{r}, \mathbf{r}')|^2 - n_0(\mathbf{r})n_0(\mathbf{r}')}{G^{(1)}(\mathbf{r}, \mathbf{r})G^{(1}(\mathbf{r}', \mathbf{r}')}. \tag{13.43}$$

where $n_0(\mathbf{r})$ is given by Eq.13.27. With $Z \sim 1$ and $T \sim 0$, τ_α goes to infinity and, in the expression of Eq.13.25, the hyperbolic functions tend to one and the square-root in the pre-factor to σ_α^{-1}. In this case, $G^{(1)}(\mathbf{r}, \mathbf{r}')$ factorizes as

$$G^{(1)}(\mathbf{r}, \mathbf{r}') \sim n_0(\mathbf{r})n_0(\mathbf{r}').$$

The expression in Eq.13.43 takes on its limiting value $g^{(2)}(\mathbf{r}, \mathbf{r}') = 1$ for any $\mathbf{r}$ and $\mathbf{r}'$ and the correlation length tends to infinity.

The behavior of $g^{(2)}(\mathbf{r}, \mathbf{r}')$ for cloud temperatures near $T = T_c$ is traced in Fig.13.8. The critical temperature is computed following the same criteria that led to Eq.13.40.[84] This Figure shows that, close to the BEC transition, the bunching is already significantly different from 2 near the center of the trap.

13.5.4. *Integrated signals*

As referred above, for temperatures close to the critical temperature, the particle's correlation function is different at different locations within the atomic cloud. To fully characterize the correlation function we would need to make *local* measurements of this function. However, from the experimental point of view, this is very difficult since the number of atoms in a cloud is quite small, resulting in a poor sinal to noise ratio. Such a local measurement would only be feasible by averaging over many different clouds with similar characteristics, namely equal temperatures and fugacities[n].

One way to avoid this problem is to average the correlation function over all locations within each cloud. This leads to an averaged second order correlation function, which we define as

$$g_m^{(2)}(\delta\mathbf{r}) = \frac{\int d\mathbf{R}\, G^{(2)}(\mathbf{R} - \frac{1}{2}\delta\mathbf{r}, \mathbf{R} + \frac{1}{2}\delta\mathbf{r})}{\int d\mathbf{R}\, n(\mathbf{R} - \frac{1}{2}\delta\mathbf{r}) n(\mathbf{R} + \frac{1}{2}\delta\mathbf{r})}, \tag{13.44}$$

a function that depends only in the relative distance $\delta\mathbf{r} = |\mathbf{r} - \mathbf{r}'|$.

Although it is easier to obtain experimentally, the averaged correlation function hides the quantum behavior of $g^{(2)}(\mathbf{r}, \mathbf{r}')$ close to quantum degeneracy, which is washed out by the integration. This can be seen in graph (a) of Fig.13.9, which plots $g_m^{(2)}(\delta\mathbf{r}) - 1$ for the same situations as in Fig.13.8. The decrease of bunching for smaller temperatures is much less pronounced than in those of Fig.13.8. Moreover, the curves of Fig.13.9(a) are more reminiscent of a gaussian, the simple behavior expected at the high temperature limit (cf. Fig.13.7).

A simple way of interpreting these curves is to consider that locally, at a given location $\mathbf{r}$, the *effective* chemical potential is equal to $\mu - V(\mathbf{r}_0)$. This corresponds to considering the gas as locally homogeneous and is known as the *local density approximation*.[41] Within this approximation, even when the cloud's fugacity is close to one, the off-center effective fugacity is always much smaller. The correlation function then approaches that of the thermal case.

The slow reduction of the bunching amplitude of $g_m^{(2)}(\delta\mathbf{r})$ is also a consequence of the integration in Eq.13.44. This amplitude is determined by the normalization which is highly influenced by the ground state occupation, which is a very localized state within a small region at the center of the cloud. In graph (b)

[n]This would be, however, a herculean task considering the number of necessary experimental runs to get a proper signal-to-noise ratio.[85]

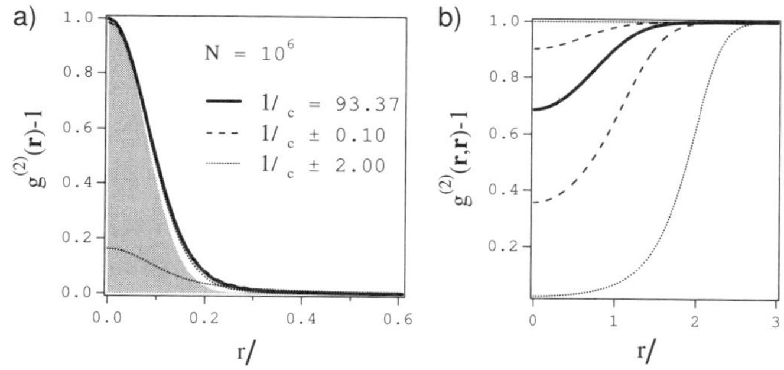

Fig. 13.9. (a) The normalized correlation function $g_m^{(2)}(\mathbf{r})$ for 10^6 atoms confined in an isotropic harmonic trap. The temperatures considered are the same as for Fig13.8. Unlike the graphs in Fig.13.8, the shape is always almost Gaussian. The transition to a flat correlation, at low temperatures, occurs less rapidly. (b) The corresponding non-integrated correlation function $g^{(2)}(\mathbf{r},\mathbf{r})$. Due to the finite spatial extent of the condensate, even for $T < T_c$ the correlation approaches 2 far from the center. This can be understood in terms of the chemical potential $\mu(r)$ which, in a local density approximation, decreases as r increases and thus the correlation is equivalent to that of a hotter cloud.

of Fig.13.9, we represent the *bunching amplitude* $g^{(2)}(\mathbf{r},\mathbf{r}) - 1$ as a function of $\mathbf{r}$. This Figure shows that the local correlation function at distances far from the center can be equal to two, even for clouds with $T < T_c$.

13.6. The atomic *time of flight*

A trapped atomic cloud is, in general, to small to be imaged *in situ*. For example, a He* cloud at 1 μK (close to critical temperature) trapped in a harmonic potential with an oscillation frequency of 500 Hz (typical value for most cold atom experiments) has a size of around 15 μm. Similar clouds made of ^{23}Na or ^{87}Rb would have half and one fifth of this value respectively. These values are close to the diffraction limit for optical imaging and are much smaller than the resolution of the detector used in the He* experiment. To be able to properly image the cloud, it is first necessary to release it from the trap and let it expand for a while [o].

13.6.0.1. *Time evolution of a h.o. wave function in free fall*

The ideal gas cloud's expansion can be derived by computing the time evolution, under the influence of gravity, of each h.o. wavefunction. After a certain period

[o]Ref. 21 reports one of the few exceptions to this.

of time, the cloud's density distribution can be worked out by just averaging over all the wavefunctions, with the statistical weight corresponding to the thermal equilibrium inside the trap. The time evolution of each wavefunction can be easily described using the appropriate Green's function.

The free fall Green's function

The time evolution of the wave function of a particle of mass M falling in the gravitational field $-Mgz$ can be workout using the Green's function [p],

$$K(\mathbf{r},t;\mathbf{r}_0)=\left(\tfrac{M}{2\pi i\hbar t}\right)^{3/2}\exp\left\{\tfrac{iM}{2\hbar t}\left[(\mathbf{r}-\mathbf{r}_0)^2+2(z+z_0)\eta(t)-\tfrac{1}{3}\left(\tfrac{1}{2}gt^2\right)\right]\right\}.$$

After a certain fall time t, the wavefunction $\psi_{\mathbf{m}}(\mathbf{r}_0)$ describing the particles of the h.o. level $\mathbf{m}$ (cf. Eqs.13.17) evolves to [q]

$$\begin{aligned}\psi_{\mathbf{m}}(\mathbf{r},t) &= \int d\mathbf{r}_0\; K(\mathbf{r},t;\mathbf{r}_0)\;\psi_{\mathbf{m}}(\mathbf{r}_0)\\ &= \prod_\alpha i^{m_\alpha}\frac{\exp\left[i(m_\alpha\delta_\alpha+\phi_\alpha)\right]}{\sqrt{\omega_\alpha t-i}}\;\psi_{m_\alpha}(\tilde{r}_\alpha)\end{aligned} \qquad (13.45)$$

with $\delta_\alpha=\arctan(1/\omega_\alpha t)$. Here $\tilde{r}_\alpha$ represents re-scaled coordinates, defined assuming that the detector is located at a distance H below the center of the trap, as

$$\tilde{x}=\frac{x}{\sqrt{1+\omega_x^2t^2}}\;,\quad \tilde{y}=\frac{y}{\sqrt{1+\omega_y^2t^2}}\quad\text{and}\quad \tilde{z}=\frac{H-\frac{1}{2}gt^2}{\sqrt{1+\omega_z^2t^2}} \qquad (13.46)$$

with $\phi_\alpha\equiv\phi_\alpha(t)$ a global phase that depends on α and t but not on the index m_α. This phase disappears in the calculation of the atomic flux or in the particles' correlation function. However, it gives rise to fringes in the first order correlation function.

The generalization of Eq.13.18 for the time dependent case is given for a non-interacting gas by

$$n(\mathbf{r},t)=\sum_{\mathbf{m}=\mathbf{0}}^{\infty}\psi_{\mathbf{m}}(\mathbf{r},t)\psi^*_{\mathbf{m}}(\mathbf{r},t)\;\langle n_{\mathbf{m}}\rangle.$$

Substituting Eq.13.45 we obtain

$$n(\mathbf{r},t_0)=\frac{n(\tilde{\mathbf{r}})}{\prod_\alpha\sqrt{1+\omega_\alpha^2t_0^2}}, \qquad (13.47)$$

with $t_0=\sqrt{2H/g}$ the mean fall time.

[p]This Green's function is given by $K(z,t;z_0)=\int\varphi^*_E(z_0)\varphi_E(z)\exp(-iEt/\hbar)\,dE$ where φ_E is the solution of the time independent Schrödinger equation $[-(\hbar^2/2M)(d^2/dz^2)-Mgz]\varphi_E(z)=E\varphi_E(z)$.

[q]Note that,[86] $\int e^{-(x-y)^2}H_n(u\,x)dx=\sqrt{\pi}(1-u^2)^{n/2}H_n[u\,y/\sqrt{1-u^2}]$.

Ballistic expansion. The *far field* approximation

If $\omega_\alpha t_0 \gg 1$, $1+\omega_\alpha^2 t_0^2$ may be replaced by $\omega_\alpha^2 t_0^2$ in all the previous expressions. This approximation, which we shall call the *far field approximation* assumes that t_0 is much larger than the period of one oscillation inside the trap. In most experiments and, particularly in the He* one, its an excellent approximation (in this latter experiment, $\omega t \sim 125$ in the axial axis and more than 3000 in the radial one). Disregarding gravity, we can obtain an expression for the cloud density under ballistic expansion after being released from the trap. Invoking the far field and the semi-classical approximations, the atomic density given by Eq.13.47 becomes

$$n(\mathbf{r},t_0) = (\sqrt{2\pi}\overline{\tau} v_T t_0))^{-3} g_{3/2}\left[Z\exp\left(-\tfrac{\mathbf{r}^2}{(v_T t_0)^2}\right)\right]. \qquad (13.48)$$

In the semi-classical approximation, the momentum distribution of the trapped cloud is [r],

$$\tilde{n}(\mathbf{p}) = \left(\sqrt{2\pi}\overline{\tau}(M v_T)\right)^3 g_{3/2}\left[Z\exp\left(-\tfrac{p_\alpha^2}{2(M v_T)^2}\right)\right]. \qquad (13.49)$$

Thus, $n(\mathbf{r},t_0) = (M/t_0)^3\tilde{n}(\mathbf{p} = M\mathbf{r}/t_0)$, the atomic density after expansion is just the initial momentum distribution of the cloud, which expands isotropically and linearly in time. Its size, at any time t, is given by

$$s_\alpha(t) = s_\alpha(t=0)\omega_\alpha t = v_T t. \qquad (13.50)$$

13.6.1. *Quantum mechanical flux*

For interpreting the results obtained in the He* experiment we need to compute the atomic flux passing through the MCP. Its quantum mechanical definition is

$$\hat{I}(\mathbf{r},t) = \frac{\hbar}{M}\mathbf{Im}\left[\hat{\Psi}^\dagger(\mathbf{r},t)\frac{\partial}{\partial_z}\hat{\Psi}(\mathbf{r},t)\right]$$

where $\mathbf{Im}(\cdot)$ stands for the imaginary part of the expression. If we assume that there is no scattering of particles between different quantum states, we can substitute the wavefunctions of Eq.13.45 into the expression 1.7 to arrive at

$$\left\langle \hat{I}(\mathbf{r},t)\right\rangle = \frac{i\hbar}{2M}\sum_{\mathbf{m}}\left\{[\tfrac{\partial}{\partial z}\psi_{\mathbf{m}}(\mathbf{r},t)]^*\psi_{\mathbf{m}}(\mathbf{r},t) - c.c.\right\}\langle\hat{n}_{\mathbf{m}}\rangle. \qquad (13.51)$$

Using the identity $\partial_z H_n(z) = 2nH_{n-1}$, the partial spatial derivative of the wave function can be carried out

$$\frac{\partial}{\partial_z}\psi_{\mathbf{m}}(\mathbf{r},t) = \frac{M}{\hbar}\left\{[-v_1 + iv_2]\psi_{m_z}(z,t) - i\, v_3\, e^{i\delta_z}\sqrt{m_z}\,\psi_{m_z-1}(z,t)\right\}\times$$
$$\times\psi_{m_x}(x,t)\psi_{m_y}(y,t), \qquad (13.52)$$

[r]This quantity is just $\tilde{n}(\mathbf{p}) = \sum_{l=1}^{\infty} Z^l \iint d\mathbf{r}\, d\mathbf{r}'\, \langle\mathbf{p}|\mathbf{r}\rangle\mathcal{G}(\mathbf{r},\mathbf{r}';l\tau)\langle\mathbf{r}'|\mathbf{p}'\rangle$.

where the *velocities* v_1, v_2 and v_3 are given by

$$v_1 = \omega_z \frac{H - \frac{1}{2}gt^2}{1+\omega_z^2 t^2},\ v_2 = \frac{1}{t}\left[H + \frac{1}{2}gt^2 - \frac{H-\frac{1}{2}gt^2}{1+\omega_z^2 t^2}\right] \text{ and } v_3 = \frac{\sqrt{2}\omega_z \sigma_z}{\sqrt{1+\omega_z^2 t^2}}. \quad (13.53)$$

Substituting this expression in Eq.13.51, we see that the term depending on v_1 disappears when subtracted with its complex conjugate. Moreover, due to the fact that the h.o. wavefunctions are real, the terms containing v_3 also cancel out. The flux expression simplifies to [s]

$$\langle \hat{I}(\mathbf{r},t)\rangle = \frac{v_2}{\prod_\alpha (1+\omega_\alpha^2 t^2)^{1/2}}\, n(\tilde{\mathbf{r}}), \quad (13.54)$$

where $n(\tilde{\mathbf{r}})$ is the cloud's density in thermal equilibrium for the re-scaled coordinates defined in Eqs.13.46. During the expansion/fall, the atomic cloud maintains its original density distribution but in rescaled coordinates. The atomic flux is proportional to a certain velocity v_2 (which is, as we show later, just the classical velocity of the center of mass). Also, there is an additional pre-factor $\prod_\alpha (1+\omega_\alpha^2 t^2)^{-1/2}$, which accounts for the overall decrease in the cloud's density due to its expansion.

The *long fall approximation*

The expressions in Eqs13.47 and 13.54 can be further simplified assuming that the fall time t_0 is much larger than the cloud's temporal extension t_{cl} as it passes through the plane of the detector. This latter quantity is approximatively given by $s_\alpha(t_0)/gt \simeq v_T/g$. We may define a *long fall* whenever $t_0 \gg t_{cl}$ or, equivalently, when $v_T \ll v_G$, with $v_G = gt_0$ the final velocity of the cloud due to the gravitational acceleration. For most experiments this condition is very well respected. In the He* experiment, for example, $v_G \sim 1$ m/s ($t_0 \sim 0.1$ s), whereas for a cloud with $T \sim 3\ \mu$K, the thermal velocity is $v_T \sim 0.08$ m/s.

Using this *long fall approximation*, v_2 in Eq.13.53 simplifies to v_G and, the semi-classical expression for the atomic flux is

$$I(x,y,\delta t) \sim \frac{v_G}{(\overline{\omega}t)^3}\frac{1}{\lambda_T^3}\, g_{3/2}\Big[Z \exp\Big(-\frac{x^2+y^2}{2(v_T t)^2}\Big)\exp\Big(-\frac{g^2}{2v_T^2}(t-t_0)^2\Big)\Big].$$

If the atoms are detected by a `MCP` with an transversal area larger than the cloud itself, the *tof* signal is obtained from a simple integration of the previous expression over the xOy plane, yielding

$$I(t) \sim \frac{\overline{T}^{-2}}{\overline{\omega}t_0}\frac{v_G}{\lambda_T}\, g_{5/2}\Big[Z \exp\Big(-\frac{(t-t_0)^2}{t_{cl}^2}\Big)\Big]. \quad (13.55)$$

[s]Assuming an ideal gas, this is an exact result.

This expression doesn't take into account the finite size effect term (the second term in the right hand side of Eq.13.37). Its inclusion is nonetheless trivial, resulting in an extra $g_{3/2}(\cdot)$ term in the expression of Eq.13.55. Such an expression, also corrected for fact that the MCP is finite, was used to fit the curves shown in Fig.13.10.

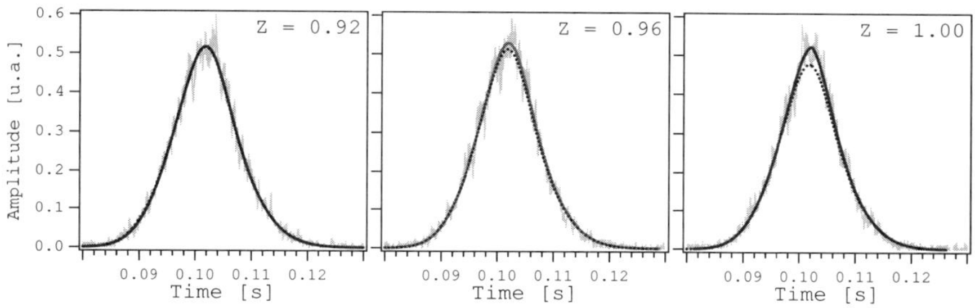

Fig. 13.10. Three *tof* signals (light curve) with similar temperatures but different fugacities fitted with expressions corresponding to the semi-classical (solid line) and the high temperature limit (dotted line) models.

13.6.2. *Intensity-intensity correlation function of a expanded cloud*

We extend now the results of the previous section to calculate the atomic flux correlation function. We are interested in computing the function

$$G^{(2)}_{fl.}(\mathbf{r},t;\mathbf{r}',t') = \langle \hat{I}(\mathbf{r},t)\hat{I}(\mathbf{r}',t')\rangle$$

where $\mathbf{r} = \{x, y, z = H\}$, $\mathbf{r}' = \{x', y', z' = H\}$ and $\hat{I}$ is the flux operator defined above in Eq.13.51. This quantity is the second order correlation function if the shot-noise term is neglected. Our aim is to show that, within the far field and the long fall approximations, the normalized version of this correlation function is equivalent to the one for the trapped gas (cf. Eq.13.43) with the coordinates rescaled according to Eqs.13.46. In this case the measurements obtained in the He* experiment can be directly related to the correlation functions of the atoms within the trap before being released. Using the shorthand definitions $\psi_{\mathbf{j}}(\mathbf{r}) \equiv \psi_{\mathbf{j}}$, $\psi_{\mathbf{j}}(\mathbf{r}') \equiv \psi'_{\mathbf{j}}$ and $\partial z \equiv \frac{\partial}{\partial z}$, the above expression may be written as

$$\langle \hat{I}(\mathbf{r},t)\hat{I}(\mathbf{r},t')\rangle = -\left(\tfrac{\hbar}{2M}\right)^2 \sum_{\mathbf{j},\mathbf{k},\mathbf{l},\mathbf{m}} [\psi^*_{\mathbf{j}}(\partial_z\psi_{\mathbf{k}}) - (\partial_z\psi^*_{\mathbf{j}})\psi_{\mathbf{k}}] \times$$
$$\times [\psi'^*_{\mathbf{l}}(\partial_z\psi'_{\mathbf{m}}) - (\partial_z\psi'^*_{\mathbf{l}})\psi'_{\mathbf{m}}]\, \langle \hat{a}^\dagger_{\mathbf{j}}\hat{a}_{\mathbf{k}}\hat{a}^\dagger_{\mathbf{l}}\hat{a}_{\mathbf{m}}\rangle.$$

Neglecting for now the ground-state contribution, this expression further simplifies to

$$\langle \hat{I}(\mathbf{r},t)\hat{I}(\mathbf{r}',t')\rangle = \langle \hat{I}(\mathbf{r},t)\rangle\langle \hat{I}(\mathbf{r}',t')\rangle + Re(A) \tag{13.56}$$

where

$$A \equiv \sum_{\mathbf{j},\mathbf{l}} \left\{ \begin{array}{c} v_2 v_2' \, \psi_{\mathbf{j}}^* \psi_{\mathbf{j}}' \psi_{\mathbf{l}} \psi'^*_{\mathbf{l}} \\ +\frac{1}{2} v_3 v_3' \sqrt{j_z l_z} \, \psi_{\mathbf{j}}^* \psi'_{\mathbf{j}-\mathbf{1_z}} \psi_{\mathbf{l}-\mathbf{1_z}} \psi'^*_{\mathbf{l}} \\ +\frac{1}{2} v_3 v'^*_3 \, l_z \, \psi_{\mathbf{j}}^* \psi_{\mathbf{j}}' \psi_{\mathbf{l}-\mathbf{1_z}} \psi'^*_{\mathbf{l}-\mathbf{1_z}} \\ -v_2 v_3' \sqrt{j_z} \, \psi_{\mathbf{j}}^* \psi'_{\mathbf{j}-\mathbf{1_z}} \psi_{\mathbf{l}} \psi'^*_{\mathbf{l}} \\ -v_2' v_3 \sqrt{l_z} \, \psi_{\mathbf{j}}^* \psi_{\mathbf{j}}' \psi_{\mathbf{l}-\mathbf{1_z}} \psi'^*_{\mathbf{l}} \end{array} \right\} \times \langle \hat{a}_{\mathbf{j}}^\dagger \hat{a}_{\mathbf{j}} \rangle \langle \hat{a}_{\mathbf{l}}^\dagger \hat{a}_{\mathbf{l}} \rangle. \tag{13.57}$$

with $\mathbf{j} - \mathbf{1_z}$ standing for the vector $(j_x, j_y, j_z - 1)$.

Two major differences appear compared to the mean flux calculation: in this latter expression, the terms in v_3, which comes from the derivative of Eq.13.52 and the phase factor [t] $\delta_\alpha + 3\pi/2$ in Eq.(13.45) don't cancel, making the exact calculation non trivial. However, if the conditions of validity of the far field and long fall approximations are fulfilled, we see that $v_3/v_2 = 2^{-1/2}\sigma_z/H \ll 1$, and only the the first term of Eq.13.57 has a non negligible contribution to the correlation function. We have shown elsewhere[85] that keeping only this term leads to an expression for $\langle \hat{I}(\mathbf{r},t)\hat{I}(\mathbf{r}',t')\rangle$ which is sufficiently accurate to interpret the experimental data we obtain within the He* experiment [u].

Explicit calculation of the flux correlation function within the far field and long fall approximations

In the following we will retain only the first term of Eq.13.57. This one is

$$\begin{aligned} A &\simeq v_2 v_2' \sum_{\mathbf{j},\mathbf{l}} \prod_{\alpha} \frac{e^{i(j_\alpha - l_\alpha)(\delta'_\alpha - \delta_\alpha)}}{\omega_\alpha^2 t t'} \, \tilde{\psi}_{\mathbf{j}} \tilde{\psi}'_{\mathbf{j}} \tilde{\psi}_{\mathbf{l}} \tilde{\psi}'_{\mathbf{l}} \, \langle \hat{a}_{\mathbf{j}}^\dagger \hat{a}_{\mathbf{j}} \rangle \langle \hat{a}_{\mathbf{l}}^\dagger \hat{a}_{\mathbf{l}} \rangle \\ &= \frac{v_2 v_2'}{(\overline{\omega}^2 t t')^3} \left| \sum_{\mathbf{j}} \tilde{\psi}_{\mathbf{j}} \tilde{\psi}'_{\mathbf{j}} \, e^{i\mathbf{j}.\boldsymbol{\Delta}} \langle \hat{a}_{\mathbf{j}}^\dagger \hat{a}_{\mathbf{j}} \rangle \right|^2 , \end{aligned} \tag{13.58}$$

where $\tilde{\psi}_{\mathbf{m}} \equiv \psi_{\mathbf{m}}(\tilde{\mathbf{r}})$ is the harmonic oscillator wave function with re-scaled coordinates and also, in the last line we have used the definitions $\Delta_\alpha = \delta'_\alpha - \delta_\alpha$ and $\sum_\alpha j_\alpha(\delta'_\alpha - \delta_\alpha) = \mathbf{j}.\boldsymbol{\Delta}$.

[t] Recall that $\delta_\alpha = \arctan(1/\omega_\alpha t)$.

[u] We note that although algebraically nontrivial, these calculations could be carried out numerically from Eq.13.56 without any approximation. This would be necessary to interpret data of experiments where, the fall time is small as for example the one of Ref. 17.

The term inside the modulus squared in the second line of Eq.13.58 is just the modified version of the first order correlation function with an extra phase factor. It can be written down using the temperature Green's function as

$$\sum_{\mathbf{j}} \psi_{\mathbf{j}} \psi'_{\mathbf{j}} e^{i\mathbf{j}.\boldsymbol{\Delta}} \langle \hat{a}^{\dagger}_{\mathbf{j}} \hat{a}_{\mathbf{j}} \rangle = \sum_{l} Z^{l} \, G^{ho}(\tilde{\mathbf{r}}, \tilde{\mathbf{r}}', \{l\tau_{\alpha} - i\Delta_{\alpha}\}), \tag{13.59}$$

where the *temperature* parameter is now a complex number.

Further simplification

Again, an exact evaluation of the expression in Eq.13.59 can only be obtained numerically. Fortunately, a subsequent use of the far field and long fall approximations permit us to obtain an analytical result. In fact, approximating $\delta_{\alpha} \approx 1/\omega_{\alpha} t$ and $\delta'_{\alpha} \approx 1/\omega_{\alpha} t'$ we may write

$$\Delta_{\alpha} \simeq \frac{1}{\omega t'} - \frac{1}{\omega t} \simeq \frac{1}{\omega_{\alpha} t_0} \frac{\delta t}{t_0},$$

with also $\delta t = t - t'$. This quantity is of the order of the time of coherence t^{coh} which, as we will shown in the following, is equal to

$$t^{coh} = \frac{l^{coh} \omega_z}{g}. \tag{13.60}$$

with ω_z the trap oscillation frequency in the Oz fall direction and $l^{coh} = \lambda_T / \sqrt{2\pi}$. Using the last two expressions we may conclude that,

$$\frac{\Delta_{\alpha}}{\tau_{\alpha}} \lesssim \frac{\sigma_{\alpha}}{H} \ll 1.$$

For most experiments this ratio is vanishingly small (in the He* experiment it is smaller than 10^{-5}) and the phase term proportional to Δ_{α} in Eq.13.59 can be neglected.

The normalized intensity correlation function

Adding the ground state contribution and after invoking the far field and long fall approximations, we obtain for the intensity correlation function the expression

$$\langle I(\mathbf{r}, t) \rangle \langle I(\mathbf{r}', t') \rangle = \frac{v_2 v'_2}{(\overline{\omega}^2 t t')^3} \left[n(\tilde{\mathbf{r}}) n(\tilde{\mathbf{r}}') + |G^{(1)}(\tilde{\mathbf{r}}, \tilde{\mathbf{r}}')|^2 - n_0(\tilde{\mathbf{r}}) n_0(\tilde{\mathbf{r}}') \right]. \tag{13.61}$$

This is essentially the same correlation function that describes the trapped atomic cloud but with the coordinates rescaled and also a scaling pre-factor that reflects the expansion of the cloud between the times t and t'. The corresponding normalized correlation function is

$$g^{(2)}(\mathbf{r},t;\mathbf{r}',t') = \frac{\langle I(\mathbf{r},t;\mathbf{r}',t')\rangle}{\langle I(\mathbf{r},t)\rangle\langle I(\mathbf{r}',t')\rangle} = 1 + \frac{|G^{(1)}(\tilde{\mathbf{r}},\tilde{\mathbf{r}}')|^2 - n_0(\tilde{\mathbf{r}})n_0(\tilde{\mathbf{r}}')}{n(\tilde{\mathbf{r}})n(\tilde{\mathbf{r}}')}, \quad (13.62)$$

which, as we have claimed earlier, is equivalent to the expression of Eq.13.43 for the density correlation function of the trapped gas. Thus, if the approximations we used to derive the Eq.13.62 are valid, an intensity correlation measurement of an ideal gas falling under the effect of the gravity gives the same information as a local measure of the particles' density correlation within the trap.

The cloud's temporal coherence at the detector. The coherence volume

In the high temperature limit, the volume of coherence, within which the atoms are correlated, can be easily derived from Eq.13.62 using Eqs.13.41 and 13.46. It takes the familiar form of Eq.13.42,

$$g^{(2)}(\delta\tilde{\mathbf{r}}) = 1 + \exp\left[-\delta\tilde{\mathbf{r}}^2/\mathbf{x}^{coh}(t)^2\right], \quad (13.63)$$

where the coherence volume is defined through $\mathbf{x}^{coh}(t)$ which, along each axis, is equal to

$$x_\alpha^{coh}(t) = l^{coh}\omega_\alpha t = \tau_\alpha s_\alpha(t), \quad (13.64)$$

with $s_\alpha(t)$ the expanding cloud's size along the corresponding axis (cf. Eq.13.50). Thus, the ratio between the coherence and the cloud volumes remains constant over time being equal to $\tau_\alpha \equiv \hbar\omega_\alpha/k_BT$.

The cloud's temporal coherence seen by the detector is just x_z^{coh}/gt_0, the time independent quantity given before in Eq.13.60. The cloud's coherence time remains independent of the propagation time as long as the far field and the long fall approximations are valid.

Final remarks on the intensity correlation function on an expanding atomic cloud

The correlation length increases linearly with the time of flight. A simple way to understand this result is to consider the analogy with optical speckle. Increasing the time of flight corresponds to increasing the propagation distance to the observation plane in the optical analog. The speckle size, *i.e.* the correlation length, obviously increases linearly with the propagation distance. Another way to understand the time dependence is to notice that after release, the atomic cloud is free and the phase space density should be constant. Since the density decreases with time as $\prod_{\alpha}(\omega_\alpha t)$ and the spread of the velocity distribution is constant, the correlation volume must increase by the same factor.

13.7. Brief description of experimental results obtained with the He* experiment

We conclude this Chapter by briefly describing the *HBT* experiment made with the He* setup.[87] As already emphasized this experiment is very well suited for carrying out particle correlation measurements since it is possible to detect single atoms.

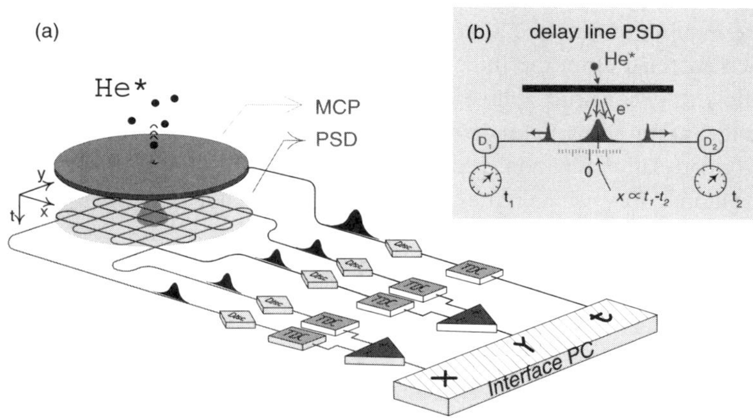

Fig. 13.11. Schematic of the detection apparatus. Single particle detection of the He* is possible due to its 20 eV of internal energy that is released at contact with the MCP. The position sensitivity is obtained through a delay-line anode (PSD) at the rear side of the MCP. The PSD is made of two long wires displaced as shown in (a). Its working principle is summarized in the inset Figure. The amplified charge generated by the MCP when it detects an atom is collected by two wires (the *delay lines*) placed below the MCP. This charge propagates to both ends of these wires and, depending on the location of the incident atom, the resulting signals arrive at the *time-to-digital converters* (TDC), at different times. By computing this time difference it is possible to infer the location where the charge was generated. The system with two delay lines use four discriminators(Disc) and TDCs. This is equivalent to a discrete anode detector with 10^5 *pixels* with a spatial resolution of about 250 μm. Another chain with a Disc and a TDC is used to get the absolute time arrival of the atoms, with a resolution better than 1 ns.

The atomic source is a magnetically trapped cloud of He* that is evaporatively cooled close to the BEC transition temperature (about 0.5 μK).

To measure the correlation, we switch off the trap, let the cloud expand and fall toward the detector placed 47 cm below the center of the magnetic trap. During its free fall toward the detector, the atomic cloud acquires a spherical shape [v] with a root mean square (*rms*) radius of a few centimeters. At the location of the detector (after 308 ms of flight) the correlation lengths are typically 800 μm in the fall direction and around 30 μm by 800 μm in the transverse directions[w].

This detector is an 8−cm diameter MCP. A delay line anode permits the position-sensitive detection of individual particles.[88,89] This device allows to detect around 10% of the incoming He*and to give for each detected particle its arrival time and in-plane position (x, y coordinates). Refer to Fig.13.11 where we sketch this apparatus and further explain its working principle in the caption. The *rms* resolution is of about 250 μm in the transversal direction. The atoms hit the detector at 3 m/s with a velocity spread below 1%, and so we convert t into a vertical position z. In this axis the *rms* resolution is then of only 2 nm.

The number of detected atoms is typically of a few thousands per shot and for each detected atom the three coordinates x, y and z are registered. These data allow us to construct a three-dimensional histogram of pair separations (Δx, Δy, and Δz) for all particles detected in a single cloud. This corresponds to the intensity correlation function, as can be seen by the following argument. Suppose that the atomic flux can be represented as a sum of N delta functions at different locations $\mathbf{r}$,

$$I(\mathbf{r}) = \sum_{i=1}^{N} \delta(\mathbf{r} - \mathbf{r}_i).$$

The correlation function of this quantity is, by definition, equal to

$$h(\delta r) = \int I(\mathbf{r}) I(\mathbf{r} + \delta\mathbf{r})\, d\mathbf{r} = 2 \sum_{i>j} \delta(|\mathbf{r}_i - \mathbf{r}_j| - \delta\mathbf{r}).$$

This expression is nothing else than a count of the pairs of atoms separated by a "distance" δr, which amounts to the calculation of the respective histogram. In the experiment of Ref. 15, this histogram was built over the entireatomic distribution and then summed over many shots, typically 1000.

This histogram is un-normalized and has a double peak structure: a broad shape corresponding to the auto-convolution of the atomic cloud and a narrow peak corresponding to the bosonic bunching behavior. To highlight the bunching behavior the above histogram is normalized by a second "fictitious" histogram. This fictitious histogram is computed by exactly the same algorithm, but applied to

[v]This is not the case of a *BEC*, where interatomic interactions force an inversion of the cloud's geometry from an initially cigar shaped to a disk shaped one.

[w]Note that this anisotropy in the transversal area of coherence follows from the different trapping oscillation frequencies in the Ox and Oy directions.

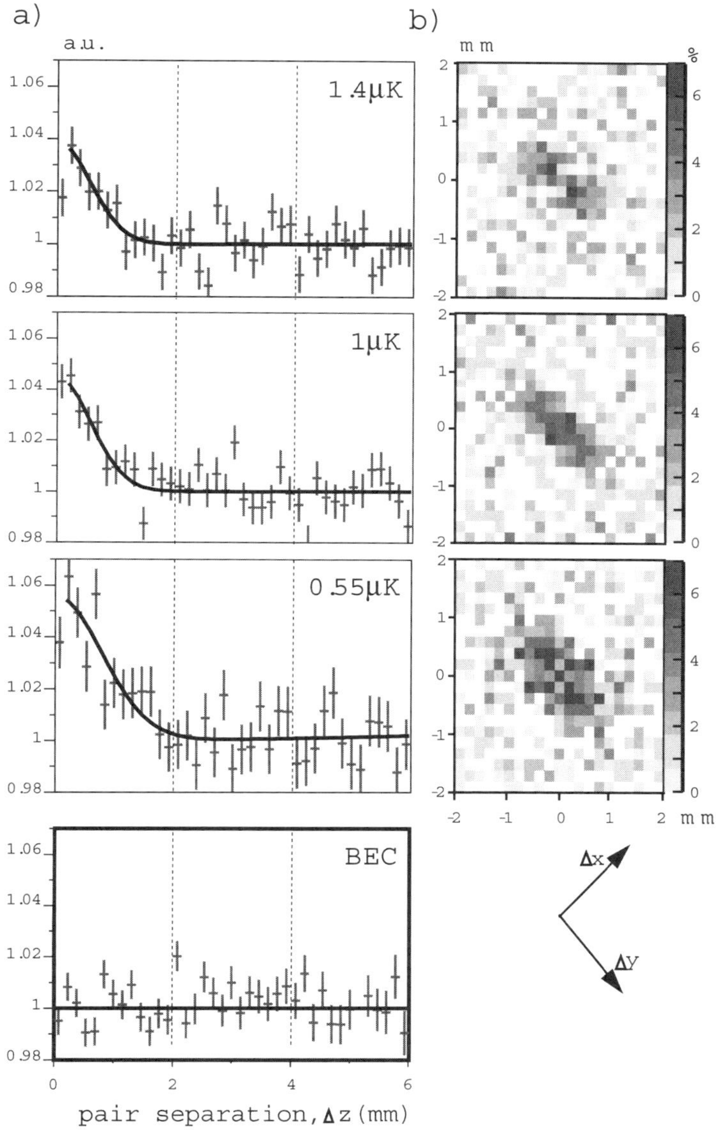

Fig. 13.12. a) normalized correlation functions along the vertical (z) axis for thermal gases at three different temperatures and for a BEC. For the thermal clouds, each plot corresponds to the average of a large number of clouds at the same temperature. Error bars correspond to the square root of the number of pairs; b) Normalized correlation functions in the $\Delta x - \Delta y$ plane for the three thermal cloud cases. The arrows at the lower right show the 45° rotation of the reference coordinate system with respect to the axes of the detector. The inverted ellipticity of the correlation function relative to the trapped cloud is clearly visible. From Ref. 15. Reprinted with permission from AAAS.

a single fictitious cloud which is the sum of all clouds. Since the individual clouds are independent, this fictitious cloud does not display a bunching behavior[x].

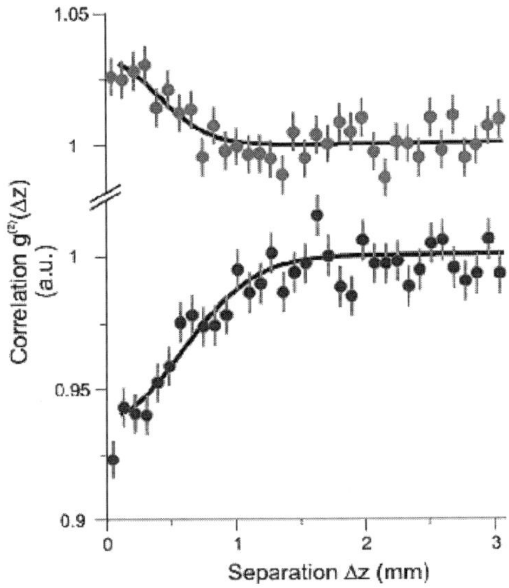

Fig. 13.13. Normalized correlation functions for ^{4}He*(bosons) in the upper graph, and ^{3}He*(fermions) in the lower graph. Both functions are measured at the same cloud temperature (0.5 μK), and with identical trap parameters. Error bars correspond to the root of the number of pairs in each bin. The line is a fit to a Gaussian function. The bosons show a bunching effect; the fermions anti-bunching. The correlation length for ^{3}He* is expected to be 33% larger than that for ^{4}He* due to the smaller mass. Reproduced from Ref. 22 published in *Nature*, where further details can be found.

The normalized correlation function, $g^{(2)}(\Delta z)$, is shown for various experimental conditions in Fig.13.12(a). The *HBT* bunching effect corresponds to the bump in the top three graphs of these figures. In Fig.13.12(b) we show the normalized correlation functions in the xOy plane and for $\Delta z = 0$ for the same three data sets displayed in the graphs on the left hand side. These plots show the asymmetry in the correlation function arising from the difference in the two transverse oscillation frequencies of the trap. The long axis of the correlation function is orthogonal to the less confined axis. The measured correlation lengths are in very good agreement with our predictions using an ideal gas model.[15,85]

The fourth graph in Fig.13.12(a) shows the result for a BEC. As expected, no correlation is observed for this latter one.

[x]The normalization histogram is nothing else than the autoconvolution of the average single particle distribution.

Recently this experiment was repeated in collaboration with the He* group of W. Vassen in Amsterdam [y], on a ultra-cold gas of ^{3}He* atoms.[22] This isotope is a Fermion. As expected, a dip was observed in the correlation function instead of a bump, due to the Pauli exclusion principle. The contrasting behavior between the fermionic and bosonic particles is shown in Fig.13.13.

13.8. Conclusion

We have shown in this chapter the profound analogy between light and matter. The ideal gas model that we developed explains qualitatively and quantitatively the experimental results obtained with clouds of metastable helium atoms. But the true atomic correlation functions are richer than their optical counterpart. In particular the interaction between particles and the dimensionality of the physical system leads to new physics. The correlation of a N-body system is largely an open question and the "purity" and control over a ultra-cold atomic sample make such systems good candidates for this study. For example ultra-cold atoms could be confined so tightly in two dimensions that their oscillations are totally frozen out in these two directions, meaning that this atomic sample behaves as a one-dimensional gas. Under some conditions this cloud has a fermionic behavior (Tonks-Girardeau gas[90,91]) that should be revealed by a measurement of the density correlation function.

Acknowledgments

This text is partially based on the Ph.D dissertation of JVG, which was done in co-tutelle between the *Universidade do Minho* and the *Université Paris-sud* under the supervision of MB and C. I. Westbrook. JVG acknowledges the financial support provided by the *Fundação para a Ciência e a Tecnologia*(FCT). The Atom Optics group of LCFIO is member of the *Institut Francilien de Recherche sur les Atomes Froids* (IFRAF) and of the *Fédération LUMAT* of the CNRS (FR2764). This work is supported by the PESSOA program 07988NJ, by the *Atom Chips* network MCRTN-CT-2003-505032, and the ANR under contract 05-NANO-008-01.

References

1. A. Einstein, Quantentheorie des einatomigen idealen gases. zweite abhandlung, *Sitzungber. Preuss. Akad. Wiss.* **1925**, 3, (1925).

[y] The detection apparatus of the He* experiment was taken to Amsterdam and used in the ^{3}He* fermionic experiment during two months in the summer of 2006.

2. S. Bose, Plancks Gesetz und Lichtquantenhypothese, *Z. Phys.* **26**(3), 178, (1924).
3. K. Huang, *Statistical mechanics*. (Wiley, New York, 1990).
4. M. R. Andrews, C. Townsend, H.-J. Miesner, D. Durfee, D. Kurn, , and W. Ketterle, Observation of interference between two bose condensates., *Science*. **275**, 637, (1997).
5. B. P. Anderson and M. A. Kasevich, Macroscopic quantum interference from atomic tunnel arrays, *Science*. **282**, 1686, (1998).
6. M.-O. Mewes, M. R. Andrews, D. M. Kurn, D. S. Durfee, C. G. Townsend, and W. Ketterle, Output coupler for bose-einstein condensed atoms, *Phys. Rev. Lett.* **78** (4), 582, (1997).
7. E. Hagley, L. Deng, M. Kozuma, J. Wen, K. Helmerson, S. Rolston, and W. Phillips, A well-collimated quasi-continuous atom laser, *Science*. **283**, 1709, (1999).
8. I. Bloch, T. W. Hänsch, and T. Esslinger, Atom laser with a cw output coupler, *Phys. Rev. Lett.* **82**(15), 3008, (1999).
9. I. Bloch, T. W. Hänsch, and T. Esslinger, Measurement of the spatial coherence of a trapped bose gas at the phase transition, *Nature*. **403**, 166, (2000).
10. P. Meystre, *Atom Optics*. (Springer Verlag, New York, 2001).
11. S. L. Rolston and W. D. Phillips, Nonlinear and quantum atom optics, *Nature*. **416**, 219, (2002).
12. R. Hanbury and R. Q. Twiss, Correlation between photons in two coherent beams of light, *Nature*. **177**, 27, (1956).
13. R. Hanbury Brown and R. Q. Twiss, A test of a new stellar interferometer on sirius, *Nature*. **178**, 1046, (1956).
14. M. Yasuda and F. Shimizu, Observation of two-atom correlation of an ultracold neon atomic beam, *Phys. Rev. Lett.* **77**(15), 3090, (1996).
15. M. Schellekens, R. Hoppeler, A. Perrin, J. V. Gomes, D. Boiron, A. Aspect, and C. I. Westbrook, Hanbury brown twiss effect for ultracold quantum gases, *Science*. **310**, 648, (2005).
16. F. T. Arecchi, Measurement of the statistical distribution of gaussian and laser sources, *Phys. Rev. Lett.* **15**(24), 912, (1965).
17. A. Ottl, S. Ritter, M. Kohl, and T. Esslinger, Correlations and counting statistics of an atom laser, *Phys. Rev. Lett.* **95**(9), 090404, (2005).
18. D. Hellweg, L. Cacciapuoti, M. Kottke, T. Schulte, K. Sengstock, W. Ertmer, and J. J. Arlt, Measurement of the spatial correlation function of phase fluctuating bose-einstein condensates, *Phys. Rev. Lett.* **91**(1), 010406, (2003).
19. M. Greiner, C. A. Regal, J. T. Stewart, and D. S. Jin, Probing pair-correlated fermionic atoms through correlations in atom shot noise, *Phys. Rev. Lett.* **94**(11), 110401, (2005).
20. S. Fölling, F. Gerbier, A. Widera, O. Mandel, T. Gericke, and I. Bloch, Spatial quantum noise interferometry in expanding ultracold atom clouds, *Nature*. **434**, 481, (2005).
21. J. Esteve, J.-B. Trebbia, T. Schumm, A. Aspect, C. I. Westbrook, and I. Bouchoule, Observations of density fluctuations in an elongated bose gas: Ideal gas and quasicondensate regimes, *Phys. Rev. Lett.* **96**(13), 130403, (2006).
22. T. Jeltes, J. McNamara, W. Hogervorst, W. Vassen, V. Krachmalnicoff, M. Schellekens, A. Perrin, H. Chang, D. Boiron, A. Aspect, and C. I.Westbrook, Comparison of the Hanbury Brown-Twiss effect for bosons and fermions, *Nature*. **445**, 402, (2007). Available in cond-mat/0612278.

23. T. Rom, T. Best, D. van Oosten, U. Schneider, S. Fölling, B. Paredes, and I. Bloch, Free fermion antibunching in a degenerate atomic fermi gas released from an optical lattice, *Nature*. **444**, 733, (2006).
24. G. Baym, The physics of Hanbury Brown-Twiss intensity interferometry: from stars to nuclear collisions, *ACTA PHYS.POLON.B*. **29**, 1839, (1998).
25. D. H. Boal, C.-K. Gelbke, and B. K. Jennings, Intensity interferometry in subatomic physics, *Rev. Mod. Phys.* **62**(3), 553, (1990).
26. U. Heinz and B. V. Jacak, Two-particle correlations in relativistic heavy-ion collisions, *Ann. Rev. Nucl. Part. Sci.* **49**, 529, (1999).
27. C. Y. Wong, *Introduction to High-Energy Heavy-Ion Collisions*. (World Scientific, 1994).
28. M. Iannuzzi, A. Orecchini, F. Sacchetti, P. Facchi, and S. Pascazio, Direct experimental evidence of free-fermion antibunching, *Phys. Rev. Lett.* **96**(8), 080402, (2006).
29. M. Henny, S. Oberholzer, C. Strunk, T. Heinzel, K. Ensslin, M. Holland, and C. Schonenberger, The fermionic hanbury brown and twiss experiment, *Science*. **284**, 296, (1999).
30. W. D. Oliver, J. Kim, R. C. Liu, and Y. Yamamoto, Hanbury brown and twiss-type experiment with electrons, *Science*. **284**, 299, (1999).
31. E. A. Donley, N. R. Claussen, S. T. Thompson, and C. E. Wieman, Atom-molecule coherence in a bose-einstein condensate, *Nature*. **417**, 529, (2002).
32. C. A. Regal, C. Ticknor, J. L. Bohn, and D. S. Jin, Creation of ultracold molecules from a fermi gas of atoms, *Nature*. **424**, 47, (2003).
33. M. Born and E. Wolf, *Optics*. (Pergamon, Oxford, 1980).
34. Goodman, *Statistical optics*. (John Wiley & Sons, New York, 1985).
35. R. J. Glauber, Photon correlations, *Phys. Rev. Lett.* **10**(3), 84, (1963).
36. R. J. Glauber, The quantum theory of optical coherence, *Phys. Rev.* **130**(6), 2529, (1963).
37. M. Scully and M. Zubairy, *Quantum Optics*. (Cambridge University Press, Cambridge, England, 1997).
38. L. D. Landau and E. M. Lifshitz, *Statistical Physics, part 1*. (Butterworth-Heynemann Ltd., London, 1980).
39. L. Mandel and E. Wolf, *Optical coherence and quantum optics*. (Cambridge University Press, Cambridge, MA, 1990).
40. H. D. Politzer, Condensate fluctuations of a trapped, ideal bose gas, *Phys. Rev. A*. **54** (6), 5048, (1996).
41. M.Naraschewski and R. Glauber, Spatial coherence and density correlations of trapped bose gases, *Phys. Rev. A*. **59**(6), 4595, (1999).
42. F. Dalfovo, S. Giorgini, L. P. Pitaevskii, and S. Stringari, Theory of bose-einstein condensation in trapped gases, *Rev. Mod. Phys.* **71**(3), 463, (1999).
43. A. Ashkin, Acceleration and trapping of particles by radiation pressure, *Phys. Rev. Lett.* **24**(4), 156, (1970).
44. J. E. Bjorkholm, R. R. Freeman, A. Ashkin, and D. B. Pearson, Observation of focusing of neutral atoms by the dipole forces of resonance-radiation pressure, *Phys. Rev. Lett.* **41**(20), 1361, (1978).
45. S. Chu, J. E. Bjorkholm, A. Ashkin, and A. Cable, Experimental observation of optically trapped atoms, *Phys. Rev. Lett.* **57**(3), 314, (1986).

46. D. J. Wineland, R. E. Drullinger, and F. L. Walls, Radiation-pressure cooling of bound resonant absorbers, *Phys. Rev. Lett.* **40**, 1639, (1978).
47. P. D. Lett, R. N. Watts, C. I. Westbrook, W. D. Phillips, P. L. Gould, and H. J. Metcalf, Observation of atoms laser cooled below the doppler limit, *Phys. Rev. Lett.* **61**, 169, (1988).
48. J. Dalibard, J.-M. Raimond, and J. Zinn-Justin, Eds., *Fundamental Systems in Quantum Optics*. (North-Holland, 1990).
49. A. L. Migdall, J. V. Prodan, W. D. Phillips, T. H. Bergeman, and H. J. Metcalf, First observation of magnetically trapped neutral atoms, *Phys. Rev. Lett.* **54**, 2596, (1985).
50. H. F. Hess, G. P. Kochanski, J. M. Doyle, N. Masuhara, D. Kleppner, and T. J. Greytak, Magnetic trapping of spin-polarized atomic hydrogen, *Phys. Rev. Lett.* **59**, 672, (1987).
51. N. Masuhara, J. M. Doyle, J. C. Sandberg, D. Kleppner, T. J. Greytak, H. F. Hess, and G. P. Kochanski, Evaporative cooling of spin-polarized atomic hydrogen, *Phys. Rev. Lett.* **61**, 935, (1988).
52. M. H. Anderson, J. R. Ensher, M. R. Matthews, C. E. Wieman, and E. A. Cornell, Observation of Bose-Einstein condensation in a dilute atomic vapor, *Science*. **269**(0), 198, (1995).
53. K. B. Davis, M.-O. Mewes, M. R. Andrews, N. J. van Druten, D. S. Durfee, D. M. Kurn, and W. Ketterle, Bose-einstein condensation in a gas of sodium atoms, *Phys. Rev. Lett.* **75**, 3969, (1995).
54. C. C. Bradley, C. A. Sackett, J. J. Tollet, and R. G. Hulet, Evidence of bose-einstein condensation in an atomic gas with attractive interactions, *Phys. Rev. Lett.* **75**, 1687, (1995).
55. G. Modugno, G. Ferrari, G. Roati, R. J. Brecha, A. Simoni, and M. Inguscio, Bose-einstein condensation of potassium atoms by sympathetic cooling, *Science*. **294**(5545), 1320, (2001).
56. S. L. Cornish, N. R. Claussen, J. L. Roberts, E. A. Cornell, and C. E. Wieman, Stable ^{85}Rb bose-einstein condensates with widely tunable interactions, *Phys. Rev. Lett.* **85** (9), 1795, (2000).
57. T. Weber, J. Herbig, M. Mark, H.-C. Nägerl, and R. Grimm, Bose-einstein condensation of cesium, *Science*. **299**(5604), 232, (2002).
58. R. Grimm, M. Weidemuller, and Y. B. Ovchinnikov, Optical dipole traps for neutral atoms, *Advances in Atomic, Molecular and Optical Physics*. **42**, 95, (2000). Also available in physics/9902072.
59. Y. Takasu, K. Maki, K. Komori, T. Takano, K. Honda, M. Kumakura, T. Yabuzaki, and Y. Takahashi, Spin-singlet bose-einstein condensation of two-electron atoms, *Phys. Rev. Lett.* **91**(4), 040404, (2003).
60. A. Griesmaier, J. Werner, S. Hensler, J. Stuhler, and T. Pfau, Bose-einstein condensation of chromium, *Phys. Rev. Lett.* **94**(16), 160401, (2005).
61. D. G. Fried, T. C. Killian, L. Willmann, D. Landhuis, S. C. Moss, D. Kleppner, and T. J. Greytak, Bose-einstein condensation of atomic hydrogen, *Phys. Rev. Lett.* **81**(18), 3811, (1998).
62. A. Robert, O. Sirjean, A. Browaeys, J. Poupard, S. Nowak, D. Boiron, C. I. Westbrook, and A. Aspect, A bose-einstein condensate of metastable atoms, *Science*. **292**, 461, (2001).

63. F. P. D. Santos, J. Léonard, J. Wang, C. J. Barrelet, F. Perales, E. Rasel, C. S. Unnikrishnan, M. Leduc, and C. Cohen-Tannoudji, Bose-einstein condensation of metastable helium, *Phys. Rev. Lett.* **86**(16), 3459, (2001).
64. A. S. Tychkov, T. Jeltes, J. M. McNamara, P. J. J. Tol, N. Herschbach, W. Hogervorst, and W. Vassen, Metastable helium bose-einstein condensate with a large number of atoms, *Phys. Rev. A.* **73**(3), 031603, (2006).
65. J. A. Swansson, R. G. Dall, and A. G. Truscott, Efficient loading of a He* magneto-optic trap using a liquid he cooled source, *Rev. of Sci. Instrum.* **77**(4), 046103, (2006).
66. R. J. W. Stas, J. M. McNamara, W. Hogervorst, and W. Vassen, Simultaneous magneto-optical trapping of a boson-fermion mixture of metastable helium atoms, *Phys. Rev. Lett.* **93**(5), 053001, (2004).
67. T. W. Hänsch and A. L. Schawlow, Cooling of gases by laser radiation, *Opt. Comm.* **13**, 68, (1975).
68. W. D. Phillips and H. Metcalf, Laser deceleration of an atomic beam, *Phys. Rev. Lett.* **48**(9), 596, (1982).
69. W. Ketterle, D. S. Durfee, and D. M. Stamper-Kurn. Making, probing and understanding bose-einstein condensates. In eds. M. Inguscio, S. Stringari, and C. Wieman, *Bose-Einstein Condensation in Atomic Gases*, Amsterdam, (1999). International School of Physics "Enrico Fermi"-Course CXL, IOS. Also available in cond-mat/9904034.
70. J. L. Wiza, Microchannel plate detectors, *Nucl. Instr. and Meth.* **162**, 587, (1979).
71. G. V. Shlyapnikov, J. T. M. Walraven, U. M. Rahmanov, and M. W. Reynolds, Decay kinetics and bose condensation in a gas of spin-polarized triplet helium, *Phys. Rev.Lett.* **73**(24), 3247, (1994).
72. A. Browaeys. *Piégeage magnétique d'un gaz d'Hélium métastable : vers la condensation de Bose-Einstein.* Thèse de doctorat, Université de Paris VI, (2000). Available at http://tel.ccsd.cnrs.fr.
73. W. H. Wing, On neutral particle trapping in quasistatic electromagnetic fileds, *Prog. Quant. Electr.* **8**, 181, (1984).
74. M.-O. Mewes, M. R. Andrews, N. J. van Druten, D. M. Kurn, D. S. Durfee, and W. Ketterle, Bose-einstein condensation in a tightly confining dc magnetic trap, *Phys. Rev. Lett.* **77**(3), 416, (1996).
75. H. F. Hess, G. P. Kochanski, J. M. Doyle, N. Masuhara, D. Kleppner, and T. J. Greytak, Magnetic trapping of spin-polarized atomic hydrogen, *Phys. Rev. Lett.* **59**(6), 672, (1987).
76. D. E. Pritchard, Cooling neutral atoms in a magnetic trap for precision spectroscopy, *Phys. Rev. Lett.* **51**(15), 1336, (1983).
77. W. Petrich, M. H. Anderson, J. R. Ensher, and E. A. Cornell, Stable, tightly confining magnetic trap for evaporative cooling of neutral atoms, *Phys. Rev. Lett.* **74**(17), 3352, (1995).
78. O. Sirjean, S. Seidelin, J. V. Gomes, D. Boiron, C. I. Westbrook, A. Aspect, and G. V. Shlyapnikov, Ionization rates in a bose-einstein condensate of metastable helium, *Phys. Rev. Lett.* **89**(22), 220406, (2002).
79. S. Seidelin, J. V. Gomes, R. Hoppeler, O. Sirjean, D. Boiron, A. Aspect, and C. I. Westbrook, Getting the elastic scattering length by observing inelastic collisions in ultracold metastable helium atoms, *Phys. Rev. Lett.* **93**, 090409, (2004).

80. J. V. Gomes. *Thermometry and coherence properties of a ultracold quantum gas of metastable Helium.* Thèse de doctorat, Université de Paris XI.
81. S. Seidelin, O. Sirjean, J. V. Gomes, D. Boiron, C. I. Westbrook, and A. Aspect, Using ion production to monitor the birth and death of a metastable helium bose eintein condensate., *J. Opt. B: Quantum Semiclass. Opt.* **5**(5), S112, (2003).
82. R. P. Feynman, *Statistical Mechanics: A Set of Lectures*. (Perseus Books, 1972).
83. R. P. Feynman and A. Hibbs, *Quantum mechanics and path integrals*. (Mc-Graw Hill, New-York, 1965).
84. R. Hoppeler, J. Viana Gomes, and D. Boiron, Atomic density of an harmonically trapped ideal gas near bose-einstein transition temperature, *Eur. Phys. J. D.* **41**, 157, (2006).
85. J. V. Gomes, A. Perrin, M. Schellekens, D. Boiron, C. I. Westbrook, and M. Belsley, Theory for a hanbury brown twiss experiment with a ballistically expanding cloud of cold atoms, *Phys. Rev. A.* **74**(5), 053607, (2006).
86. I. S. Gradshteyn and I. M. Ryzhik, *Table of series, integrals and products* $(7.374-8)$. (Academic Press, London, 1980).
87. http://www.atomoptic.fr.
88. RoentDek. Manual of the *MCP-Delay line* model dld80. Available at the manufacturer's web page: http://www.roentdek.com/.
89. O. Jagutzki, V. Mergel, K. Ullmann-Pfleger, L. Spielberger, U. Spillmann, R. Dörner, and H. Schmidt-Böcking, A broad-application microchannel-plate detector system for advanced particle or photon detection tasks: Large area imaging, precise multi-hit timing information and high detection rate, *Nucl. Instr. and Meth. in Phys. Res. A.* **477**, 244, (2002).
90. M. Girardeau, Relationship between systems of impenetrable bosons and fermions in one dimension., *J. Math. Phys.* **1**, 516, (1960).
91. M. Olshanii, Atomic scattering in the presence of an external confinement and a gas of impenetrable bosons., *Phys. Rev. Lett.* **81**, 938, (1998).

Chapter 14

Atomic Bose-Einstein condensation: Beyond mean-field theory

Guilherme S. Nunes
Instituto Superior de Ciências do Trabalho e da Empresa (ISCTE), Av. das Forças Armadas, 1600 Lisboa, Portugal
Guilherme.Nunes@iscte.pt

The hamiltonian describing the physics of atomic Bose gases is a many-body hamiltonian with a confining potential and two-particle repulsive interactions. Dealing with such a hamiltonian is difficult, even for dilute gases, at the level of mean-field theory, because of the very repulsive nature of the potential. It becomes even more difficult when the atomic density is increased and many-particle correlations need to be considered. In this article we overview the way in which density functional theory deals with both these problems.

Contents

14.1. Introduction to density functional theory

Bose-Einstein condensation, has been observed since 1995 for a number of atomic systems.[1–7] In these experiments, a number of atoms, which typically form a fairly dilute gas, are spin polarized and trapped by a spatially varying magnetic field. The interaction between the atoms and the confining trap can therefore be described by an external potential V_{ext}. The atomic condensates, which are formed when the gas is cooled, by laser beams and evaporation, are only metastable. In the long run such a gas should solidify due to three body interactions. However, if such processes are sufficiently rare, the condensate can survive for more than a few seconds, and be observed in the laboratory. Therefore, only two-body collisions are of relevance during the condensates's lifetime and the attractive tail of the interaction potential $V(r_1 - r_2)$ can be discarded because no atomic bound states can be formed in the two-body scattering events. $V(r_1 - r_2)$ can therefore be considered repulsive, in all generality.

The hamiltonian for such a system reads:

$$\begin{aligned} H &= \sum_{i=1}^{N} -\frac{\hbar^2}{2m} \nabla^2{}_{r_i} + \frac{1}{2} \sum_{i \neq j}^{N} V(r_j - r_i) + \sum_{i=1}^{N} V_{\text{ext}}(r_i) \\ &= T + V + V_{\text{ext}} \ . \end{aligned} \tag{14.1}$$

The lowest eigenvalue of this operator is, of course, the ground state energy E_0, and the corresponding eigenstate $| \Psi_0 >$ is associated with the ground state wavefunction $< r_1..r_N | \Psi_0 >$. And, of course, $| \Psi_0 >$ minimizes the expectation value $< \Psi | H | \Psi >$, which can be viewed as a functional of the many-body wavefunction Ψ. The main idea of density functional theory is that this minimization can be carried out in two stages. First, choose an arbitrary density $\rho(r)$ and then consider only those many-body wavefunctions Ψ from which ρ can be derived. If one minimizes the expectation of H in this subset, the result is obviously a functional of ρ :

$$\bar{F}(\rho) = F(\rho) + \int V_{\text{ext}} \, \rho \quad , \tag{14.2}$$

where

$$F(\rho) \equiv \min < \Psi \,|\, T + V \,|\, \Psi > |_{\rho} \ , \tag{14.3}$$

because $\rho(r) = \int | \Psi(r, r_2, .., r_N) |^2 \ d^3r_2..d^3r_N$. Secondly, one minimizes $\bar{F} = F + \int V_{\text{ext}} \ \rho$ with respect to ρ. That is:

$$E_0 = \min \bar{F}(\rho) \ . \tag{14.4}$$

So that E_0 is the minimum of a certain density functional $\bar{F}$. This can be accomplished, in principle, by solving the equation

$$\frac{\delta \bar{F}}{\delta \rho(r)} = 0 \tag{14.5}$$

for the ground state density $\rho_0(r)$. Then, $E_0 = \bar{F}(\rho_0)$. Notice that the above derivation is completely independent of statistics. It is equally valid for Fermi-Dirac and Bose-Einstein statistics. And it is also independent of any specific properties of V.

Formally, it is possible to gain some understanding of F and $\bar{F}$ in the following way: The functional F is obtained by minimizing the expectation value of $T+V$, under the conditions $< \Psi \mid \hat{\rho}(r) \mid \Psi > = \rho(r)$ and $< \Psi \mid \Psi > = 1$. So that one can associate a lagrange multiplier $\lambda(r)$ with the first condition and a lagrange multiplier ϵ with the second. Here we distinguish the density operator $\hat{\rho}(r)$ — which should not be confused with an ordinary function, obviously. So that the minimizing Ψ, which leads to $F(\rho)$ satisfies the Schrodinger equation:

$$(T+V) \mid \Psi > = \int \lambda(r)\, \hat{\rho}(r)\, d^3r \mid \Psi > + \epsilon \mid \Psi > , \tag{14.6}$$

and, therefore:

$$< \Psi \mid T+V \mid \Psi > = \int \lambda(r)\, \rho(r)\, d^3r \; + \; \epsilon\, . \tag{14.7}$$

Now, it is possible to interpret $-\lambda(r)$ as an external potential, $V_\rho(r)$, which leads to a ground state density $\rho(r)$, and $\epsilon(\rho)$ as the corresponding ground state energy (recall that ρ is not the ground state density ρ_0). This is because the solution of the Schrodinger equation

$$(T+V+V_\rho) \mid \Psi > = \epsilon \mid \Psi > , \tag{14.8}$$

is necessarily a state of density $\rho(r)$, by construction. Therefore

$$F(\rho) \; = < \Psi \mid (T+V) \mid \Psi > = \epsilon(\rho) - \int V_\rho(r)\, \rho(r)\, d^3r \quad . \tag{14.9}$$

And

$$\bar{F}(\rho) \; = \int V_{\text{ext}}\, \rho\, d^3r \; - \int V_\rho\, \rho\, d^3r \; + \; \epsilon\, (V_\rho) \quad . \tag{14.10}$$

Clearly, if ρ is the ground state density, ρ_0, for the original external potential, $V_{\text{ext}}(r)$, equation (14.10) reproduces the corresponding ground state energy, $E_0 = \epsilon\, (V_{\text{ext}}) = \epsilon(\rho_0)$, as it should.

However, even if this formalism is exact, we still have not been able to find an exact, explicit, expression for F or $\bar{F}$. To continue, an approximation must be sought.

14.2. Approximate functionals

Let us see what kind of sensible approximations can be made. First, let us consider that, at least as a useful model, the two-particle interaction potential V is sufficiently well behaved that the Hartree-Fock (Hartree-Bose) ground state of the Bose hamiltonian is well defined. That is, if we take

$$\Psi_H(r_1, ..., r_N) = \psi(r_1)\,\psi(r_2)\,...\,\psi(r_N) \ , \tag{14.11}$$

the expectation value $< \Psi_H \mid T + V + V_{\text{ext}} \mid \Psi_H >$ is well defined and equal to

$$F_H(\rho(r)) = -N\,\frac{\hbar^2}{2m}\int \psi(r)\,\nabla^2\,\psi(r)\,d^3r + \frac{1}{2}\int\!\!\int \rho(r)\,V(r-r')\,\rho(r')\,d^3r\,d^3r' + \int V_{\text{ext}}(r)\,\rho(r)\,d^3r \ , \tag{14.12}$$

where, obviously,

$$\rho(r) = N \mid \psi(r)|^2 \ . \tag{14.13}$$

Therefore, we must require that the integral

$$\int\!\!\int \rho(r)\,V(r-r')\,\rho(r')\,d^3r\,d^3r' \tag{14.14}$$

is well defined. This is certainly true if the interaction potential is integrable. Again, a rigorous model for interatomic interactions should be very repulsive and not integrable, and the Hartree integral (14.14) should not be well defined. However, it may be useful to consider, as a rough model or for some other reason, interaction potentials which are better behaved. We will, later in the text, deal with more realistic interatomic potentials.

Notice, here, that the Hartree ground state wave function for a Bose gas is automatically a Hartree-Fock (Hartree-Bose), symmetric, state, because the same orbital can be occupied for each particle (not so for excited states).

Equation 14.12 can alternatively be written as:

$$F_H(\rho(r)) = \int -\frac{\hbar^2}{2m}\sqrt{\rho(r)}\,\nabla^2\sqrt{\rho(r)}\,d^3r + \frac{1}{2}\int\!\!\int \rho(r)\,V(r-r')\,\rho(r')\,d^3r\,d^3r' + \int V_{\text{ext}}(r)\,\rho(r)\,d^3r \ . \tag{14.15}$$

Minimization of (14.12) with respect to $\psi(r)$, to find the ground state density, leads to the Hartree-Bose equation:

$$-\frac{\hbar^2}{2m}\nabla^2\psi(r) + \int \rho(r')\,V(r'-r)\,\psi(r)\,d^3r' + V_{\text{ext}}(r)\,\psi(r) = \epsilon_0\,\psi(r)\ . \tag{14.16}$$

Or,

$$-\frac{\hbar^2}{2m}\nabla^2\sqrt{\rho(r)} + \int \rho(r')\,V(r'-r)\sqrt{\rho(r)}\,d^3r' + V_{\text{ext}}\sqrt{\rho(r)} = \epsilon_0\sqrt{\rho(r)}\ . \tag{14.17}$$

This is also known as the Gross-Pitaevskii equation.[8,9] Do not mistake the eigenvalue ϵ_0 for the ground state energy E_0. In the Hartree approximation, $E_0 = N\epsilon_0 - \frac{1}{2}\int \rho(r)\,V(r-r')\,\rho(r')\,d^3r\,d^3r'$.

Recall that the Hartree functional is only approximate. This should be quite obvious since the true many-body wavefunction contains correlations between different atoms, whereas in the Hartree approximation, all the atoms are uncorrelated. Therefore the true ground state energy for this many-particle system, E_0 (true), differs from the Hartree ground state energy, E_0 (Hartree), by what is known as the correlation energy E_c.

To improve on the Hartree functional, this energy must be somehow taken into account. The way in which this is done, for electrons, is to first consider the homogeneous many-body problem. The homogeneous energy of the electron gas can be calculated in various ways, analytically[10] and numerically. The best results are obtained using diffusion Monte-Carlo.[11] Therefore, numerically, the ground state energy of the electron gas is quite well known as a function of the (constant) density. A very common and good approximation for the electronic density functional was first proposed by Kohn and Sham[12] soon after density functional theory was invented, by Kohn and Hohenberg.[13] Their main idea was to approximate the true density functional by

$$F_{KS} = F_H + \int \epsilon_{xc}(\rho)\,\rho\,d^3r\ , \tag{14.18}$$

where $\epsilon_{xc} = E_{xc}/N$, is the correlation energy per particle calculated (numerically) for the homogeneous gas. More exactly, for a homogeneous Fermi gas, the correct ground state energy, minus the Hartree energy, is usually known as the exchange-correlation energy, and usually denoted by E_{xc}. For a Fermi gas, one also defines the exchange energy as the difference between the Hartree-Fock energy, E_{HF}, calculated with a Slater determinant of atomic orbitals, which is therefore antisymmetric, and the Hartree energy, E_H: $E_x = E_{HF} - E_H$. The approximation (14.18) is called the local density approximation (LDA) because it is valid locally, in an area of space for which the density may be considered constant. Globally, it is exact if the gas is homogeneous (ρ is then constant), and it is best when ρ varies in a fairly smooth way. Roughly speaking, $\rho / \mid \nabla\rho \mid$ should be large compared to the mean distance between atoms.

The Kohn-Sham local density approximation is also feasible for bosons, by calculating numerically, with diffusion Monte-Carlo, the correlation energy per particle, ϵ_c, as a function of the density ρ for the homogeneous Bose gas. This can be done for any interatomic potential for which the Hartree energy is well defined. A well known recent calculation appeared in.[14] The Kohn-Sham functional is then:

$$F_{KS}(\rho) = \int -\frac{\hbar^2}{2m}\sqrt{\rho}\,\nabla^2\sqrt{\rho}\,d^3r + \frac{1}{2}\int\!\!\int \rho(r)V(r-r')\rho(r')\,d^3r d^3r' + \int \epsilon_c(\rho)\,\rho\,d^3r + \int V_{\text{ext}}(r)\,\rho(r)\,d^3r \,. \tag{14.19}$$

And the corresponding equation (obtained by minimizing the KS functional) from which the ground state density can be calculated is:

$$-\frac{\hbar^2}{2m}\nabla^2\sqrt{\rho(r)} + \int \rho(r')\,V(r'-r)\,\sqrt{\rho(r)}\,d^3r + \frac{\partial}{\partial\rho}(\epsilon_c\,\rho)\sqrt{\rho(r)} + V_{\text{ext}}\sqrt{\rho(r)} = \epsilon_0\sqrt{\rho(r)} \,. \tag{14.20}$$

This can be called a generalized Gross-Pitaevskii equation.

What can be done if the interatomic potential is such that the Hartree integral is not well defined ? As previously mentioned, this is an important question because realistic interatomic potentials or commonly used model potentials fall into this category (they are very repulsive at short distances). In fact, in the Bose-Einstein literature, the interaction potential which is almost always used is a hard sphere potential of finite range a, equal to the s-wave scattering length:

$$V(r) = \begin{cases} +\infty \Leftarrow & r < a \\ 0 \Leftarrow & r \geq a \end{cases} \tag{14.21}$$

In such cases, the correlation energy per particle, ϵ_c, for the homogeneous Bose gas, is clearly not defined. However, the total energy per particle, for the homogeneous Bose gas, ϵ, is still well defined, and it is this energy that can be used to construct a density functional (in such cases) with a local density approximation:[15,16]

$$F = \int - \frac{\hbar^2}{2m} \sqrt{\rho}\ \nabla^2 \sqrt{\rho}\, d^3 r \ + \ \int \epsilon\,(\rho)\,\rho\, d^3 r \tag{14.22}$$

Or,

$$F = N \int - \frac{\hbar^2}{2m} \psi\, \nabla^2\, \psi\, d^3 r \ + \ \int \epsilon\,(\rho)\,\rho\, d^3 r \ . \tag{14.23}$$

The corresponding Kohn-Sham equation is then:[15,16]

$$- \frac{\hbar^2}{2m} \nabla^2 \psi_0 + \frac{\partial}{\partial \rho} (\epsilon\, \rho)\, \psi_0 + V_{\text{ext}}\, \psi_0 = \epsilon_0\, \psi_0 \ . \tag{14.24}$$

The hard sphere potential was mentioned because it is a very commonly employed model potential but the above formalism is valid for any potential which we might wish to consider. This is a distinct advantage of the Kohn-Sham equation over the usual Gross-Pitaevskii equation, which always assumes the potential (14.21). That is, the functional (14.22) is defined for any non-integrable interaction potential and not just the hard-sphere potential. All we need is the energy of the homogeneous Bose gas, ϵ, as a function of the constant density ρ, for a particular interaction V. The local density approximation to the exact density functional can then be defined and applied to any problem in which the Bose gas is inhomogeneous, due to a confining potential.

However, we shall now focus, almost exclusively, on the hard-sphere interaction and see exactly what the functional (14.22) is like (for such an interaction) since this is the case for which the energy of the homogeneous Bose gas has been most studied and is best known. Indeed, the total energy per particle for the homogeneous Bose gas, with hard-sphere interactions, has been obtained analytically.[17–23] The result is:

$$\epsilon = \frac{2\pi\hbar^2 a}{m} \rho \left[1 + g\left(\rho a^3\right)\right] \ , \tag{14.25}$$

with

$$g = \frac{128}{15\sqrt{\pi}} \sqrt{\rho a^3} + \cdots \quad . \tag{14.26}$$

The same calculation has been done numerically,[14] with diffusion Monte-Carlo and agrees nicely with the previous equations. For dilute gases, such as the ones which are usually produced in Bose-Einstein experiments, like the pioneering experiments of the second half of the 1990's, $\rho a^3 \ll 1$. And so ϵ is proportional to the density, $\epsilon = 2\pi\hbar^2 a\rho/m$, which leads to the usual Gross-Pitaevskii equation.[8,9] That is, equation (14.24) becomes:

$$-\frac{\hbar^2}{2m}\nabla^2\psi_0 \; + \; V_{\rm ext}(r)\; \psi_0 \; + \; \frac{4\pi\hbar^2 a}{m}\rho\, \psi_0 \; = \; \epsilon_0\, \psi_0 \; . \tag{14.27}$$

So density functional theory leads to the usual GP equation, without the need to discuss pseudopotentials, which is the way in which the difficulties with very repulsive potentials are traditionally overcome.[17–19] For Bose gases which are not so dilute, $\rho a^3 > 10^{-4}$, the extra term in (14.25) becomes non-negligible and should be included. Equation (14.24) then becomes:[15]

$$\begin{aligned} -\frac{\hbar^2}{2m}\nabla^2\psi_0 \; + \; V_{\rm ext}(r)\; \psi_0 \; + \; \frac{4\pi\hbar^2 a}{m}\rho\,(1+g \; + \; \frac{1}{2}\,(\rho\, a^3)\; g' \;)\; \psi_0 \\ = \epsilon_0\; \psi_0 \;\; , \end{aligned} \tag{14.28}$$

or,

$$\begin{aligned} -\frac{\hbar^2}{2m}\nabla^2\psi_0 \; + \; V_{\rm ext}(r)\; \psi_0 \; + \; \frac{4\pi\hbar^2 a}{m}\rho\,(1+\frac{5}{4}\,\frac{128}{15\sqrt{\pi}}\,\sqrt{\rho\, a^3})\; \psi_0 \\ = \; \epsilon_0\; \psi_0 \;\; . \end{aligned} \tag{14.29}$$

This is also an extended Gross-Pitaevskii equation, although different from (14.20). Of course, in the dilute limit, this equation reduces to the usual GP form (14.27). In the literature, it is frequently referred to as the modified Gross-Pitaevskii equation (MGP).[24–26] The GP theory is also commonly referred to as mean-field theory, so the modified equation is an extension beyond mean-field theory. The question of experimentally testing such an extension, in the laboratory, now arises.

14.3. Some numerical results

In the laboratory, Bose condensates are trapped by external potentials which may be considered harmonic: $V_{\rm ext} = \frac{1}{2}m\omega^2 r^2$. The MGP becomes, in harmonic oscillator units:

$$\begin{aligned} -\nabla^2\psi_0 \; + \; r^2\; \psi_0 + 8\pi\, \frac{a}{l_\omega}\; \rho\; (1 \; + g \; + \; \frac{1}{2}\rho\; (\frac{a}{l_\omega})^3\; g' \;)\; \psi_0 \\ = \; \epsilon_0\, \psi_0 . \end{aligned} \tag{14.30}$$

Here, $l_\omega \equiv \frac{\hbar}{m\omega}$, is the harmonic length. The other important lengthscale is the scattering length a. Remember that we are discussing only the ground state of the inhomogenous Bose gas. That is, $T = 0$. There are two important combinations of these scales in the above equation. The combination $N\, a^3/{l_\omega}^3$, which contains the number of trapped atoms, N, determines the importance of the extra terms in the Gross-Pitaevskii equation. For typical experimental values, this term is quite small. It can be increased, experimentally, by increasing a, which has recently become possible by exploiting Feshbach ressonances,[27–31] or by increasing the value of N. The other combination, $N\, a/l_\omega$, determines the importance of the interactions, and is generally larger than 1, for typical experimental values. The energy scale is $\hbar\omega/2$ and ϵ_0 is usually denoted as the chemical potential μ.

Table 14.1. Chemical potential μ, and total energy per particle E_0/N in units of $\hbar\omega$, as a function of the number N of trapped ^{133}Cs atoms, in a 10 Hz trap, for a scattering length $a_F = 2500\mathring{A}$, without the Thomas-Fermi approximation.

N	μ (GP)	μ (GGP)	E_0/N (GP)	E_0/N (GGP)	$\rho_{\max}a^3$ (GP)	$\rho_{\max}\, a^3$ (GGP)
10	2.0246	2.0707	1.7871	1.8000	7.8144×10^{-4}	7.3584×10^{-4}
100	3.8822	4.1350	2.9817	3.1324	2.3854×10^{-3}	2.0293×10^{-3}
500	6.9862	7.6568	5.1258	5.5439	4.4957×10^{-3}	3.6319×10^{-3}
1000	9.1190	10.1236	6.6251	7.2571	5.9086×10^{-3}	4.6619×10^{-3}

Table 14.2. The chemical potential μ, the total energy per particle, E_0/N, and the atomic density, calculated with the GP equation and the GGP equation for various numbers of ^{87}Rb atoms, with a scattering length of $1000\, a_0$, in a 77.87 Hz trap. All energies are in units of $\hbar\omega$.

N	μ (GP)	μ (GGP)	E_0/N (GP)	E_0/N (GGP)	$\rho_{\max}a^3$ (GP)	$\rho_{\max}\, a^3$ (GGP)
10	1.7875	1.7982	1.6520	1.6569	1.0689×10^{-4}	1.0526×10^{-4}
100	3.0444	3.1224	2.4246	2.4689	4.0064×10^{-4}	3.7388×10^{-4}
1000	6.8654	7.2014	5.0412	5.2490	1.0070×10^{-3}	8.9736×10^{-4}
5000	12.8136	13.6840	9.2394	9.7863	1.9023×10^{-3}	1.6354×10^{-3}

We shall first consider large values of a, artificially induced by a magnetic field. for Cs and Rb atoms. Tables 14.1, 14.2, 14.3 and 14.4, clearly show that the modified theory predicts a non-negligible correction to mean-field theory for large values of the scattering length, even if the number of atoms confined is rather small. For 1000 Cs atoms, such corrections can reach about 11% for the chemical potential, 9.5% for the total energy and 21% for the peak density. For 5000 ^{87}Rb

Table 14.3. Chemical potential μ, total energy per particle, E_0/N, and maximum condensate density $\rho_{\max}\, a^3$, for $N = 10, 100, 1000$ and 5000 ^{85}Rb atoms, with a scattering length of $a = 1000\, a_0$, in a 77.87 Hz trap. All energies are in units of $\hbar\omega$.

N	μ (GP)	μ (GGP)	E_0/N (GP)	E_0/N (GGP)	$\rho_{\max}\, a^3$ (GP)	$\rho_{\max}\, a^3$ (GGP)
10	1.7847	1.7951	1.6504	1.6552	1.0352×10^{-4}	1.0198×10^{-4}
100	3.0334	3.1099	2.4174	2.4608	3.8946×10^{-4}	3.6382×10^{-4}
1000	6.8352	7.1652	5.0201	5.2242	9.7936×10^{-4}	8.7395×10^{-4}
5000	12.7551	13.6111	9.1978	9.7351	1.8500×10^{-3}	1.5933×10^{-3}

Table 14.4. Up to $N = 10000$ Rubidium 85 atoms, confined in a 77.87 Hz magnetic trap, with scattering length $a = 500\, a_0$. All energies are in units of $\hbar\omega$.

N	μ (GP)	μ (GGP)	E_0/N (GP)	E_0/N (GGP)	$\rho_{\max} a^3$ (GP)	$\rho_{\max}\, a^3$ (GGP)
1000	5.2740	5.3861	3.9337	4.002	1.8652×10^{-4}	1.7680×10^{-4}
5000	9.7249	10.0230	7.0527	7.2383	3.5142×10^{-4}	3.2692×10^{-4}
10000	12.7551	13.2055	9.1977	9.4794	4.6251×10^{-4}	4.2616×10^{-4}

atoms in a 77.87 Hz trap, with a scattering length of 1000 Bohr units, we find 6.8% for chemical potential, 5.9% for the total energy and about 14% for the peak density. For ^{85}Rb, in the same trap, with the same induced scattering length, these corrections are similar, as shown in table 14.3.

Next, consider the possibility of testing these corrections with natural (no Feshbach ressonances) scattering lenghts, of the order of a few nanometers, for very large numbers of confined atoms.

14.4. The Thomas-Fermi approximation

Let us relate the two adimensional parameters $\alpha = N\,(a/l_\omega)^3$ and $\beta = N\,(a/l_\omega)$. Obviously $\beta = N^{2/3}\,\alpha^{1/3}$. So, for the same value of α, the parameter which measures the strength of the interatomic interactions, β, increases with N. In the previous section, the modified Gross-Pitaevskii equation (which extends the usual mean-field theory of the Bose-Einstein condensate), was solved for relatively large values of α, but relatively small values of N. We shall now consider similar values of α, with N large. That is, we shall consider small values of the ratio a/l_ω, as it commonly occurs in experiments where the scattering length, a, is not artificially enhanced. One then finds that β can be very large. For example, for ^{87}Rb, if $a = 7$ nm and $f = 10$ Hz, then $a/l_\omega \simeq 2 \times 10^{-3}$, and this means that, for $N = 10^7$, $\beta = 2 \times 10^4$! Now consider that the same value of α is obtained with $N = 1000$. Then $a/l_\omega \simeq 0.043$ and $\beta = 43$! So the two values of β are indeed quite different for the same value of α. >From a mathematical point of view, β measures the strength of the nonlinearities in the MGP equation. Large β means a very nonlinear equation and such a problem is much more difficult, numerically. However, it also means that the relative importance of the kinetic energy is decreased. A popular approximation, which has been found to work quite well for large β, is to completely obliterate the kinetic energy. The Kohn-Sham functional is then approximated by the Thomas-Fermi functional:[16,32,33]

$$F_{TF} = \int [V_{\text{ext}}(r)\, \rho + \epsilon(\rho)\rho]\, d^3 r \quad . \tag{14.31}$$

The MGP equation becomes:

$$V_{\text{ext}} + \frac{\partial \epsilon}{\partial \rho} \rho + \epsilon = \epsilon_0 \quad . \tag{14.32}$$

Or,

$$r^2 + 8\,\pi \frac{a}{l_\omega} \rho\, (1 + \frac{32}{3\sqrt{\pi}} \sqrt{\rho a^3}) = \mu \quad . \tag{14.33}$$

This is the Thomas-Fermi equation.[32,33] It will be convenient in what follows to define c_1 and c_2 as:

$$c_1 = 8\,\pi \frac{a}{l_\omega}\, N \tag{14.34}$$

$$c_2 = c_1 \frac{32}{3\sqrt{\pi}} (\frac{a}{l_\omega})^{\frac{3}{2}}\, N^{\frac{1}{2}} \quad . \tag{14.35}$$

The TF equation then reads:

$$r^2 + c_1\, n + c_2\, n^{\frac{3}{2}} = \mu \quad , \tag{14.36}$$

where n is the normalized density: $n = \rho/N$. This equation can be solved numerically and the value of μ can also be determined, numerically, by verifying the condition $\int n\, d^3 r = 1$. Analytically, it can be solved as a function of μ. But the latter integral is hard to do. If c_2 is ignored, in (14.36), we obtain

$$r^2 + c_1\, n = \mu \quad . \tag{14.37}$$

which is the GP equation with the kinetic energy term dropped out. The solution to this is absolutely trivial:

$$n = (\mu - r^2)/c_1 \quad . \tag{14.38}$$

The normalization condition for n then yields:

$$\mu = (\frac{15\,c_1}{8\,\pi})^{\frac{2}{5}} \quad . \tag{14.39}$$

And the total energy is:

$$E_0 = \frac{5}{7}\,N\,\mu \quad . \tag{14.40}$$

So E_0 scales with $N^{7/5}$ (for large N). It is easy to see that the total energy per particle must always be smaller than the chemical potential.

The following tables show our results, in the Thomas-Fermi approximation, for a few special cases, of experimental interest.

Table 14.5. Chemical potential μ, and total energy per particle E_0/N in units of $\hbar\omega$, as a function of the number N of trapped Cs atoms, in a 10 Hz trap, in the Thomas-Fermi approximation. The scattering length is $a = 3.20$ nm.

N	μ (GP)	μ (GGP)	E_0/N (GP)	E_0/N (GGP)	$\rho_{\max}\,a^3$ (GP)	$\rho_{\max}\,a^3$ (GGP)
10^5	9.8911	9.9091	7.0651	7.0763	1.0598×10^{-6}	1.0552×10^{-6}
10^6	24.8453	24.9168	17.7466	17.7913	2.6621×10^{-6}	2.6439×10^{-6}
10^7	62.4085	62.6924	44.5775	44.7551	6.6868×10^{-6}	6.6149×10^{-6}
10^8	156.7631	157.8812	111.9737	112.6777	1.6797×10^{-5}	1.6513×10^{-5}
10^9	393.7712	398.2189	281.2651	284.0507	4.2191×10^{-5}	4.1083×10^{-5}
10^{10}	989.1085	1006.6261	706.5061	717.4898	1.0598×10^{-4}	1.0169×10^{-4}

Table 14.6. Chemical potential μ, and total energy per particle E_0/N in units of $\hbar\omega$, as a function of the number N of trapped ^{85}Rb atoms, in a 77.78 Hz trap, in the Thomas-Fermi approximation, with a scattering length of $a = 100\,a_0$, where $a_0 = 0.529 \times 10^{-10}$ m.

N	μ (GP)	μ (GGP)	E_0/N (GP)	E_0/N (GGP)	$\rho_{\max}\,a^3$ (GP)	$\rho_{\max}\,a^3$ (GGP)
10^5	16.6668	16.8102	11.9049	11.9947	2.4259×10^{-5}	2.3770×10^{-5}
10^6	41.8652	42.4314	29.9037	30.2585	6.0936×10^{-5}	5.9031×10^{-5}
10^7	105.1607	107.3862	75.1147	76.5110	1.5307×10^{-4}	1.4572×10^{-4}
10^8	264.1517	272.8416	188.6797	194.1423	3.8448×10^{-4}	3.5661×10^{-4}
10^9	663.5192	697.1318	473.9420	495.1305	9.6578×10^{-4}	8.6231×10^{-4}
10^{10}	1666.6848	1795.0267	1190.4885	1271.6980	2.4259×10^{-3}	2.0529×10^{-3}

Table 14.5 shows that the "beyond mean-field" effects, for a large number of Caesium atoms, can become 1.77% for the chemical potential, 1.55% for the ground state energy and about 4% for the maximum density. May it be reminded

Table 14.7. Chemical potential μ, and total energy per particle E_0/N in units of $\hbar\omega$, as a function of the number N of trapped ^{87}Rb atoms, in a 20 Hz trap, in the Thomas-Fermi approximation, with a scattering length of $a = 7$ nm.

N	μ (GP)	μ (GGP)	E_0/N (GP)	E_0/N (GGP)	$\rho_{\max}\, a^3$ (GP)	$\rho_{\max}\, a^3$ (GGP)
10^5	14.2746	14.3521	10.1961	10.2446	9.5749×10^{-6}	9.4520×10^{-6}
10^6	35.8561	36.1633	25.6115	25.8038	2.4051×10^{-5}	2.3568×10^{-5}
10^7	90.0664	91.2793	64.3331	65.0931	6.0413×10^{-5}	5.8532×10^{-5}
10^8	226.2366	231.0044	161.5975	164.5887	1.5175×10^{-4}	1.4450×10^{-4}
10^9	568.2807	586.8993	405.9146	417.6183	3.8118×10^{-4}	3.5365×10^{-4}
10^{10}	1427.4565	1499.4811	1019.6113	1065.0122	9.5749×10^{-4}	8.5527×10^{-4}

that the total energy and the atomic density are directly measurable quantities. Table 14.6 shows that the "beyond mean-field" effects, for $N = 10^9$ ^{85}Rb atoms, can reach 5.07% for the chemical potential, 4.47% for the condensate energy and 10.71% for the maximum atomic density. I also want to present some calculations, done for a ^{87}Rb gas, in a 20 Hz trap, with a scattering length of $a = 7$ nm. The results are shown in table 14.7. A relative difference in the ground state peak density of about 4.8%, for $N = 10^8$, and 7.2% for $N = 10^9$ ^{87}Rb atoms, is found. For the ground state energy, relative differences of 1.85% and 2.88% are found for $N = 10^8$ and $N = 10^9$ atoms, respectively.

Notice that, in spite of the very large number of atoms confined, the maximum atomic density is still low enough for the Huang-Lee-Yang expression for the energy of the homogeneous Bose gas to be valid, locally. Indeed, according to the simple TF version of the GP theory, we would obtain $\rho_{\max}\, a^3 = 1$, exactly, for

$$\mu = 8\pi \frac{l^2}{a^2} \quad , \tag{14.41}$$

which means

$$N_{\max} = \frac{(8\pi)^{\frac{5}{2}}}{15} (l/a)^6 \quad . \tag{14.42}$$

For Caesium atoms this yields $N_{\max} \simeq 10^{20}$. So the values of N used in the above calculations are well below this limit.

Let us now return to the TF approximation for the extended GP theory of the BE condensate (14.36). As previously stated, this is a cubic equation which can be solved exactly. That is, ρ can be found exactly as a function of r and μ. The problem is that it is then difficult to find μ. In this section I wish to consider a simple analytical approximation to the solution of (14.36), which also provides an expression for μ and E_0. The main point is that $c_2 \ll c_1$ and, therefore, the extra term, due to $c_2 \neq 0$, in (14.36), can treated as a perturbation. That is, the density distribution changes only slightly, due to $c_2 \neq 0$. How does it change ? It changes so as to decrease the energy due to the interatomic interactions, so that the peak density decreases slightly and the size of the condensate, R, increases slightly — R is defined by the condition $\rho(R) = 0$. That is, $\rho(0)$ decreases to $\rho'(0)$ and R increases to R'. But this change is subject to the normalization constraint: $\int \delta\,\rho\, d^3\, r = 0$, where $\delta\rho \equiv \rho' - \rho$. It is also obvious that μ increases with c_2, because $\mu' = R'^2$. So $\mu' > \mu$. Now, we know that $\delta\, n(0) = -k_1$, where $k_1 > 0$. That is, $n'(0) = n(0) - k_1$. And we know that $\delta\, n$ is increasing. We also know that that δn is minimum at $r = 0$, so the simplest approximation to δn is the quadratic form $\delta n = -k_1 + k_2\, r^2$.

What we are proposing, then, is the approximation:

$$n' \;=\; (\mu - r^2)/c_1 \;-\; k_1 + k_2\, r^2 \tag{14.43}$$

Therefore:

$$0 \;=\; (\mu - \mu')/c_1 \;-\; k_1 + k_2\, \mu' \quad . \tag{14.44}$$

The normalization condition, $\int \delta\, n\, d^3\, r \;=\; 0$, can be well approximated by $\int_0^{R'} (n' - n)\; r^2\; dr = 0$. We then obtain:

$$-k_1\, R'^3/3 \;+\; k_2\, R'^5/5 \;=\; 0 \quad . \tag{14.45}$$

Or,

$$-k_1/3 \;+\; k_2\, \mu'/5 \;=\; 0 \quad . \tag{14.46}$$

Now, recall that the defining equation for n, in the TF approximation, is:

$$-\mu' + r^2 + c_1\, n' + c_2\, n'^{3/2} \;=\; 0 \quad . \tag{14.47}$$

But, at $r = 0$, we have $n \;=\; \mu/c_1 - k_1$, where k_1 is relatively small. We therefore approximate $n'(0)^{3/2}$ by $(\mu/c_1)^{3/2} - \frac{3}{2}\,(\mu/c_1)^{1/2}\, k_1$. Then, substituting this and $r = 0$ in (14.45), we get a third equation:

$$-\mu' + \mu - c_1\, k_1 + c_2\, (\mu/c_1)^{3/2} \;-\; \frac{3}{2}\, c_2\, (\mu/c_1)^{1/2}\, k_1 \;=\; 0 \quad . \tag{14.48}$$

So (14.40), (14.44) and (14.46) are three linear equations in the three unknowns k_1, k_2 and μ'. Remember that μ is given by equation (14.38). Here are the solutions:

$$k_1 \;=\; \frac{c_2\,(\frac{\mu}{c_1})^{3/2}}{\frac{5}{3}c_1 \;+\; \frac{3}{2}\, c_2\,(\frac{\mu}{c_1})^{1/2}} \tag{14.49}$$

$$\mu' = \mu + \frac{\frac{2}{3}\,\mu\,c_2\left(\frac{\mu}{c_1}\right)^{1/2}}{\frac{5}{3}\,c_1 + \frac{3}{2}\,c_2\left(\frac{\mu}{c_1}\right)^{1/2}} \tag{14.50}$$

$$k_2 = \frac{\frac{5}{3}\frac{c_2}{c_1}\left(\frac{\mu}{c_1}\right)^{1/2}}{\frac{5}{3}\,c_1 + \frac{13}{6}\,c_2\left(\frac{\mu}{c_1}\right)^{1/2}} \quad . \tag{14.51}$$

We shall next find out how well this ATF works, for the cases already studied, in the previous section — tables 14.8-14.13.

Table 14.8. Chemical potential μ as a function of the number N of trapped Cs atoms, in a 10 Hz trap, in different TF approximations. We take $a = 3.20$ nm.

N	μ (TF-GP)	μ' (TF)	μ' (ATF)
10^5	9.8911	9.9091	9.9155
10^6	24.8453	24.9168	24.9420
10^7	62.4085	62.6924	62.7916
10^8	156.7631	157.8881	158.2761
10^9	393.7712	398.2189	399.7189
10^{10}	989.1085	1006.6261	1012.3254

Table 14.9. Chemical potential μ as a function of the number N of trapped ^{85}Rb atoms, in a 77.78 Hz trap, in different TF approximations. We take $a = 100 \times a_0$.

N	μ (TF-GP)	μ' (TF)	μ' (ATF)
10^5	16.6668	16.8102	16.8593
10^6	41.8652	42.4314	42.6200
10^7	105.1606	107.3862	108.0958
10^8	264.1516	272.8416	275.4228
10^9	663.5188	697.1318	706.0046
10^{10}	1666.6839	1795.0267	1822.6781

We now turn to the condensate energy per particle (in the ATF). This is:

Table 14.10. Chemical potential μ as a function of the number N of trapped ^{87}Rb atoms, in a 20.00 Hz trap, in different TF approximations. $a = 7$ nm.

N	μ (TF-GP)	μ' (TF)	μ' (ATF)
10^5	14.2746	14.3521	14.3791
10^6	35.8561	36.1633	36.2684
10^7	90.0664	91.2793	91.6835
10^8	226.2365	231.0044	232.5257
10^9	568.2804	586.8993	592.4343
10^{10}	1427.4558	1499.4812	1518.5207

Table 14.11. Maximum atomic density as a function of the number N of trapped ^{133}Cs atoms, in a 10.00 Hz trap, in different TF approximations. $a = 3.20$ nm.

N	$\rho_{\max}\, a^3$ (TF-GP)	$\rho'_{\max}\, a^3$ (TF)	$\rho'_{\max}\, a^3$ (ATF)
10^5	1.0598×10^{-6}	1.0552×10^{-6}	1.0559×10^{-6}
10^6	2.6621×10^{-6}	2.6439×10^{-6}	2.6465×10^{-6}
10^7	6.6868×10^{-6}	6.6149×10^{-6}	6.6253×10^{-6}
10^8	1.6797×10^{-5}	1.6513×10^{-5}	1.6553×10^{-5}
10^9	4.2191×10^{-5}	4.1083×10^{-5}	4.1235×10^{-5}
10^{10}	1.0598×10^{-4}	1.0169×10^{-4}	1.0225×10^{-4}

$$\begin{aligned} E_0'/N \;=\; & \int_0^{R'} \frac{1}{2}\, c_1\, n'^2\, 4\pi r^2\, dr \;+\; \int_0^{R'} r^2\, n'\, 4\pi r^2\, dr \\ & + \int_0^{R'} \frac{1}{2}\, \frac{4}{5}\, c_2\, n'^{5/2}\, 4\pi r^2\, dr \quad , \end{aligned} \tag{14.52}$$

where

$$n' \;=\; a - b\, r^2 \tag{14.53}$$

and $a = \frac{\mu}{c_1} - k_1$ and $b = \frac{1}{c_1} - k_2$. This results in:

$$\begin{aligned} E_0'/N \;=\; & c_1\, a^2\, \mu'^{3/2}\, \frac{16\pi}{105} \;+\; a\, \mu'^{5/2}\, \frac{8\pi}{35} \\ & + \; c_2\, a^{5/2}\, \mu'^{3/2}\, \frac{\pi^2}{32} \quad . \end{aligned} \tag{14.54}$$

Table 14.14 contains some results.

Table 14.12. Maximum atomic density as a function of the number N of trapped ^{85}Rb atoms, in a 77.78 Hz trap, in different TF approximations. $a = 100\, a_0$.

N	$\rho_{\max}\, a^3$ (TF-GP)	$\rho'_{\max}\, a^3$ (TF)	$\rho'_{\max}\, a^3$ (ATF)
10^5	2.4260×10^{-5}	2.3771×10^{-5}	2.3839×10^{-5}
10^6	6.0936×10^{-5}	5.9031×10^{-5}	5.9289×10^{-5}
10^7	1.5307×10^{-4}	1.4572×10^{-4}	1.4666×10^{-4}
10^8	3.8448×10^{-4}	3.5661×10^{-4}	3.5987×10^{-4}
10^9	9.6578×10^{-4}	8.6231×10^{-4}	8.7302×10^{-4}
10^{10}	2.4259×10^{-3}	2.0529×10^{-3}	2.0853×10^{-3}

Table 14.13. Maximum atomic density as a function of the number N of trapped ^{87}Rb atoms, in a 20.00 Hz trap, in different TF approximations. $a = 7$ nm.

N	$\rho_{\max}\, a^3$ (TF-GP)	$\rho'_{\max}\, a^3$ (TF)	$\rho'_{\max}\, a^3$ (ATF)
10^5	9.5749×10^{-6}	9.4520×10^{-6}	9.4697×10^{-6}
10^6	2.4051×10^{-5}	2.3569×10^{-5}	2.3636×10^{-5}
10^7	6.0413×10^{-5}	5.8532×10^{-5}	5.8786×10^{-5}
10^8	1.5175×10^{-4}	1.4450×10^{-4}	1.4542×10^{-4}
10^9	3.8118×10^{-4}	3.5365×10^{-4}	3.5688×10^{-4}
10^{10}	9.5749×10^{-4}	8.5527×10^{-4}	8.6586×10^{-4}

So, this analytical approximation to the Thomas-Fermi approximation is quite good. For example, for $N = 10^{10}$ Cs atoms, one finds a relative difference, between the exact TF and the ATF, of 5.66×10^{-3} for the chemical potential, 5.51×10^{-3} for the maximum density and 1.45×10^{-3} for the condensate energy. These are much smaller than the corresponding relative differences between the Thomas-Fermi approximation to the GP equation, TF-GP, and the exact TF results.

Finally, one should beware that this approximation yields a density per particle, n', which is not exactly normalized. This follows from the approximation $\int_0^{R'} (n' - n)\, r^2\, dr = 0$. Indeed, the above approximation implies that

Table 14.14. Chemical potential μ as a function of the number N of trapped Cs atoms, in a 10 Hz trap, in different TF approximations: The TF-GP approximation, the exact TF approximation, the ATF and the renormalized ATF. We take $a = 3.20$ nm.

N	E_0/N (TF-GP)	E_0'/N (TF)	E_0'/N (ATF)	E_0'/N (RATF)
10^5	7.0651	7.0763	7.0762	7.0763
10^6	17.7466	17.7913	17.7906	17.7913
10^7	44.5775	44.7551	44.7507	44.7551
10^8	111.9737	112.6777	112.6502	112.6779
10^9	281.2651	284.0507	283.8807	284.0519
10^{10}	706.5061	717.4898	716.4489	717.4965

$$\begin{aligned} I &= \int_0^{R'} n' \, 4\pi r^2 \, dr \\ &= 1 + \frac{4\pi\mu}{c_1} (R'^3 - R^3)/3 - \frac{4\pi}{c_1} (R'^5 - R^5)/5 \quad . \end{aligned} \tag{14.55}$$

So $I = \int_0^{R'} n' \, 4\pi r^2 \, dr$ is very slightly less than 1. One can improve our approximation by renormalizing $n' : n' \to n'/I$. This raises very slightly the maximum density, which increases the corresponding discrepency between the TF value and the (R)ATF value. But for the total energy, we find an excellent improvement (table 14.14). The new expression for the energy is:

$$\begin{aligned} E_0'/N &= c_1 \, a^2 \, \mu'^{3/2} \frac{16\pi}{105} \frac{1}{I^2} + a \, \mu'^{5/2} \frac{8\pi}{35} \frac{1}{I} \\ &+ c_2 \, a^{5/2} \, \mu'^{3/2} \frac{\pi^2}{32} \frac{1}{I^{5/2}} \quad . \end{aligned} \tag{14.56}$$

References

1. M. H. Anderson, J. R. Ensher, M. R. Matthews, C. E. Wieman, and E. A. Cornell, Observation of Bose-Einstein condensation in a dilute atomic vapor, *Science*. **269**, 198, (1995).
2. C. C. Bradley, C. A. Sackett, J. J. Tollet, and R. G. Hulet, Evidence of Bose-Einstein condensation in an atomic gas with attractive interactions, *Phys. Rev. Lett.* **75**, 1687, (1995).
3. K.B.Davis, M.O.Mewes, M.R.Andrews, N. Druten, D.S.Durfee, D.M.Kurn, and W.Ketterle, Bose-Einstein condensation in a gas of sodium atoms, *Phys. Rev. Lett.* **75**, 3969, (1995).

4. D.G.Fried, T.C.Killian, L.Willmann, D. Landhuis, S. C. Moss, D. Kleppner, and T. J. Greytak, Bose-Einstein condensation of atomic hydrogen, *Phys. Rev. Lett.* **81**, 3811, (1998).
5. A. Robert, O. Sirjean, A. Browaeys, J. Poupard, S. Nowak, D. Boiron, C. Westbrook, and A. Aspect, A Bose-Einstein condensate of metastable atoms, *Science.* **292**, 461, (2001).
6. F. P. dos Santos, J. Léonard, J. Wang, C. J. Barrelet, F. Perales, E. Rasel, C. S. Unnikrishnan, M. Leduc, and C. Cohen-Tannoudji, Bose-Einstein condensation of metastable helium, *Phys. Rev. Lett.* **86**, 3459, (2001).
7. M. W. Zwierlein, C. A. Stan, C. H. Schunck, S. M. F. Raupach, S. Gupta, Z. Hadzibabic, and W. Ketterle, Observation of Bose-Einstein condensation of molecules, *Phys. Rev. Lett.* **91**, 250401, (2003).
8. E. P. Gross, Structure of a quantized vortex in boson systems, *Nuovo Cimento.* **20**, 454, (1961).
9. L. P. Pitaevskii, Vortex lines in an imperfect bose gas, *Zh. Eksp. Teor. Fiz.* **40**, 646, (1961).
10. A.L.Fetter and J.D.Walecka, *Quantum theory of many-particle systems.* (McGraw-Hill, 1980).
11. D.M.Ceperley and B.J.Alder, Ground state of the electron gas by a stochastic method, *Phys. Rev. Lett.* **45**, 566, (1980).
12. W.Kohn and L.J.Sham, Self-consistent equations including exchange and correlation effects, *Phys. Rev.* **140**, A1133, (1965).
13. P.Hohenberg and W.Kohn, Inhomogeneous electron gas, *Phys. Rev.* **136**, B864, (1964).
14. S.Giorgini, J.Boronat, and J.Casulleras, Ground state of a homogeneous Bose gas: A diffusion Monte Carlo calculation, *Phys. Rev. A.* **60**, 5129, (1999).
15. G. S. Nunes, Density functional theory of the inhomogeneous Bose-Einstein condensate, *J. Phys. B: At. Mol. Opt. Phys.* **32**, 4293, (1999).
16. G.S.Nunes, *Structural properties and electronic structure of crystalline silver halides — PhD Thesis.* (SUNY at Stony Brook, NY, USA, 1997).
17. K.Huang and C.N.Yang, Quantum-mechanical many-body problem with hard-sphere interaction, *Phys. Rev.* **105**, 767, (1957).
18. K.Huang, C.N.Yang, and J.M.Luttinger, Imperfect Bose gas with hard sphere interaction, *Phys. Rev.* **105**, 776, (1957).
19. T.D.Lee, K.Huang, and C.N.Yang, Eigenvalues and eigenfunctions of a Bose system of hard spheres and its low-temperature properties, *Phys. Rev.* **106**, 1135, (1957).
20. T.T.Wu, Ground state of a Bose system of hard spheres, *Phys. Rev.* **115**, 1390, (1959).
21. N.Hugenholtz and D.Pines, Ground state and excitation spectrum of a system of interacting bosons, *Phys. Rev.* **116**, 489, (1959).
22. K.Sawada, Ground-state of a Bose-Einstein gas with repulsive interaction, *Phys. Rev.* **116**, 1344, (1959).
23. E.Braaten and A.Nieto, Renormalization effects in a dilute Bose gas, *Phys. Rev. B.* **55**, 8090, (1997).
24. A.Banerjee and M.P.Singh, Ground-state properties of a trapped Bose gas beyond the mean-field approximation, *Phys. Rev. A.* **64**, 063604, (2001).
25. E. Erdemir and B. Tanatar, q-gaussian trial function in high density Bose-Einstein condensates, *Physica A.* **322**, 449, (2003).

26. B.A.McKinney, M.Dunn, and D.K.Watson, Beyond-mean-field results for atomic Bose-Einstein condensates at interaction strengths near Feshbach resonances: A many-body perturbation-theory calculation, *Phys. Rev. A.* **69**, 053611, (2004).
27. S.Inouye, M.R.Andrews, J.Stenger, H.J.Miesner, D.M.Stamper-Kurn, and W.Ketterle, Observation of Feshbach resonances in a Bose-Einstein condensate, *Nature.* **392**, 151, (1998).
28. P.Courteille, R.S.Freeland, D.J.Heinzen, F.A.Abeelen, and B.J.Verhaar, Observation of a Feshbach resonance in cold atom scattering, *Phys. Rev. Lett.* **81**, 69, (1998).
29. J.L.Roberts, N.R.Claussen, J.P.Burke, C.H.Greene, E.A.Cornell, and C.E.Wieman, Resonant magnetic field control of elastic scattering in cold ^{85}rb, *Phys. Rev. Lett.* **81**, 5109, (1998).
30. S.L.Cornish, N.R.Claussen, J.L.Roberts, E.A.Cornell, and C.E.Wieman, Stable ^{85}rb Bose-Einstein condensates with widely tunable interactions, *Phys. Rev. Lett.* **85**, 1795, (2000).
31. V. Vuletić, A. J. Kerman, , C. Chin, and S. Chu, Observation of low-field Feshbach resonances in collisions of cesium atoms, *Phys. Rev. Lett.* **82**, 1406, (1999).
32. G.Baym and C.J.Pethick, Ground-state properties of magnetically trapped Bose-condensed rubidium gas, *Phys. Rev. Lett.* **76**, 6, (1996).
33. P.Schuck and X.Vinas, Thomas-Fermi approximation for Bose-Einstein condensates in traps, *Phys. Rev. A.* **61**, 043603, (2000).

Chapter 15

Wave kinetic description of Bose Einstein condensates

J. T. Mendonça

CFIF and CFP, Instituto Superior Técnico, 1049-001 Lisboa, Portugal,
titomend@ist.utl.pt

A kinetic approach to cold atoms and Bose-Einstein condensates is explored. This approach is based on the Wigner transformation, which allows for a classical phase space representation of a quantum system. Wave kinetic equations exactly equivalent to the Gross Pitaevskii equation are considered, and various approximations are discussed. In the quasi-classical limit, we obtain the particle number conservation equation. Several different examples of application of this method are given. They include, self-phase modulation of a BE condensate, modulational instability and wakefield generation by a cold atom beam in a thermal background, and kinetic dispersion relation of Bogoliubov oscillations with collisionless Landau damping.

Contents

15.1. Introduction

In recent years, research in cold atom beams and Bose-Einstein (BE) condensates has become extremely attractive (see the recent reviews[1,2]). Here we explore the possible use of wave kinetics to describe the physical properties of these systems.

Bose Einstein condensates are ensembles of highly correlated atoms that all are in the same low energy state and can be described by a single collective wavefunction . The evolution of this collective state is determined by the Gross-Pitaevskii (GP) equation,[3,4] which takes the form of a nonlinear Schrödinger equation where the nonlinear term describes the atomic collisions inside the condensate. These collisions are the ultimate reason for the existence of strong correlations between the atomic states and for the use of a single collective wavefuntion. In a sense, atomic collisions can built-up an effective potential that affects the evolution of the collective system.

In contrast with the traditional theoretical approach based on the GP equation, we have developed in recent years a wave kinetic theory, which corresponds to a distinct but exactly equivalent approach to BE condensates. One can question the interest of using a new approach. But Physics has shown in the past that alternative models can shine new light on already known problems and reveal the existence of new aspects of the physical reality. Our aim is therefore to provide an independent view on BE condensates. And we will show, through a series of examples, that our new approach is particularly well adapted to the study of dynamical processes in the condensates.

Wave kinetic theory emerged from Quantum Mechanics, in the early 1930's when E. Wigner and H. Weyl introduce the classical phase space representation of quantum states.[5,6] The instrument for such a representation is the so called Wigner function, or Wigner quasi-probability.[7] An evolution equation for this quasi-probability was later derived by Moyal,[8] starting from the Schrödinger equation. The Moyal equation is an exact kinetic equation, equivalent to the Schrödinger equation, which can be generalized to the GP equation as discussed below. The Wigner approach has been widely used in many different areas of physics and engineering.

In more recent times we have been able to show that the wave kinetic approach, based on the Wigner-Moyal (WM) theory, is an important theoretical instrument for the study of collective plasma phenomena (see[9] and the recent review[10]). Plasma turbulence is a particularly challenging problem, but it can be considerably simplified if part of the complex turbulent fields are described as a distribution of quasi-particles. The same methods can be applied to nonlinear optics and photonics, where radiation evolving in a given optical medium can be seen as a gas of photons described by a quasi-distribution and obeying an appropriate kinetic equation.[11] More recent applications of this approach range from nonlinear optics in fibers[12] and in crystals,[13] to neutrino physics (see[14] and for a recent review[15]). In most of these areas, the original WM formulation had to be generalized in order to account for different wave equations and to incorporate self-consistent nonlinear coupling with other particle and field distributions which are, apart from quasi-particles, the other basic components of the medium. From all this work, a self-consistent description of nonlinear collective phenomena based on the use of wave kinetic equations of the WM type has been established.

The key point of our present approach will be the use of a WM equation for the BE condensates, describing the spatio-temporal evolution of the appropriate Wigner function. Wigner functions for the BECs were discussed in the past[16,17] and the WM equation has also been sporadically used.[18] But no systematic application of the WM equation has previously been considered. This equation can be used both in exact form and in the quasi-classical limit. In the quasi-classical limit, the wave kinetic equation reduces to the particle number conservation equation, which is a kinetic equation formally analogous to the one-particle Liouville equation, also called a Vlasov equation because it incorporates a collective mean field potential. A description of the BE condensates in terms of both the exact and the quasi-classical kinetic equation is adequate to deal with a large variety of problems, as exemplified here. The possible existence of a thermal (or non-condensate) background gas is also considered. The non-condensate gas can be described by fluid equations.

The manuscript is organized in the following fashion. In Section 2, we establish the wave kinetic equation and discuss its exact and approximate versions. We then apply them to several different physical problems. The first one, considered in Section 3, is the self-phase modulation of a cold atom or a BE condensate beam. A similar problem has been studied numerically in the past,[19] but the present formulation leads to explicit analytical results. These results show that a part of the BE condensate beam is decelerated and eventually comes to a complete halt, as a result of the collective forces acting on the condensate. Another example is considered in Section 4, where we establish the kinetic dispersion relation for sound waves in the BE condensates, giving a kinetic correction to the usual Bogoliubov sound speed,[20,21] and predicting the occurrence of Landau damping.[22,23] Our description of Landau damping is significantly different from that previously considered for transverse oscillations of BECs.[24]

In Section 5, we consider the excitation of a wakefield left behind the condensate, when it moves across the non-condensate gas and, in Section 6, we study the modulational instability of the BE condensate and discuss its relation with the wakefield. The aim of Section 7, is to improve on the kinetic description of the BE condensate, by using an exact form of wave kinetic equation, which is therefore able to describe all the quantum features of the condensate. We illustrate this improved kinetic treatment by applying it to the Bogoliubov oscillations. An exact description of Landau damping is also considered. New quantum features can be identified both in the dispersion relation, and in the exact damping coefficients. The range of validity of the quasi-classical approach used in Section 4 is clarified. Finally, in Section 8, the virtues and limitations of the kinetic approach are discussed and some conclusions are stated.

15.2. Kinetic equation for the condensate

It is known that for an ultra-cold atomic ensemble, and in particular for BE condensates, the ground state atomic quantum field can be replaced by a macroscopic atomic wave function ψ. In a large variety of situations, the evolution of ψ is determined by the GP equation

$$i\hbar\partial_t\psi = -\frac{\hbar^2}{2m}\nabla^2\psi + (V_0 + V_{NL})\psi \tag{15.1}$$

where $\partial_t \equiv \partial/\partial t$, $\psi \equiv \psi(\vec{r}, t)$ is the macroscopic wave function of the ground state atomic field, $V_0 = V(\vec{r})$ is the external confining potential, and the nonlinear potential V_{NL} is defined as

$$V_{NL}(\vec{r}, t) = g\,|\psi(\vec{r}, t)|^2 \tag{15.2}$$

Here g is the interaction strength[3,4] defined by $g = 4\pi a\hbar^2/m$, with a, the s-wave scattering length, and m the mass of each boson. In order to built up our alternative kinetic approach we first define the auto-correlation function for the BE condensate wave function, $K \equiv \psi(\vec{r}_1, t_1)\psi^*(\vec{r}_2, t_2)$, which can also be written in the form

$$K(\vec{r}, \vec{s}) = \psi(\vec{r} + \vec{s}/2, t + \tau/2)\;\psi^*(\vec{r} - \vec{s}/2, t - \tau/2) \tag{15.3}$$

where the two vector positions are $\vec{r} = (\vec{r}_1 + \vec{r}_2)/2$ and $\vec{s} = \vec{r}_1 - \vec{r}_2$, and the two time variables are $t = (t_1 + t_2)/2$ and $\tau = t_1 - t_2$. We now define the double Fourier transformation of this auto-correlation function, by integrating over the fast and short scale variables $\vec{s}$ and τ, as shown by

$$W(\omega, \vec{k}; \vec{r}, t) = \int d\vec{s} \int d\tau\; K(\vec{r}, \vec{s}) \exp(-i\vec{k}\cdot\vec{s} + i\omega\tau) \tag{15.4}$$

This is a straightforward generalization of the usual Wigner function,[7] which is useful here for a consistent description of the BE condensate, as shown below. Going back to the original GP equation (15.1)-(15.2), it is then possible to derive the following exact evolution equation for the function $W(\omega, \vec{k}; \vec{r}, t)$, in the form

$$i\hbar\left(\partial_t + \frac{\hbar}{m}\vec{k}\cdot\nabla\right)W = \int \frac{d\vec{q}}{(2\pi)^3}\int \frac{d\Omega}{2\pi}\, V(\vec{q},\Omega)\,[W_- - W_+]\exp(i\vec{q}\cdot\vec{r} - i\Omega t) \tag{15.5}$$

where $V(\vec{q},\Omega)$ is the double Fourier transformation of the total potential $V = V_0 + V_{NL}$, as determined by

$$V(\vec{r},t) = \int \frac{d\vec{q}}{(2\pi)^3}\int \frac{d\Omega}{d2\pi}\, V(\vec{q},\Omega)\exp(i\vec{q}\cdot\vec{r} - i\Omega t) \tag{15.6}$$

and $W_\pm$ are determined by

$$W_\pm \equiv W(\omega \pm \Omega/2, \vec{k} \pm \vec{q}/2) \tag{15.7}$$

Equation (15.3) can be seen as the WM equation describing the space and time evolution of the BE condensates, and it is exactly equivalent to the GP equation (15.1). However, it is sometimes more convenient to use a quasi-classical approximation. This is justified for the important case of slowly varying potentials. In this case, we can neglect the higher order space and time derivatives. This corresponds to the quasi-classical approximation. Introducing the simplifying assumptions that lead to the quasi-classical limit, we reduce the above WM equation to a much simpler form

$$\left(\frac{\partial}{\partial t} + \vec{v}\cdot\nabla + \vec{F}\cdot\frac{\partial}{\partial\vec{k}}\right)W = 0, \tag{15.8}$$

where $\vec{v} = \hbar\vec{k}/m$ is the velocity of the condensate atoms corresponding to the wavevector state $\vec{k}$, and $\vec{F} = \nabla V$ is a force associated with the inhomogeneity of the condensate self-potential. The nonlinear term in the GP equation (15.1) is hidden in this force $\vec{F}$. As will be seen, this nonlinear term looks very much like a ponderomotive force, similar to radiation pressure.

In this quasi-classical limit, W is nothing but the particle occupation number for translational states with momentum $\vec{p} = \hbar\vec{k}$. Equation (15.8) is equivalent to a conservation equation, stating the conservation of the quasi-probability W in the six-dimensional classical phase space $(\vec{r},\vec{k})$, and can also be written as

$$\frac{d}{dt}W(\vec{r},\vec{k},t) = 0. \tag{15.9}$$

This kinetic equation can then be used to describe physical processes occurring in a BE condensate, as long as the quasi-classical approximation of slowly varying potentials is justified. The interest of such a kinetic descriptions will be illustrated with the aid of two simple and different examples, to be presented in the next sections. Many other applications can be envisaged, and will be explored in future publications.

In the presence of a non-condensate background, the BE condensate can be described by a generalized Gross-Pitaevskii equation where the nonlinear potential V_{NL} is now determined by

$$V_{NL}(\vec{r},t) = g|\psi(\vec{r},t)|^2 + 2gn(\vec{r},t) \tag{15.10}$$

where g was defined above and $n(\vec{r},t)$ is the density of the non-condensate gas. In the kinetic equation for the BE condensate, the force is determined by $\vec{F} = \nabla(V_0 + V_{NL})$, which means that local inhomogeneities in the non-condensate gas will contribute to the force acting on the condensate, therefore coupling the evolution of the two gaseous phases. In order to describe the evolution of the background thermal gas we can use a similar kinetic equation for its distribution function $f_b(\vec{r},\vec{v_b},t)$. In the general case, due to collisions, the number of particles in the BE condensate and in the background thermal gas will not be constant, because some of the atoms will eventually be capture (or emitted) by the condensate phase. This means that a general kinetic description of the two gaseous phases wold require the inclusion of source terms in both kinetic equations, according to the expressions

$$\left(\frac{\partial}{\partial t} + \vec{v}\cdot\nabla + \vec{F}\cdot\frac{\partial}{\partial \vec{k}}\right) W = \left(\frac{\partial W}{\partial t}\right)_{coll},$$

$$\left(\frac{\partial}{\partial t} + \vec{v_b}\cdot\nabla + \vec{F_b}\cdot\frac{\partial}{\partial \vec{k}}\right) f_b = \left(\frac{\partial f_b}{\partial t}\right)_{coll}, \tag{15.11}$$

where F_b is the force acting on the background atoms. However, for phenomena with short time scales such as the waves and instabilities considered here, the collision terms can usually be neglected. Furthermore, if we are not interested in the kinetic effects associated with the background gas, but only with those referring to the BE condensate, we can take the momenta of the second kinetic equation an derive a set of fluid equations that can be written as[21]

$$\frac{\partial n}{\partial t} + \nabla\cdot(n\vec{v_n}) = 0$$

$$mn\left[\frac{\partial \vec{v}_n}{\partial t} + (\vec{v}_n \cdot \nabla)\vec{v}_n\right] = -\nabla P - n\nabla U \tag{15.12}$$

$$\frac{\partial \epsilon}{\partial t} + \frac{5}{3}\nabla \cdot (\epsilon \vec{v}_n) = \vec{v}_n \cdot \nabla P$$

where $\vec{v}_n, P$ and ϵ are the mean velocity, the pressure and the (non-convective) energy density of the thermal gas. The two last quantities are simply related by $P = 2\epsilon/3$. The potential U includes the self-consistent Hartree-Fock mean field of the condensate and can be written as $U(\vec{r},t) = V_0(\vec{r}) + 2g[n(\vec{r},t) + N(\vec{r},t)]$ with $N(\vec{r},t) = \int W(\vec{r},\vec{k},t)(d\vec{k}/2\pi)^3$. In the classical limit, this quantity N can be identified with the density of the BE condensate.

15.3. Self-phase modulation

We first consider self-phase modulation of a BE condensate, moving with respect to the confining potential $V_0(\vec{r})$, in the absence of a background gas $n = 0$. This is analogous to the well known problem of self-phase modulation of short laser pulses moving in a nonlinear optical medium.[12] We consider the one-dimensional problem of a beam moving along the z-axis. The kinetic equation (15.8) can then be written as

$$\left(\frac{\partial}{\partial t} + v_z\frac{\partial}{\partial z} + F_z\frac{\partial}{\partial k}\right) W(z,k,t) = 0, \tag{15.13}$$

with v_z and F_z given by, respectively,

$$v_z = \frac{\hbar k}{m} + g\frac{\partial}{\partial t}N(z,t) \quad , \quad F_z = \frac{dk}{dt} = -g\frac{\partial}{\partial z}N(z,t), \tag{15.14}$$

where the axial density (or intensity) of the beam condensate, is defined as $N(z,t) = \int W(z,k,t)\ dk/2\pi$. Let us assume that the ultra-cold atomic beam has a mean velocity $v_0 = \hbar k_0/m$. This suggests the use of a new space coordinate $\eta = z - v_0 t$. In terms of this new coordinate, the quasi-classical equations of motion of a cold atom in the beam can be written in canonical form

$$\frac{d\eta}{dt} = \frac{\partial h}{\partial k} = \frac{1}{m}(k - k_0), \tag{15.15}$$
$$\frac{dk}{dt} = -\frac{\partial h}{\partial \eta} = -\frac{g}{\hbar}\frac{\partial}{\partial \eta}N(\eta,t),$$

where we have introduced the Hamiltonian function

$$h(\eta,k,t)=\omega(\eta,k,t)-kv_0=\frac{k}{m}\left(\frac{k}{2}-k_0\right)+\frac{g}{\hbar}N(\eta,t). \tag{15.16}$$

Here $\omega(\eta,k,t)$ is the Hamiltonian in the rest frame expressed in the new coordinate. A straightforward integration of the equations of motion leads to

$$k(t)=k_0-\frac{g}{\hbar}\int_0^t\frac{\partial}{\partial\eta}N(\eta,t')dt'. \tag{15.17}$$

Let us introduce the concept of beam energy chirp, $<\epsilon(\eta,t)>$, in analogy with the frequency chirp of a laser pulse.[12] It determines the beam mean energy, at a given position and at a given time

$$<\epsilon(\eta,t)>=\hbar\int W(\eta,k,t)\,\omega(\eta,k,t)\frac{dk}{2\pi}, \tag{15.18}$$

where the weighting function $W(\eta,k,t)$ is the solution of the kinetic one-dimensional equation (15.13). It is possible to show that[29]

$$<\epsilon(\eta,t)>=<\epsilon(0)>-\frac{k_0}{m}g\int_0^t\frac{\partial}{\partial\eta}N(\eta,t')dt', \tag{15.19}$$

where $<\epsilon(0)>\equiv<\epsilon(\eta_0,t_0)>$ is the initial beam energy chirp. If the beam density profile $N(\eta)$ is independent of time this reduces to

$$<\epsilon(\eta,t)>=<\epsilon(0)>-\hbar v_0 g\left(\frac{\partial N}{\partial\eta}\right)t. \tag{15.20}$$

The maximum energy shift will be attained at some position inside the beam profile, $\eta=\eta_{max}$, determined by the stationarity condition

$$\frac{\partial}{\partial\eta}<\epsilon(\eta,t)>=\left(\frac{\partial^2N}{\partial\eta^2}\right)=0. \tag{15.21}$$

For a Gaussian beam profile, $N(\eta)=N_0\exp(-\eta^2/\sigma^2)$, where σ determines the beam width, we have $\eta_{max}=\pm\sigma/\sqrt{2}$. This leads to the following value of the maximum energy shift

$$\Delta\epsilon(t)\equiv<\epsilon(t)>_{max}-<\epsilon(0)>=\pm\frac{\hbar\sqrt{2}}{\sigma}gv_0N_0e^{-1/2}t. \tag{15.22}$$

This shows that the maximum energy chirp due to self-phase modulation is proportional to the distance travelled by the beam, $d = v_0 t$. Due to the plus and minus signs, the initial beam will split into two, one part of the beam being accelerated to higher speeds, while the other is decelerated. This would correspond to the red-shift and to the blue-shift observed in nonlinear optics. The decelerated beam will eventually stop at a time $t \simeq \tau$, such that $\Delta\epsilon(\tau) =< \epsilon(0) >$. We therefore established the condition for beam freezing.

15.4. Bogoliubov oscillations

Our second example deals with the dispersion relation of sound waves. Again, we assume a given equilibrium distribution $W_0(z, k, t)$, corresponding, for instance, to the Thomas-Fermi equilibrium for a given confining potential $V_0(\vec{r}_\perp, z)$.[25] Linearizing the one-dimensional kinetic equation with respect to the perturbation $\tilde{W}$, we obtain

$$\left(\frac{\partial}{\partial t} + v_z \frac{\partial}{\partial z}\right) \tilde{W}(z, k, t) + \tilde{F} \frac{\partial}{\partial k} W_0(z, k, t) = 0, \tag{15.23}$$

where the perturbed force is determined by

$$\tilde{F} = -\frac{g}{\hbar} \frac{\partial}{\partial z} \tilde{N}(z, t) = -\frac{g}{\hbar} \frac{\partial}{\partial z} \int \tilde{W}(z, k, t) \frac{dk}{2\pi}. \tag{15.24}$$

Let us now assume perturbations of the form $\tilde{W}, \tilde{N} \sim \exp(ikz - i\omega t)$. Integration over the momentum spectrum of the particle condensate then leads to the following expression

$$1 + \frac{g}{\hbar} k \int \frac{\partial W_0(k')/\partial k'}{(\omega - \hbar k k'/m)} \frac{dk'}{2\pi} = 0. \tag{15.25}$$

This is the kinetic dispersion relation for axial perturbations in the BE condensate. Let us illustrate this for a condensate beam with no translational dispersion. The equilibrium state of the beam can then be described by $W_0(k') = 2\pi N_0 \delta(k' - k_0')$, where $N_0 = \int W_0(k') dk'/2\pi$ is the particle number density in the condensate. Replacing this in the dispersion relation (15.25). we have

$$1 - \frac{g k^2}{m} \frac{N_0}{(\omega - k v_0')^2} = 0, \tag{15.26}$$

where $v_0' = \hbar k_0'/m = p_0'/m$ is the beam velocity. This can also be written as $(\omega - k v_0')^2 = k^2 c_s^2$, where $c_s = \sqrt{g N_0/m}$ is the Bogoliubov sound speed.

Obviously, equation (15.35) is the Doppler shifted dispersion relation of sound waves in the BEC gas. In its reference frame it reduces to $\omega = kc_s$.

Let us now consider a situation where, instead of a cold beam, we have a beam with a finite translational velocity spread. In this case, the dispersion relation written in the condensate frame of reference gives $\omega = kc_s + i\gamma$, where $|\gamma| \ll kc_s$, is the damping coefficient

$$\gamma = \frac{\omega}{4}\frac{gm}{\hbar^2}\left(\frac{\partial W_0}{\partial k'}\right)_{k'=k'_s} . \tag{15.27}$$

This corresponds to the non-collisional Landau damping of Bogoliubov oscillations in the BE condensate.

15.5. Wakefield excitation

In this Section, we consider the role of the background on the BE condensate oscillations.[30] We now assume that the relative motion of the condensate and the thermal gas produces density perturbations $\tilde{N}$ and $\tilde{n}$ around their equilibrium values N_0 and n_0. Similarly, we take the pressure as $P = P_0 + \tilde{P}$. Linearizing equations (15.12) with respect to the perturbations, we obtain

$$\left(\frac{\partial^2}{\partial t^2} - u_s^2 \nabla^2\right)\tilde{n} = 2n_0 \frac{g}{m}\nabla^2(N_0 + \tilde{N}) \tag{15.28}$$

where we have used the sound speed u_s in the thermal gas

$$u_s^2 = \frac{5}{3}\frac{P_0}{mn_0} + \frac{2g}{m}n_0 \tag{15.29}$$

Let us first consider the case of a very short BE condensate beam moving with velocity $\vec{u}_0 = u_0\vec{e}_x$, across the non-condensate background gas. In such conditions, the gradient of N_0 in equation (15.28) dominates, and we can use $\tilde{N} = 0$. We retain the possible existence of a wave structure in the perpendicular direction by introducing a cuf-off frequency ω_0, which is determined by the transverse part of the ∇^2 operator. We then get

$$\left(\frac{\partial^2}{\partial t^2} - u_s^2\frac{\partial^2}{\partial x^2}\right)\tilde{n} + \omega_0^2\tilde{n} = 2n_0\frac{g}{m}\frac{\partial^2}{\partial x^2}N_0(x - u_0 t) \tag{15.30}$$

In the quasi-static approximation, this equation can be reduced to

$$\left(\frac{d^2}{d\eta^2} + 1\right)\tilde{n} = f(\eta) \tag{15.31}$$

with $\eta = k_0(x - u_0 t)$. This is a driven harmonic oscillator with unit frequency, where the force is determined by

$$f(\eta) = 2n_0 \frac{g}{m} \frac{1}{(u_0^2 - u_s^2)} \frac{\partial^2 N_0}{\partial \eta^2} \tag{15.32}$$

Using appropriate initial conditions, we obtain the following solution

$$\tilde{n}(\eta) = 2n_0 \frac{g}{m} \frac{1}{(u_0^2 - u_s^2)} \left[N_0(\eta) - \int_{\infty}^{\eta} N_0(\eta') \sin(\eta - \eta') d\eta' \right] \tag{15.33}$$

Here, the first term is just a local perturbation of the background gas that occurs where the BE condensate is located at a given time. And the second term is the wakefield, which corresponds to an acoustic wave propagating in the background gas, with a phase velocity equal to the velocity of the BE condensate beam. By using the linear dispersion relation of the acoustic waves, $\omega^2 = \omega_0^2 + k^2 u_s^2$, we get the wake frequency in the laboratory frame $\omega = \omega_0 \, (u_0^2 - u_s^2)^{1/2}/u_0$. This shows that the wakefields can only be generated if the acoustic wave has some transverse structure, which means $\omega_0 \neq 0$, and also if the BE condensate moves faster than the sound speed, $u_0 > u_s$.

15.6. Modulational instability

We now go back to equation (15.28) but neglect the driving term in N_0, and retain $\tilde{N}$. It can then be shown that, even in the absence of the driving term, the system is unstable. In order to solve equation (15.28) we now need an additional relation between $\tilde{N}$ and $\tilde{n}$, which is determined by the linearized kinetic equation. We then get two coupled equations for the perturbed quantities $\tilde{n}$, $\tilde{W}$ and $\tilde{N}$. For perturbations evolving as $\exp(i\vec{k} \cdot \vec{r} - i\omega t)$, these equations are

$$(\omega^2 - u_s^2 k^2)\tilde{n} = 2n_0 \frac{g}{m} k^2 \tilde{N}$$

and

$$(\omega - \vec{k} \cdot \vec{v})\tilde{W} = -g(\tilde{N} + 2\tilde{n})\vec{k} \cdot \frac{\partial W_0}{\partial \vec{p}} \tag{15.34}$$

This leads to the following dispersion relation of the system BE condensate plus the thermal gas background

$$(\omega^2 - u_s^2 k^2) = -4n_0 \frac{g}{m} k^2 \frac{G}{(1 + G)} \tag{15.35}$$

with

$$G = g \int \frac{\vec{k} \cdot \partial W_0/\partial \vec{p}}{(\omega - \vec{k} \cdot \vec{p}/m)} \frac{d\vec{p}}{(2\pi\hbar)^3} \tag{15.36}$$

For a nearly mono-kinetic BE condensate beam, such that $W_0(\vec{p}) = (2\pi\hbar)^3 N_0 \delta(\vec{p} - \vec{p}_0)$, this reduces to

$$(\omega^2 - u_s^2 k^2)(\omega - k u_0)^2 = 4\frac{n_0}{N_0}(k c_s)^4 \tag{15.37}$$

where $u_0 = p_0/m$, and the Bogolioubov sound speed is $c_s = \sqrt{g N_0/m}$. In the absence of coupling between the two gaseous phase (condensate and background), this expression splits into two separate dispersion relation, $\omega = k u_s$ for sound waves in the thermal gas, and $\omega - k u_0 = k c_s$, for Doppler shifted sound waves in the BE condensate. The coupling between the two fluids, implied by G, will produce beam-like instabilities. To illustrate this important question, we consider the resonant condition $u_s = u_0$. The instability growth rate is then given by

$$\gamma = \frac{\sqrt{3}}{2^{2/3}} \left(\frac{n_0}{N_0}\right)^{1/3} \left(\frac{c_s}{u_0}\right)^{4/3} \omega_r \tag{15.38}$$

This is valid for a cold beam. Inclusion of a finite translational beam temperature has also been considered.[30]

15.7. Purely quantum effects

Until now we have only applied the wave kinetic equation (15.8), valid in the quasi-classical approximation. In this Section we will use the exact kinetic equation, which will allow us to discuss purely quantum effects. In this case, the dispersion relation for Bogoliubov oscillations (15.25) is no longer valid, and is replaced by[32]

$$1 + \frac{k}{\hbar} \int W_0(k_z') \left[\frac{1}{(\omega_+ - k v_z)} - \frac{1}{(\omega_- - k v_z)}\right] \frac{dk_z'}{2\pi} = 0 \tag{15.39}$$

where $\omega_\pm = \omega \pm \hbar k^2/2m$. For the simple case of a condensate beam with no translactional temperature, e $W_0(k_z') = 2\pi W_0 \delta(k_z' - k_0')$, this reduces to

$$(\omega - k v_0)^2 = k^2 c_s^2 + k^4 \frac{\hbar^2}{4m^2} \tag{15.40}$$

where $c_s = \sqrt{kW_0/m}$ is again the Bogoliubov sound speed. In the classical limit, $\hbar \to 0$, and for a BE condensate at rest $v_0 = 0$, we get the classical expression $\omega = kc_s$, as discussed before. The new dispersion relation shows that finite quantum effects introduce a dependence of the wave phase velocity on the wavenumber k.

On the other hand, a finite beam temperature leads to wave damping, determined by the coefficient[32]

$$\gamma = \frac{g^2 W_0}{4\hbar^2} \frac{k}{\omega} \left[W_0(\omega_+) - W_0(\omega_-) \right] \tag{15.41}$$

where

$$W_0(\omega_\pm) = W_0 \left(v_z = \omega/k \pm \frac{\hbar k}{2m} \right) \tag{15.42}$$

Equation (15.41) describes the kinetic non-dissipation wave attenuation, or Landau damping. If the lower translactional energy level $\hbar\omega_-$ is more populated than the higher energy level $\hbar\omega_+$, we have $\gamma < 0$ and wave attenuation occurs. In the opposite case of inversion of population we have a wave instability, or wave growth. In the case of $\hbar k/m \ll \omega/k$, corresponding to the quasi-classical approximation the Landau damping coefficient takes the form

$$\gamma \simeq \frac{g\omega W_0}{4\hbar^2} \left(\frac{\partial W_0}{\partial k'_z} \right)_{res} \tag{15.43}$$

This expression could also be derived directly from the quasi-classical wave kinetic equation,[29] as indicated in Section 4.

15.8. Conclusions

The wave kinetic description of Bose-Einstein condensates was considered. This is based on the Wigner-Moyal equation for the quantum system, which is exactly equivalent to the Gross-Pitaevskii equation. We have used the wave kinetic approach, in both the exact and the quasi-classical formulations. The quasi-classical approach to BE condensates can be seen as an intermediate step between the Gross-Pitaievskii equation and the hydrodynamical equations for the condensate gas, often found in the literature.

A number of different physical problems was considered, in order to illustrate the versatility of the kinetic theory. One is self-phase modulation of a BE condensate beam. This first example shows that, due to the influence of its own inhomogeneous self-potential, nearly half of the beam is accelerated while the other half is decelerated. After some time, the decelerated part of the beam will tend to a state of complete halt. The second example concerns Bogolioubov or sound-like oscillations of the condensate. A kinetic dispersion relation for sound waves in BE condensates could be established.

We have also considered the wakefield excitation by BE condensate beams moving in a non-condensate gas background. The kinetic equation for the BE condensate was modified by the introduction of an additional force term, and the non-condensate background gas was described by fluid equations including the condensate potential, and coupled with the wave kinetic equation. Using these two coupled equations, we were able to study the modulational instabilities of the system. Instability growth rates were derived. A general dispersion relation for the system of coupled BE condensate and non-condensate gas was also established.

An exact wave kinetic description was also used to describe the Bogoliubov oscillations. We have obtained an exact dispersion relation, where quantum dispersion effects could be identified. These effects lead to the existence of dispersion (sound speed dependence on the phonon wavenumber), due to atom recoil during emission or absorption of phonons. Our dispersion relation also contains a wave attenuation term of a kinetic nature, the non-dissipative type of attenuation known as Landau damping. Conditions for sound wave instability correspond to negative Landau damping and can occur for inversion of population in the kinetic energy quantum states. Comparison with the quasi-classical approach of Section 4 shows the influence of atom recoil and clarifies the quasi-classical approximation.

Several other different problems relevant to BECs can also be considered in the frame of the wave kinetic theory, such as acoustic oscillations with a transverse structure. This indicates that the kinetic theory is a very promising approach to the physics of BE condensates, which will eventually allow us to introduce new ideas in this stimulating area of research, and to suggest new configurations to the experimentalists. The examples discussed here show that this theory is particularly well adapted to study dynamical effects associated with the BE condensates.

References

1. A. J. Legget, Bose-Einstein condensation in the alkali gases: Some fundamental concepts, *Rev. Mod. Phys.* **73**, 307 (2001).
2. F. Dalfovo *et al.*, Theory of Bose-Einstein condensation in trapped gases, *Rev. Mod. Phys.* **71**, 463 (1999).
3. E. P. Gross, Structure of a quantized vortex in boson systems, *il Nuovo Cimento* **20**, 454 (1961); Hydrodynamics of a superfluid condensate, *J. Math. Phys.* **4**, 195 (1963).
4. L. P. Pitaevskii, Vortex lines in an imperfect bose gas, *Zh. Eksp. Teor. Fiz* bf 40, 646 (1961) [*Sov. Phys. JETP* **13**, 431 (1961)].

5. E.P. Wigner, On the quantum correction for thermodynamic equilibrium, *Phys. Rev.*, **40**, 749 (1932).
6. H. Weyl, Quantenmechanik und Gruppentheorie *Z. Phys.*, **46**, 1 (1927).
7. M. Hillary, R. F. O' Connel, M. O. Scully and E. P. Wigner, Distribution functions in physics: fundamentals, *Phys. Rep.* **106**, 121 (1984).
8. J.E. Moyal, Quantum mechanics as a statistical theory, *Proc. Cambr. Phil. Soc.*, **45**, 99 (1949).
9. J.T. Mendonça, R. Bingham and P.K. Shukla, Resonant quasiparticles in plasma turbulence, *Phys. Rev. E*, **68**, 016406 (2003).
10. J.T. Mendonça, Wave kinetics and photon acceleration, *Phys. Scripta*, **74**, C61 (2006).
11. J.T. Mendonça, *Theory of Photon Acceleration*, Institute of Physics Publishing, Bristol (2001).
12. L.O. Silva and J.T. Mendonça, Photon kinetic theory of self-phase modulation, *Opt. Commun.*, **196**, 285 (2001).
13. J.T. Mendonça, Photon acceleration and polariton wakefields in dielectric crystals, *New J. of Phys.*, **8**, 185 (2006).
14. R. Bingham et al., Collective interactions between neutrinos and dense-plasmas, *Phys. Lett. A*, **193**, 279 (1994); Interaction between neutrinos and nonstationary plasmas, J.T. Mendonça et al., *Phys. Lett. A*, **209**, 78 (1995); Neutrinos generating inhomogeneities and magnetic fields in the early universe, P.K. Shukla et al., *Phys. Plasmas*, **5**, 2815 (1998); A. Serbeto, Solitons and shock fronts in the scattering of neutrinos by plasmas, *Phys. Plasmas*, **6**, 2943 (1999).
15. R. Bingham et al., Neutrino plasma coupling in dense astrophysical plasmas, *Plasma Phys. Control. Fusion*, **46**, B327 (2005).
16. M. J. Steel *et al.*, Dynamical quantum noise in trapped Bose-Einstein condensates, *Phys. Rev. A* **58**, 4824 (1998).
17. C. W. Gardiner and M. J. Davis, The stochastic Gross-Pitaevskii equation: II, *J. Phys. B: At. Mol. Opt. Phys* **36**, 4731 (2003).
18. S.A. Gardiner et al., Nonlinear matter wave dynamics with a chaotic potential, *Phys. Rev. A*, **62**, 023612 (2000).
19. W. Zhang and D. F. Walls, Quantum field theory of interaction of ultracold atoms with a light wave: Bragg scattering in nonlinear atom optics, *Phys. Rev. A* **49**, 3799 (1994).
20. M. R. Andrews *et al.*, Propagation of sound in a Bose-Einstein condensate, *Phys. Rev. Lett.* **79**, 553 (1997).
21. E. Zaremba, Sound propagation in a cylindrical Bose-condensed gas, *Phys. Rev. A* **57**, 518 (1998).
22. L. P. Pitaevskii and S. Stringari, Landau damping in dilute Bose gases, Phys. Lett. A **235**, 398 (1997).
23. B. Jackson and E. Zaremba, Landau damping in trapped Bose condensed gases, *New J. Phys.* **5**, 88 (2003).
24. M. Guilleumas and L. P. Pitaevskii, Landau damping of transverse quadrupole oscillations of an elongated Bose-Einstein condensate, *Phys. Rev. A* **67**, 053607 (2003).
25. G. Byam and C. J. Pethick, Ground-state properties of magnetically trapped Bose-Condensed rubidium gas, *Phys. Rev. Lett.* **76**, 6 (1996).
26. S. Stringari, Collective excitations of a trapped Bose-condensed gas, *Phys. Rev. Lett.* **77**, 2360 (1996).

27. V.V. Konotop and M. Salerno, Modulational instability in Bose-Einstein condensates in optical lattices, *Phys. Rev. A*, **65**, 021602 (2002).
28. J. T. Mendonça and N. L. Tsintsadze, Analog of the Wigner-Moyal equation for the electromagnetic field, *Phys. Rev. E* **62**, 4276 (2000).
29. J.T. Mendonça, P.K. Shukla and R. Bingham, Wakefield of Bose-Einstein condensates in a background thermal gas, *Phys. Lett. A*, **340**, 355 (2005).
30. J.T. Mendonça, R. Bingham and P.K. Shukla, A kinetic approach to Bose-Einstein condensates: self-phase modulation and Bogoliubov oscillations *JETP*, **201**, 942-948 (2005).
31. E. Zaremba, Sound propagation in a cylindrical Bose-condensed gas, *Phys. Rev. A*, **57**, 518 (1998).
32. J.T. Mendonça and A. Serbeto, Wave kinetic description of Bogoliubov oscillations in a Bose Einstein condensate, (2006) submitted for publication.

Chapter 16

Critical magnetic fields in superconductors with singular density of states

R. G. Dias

Departamento de Física, Universidade de Aveiro,
3810 Aveiro, Portugal,
rdias@fis.ua.pt

We present a review of the temperature-magnetic field phase diagram of homogeneous and inhomogeneous superconductivity in the case of a clean quasi-two-dimensional superconductor with singular density of states. For transverse magnetic field, the superconducting pairing susceptibility $K_T(r)$ displays anomalous short-range behavior which leads to positive curvature in the upper critical field. The Pauli limit (H_p) is strongly enhanced and a huge metastability region appears when the magnetic field is applied parallely to the conducting planes. A non-uniform superconducting FFLO state is not favored by the presence of the van Hove singularity.

Contents

16.1. Introduction

Dimensionality plays a determinant role in the behavior of interacting electron systems.[1,2] For example, one of the key features behind the strange normal state behavior of the high-T_c superconductors is certainly the quasi-two dimensionality of these materials.[1] A common characteristic of two dimensional models of interacting electrons in a periodic lattice potential is the presence of a van Hove singularity (vHS) in the density of states. Such singularities may induce anomalous behavior if in the proximity of the Fermi level. In this chapter, we review the effects of such singularities in the temperature-magnetic field phase diagram of homogeneous and inhomogeneous superconductivity in the case of a quasi-two-dimensional superconductor.[3–5]

The superconducting transition under magnetic field (usually designated by upper critical field, H_{c2}, in the case of a type-II superconductor), is directly related to a particular correlation function, the pair propagator or pairing susceptibility.[6] Anomalous behavior of the pair propagator reflects itself in a unusual form for the upper critical field and therefore, the H_{c2} curve is a probe of the spatial and temperature dependence of the normal state pair propagator.

In the case of quasi-two-dimensional superconductors, under parallel magnetic field, the magnetic orbital frustration is greatly reduced and the superconducting transition is determined by Zeeman pair breaking.[7,8] For such magnetic field configuration, the temperature-magnetic field phase diagram displays a low temperature first-order transition line.[9,10] The zero-temperature critical field associated with this first-order transition is usually denominated by Pauli limit or Chandrasekhar-Clogston limit.[7,8] Superconductivity may however persist at magnetic fields higher than the Pauli limit if the formation of an inhomogeneous superconducting phase, i.e., the so-called Fulde-Ferrel-Larkin-Ovchinnikov (FFLO) phase,[11,12] occurs.

16.1.1. *Van Hove singularities*

In a two-dimensional metal, a vHS in the density of states results usually from the presence of a saddle point in the energy dispersion $\epsilon_{\boldsymbol{k}}$. In the proximity of a simple saddle point, $\epsilon_{\boldsymbol{k}} \sim q_x^2 - q_y^2$, or with an axis rotation, $\epsilon_{\boldsymbol{k}} \sim q_x q_y$,, where

$\boldsymbol{q} = \boldsymbol{k} - \boldsymbol{k}_{vh}$, and therefore a logarithmic singularity appears in the density of states.

Saddle points in the energy dispersion can be directly probed using the angle-resolved photoemission technique. For instance, numerous reports have provided evidence of the existence of saddle points in the case of the Copper-Oxide superconductors.[13] Furthermore, in some cases, extended saddle points have also been found.[14] The existence of extended saddle points is also indicated by numerical work on the two-dimensional Hubbard model[15,16] which is believed to describe the physics of the copper-oxide planes.[1]

In the case of an extended saddle point, $\epsilon_{\boldsymbol{k}} \sim |q_x|^n - |q_y|^m$, with higher powers and this leads to a power-law divergence in the density of states $N(\epsilon) \sim (\epsilon - \epsilon_{vh})^{-\alpha}$ with $\alpha = 1 - \frac{1}{n} - \frac{1}{m}$. Power-law divergences are also present in one-dimensional systems. In a one-dimensional system, a q^2 dispersion leads to a inverse square root divergence in the density of states. In order to eliminate the anisotropy effects associated with the vHS, it is convenient to introduce an isotropic dispersion relation that reproduces the vHS in the density of states, $\epsilon_{\boldsymbol{k}} - \epsilon_{vh} = a \cdot \text{sign}(q)|q|^b$ where $q = k - k_{vh}$. Using this form for the energy dispersion relation eliminates the effects of anisotropy that are inevitably associated with a saddle point. The density of states for the above model is $N(\epsilon) \sim a^{-1/b} b^{-1} (\epsilon - \epsilon_{vh})^{\frac{1}{b}-1}$. The vHS must be pinned at or in the proximity of the Fermi level in order for its effects to be important.

The remaining part of this chapter gives an exposition of the effects of a vHS pinned at the Fermi level in the temperature-magnetic field phase diagram of a 2D weak-coupling superconductor. Section 16.2 introduces a brief derivation of the superconducting gap equation in the presence of magnetic field. In section 16.3, a description of vHS effects in the mean-field superconducting instability for a transverse orientation of magnetic field is presented. The Pauli limit and the possibility of a inhomogeneous superconducting phase in this van Hove scenario are addressed in Section 16.4. Section 16.5 contains brief discussions of how other features such as anisotropy or non-s-wave pairing symmetry may modify the results presented in sections 16.3 and 16.4. In section 16.6, the relevance of these results for the cuprates is discussed. Finally, a short conclusion is presented.

16.2. The gap equation

The BCS pairing Hamiltonian in the absence of magnetic field is

$$H = \sum_{\boldsymbol{k},\sigma=\uparrow,\downarrow} \xi_{\boldsymbol{k}} a^\dagger_{\boldsymbol{k}\sigma} a_{\boldsymbol{k}\sigma} + \sum_{\boldsymbol{k},\boldsymbol{k}'} V_{\boldsymbol{k},\boldsymbol{k}'} a^\dagger_{\boldsymbol{k},\uparrow} a^\dagger_{-\boldsymbol{k}+\boldsymbol{q},\downarrow} a_{-\boldsymbol{k}'+\boldsymbol{q},\downarrow} a_{\boldsymbol{k}',\uparrow} \tag{16.1}$$

with $\xi_{\boldsymbol{k}} = \epsilon_{\boldsymbol{k}} - \mu$ and where $V_{\boldsymbol{k},\boldsymbol{k}'}$ and μ are respectively the attractive interaction constant and the chemical potential. The pairing interaction was written in a more general form to allow Cooper pairing with finite total momentum and the usual approximation for the pairing potential will be adopted, $V_{\boldsymbol{k},\boldsymbol{k}'} = -V$. The $\boldsymbol{k}$ sums in the interaction term also follow the usual BCS restrictions.

Minimizing the superconducting free energy, one obtains the so-called gap equation.[17] This equation plays a central role in the theory of superconductivity and in particular, it allows the determination of the superconducting transition temperature under magnetic field in the case of a second-order transition. The determination of the first-order phase transition points requires, besides the gap equation, the evaluation of the free energy at the local minima.

16.2.1. *Out-of-plane magnetic field*

For magnetic fields perpendicular to the conducting planes, the orbital effects dominate and the critical fields are well below the Pauli limit. Zeeman pair breaking terms can be neglected and the effect of the magnetic field may be introduced in a semi-classical way.[6]

The superconducting second-order transition is characterized by the vanishing of the gap function $\Delta(\boldsymbol{r},\boldsymbol{r}')$, defined in real space as

$$\Delta(\boldsymbol{r},\boldsymbol{r}') = \frac{V(\boldsymbol{r}-\boldsymbol{r}')}{\beta} \sum_{\omega} \mathcal{F}_{\omega}(\boldsymbol{r},\boldsymbol{r}'), \tag{16.2}$$

where the anomalous temperature Green's function $\mathcal{F}_{\omega}(\boldsymbol{r},\boldsymbol{r}')$ is the Fourier transform of $\mathcal{F}(\boldsymbol{r},\boldsymbol{r}';\tau-\tau') = \langle T_{\tau}\psi_{\downarrow}(\boldsymbol{r}\ \tau)\psi_{\uparrow}(\boldsymbol{r}'\ \tau')\rangle$.[18,19] In the presence of magnetic field, $\Delta(\boldsymbol{r},\boldsymbol{r}') = \Delta(\boldsymbol{r})\delta(\boldsymbol{r}-\boldsymbol{r}')$, where the Dirac delta function reflects the s-wave local pairing, $V(\boldsymbol{r}-\boldsymbol{r}') = -V\delta(\boldsymbol{r}-\boldsymbol{r}')$.

In the vicinity of the upper critical transition curve, the gap parameter is small and one can carry out a perturbation expansion in powers of Δ. This leads to a linearized gap equation,[6]

$$\Delta(\boldsymbol{r}) = V \int d\boldsymbol{r}' K_{\beta}^{H}(\boldsymbol{r}',\boldsymbol{r})\Delta(\boldsymbol{r}'), \tag{16.3}$$

which is an homogeneous integral equation with a kernel $K_{\beta}^{H}(\boldsymbol{r}',\boldsymbol{r})$ which represents the normal state electron pair propagator in a magnetic field H,

$$K_{\beta}^{H}(\boldsymbol{r}',\boldsymbol{r}) = \frac{1}{\beta}\sum_{\omega} \mathcal{G}_{-\omega}^{H}(\boldsymbol{r}',\boldsymbol{r})\mathcal{G}_{\omega}^{H}(\boldsymbol{r}',\boldsymbol{r}). \tag{16.4}$$

The Green's function $\mathcal{G}_{\omega}^{H}$ describes the normal state under magnetic field.

The magnetic field effects can be incorporated into the pair propagator in a semi-classical way.[6,18,19] The semi-classical approximation assumes that the magnetic field provides an slowly varying envelope function for the oscillations of the Green's function and therefore $K_\beta^H(\boldsymbol{r}',\boldsymbol{r}) = K_\beta(\boldsymbol{r}'-\boldsymbol{r})\exp[i\phi(\boldsymbol{r}'-\boldsymbol{r})]$ with $\phi(\boldsymbol{r}) = \frac{2e}{\hbar c}\boldsymbol{A}(\boldsymbol{r})\cdot\boldsymbol{r}$, where $K_\beta(\boldsymbol{r})$ is the fermion pair propagator in the absence of the external field and $\boldsymbol{A}(\boldsymbol{r})$ is the vector potential. In the following, $\hbar = c = e = 1$. Using Kramers-Kronig relations, $K_\beta(\boldsymbol{r})$ can be written as

$$K_\beta(\boldsymbol{r}) = \frac{2}{\pi}\int d\omega \tanh(\beta\omega/2)A(\boldsymbol{r},\omega)B(\boldsymbol{r},-\omega), \tag{16.5}$$

with $A(\boldsymbol{k},\omega) = \mathrm{Im}G^R(\boldsymbol{k},\omega)$ and $B(\boldsymbol{k},\omega) = \mathrm{Re}G^R(\boldsymbol{k},\omega)$ where $G^R(\boldsymbol{k},\omega)$ is the retarded Green's function in the absence of magnetic field and pairing potential.

The linearized gap equation in the symmetric gauge $\mathbf{A} = \frac{1}{2}\mathbf{H}\times\mathbf{r}$ becomes

$$\Delta(\mathbf{r}) = V\int d^2\mathbf{r}' K_\beta(\mathbf{r}')\exp\left(i\frac{\mathbf{r}\times\mathbf{r}'}{l^2}\right)\Delta(\mathbf{r}+\mathbf{r}'), \tag{16.6}$$

where β is the inverse temperature, the magnetic length l is related to the applied field by $H = \phi_0(2\pi l^2)^{-1}$, and ϕ_0 is the flux quantum hc/e. The highest eigenvalue of this linear equation determines the upper critical field. In the case of an isotropic energy dispersion,[20] one finds a solution of the form $\Delta(\mathbf{r}) = \Delta_0 e^{-\frac{1}{2}\left(\frac{r}{l}\right)^2}$ which, substituted into Eq.16.6, leads to the mean-field pairing instability condition

$$\frac{1}{2\pi V} = \int_{r_0}^{\infty} K_\beta(r)\exp\left(-\frac{r^2}{2l^2}\right) r dr \tag{16.7}$$

where r_0 is a lower cutoff. This equation determines the critical magnetic length $l_c(\beta)$, or in other words, the upper critical field $H_{c2}(T)$.

16.2.2. *Parallel magnetic field*

In the case of a quasi-two-dimensional superconductor under in-plane magnetic fields, orbital frustration can be neglected and the superconducting critical field is dominated by Zeeman pair breaking.[7,8] Therefore, one must replace the energy dispersion $\xi_{\boldsymbol{k}}$ in the Hamiltonian given by Eq. 16.1 by $\xi_{\boldsymbol{k}\sigma} = \xi_{\boldsymbol{k}} - \sigma h$ with $h = \mu_B H$, and where H and μ_B are respectively the magnetic field and the Bohr magneton.

The free energy minimization[10,17] in the case of an homogeneous gap function, $\Delta = -V\sum_{\boldsymbol{k}}\langle a_{-\boldsymbol{k}\downarrow}a_{\boldsymbol{k}\uparrow}\rangle$ now leads to the gap equation

$$1 = V\sum_{\boldsymbol{p}}\frac{1 - f(\xi_{\boldsymbol{p}\uparrow}^\Delta) - f(\xi_{\boldsymbol{p}\downarrow}^\Delta)}{2\xi_{\boldsymbol{p}\uparrow}^\Delta}, \tag{16.8}$$

with $\xi^{\Delta}_{\boldsymbol{k}\sigma} = \xi^{\Delta}_{\boldsymbol{k}} - \sigma h$, $\xi^{\Delta}_{\boldsymbol{k}} = \sqrt{\xi^2_{\boldsymbol{k}} + \Delta^2}$ and where $f(x)$ is the Fermi distribution function.

16.3. Effects of a vHS on the upper critical field

If sufficiently close to the Fermi level, a vHS in the density of states is known to modify the behavior of physical quantities such as the specific heat or the resistivity.[21,22] In the particular case of superconducting properties, the effects of such singularity can be understood from the fact that the higher the density of states becomes, the larger the superconducting condensation energy is.[23] Therefore the superconducting features are enhanced by the presence of a vHS in the density of states. In particular, strong modifications of the BCS predicted behavior of the upper critical field should be expected if the density of states has a strong energy dependence at the Fermi level. Weak coupling BCS theory[6] predicts an approximately parabolic shape for the H_{c2} curve, which at low temperature saturates as T^2, and has linear temperature dependence near the zero-field critical temperature T_{co}. Furthermore, normalized H_{c2} curves fall onto an universal line.

16.3.1. *The spectral function*

An important quantity in the calculation of the superconducting critical magnetic fields is the spectral function. For a one-dimensional model of free fermions with a well defined Fermi velocity v_F, the spectral function, $\rho^{1\mathrm{D}}(\omega, x) = -A^{1\mathrm{D}}(\omega, x)/\pi$, is given by

$$\rho^{1\mathrm{D}}(\omega, x) = \frac{1}{\pi v_F} \cos\left[\left(\frac{\omega}{v_F} + k_F\right) x\right]. \tag{16.9}$$

For a circular Fermi surface, one has

$$\rho^{2\mathrm{D}}(\mathbf{r}, \omega) \sim \frac{1}{4\pi v_F}\sqrt{\frac{2k_F}{\pi r}} \cos\left[r\left(\frac{\omega}{k_F} + k_F\right) - \frac{\pi}{4}\right]. \tag{16.10}$$

In the case of a logarithmic vHS pinned at the Fermi level, and considering only the contribution of a small region around the saddle point, the spectral function becomes

$$\rho(x, q_y, \omega) = \frac{1}{2\pi|q_y|} e^{i\frac{x\omega}{q_y}} e^{i\boldsymbol{k}_F x}, \tag{16.11}$$

where the energy dispersion $\epsilon_{\boldsymbol{k}} - \epsilon_{vh} = q_x q_y$ was considered with $\boldsymbol{q} = \boldsymbol{k} - \boldsymbol{k}_{vh}$. It is convenient to write the spectral function in a mixed representation, when the energy dispersion is anisotropic. For an "extended" vHS, one has

$$\rho(r,\omega) = \frac{1}{2\pi ab}\left(\frac{|\omega|}{a}\right)^{\frac{1}{b}-1} \cdot \sqrt{\frac{2k_F}{\pi r}}$$
$$\cos\left[r\left(\left(\frac{|\omega|}{a}\right)^{\frac{1}{b}} \mathrm{sign}(\omega) + k_f\right) - \frac{\pi}{4}\right] \qquad (16.12)$$

where the energy dispersion $\epsilon_{\mathbf{k}} - \epsilon_{vh} = a \cdot \mathrm{sign}(q)|q|^b$ was considered, a is a constant and $q = k - k_{vh}$. These two expressions are similar to those of the one-dimensional and two-dimensional models with linearized energy dispersion, but with the role of the Fermi velocity being played by a function of q_y, in the case of the logarithmic vHS and by a function of ω in the case of the "extended" vHS.

16.3.2. *The pair propagator*

The upper critical field H_{c2} curve is a probe of the spatial and temperature dependence of the normal state pair propagator. The thermal and magnetic lengths define regions of different behavior of $K_\beta^H(r)$ which are integrated over in the gap equation.[6,18] The value of the pairing interaction defines a contour of solutions in the temperature-magnetic field plane.

Starting from the spectral functions given above and using Eq. 16.5 for the pair propagator, one obtains, in the case of free fermions,

$$K_\beta^{\mathrm{free}}(r) = \left(\frac{k_F}{2\pi r}\right)^{D-1} \frac{1}{v_F^2\beta} \frac{1}{\sinh\left(\dfrac{2\pi r}{\beta v_F}\right)}. \qquad (16.13)$$

When $r < \beta v_F = \xi_\beta$, the real space pair susceptibility decays as a power law $K(r) \propto r^{-2}$ in 2D. At distances longer than the thermal length ξ_β, the pair propagator is exponentially small.

For a logarithmic vHS, the pair propagator is

$$\widetilde{K}_\beta(x, 2q_y) = \frac{1}{2}\int_{|q_y|}^{p_o} dp \frac{2\pi/\beta}{(p^2 - q_y^2)} \frac{1}{\sinh\left[\dfrac{2\pi}{\beta}|x|\dfrac{p}{p^2 - q_y^2}\right]}. \qquad (16.14)$$

where p_o is a cutoff for the y component of the momentum.[5] Again, it is convenient to work in a mixed representation. At zero temperature, the asymptotic decay of Eq. 16.14 in real space is $K \sim |x|^{-1}|y|^{-1}$. This is a slow decay in comparison with the 2D free fermions result given above.

In the case of the extended vHS, one obtains the following scaling for the pair propagator,[3] $K_\beta(r) = r^{b-3}F[(\beta a/2)^{1/b}/r]$ with $F[X] \sim X^{b-2}$ when $X \gg 1$

and $b > 2$. Note that the thermal length is given by $\xi_T \sim (a/T)^{1/b}$. The pair propagator for distances smaller that the thermal length is approximately given by $K_T(r) \sim r^{-1}T^{-1+\frac{2}{b}}$ and therefore, it diverges as the temperature goes to zero. For $1 \leq b < 2$, $F[X] \sim$ const, if $X \ll 1$ and therefore, the pair propagator has a different short range dependence, $K_T(r) \sim r^{b-3}$. For $X \ll 1$, the function is exponentially small.

16.3.3. *Zero-field critical temperature*

The influence of the vHS on the zero-field critical temperature and the role of the frequency cutoff can be qualitatively understood if, in the usual BCS expression for the critical temperature, one considers an effective density of states, $T_c \sim T_D e^{-1/[\langle N(\epsilon)\rangle_{T_c} g]}$, where $\langle N(\epsilon)\rangle_{T_c}$ is the thermal averaged density of states, $\langle N(\epsilon_F)\rangle_T \sim \int_{-\infty}^{\infty} d\epsilon(\partial f/\partial\epsilon)N(\epsilon)$ and T_D is a constant determined by the short distance cutoff in Eq. 16.7. In the case of a logarithmic singularity, $\langle N(\epsilon)\rangle_T \sim \log(Tx_o/p_o)$. Solving the self-consistent equation, one obtains the weak-coupling critical temperature $T_c \propto \exp(-\sqrt{2/V})$.[24] For the "extended" vHS, $\langle N(\epsilon)\rangle_{T_{co}} \sim a^{-1/b}b^{-1}T_{co}^{\frac{1}{b}-1}$ which implies in the weak coupling limit, $T_{co}^{\frac{1}{b}-1} \sim 1/V$. For the case of an 1D vHS, $N(\epsilon) \sim \epsilon^{-1/2}$, this broadening argument leads correctly to the transition temperature $T_{co} \propto V^2$.[14,25]

The exact expression for the zero-field critical temperature in the case of a logarithmic vHS is obtained from the uniform pair propagator, $\mathcal{K}_\beta(0,0) = \int_{x_0}^{\infty} dx\widetilde{K}_\beta(x, q_y = 0)$, where x_0 is a short distance cutoff in the x-direction. The zero-field critical temperature is determined by the relation $1 = V\mathcal{K}_{\beta_c}(0,0)$ and for the logarithmic vHS, one obtains $T_c = p_o/(\pi x_o)\exp(-\sqrt{2/V})$. In the case of the "extended" vHS, the zero-field critical temperature is similarly obtained and one has $T_{co} \sim \{[(k_F/\pi)a^{-\frac{1}{b}}/(b-1)]V\}^{b/(b-1)}$ in agreement with the qualitative argument.

16.3.4. *Zero-temperature critical field*

In the case of a logarithmic vHS, the low temperature behavior of H_{c2} can be obtained using a variational method.[5] The zero-temperature pair propagator has a simple form, $\widetilde{K}_\beta(x, q_y) = 1/(2|x|)\log[2p_o/|q_y|]$ and the gap equation can be written as

$$\widetilde{\Delta}(x) = g\int dx'\widetilde{K}_\beta[x' - x, -H(x + x')]\widetilde{\Delta}(x'). \tag{16.15}$$

Assuming a variational form for the gap function, $\widetilde{\Delta}(x) = \Theta(a - |x|)/\sqrt{2a}$, one obtains $a = e\sqrt{x_o p_o/2H}$ which leads to $1/V \gtrsim 1/4\log^2[e^2 p_o/(2x_o H_{co})]$.

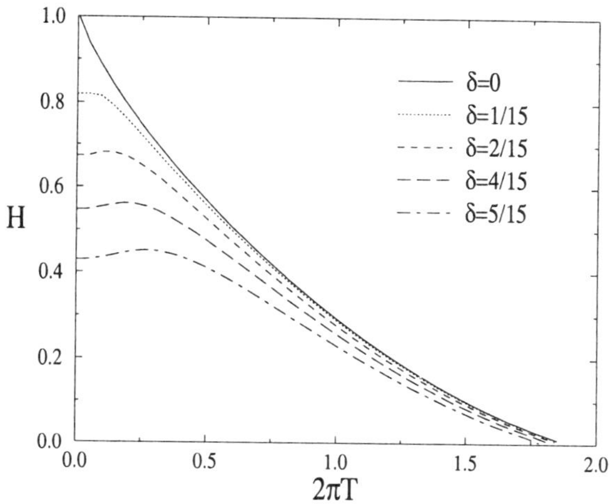

Fig. 16.1. H_{c2} curves for $p_o/x_0 = 20$, $V = 0.2$ and several values of $\delta = \epsilon_F - \epsilon_{vh}$ in the case of the logarithmic vHS. $H_{c2}(T)$ shows saturation at temperatures below δ and upward curvature at higher temperatures. A small increase of H_{c2} with temperature is observed in the curves with higher δ. Reproduced from Ref. 5.

Therefore

$$H_{co} \sim \frac{e^2 p_o}{2x_o} \exp\left(-\frac{2}{\sqrt{V}}\right). \tag{16.16}$$

This implies $x_o H_{co}/p_o \sim (x_o T_{co}/p_o)^{\sqrt{2}}$ in contrast with the usual Fermi liquid BCS scaling $H_{co} \sim T_{co}^2$. By expanding the kernel [Eq. (16.14)] around the zero-temperature critical point, the low temperature behavior of H_{c2} can be obtained and one finds that $H_c(T)$ does not saturate as T goes to zero. Instead it decreases linearly with temperature.

For the power-law vHS, the low temperature scaling of H_{c2} can be obtained rewriting the gap equation in magnetic length units.[3] Then, the gap solutions fall into an universal Gaussian curve (as in the usual BCS case) and become independent of the magnetic field. The magnetic phase acquired by the Cooper pair also becomes independent of H and the pair propagator scales as $K_\beta(r) = H^{-1/2}G(r,\beta)$. This leads to the low temperature scaling of the upper critical field, $H_{c2}(T) \sim T^{-2+\frac{4}{b}}$, for $b > 2$. Note that for $b = 1$, with the introduction of a cutoff, one recovers the usual BCS results and, in particular, $H_{co} \sim T_{co}^2$. For $1 \leq b < 2$, the pair propagator shows a different short range dependence, $K_T(r) \sim r^{b-3}$. The pair propagator does not diverge as we decrease the temperature, and with a scaling argument, one can show that now the zero-temperature

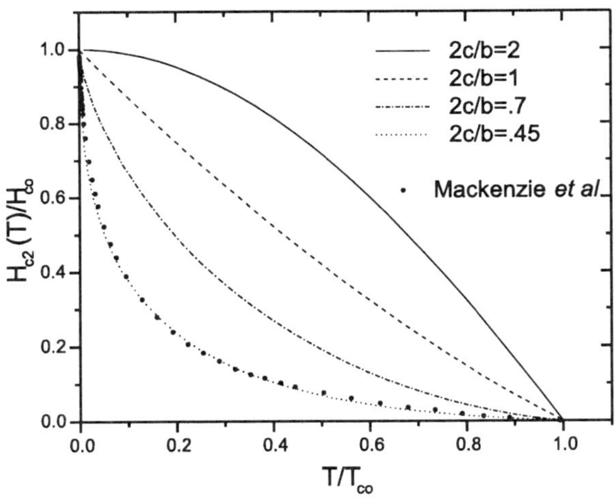

Fig. 16.2. Normalized H_{c2} curves for the "extended" vHS with $1 < b < 2$. As the dispersion relation changes from linear to quadratic, the upper critical curve changes from the usual BCS curve to a curve with strong positive curvature. Reproduced from Ref. 3.

critical field is finite, $1/g \sim H_{co}^{(1-b)/2}$ and thus, $H_{co} \sim T_{co}^{2/b}$.

The low temperature dependence of H_{c2} can be obtained expanding the pair propagator in powers of T, $[K_T(r) - K_0(r)]/r^{b-3} \sim -(rT^{1/b})^c$ and following Gorkov,[6] one obtains $H_{c2}(T) - H_{c2}(0) \sim -T^{2c/b}$, where c is a function of b such that $c = (2-b)/2$, when $b \sim 2$ and $c = (3-b)/2$, when $b \to 1$. As in the case of the logarithmic vHS, $H_c(T)$ does not saturate as T goes to zero.

16.3.5. *Numerical* H_{c2}

The full temperature dependence of the upper critical field requires a numerical approach.[3,5] The numerical H_{c2} curves are displayed in Fig. 16.1 in the case of an logarithmic vHS for several values of doping $\delta = \epsilon_F - \epsilon_{vh}$ and fixed coupling V. When $\delta = 0$, that is, when the Fermi energy coincides with the vHS, there is no saturation region and the H_{c2} curve shows upward curvature throughout the complete temperature range.

The upper critical field curves for the "extended" vHS are characterized by a strong divergence as $T \to 0$ when $b > 2$ and linear behavior close to T_{co}. For $1 \leq b < 2$, H_{c2} does not diverge as $T \to 0$. In Fig. 16.2, H_{c2} curves obtained for $1 \leq b < 2$ are displayed. A drastic transformation from conventional parabolic-like curves ($b \sim 1$) to curves with strong positive curvature ($b \sim 2$) is observed.

Note that for a given vHS exponent α, the reduced vHS curves are independent of the coupling constant in the weak coupling limit. Such universal H_{c2} behavior is not observed in the case of the logarithmic vHS.[3,5] In this case, the upward curvature becomes stronger as V is decreased.

16.4. Paramagnetic critical fields

Superconductivity due to spin-singlet Cooper pairing is suppressed by Zeeman spin-splitting and consequently, the upper critical field cannot exceed the paramagnetic limit H_p, also designated as Pauli limit or Chandrasekhar-Clogston limit.[7,8] H_p is the zero-temperature critical field associated with a first-order transition from the homogeneous superconducting phase to the normal phase when only the effects of magnetic field coupling to electronic spins are considered. Its value is determined by the energy balance between the magnetic energy density gained by the difference in susceptibilities of the normal and superconducting states and the superconducting condensation energy density U_c. The normal phase has a finite Pauli susceptibility χ_p while the susceptibility of the spin-singlet superconducting phase vanishes at zero temperature and therefore, at zero temperature, $F_S - \chi_p H^2/2 = F_n$. In BCS theory, the condensation energy density is $U_c = N(\epsilon_F)\Delta_0^2/2$, where $N(\epsilon_F)$ is the density of states at the Fermi energy and Δ_0 is the zero-temperature energy gap. Making use of these relations and of $\chi_p = 2\mu_B^2 N(\epsilon_F)$, one obtains $\mu_B H_p = \Delta_0/\sqrt{2}$.[7,8]

16.4.1. *Supercooling field*

In the case of a quasi-two-dimensional superconductor under parallel magnetic fields, the low temperature first-order transition line ends in a tricritical point where a high temperature second-order phase transition line begins.[9,10] Associated with the first-order transition line, there is a region of metastability limited below by a supercooling field h_{sc} and above by a superheating field h_{sh}. The superheating fields are the highest magnetic fields associated with finite gap solution branches of the coupled gap equations.[26]

The temperature dependence of the critical field (h_{sc}) that induces the second-order phase transition from the homogeneous superconducting state to the normal state is obtained taking the limit $\Delta \to 0$ in Eq. 16.8,

$$\frac{1}{V} = \int_0^{\omega_D} d\xi \frac{N(\xi)}{2\xi} \left[\tanh \frac{\xi - h}{2t} + \tanh \frac{\xi + h}{2t} \right] \tag{16.17}$$

where ω_D is the usual frequency cutoff and $t = k_B T$. Below the tricritical point temperature, the phase transition becomes of first-order and the field given by the

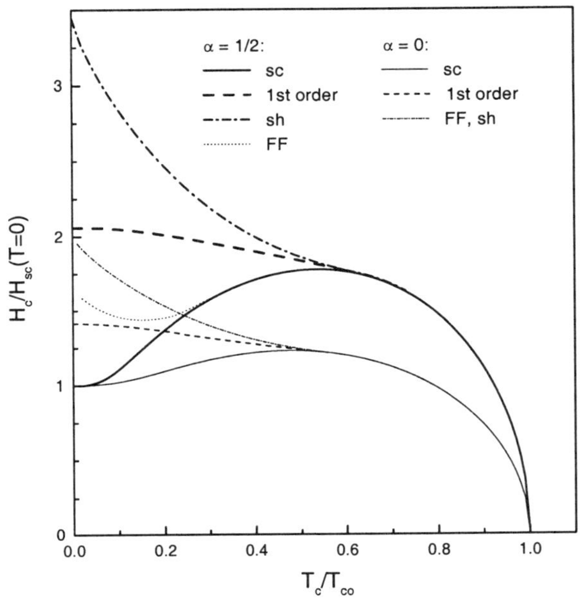

Fig. 16.3. The phase diagram of a paramagnetically limited two-dimensional superconductor with constant density of states ($\alpha = 0$) and with a power-law divergence in the density of states ($\alpha = 1/2$). The reentrant behavior of the supercooling field H_{sc} is strongly enhanced if a vHS is present in the DOS. Reproduced from Ref. 4.

previous equation becomes a supercooling field.

Considering the expression for the density of states of the "extended" vHS, $N(\epsilon) = N_o|\epsilon - \epsilon_{vh}|^{-\alpha}$, the zero-field critical temperature is given by $2t_{co} \sim [g/(\alpha - \alpha^2)]^{1/\alpha}$. The zero-temperature supercooling field ($h_{sc,0}$) is given by $h_{sc,0} = (2g/\alpha)^{1/\alpha}$. Therefore, $h_{sc,0} \sim (1-\alpha)^{1/\alpha} t_{c0}$ and one has a much larger enhancement of t_{co} than that of $h_{sc,0}$ in the limit $\alpha \rightarrow 1$.[4]

In Fig. 16.3, the temperature dependence of the reduced upper critical fields (or supercooling fields) of a superconductor with a vHS and a superconductor with constant density of states are displayed. Note that the second-order critical field reaches its maximal value at finite temperature.[9] This reentrant behavior for the supercooling field has been recently observed in thin aluminum films.[27] In the case of a vHS superconductor, this maximum is enhanced relatively to the zero-temperature supercooling field and the reentrance becomes more pronounced.[4]

16.4.2. *Blocking region*

The reentrant behavior of the supercooling field is due to the existence of a blocking region for Cooper pairing at zero temperature. In order to contribute to the formation of a Cooper pair of total momentum $\boldsymbol{q} = 0$, two states with opposite spins and of momenta $\boldsymbol{k}$ and $-\boldsymbol{k}$ must be either both empty or both occupied at zero temperature. Therefore the Zeeman splitting of the free fermion Fermi surface creates a no-pairing or blocking region around the zero field Fermi surface. At finite temperature, thermal excitations provide low energy pairing possibilities within this no-pairing region and therefore, the pairing susceptibility grows with temperature. When a vHS is pinned at the Fermi level, the number of these thermal excitations is much larger and the reentrant behavior becomes more pronounced.

16.4.3. *Free energy*

The free energy difference between the superconducting and the normal state can be determined from[9]

$$F_s(T,H) - F_n(T,H) = \int_0^{\Delta} \frac{d(1/V)}{d\Delta'} \Delta'^2 d\Delta'. \tag{16.18}$$

When the zero-gap local extreme of the free energy becomes the absolute minimum, the transition to the normal state occurs. If the local minimum of the superconducting phase converges to the normal state local extreme, the transition is of second-order, otherwise it is of first-order. Associated with the first-order transition line, there is a region of metastability limited below by a supercooling field h_{sc} and above by a superheating field h_{sh}.

16.4.4. *Superheating field*

The superheating field is the highest field for which there is still a finite gap solution of Eq. 16.17.[9] The first-order critical field h_1 and the superheating field h_{sh} determined numerically from Eqs. 16.17 and 16.18 are shown in Fig. 16.3. The zero-temperature superheating field coincides with the zero-temperature gap function, $h_{sh,0} = \Delta_0 = [A(\alpha)g]^{1/\alpha} \sim t_{c0}$, with $A(\alpha) = 2\pi \text{cosec}[(1-\alpha)\pi] P_{-\alpha}(0)$ where $P_\nu(x)$ is the Legendre function of the first kind. This implies that the superheating field is enhanced relatively to the supercooling field, $h_{sc,0} \sim (1-\alpha)^{1/\alpha} h_{sh,0}$, and therefore a huge metastability region is present in the phase diagram.[4]

16.4.5. *Pauli limit*

At zero temperature, the free energy difference is given by $\Delta F(0,H) = N_o/(4-2\alpha)[4/(1-\alpha)\,h^{2-\alpha} - \alpha A(\alpha)\,\Delta^{2-\alpha}]$. Setting this diference equal to zero, one obtains a strongly enhanced Pauli limit, $h_p = \{\alpha(1-\alpha)A(\alpha)/4\}^{\frac{1}{2-\alpha}}\Delta_0$. For $\alpha = 0$, one recovers the well known result, $h_p = \Delta_0/\sqrt{2}$.[9] In the limit $\alpha \to 1$, $h_p \to \Delta_0/2$.

16.4.6. *Fulde-Ferrel phase*

In a quasi-2D superconductor under parallel magnetic field, superconductivity may persist at magnetic fields higher than the Pauli limit if the formation of Cooper pairs with nonzero momentum is considered,[10–12] i.e., if the FFLO phase occurs. As a consequence of the finite total momentum, the superconducting order parameter oscillates in space.

In order to infer the possible existence of FFLO phase in the phase diagram. one has to consider the BCS mean-field approach for the case of an inhomogeneous gap function, $\Delta_{\boldsymbol{q}} = -VS^{-1}\sum_{\boldsymbol{k}}\langle a_{-\boldsymbol{k}-\boldsymbol{q}\downarrow}a_{\boldsymbol{k}\uparrow}\rangle$. Since the phase transition from the FFLO state to the normal phase is of second-order, taking the limit $\Delta_{\boldsymbol{q}} \to 0$, one obtains the following two-dimensional gap equation[10]

$$1 = \frac{V}{S}\sum_{\boldsymbol{p}} \frac{1 - f(\xi_{\boldsymbol{p}+\boldsymbol{q}/2\uparrow}) - f(\xi_{\boldsymbol{p}-\boldsymbol{q}/2\downarrow})}{\xi_{\boldsymbol{p}+\boldsymbol{q}/2\uparrow} + \xi_{\boldsymbol{p}-\boldsymbol{q}/2\downarrow}}. \tag{16.19}$$

Considering the isotropic energy dispersion for the "extended" vHS, the inhomogeneous gap equation can be rewritten as

$$1 = V\int_0^{\omega_D} d\xi N(\xi)\int_0^{\pi}\frac{d\theta}{\pi}\frac{\tanh\left(\frac{\xi^+}{2t}\right) + \tanh\left(\frac{\xi^-}{2t}\right)}{\xi^+ + \xi^-}, \tag{16.20}$$

with $\xi^{\pm} = [|\xi|^{1/b}\mathrm{sgn}(\xi) \pm 1/2q\cos\theta]^b \pm h$ where a q^2/k term has been neglected since $q \ll k_F$ and where $[x]^b$ should be understood as $\mathrm{sgn}(x)|x|^b$. The zero-temperature critical field is the maximum $H_c(q)$ which is reached when $\tilde{q} = q/(2h^{1/b}) = 1$. This field is significantly enhanced by the vHS since it is proportional to $V^{\frac{1}{\alpha}}$, with $\alpha = 1 - \frac{1}{b}$. However, there is only a weak enhancement in comparison with the $q = 0$ supercooling field as observed in Fig. 16.3, in the case of $\alpha = 1/2$. The FFLO phase boundary lies clearly below the first-order critical field h_1 and therefore, this phase will not be observed.

In the case of the energy dispersion with saddle points, the zero gap pairing susceptibility has its maximum value at $q = 0$ even for a simple quadratic saddle point as can be observed in Fig. 16.4. For a weaker saddle point with exponents smaller than two, the maximum shifts to finite q, until it becomes fixed at $\tilde{q} = 1$.

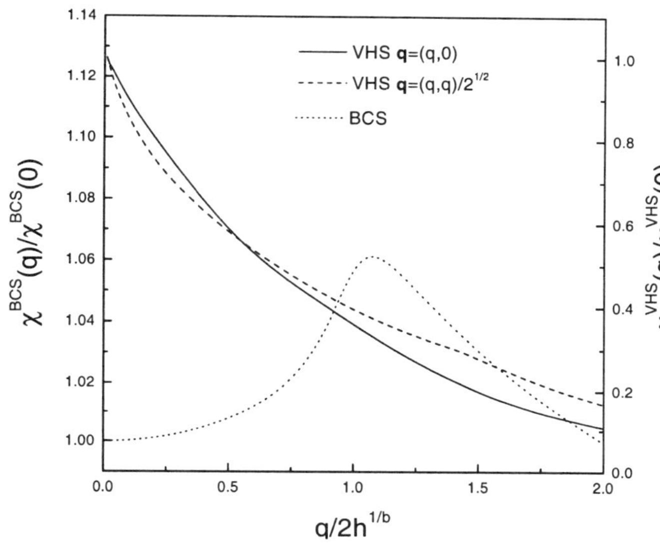

Fig. 16.4. The zero-gap pairing susceptibility as a function of the renormalized pair momentum $q/(2h^{1/b})$ for a BCS superconductor and a superconductor with an an extended saddle point with $m = n = 4$ at a small, but finite temperature. Reproduced from Ref. 4.

16.5. Other influences in the magnetic critical fields

The effects of the vHS on the superconducting magnetic critical fields described in the previous sections assumed a simple picture of singlet superconductivity in a clean two-dimensional metal with s-wave pairing symmetry and the vHS pinned at the Fermi level. For transverse magnetic fields, the semi-classical approximation was used. In a more general picture, other factors have to be considered in conjunction with the vHS.

16.5.1. *Pairing symmetry*

If the pairing potential is non-local, the possibility of pairing with d-wave symmetry has to be considered. Unless the saddle points are in the proximity of the nodal lines, the results presented in section 16.3 for the upper critical field remain valid.

The Pauli limit of a superconductor with constant density of states and d-wave pairing symmetry has been addressed by several authors.[28,29] If only homogeneous superconductivity is considered, one obtains again the typical phase diagram of a s-wave superconductor and a vHS pinned at the Fermi level should lead to the effects described in section 16.4. D-wave symmetry leads however to characteristic modifications in the phase diagram for inhomogeneous superconductivity with, in particular, the appearance of a low-temperature kink in the phase boundary between the FFLO phase and the normal phase. This kink is associated with a modification in the direction of the pairing momentum from the gap maxima to the gap minima. However, since d-wave pairing does not leads to an enhancement of the FFLO features in the phase diagram, the conclusion that homogeneous superconductivity becomes dominant in the phase diagram when a vHS is pinned at the Fermi level can still be extracted.

16.5.2. *Anisotropy*

In the case of transverse fields, unless a dimensional crossover is present, the reduced upper critical field (H_{c2}/H_{co} as a function of T/T_{co}) shows very little sensitivity to anisotropy.[30–32] For a system with an elliptical Fermi surface, the reduced upper critical field follows the universal parabolic-like BCS curve. This is also the case for an open warped Fermi surface, if T_{co} is smaller than the t_y modulation of the Fermi surface. If $t_y \ll T_{co}$, a reduction of the effective dimension of our system will occur induced by the magnetic field and H_{c2} will diverge at a finite temperature according to the mean-field analysis.[30–32] Saturation should however arise at low temperatures due to fluctuations and Pauli pair breaking.[32]

Concerning the paramagnetic critical fields, as one can easily conclude from the equations of section 16.4, the phase diagram for homogeneous superconductivity has no dependence on the Fermi surface shape, only density of states dependence. However, the FFLO state is extremely sensitive to nesting properties of the Fermi surface. Note that, in section 16.4, only the saddle point contribution to the formation of this state has been considered and the rest of the Fermi surface has been ignored. This procedure is justified for homogeneous superconductivity, since the saddle point contribution clearly dominates. However, in the case of FFLO superconductivity, if some portion of the Fermi surface is strongly nested, the respective critical field may be higher than that of homogeneous superconductivity and therefore, a region of existence of FFLO phase would appear in the phase diagram presented in Fig. 16.2.

16.5.3. *Doping effects*

In sections 16.3 and 16.4, the vHS was assumed to be pinned at the Fermi level. Doping implies, in an independent electron picture, a shift δ of the Fermi energy from the vHS, with $\delta = \epsilon_F - \epsilon_{vh}$. Since the BCS critical temperature has strong density of states dependence, one would expect that this would lead to a strong reduction of T_c. This doping dependence is however partially reduced because the critical temperature is dependent on the temperature averaged density of states.

The zero-temperature upper critical field H_{co} has a much stronger dependence on δ that T_{co}.[5] The effects of temperature and doping in the gap equation are similar as they provide a cutoff for large distances. Consequently, $H_{c2}(T)$ saturates at a temperature of the order of δ. For $T > \delta$, $H_{c2}(T)$ is independent of the doping. This behavior can be observed in Fig. 16.1.

16.6. Van Hove singularities and high-T_c superconductivity

Angle resolved photoemission experiments clearly indicate that the energy dispersion of the CuO_2 planes of the high-T_c superconductors contains saddle points, in some cases extended.[13,14,33,34] It has been claimed that the high superconducting critical temperatures of the cuprates could be explained taking into account the existence of these saddle points in the proximity of the Fermi surface.[24,35] This is the so-called van Hove scenario for the high-T_c superconductors. Extended saddle points have also been found in the case of the two-dimensional Hubbard model which is believed to describe the physics of the CuO_2 planes.[15,16]

The van Hove scenario offers a simple explanation for the very small isotope effect of the copper-oxides. The isotope effect is reduced in this scenario since an electronic energy has replaced the Debye temperature in the expressions for T_c as shown in section 16.3 and consequently, T_c is independent of the isotope mass. The corrections to the weak coupling limit expressions give a finite but small isotope mass exponent.[22] Besides the critical temperature enhancement, the van Hove scenario explains anomalous normal state properties of the copper-oxides as for example the linear-T resistivity.[21]

For the electron doped materials, photoemission experiments[34,36] indicate that the saddle points sit considerably far from the Fermi surface and therefore, these materials have lower critical temperatures. Note that ARPES experiments in copper-oxides compounds as, for example, Bi2201[34] and non-cuprate materials as Sr_2RuO_4[37] have also found saddle points very close to the Fermi surface, but these materials have low critical temperatures. In these cases, the low T_cs might be due to the experimentally found smaller extent of the saddle points in these materials.[34,37] Note that photoemission experiments on $YBa_2Cu_3O_{6.9}$,[33] $YBa_2Cu_3O_{6.5}$, and $YBa_2Cu_3O_{6.3}$[38] report a clear doping independence of the pinning of the Fermi level at the vHS. Moreover, this doping independence of the pinning is predicted by many numerical studies, from slave-boson calculations[39,40] to renormalization group calculations.[41]

The upper critical field of overdoped $Tl_2Ba_2CuO_{6+\delta}$[42]and $Bi_2Sr_2CuO_y$[43] and underdoped $YBa_2Cu_3O_{7-\delta}$[44] obtained in magnetoresistance experiments down to very low temperatures has very strong positive curvature and no evidence of saturation at low temperatures. This behavior contrasts strongly with the weak-coupling BCS result[6] which predicts an approximately parabolic shape for the H_{c2} curve. According to the results of section 16.3, these curves may reflect the strong energy dependence of the density of states which results from the presence of saddle points in the proximity of the Fermi surface. The upper critical field of the electron doped cuprate material $Nd_{2-x}Ce_xCuO_4$ with $x \sim .15$ has been determined by Hidaka and Suzuki[45] and a H_{c2} curve with saturation at low temperatures but upward curvature close to T_c was observed in agreement with Fig. 16.1. Anomalous H_{c2} curves have also been obtained in layered non-cuprate superconductors[46] such as $k-$(BEDT-TTF)$_2$Cu(NCS)$_2$.[47] The common feature to the H_{c2} curves of these materials is that they all show upward curvature, extending to $T \sim 0$ in some cases. In Fig. 16.2, the experimental H_{c2} points for $Tl_2Ba_2CuO_{6+\delta}$[42] are also displayed. A good fit is observed for $2c/b = .45$, which according to the picture presented in section 16.3, implies that the density of states diverges as $N(\epsilon) \sim \epsilon^{-.28}$ in agreement with the saddle point energy dispersion observed in photoemission experiments.[14]

Under in-plane magnetic fields, if the van Hove scenario applies to copper-oxides, a huge metastability region should appear at low temperature in the phase diagram as described in section 16.3. This region can be probed by resistive critical field measurements[48,49] or tunnelling measurements of density of states (as recently in thin Al films[27]). Unfortunately, in-plane critical fields of the high-T_c superconductors are presently outside the experimental magnetic field range. However, experiments with explosive-driven magnetic fields indicate that the in-plane upper critical field in the copper-oxides exceeds considerably the BCS Pauli limit.[50]

16.7. Conclusion

The vHS provides a simple example of a system where unusual normal state correlations show up strongly in the temperature dependence of the magnetic critical field of a clean weak-coupling superconductor. The anomalous H_{c2} behavior reflects the short-range enhancement of the pair propagator and the unusual temperature dependence of the thermal length which result from the presence of a density of states divergence at the Fermi level. Upward curvature is observed in the mean-field upper critical field $H_{c2}(T)$. When the Fermi energy coincides with the vHS, this upward curvature extends to $T \sim 0$. Under in-plane magnetic fields, at low temperature, a huge metastability region appears in the temperature-magnetic field phase diagram and the Pauli limit, H_p, is strongly enhanced. FFLO superconductivity is absent from the phase diagram unless there are Fermi surface sections (away from the saddle point region) with very good nesting properties.

Despite intense theoretical activity over the past 20 years, there is still no consensual theory for the origin of high-T_c superconductivity. Recently, experiments have again raised the possibility that the electron-phonon mechanism plays an important role in the cuprates.[51] Even if a novel mechanism is the key to the superconductivity in the copper-oxides, it is quite unlikely that the presence of saddle points in the energy dispersion of the CuO_2 planes in the proximity of the Fermi surface has no effect on the superconducting phase diagram.

References

1. P. W. Anderson, The resonating valence bond state in La_2CuO_4 and superconductivity, *Science*. **235**(4793), 1196–1198, (1987).
2. F. D. M. Haldane, Luttinger liquid theory of one-dimensional quantum fluids: 1. properties of the Luttinger model and their extension to the general 1D interacting spinless fermi gas, *J. Phys. C*. **14**(19), 2585–2609, (1981).
3. R. G. Dias, Effects of van Hove singularities on the upper critical field, *J. Phys.-Condens. Matter*. **12**(42), 9053–9060, (2000).
4. R. G. Dias and J. A. Silva, Huge metastability in high-T_c superconductors induced by parallel magnetic field, *Phys. Rev. B*. **67**(9), 092511, (2003).
5. R. G. Dias and J. M. Wheatley, Superconducting upper critical field near a 2D van Hove singularity, *Solid State Commun*. **98**(10), 859–862, (1996).
6. L. P. Gorkov, The critical supercooling field in superconductivity theory, *JEPT*. **10**(3), 593–599, (1960).
7. B. S. Chandrasekhar, A note on the maximum critical field of high field superconductors, *Appl. Phys. Lett*. **1**(1), 7–8, (1962).
8. A. M. Clogston, Upper limit for critical field in hard superconductors, *Phys. Rev. Lett*. **9**(6), 266–267, (1962).
9. K. Maki and T. Tsuneto, Pauli paramagnetism and superconducting state, *Prog. Theor. Phys*. **31**(6), 945–956, (1964).
10. H. Shimahara, Fulde-Ferrell state in quasi-2-dimensional superconductors, *Phys. Rev. B*. **50**(17), 12760–12765, (1994).
11. P. Fulde and R. A. Ferrell, Superconductivity in strong spin exchange field, *Phys. Rev*. **135**(3A), A550, (1964).
12. A. I. Larkin and Y. N. Ovchinnikov, Inhomogeneous state of superconductors. **20**(3), 762–769, (1965).
13. K. Gofron, J. C. Campuzano, A. A. Abrikosov, M. Lindroos, A. Bansil, H. Ding, D. Koelling, and B. Dabrowski, Observation of an extended Van Hove singularity in $YBa_2Cu_4O_8$ by ultrahigh-energy resolution angle-resolved photoemission, *Phys. Rev. Lett*. **73**(24), 3302–3305, (1994).

14. A. A. Abrikosov, J. C. Campuzano, and K. Gofron, Experimentally observed extended saddle point singularity in the energy spectrum of $YBa_2Cu_3O_{6.9}$ and $YBa_2Cu_4O_8$ and some of the consequences, *Physica C.* **214**(1-2), 73–79, (1993).
15. F. F. Assaad and M. Imada, Unusually flat hole dispersion relation in the two-dimensional Hubbard model and restoration of coherence by addition of pair hopping processes, *Eur. Phys. Journal B.* **10**(4), 595–598, (1999).
16. M. Imada, A. Fujimori, and Y. Tokura, Metal-insulator transitions, *Rev. Mod. Phys.* **70** (4), 1039–1263, (1998).
17. G. Rickayzen, *Theory of Superconductivity*. (John Wiley & Sons, New York, 1965).
18. A. A. Abrikosov, L. P. Gorkov, and I. E. Dzyaloshinskii, *Methods of Quantum Field Theory in Statistical Physics*. (Dover Publications, New York, 1963).
19. A. L. Fetter and J. D. Walecka, *Quantum Theory of Many-Particle Systems*. (McGraw-Hill, New York, 1971).
20. A. K. Rajagopal and R. Vasudevan, De Haas-Van Alphen oscillations in critical temperature of type 2 superconductors, *Phys. Letts.* **23**(9), 539, (1966).
21. D. M. Newns, C. C. Tsuei, R. P. Huebener, P. J. M. Vanbentum, P. C. Pattnaik, and C. C. Chi, Quasiclassical transport at a van Hove singularity in cuprate superconductors, *Phys. Rev. Lett.* **73**(12), 1695–1698, (1994).
22. C. C. Tsuei, D. M. Newns, C. C. Chi, and P. C. Pattnaik, Anomalous isotope effect and Van Hove singularity in superconducting Cu oxides, *Phys. Rev. Lett.* **65**(21), 2724–2727, (1990).
23. J. Bardeen, L. N. Cooper, and J. R. Schrieffer, Theory of superconductivity, *Phys. Rev.* **108**(5), 1175–1204, (1957).
24. J. Labbe and J. Bok, Superconductivity in alkaline-earth-substituted La_2CuO_4: a theoretical model, *Europhys. Lett.* **3**(11), 1225–1230, (1987).
25. J. Labbe, S. Barisic, and J. Friedel, Strong coupling superconductivity in V_3X type of compounds, *Phys. Rev. Lett.* **19**(18), 1039–1041, (1967).
26. R. G. Dias, Zeeman splitting in multiple-band superconductors, *Phys. Rev. B.* **72**(1), (2005).
27. V. Y. Butko, P. W. Adams, and E. I. Meletis, State memory and reentrance in a paramagnetically limited superconductor, *Phys. Rev. Lett.* **83**(18), 3725–3728, (1999).
28. K. Yang and S. L. Sondhi, Response of a $d_{x^2-y^2}$ superconductor to a Zeeman magnetic field, *Phys. Rev. B.* **57**(14), 8566–8570, (1998).
29. K. Maki and H. Won, The sine-wave like d-wave superconductivity in high magnetic fields, *Czech. J. Phys.* **46**, 1035–1036, (1996).
30. R. A. Klemm, M. R. Beasley, and A. Luther, Upper critical field of layered superconductors, *J. Low Temp. Phys.* **16**(5-6), 607–613, (1974).
31. R. A. Klemm, A. Luther, and M. R. Beasley, Theory of upper critical field in layered superconductors, *Phys. Rev. B.* **12**(3), 877–891, (1975).
32. A. G. Lebed and K. Yamaji, Restoration of superconductivity in high parallel magnetic fields in layered superconductors, *Phys. Rev. Lett.* **80**(12), 2697–2700, (1998).
33. J. C. Campuzano, K. Gofron, H. Ding, R. Liu, B. Dabrowski, and B. J. W. Veal, Photoemission from the high-T_c superconductors. **95**(1-2), 245–250, (1994).
34. D. M. King, Z. X. Shen, D. S. Dessau, D. S. Marshall, C. H. Park, W. E. Spicer, J. L. Peng, Z. Y. Li, and R. L. Greene, Observation of a saddle point singularity in

$Bi_2(Sr_{0.97}Pr_{0.03})_2CuO_{6+\delta}$ and its implications for normal and superconducting state properties, *Phys. Rev. Lett.* **73**(24), 3298–3301, (1994).

35. J. Friedel, On quasi one-dimensional or two-dimensional superconductors, *J. Phys. (Paris).* **48**(10), 1787–1797, (1987).
36. D. S. Dessau, Z. X. Shen, D. M. King, D. S. Marshall, L. W. Lombardo, P. H. Dickinson, A. G. Loeser, J. Dicarlo, C. H. Park, A. Kapitulnik, and W. E. Spicer, Key features in the measured band structure of $Bi_2Sr_2CaCu_2O_{8+\delta}$: Flat bands at e_f and Fermi surface nesting, *Phys. Rev. Lett.* **71**(17), 2781–2784, (1993).
37. D. H. Lu, M. Schmidt, T. R. Cummins, S. Schuppler, F. Lichtenberg, and J. G. Bednorz, Fermi surface and extended van Hove singularity in the noncuprate superconductor Sr_2RuO_4, *Phys. Rev. Lett.* **76**(25), 4845–4848, (1996).
38. R. Liu, B. W. Veal, C. Gu, A. P. Paulikas, P. Kostic, and C. G. Olson, Electronic structure as a function of doping in $YBa_2Cu_3O_x$ for $6.2 \leq x \leq 6.9$ studied by angle-resolved photoemission, *Phys. Rev. B.* **52**(1), 553–558, (1995).
39. R. S. Markiewicz, Phase-separation near the Mott transition in $La_{2-x}Sr_xCuO_4$, *Journal of Physics-Condensed Matter.* **2**(3), 665–676, (1990).
40. D. M. Newns, P. C. Pattnaik, and C. C. Tsuei, Role of Van Hove singularity in high temperature superconductors: Mean field, *Phys. Rev. B.* **43**(4), 3075–3084, (1991).
41. J. Gonzalez, F. Guinea, and M. A. H. Vozmediano, Renormalization group analysis of electrons near a van Hove singularity, *Europhys. Letts.* **34**(9), 711–716, (1996).
42. A. P. Mackenzie, S. R. Julian, G. G. Lonzarich, A. Carrington, S. D. Hughes, R. S. Liu, and D. C. Sinclair, Resistive upper critical field of $Tl_2Ba_2CuO_6$ at low temperature and high magnetic fields, *Phys. Rev. Lett.* **71**(8), 1238–1241, (1993).
43. M. S. Osofsky, R. J. Soulen, S. A. Wolf, J. M. Broto, H. Rakoto, J. C. Ousset, G. Coffe, S. Askenazy, P. Pari, I. Bozovic, J. N. Eckstein, and G. F. Virshup, Anomalous temperature dependence of the upper critical magnetic field in Bi-Sr-Cu-O, *Phys. Rev. Lett.* **71**(14), 2315–2318, (1993).
44. D. J. C. Walker, O. Laborde, A. P. Mackenzie, S. R. Julian, A. Carrington, J. W. Loram, and J. R. Cooper, Resistive upper critical field of thin films of underdoped $YBa_2(Cu_{0.97}Zn_{0.03})_3O_{7-\delta}$, *Phys. Rev. B.* **51**(14), 9375–9378, (1995).
45. Y. Hidaka and M. Suzuki, Growth and anisotropic superconducting properties of $Nd_{2-x}Ce_xCuO_{4-y}$ single crystals, *Nature.* **338**(6217), 635–637, (1989).
46. D. E. Prober, R. E. Schwall, and M. R. Beasley, Upper critical fields and reduced dimensionality of the superconducting layered compounds, *Phys. Rev. B.* **21**(7), 2717–2733, (1980).
47. K. Murata, Y. Honda, H. Anzai, M. Tokumoto, K. Takahashi, N. Kinoshita, T. Ishiguro, N. Toyota, T. Sasaki, and Y. Muto, Transport properties of $\kappa-$(BEDT-TTF)$_2$Cu(NCS)$_2$: H_{c2}, its anisotropy and their pressure dependence, *Synth. Met. A.* **27**(1-2), A341–A346, (1988).
48. P. W. Adams, P. Herron, and E. I. Meletis, First-order spin-paramagnetic transition and tricritical point in ultrathin Be films, *Phys. Rev. B.* **58**(6), R2952–R2955, (1998).
49. W. H. Wu, R. G. Goodrich, and P. W. Adams, Spin-paramagnetic transition of ultrathin granular Al films in a tilted magnetic field, *Phys. Rev. B.* **51**(2), 1378–1380, (1995).
50. A. S. Dzurak, B. E. Kane, R. G. Clark, N. E. Lumpkin, J. O'Brien, G. R. Facer, R. P. Starrett, A. Skougarevsky, H. Nakagawa, N. Miura, Y. Enomoto, D. G. Rickel, J. D. Goettee, L. J. Campbell, C. M. Fowler, C. Mielke, J. C. King, W. D. Zerwekh,

D. Clark, B. D. Bartram, A. I. Bykov, O. M. Tatsenko, V. V. Platonov, E. E. Mitchell, J. Herrmann, and K. H. Muller, Transport measurements of in-plane critical fields in $YBa_2Cu_3O_{7-\delta}$ to 300T, *Phys. Rev. B*. **57**(22), R14084–R14087, (1998).
51. J. Lee, K. Fujita, K. McElroy, J. A. Slezak, M. Wang, Y. Aiura, H. Bando, M. Ishikado, T. Masui, J. X. Zhu, A. V. Balatsky, H. Eisaki, S. Uchida, and J. C. Davis, Interplay of electron lattice interactions and superconductivity in $Bi_2Sr_2CaCu_2O_{8+\delta}$, *Nature*. **442** (7102), 546–550, (2006).

Chapter 17

Green function study of impurity effects in high-T_c superconductors

Yu.G. Pogorelov,[a] M.C. Santos,[b] and V.M. Loktev[c]

[a]*Departamento de Física, Universidade do Porto, Rua do Campo Alegre 687, 4169-007 Porto, Portugal; ypogorel@fc.up.pt*

[b]*Departamento de Física, Universidade de Coimbra, R. Larga, Coimbra, 3004-535, Portugal; mario.santos@ci.uc.pt*

[c]*N.N. Bogolyubov Institute for Theoretical Physics, NAN of Ukraine, Metrologichna 14b, 03134 Kiev, Ukraine; vloktev@bitp.kiev.ua*

The revision is made of Green function methods that describe the dynamics of electronic quasiparticles in disordered superconducting systems with d-wave symmetry of order parameter. Various types of impurity perturbations are analyzed within the simplest T-matrix approximation. The extension of the common self-consistent T-matrix approximation (SCTMA) to the so-called group expansions in clusters of interacting impurity centers is discussed and hence the validity criteria for SCTMA are established. A special attention is payed to the formation of impurity resonance states and localized states near the characteristic points of energy spectrum, corresponding to nodal points on the Fermi surface.

Contents

17.1. Introduction

The two decades of intensive study after the discovery of high-T_c superconductivity in perovskite metal-oxide materials[1,2] provided a solid experimental base on their electronic structure in normal and superconducting (SC) state, in particular on the specific type of SC order and its parameters.[3] However there still exist a lot of problems in theoretical understanding of fundamental physics behind the observable properties of high-T_c materials, and many of them are related to the effects of disorder by presence of impurities.[4] The impurity effects are known to be of less importance for quasiparticle dynamics and SC properties in the traditional metals and alloys with s-wave type of SC pairing,[5,6] except for strong pair-breaking effect from paramagnetic impurities[7] and related localized impurity levels within the SC gap.[8–10] However, in metal-oxide materials, seen as strongly doped semiconductors in the normal state[11] and establishing d-wave SC order[12] below T_c, both magnetic and non-magnetic impurities can act as effective pair-breakers[13] and produce quasiparticle resonances in the finite density of states (DOS) between the d-wave coherence peaks.[14,15] Such effects were already noted, though not properly recognized, in early point-contact experiments[16] but were fully verified later on by spectacular observations in scanning tunneling microscopy (STM).[17–19] The further interest to the impurity effects in high-T_c materials is stimulated by their possible influence on fundamental physical properties as infrared quasiparticle conductivity,[20,21] dynamics of magnetic vortices,[22,23] low temperature heat conduction,[24,25] etc.

The theoretical metods for study of impurity effects in SC systems are widely adopted from the well developed general field of elementary excitations in disordered solids, beginning from the classical works by Lifshitz,[26] Mott,[27] and Anderson.[28] The most effective approach to the quasiparticle dynamics is provided by the Green function (GF) method,[29,30,32] modified especially for the SC quasiparticles by Gorkov.[31] The GF analysis of the disorder effects in superconductors with non-trivial symmetry of order parameter was first developed yet before the discovery of high-T_c materials, mostly based on the concept of self-consistent T-matrix approximation (SCTMA)[33–35] for the quasiparticle self-energy that describes modification of their dispersion law and lifetime under disorder. The following investigations of the effects of disorder on metal-oxide SC systems with nodal points on the quasi-2D Fermi surface lead to the conclusions on a great universality of their transport properties,[36–38] and these conclusions could be also important for other Fermi systems with similar structure of excitation spectrum, as recently discovered graphene sheets.[39,40] Nevertheless, the existing dicrepancies between the predicted universal transport behavior and available experimental data,[41,42] indicate a need in critical revision of the used theoretical approach. In particular, the fact that SCTMA is essentially a single-impurity approximation makes it possible that the omitted effects of inter-impurity correlations can in principle introduce an important modification into dynamics and kinetics of quasiparticles.

The main purpose of this Chapter is to review the corresponding theoretical work made by the authors during last five years on validity of different approximations for GF's in disordered d-wave superconductors, for different types of impurity perturbations. The topic of our particular interest is the extension of the SCTMA approach to a more general form of the so-called group expansions (GE's) of self-energy,[44] where the first SCTMA term is followed by a group series in increasing numbers of interacting impurity centers. They are alike the classical Ursell-Mayer group expansions in the theory of non-ideal gases[43] where the particular terms (the group integrals) include physical interactions between the particles. In our case, these expansions include indirect (and, what is important, dependent on ε) interactions between the impurity centers, through the exchange by virtual excitations from (admittedly renormalized) band spectrum, so that each term corresponds to summation of a certain infinite series of diagrams. Actually, there are possible different types of GE's for particular regions of energy spectrum, one of them, called fully renormalized GE, is more adequate to extended (band-like) states and the other, non-renormalized GE, to localized states.[44,45] Then the issue of SCTMA validity is defined by the convergence of fully renormalized GE, otherwise new important impurity effects *beyond* SCTMA can be obtained from the non-renormalized GE as discussed in the following sections.

17.2. General formalism

Below we use the particular type of two-time GF's,[30] since they are more adapted to the systems with intrinsic disorder than the Matsubara functions, commonly used in the field-theoretical approaches for uniform systems.[32] The Fourier transformed two-time (advanced) GF is defined as

$$\langle\langle a|b\rangle\rangle_\varepsilon = i\int_{-\infty}^{0} e^{i(\varepsilon - i0)t}\langle\{a(t), b(0)\}\rangle dt, \qquad (17.1)$$

where a and b are Heisenberg fermionic operators, $\langle\ldots\rangle$ is the quantum-statistical average with the corresponding Hamiltonian, and $\{.,.\}$ is the anticommutator.

Various observable quantities at given temperature $T = (k_B\beta)^{-1}$ are obtained after these functions through the known spectral formula[30] for an average of operator product

$$\langle ba \rangle = \frac{1}{\pi} \int_{-\infty}^{\infty} \frac{d\varepsilon}{\mathrm{e}^{\beta(\varepsilon-\mu)}+1} \langle\langle a|b \rangle\rangle_\varepsilon , \tag{17.2}$$

including the chemical potential μ. The energy argument ε (in units where $\hbar = 1$) beside a GF will be dropped in what follows, unless necessary.

The explicit forms of two-time GF's can be found from the Heisenberg equation of motion for operators:[30]

$$i\frac{d}{dt}a(t) = [a(t), H],$$

where $[.,.]$ is the commutator. Then we have for the Fourier transformed GF's:

$$\varepsilon \langle\langle a|b \rangle\rangle = \langle \{a, b\} \rangle + \langle\langle [a, H] |b \rangle\rangle .$$

In particular, for the operators of creation $a_{\mathbf{k}}^\dagger$ and annihilation $a_{\mathbf{k}}$ of free quasiparticles with quasimomentum $\mathbf{k}$ and eigen-energy $\varepsilon_{\mathbf{k}}$, obeying the Hamiltonian

$$H = \sum_{\mathbf{k}} \varepsilon_{\mathbf{k}} a_{\mathbf{k}}^\dagger a_{\mathbf{k}},$$

the diagonal GF $\langle\langle a_{\mathbf{k}}|a_{\mathbf{k}}^\dagger \rangle\rangle$ is simply

$$\langle\langle a_{\mathbf{k}}|a_{\mathbf{k}}^\dagger \rangle\rangle = \frac{1}{\varepsilon - \varepsilon_{\mathbf{k}}}.$$

A much more complicated case of interacting quasiparticles (but in a uniform system) has a general solution

$$\langle\langle a_{\mathbf{k}}|a_{\mathbf{k}}^\dagger \rangle\rangle = \frac{1}{\varepsilon - \varepsilon_{\mathbf{k}} - \Sigma_{\mathbf{k}}(\varepsilon)},$$

where the complex self-energy $\Sigma_{\mathbf{k}}(\varepsilon)$ is usually realized through the diagrammatic series.[32]

17.3. Green functions for *d*-wave superconductor

We start from the simplest single-band model for a *d*-wave superconductor (in absence of impurities), formed by 2D hopping of holes between nearest neighbor oxygen sites in CuO_2 planes[46–48] (Fig. 17.1). Taking explicit account of spin indices $\sigma = \uparrow, \downarrow$ and of the mean-field anomalous coupling, the Hamiltonian can be presented in the compact matrix form with use of *Nambu spinors*, the row-spinor $\psi_{\mathbf{k}}^{\dagger} = (a_{\mathbf{k},\uparrow}^{\dagger}, a_{-\mathbf{k},\downarrow})$ and respective column-spinor $\psi_{\mathbf{k}}$:

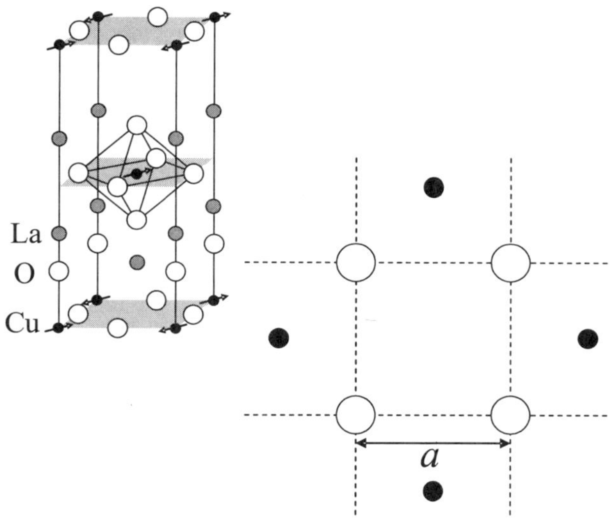

Fig. 17.1. The exemplary metal-oxide perovskite structure of La_2CuO_4 with conducting CuO_2 planes (shadowed) and a fragment of square lattice in such a plane. Arrows indicate AFM spin order at Cu sites.

$$H = \sum_{\mathbf{k}} \left[\psi_{\mathbf{k}}^{\dagger} \left(\xi_{\mathbf{k}} \widehat{\tau}_3 + \Delta_{\mathbf{k}} \widehat{\tau}_1 \right) \psi_{\mathbf{k}} \right]. \tag{17.3}$$

Here $\xi_{\mathbf{k}} = \varepsilon_{\mathbf{k}} - \mu$ is the energy of normal quasiparticle with quasimomentum $\mathbf{k}$ referred to the chemical potential μ (this will be also the reference for the energy variable ε) and $\widehat{\tau}_j$ $(j = 1, 2, 3)$ are the Pauli matrices in Nambu indices. The SC pairing parameter $\Delta_{\mathbf{k}}$ satisfies the BCS gap equation:[6]

$$\Delta_{\mathbf{k}} = \frac{1}{N} \sum_{\mathbf{k}'} V_{\mathbf{k},\mathbf{k}'} \frac{\Delta_{\mathbf{k}'}}{E_{\mathbf{k}'}} \tanh\left(\frac{\beta E_{\mathbf{k}'}}{2}\right) \tag{17.4}$$

where $E_{\mathbf{k}} = \sqrt{\xi_{\mathbf{k}}^2 + \Delta_{\mathbf{k}}^2}$ is the SC quasiparticle energy and the Cooper separable ansatz is used for the SC coupling function $V_{\mathbf{k},\mathbf{k}'} = V_{SC}\gamma_{\mathbf{k}}\gamma_{\mathbf{k}'}$ with the SC coupling constant V_{SC}. The coupling function $\gamma_{\mathbf{k}} = \theta(\varepsilon_{\mathrm{D}}^2 - \xi_{\mathbf{k}}^2)\cos 2\varphi_{\mathbf{k}}$ includes the restriction to the BCS shell of width ε_{D} (the "Debye energy") around the Fermi level and the d-wave symmetry cosine factor with the angular variable $\varphi_{\mathbf{k}} = \arctan k_y/k_x$ for the 2D Brillouin zone. Then Eq. 17.4 yields in the gap function $\Delta_{\mathbf{k}} = \Delta\cos 2\varphi_{\mathbf{k}}$ where the parameter Δ is found from the specific d-wave gap equation:

$$1 = \frac{V_{SC}}{N}\sum_{\mathbf{k}} \frac{\theta(\varepsilon_{\mathrm{D}}^2 - \xi_{\mathbf{k}}^2)\cos^2 2\varphi_{\mathbf{k}}}{\sqrt{\xi_{\mathbf{k}}^2 + \Delta^2\cos^2 2\varphi_{\mathbf{k}}}} \tanh\left(\beta\frac{\sqrt{\xi_{\mathbf{k}}^2 + \Delta^2\cos^2 2\varphi_{\mathbf{k}}}}{2}\right). \qquad (17.5)$$

Next we define the 2×2 Nambu matrix of GF's:

$$\widehat{G}_{\mathbf{k},\mathbf{k}'} = \langle\langle \psi_{\mathbf{k}} | \psi_{\mathbf{k}'}^{\dagger} \rangle\rangle, \qquad (17.6)$$

which matrix elements are the well-known Gor'kov normal and anomalous functions.[31] In what follows we distinguish between Nambu indices (N-indices) and quasi-momentum indices (m-indices) in this matrix and in related (more complicated) matrices. The exact GF matrix for the uniform system, Eq. 17.3, is m-diagonal, $\widehat{G}_{\mathbf{k},\mathbf{k}'} = \delta_{\mathbf{k},\mathbf{k}'}\widehat{G}_{\mathbf{k}}^0$, with

$$\widehat{G}_{\mathbf{k}}^0 = \frac{\varepsilon + \xi_{\mathbf{k}}\widehat{\tau}_3 + \Delta_{\mathbf{k}}\widehat{\tau}_1}{\varepsilon^2 - E_{\mathbf{k}}^2}. \qquad (17.7)$$

The physical properties of SC state are suitably given by these GF's. Thus, the global single-particle DOS, which defines, e.g., the electronic specific heat, is defined straightforwardly by

$$\rho(\varepsilon) = \pi^{-1}\mathrm{Im}\,\mathrm{Tr}\,\widehat{G} \qquad (17.8)$$

where $\widehat{G} = N^{-1}\sum_{\mathbf{k}}\widehat{G}_{\mathbf{k}}$ is the local GF matrix with $\widehat{G}_{\mathbf{k}} \equiv \widehat{G}_{\mathbf{k},\mathbf{k}}$. The local DOS (LDOS) at $\mathbf{n}$th lattice site:

$$\rho_{\mathbf{n}}(\varepsilon) = \frac{1}{\pi N}\sum_{\mathbf{k},\mathbf{k}'} \mathrm{e}^{i(\mathbf{k}-\mathbf{k}')\mathbf{n}}\mathrm{Im}\,\mathrm{Tr}\,\widehat{G}_{\mathbf{k},\mathbf{k}'}, \qquad (17.9)$$

is relevant for interpretation of topography STM data.[18] Other expressions for observable characteristics through GF's, are obtained from the spectral formula, Eq. 17.2, as discussed in what follows.

For the 2D lattice sums, like Eqs. 17.8, 17.9 at given d-wave symmetry, it is suitable to use the "polar coordinates" $\xi \equiv \xi_{\mathbf{k}}$ and $\varphi \equiv \varphi_{\mathbf{k}}$, accordingly to the integration rule:

$$\frac{1}{N}\sum_{\mathbf{k}} f_{\mathbf{k}} = \left(\frac{a}{2\pi}\right)^2 \int d\mathbf{k} f\left(k_x, k_y\right)$$
$$\approx \frac{\rho_N}{4\pi}\int_{-\mu}^{W-\mu} d\xi \int_0^{2\pi} d\varphi f\left(\xi,\varphi\right). \qquad (17.10)$$

Here a is the square lattice constant, $\rho_N \approx 4/(\pi W)$ is the normal state DOS, and the limits for "radial" integration (including the bandwidth W) are rather qualitative, however they only define some less sensitive logarithmic factors.

Thus the local GF matrix for uniform system

$$\widehat{G}^0 = \frac{1}{N}\sum_{\mathbf{k}} \widehat{G}^0_{\mathbf{k}} = \rho_N\left(g_0 - g_{as}\widehat{\tau}_3\right), \qquad (17.11)$$

contains the energy dependence mainly in the function $g_0\left(\varepsilon\right) = \left(\varepsilon/N\right)\sum_{\mathbf{k}} \left(\varepsilon^2 - E^2_{\mathbf{k}}\right)^{-1}$, and from Eq. 17.10 (within accuracy to $O\left(\Delta^3/\mu^3\right)$) the latter is:

$$g_0\left(\varepsilon\right) \approx -\mathrm{Im}\,\mathrm{K}\left(\frac{\Delta^2}{\varepsilon^2}\right)\mathrm{sign}\left(\varepsilon\right) + i\mathrm{Re}\,\mathrm{K}\left(\frac{\Delta^2}{\varepsilon^2}\right) + \frac{\varepsilon}{2\widetilde{\mu}}, \qquad (17.12)$$

where $\widetilde{\mu} = \mu(1-\mu/W) \approx \mu$ and the complete elliptic integral of 1st kind $\mathrm{K}\left(k\right)$ behaves as $\approx \pi\left(1 + k/4\right)/2$ at $k \ll 1$, as $\approx \ln\left(4/\sqrt{k-1}\right)$ at $|k-1| \ll 1$, and as $\approx -i\ln\left(4i\sqrt{k}\right)/\sqrt{k}$ at $k \gg 1$.[49] It should be noted that the analytic result, Eq. 17.12, reflects the fact that $\mathrm{Re}\,g_0(\varepsilon)$ is odd and $\mathrm{Im}\,g_0(\varepsilon)$ is even in energy (understood as $\varepsilon - i0$). Correspondingly, DOS for a uniform d-wave SC crystal:

$$\rho\left(\varepsilon\right) = \frac{1}{\pi}\mathrm{Im}\,\mathrm{Tr}\,\widehat{G}^0 = \frac{2}{\pi}\rho_N\,\mathrm{Im}\,g_0\left(\varepsilon\right),$$

displays sharp SC coherence peaks: $\rho(\varepsilon) \approx (2/\pi)\rho_N \ln\left(4|\varepsilon|/\sqrt{|\varepsilon^2-\Delta^2|}\right)$ at $|\varepsilon| \to \Delta$, decays linearly as $\rho\left(\varepsilon\right) \approx |\varepsilon|\rho_N/\Delta$ at $|\varepsilon| \ll \Delta$, and tends to the normal state constant DOS value ρ_N at $|\varepsilon| \gg \Delta$ (Fig. 17.2).

The asymmetry factor beside $\widehat{\tau}_3$ in Eq. 17.11 is almost constant: $g_{as} = -N^{-1}\sum_{\mathbf{k}} \xi_{\mathbf{k}}/\left(\varepsilon^2 - E^2_{\mathbf{k}}\right) \approx \ln\sqrt{W/\mu - 1}$ (until $|\varepsilon| \ll \mu$, W) and only turns zero at exact half-filling, $\mu = W/2$. As will be seen below, this non-zero value has an important role for impurity perturbations on the d-wave spectrum.

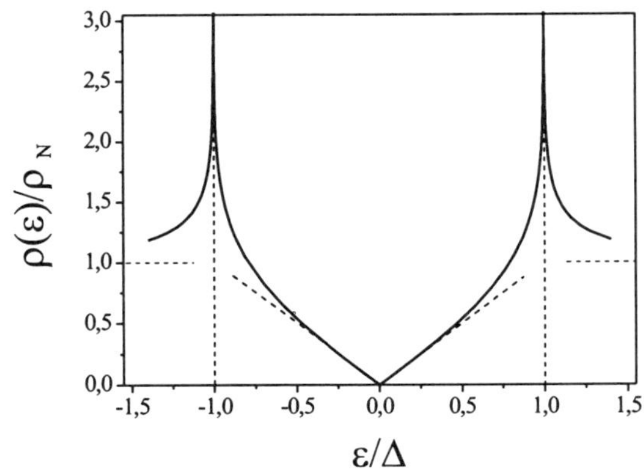

Fig. 17.2. DOS in a clean d-wave SC system (solid line). Dashed lines indicate the linear low energy asymptotics, the logarithmic divergence at $|\varepsilon| \to \Delta$, and the tendency to constant value ρ_N at $|\varepsilon| \gg \Delta$.

Similarly, we can calculate the gap equation, Eq. 17.5, at $T \to 0$:

$$1 = \frac{V_{SC}\rho_N}{4\pi}\int_0^{2\pi} d\varphi \cos^2 2\varphi \int_{-\varepsilon_D}^{\varepsilon_D} \frac{d\xi}{\sqrt{\xi^2 + \Delta^2\cos^2 2\varphi}}. \qquad (17.13)$$

Integrating this first in ξ, we have

$$\int_{-\varepsilon_D}^{\varepsilon_D} \frac{d\xi}{\sqrt{\xi^2 + \Delta^2\cos^2 2\varphi}} = 2\mathrm{arcsinh}\frac{\varepsilon_D}{|\Delta\cos 2\varphi|} \approx 2\ln\frac{2\varepsilon_D}{\Delta\,|\cos 2\varphi|},$$

(since $\Delta \ll \varepsilon_D$). Doing next the angular integration:

$$\int_0^{2\pi} d\varphi \cos^2 2\varphi \ln\frac{2\varepsilon_D}{\Delta\,|\cos 2\varphi|} = \pi\left(\ln\frac{4\varepsilon_D}{\Delta} - \frac{1}{2}\right), \qquad (17.14)$$

we arrive at the gap parameter:

$$\Delta = 4\varepsilon_D e^{-1/\lambda - 1/2}. \qquad (17.15)$$

with the dimensionless pairing constant $\lambda = V_{SC}\rho_N/2$. As usually, this value can be compared with the critical temperature T_c of SC transition, found from the same gap equation, Eq. 17.5, under the condition $\Delta_{\mathbf{k}} \equiv 0$:

$$\frac{1}{\lambda} = \int_0^{\varepsilon_{\rm D}} \frac{d\xi}{\xi} \tanh \frac{\xi}{2k_{\rm B}T_c} \approx \ln \frac{2\gamma_{\rm E}\varepsilon_{\rm D}}{\pi k_{\rm B}T_c},$$

where $\gamma_{\rm E} \approx 1.781$ is the Euler constant,[49] so that

$$k_{\rm B}T_c = (2\gamma_{\rm E}\varepsilon_{\rm D}/\pi)\,{\rm e}^{-1/\lambda}. \quad (17.16)$$

From comparison of Eqs. 17.15 and 17.16 we conclude that the characteristic ratio $r = 2\Delta/k_{\rm B}T_c$ in this case is $2/\sqrt{\rm e}$ times higher than the s-wave BCS value $r_{\rm BCS} = 2\pi/\gamma_{\rm E} \approx 3.52$, reaching ≈ 4.27.

17.4. Impurity perturbations in *d*-wave superconductor and group expansions for Green functions

The simplest impurity perturbation of the Hamiltonian, Eq. 17.3, is realized by the point-like Lifshitz potential $V_{\rm L}$ on random lattice sites $\mathbf{p}$ with concentration $c \ll 1$. Its matrix form:

$$H' = \frac{1}{N} \sum_{\mathbf{p},\mathbf{k},\mathbf{k}'} {\rm e}^{i(\mathbf{k}-\mathbf{k}')\mathbf{p}} \psi^{\dagger}_{\mathbf{k}'} \widehat{V} \psi_{\mathbf{k}}, \quad (17.17)$$

includes the impurity perturbation matrix $\widehat{V} = V_{\rm L}\widehat{\tau}_3$. This is the most extensively used model for impurity effects in superconductors,[4,15,36,50–52] but below we also consider some extensions of this form, either in spatial range of perturbed sites and in spin variables.

In presence of impurities, the equation of motion for the Nambu matrix GF, related to the Hamiltonian $H + H'$ reads:

$$\widehat{G}_{\mathbf{k},\mathbf{k}'} = \delta_{\mathbf{k},\mathbf{k}'} \widehat{G}^0_{\mathbf{k}} - \frac{1}{N} \sum_{\mathbf{p},\mathbf{k}''} {\rm e}^{i(\mathbf{k}-\mathbf{k}'')\mathbf{p}} \widehat{G}^0_{\mathbf{k}} \widehat{V} \widehat{G}_{\mathbf{k}'',\mathbf{k}'}. \quad (17.18)$$

and we shall choose different routines to close the infinite chain of equations for the "scattered" GF's, like $\widehat{G}_{\mathbf{k}'',\mathbf{k}'}$ in Eq. 17.18. In particular, the routine to obtain the fully renormalized GE consists in consecutive iterations of this equation for the "scattered" GF's and in systematic separation of all those already present in the previous iterations.[44] It should be noted that the observable characteristics of a disordered system are described by the so-called *self-averaging* GF's, which values for all particular realizations of disorder are practically non-random, equal to those averaged over disorder.[26] GE's are well defined just for self-averaging quantities, so one should always try to formulate each particular problem in terms of these quantities.

Thus, considering the m-diagonal $\widehat{G}_{\mathbf{k}}$, the most important example of a self-averaging GF, we separate, among the scattering terms, the function $G_{\mathbf{k}}$ itself from those with $\widehat{G}_{\mathbf{k}',\mathbf{k}}$, $\mathbf{k}' \neq \mathbf{k}$:

$$\begin{aligned}\widehat{G}_{\mathbf{k}} &= \widehat{G}^0_{\mathbf{k}} + \frac{1}{N}\sum_{\mathbf{k}',\mathbf{p}} \mathrm{e}^{i(\mathbf{k}-\mathbf{k}')\cdot\mathbf{p}}\widehat{G}^0_{\mathbf{k}}\widehat{V}\widehat{G}_{\mathbf{k}',\mathbf{k}} \\ &= \widehat{G}^0_{\mathbf{k}} + c\widehat{G}^0_{\mathbf{k}}\widehat{V}G_{\mathbf{k}} + \frac{1}{N}\sum_{\mathbf{k}'\neq\mathbf{k},\mathbf{p}} \mathrm{e}^{i(\mathbf{k}-\mathbf{k}')\cdot\mathbf{p}}\widehat{G}^0_{\mathbf{k}}\widehat{V}\widehat{G}_{\mathbf{k}',\mathbf{k}}. \qquad (17.19)\end{aligned}$$

Then for each $\widehat{G}_{\mathbf{k}',\mathbf{k}}$, $\mathbf{k}' \neq \mathbf{k}$ we write down Eq. 17.18 again and single out the scattering terms with $\widehat{G}_{\mathbf{k}}$ and $\widehat{G}_{\mathbf{k}',\mathbf{k}}$ in its r.h.s:

$$\begin{aligned}\widehat{G}_{\mathbf{k}',\mathbf{k}} &= \frac{1}{N}\sum_{\mathbf{k}'',\mathbf{p}'} \mathrm{e}^{i(\mathbf{k}'-\mathbf{k}'')\cdot\mathbf{p}}\widehat{G}^0_{\mathbf{k}'}\widehat{V}\widehat{G}_{\mathbf{k}'',\mathbf{k}} \\ &= c\widehat{G}^0_{\mathbf{k}'}\widehat{V}\widehat{G}_{\mathbf{k}',\mathbf{k}} + \frac{1}{N}\mathrm{e}^{i(\mathbf{k}'-\mathbf{k})\cdot\mathbf{p}}\widehat{G}^0_{\mathbf{k}'}\widehat{V}\widehat{G}_{\mathbf{k}} \\ &+ \frac{1}{N}\sum_{\mathbf{p}'\neq\mathbf{p}} \mathrm{e}^{i(\mathbf{k}'-\mathbf{k})\cdot\mathbf{p}'}\widehat{G}^0_{\mathbf{k}'}\widehat{V}\widehat{G}_{\mathbf{k}} \\ &+ \frac{1}{N}\sum_{\mathbf{k}''\neq\mathbf{k},\mathbf{k}';\mathbf{p}'} \mathrm{e}^{i(\mathbf{k}'-\mathbf{k}'')\cdot\mathbf{p}'}\widehat{G}^0_{\mathbf{k}'}\widehat{V}\widehat{G}_{\mathbf{k}'',\mathbf{k}}. \qquad (17.20)\end{aligned}$$

Note that, among the terms with $\widehat{G}_{\mathbf{k}}$, the $\mathbf{p}' = \mathbf{p}$ term (the second in r.h.s. of Eq. 17.20) bears the phase factor $\mathrm{e}^{i(\mathbf{k}'-\mathbf{k})\cdot\mathbf{p}}$, so it is coherent to that already present in the last sum in Eq. 17.19. That is why this term is explicitly separated from other, incoherent ones, $\propto \mathrm{e}^{i(\mathbf{k}'-\mathbf{k})\cdot\mathbf{p}'}$, $\mathbf{p}' \neq \mathbf{p}$ (but there will be no such separation when doing 1st iteration of Eq. 17.18 for the m-non-diagonal GF $\widehat{G}_{\mathbf{k}'',\mathbf{k}}$ itself).

Continuing the sequence, we collect the terms with the initial function $\widehat{G}_{\mathbf{k}}$ which result from:

i) all multiple scatterings on the same site $\mathbf{p}$, and

ii) such processes on the same pair of sites $\mathbf{p}$ and $\mathbf{p}' \neq \mathbf{p}$, and so on.

Then summation in $\mathbf{p}$ of the i)-terms gives rise to the first term of GE as $c\widehat{T}$ where:

$$\widehat{T} = \widehat{V}\left(1 - \widehat{G}\widehat{V}\right)^{-1}, \qquad (17.21)$$

and, if the impurity cluster processes were neglected, this term would be just the self-consistent T-matrix.[33,53] The second term of GE, obtained by summation of the ii)-terms in $\mathbf{p}, \mathbf{p}' \neq \mathbf{p}$, contains the interaction matrices[15] $\widehat{A}_{\mathbf{p}',\mathbf{p}} = N^{-1}\sum_{\mathbf{k}'} e^{i\mathbf{k}'\cdot\mathbf{p}}\widehat{G}_{\mathbf{k}'}\widehat{T}$ generated by the multiply scattered GF's $\widehat{G}_{\mathbf{k}',\mathbf{k}}$, $\mathbf{k}' \neq \mathbf{k}$, etc., (including their own renormalization). For instance, the iterated equation of motion for $\widehat{G}_{\mathbf{k}'',\mathbf{k}}$ with $\mathbf{k}'' \neq \mathbf{k}, \mathbf{k}'$ in the last term of Eq. 17.20 will produce:

$$
\begin{aligned}
\widehat{G}_{\mathbf{k}'',\mathbf{k}} &= \frac{1}{N}\sum_{\mathbf{k}''',\mathbf{p}''} e^{i(\mathbf{k}''-\mathbf{k}''')\cdot\mathbf{p}''}\widehat{G}^0_{\mathbf{k}''}\widehat{V}\widehat{G}_{\mathbf{k}''',\mathbf{k}} \\
&= \frac{1}{N}e^{i(\mathbf{k}''-\mathbf{k})\cdot\mathbf{p}}\widehat{G}^0_{\mathbf{k}''}\widehat{V}\widehat{G}_{\mathbf{k}} + \frac{1}{N}e^{i(\mathbf{k}''-\mathbf{k})\cdot\mathbf{p}'}\widehat{G}^0_{\mathbf{k}''}\widehat{V}\widehat{G}_{\mathbf{k}} \\
&+ \quad \text{terms} \quad \text{with} \quad \widehat{G}_{\mathbf{k}',\mathbf{k}} \quad \text{and} \quad \widehat{G}_{\mathbf{k}'',\mathbf{k}} \\
&+ \quad \text{terms} \quad \text{with} \quad \widehat{G}_{\mathbf{k}''',\mathbf{k}}\ (\mathbf{k}''' \neq \mathbf{k}, \mathbf{k}', \mathbf{k}''). \qquad (17.22)
\end{aligned}
$$

Finally we arrive at the fully renormalized representation for the m-diagonal GF as

$$
\widehat{G}_{\mathbf{k}} = \widehat{G}_{\mathbf{k},\mathbf{k}} = \left[\left(\widehat{G}^0_{\mathbf{k}}\right)^{-1} - \widehat{\Sigma}_{\mathbf{k}}\right]^{-1}, \qquad (17.23)
$$

where the renormalized self-energy matrix is presented by the GE

$$
\widehat{\Sigma}_{\mathbf{k}} = c\widehat{T}\left(1 - c\widehat{A}_0 - c\widehat{A}_0^2 + c\widehat{B}_{\mathbf{k}} + \cdots\right) \qquad (17.24)
$$

with the pair term

$$
\widehat{B}_{\mathbf{k}} = \sum_{\mathbf{n}\neq 0}\left(\widehat{A}^3_{\mathbf{n}}e^{-i\mathbf{k}\mathbf{n}} + \widehat{A}^4_{\mathbf{n}}\right)\left(1 - \widehat{A}^2_{\mathbf{n}}\right)^{-1}, \qquad (17.25)
$$

An alternative routine consists in iterations of the equation of motion for *all* the terms $\widehat{G}_{\mathbf{k}'',\mathbf{k}}$ in Eq. 17.18 and summing the contributions $\propto \widehat{G}^0_{\mathbf{k}}$, like the first term in the r.h.s. This finally leads to the solution of form:

$$
\widehat{G}_{\mathbf{k}} = \widehat{G}^0_{\mathbf{k}} + \widehat{G}^0_{\mathbf{k}}\widehat{\Sigma}^0_{\mathbf{k}}\widehat{G}^0_{\mathbf{k}}, \qquad (17.26)
$$

where the non-renormalized self-energy

$$
\widehat{\Sigma}^0_{\mathbf{k}} = c\widehat{T}^0\left(1 + c\widehat{B}^0_{\mathbf{k}} + \cdots\right) \qquad (17.27)
$$

includes the non-renormalized T-matrix:

$$\widehat{T}^0 = \widehat{V}\left(1 - \widehat{G}^0\widehat{V}\right)^{-1}, \tag{17.28}$$

and the corresponding pair term:

$$\widehat{B}^0_{\mathbf{k}} = \left(\widehat{A}^0_{\mathbf{n}}\mathrm{e}^{-i\mathbf{k}\cdot\mathbf{n}} + \widehat{A}^0_{\mathbf{n}}\widehat{A}^0_{-\mathbf{n}}\right)\left(1 - \widehat{A}^0_{\mathbf{n}}\widehat{A}^0_{-\mathbf{n}}\right)^{-1}, \tag{17.29}$$

with the non-renormalized interaction matrices given by:

$$\widehat{A}^0_{\mathbf{n}} = \widehat{G}^0_{\mathbf{n}}\widehat{T}^0, \quad \widehat{G}^0_{\mathbf{n}} = \frac{1}{N}\sum_{\mathbf{k}}\mathrm{e}^{i\mathbf{k}\cdot\mathbf{n}}\widehat{G}^0_{\mathbf{k}}, \quad \widehat{G}^0 \equiv \widehat{G}^0_0. \tag{17.30}$$

Like the previous Eq. 17.23, the next to unity term in the brackets of Eq. 17.27 describes the contribution from all possible clusters of two impurities and the dropped terms are for clusters of three and more impurities. This permits to describe, in principle, the hierarchical structure of quasi-continuous spectrum of localized states in the crystal with impurities.[26] Besides the two above presented, other types of GE's (with different grades of their renormalization) are also possible after specific routines for the equations of motion.[54]

The crucial issue for group expansions is their convergence. Strictly speaking, it can be only asymptotic, moreover, it essentially depends on the chosen value of ε. In practice, we simply consider the group series converging, at a given ε, if the contribution to the self-energy $\widehat{\Sigma}$ from the 1st (single-impurity) term of GE dominates over that from the 2nd term (impurity pairs). Then it is believed that the pair term and all the rest of the series can be dropped, as is proved in some simplest model cases.[45] But with varying ε, a condition can be reached that the first two GE terms turn to be of the same order (and, supposedly, all the rest too), this is expected to define a limit of convergence for the given type of GE. Such limits for different GE types are different,[45] and we can combine between them to cover the maximum energy range. Finally, the areas of the spectrum, where no GE is convergent, define special regions, like the regions of concentration broadening around localized levels or the mobility edges (dividing band-like states from localized states). The quantitative analysis of GE convergence and of impurity cluster effects begins from the simplest single-impurity level, providing the elements for all the following steps.

17.5. Single-impurity approximations

Using Eqs. 17.11, 17.28, we express the non-renormalized T-matrix as

$$\widehat{T}^0(\varepsilon) = \frac{v}{\rho_N}\frac{vg_0 - \widehat{\tau}_3}{1 - v^2 g_0^2}, \tag{17.31}$$

through the function $g_0(\varepsilon)$, Eq. 17.12, and the dimensionless perturbation parameter $v = V_{\mathrm{L}}\rho_N/\left(1 - V_{\mathrm{L}}\rho_N g_{as}\right)^{-1}$.[55] This T-matrix permits existence of low energy resonances,[14,15] at symmetric points $\varepsilon = \pm\varepsilon_{res}$, found from the condition $\mathrm{Re}\left[g_{0d}(\varepsilon_{res})\right]^2 = 1/v^2$, analogous to the known Lifshitz equation in normal metals and semiconductors.[26]

Since $\mathrm{Re}\left[g_0(\varepsilon)\right]^2 \approx \left[\mathrm{Im}\,\mathrm{K}(\Delta^2/\varepsilon^2)\right]^2 - \left[\mathrm{Re}\,\mathrm{K}(\Delta^2/\varepsilon^2)\right]^2$ reaches its highest value ≈ 0.47 at $|\varepsilon| \approx 0.443\Delta$, the formal solution to the Lifshitz equation first appears just at these points when v reaches ≈ 1.46. However, this formal solution may not yet correspond to a true resonance because its broadening $\Gamma_{res} \approx \mathrm{Im}g_0(\varepsilon_{res})/\left(d\mathrm{Re}\,g_0/d\varepsilon\right)_{\varepsilon_{res}}$ turns out to be $\sim \Delta\ln\left(\Delta/\sqrt{|\varepsilon_{res}^2 - \Delta^2|}\right)$, that is large compared to ε_{res} itself. But for strong enough perturbation: $v \gg 1$, the resonance energy is low, $\varepsilon_{res} \ll \Delta$, and from the logarithmic asymptotics: $\mathrm{Re}\,g_0(\varepsilon) \approx (\varepsilon/\Delta)\ln(4\Delta/\varepsilon)$, it estimates as $\varepsilon_{res} \approx \Delta/\left[v\ln(4v)\right]$.[14,15] Then also the broadening $\Gamma_{res} \approx \pi\varepsilon_{res}/\left[2\ln(4v)\right]$ is smaller than ε_{res} (though not very much) already for $v \gtrsim 1.7$, as seen in Fig. 17.3.

It is important to notice that the perturbation parameter v can turn very large, $|v| \gg 1$, even for relatively moderate perturbation potential $V_{\mathrm{L}} \sim 1/\rho_N$, due to presence of the finite asymmetry factor $g_{as} \sim 1$, and this can open possibility for the unitary scattering regime[36] in superconductors with rather common impurity substitutes as transition metals or Zn for Cu. This specific property of impurity scattering in high-T_c systems with asymmetric bands was first indicated by Hirschfeld *et al.*[56]

The local SC order parameter in the d-wave system is constructed from the off-diagonal correlators between nearest neighbor sites in the lattice, $\langle a_{\mathbf{n}+\delta,\uparrow}a_{\mathbf{n},\downarrow}\rangle$, where $\delta = (\pm a/2, \pm a/2)$, taking the GF form (in agreement with Eq. 17.2)

$$\begin{aligned}\Delta_{\mathbf{n}} = {} & \frac{V_{SC}}{2\pi N}\sum_{\mathbf{k},\mathbf{k}'} \mathrm{e}^{i\left[\mathbf{k}(\mathbf{n}+\delta)-\mathbf{k}'\mathbf{n}\right]}\theta\left(\varepsilon_{\mathrm{D}}^2 - \xi_{\mathbf{k}}^2\right)\theta\left(\varepsilon_{\mathrm{D}}^2 - \xi_{\mathbf{k}'}^2\right) \\ & \times \int_{-\infty}^{\infty}\frac{d\varepsilon}{\mathrm{e}^{\varepsilon/T}+1}\mathrm{Im}\;\mathrm{Tr}\widehat{G}_{\mathbf{k},\mathbf{k}'}\widehat{\tau}_1.\end{aligned} \tag{17.32}$$

Its average value over all the sites $\mathbf{n}$ is presented in a specific GF form as

$$\Delta = \frac{V_{SC}}{2\pi}\int_{-\infty}^{\infty}\frac{d\varepsilon}{\mathrm{e}^{\varepsilon/T}+1}\mathrm{Im}\;\mathrm{Tr}\widehat{F}_{\delta}^{0}\widehat{\tau}_1. \tag{17.33}$$

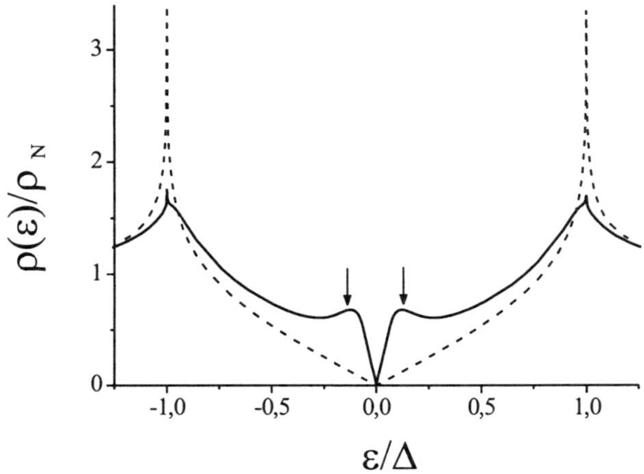

Fig. 17.3. Low energy resonance levels $\pm\varepsilon_{res}$ (arrows) in the DOS of d-wave superconductor containing a finite concentration $c = 0.2\rho_N\Delta$ of impurity scatterers with $V_{\rm L} \approx 0.7/\rho_N$ (corresponding to $v \approx 2.3$). Other distinctions from the pure crystal DOS in Fig. 17.2 (shown here by the dashed line) are the finite spikes at $|\varepsilon| = \Delta$ and the enhanced slope at $|\varepsilon| < \varepsilon_{res}$.

Here the matrix

$$\widehat{F}_\delta^0 = \frac{1}{N}\sum_{\mathbf{k}} {\rm e}^{i\mathbf{k}\cdot\delta}\theta\left(\varepsilon_{\rm D}^2 - \xi_{\mathbf{k}}^2\right)\widehat{G}_{\mathbf{k}}^0, \tag{17.34}$$

is calculated accordingly to the rules:

$$\sum_{\mathbf{k}} f(k)\,{\rm e}^{i\mathbf{k}\delta} \approx \sum_{\mathbf{k}} f(k),$$
$$\sum_{\mathbf{k}} f(k)\,{\rm e}^{i\mathbf{k}\delta}\cos 2\varphi_{\mathbf{k}} \approx \frac{\pi\mu\rho_N}{2}\sum_{\mathbf{k}} f(k)\cos^2 2\varphi_{\mathbf{k}},$$
$$\sum_{\mathbf{k}} \xi_{\mathbf{k}}\theta\left(\varepsilon_{\rm D}^2 - \xi_{\mathbf{k}}^2\right) f\left(\xi_{\mathbf{k}}^2\right) = 0,$$

to result in:

$$\widehat{F}_\delta^0 \approx \rho_N\left(f_0 + \frac{\pi\mu\rho_N}{2} g_1\widehat{\tau}_1\right). \tag{17.35}$$

This includes other energy function

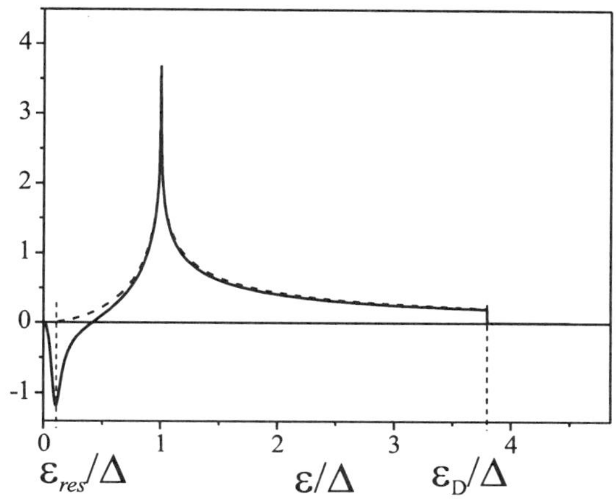

Fig. 17.4. Comparison of the integrand functions in the numerator (solid line) and in the denominator (dashed line) of Eq. 17.39 at the same value of perturbation parameter $v = 2.3$ as used in Fig. 17.3. Note the sizeable negative effect of the resonance level at ε_{res}, which reduces the value $\eta \approx 0.83$, compared to that value in the s-wave case $\eta_s \approx 0.98$ for the same v.

$$\begin{aligned} g_1(\varepsilon) &= \frac{\Delta}{2\pi}\int_0^{2\pi}\cos^2 2\varphi\, d\varphi \int_0^{\varepsilon_D}\frac{d\xi}{\varepsilon^2-\xi^2-\Delta^2\cos^2 2\varphi} \\ &\approx \frac{\Delta}{2\varepsilon_D} - i\frac{\varepsilon}{\Delta}\left[\mathrm{K}\left(\frac{\Delta^2}{\varepsilon^2}\right)-\mathrm{E}\left(\frac{\Delta^2}{\varepsilon^2}\right)\right], \end{aligned} \tag{17.36}$$

with the full elliptic integral of 2nd kind E,[49] and the function f_0 which only differs from g_0, Eq. 17.12, by its last term $\varepsilon/\varepsilon_D$. The absence of asymmetry term $\propto \widehat{\tau}_3$ is due to the BCS shell factor $\theta\left(\varepsilon_D^2-\xi_{\mathbf{k}}^2\right)$ in Eq. 17.34.

The local variation of SC order on **n**th site from its average value is only defined by m-non-diagonal contributions into Eq. 17.32, which are easily calculated for the simplest case of a single impurity center at $\mathbf{p}=0$:

$$\widehat{G}_{\mathbf{k},\mathbf{k}'} = \frac{1}{N}\widehat{G}^0_{\mathbf{k}}\widehat{T}^0\widehat{G}^0_{\mathbf{k}'}, \tag{17.37}$$

expressing this variation in terms of matrices $\widehat{F}$ as

$$\Delta_{\mathbf{n}} - \Delta = \frac{V_{SC}}{2\pi}\int_{-\infty}^{\infty}\frac{d\varepsilon}{e^{\varepsilon/T}+1}\mathrm{Im}\,\mathrm{Tr}\widehat{F}^0_{\mathbf{n}+\delta}\widehat{T}^0\widehat{F}^0_{\mathbf{n}}\widehat{\tau}_1. \tag{17.38}$$

Taking into account that $\widehat{F}_0^0$ is simply a scalar $\rho_N f_0$ and considering zero temperature, we obtain the parameter $\eta = 1 - \Delta_0/\Delta$ of relative suppression of SC order on the very impurity site $\mathbf{n} = 0$ as:

$$\eta = -\frac{\int_0^{\varepsilon_D} d\varepsilon \mathrm{Im}\,\mathrm{Tr}\,\widehat{F}_\delta^0 \widehat{T}^0 \widehat{F}_0^0 \widehat{\tau}_1}{\int_0^{\varepsilon_D} d\varepsilon \mathrm{Im}\,\mathrm{Tr}\,\widehat{F}_\delta^{(0)} \widehat{\tau}_1} = -\frac{v^2 \int_0^{\varepsilon_D} d\varepsilon \mathrm{Im}\left[g_0 f_0 g_1 / \left(1 - v^2 g_0^2\right)\right]}{\int_0^{\varepsilon_D} d\varepsilon \mathrm{Im}\, g_1}. \tag{17.39}$$

as a function of the perturbation parameter v. The numerically calculated function $\eta(v)$ turns to be quite similar to the analytic result for impurity in an s-wave superconductor:[15] $\eta_s(v) = \pi^2 v^2 / \left(1 + \pi^2 v^2\right)$, at weak perturbations, $v \ll 1$, but its saturation to unity at stronger perturbations, $v \gtrsim 1$, is sensibly delayed. This is due to the pronounced negative effect of the low energy resonance level ε_{res} on the integrand function in numerator of Eq. 17.37, compared to the integrand in its denominator in Fig. 17.4, whereas in the s-wave case, at absence of impurity resonance, the two integrands (in numerator and denominator) are simply proportional to each other.

17.5.1. *Extended impurity center*

The predicted effects of a point-like scatterer on the local SC order are perhaps too strong to be adequate to the observed stability of SC state under disorder. But in reality the impurity perturbations in high-T_c materials are not exactly point-like but rather extended to a finite number of neighbor sites to the impurity center. This raises an important question on how robust are the results of point-like model with respect to the spatial extent and geometry of impurity perturbation. The opposite limit to the point-like perturbation, when the defect is much bigger of the Fermi wavelength and can be treated quasiclassically,[57] hardly applies to real atomic substitutes in high-T_c systems with perturbation limited to few nearest neighbors of the impurity site. To model the latter situation, we extend the perturbation Hamiltonian, Eq. 17.17, to the form:[58]

$$H_{ext} = -\frac{1}{N} \sum_{\mathbf{k},\mathbf{k}',\mathbf{p}} \mathrm{e}^{i(\mathbf{k}'-\mathbf{k})\mathbf{p}} \sum_{\delta} \mathrm{e}^{i(\mathbf{k}'-\mathbf{k})\delta} \Psi_{\mathbf{k}'}^{\dagger} \widehat{V} \Psi_{\mathbf{k}}. \tag{17.40}$$

It contains formally the same perturbation matrix $\widehat{V} = V_{\mathrm{L}} \widehat{\tau}_3$, but takes an explicit account of the phase shifts $\mathrm{e}^{i(\mathbf{k}'-\mathbf{k})\delta}$ at quasiparticle scattering by extended perturbation on the nearest neighbor lattice sites δ to the impurity center $\mathbf{p}$ (which

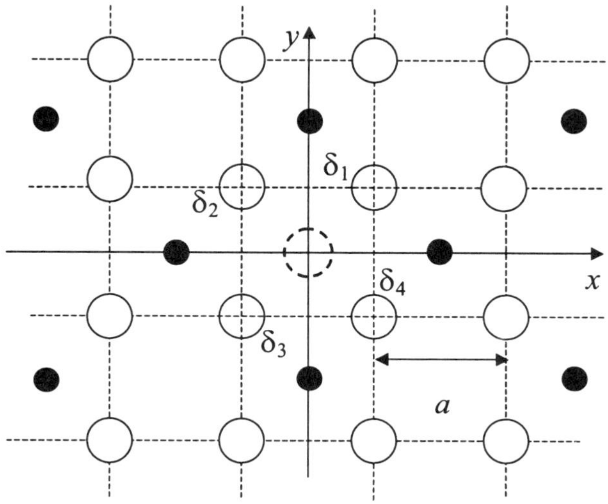

Fig. 17.5. Extended perturbation over four nearest oxygen sites δ_i to the impurity ion (its projection onto the CuO_2 plane is shown by the dashed circle at the origin).

itself does not pertain in this case to the lattice, see Fig. 17.5). Then the equation of motion, Eq. 17.18, is modified to:

$$\begin{aligned} \widehat{G}_{\mathbf{k},\mathbf{k}'} &= \widehat{G}^0_{\mathbf{k}}\delta_{\mathbf{k},\mathbf{k}'} - \\ &+ \frac{1}{N}\sum_{\mathbf{k}'',\mathbf{p},j} \mathrm{e}^{i(\mathbf{k}-\mathbf{k}'')\mathbf{p}}\alpha_{j\mathbf{k}}\alpha_{j\mathbf{k}''}\widehat{G}^0_{\mathbf{k}}\widehat{V}\widehat{G}_{\mathbf{k}'',\mathbf{k}'}, \end{aligned} \tag{17.41}$$

where the functions

$$\begin{aligned} \alpha_{1,\mathbf{k}} &= 2\cos\frac{ak_x}{2}\cos\frac{ak_y}{2}, \qquad & \alpha_{2,\mathbf{k}} &= 2\cos\frac{ak_x}{2}\sin\frac{ak_y}{2}, \\ \alpha_{3,\mathbf{k}} &= 2\sin\frac{ak_x}{2}\cos\frac{ak_y}{2}, \qquad & \alpha_{4,\mathbf{k}} &= 2\sin\frac{ak_x}{2}\sin\frac{ak_y}{2}, \end{aligned} \tag{17.42}$$

realize irreducible representations of the C_4 point group ($j = 1$ being related to A-, $j = 2, 3$ to E-, and $j = 4$ to B-representations[59]) and thus satisfy the orthogonality condition

$$\frac{1}{N}\sum_{\mathbf{k}}\alpha_{j,\mathbf{k}}\alpha_{j'\mathbf{k}} = \delta_{jj'}. \tag{17.43}$$

The impurity effects on quasiparticle spectrum are naturally classified along these representations, alike the known effects of magnetic impurities in ferro- and antiferromagnetic crystals.[45,60]

The orthogonality of the $\alpha_{j,\mathbf{k}}$ functions results in that Eq. 17.41 has a solution formally coinciding with Eq. 17.26, but with the T-matrix additive in these representations: $\widehat{T}^0 = \sum_j \widehat{T}_j^0$, where each partial T-matrix $\widehat{T}_j^0 = \widehat{V}(1-\widehat{G}_j^0\widehat{V})^{-1}$ includes the specific local GF matrix: $\widehat{G}_j^0 = N^{-1}\sum_{\mathbf{k}}\alpha_{j,\mathbf{k}}^2\widehat{G}_{\mathbf{k}}^0$. Alike Eq. 17.31, this matrix can be expanded in the basis of Pauli matrices

$$\widehat{G}_j^0 = \rho_N(g_{j0} + g_{j1}\widehat{\tau}_1 - g_{j3}\widehat{\tau}_3). \tag{17.44}$$

The functions g_{ji} are calculated by using Eq. 17.10; some of them are zero by the symmetry reasons: $g_{11} = g_{41} = 0$, and the rest can be approximated as:

$$g_{j0} \approx \overline{\alpha_j^2}g_0, \qquad g_{j3} \approx \overline{\alpha_j^2}g_{as}, \qquad g_{21} = -g_{31} \approx \overline{\alpha_2^2}g_1. \tag{17.45}$$

Here $\overline{\alpha_j^2}$ are the average values of $\alpha_{j\mathbf{k}}^2$ over the Fermi surface: $\overline{\alpha_1^2} \approx 4(1-\mu/W)$, $\overline{\alpha_{2,3}^2} \approx 4\mu/W$, $\overline{\alpha_4^2} \approx 2(\mu/W)^2$, where the band occupation parameter μ/W is supposedly small. Then the most important contribution to $\widehat{T}^0$ comes from the $j = 1$ term (A-representation):

$$\widehat{T}_1^0 = \frac{v_A}{\overline{\alpha_1^2}\rho_N}\frac{v_A g_0 - \widehat{\tau}_3}{D_A}. \tag{17.46}$$

with the A-channel perturbation parameter $v_A = \overline{\alpha_1^2}V_{\mathrm{L}}\rho_N/(1-\overline{\alpha_1^2}V_{\mathrm{L}}\rho_N g_{as})$ and the denominator $D_A(\varepsilon) = 1 - v_A^2 g_0^2(\varepsilon)$. It can produce low energy resonances at $\varepsilon = \pm\varepsilon_{res}$ (such that $\mathrm{Re}\,D_A(\varepsilon_{res}) = 0$), similar to the above mentioned resonances for point-like impurity center. This again requires that v_A exceeds a critical value $v_{A,cr} \approx 2/\pi$.

The contributions from $j = 2, 3$ (E-representation) are:

$$\widehat{T}_{2,3}^0 = \frac{v_E}{\overline{\alpha_2^2}\rho_N}\frac{v_E(g_0 \mp g_1\widehat{\tau}_1) - \widehat{\tau}_3}{D_E}, \tag{17.47}$$

with $v_E = \overline{\alpha_2^2}V_{\mathrm{L}}\rho_N/(1-\overline{\alpha_2^2}V_{\mathrm{L}}\rho_N g_{as})$ and $D_E = 1 - v_E^2(g_0^2 - g_1^2)$. It is less probable to have a resonance effect in this channel at low occupation $\mu/W \ll 1$, since the reduced parameter v_E compared to v_A and competition between Re g_0^2 and Re g_1^2 in D_E.

The B-channel $\widehat{T}_4^0$ has the same structure as the A-channel term, Eq. 17.48, but with v_A replaced by a strongly reduced value $v_B = \overline{\alpha_4^2} V_{\rm L} \rho_N / (1 - \overline{\alpha_4^2} V_{\rm L} \rho_N g_{as})$, hence it turns even less important than the E-channel terms.

Now, using Eqs. 17.46, 17.47 in Eq. 17.8, the global DOS is obtained as

$$\rho(\varepsilon) \approx \frac{\rho_N}{\pi} \text{Im } g_0(\varepsilon - \Sigma_0), \tag{17.48}$$

where the scalar self-energy

$$\Sigma_0 = \frac{\pi c W g_0(\varepsilon)}{4} \left(\frac{v_A^2}{\overline{\alpha_1^2} D_A} + \frac{2 v_E^2}{\overline{\alpha_2^2} D_E} + \frac{v_B^2}{\overline{\alpha_4^2} D_B} \right) \tag{17.49}$$

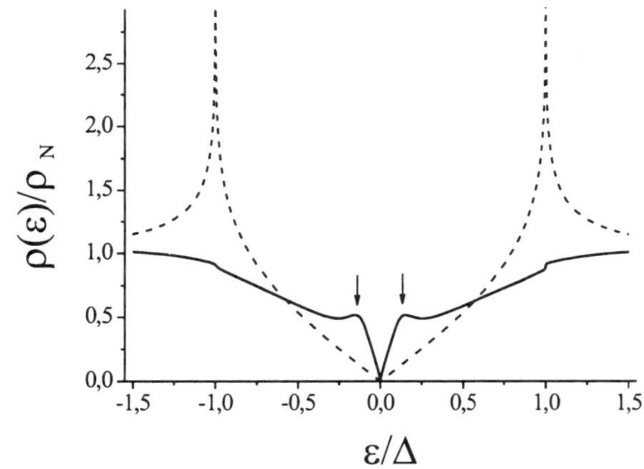

Fig. 17.6. DOS in the d-wave superconductor with extended impurity centers (the solid line), for the choice of parameters $W = 2$ eV, $\mu = 0.3$ eV, $\varepsilon_{\rm D} = 0.15$ eV, $V_{\rm L} = 0.3$ eV, $c = 0.1$. The arrow indicates the low-energy resonance by the A-channel impurity effect and the dashed line represents the pure d-wave DOS.

includes the effects of all three channels. Fig. 17.5 presents the results of direct calculation from Eq. 17.48 with use of Eq. 17.49 at a typical choice of parameters, $W = 2$ eV, $\mu = 0.3$ eV, $\varepsilon_{\rm D} = 0.15$ eV, $V_{\rm L} = 0.3$ eV (this gives for particular channels: $v_A \approx 1.763$, $v_E \approx 0.129$, and $v_B \approx 0.009$), and $c = 0.1$. They are similar to the results for point-like impurities,[4,15] showing low-energy resonances at $\pm\varepsilon_{res}$, mainly due to the A-channel effect. But the reduction of coherence peaks at $|\varepsilon| = \Delta$ is much stronger, due to additional effect from the E-channel, while the B-channel has no appreciable effect at all.

In analogy with Eq. 17.31 for variation of the SC order parameter, variation of LDOS $\rho_{\mathbf{n}}(\varepsilon)$, Eq. 17.9, near the impurity site, compared to the average value $\rho(\varepsilon)$, is only given by the m-non-diagonal GF's:

$$\delta\rho_{\mathbf{n}}(\varepsilon) = \frac{1}{\pi N} \sum_{\mathbf{k},\mathbf{k}' \neq \mathbf{k}} \mathrm{e}^{i(\mathbf{k}-\mathbf{k}')\cdot\mathbf{n}} \mathrm{Im\,Tr}\,\widehat{G}_{\mathbf{k},\mathbf{k}'}. \tag{17.50}$$

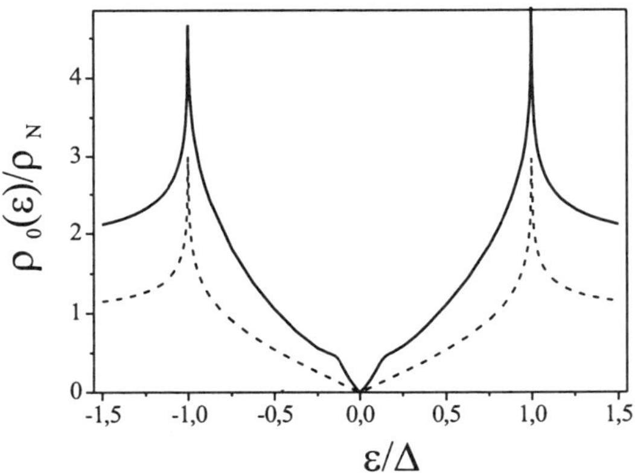

Fig. 17.7. Local density of states on the nearest neighbor site to an extended impurity center, for the same choice of parameters as in Fig. 17.6 (but supposing $c \to 0$). Note the overall enhancement of electronic density compared to that on remote sites from impurity (dashed line).

Here $\widehat{G}_{\mathbf{k},\mathbf{k}'} = N^{-1}\sum_j \alpha_{j,\mathbf{k}} \widehat{G}^0_{\mathbf{k}} \widehat{T}^0_j \widehat{G}^0_{\mathbf{k}'} \alpha_{j,\mathbf{k}'}$, and the strongest variation is attained at $\mathbf{n} = \delta$, the nearest neighbor sites to the impurity. Using Eq. 17.50 and the orthogonality relations, we expand this value into a sum:

$$\begin{aligned}\delta\rho_{\mathbf{n}=\delta}(\varepsilon) &= \frac{1}{\pi N^2} \sum_{\mathbf{k},\mathbf{k}',j} \mathrm{Im\,Tr}\,\mathrm{e}^{i\mathbf{k}\cdot\delta} \alpha_{j,\mathbf{k}} \widehat{G}^0_{\mathbf{k}} \widehat{T}^0_j \widehat{G}^0_{\mathbf{k}'} \alpha_{j,\mathbf{k}'} \mathrm{e}^{-i\mathbf{k}'\cdot\delta} \\ &= \frac{1}{\pi} \sum_j \mathrm{Im\,Tr}\,\widehat{G}^0_j \widehat{T}^0_j \widehat{G}^0_j,\end{aligned}$$

and present the overall maximum LDOS as:

$$\rho_{\mathbf{n}=\delta}(\varepsilon) = \frac{2\rho_N}{\pi}\mathrm{Im}\left[g_0(\varepsilon)\left(1 + \frac{v_A N_A}{\alpha_1^2 D_A} + 2\frac{v_E N_E}{\alpha_2^2 D_E} + \frac{v_B N_B}{\alpha_4^2 D_B}\right)\right]. \quad (17.51)$$

Alike Eq. 17.48 for global DOS, the resonance contribution to Eq. 17.51 at low energies $\varepsilon \sim \varepsilon_{res}$ comes from the A-term with the numerator $N_A = 2g_3 + v_A(g_0^2 + g_3^2)$. Other channels, with $N_E = 2g_3 + v_E(g_0^2 - g_1^2 - g_3^2)$ and $N_B = 2g_3 + v_B(g_0^2 + g_3^2)$, mainly contribute to modification of the pure d-wave DOS $\rho_d(\varepsilon) = 2/\pi \mathrm{Im}\, g_0(\varepsilon)$ far from the resonance. LDOS on nearest neighbor sites to the impurity, calculated from Eq. 17.51 (solid line in Fig. 17.7), displays less pronounced low energy resonances than those in the global DOS, Fig. 17.6, but an overall enhancement compared to the LDOS curve for remote sites from impurity $\rho_{\mathbf{n}\to\infty}(\varepsilon) = \rho(\varepsilon)$ (the dashed line). This picture resembles the direct experimental measurements of differential conductance through the STM tip positioned close to and far from an impurity center.[18]

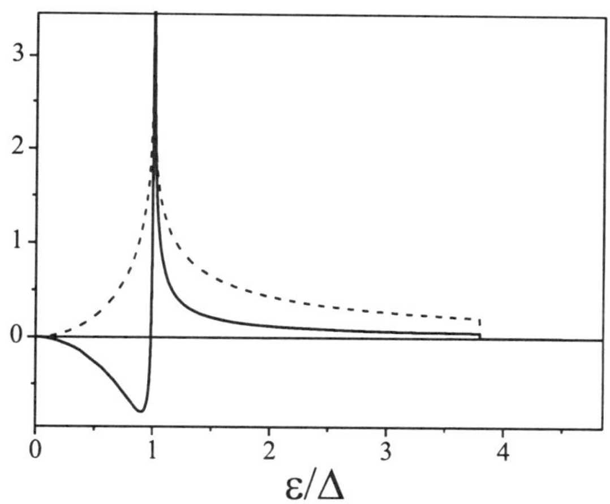

Fig. 17.8. The dimensionless function $F(\varepsilon)$ (solid line) used in the numerator of Eq. 17.52, compared to the integrand in the denominator (dashed line) to calculate the suppression parameter η_{sup} at the same conditions as in Fig. 17.4.

Similarly, the local perturbation of SC order parameter, Eq. 17.37, can be considered. The local d-wave SC order in the unit cell containing the impurity (see Fig. 17.5) is given by the average $\Delta_{\mathbf{n}} = V_{SC}\langle a_{\mathbf{n}+\delta_1,\downarrow} a_{\mathbf{n}+\delta_2,\uparrow}\rangle$.[a] Again, the

[a] Of course, this definition admits the choice of any pair of nearest neighbor sites closest to impurity, instead of δ_1 and δ_2.

suppression parameter for this extended defect is defined as $\eta_{ext} = 1 - \Delta_0/\Delta$, and it is only contributed by the non-diagonal GF's:

$$\begin{aligned}\eta_{ext} &= -\frac{V_{SC}W}{4\mu N\Delta}\sum_{\mathbf{k},\mathbf{k}'\neq\mathbf{k}} \mathrm{e}^{i(\mathbf{k}\cdot\delta_2-\mathbf{k}'\cdot\delta_3)} \\ &\times\ \theta\left(\varepsilon_{\mathrm{D}}^2-\xi_{\mathbf{k}}^2\right)\theta\left(\varepsilon_{\mathrm{D}}^2-\xi_{\mathbf{k}'}^2\right)\langle a_{-\mathbf{k},\downarrow}a_{\mathbf{k}',\uparrow}\rangle \\ &= -\frac{\sum_j(-1)^j\int_{-\infty}^0 d\varepsilon \mathrm{Im\,Tr}\,\widehat{F}_j^0\widehat{T}_j^0\widehat{F}_j^0\widehat{\tau}_1}{\int_{-\infty}^0 d\varepsilon \mathrm{Im\,Tr}\,\widehat{F}_\delta^0\widehat{\tau}_1},\end{aligned} \qquad (17.52)$$

where the matrices $\widehat{F}_j^0 = N^{-1}\sum_{\mathbf{k}}\alpha_{j,\mathbf{k}}^2\theta\left(\varepsilon_{\mathrm{D}}^2-\xi_{\mathbf{k}}^2\right)\widehat{G}_{\mathbf{k}}^{(0)}$ mainly differ from $\widehat{G}_j^0$ by the absence of the $\propto\widehat{\tau}_3$ term, like Eq. 17.36, and $\widehat{F}_\delta^0$ is given by Eq. 17.35. Using here Eqs. 17.52, 17.44, one arrives at the expression:

$$\eta_{ext} = -2v_E^2\frac{\int_0^{\varepsilon_{\mathrm{D}}}\mathrm{Im}\left[g_1(2f_0g_0+f_0^2+g_1^2)/D_E\right]d\varepsilon}{\int_0^{\varepsilon_{\mathrm{D}}}\mathrm{Im}\,g_1(\varepsilon)d\varepsilon}, \qquad (17.53)$$

where only the E-channel terms contribute to the numerator (see Fig. 17.8 to compare it with the denominator).

Numeric analysis of this expression for the above chosen perturbation parameters results in $\eta_{ext}\approx 0.132$. This value is much smaller than that for the point-like impurity (assuming v equal to v_A, it would be $\eta\approx 0.763$), and, in view of the said in the beginning of this Section, the present model looks more plausible for disordered high-T_c systems. The most evident physical reason for so drastic reducing is the separate action of the extended impurity center in different symmetry channels, so that the stronger perturbation, v_A, is effective for the N-diagonal characteristics (DOS and LDOS) while the N-non-diagonal ones (as SC order) are only defined by the weak perturbation, v_E. As will be shown below, even more diverse effects on local DOS and d-wave order parameter are produced by the extended impurity perturbation if it is spin-dependent.

17.5.2. *Magnetic perturbation from non-magnetic impurity*

Unlike the above mentioned distinct effects of magnetic and non-magnetic impurities in traditional SC materials, introduction of non-magnetic Zn^{2+} ions instead of Cu^{2+} into the cuprate planes has a suppression effect on HTSC not weaker but rather stronger than that by magnetic Ni^{2+} ions.[13] Therefore non-magnetic impurity ions in HTSC are seen as extremely strong scatterers,[61] treated in the unitary limit.[36,62,63] However, the impurity centers formed in the CuO_2 plane by *homovalent* substitution (as Zn^{2+} or Ni^{2+} for Cu^{2+}), hardly could produce such a strong perturbation potential. Also the predicted symmetric resonances by non-magnetic impurities in LDOS are not observed in experimental STM spectra taken near Zn sites in $Bi_2Sr_2CaCu_2O_{8+}$.[18]

This Section presents an alternative approach to the problem of Cu-substituting impurities. It will be shown that, irrespectively of their type (magnetic or non-magnetic), the resulting center generally acts on charge carriers as *magnetic*. In accordance with the general concept of time-reversal symmetry breaking, such center should in fact strongly suppress SC order either of *s*- or *d*-type, as was first qualitatively stated yet by Mahajan *et al.*[64] Similar views on the effect of Zn impurities in high-T_c cuprates were expressed more recently,[65–67] though still focusing on unitary scattering. As seen below, the spin-dependent effect of isolated (non-magnetic) impurity in a CuO_2 plane, not reaching the unitary limit of perturbation but changing its Nambu structure, can become really strong.[68]

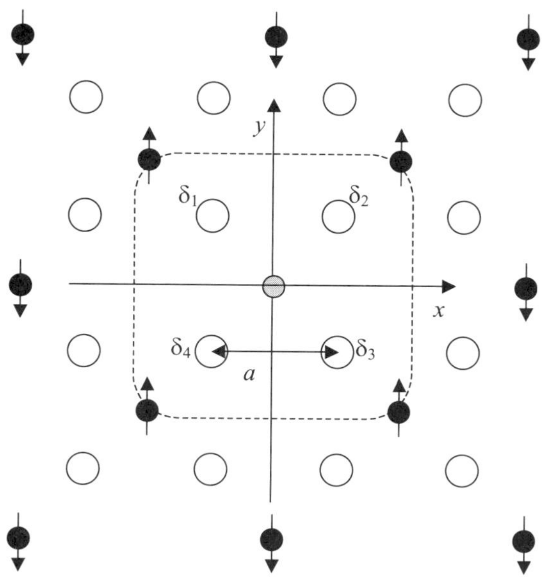

Fig. 17.9. Effective magnetic perturbation for charge carriers on nearest neighbor sites to the non-magnetic impurity substitute for Cu^{2+} in CuO_2 plane.

Fig. 17.8 shows a cathion impurity substitute for Cu in a CuO_2 plane, like real Zn, Fe, or Ni impurities in high-T_c compounds, and this center presents a notable geometric similarity to the extended center, Fig. 17.5 from the previous Section. However, there is also a notable difference in the mechanism of perturbation by these two centers. The main perturbation on O^- holes by the present type of impurity (regardless of being magnetic or non-magnetic) is due to the fact that its neighbor O sites occur in a non-zero exchange field by Cu^{2+} ions,[69,70] which is equivalent to the effect of magnetic impurity in a common superconductor. On the other hand, there are no reasons to consider any sizeable spin-independent perturbation from such isovalent impurity.

The respective model Hamiltonian consists in three terms: $H + H_c + H_{int}$, where H is given by Eq. 17.3. The first perturbation term $H_c = -hS_z$ models the (AFM) correlation between the impurity center and its environment, where $h \sim J_{dd}$, the Cu-Cu exchange constant, and $\mathbf{S}$ is the spin of a fictitious "magnetic impurity". It can be seen as a cluster of four 1/2 spins of Cu nearest neighbors to real non-magnetic impurity. In reality, its quantization axis z is only defined over time periods no longer than $\tau_s \sim \hbar\xi_s/(aJ_{dd}) \sim 10^{-13}$ s for experimentally measured spin correlation length $\xi_s \sim a/\sqrt{x}$[71] and doping levels $x \sim 0.1$ (this also agrees with the NMR data[65]). However this τ_s is much longer than typical electronic times $\sim \hbar/\mu \sim 10^{-15}$ s for HTSC compounds. For $h > 0$ we have $\langle S_z \rangle \equiv s$ and $0 < s < S$, which accounts for the short-range AFM order, whereas $s \to 0$ in the paramagnetic limit $h \ll k_B T$. The spin-dependent interaction between charge carriers and impurity is in its turn separated into three parts:

$$H_{int} = H_{int}^{\mathrm{MF}} + H_{int}^{\parallel} + H_{int}^{\perp}, \tag{17.54}$$

where

$$H_{int}^{\mathrm{MF}} = \frac{Js}{N} \sum_{\mathbf{k},\mathbf{k}'} \sum_{\sigma=\pm} \alpha_{j,\mathbf{k}} \alpha_{j,\mathbf{k}'} \sigma a_{\mathbf{k}',\sigma}^{\dagger} a_{-\mathbf{k},\sigma}$$

is the "mean-field" (MF) polarization of carrier spins by the impurity center, and

$$H_{int}^{\parallel} = \frac{J}{N} \sum_{\mathbf{k},\mathbf{k}'} \sum_{\sigma=\pm} \alpha_{j,\mathbf{k}} \alpha_{j,\mathbf{k}'} \sigma \left(S_z - s\right) a_{\mathbf{k}',\sigma}^{\dagger} a_{\mathbf{k},\sigma},$$

$$H_{int}^{\mathrm{MF}} = \frac{J}{N} \sum_{\mathbf{k},\mathbf{k}'} \sum_{\sigma=\pm} \alpha_{j,\mathbf{k}} \alpha_{j,\mathbf{k}'} S_\sigma a_{\mathbf{k}',-\sigma}^{\dagger} a_{\mathbf{k},\sigma},$$

are their interactions with longitudinal and transversal fluctuations of $\mathbf{S}$. In the paramagnetic limit: $s \to 0$, Eq. 17.54 is reduced to the common Kondo interaction.[72,73] For definiteness, the Cu-O p-d exchange parameter J is considered positive. The functions $\alpha_{j,\mathbf{k}}$ are formally the same as given by Eq. 17.42 for extended impurity center in the previous Section, but the distinctive features of the perturbation, Eq. 17.54, are: i) additional degrees of freedom by spin $\mathbf{S}$, and ii) coupling of $\mathbf{S}$ to the local AFM correlations.

To describe the local effects by this impurity center, we look for m-non-diagonal GF's $\widehat{G}_{\mathbf{k},\mathbf{k}'}$ from the equation of motion

$$\widehat{G}_{\mathbf{k},\mathbf{k}'} = JN^{-1}\sum_{\mathbf{k}'',j} \alpha_{j,\mathbf{k}}\widehat{G}_{\mathbf{k}}\left(s\widehat{G}_{\mathbf{k}'',\mathbf{k}'} + \widehat{G}^{(z)}_{\mathbf{k},\mathbf{k}'} + \widehat{G}^{(-)}_{\mathbf{k},\mathbf{k}'}\right)\alpha_{j,\mathbf{k}''}$$

including three scattered GF's: the MF one $\widehat{G}_{\mathbf{k}'',\mathbf{k}'}$, the longitudinal $\widehat{G}^{(z)}_{\mathbf{k}'',\mathbf{k}'} = \langle\langle\psi_{\mathbf{k}''}(S_z - s)|\psi^{\dagger}_{\mathbf{k}'}\rangle\rangle$ and the transversal $\widehat{G}^{(-)}_{\mathbf{k}'',\mathbf{k}'} = \langle\langle\overline{\psi}_{\mathbf{k}''}S_{-}|\psi^{\dagger}_{\mathbf{k}'}\rangle\rangle$ with the "spin-inverted" spinor $\overline{\psi}^{\dagger}_{\mathbf{k}} = \left(a^{\dagger}_{\mathbf{k},\downarrow}, a_{-\mathbf{k},\uparrow}\right)$. The two last terms are analogous to the well known Nagaoka's Γ-term[73,74] and treating them with a similar decoupling procedure leads to the solution of type Eq. 17.48:

$$\widehat{G}_{\mathbf{k},\mathbf{k}'} = \frac{1}{N}\sum_{j} \alpha_{j,\mathbf{k}}\widehat{G}_{\mathbf{k}}\widehat{T}_j\widehat{G}_{\mathbf{k}'}\alpha_{j,\mathbf{k}'}, \tag{17.55}$$

but with more complicated partial T-matrices:

$$\widehat{T}_j = \left[Js + J^2\left(\Sigma^2\widehat{G}_j + \widehat{X}_j\right)\right]\left[1 - Js - J^2\left(\Sigma^2\widehat{G}_j + \widehat{X}_j\right)\right]^{-1},$$
$$\Sigma^2 = \langle S_z^2\rangle - s^2, \quad \widehat{X}_j = \frac{1}{N}\sum_{\mathbf{k}} \alpha^2_{j,\mathbf{k}}\widehat{G}_{\mathbf{k}}(\varepsilon + h)\widehat{X}_{\mathbf{k}},$$
$$\widehat{X}_{\mathbf{k}} = S(S+1) - s(s+1) - \Sigma^2 + \left(1 + 2\frac{\xi_{\mathbf{k}}}{E_{\mathbf{k}}}\right)\widehat{\tau}_3,$$

where the shift of energy argument, $\varepsilon \to \varepsilon + h$, is due to the non-elastic scattering effect of AFM stiffness. By the definition of this model, Js is positive.

It is interesting to trace the behavior of $\widehat{T}_j$ in the two characteristic limits for AFM correlations between Cu^{2+}spins. In the paramagnetic limit: $h \to 0$, $s \to 0$, we have $\Sigma^2 \to S(S+1)/3$ and $\widehat{X}_j \to 2S(S+1)/3$, so that:

$$\widehat{T}_j \to J^2S(S+1)\widehat{G}_j\left[1 - Js - J^2\left(\Sigma^2\widehat{G}_j + \widehat{X}_j\right)\right]^{-1},$$

generalizing the known results[7,74] for the case of extended impurity center.

Another limit, fully polarized, $h \to \infty$, $s \to S$, corresponds to $\Sigma^2 \to 0$, $\widehat{X}_j \to 0$ and results in

$$\widehat{T}_j \to JS\left(1 - JS\widehat{G}_j\right)^{-1}, \tag{17.56}$$

which is only due to MF magnetic scattering and similar to the simple forms, Eqs. 17.46, 17.47. This limit is justified at $JS \gg k_{\rm B}T$, hence it well applies in the SC phase at $T < T_c \sim \Delta/k_{\rm B}$ and will be used for the T-matrices below. The important change of the perturbation matrix structure, from $\widehat{V} \propto \widehat{\tau}_3$ to $\widehat{V} \propto \widehat{\tau}_0$, causes a strong modification of the impurity effects.

Thus, the variation of LDOS, Eq. 17.9, compared to the uniform value $\rho(\varepsilon)$, Eq. 17.8, is here presented as:

$$\begin{aligned}\rho_{\mathbf{n}}(\varepsilon) - \rho(\varepsilon) &= \frac{1}{\pi N}\sum_{\mathbf{k},\mathbf{k}'\neq\mathbf{k}} \operatorname{Im}\operatorname{Tr} e^{i(\mathbf{k}-\mathbf{k}')\mathbf{n}}\widehat{G}_{\mathbf{k},\mathbf{k}'} \\ &= \sum_j \operatorname{Im}\operatorname{Tr}\widehat{G}_j(\mathbf{n})\,\widehat{T}_j\widehat{G}_j(\mathbf{n}),\end{aligned}$$

where the matrices $\widehat{G}_j(\mathbf{n}) = N^{-1}\sum_{\mathbf{k}} e^{i\mathbf{kn}}\alpha_{j,\mathbf{k}}^2\widehat{G}_{\mathbf{k}}^{(0)}$ appear by virtue of Eq. 17.55. The maximum LDOS variation is attained at $\mathbf{n} = \delta$, the nearest neighbor sites to the impurity, mainly contributed by $j = 1$:

$$\rho_\delta(\varepsilon) - \rho(\varepsilon) \approx \operatorname{Im}\operatorname{Tr}\widehat{G}_1(\delta)\,\widehat{T}_1\widehat{G}_1(\delta). \tag{17.57}$$

The relevant GF matrices are obtained in similarity with Eq. 17.44:

$$\widehat{G}_1(\delta) \approx \overline{\alpha_1^2}\rho_N\left(g_0 + g_{as}\widehat{\tau}_3\right),$$

so that resulting $\rho_\delta(\varepsilon)$ can display a resonance at ε_{res}, defined by the denominator of $\hat{T}_1$:

$$\operatorname{Re}\left\{\left[1 - u_A g_0(\varepsilon_{res})\right]^2 - u_A^2 g_{as}^2\right\} = 0, \tag{17.58}$$

with the spin-dependent perturbation parameter $u_A = JS\rho_N\overline{\alpha_1^2}$, for the A-channel. Alike the case of non-magnetic impurity, the finite asymmetry factor g_{as} makes it possible to have ε_{res} close to zero at rather moderate perturbation parameter J, close to $J_{cr} = 1/(S\overline{\alpha_1^2}g_{as}\rho_N)$, then the peak in LDOS becomes very sharp, as shown in Fig. 17.10 for $\rho_\delta(\varepsilon)$ at the choice $J = 0.3$ eV $\approx 1.15 J_{cr}$. The most notable distinction from the non-magnetic perturbation consists in that the resonance condition $\operatorname{Re} g_0(\varepsilon_{res}) = u_A^{-1} \pm g_{as}$ for *odd* function $\operatorname{Re} g_0(\varepsilon)$ leads to a *single* sharp peak on one side from the Fermi level, instead of two symmetric peaks as in Figs. 17.3, 17.6. This is just the observed behavior for LDOS near Zn sites in $Bi_2Sr_2CaCu_2O_8$[18] (as seen in the right panel of Fig. 17.10), which

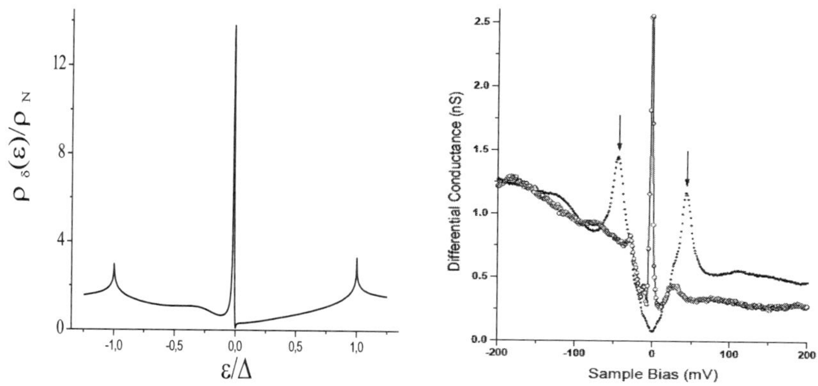

Fig. 17.10. (a) Local density of states near magnetic impurity $\rho_\delta(\varepsilon)$, given by Eqs. 17.57, 17.58, at the choice of perurbation parameter J = 0.3 eV, slighly above the critical value $J_{cr} = 1/(S\overline{\alpha_1^2} g_{as}\rho_N) \approx 0.26$ eV, presents a sharp resonance just below the Fermi level, similar to that observed in (b) the STM spectrum on Zn site in $Bi_2Sr_2CaCu_2O_8$ at ≈ -1.5 meV (Reproduced with permission of Nature from Ref. 18).

is a strong argument in favor of the proposed perturbation mechanism for these impurities in CuO_2 planes.

The local effect on SC correlation is characterized by the same average $\Delta_\mathbf{n} = (V_{SC})\langle a_{\mathbf{n}+\delta_1,\downarrow} a_{\mathbf{n}+\delta_2,\uparrow}\rangle$ as in the preceding Section, and the suppression parameter $\eta_S = 1 - \Delta_0/\Delta$ is formally given by the same Eq. 17.52, but with the partial T-matrices defined by Eq. 17.56. Again, it is only contributed by the E-channel terms:

$$\begin{aligned}\eta_S &= -\frac{2\int_{-\infty}^{0} d\varepsilon \mathrm{Im\,Tr}\,\widehat{F}_2^0\widehat{T}_2^0\widehat{F}_2^0\widehat{\tau}_1}{\int_{-\infty}^{0} d\varepsilon \mathrm{Im\,Tr}\,\widehat{F}_\delta^0\widehat{\tau}_1}\\ &= -2u_E\frac{\int_0^{\varepsilon_D}\mathrm{Im}\left\{g_1\left[2f_0 - u_E\left(2f_0g_0 + f_0^2 + g_1^2\right)\right]/D_E\right\}d\varepsilon}{\int_0^{\varepsilon_D}\mathrm{Im}\,g_1(\varepsilon)\,d\varepsilon},\end{aligned} \quad (17.59)$$

where $D_E = (1-u_E g_0)^2 - u_E^2\left(g_1^2 + g_3^2\right)$ and the spin-dependent perturbation parameter for E-channel $u_E = JS\rho_N\overline{\alpha_2^2}$. Comparing this function (see Fig. 17.11) to its analogue, Eq. 17.52, for the spin-independent perturbation, shows a much more pronounced suppression effect. The evident reason for this is the different structure of D_E in Eq. 17.59 compared to that in Eq. 17.47, directly related to the above mentioned change of the perturbation matrix structure.

In fact, numeric integration in Eq. 17.59 with the same set of parameters as used above for LDOS (corresponding to $u_E \approx 0.15$) shows a considerable suppression of local SC order: $\eta_{sup} \approx 0.349$, almost triple of than for equal

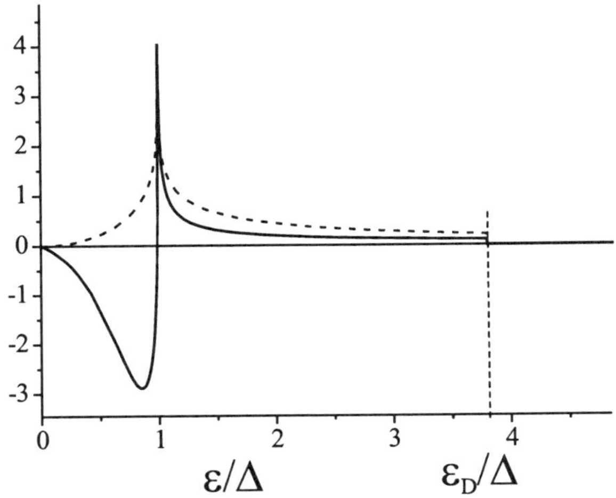

Fig. 17.11. The comparison between the behavior of numerator (solid line) and denominator (dashed line) in Eq. 17.59.

spin-independent perturbation in the previous Section. The dependence $\eta_S(J)$ is generally non-monotonous, anyhow it should be stressed that no unitary limit $JS\rho_N \gg 1$ is needed to get such a strong effect.

17.6. Self-consistent approximation and its validity

At higher concentrations of impurity scatterers the non-renormalized T-matrix approximation is no more reliable, and is usually changed for its self-consistent version. In the simplest case of point-like scatterers, Eq. 17.3, we define the self-consistent approximation for m-diagonal SPGF as:

$$\widehat{G}_{\mathbf{k}}^{(sc)} = \left\{ \left[\widehat{G}_{\mathbf{k}}^{(0)}\right]^{-1} - \widehat{\Sigma}^{(sc)} \right\}^{-1}, \tag{17.60}$$

including the self-consistent matrices for self-energy:

$$\widehat{\Sigma}^{(sc)} = -c\widehat{V}\left[1 + \widehat{G}^{(sc)}\widehat{V}\right]^{-1} \tag{17.61}$$

and for local GF:

$$\widehat{G}^{(sc)} = \frac{1}{N}\sum_{\mathbf{k}} \widehat{G}_{\mathbf{k}}^{(sc)}. \tag{17.62}$$

The self-energy matrix, Eq. 17.61, is expanded in Pauli matrices:

$$\widehat{\Sigma}^{(sc)} = \Sigma_0 + \Sigma_1 \widehat{\tau}_1 + \Sigma_3 \widehat{\tau}_3, \tag{17.63}$$

where Σ_i are generally some complex-valued functions of energy.[55] Then integration in Eq. 17.62, using Eq. 17.10, results in a similar expansion for the self-consistent GF matrix:

$$\widehat{G}^{(sc)} = G_0 - G_1 \widehat{\tau}_1 - G_3 \widehat{\tau}_3. \tag{17.64}$$

It can be easily shown that the self-consistent value of $G_1(\varepsilon)$ is actually zero,[55] and G_3 is practically constant: $G_3 \approx \rho_N g_{as}$, hence the self-consistency problem for point-like impurities in the *d*-wave SC is reduced to a single complex function $G_0(\varepsilon) = \rho_N g_0(\varepsilon - \Sigma_0)$ defined from the equation for scalar self-energy $\Sigma_0(\varepsilon)$:

$$F_\varepsilon(\Sigma_0) = 1 - v^2 g_0(\varepsilon - \Sigma_0)\left[g_0(\varepsilon - \Sigma_0) - \frac{c}{\Sigma_0 \rho_N}\right] = 0. \tag{17.65}$$

In the most important region of low energies, $|\varepsilon| \ll \Delta$, this equation can be simplified using for g_0 the arcsine approximation,[55] instead of exact elliptic functions, Eq. 17.12:

$$g_0(\varepsilon) \approx \varepsilon\left(\frac{1}{\tilde{\mu}} + \frac{i\pi}{\Delta}\arcsin\frac{\Delta}{\varepsilon}\right). \tag{17.66}$$

Considering such form of Eq. 17.65, P. Lee supposed that in *unitary* limit, $v \to \infty$, the unity term can be dropped,[36] then the self-energy gets related to the local GF as: $\Sigma_0 = -c/G_0$ (inverse to the common relation $\Sigma_0 \approx cv^2 G_0/\rho_N^2$ in the Born limit, $v \ll 1$). This surprising relation leads straightforwardly to the conclusion that, in a *d*-wave superconductor with unitary scatterers, the self-energy should tend to a *finite* limiting value: $\Sigma_0(\varepsilon \to 0) \to -i\gamma_0$, and so the DOS: $\rho(\varepsilon \to 0) \to \rho_0 = c/(\pi\gamma_0)$, or in the present notations: $\rho_0 \approx \sqrt{\pi c \rho_N/\Delta}$ (within to some logarithmic corrections and neglecting the terms $\sim \Delta/\tilde{\mu}$ beside unity). A similar conclusion for the case of Born scatterers was made earlier by Gor'kov and Kalugin,[75] and their predicted finite DOS reads in these notations as $\rho_0 \approx (2\rho_N/\pi\alpha)\,\mathrm{e}^{-1/\alpha}$ with $\alpha = cv^2/(\rho_N\Delta) \ll 1$.

These finite limits should mean a spontaneous breakdown of the d-wave symmetry in presence of scatterers and a qualitative rearrangement of the low energy excitation spectrum, including appearance of strongly localized quasiparticle states (in spite of absence of such localization in the simple T-matrix treatment[14,15]). The last decade produced an extensive theoretical discussion on reality of such SCTMA behavior, and astonishing variety of results was obtained, including power law convergence to zero: $\rho(\varepsilon \to 0) \propto \varepsilon^{\alpha}$, with universal[76] or non-universal,[7751] values of the exponent α, different finite limits,[75],[3678] and even divergence $\rho(\varepsilon \to 0) \propto \varepsilon \ln(1/\varepsilon)$.[79] On the other hand, numerous experimental studies have been done to check the principal conclusion from existence of finite ρ_0 in the unitary limit, the so-called *universal* values of quasiparticle electrical conductivity $\sigma_0 = \left(e^2/\pi^2\hbar\right) v_{\rm F}/v_{\Delta}$[36] and heat conductivity $\kappa_0/T = \left(k_{\rm B}^2/3\right)\left(v_{\rm F}^2 + v_{\Delta}^2\right)/v_{\rm F}v_{\Delta}$,[38] and also the results of these measurements are still contradictory.

It should be stressed that the above mentioned theoretical construction uses two main assumptions: *i*) that certain impurities in high-T_c superconductors are extremely strong scatterers (even as high values as $v \sim 10^2 \div 10^3$ are used sometimes to adujst the theoretical predictions to observable data); *ii*) that solutions of self-consistent equations (linear in c) can apply either to extended and localized states (since the finite ρ_0 is associated with localized states).

However, even for possibly high value of v (enhanced by the above discussed asymmetry factor, though less probable to exceed few units), the neglect of the unity term in Eq. 17.65 may be unjustified at very low energies. This rises a technical question about existence of different formal solutions of this equation. But a more fundamental issue is that the self-consistency procedure is only well defined for *extended* electronic states,[81] which assure effective averaging of effects of different impurity scatterers, say, along the mean free path. This suggests to additionally check whether the obtained solutions are compatible with the self-consistency, and, if they do not, to look for alternative solutions, beyond the framework of SCTMA. Below we consider in more detail whether the finite DOS at zero energy necessarily follows from the SCTMA solution and what alternatives it can have.[80]

Eq. 17.65 can be formally solved with respect to $g_0(\varepsilon - \Sigma_0)$:

$$g_0(\varepsilon - \Sigma_0) = \frac{-c \pm \sqrt{c^2 + (2\Sigma_0\rho_N/v)^2}}{2\Sigma_0\rho_N}, \tag{17.67}$$

so that the Lee's choice in the unitary limit is related to the *minus* sign while the Gor'kov and Kalugin's choice in the Born limit to the *plus* sign. Keeping *both*

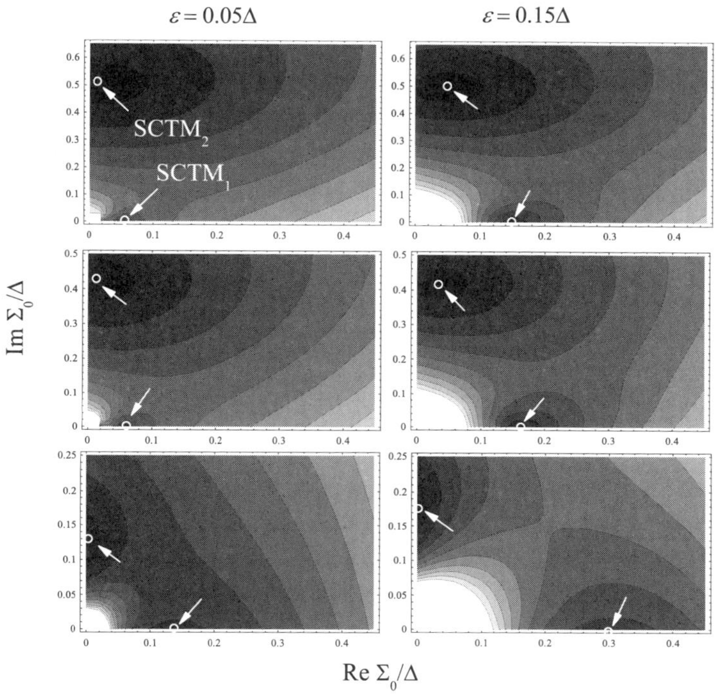

Fig. 17.12. Contour plots of $|F_\varepsilon(\Sigma_0)|$, Eq. 17.65, in function of complex self-energy Σ_0 for two different energies ε at the choice of $c = \rho_N \Delta$ and three different perturbation parameters: unitary limit $v = 10$ (upper row), intermediate regime $v = 1$ (middle row), and Born limit $v = 0.35$ (bottom row). There are always two roots shown by white circles and denoted $SCTMA_1$ and $SCTMA_2$, and at $\varepsilon \to 0$ the first of them tends to zero, close to the real axis, while the other tends to a finite imaginary limit.

these options for any value of perturbation parameter v, multiple solutions for $\Sigma_0(\varepsilon)$ are yet possible due to essential non-linearity of $g_0(\varepsilon - \Sigma_0)$ as a function of Σ_0. Finally, a single physical solution for any given energy ε should be selected on the basis of SCTMA validity criterion (this analysis will be done in the next Section). The numerical solutions of Eq. 17.65 (using the arcsine form Eq. 17.66) in the complex plane of self-energy Σ_0 for different values of energy ε and perturbation parameter v are summarized in Fig. 17.12. It is seen that there are two roots in each case, denoted $SCTMA_1$ and $SCTMA_2$. The $SCTMA_2$ root tends to a finite and imaginary value at $\varepsilon \to 0$, and, and passing to the unitary or Born limits in v, one reproduces respectively the Lee's and Gor'kov and Kalugin's predictions in a unified way. At the same time, the $SCTMA_1$ root tends to *zero* at any value of v, suggesting a zero limit for the DOS. This general behavior is essentially the same for both functional forms of g_0 in Eq. 17.65. Hence, at any regime of impurity perturbation, there is an alternative to the finite limit of low-energy DOS, consistently calculated within the SCTMA framework!

One additional comment is in order to the self-consistent equation with multiple solutions, like that in Fig. 17.12. It is seen there that the SCTMA_2 root have a much wider "attraction basin" than SCTMA_1, especially at very low energies. This can hinter detection of the alternative solution when running a numeric routine, as probably was the case for several numerical SCTMA studies which found finite DOS at zero energy,[78].[51]

It will be seen below that each of the two suggested SCTMA solutions has its specific validity domain, beyond the area of impurity resonance, while no one of them is a good approximation within this area. Let us specify the low energy behavior of each solution and try to build a "pragmatic" combination of the two, in order to obtain a correctly normalized quasiparticle DOS.

The low energy limit for the SCTMA_2 solution, $\Sigma_0\left(\varepsilon \longrightarrow 0\right) = -i\gamma_0$, is obtained accordingly to Eq. 17.65 as a root of:

$$1 + v^2 \left[2\mathrm{K}\left(-\frac{\Delta^2}{\gamma_0^2}\right) + \frac{\gamma_0}{2\tilde{\mu}} \right]^2 + \frac{cv^2}{\rho_N \gamma_0} \left[2\mathrm{K}\left(-\frac{\Delta^2}{\gamma_0^2}\right) + \frac{\gamma_0}{2\tilde{\mu}} \right] = 0, \qquad (17.68)$$

in "elliptic" form, or

$$1 + v^2\gamma_0^2 \left(\frac{\pi}{\Delta}\mathrm{arcsinh}\frac{\Delta}{\gamma_0} - \frac{1}{\tilde{\mu}} \right)^2 - \frac{cv^2}{\rho_N} \left(\frac{\pi}{\Delta}\mathrm{arcsinh}\frac{\Delta}{\gamma_0} - \frac{1}{\tilde{\mu}} \right) = 0, \qquad (17.69)$$

in "arcsine" form. The numeric solution of Eq. 17.69 for γ_0 in function of perturbation parameter v (shown in Fig. 17.13 for the choice of $c = \rho_N \Delta$) reproduces the Lee's limit already for $v \gtrsim 3$ and the Gor'kov and Kalugin's limit for $v \lesssim 0.5$ and thus justifies the attribution of regimes in Fig. 17.12. The result for Eq. 17.68 is essentially the same.

The behavior of the SCTMA_2 solution at finite energies ε can be obtained from Eq. 17.65 only numerically, and the related DOS, as shown in Fig. 17.14, grows slowly from the residual value ρ_0 at $\varepsilon \lesssim \varepsilon_{res}$ and then at $\varepsilon > \varepsilon_{res}$ goes closely to the result of simple T-matrix approximation of Sec. 17.5, which suggests reliability of SCTMA_2 in this energy range.

For the alternative SCTMA_1 solution, Eq. 17.65 admits an analytical approximation[b] by using the logarithmic asymptotics for the elliptic K-integral (or $\arcsin x \approx -i \ln\left(2ix\right)$ at $|x| \gg 1$[55] for the arcsine form):

[b]But only valid at extremely low energies.

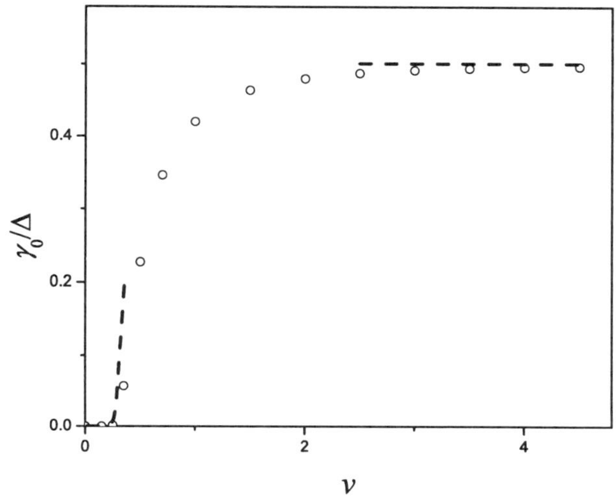

Fig. 17.13. Residual self-energy $\gamma_0 = \lim_{\varepsilon \to 0} i\Sigma(\varepsilon)$ for the SCTMA$_2$ solution, calculated from Eq. 17.69 with the choice of $\tilde{\mu} = 10\Delta$ in function of the perturbation parameter v (open circles). The dashed lines show the limiting behaviors: exponential in the Born limit, $\gamma_0/\Delta \approx 4\exp(-1/v^2)$, and a constant value in the unitary limit.

$$G_0(\varepsilon) \approx \frac{\rho_N^2 \varepsilon}{cv^2 \ln(4icv^2/\rho_N\varepsilon)}. \tag{17.70}$$

The corresponding analytic function for the low-energy DOS is:

$$\rho(\varepsilon) \approx \frac{\rho_N^2 \varepsilon}{cv^2 \ln^2(4cv^2/\rho_N\varepsilon)}, \tag{17.71}$$

though, for the instance in Fig. 17.14, the numerical SCTMA$_1$ solution attains this behavior only at $\varepsilon \lesssim 10^{-3}\Delta$. However, it is just this function that describes the asymptotic vanishing of DOS, even faster than the non-perturbed function, Eqs. 17.13, or the simple T-matrix function, Fig. 17.3. Also it vanishes faster than the power laws, $\rho(\varepsilon) \propto \varepsilon^\alpha$ with $\alpha \leq 1$,[51,77] or with $\alpha = 1$,[76] using other than SCTMA approaches. That fast vanishing can be seen as a certain narrow "quasi-gap" near the Fermi energy, and beyond this quasi-gap a plausible matching between the two SCTMA solutions, over the interval of broadening Γ_{res} of the resonance ε_{res}, can be done by the simple T-matrix function, in order to preserve the overall normalization of DOS[68]

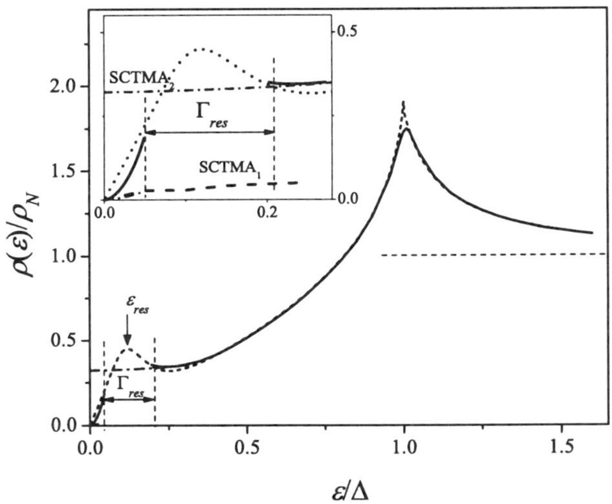

Fig. 17.14. Construction of the self-consistent DOS (solid line) adjusted to the two different SCTMA solutions beyond the region of impurity resonance ε_{res} of width Γ_{res}. The impurity parameters are chosen as $v = 1$ and $c = 0.2\rho_N\Delta$. The SCTMA$_1$ solution is shown by the dashed line, the SCTMA$_2$ solution by the dash-dotted line, and the short-dash line shows the common T-matrix solution from Fig. 17.3.

$$\int d\varepsilon \left[\rho\left(\varepsilon\right) - \rho_N\right] = 0. \tag{17.72}$$

Notice that Eq. 17.72 is already satisfied if $\rho\left(\varepsilon\right)$ is chosen in the simple T-matrix form (short-dash line in Fig. 17.14). Hence it is also satisfied if the positive and negative areas between that and SCTMA (solid line) curves in the energy intervals beyond the Γ_{res} range are equal, as approximately realized by the construction in Fig. 17.14. This provides the sought "compromise" SCTMA solution for DOS.

17.6.1. *Ioffe-Regel-Mott criterion and validity of SCTMA solutions*

A criterion for a quasiparticle state with excitation energy ε and momentum p to be of extended type, as necessary for the SCTMA solutions, was first proposed by Ioffe and Regel[82] and then substantiated by Mott,[27] consisting in that the quasiparticle mean free path ℓ be longer than its wavelength $\lambda = 2\pi\hbar/p$, or else that its lifetime τ be longer than the oscillation period $\hbar/\varepsilon$. This criterion is widely used for analysis of normal excitation spectra in disordered systems,[81] however, when applying it to disordered superconductors, one has first to redefine the quasiparticle basic characteristics. Thus, excitation of a Bogolyubov quasiparticle with the nominal wave vector $\mathbf{k}$ over the BCS ground state changes the system energy by $E_{\mathbf{k}}$ and hence its momentum by $p_{\mathbf{k}} = \hbar E_{\mathbf{k}} / \left|\nabla_{\mathbf{k}} E_{\mathbf{k}}\right|$. Then the related wavelength is $\lambda_{\mathbf{k}} = 2\pi \left|\nabla_{\mathbf{k}} E_{\mathbf{k}}\right| / E_{\mathbf{k}}$, generally different from the free particle value $2\pi/k$. Next, the mean free path $\ell_{\mathbf{k}}$ is defined as the group velocity $\left|\nabla_{\mathbf{k}} E_{\mathbf{k}}\right| / \hbar$ times the lifetime $\hbar / \mathrm{Im}\,\Sigma\left(E_{\mathbf{k}}\right)$, so that the Ioffe-Regel-Mott (IRM) criterion $\ell_{\mathbf{k}} \gg \lambda_{\mathbf{k}}$ can be presented as

$$E_{\mathbf{k}} \gg \mathrm{Im}\,\Sigma\left(E_{\mathbf{k}}\right). \tag{17.73}$$

In fact, the dispersion law is renormalized due to impurity scattering from $E_{\mathbf{k}} = \sqrt{\xi_{\mathbf{k}}^2 + \Delta_{\mathbf{k}}^2}$ to $\widetilde{E}_{\mathbf{k}}$, defined by

$$\widetilde{E}_{\mathbf{k}} - \mathrm{Re}\,\Sigma\left(\widetilde{E}_{\mathbf{k}}\right) = E_{\mathbf{k}}. \tag{17.74}$$

Using the simple T-matrix solution, $\Sigma = cv^2 \rho_N^{-1} g_0 / \left(1 - v^2 g_0^2\right)$, we have in the long-wave limit:

$$\widetilde{E}_{\mathbf{k}} \approx \frac{\rho_N \Delta}{cv^2 \ln\left(4cv^2/\rho_N E_{\mathbf{k}}\right)} E_{\mathbf{k}},$$

$$\mathrm{Im}\left(\widetilde{E}_{\mathbf{k}}\right) \approx \frac{cv^2}{\rho_N \Delta} \widetilde{E}_{\mathbf{k}} \approx \frac{E_{\mathbf{k}}}{\ln\left(4cv^2/\rho_N E_{\mathbf{k}}\right)}. \tag{17.75}$$

Thus the criterion 17.73 is only fulfilled for low enough concentration of scatterers: $c < c_\Delta \equiv \rho_N \Delta / v^2$, and this can be considered the validity condition for simple T-matrix approximation.

At higher impurity concentrations, $c \gg c_\Delta$, we need to pass to the SCTMA solutions of the preceding Section and to renormalize the dispersion law $\widetilde{E}_{\mathbf{k}}$ and self-energy $\Sigma\left(\widetilde{E}_{\mathbf{k}}\right)$ in a way specific for each solution.[83] Since the SCTMA self-energy only depends on energy, not on momentum, Eq. 17.74 holds for any relation between the radial and tangential components, $\xi_{\mathbf{k}}$ and $\Delta_{\mathbf{k}}$, of excitation energy $E_{\mathbf{k}}$. In particular limits: *i)* $\Delta_{\mathbf{k}} \to 0$, $\xi_{\mathbf{k}} \to E_{\mathbf{k}}$ and *ii)* $\xi_{\mathbf{k}} \to 0$, $\Delta_{\mathbf{k}} \to E_{\mathbf{k}}$, they are renormalized as

$$i)\quad \widetilde{\xi}_{\mathbf{k}} - \mathrm{Re}\,\Sigma\left(\widetilde{\xi}_{\mathbf{k}}\right) = \xi_{\mathbf{k}}, \quad ii)\quad \widetilde{\Delta}_{\mathbf{k}} - \mathrm{Re}\,\Sigma\left(\widetilde{\Delta}_{\mathbf{k}}\right) = \Delta_{\mathbf{k}}.$$

Then, using the SCTMA$_1$ solution[c] in Eq. 17.74 gives the long-wave dispersion law within logarithmic accuracy as:

[c] In elliptic K-form.

$$\widetilde{E}_{\mathbf{k}}^{(1)} \approx \frac{c}{c_\Delta} E_{\mathbf{k}} \ln \frac{4\Delta}{E_{\mathbf{k}}}$$

(note the difference with Eq. 17.75 for simple T-matrix). Respectively, its renormalized components are:

$$\widetilde{\xi}_{\mathbf{k}}^{(1)} \approx \frac{c}{c_\Delta} \xi_{\mathbf{k}} \ln \frac{4\Delta}{E_{\mathbf{k}}}, \quad \widetilde{\Delta}_{\mathbf{k}}^{(1)} \approx \frac{c}{c_\Delta} \Delta_{\mathbf{k}} \ln \frac{4\Delta}{E_{\mathbf{k}}},$$

and the related damping:

$$\Gamma_{\mathbf{k}}^{(1)} = \operatorname{Im} \Sigma^{(1)} \left(\widetilde{E}_{\mathbf{k}}^{(1)} \right) \approx \frac{\pi E_{\mathbf{k}}}{2 \ln \left(4\Delta / E_{\mathbf{k}} \right)}.$$

Using this in the IRM criterion, Eq. 17.73, leads to the SCTMA$_1$ validity condition:

$$E_{\mathbf{k}} \ll \Delta \exp \left(-\sqrt{\frac{\pi c_\Delta}{2c}} \right), \tag{17.76}$$

which holds in a very narrow vicinity of the Fermi energy at $c < c_\Delta$. At $c < c_\Delta$, this vicinity is defined by the condition $\Gamma_{\mathbf{k}}^{(1)} \ll E_{\mathbf{k}}$, holding at $E_{\mathbf{k}} \ll \Delta$ (for instance, it can be as small numerically as $\sim 10^{-2}\Delta$).

Applying the same treatment to the SCTMA$_2$ solution, which formally defines the low energy dispersion law $\widetilde{E}_{\mathbf{k}}^{(2)} \approx E_{\mathbf{k}}$ and the damping $\Gamma_{\mathbf{k}}^{(2)} = \operatorname{Im} \Sigma^{(2)} \left(\widetilde{E}_{\mathbf{k}}^{(2)} \right) \approx cv^2 \rho_N^{-1} \operatorname{Im} g_0$, we obtain the condition:

$$E_{\mathbf{k}} \gg \frac{c}{c_\Delta} \Delta \operatorname{Im} g_0 \left(E_{\mathbf{k}} \right), \tag{17.77}$$

so that this solution is valid only far enough from the nodal points, where it provides the correct limit of pure *d*-wave DOS. However, this solution is clearly eliminated near the nodal point. Thus, the only SCTMA solution, which may be valid in a close vicinity of the Fermi energy, is the SCTMA$_1$ solution, Eq. 17.71.

Notably, the two estimates, Eqs. 17.76, 17.77, do not necessarily assure the overlap between the two validity regions, so that for $c \gg c_\Delta$ there can exist some intermediate energy range where neither of SCTMA solutions applies. This range roughly corresponds to the broad linewidth of the impurity resonance ε_{res} where DOS cannot be rigorously obtained even with use of the next GE terms, and where it was interpolated by the simple T-matrix form between the two SCTMA asymptotics in the previous Section.

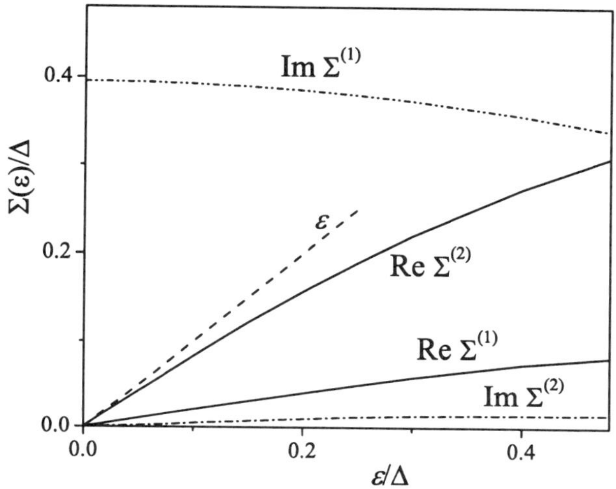

Fig. 17.15. Real and imaginary parts of the self-energy $\Sigma\left(\varepsilon\right)$ (in units of Δ) obtained for two different SCTMA solutions at the choice of perturbation parameters $c = \rho_N \Delta$, $v = 1$. Note the tendency of $\operatorname{Re}\Sigma^{(1)}\left(\varepsilon\right)$ to ε (dashed line) at $\varepsilon \to 0$.

Finally, we notice that other known non-perturbative solutions for d-wave disordered systems with DOS vanishing at $\varepsilon \to 0$ as a certain power law: $\rho\left(\varepsilon\right) \sim \varepsilon^{\alpha}$,[77],[76] also have to satisfy IRM criterion since they use field theoretic approach, only compatible with band-like states. But it can be easily shown to be only fulfilled for such DOS if the power is $\alpha > 1$, while the reported values are $\alpha = 1/7$[77] and $\alpha = 1$.[76] In fact, let the renormalized components of dispersion law (in the low energy limit) behave as $\widetilde{\xi}_k \sim (k - k_{\mathrm{F}})^{\nu} \propto \xi^{\nu}$, $\widetilde{\Delta}_k \propto \Delta^{\nu}$ with a certain $\nu > 0$, then the simplest estimate for d-wave DOS is

$$\begin{aligned}\rho\left(\varepsilon\right) &\propto \varepsilon \int d\Delta \int d\xi \delta\left(\varepsilon^2 - \widetilde{\xi}^2 - \widetilde{\Delta}^2\right)\\ &\propto \varepsilon \int E\delta\left(\varepsilon^2 - \widetilde{E}^2\right) dE = \varepsilon \int E\delta\left(\varepsilon^2 - E^{2\nu}\right) dE\\ &\propto \varepsilon^{(2-\nu)/v},\end{aligned}$$

that is the DOS exponent $\alpha = (2 - \nu)/v$. In the considered field models, the quasiparticle broadening is defined through the DOS: $\Gamma_{\mathbf{k}} = u^2 \rho\left(\widetilde{E}_{\mathbf{k}}\right)$, with the disorder parameter u of Anderson model.[28] Then the criterion, Eq. 17.73, is reformulated as

$$\widetilde{E} \gg u^2 \rho \left(\widetilde{E} \right),$$

and leads to the condition $\widetilde{E} \gg \mathrm{const} \cdot \widetilde{E}^{(2-\nu)/v}$. In the limit $\widetilde{E} \to 0$, this is only possible if $(2-\nu)/v > 1$, that is $\alpha > 1$.

17.7. Group expansions and localization of nodal quasiparticles

The above considerations essentially restrict possible candidate solutions for quasiparticle spectrum in the disordered d-wave superconductor and may suggest Eq. 17.71 as the only consistent low energy solution for the problem.

However the SCTMA (or field-theoretical) analysis can not be considered fully comprehensive for the real quasiparticle spectrum in a disordered system, if the localized states are also admitted. In accordance with the above IRM check, the single-impurity scattering processes by SCTMA can not produce localized states near zero energy. But if the SCTMA contribution to DOS vanishes in this limit, the importance can pass to the next GE terms, related to scattering (and possible localization) of quasiparticles on random groups (clusters) of impurities. The essential point is that their contribution to DOS is mostly defined by the *real* parts of GF's which do not need self-consistency corrections and thus remain valid for the energy range of localized states (where the IRM criterion no more holds).

The known approaches to impurity cluster effects in disordered d-wave superconductors, either numerical[56,84] and analytical,[79] were contradictory about DOS and did not conclude definitely on localization. A practical analysis of such effects within the GE framework for an s-wave superconductor was proposed recently,[85] using a special algebraic technique in the limit $\varepsilon \to \Delta$. A similar technique for disordered d-wave systems in the limit $\varepsilon \to 0$ is presented below for the cases of non-magnetic (NM) and magnetic (M), impurities, showing that only M-impurities can provide a finite DOS, replacing the SCTMA$_1$ solution (but different from SCTMA$_2$) in a narrow vicinity of the nodal energy and manifesting onset of localization there. In particular, this should produce, instead of universal conductivity, its exponential suppression at sufficiently low temperatures.

17.7.1. *Interaction matrices and DOS at nodal points*

Let us pass to consideration of the states near the Fermi level, $\varepsilon \to 0$, with use of the non-renormalized GE, Eq. 17.27 (expected to be more adequate for localized states). Then the limit for DOS $\rho(\varepsilon \to 0)$ is related to the imaginary and traceful part of the self-energy matrix: $\gamma = \mathrm{Tr}\,\mathrm{Im}\widehat{\Sigma}_{\mathbf{k}}(0)/2$ (for simplicity we drop the superindices at all non-renormalized GE matrices). If the main contribution to γ comes from the GE pair term, where the share of $\widehat{A}_{\mathbf{n}}\mathrm{e}^{-i\mathbf{k}\cdot\mathbf{n}}$ is negligible beside that of $\widehat{A}^2_{\mathbf{n}}$, γ can be considered momentum-independent. In this approximation, supposing also $\gamma \ll \Delta$, one obtains[36]

$$\rho(0) \approx \rho_N \frac{2\gamma}{\pi\Delta} \ln \frac{\Delta}{\gamma}, \tag{17.78}$$

and the following task is reduced to proper calculation of γ in function of the impurity perturbation parameters. Then we present the interaction matrices at $\varepsilon \to 0$ as:

$$\widehat{A}_{\mathbf{n}}(0) = (g_{\mathbf{n}}\hat{\tau}_3 + f_{\mathbf{n}}\hat{\tau}_1)\,\widehat{T}(0), \tag{17.79}$$

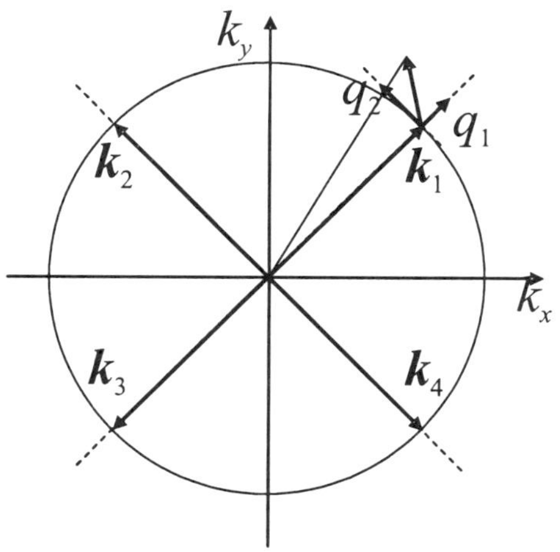

Fig. 17.16. Schematic of local coordinates near nodal points in the Brillouin zone of a *d*-wave superconductor.

where the real coefficient functions

$$g_{\mathbf{n}} = -\frac{\hbar v_{\mathrm{F}}}{N} \sum_{\mathbf{k}} \frac{\mathrm{e}^{i\mathbf{k}\cdot\mathbf{n}} q_1}{E^2_{\mathbf{k}}}, \quad f_{\mathbf{n}} = -\frac{\hbar v_{\Delta}}{N} \sum_{\mathbf{k}} \frac{\mathrm{e}^{i\mathbf{k}\cdot\mathbf{n}} q_2}{E^2_{\mathbf{k}}}$$

are obtained with linearization of the dispersion law near the nodal points as $\xi_{\mathbf{k}} \approx \hbar v_{\mathrm{F}} q_1$, $\Delta_{\mathbf{k}} \approx \hbar v_{\Delta} k_2$ (see Fig. 17.16). Since the above mentioned matrices $\hat{T}(0)$ and $\hat{A}_{\mathbf{n}}$ are real, the imaginary part of the GE pair term is generated by the poles of $\left(1-\hat{A}_{\mathbf{n}}^2\right)^{-1}$. If there is no such poles, one has to search for contributions to γ from next order GE terms. For an lth order GE term, the imaginary part is related to the poles of the inverse of a certain lth degree polynomial in $\hat{A}_{\mathbf{n}},\ldots,\hat{A}_{\mathbf{n}_{l(l-1)/2}}$ (where $\mathbf{n}_1,\ldots,\mathbf{n}_{l(l-1)/2}$ are all possible separations between l impurities).

Generally, in the energy spectrum of a crystal with impurities, one can distinguish certain intervals where DOS is dominated by contributions from band-like states, single impurities, impurity pairs, triples, etc.[26] Then, for instance, in the pair-dominated energy interval, each discrete peak (by an impurity pair at given separation $\mathbf{n}$ in the lattice) experiences small shifts, due to the effects by neighbor impurities of such a pair, different in different parts of the system. These shifts produce a broadening of pair peaks, and if it is wider than the distance between the peaks, the resulting continuous pair-dominated spectrum can be effectively described, passing from summation in discrete $\mathbf{n} \neq 0$ in Eq. 17.27 to integration in continuous $\mathbf{r}$ (for $r > r_0$ where $r_0 \sim a$). Such possibility was shown long ago for normal electron spectrum,[44] and it is even more pronounced for the superconducting system where the pair contribution to DOS at a given energy can come from multiple pair configurations (see Figs. 17.17,17.19 below) and this multiplicity turns yet much greater at involving neighbor impurities into each configuration.

17.7.2. *Non-magnetic impurities*

To treat matrix GE in s-wave superconductors, the algebraic isomorphism was used between the interaction Nambu matrices and common complex numbers.[85] Remarkably, the same isomorphism is also found for matrices of zero energy interaction $\hat{A}_{\mathbf{n}}$ between NM impurities in the d-wave system.[86] In this case, we have explicitly: $\hat{T}(0) = \tilde{V}\hat{\tau}_3$, where $\tilde{V} = v/\rho_N$, and the interaction matrices, Eq. 17.79, are presented as $\hat{A}_{\mathbf{n}} = \tilde{V}g_{\mathbf{n}} + i\tilde{V}f_{\mathbf{n}}\hat{\tau}_2$, thus pertaining to the general two-parametric family $\hat{C}(x,y) = x + iy\hat{\tau}_2$ with real x, y. This family forms an algebra with the product $\hat{C}(x,y)\hat{C}(x',y') = \hat{C}(xx'-yy', yx'+xy')$, isomorphic to that considered in Ref.[85] and to the algebra $\mathbb{C}$ of common complex numbers: $(x+iy)(x'+iy') = xx'-yy'+i(yx'+xy')$. By this isomorphism, the real matrix $\hat{A}_{\mathbf{n}}$ is related to a "complex number" $A_{\mathbf{n}} = \tilde{V}g_{\mathbf{n}} + \hat{\imath}\tilde{V}f_{\mathbf{n}}$, where the "imaginary unity" $\hat{\imath}$ corresponds to the *real* matrix $\hat{\imath} \equiv i\hat{\tau}_2$. Using such "complex" representation and the above mentioned passage from summation in $\mathbf{n}$ to integration in $\mathbf{r}$, we can write the pair contribution to γ in the form:

$$c^2 \mathrm{Im} \int_{r>r_0} \frac{d\mathbf{r}}{a^2} \Re \frac{p(\mathbf{r}) + \hat{\imath} q(\mathbf{r})}{s(\mathbf{r}) + \hat{\imath} t(\mathbf{r})}, \tag{17.80}$$

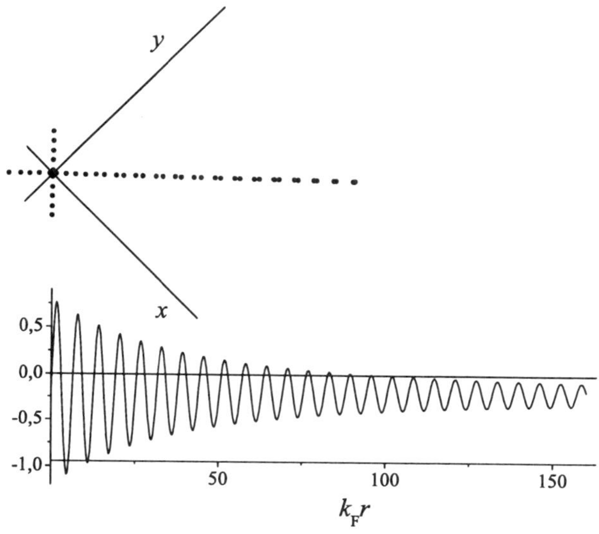

Fig. 17.17. The poles of the integrand in Eq. 17.8 in function of pair separation vector $\mathbf{r} = (x, y)$, are located along the nodal axis in the direct space, where one of the pole conditions, $f_{\mathbf{r}} = 0$, holds identically. Another pole condition, $|g_{\mathbf{r}}| = 1/|\tilde{V}| = |g_{as} - 1/V|$, is reached at discrete points (here at the choice of parameters $\beta = 0.05$, $\tilde{V} g_{as} = 5$). The plot below shows the related behavior of $g_{\mathbf{r}} - 1/\tilde{V}$ for $\mathbf{r}$ along the nodal direction.

with $s = 1 - \tilde{V}^2(g_{\mathbf{r}}^2 - f_{\mathbf{r}}^2)$, $t = -2\tilde{V}^2 g_{\mathbf{r}} f_{\mathbf{r}}$ and some continuous functions $p(\mathbf{r}), q(\mathbf{r})$. Here the symbol $\Re$ means the "real" (traceful) part of the "complex" integrand, while the common imaginary part Im is produced by its poles. These are attained at such separations $\mathbf{r} = \mathbf{r}^*$ in an impurity pair that $s(\mathbf{r}^*) = t(\mathbf{r}^*) = 0$, which requires $|g_{\mathbf{r}^*}| = \tilde{V}^{-1}$ and $f_{\mathbf{r}^*} = 0$. Direct calculation of $g_{\mathbf{r}}$ and $f_{\mathbf{r}}$ shows that all possible $\mathbf{r}^*$ lie on nodal directions, forming identical finite series along them (like those in Fig. 17.17). Also we note that poles only exist at high enough band filling, $\mu\rho_N \geq (1 + \mathrm{e})^{-1}$ and strong enough perturbation parameter, $v \geq (1 + g_{as})^{-1}$, as chosen in Fig. 17.17. If so, it is suitable to pass in the vicinity of each $\mathbf{r}^*$ from integration in the components r_1, r_2 of vector $\mathbf{r}$ to that in the components s, t of the "complex" denominator:

$$\mathrm{Im} \sum_{\mathbf{r}^*} \int ds dt \mathcal{J}_{\mathbf{r}^*}(s,t) \frac{p(s,t)s + q(s,t)t}{s^2 + t^2}, \qquad (17.81)$$

and for any $\mathbf{r}^*$ the transformation Jacobian

$$\mathcal{J}_{\mathbf{r}^*}(s,t) = \left.\frac{\partial (r_1, r_2)}{\partial (s,t)}\right|_{\mathbf{r}=\mathbf{r}^*}$$

is real and non-singular. Then the singularity in the denominator of Eq. 17.81 is canceled by the vanishing residue in the numerator and, even at formal existence of poles in Eq. 17.8, they give no contribution to the zero energy DOS. Mathematically, this simply follows from an extra dimension at 2D integration, giving zero weight to the isolated poles.

The above conclusion can be immediately generalized for any lth order GE term ($l \geq 3$), where the integrand is again presented as $(p + \hat{i}q)/(s + \hat{i}t)$ and p, q, s, t are now continuous functions of $N_l = 2(l-1)$ independent variables (components of the vectors $\mathbf{r}_1, \ldots, \mathbf{r}_{l-1}$) in the configurational space $\mathcal{S}_l$. This integrand can have simple poles on some $(N_l - 2)$-dimensional surface $\mathcal{A}_l$ in $\mathcal{S}_l$ (under easier conditions than for $l = 2$). Then the N_l-fold integration can be done over certain coordinates $u_1, \ldots, u_{N_l-2}$ in $\mathcal{A}_l$ and over the components s, t of the "complex" denominator in the normal plane to $\mathcal{A}_l$:

$$\begin{aligned} &\operatorname{Im} \int du_1 \ldots du_{N_l-2} \mathcal{J}(u_1, \ldots, u_{N_l-2}) \\ &\times \int dsdt \frac{p(\ldots, s, t)\, s + q(\ldots, s, t)\, t}{s^2 + t^2}, \end{aligned} \tag{17.82}$$

with a non-singular Jacobian $\mathcal{J}$. The latter integral has no imaginary part by the same reasons as for Eq. 17.81. Thus, it can be concluded that perturbation by NM-impurities in a d-wave system can not produce localized quasiparticles of zero energy, and this directly follows from the indicated isomorphism of the interaction matrices to the algebra $\mathbb{C}$ of complex numbers.

Moreover, the same conclusion is also valid for yet another type of NM-perturbation, due to locally perturbed SC order by the matrix $\hat{V} = V\hat{\tau}_1$. In this case, the interaction matrix:

$$\hat{A}_{\mathbf{n}} = \frac{V}{1 - V^2 g_{as}^2} \left[f_{\mathbf{n}} + V g_{as} g_{\mathbf{n}} + i \left(g_{\mathbf{n}} - V g_{as} f_{\mathbf{n}} \right) \hat{\tau}_2 \right],$$

pertains to the same family $\hat{C}(x, y)$ as in the above case, hence leading to the same absence of contribution to the zero energy DOS.

17.7.3. *Magnetic impurities*

However, an essential cluster contribution to zero energy DOS can be produced by M impurities with the scalar local perturbation $\hat{V} = \tilde{V}$ and respective T-matrix (similar to Eq. 17.56) in the zero energy limit:

$$\hat{T}(0) = \tilde{V}(1 - \tilde{V}\hat{G})^{-1} = \tilde{V}\frac{1 + vg_{as}\hat{\tau}_3}{1 - v^2 g_{as}^2}. \tag{17.83}$$

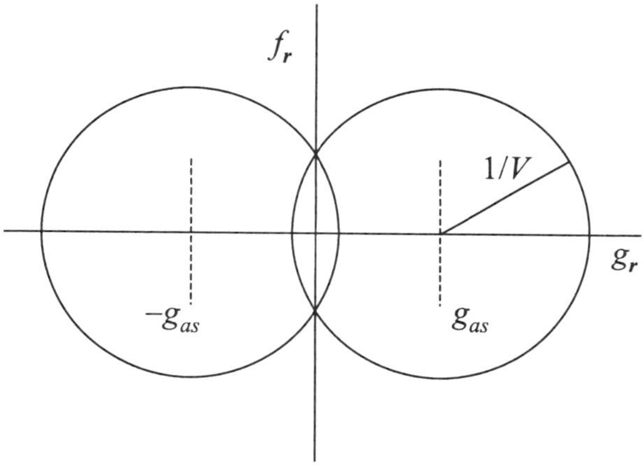

Fig. 17.18. Circular trajectories in the space of variables $g_{\mathbf{r}}$, $f_{\mathbf{r}}$, corresponding to the poles of GE denominator, Eq. 17.83, for M impurities at the choice of $v = 1$ and $g_{as} = 0.84$.

Unlike the traceless $\hat{T}(0)$ for NM-impurities, this matrix produces a finite shift of the nodal point itself, from $\varepsilon = 0$ to $\varepsilon = \varepsilon_0 = cV(1 - V^2 g_{as}^2)^{-1}$. As usual, this can be absorbed into the Fermi level position by shifting all the energy arguments of considered GF's, then DOS in T-matrix (or SCTMA) approximation will vanish at $\varepsilon \to \varepsilon_0$ (in the same way as at $\varepsilon \to 0$ for NM-impurities), fixing the relevant limit in presence of M impurities. The following treatment of higher order GE terms involves the matrix of interaction between M-impurities:

$$\hat{A}_{\mathbf{r}} = \tilde{V}\frac{(\hat{\tau}_3 + vg_{as})g_{\mathbf{r}} + (\hat{\tau}_1 - ivg_{as}\hat{\tau}_2)f_{\mathbf{r}}}{1 - v^2 g_{as}^2}. \tag{17.84}$$

Notably, it does not fit the $\mathbb{C}$ algebra, and though being harder technically, this permits M-impurities to effectively contribute to the zero energy DOS. In fact, the straightforward calculation of the corresponding GE pair term leads to the general matrix expression

$$(1 - \hat{A}_{\mathbf{r}}^2)^{-1} = \frac{N_0 + N_1\hat{\tau}_1 + iN_2\hat{\tau}_2 + N_3\hat{\tau}_3}{D_{\mathbf{r}}}. \tag{17.85}$$

Here N_j's in the numerator are certain functions of $g_{\mathbf{r}}$, $f_{\mathbf{r}}$ and the denominator D is the 4th grade polynomial:

$$D = \left[f_{\mathbf{r}}^2 + (g_{\mathbf{r}} - g_{as})^2 - \frac{1}{v^2}\right]\left[f_{\mathbf{r}}^2 + (g_{\mathbf{r}} + g_{as})^2 - \frac{1}{v^2}\right].$$

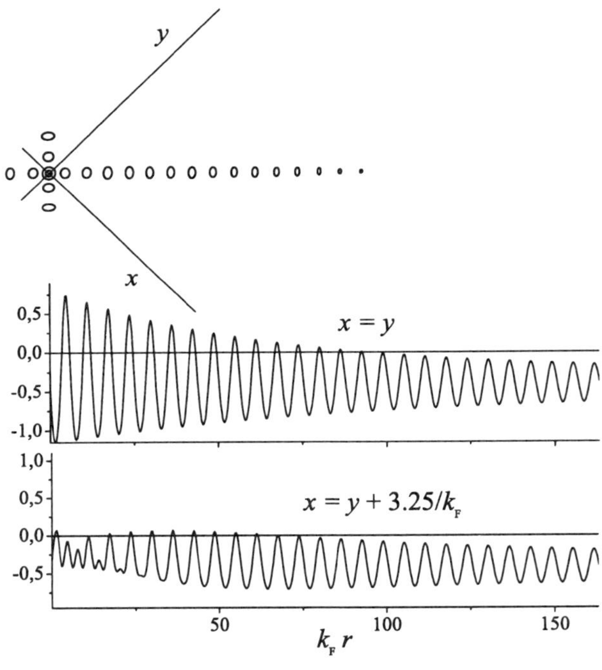

Fig. 17.19. In the $\mathbf{r}$-plane, the trajectories of zeroes of $D_{\mathbf{r}}$ form continuous loops within the stripes of $\sim v_{\mathrm{F}}/(v_{\Delta}|\delta|k_{\mathrm{F}})$ length and $\sim 1/(|\delta|k_{\mathrm{F}})$ width along the nodal axes. The plots below present the pair GE term denominator $D_{\mathbf{r}}$ (at the same choice of parameters as in Fig. 17.18), seen along the stripe axis and along the stripe border.

Zeroes of D in the space of variables $g_{\mathbf{r}}, f_{\mathbf{r}}$ form two circular trajectories of radius $1/v$ centered at $\pm g_{as}$, as shown in Fig. 17.18 (cf. to their location in the two outer points of these circles, $f_{\mathbf{r}} = 0$, $\pm g_{\mathbf{r}} = g_{as} + 1/v$, in the NM case). It is this extension of singularities, from isolated points to continuous trajectories, that allows finite imaginary part of the 2D integral

$$\gamma \approx \frac{c^2\tilde{V}}{1-v^2 g_{as}^2}\mathrm{Im}\int \frac{d\mathbf{r}}{a^2}\frac{N_0+vg_{as}N_3}{D_{\mathbf{r}}}. \tag{17.86}$$

On the other hand, it is assured by the fact that the traceful numerator $N_0+vg_{as}N_3$ in Eq. 17.86 does not vanish on these trajectories. Quantitative analysis is simplified in the case of $vg_{as} \approx 1$, when the small parameter $\delta = 1 - v^2 g_{as}^2$ (such that $|\delta| \ll 1$) defines a low energy resonance at $\varepsilon_{res} \approx \Delta\delta/[v\ln(1/\delta^2)]$.[85] In this case we have simply $N_0 + vg_{as}N_3 \approx -2g_{\mathbf{r}}^2/v^2$. Direct numeric calculation of the functions $g_{\mathbf{r}}, f_{\mathbf{r}}$ shows that the trajectories $D = 0$, when presented in variables $r_{1,2}$, form multiple loops (seen in the right panel of Fig. 17.19) of total number $\sim v_{\mathrm{F}}/(v_\Delta|\delta|)$, each contributing by $\sim \pi^2/[(ak_{\mathrm{F}})^2|\delta|]$ into $\mathrm{Im}\int g_{\mathbf{r}}^2/(V^2D)d\mathbf{r}/a^2$ in Eq. 17.86. Thus we arrive at the estimate: $\gamma \sim c^2\tilde{V}/|\delta|$, and hence to the finite residual DOS:

$$\rho(\varepsilon_0) \sim \rho_N \frac{c^2\tilde{V}}{\Delta|\delta|}\ln\frac{\Delta|\delta|}{c^2\tilde{V}}. \tag{17.87}$$

We recall that this result is impossible in the properly formulated self-consistent approximation,[55] and it is also in a striking difference to the SCTMA predictions, $\rho(0) \sim \exp(-\rho_N\Delta/cv^2)\rho_N^2\Delta/(cv^2)$ in the Born limit[75] or $\rho(0) \sim \sqrt{c\rho_N/\Delta}$ in the unitary limit.[36] The non-universality of this effect is manifested by its sensitivity to the M-perturbation parameter v, so that $\rho(\varepsilon_0)$ is defined by impurity pairs only for strong enough perturbations, $|v| \gtrsim (1+g_{as})^{-1}$. For weaker v, nodal quasiparticles will be only localized on impurity clusters of a greater number $n > 2$ (the bigger the smaller $|v|$), and the particular form of Eq. 17.87 would change to $\rho(0) \sim \rho_N c^n$ (with some logarithmic corrections). Anyhow, a finite limit of DOS at zero energy is granted by the fact that in real high-T_c systems a certain M-type perturbation can result even from nominally non-magnetic centers (as shown in Sec. 17.5.2). Moreover, the above considered condition $|v|g_{as} \approx 1$ does not seem very difficult, as testified by the observation of extremely low-energy resonance $\varepsilon_{res} \approx -1.5$ meV by Zn impurities in $Bi_2Sr_2CaCu_2O_{8+\delta}$.[18] It just fits the asymmetric M-resonance from Fig. 17.20, contrasting with the symmetric NM-resonance picture.

Then the overall impurity effect in a d-wave superconductor can be seen as a superposition (almost independent) of the above described effects from NM-impurities with perturbation parameter V_{NM} and concentration c_{NM} and from M-impurities with perturbation parameter V and concentration c (supposedly $c \ll$

$c_{\rm NM}$). An example of such situation is shown in Fig. 17.20.

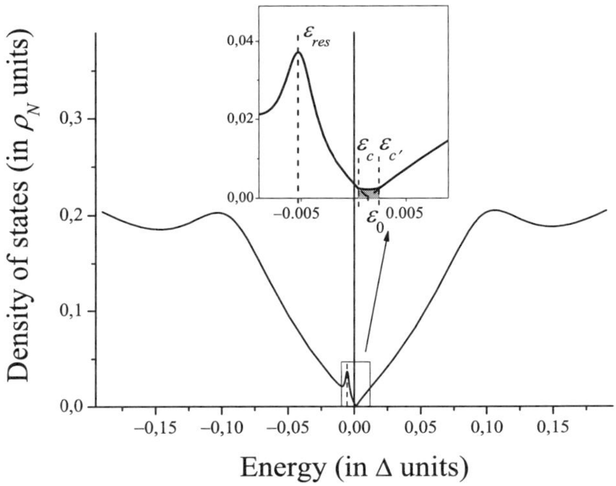

Fig. 17.20. Low-energy d-wave DOS $\rho(\varepsilon)$ in simultaneous presence of NM-impurities (with $c_{\rm NM} = 3\%$ and $v_{\rm NM} = 1$), producing two symmetric broad resonances, and M-impurities (with $c = 0.03\%$, $v = v_{\rm NM}$, and $vg_{as} = 0.9$), producing single sharp resonance at extremely low energy ε_{res}. Inset shows the mobility edges ε_c and $\varepsilon_{c'}$ around the shifted nodal point ε_0, they separate localized states (shadowed area) with almost constant DOS, $\rho(\varepsilon) \approx \rho(\varepsilon_0)$, from band-like states whose DOS is close to the T-matrix value (solid line).

The residual DOS from pair GE term prevails within a certain narrow vicinity of ε_0 where quasiparticle states are all localized on properly separated impurity pairs. Outside this vicinity, the states are extended and reasonably described by T-matrix (or SCTMA). The transition from localized to extended states occurs at the Mott mobility edges $\varepsilon_c < \varepsilon_0$ and $\varepsilon_{c'} > \varepsilon_0$, where GE and T-matrix contributions to DOS are comparable. Using the simplest approximation for the T-matrix term: $\rho(\varepsilon) \sim \rho_N|\varepsilon - \varepsilon_0|/\Delta$, we estimate the range of localized states (somewhat exaggerated in the inset of Fig. 17.20) as

$$\varepsilon_{c'} - \varepsilon_0 \sim \varepsilon_0 - \varepsilon_c \sim \delta_c \sim \frac{c^2\tilde{V}}{|\delta|} \ln \frac{\Delta|\delta|}{c^2\tilde{V}}, \tag{17.88}$$

provided it is much smaller than the distance to the M-resonance: $\delta_c \ll |\varepsilon_0 - \varepsilon_{res}|$. The same estimates for the mobility edges follow from the IRM breakdown condition: $\varepsilon - {\rm Re}\Sigma(\varepsilon) \sim {\rm Im}\Sigma(\varepsilon)$, at $\varepsilon \approx \varepsilon_0$.

The tendency to localization of quasiparticles can be generally opposed by the effects of repulsive Coulomb interaction between them[21] and this issue was also discussed for disordered d-wave superconductors.[87,88] These field theory treatments showed that localization can survive at low enough temperature. The full account of Coulomb interactions in the present GE approach is rather complicated technically, but a simple estimate follows from the overall number of (supposedly) localized particles within the energy range, Eq. 17.16, which is as small as $n_{loc} \sim \rho(\varepsilon_0)\delta_c \sim (c^4\tilde{V}/\Delta\delta^2)\ln^2|\Delta\delta/c^2\tilde{V}|$. Since the average distance between them $\sim a/\sqrt{n_{loc}}$ is much longer than the distance between charge carriers $\sim a/\sqrt{n} \sim a/\sqrt{\rho_N\varepsilon_F}$, the effects of Coulomb interaction are hopefully screened out, at least for the systems far enough from half-filling.[89]

Notably, localization turns to be yet possible near the resonance energy, $\varepsilon \approx \varepsilon_{res}$, but this requires that the concentration of M-impurities surpasses a certain characteristic value $c_{res} \sim (\varepsilon_{res}/\Delta)^2 \ln(\Delta/|\varepsilon_{res}|)$. In particular, for the choice of parameters in Fig. 17.20, we find $c_{res} \sim 3 \cdot 10^{-4}$, so that this system should be close to the onset of localization also in this spectrum range, where each localized state is associated with a *single* impurity center.

Generally, presence of localized states near the lowest excitation energies in the spectrum must influence significantly the kinetic properties of a crystal with impurities, such as electric and (electronic part of) heat conductivity at lowest temperatures. Taking in mind the above referred modification of Kubo formula for the energy range $|\varepsilon - \varepsilon_0| < \delta_c$, their temperature dependencies, instead of reaching the universal values σ_0 and $[\kappa/T](0)$, should rather tend to the exponential vanishing: $\sim \exp(-\delta_c/k_B T)$, at low enough temperatures: $T \ll \delta_c/k_B$. The latter value, at the same choice of impurity perturbation parameters and typical gap $\Delta \sim 30$ meV, is estimated as ~ 0.2 K. In this context, the intriguing sharp downturn of κ/T, recently observed at temperatures $\lesssim 0.3$ K[41] and attributed to the low-temperature decoupling of phonon heat channel,[42] can be otherwise considered as a possible experimental manifestation of the quasiparticle localization by impurity clusters. A more detailed analysis of possible non-universal behavior of transport properties of disordered d-wave superconductors will be necessary to confirm this conjecture.

17.8. Conclusions

The quasiparticle states are considered in a d-wave superconductor with impurities, extending the self-consistent T-matrix approximation (SCTMA) framework by using the techniques of group expansions of Green functions in complexes of interacting impurities. A comparative analysis of impurity effects on quasiparticle energy and lifetime is done for different types of impurity perturbations, varying their intensity, spatial extent, and matrix structure in Nambu indices. The essential difference between point-like and extended perturbation is established, consisting in separate action of different symmetry modes of perturbation on different physical properties of quasiparticles, which eventually reduces destructive effects of extended impurity centers on SC order. Using the Ioffe-Regel-Mott criterion, modified for the case of quasiparticles in d-wave superconductors, the criteria of validity are established for SCTMA and a proper combination of different self-consistent solutions for different regions of energy spectrum is suggested. Beyond the scope of SCTMA, the practical calculation of group expansions for different types of impurities is developed. It is shown that, if the impurity perturbation of magnetic type is present, the indirect interaction between impurities can essentially change the quasiparticle spectrum near nodal points, producing strongly localized states of non-universal character (depending on the perturbation strength). Experimental check for possible non-universal effects in low temperature transport properties can be done, e.g., in the Zn doped $Bi_2Sr_2CaCu_2O_{8+\delta}$ system.

References

1. J. G. Bednorz, K. A. Müller, Possible highTc superconductivity in the Ba-La-Cu-O system, *Z. Phys. B* **64** (2), 189–193, (1986).
2. M. K. Wu, J. R. Ashburn, C. J. Torng, P. H. Hor, R. L. Meng, L. Gao, Z. J. Huang, Y. Q. Wang, and C. W. Chu, Superconductivity at 93 K in a new mixed-phase Y-Ba-Cu-O compound system at ambient pressure, *Phys. Rev. Lett.* **58** (9), 908–910, (1987).
3. D. M. Ginsberg ed., *Physical Properties of High Temperature Superconductors V.* (World Scientific, Singapore, 1996).
4. A. V. Balatsky, I. Vekhter, and J. X. Zhu, Impurity-induced states in conventional and unconventional superconductors, *Rev. Mod. Phys.* **78** (2), 373–433, (2006).
5. P. W. Anderson, Theory of dirty superconductors, *J. Phys. Chem. Solids.* **11** (1-2), 26–30, (1959).
6. P. G. de Gennes, *Superconductivity of metals and alloys.* (Perseus Books Publishing, L.L.C., 1999).
7. A. A. Abrikosov and L. P. Gor'kov, Contribution to the theory of superconducting alloys with paramagnetic impurities, *Sov. Phys. JETP* **12** (6), 1243–1253, (1961).
8. M. Fowler and K. Maki, Conditions for bound states in a superconductor with a magnetic impurity, *Phys. Rev.* **164** (2), 484–488, (1967).
9. H. Shiba, Classical spins in superconductors, *Prog. Theor. Phys.* **40** (3), 435–451, (1968).
10. A. I. Rusinov, Superconductivity near a paramagnetic impurity, *JETP Lett. URSS* **9** (2), 85–88, (1969).

11. W. E. Pickett, Electronic structure of the high-temperature superconductors, *Rev. Mod. Phys.* **61** (2), 433–512, (1989).
12. C. C. Tsuei and J.R. Kirtley, Pairing symmetry in cuprate superconductors, *Rev. Mod. Phys.* **72** (4), 969–1016, (2000).
13. D. A. Bonn, S. Kamal, K. Zhang, R. Liang, D. J. Baar, E. Klein, and W. N. Hardy, Comparison of the influence of Ni and Zn impurities on the electromagnetic properties of $YBa_2Cu_3O_{6.95}$, *Phys. Rev. B* **50** (6), 4051–4063, (1994).
14. A. V. Balatsky, M. I. Salkola, and A. Rozengren, Impurity-induced virtual bound states in d-wave superconductors, *Phys. Rev. B.* **51** (21), 15547–15551, (1995).
15. Y. Pogorelov, Ground state symmetry and impurity effects in superconductors, *Solid State Commun.* **95** (4), 245–249, (1995).
16. J. M. Valles, Jr., R. C. Dynes, A. M. Cucolo, M. Gurvitch, L. F. Schneemeyer, J. P. Garno, and J. V, Waszczak, Electron tunneling into single crystals of $YBa_2Cu_3O_{7-\delta}$, *Phys. Rev. B.* **44** (21), 11986–11996, (1991).
17. A. Yazdani, C. M. Howald, C. P. Lutz, A. Kapitulnik, and D.M, Eigler, Impurity-induced bound excitations on the surface of $Bi_2Sr_2CaCu_2O_8$, *Phys. Rev. Lett.* **83** (1), 176–179, (1999).
18. E. W. Hudson, K. M. Lang, H. Eisaki, S. Ushida, and J. C. Davis, Imaging the effects of individual zinc atoms on superconductivity in $Bi_2Sr_2CaCu_2O_{8+}$, *Nature (London)* **403** (1), 746–750, (2000).
19. E. W. Hudson, K. M. Lang, V. Madhavan, S. H. Pan, H. Eisaki, S. Ushida, and J. C. Davis, Interplay of magnetism and high-T_c superconductivity at individual Ni impurity atoms in $Bi_2Sr_2CaCu_2O_{8+\delta}$, *Nature (London)* **411** (1), 920–924, (2001).
20. W. N. Hardy, S. Kamal, D. A. Bonn, K. Zhang, R. Liang, D. C. Morgan, and D. J. Baar, Microwave surface impedance of the cuprate superconductor $YBa_2Cu_3O_{6.95}$, *Physica B* **197** (1-4), 609–616, (1994).
21. D. N. Basov, A. V. Puchkov, R. A. Hughes, T. Strach, J. Preston, T. Timusk, D. A. Bonn, R. Liang, and W. N. Hardy, Disorder and superconducting-state conductivity of single crystals of $YBa_2Cu_3O_{6.95}$, *Phys. Rev. B.* **49** (17), 12165–12169, (1991).
22. S. G. Doettinger, R. P. Huebener, P. Gerdermann, A. Kühle, S. Anders, T. G. Träuble, and J. C. Villégier, Electronic instability at high flux-flow velocities in high-T_c superconducting films, *Phys. Rev. Lett.* **73** (12), 1691–1694, (1994).
23. F. Guinea, Y. Pogorelov, Vortex viscosity in superconductors with short coherence length, *Phys. Rev. Lett.* **74** (3), 462–465, (1995).
24. L. Taillefer, B. Lussier, R. Gagnon, K. Behnia, H. Aubin, Universal heat conduction in $YBa_2Cu_3O_{6.9}$, *Phys. Rev. Lett.* **79** (21), 483–486, (1997).
25. M. Sutherland, D. G. Hawthorn, R. W. Hill, F. Ronning, S. Wakimoto, H. Zhang, C. Proust, E. Boaknin, C. Lupien, L. Taillefer, R. Liang, D. A. Bonn, W. N. Hardy, R. Gagnon, N. E. Hussey, T. Kimura, M. Nohara, H. Takagi, Thermal conductivity across the phase diagram of cuprates: Low energy quasiparticles and doping dependence of the superconducting gap, *Phys. Rev. B.* **67** (17), 174520, (2003).
26. I. M. Lifshitz, Energy spectrum of disordered systems, *Adv. Phys.* **13** (52), 483–536, (1964).
27. N. F. Mott, Electrons in disordered structures, *Adv. Phys.* **16** (52), 49–144, (1967).
28. P. W. Anderson, Absence of diffusion in certain random lattices, *Phys. Rev.* **109** (52), 1492–1505, (1958).

29. T. Matsubara, A new approach to quantum-statistical mechanics, *Prog. Theor. Phys.* **14** (52), 351–376, (1955).
30. D. N. Zubarev, Double-time Green functions in statistical physics, *Sov. Phys. Uspekhi* **3** (3), 320–345, (1960).
31. L. P. Gorkov, On the energy spectrum of superconductors, *Sov. Phys. JETP* **7** (3), 505–508, (1958).
32. A. A. Abrikosov, L. P. Gorkov, and I. E. Dzyaloshinskii, *Methods of quantum field theory in statistical physics* (Dover Publications, NY, 1975).
33. C. J. Pethick, and D. Pines, Transport processes in heavy-fermion superconductors, *Phys. Rev. Lett.* **57** (1), 118–121, (1986).
34. S. Schmitt-Rink, K. Miyake, C. M. Varma, Transport and thermal properties of heavy-fermion superconductors: a unified picture, *Phys. Rev. Lett.* **57** (20), 2575–2578 (1986).
35. P.J. Hirshfeld, P. Wolfle and D. Einzel, Consequences of resonant impurity scattering in anisotropic superconductors: thermal and spin relaxation properties, *Phys. Rev. B* **37** (1), 83–97 (1988).
36. P. A. Lee, Localized states in a *d*-wave superconductor, *Phys. Rev. Lett.* **71** (12), 1887–1890 (1993).
37. M. J. Graf, S. -K. Yip, J. A. Sauls, D. Rainer, Electronic thermal conductivity and the Wiedemann-Franz law for unconventional superconductors, *Phys. Rev. B* **53** (22), 15147–15161 (1996).
38. A. C. Durst and P. A. Lee, Impurity-induced quasiparticle transport and universal-limit Wiedemann-Franz violation in *d*-wave superconductors, *Phys. Rev. B* **62** (2), 12070–12090 (2000).
39. K. S. Novoselov, A. K. Geim, S. V. Morozov, D. Jiang, Y. Zhang, S. V. Dubonos, I. V. Grigorieva, and A. A. Firsov, Electric field effect in atomically thin carbon films, *Science* **306**, 666–669 (2004).
40. N. M. R. Peres, F. Guinea, and A. H. Castro Neto, Electronic properties of disordered two-dimensional carbon, *Phys. Rev. B* **73** (12), 125411 (2006).
41. R. W. Hill, C. Proust, L. Taillefer, R. Fournier, R. L. Greene, Breakdown of Fermi-liquid theory in a copper-oxide superconductor, *Nature* **414** (6865), 711–715 (2001).
42. M. F. Smith, J. Paglione, M. B. Walker, L. Taillefer, Origin of anomalous low-temperature downturns in the thermal conductivity of cuprates, *Phys. Rev. B* **71** (1), 014506 (2005).
43. J. E. Mayer and M. Goeppert-Mayer, *Statistical mechanics* (Wiley, NY, 1940).
44. M. A. Ivanov, Dynamics of quasilocalized vibrations with a high concentration of impurity centers, *Sov. Phys. Solid State* **12** (7), 1508–1515 (1971).
45. M. A. Ivanov, V. M. Loktev, and Yu. G. Pogorelov, Long-range impurity states in magnetic crystals, *Phys. Rep.* **153** (4-5), 209–330 (1987).
46. F. C. Zhang and T. M. Rice, Effective Hamiltonian for the superconducting Cu oxides, *Phys. Rev. B* **37** (7), 3759–3761 (1988).
47. A. Fujimori, Character of doped oxygen holes in high-T_c Cu oxides, *Phys. Rev. B* **39** (1), 793–796 (1989).
48. V. A. Gavrichkov, E. V. Kuz'min, and S. G. Ovchinnikov, Electronic structure and order parameter symmetry in high-temperature superconductors *Physics Uspekhi* **43** (2), 1813–184 (2000).

49. M. Abramowitz and I. A. Stegun, *Handbook of mathematical functions* (National Bureau of Standards, NY, 1964).
50. M. Franz, C. Kallin, and A. J. Berlinsky, Impurity scattering and localization in *d*-wave superconductors, *Phys. Rev. B* **54** (4), R6897–R6900 (1996).
51. W.A. Atkinson, P. J. Hirschfeld, and A. H. McDonald, Gap inhomogeneity and the density of states in disordered *d*-wave superconductors, *Phys. Rev. Lett.* **85** (18), 3922–3925 (2000).
52. A. Polkovnikov, S. Sachdev, and M. Voyta, Impurity in a *d*-wave superconductor: Kondo effect and STM spectra, *Phys. Rev. Lett.* **86** (2), 296–299 (2001).
53. G. Baym, Self-consistent approximation in many-body systems, *Phys. Rev.* **127** (4), 1391–1401 (1962).
54. M. A. Ivanov, and Yu. G. Pogorelov, Electron-states in a crystal with impurity levels near band edge, *Sov. Phys. JETP* **72** (6), 2198–2209 (1977).
55. V. M. Loktev and Yu. G. Pogorelov, Formation of d-wave superconducting order in a randomly doped lattice, *Low. Temp. Phys.* **27** (9), 767-776 (2001).
56. L. Zhu, W. A. Atkinson, and P. J. Hirschfeld, Power spectrum of many impurities in a d-wave superconductor, *Phys. Rev. B* **69** (6), 060503(R) (2004).
57. I. Adagideli, P. M. Golgbart, A. Schnirman, and A. Yazdani, Low-energy quasiparticle states near extended scatterers in *d*-wave superconductors and their connection with SUSY quantum mechanics , *Phys. Rev. Lett.* **83** (26), 5571–5574 (2001).
58. Yu. G. Pogorelov and M. C. Santos, Effects of extended impurity perturbation in a d-wave superconductor, *Phys. Rev. B* **71** (1), 014516 (2005).
59. F. A. Cotton, *Chemical Applications of Group Theory* (Wiley NY, 1990).
60. Yu. A. Izyumov and M. V. Medvedev, *Magnetically Ordered Crystals Containing Impurities* (Consultants Bureau, NY, 1973).
61. T. R. Chien, Z. Z. Chang, and N. P. Ong, Effect of Zn impurities on the normal-state Hall angle in single-crystal $\mathrm{YBa_2Cu_{3-x}Zn_xO_{7-\delta}}$, *Phys. Rev. Lett.* **67** (15), 2088-2091 (1991).
62. P. J. Hirschfeld and N. Golgenfeld, Effect of strong scattering on the low-temperature penetration depth of a *d*-wave superconductor, *Phys. Rev. B* **48** (6), 4219-4222 (1993).
63. R. Fehrenbacher and M. R. Norman, Gap renormalization in dirty anisotropic superconductors: Implications for the order parameter of the cuprates , *Phys. Rev. B* **50** (5), 3495-3498 (1994).
64. A. V. Mahajan, H. Alloul, G. Collin, and J. F. Marucco, ^{89}Y NMR probe of Zn induced local moments in $\mathrm{YBa_2(Cu_{1-y}Zn_y)_3O_{6+x}}$, *Phys. Rev. Lett.* **72** (19), 3100-3103 (1994).
65. W. A. MacFarlane, J. Bobroff, H. Alloul, P. Mendels, and N. Blanchard, Dynamics of the Local Moment Induced by Nonmagnetic Defects in Cuprates, *Phys. Rev. Lett.* **85** (5), 11080-1111 (2000).
66. Q. Chen and R. J. Schrieffer, Pairing fluctuation theory of high-Tc superconductivity in the presence of nonmagnetic impurities, *Phys. Rev. B* **66** (1), 014512 (2002).
67. K. Park, Kondo effect of nonmagnetic impurities and coexisting charge order in cuprate superconductors, *Phys. Rev. B* **67** (9), 094513 (2003).
68. V. M. Loktev and Yu. G. Pogorelov, Magnetic effect from non-magnetic impurity in superconducting $\mathrm{CuO_2}$ plane, *Europhys. Lett.* **60** (5), 757–763 (2002).

69. R. E. Walstedt, R. F. Bell, L. F. Schneemeyer, and J. V. Waszczak, Absence of magnetic pair breaking in Zn-doped $YBa_2Cu_3O_7$, *Phys. Rev. B* **48** (14), 10646–10649 (1993).
70. M.-H. Julien, T. Fehér, M. Horvatić, C. Berthier, O. N. Bakharev, P. Ségransan, G. Collin, and J.-F. Marucco, ^{63}Cu NMR evidence for enhanced antiferromagnetic correlations around Zn impurities in $YBa_2Cu_3O_{6.7}$, *Phys. Rev. Lett.* **84** (15), 3422-3425 (2000).
71. R. J. Birgeneau, and G. Shirane, Neutron scattering studies of structural and magnetic excitations in lamellar copper oxides - a review, in: *Physical properties of high temperature superconductors I*, ed. D. M. Ginsberg (World Scientific, Singapore, 1989).
72. J. Kondo, Resistance minimum in dilute magnetic alloys, *Progr. Theor. Phys.* **32** (15), 37–49 (1964).
73. Y. Nagaoka, Self-consistent treatment of Kondo's effect in dilute alloys, *Phys. Rev.* **138** (4A), A1112–A1120 (1965).
74. J. Zittartz and E. Mueller-Hartmann, Theory of magnetic impurities in superconductors. I, *Z. Phys.* **232** (1), 11–31 (1970).
75. L. P. Gor'kov and P. A. Kalugin, Defects and an unusual superconductivity, *Sov. Phys. JETP Lett.* **41** (5), 253–255 (1985).
76. T. Senthil, M. P. A. Fisher, L. Balents, and C. Nayak, Quasiparticle transport and localization in high-Tc superconductors, *Phys. Rev. Lett.* **81** (21), 4704-4707 (1998).
77. A. A. Nersesyan, A. M. Tsvelik and F. Wenger, Disorder effects in two-dimensional Fermi systems with conical spectrum: exact results for the density of states, *Nucl. Phys. B* **438** (3), 561-588 (1995).
78. K. Ziegler, M. H. Hetter and P. J. Hirschfeld, Nonzero Fermi level density of states for a disordered *d*-wave superconductor in two dimensions, *Phys. Rev. Lett.* **77** (14), 3013-3016 (1996).
79. C. Pepin and P. A. Lee, Order from disorder: nonmagnetic impurities in the spin-gap phase of the cuprates, *Phys. Rev. Lett.* **81** (13), 2779-2782 (1998).
80. V. M Loktev and Yu. G. Pogorelov, Density of states in doped metal with *d*-wave superconducting order parameter, *Europhys. Lett.* **58** (4), 549–554 (2002)
81. R. J. Elliott, J. A. Krumhansl, and P. L. Leath, The theory and properties of randomly disordered crystals and related physical systems, *Rev. Mod. Phys.* **46** (3), 465-5432 (1974).
82. A. F. Ioffe, A. R. Regel, Noncrystalline amorphous and liquid electronic semiconductors, *Progr. Semicond.* **4** (3), 237-242 (1960).
83. V. M. Loktev and Yu. G. Pogorelov, Validity of self-consistent T-matrix for doped d-wave superconductors, *Phys. Lett. A* **320** (13), 307-311 (2004).
84. L. Zhu, W. A. Atkinson, and P. J. Hirschfeld, Two impurities in a *d*-wave superconductor: local density of states, Phys. Rev. B **67** (9), 094508 (2003).
85. Yu. G. Pogorelov and V. M. Loktev, Group expansions for impurities in superconductors, *Phys. Rev. B* **69** (21), 214508 (2004).
86. Yu. G. Pogorelov, M. C. Santos and V. M. Loktev, Impurity clusters and localization of nodal quasiparticles in *d*-wave superconductors, *cond-mat*/0605424 (2006).
87. D.V. Khveschenko, A.G. Yashenkin, I.V. Gornyj, Phys. Rev. Lett., **86**, 4668 (2001).
88. M. Fabrizio, L. Del'Anna and C. Castellani, Phys. Rev. Lett. **88**, 076603 (2002).
89. L. Boeri, E. Cappelluti, C. Grimaldi, and L. Pietronero, Phys. Rev. B **68**, 214514 (2003).

Chapter 18

Hadronica

J.E.F.T. Ribeiro
CFIF, IST-UTL,
Av.Rovisco Pais, Lisboa, Portugal,
EmilioRibeiro@netcabo.pt

The BCS theory of $^3P_0$ quark-antiquark condensation is given. A method to evaluate overlap kernels is discussed. An introduction to the derivation of Salpeter equations and its relation with RGM equations is presented. The mechanism for the spontaneous breakdown of chiral symmetry and its relation with issues like the pion Goldstne boson, chiral restoration, scalar versus vectorial confinement and hadronic decay are also treated.

Contents

18.1. Introduction

When I was invited to write a chapter dedicated to the physics of hadrons when seen from the many-body perspective, I knew I had a difficult task ahead. Partly because the lexicon of hadronic physics is somewhat remote from the usual many-body terminology and partly due to the existence, in the field of hadronic physics, of a vast literature, with some repetitions and more often than probably it should, old ideas and results redressed in new garments. Therefore, in order to introduce some concepts, I must start with a necessary introduction to the field of hadronic physics, trying as much as possible to stay close to the historical path. Within the allotted space and with the focus on the many-body aspects of hadronic physics, I cannot help leaving out fundamental aspects of Hadronic physics in such a digression. For instance, no attempt will be made at discussing the gauge aspects of QCD. Unavoidably there will also be some autobiographic contamination with such a perspective but I hope that in the end this will not hurt a faithful representation of what has been some of the problems and progress of hadronic physics.

18.2. Quarks

Quarks were born in 1964. This was the year when M.Gell-Mann[1] and, independently, G.Zweig[2] understood that the multiplicity of hadrons could be organized as combinations of more fundamental constituents having the extraordinary property of being fractionally charged: the quarks. Already in 1938 and following the experimentally observed[3] charge independence of neutron-proton and proton-proton interactions, we could have thought[4] of organizing protons and neutrons into a fundamental SU(2)doublet of constituents as follows,

$$\epsilon = \binom{p}{n}, \ \delta = (\bar{p}, \bar{n}), \ \textbf{Singlet}: \frac{1}{\sqrt{2}}(\bar{p}p + \bar{n}n), \ \textbf{Triplet}: \begin{pmatrix} \bar{n}p \\ \bar{p}n \\ \frac{1}{\sqrt{2}}(\bar{p}p - \bar{n}n) \end{pmatrix} \tag{18.1}$$

The assumption that this SU(2) group, named after Cassen and Condon[5] as Isotopic spin (Isospin for short), was (nearly) a symmetry of nature would have forced us to expect that, within each triplet, masses would be (almost) degenerate. Therefore, we could be tempted to assign the almost mass degenerate triplet of pions $\{\pi^+, \pi^0, \pi^-\}$ to the above triplet and henceforth we would be facing a choice: either these pions were bound states of nucleons or they could be thought of operational agents for the transmutation between protons and neutrons. In other words, pions, like viruses, must inject some kind of "genetic" material to transform protons into nucleons and viceversa. These "genes" are the quarks. If they were fermions, we could, at the same time, explain both the fermionic nature of nucleons and the bosonic nature of pions (see Fig.(18.1)). As the number of strongly interacting particles kept increasing from the hitherto known set of nucleons and pions, a new phenomenon was observed. A number of transitions that should have occurred, were SU(2) the symmetry group of strong interactions, did not occur (or at least they took place solely through electromagnetic or weak processes). A new conservation law, other than Isospin, must have been at work. In 1953, Gell-Mann,[6] Nakano and Nishijima[7] introduced a new additive number, known as Strangeness, which was also conserved by strong interactions. It is not

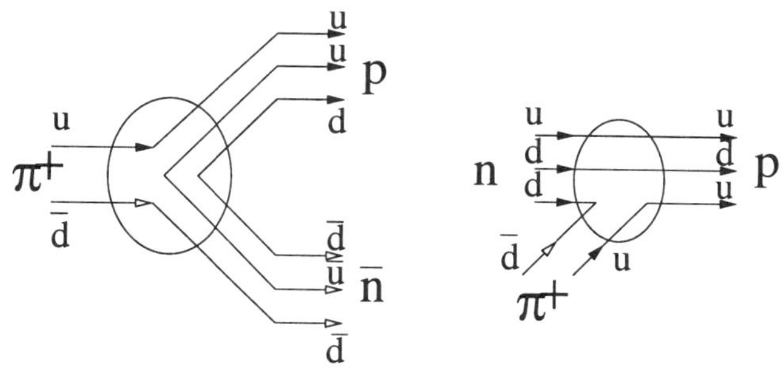

Fig. 18.1. The role of quarks: ensuring that π^+ is the transmutation agent for the transition n$\rightarrow$ p or, equivalently, the responsible agent for the creation of a nucleon-antinucleon pair, $p - \bar{n}$. The ellipsis enclosing the area where quark rearrangement takes place represents a *strong* interaction whose properties, other than flavor independence, we don't need to know at this stage.

hard to guess that in order to accommodate this new degree of freedom, we could enlarge the Isospin group SU(2) to SU(3), therefore adding a new member to the, at the time, small family of $\{u, d\}$ quarks. It came to be known as the strange quark. With the discovery of yet new particles and the observation of new conservation laws among them, more quarks were needed to take account of these and, to make a long story short, we now have come to know a large family of quark flavors:$\{u, d, s, c, b, t\}$.

18.3. Quark Models Old and New[8]

Just one year after being theoretically born, new functions were bestowed upon quarks in 1965. The obvious culprit was the Δ^{++}, a member of the symmetric multiplet 56 of the proposed final (when seen from 1964) strong interaction symmetry group $Su(3)_{flavor} \otimes SU(2)_{spin}$. The Δ^{++} was, basically, an ensemble of three u, which contained a symmetric combination of $\{u \uparrow u \uparrow u \uparrow\}$. But quarks were supposed to be fermions and they can only live together in a completely antisymmetric state, which was clearly not the case. So, in the year of 1965, Moo-Young Han with Yoichiro Nambu[9] proposed that quarks should possess an additional SU(3) degree of freedom, later called color charge. Hadronic masses will contain terms proportional to the two Casimir operators of this group C_2, C_3. For a given bound state of n quarks (3 for baryons or 2 for mesons, in which case one of the quarks is in reality an antiquark), the quadratic Casimir operator C_2 is given by $(\sum_i^n \vec{\lambda}_i) \cdot (\sum_i^n \vec{\lambda}_i)$, with the $\vec{\lambda}$. being the SU3(3) generators. In other words, C_2, and by the same token, the hadronic masses should contain terms going like $\vec{\lambda}_i \cdot \vec{\lambda}_j$. Later, these SU(3) generators were given the physical content of being the quark color charges, which, as in the case of QED, should couple to eight gauge bosons, known as gluons.

For the moment, let us follow Lipkin[10] and assume that quark-quark and quark antiquark forces have the general structure,

$$H = \frac{1}{2}\sum_{i\neq j}\vec{\lambda}_i\cdot\vec{\lambda}_j v(ij) + \sum_i (T(i) + m_i) \tag{18.2}$$

where $v(ij)$ is a given color independent potential. Then, for a state of n quarks and m antiquarks we have,

$$\langle H(n,m)\rangle = (n+m)(m_Q + \bar{T}) + \frac{1}{2}\Big((\sum_i \vec{\lambda}_i)^2 - \sum_i (\vec{\lambda}_i)^2\Big)\bar{v} \tag{18.3}$$

where $\bar{v}$ and $\bar{T}$ represent color independent averages. First, let us consider color singlets: $((\sum_i \vec{\lambda}_i)^2 = 0)$. For mesons and baryons we have respectively,

$$\langle H\rangle = 2(m_Q + \bar{T} - \frac{2}{3}\bar{v}),\ \ 3(m_Q + \bar{T} - \frac{2}{3}\bar{v}) \tag{18.4}$$

$\bar{v} > 0$. Non-singlets will have masses higher by an additional term $1/2\Big((\sum_i \vec{\lambda}_i)^2\Big)\bar{v}$. When $\bar{v} \to \infty$ we require $m_Q + \bar{T} \to \frac{2}{3}\bar{v}$ in order to have finite masses for mesonic and baryonic color singlets. In other words, if $\bar{v} = V + \hat{v}$ and $V \to \infty$ then it is sufficient that $m_Q = M + \hat{m}_Q$ with $M = \frac{2}{3}V$ to obtain finite result for singlets

$$\langle H\rangle = 2(\hat{m}_Q + \bar{T} - \frac{2}{3}\hat{v}),\ \ 3((\hat{m}_Q + \bar{T} - \frac{2}{3}\hat{v}). \tag{18.5}$$

For non-singlets, hadronic masses will diverge with $1/2\Big((\sum_i \vec{\lambda}_i)^2\Big)V$. A simple way to enforce this is to slightly change Eq.(18.2) into,

$$H = \frac{1}{2}\sum_{i,j}\vec{\lambda}_i\cdot\vec{\lambda}_j v(ij) + \sum_i (T(i) + \hat{m}_i), \tag{18.6}$$

where $\hat{m}_i$ is what we nowadays call a current quark mass. In general $\hat{m}_Q$ and $\hat{m}_i$ may come to differ by a finite quantity. $\hat{m}_Q$ is known as the valence quark mass.

So we have that the basic requirement for confinement of quarks into singlets must be that *for a singlet state, the infrared singularities of the potential must cancel against the infrared singularities of the quark masses. They do not cancel for non color singlets.*

Finally, the interaction form of Eqs.(18.2, 18.6) has the important property that the mean force experienced by a given quark, when interacting with a color singlet, is zero: $\langle singlet|\vec{\lambda}_j|siglet\rangle = 0$. This implies that any given ensemble of quark and antiquarks tends to dissolve into non-interacting color singlets.

Experimental properties other than hadronic masses, such as cross sections are of the utmost importance towards the understanding of the microscopic nature of strong interactions. The razor of Occam has it that the same theory should explain-without further additions or tunings-the largest possible body of diverse experimental facts. This logical necessity is clearly demonstrated in the case of the strong repulsive core known to exist in the strong scattering of hadrons. It cannot be explained unless the color quark interaction contains a color hyperfine term $\vec{\lambda}_i \cdot \vec{\lambda}_j \; S_i \cdot S_j$.[11] Let us see why.

Back in 1975, Neudatchin et al[12] proposed an explanation for the NN short range repulsion–a strong NN soft repulsive core–that is known to exist for small distances ($\leq 0.4fm$). In their paper they argued that the fact of ${}^1S_0$ phase shifts going through zero at $E_{lab} = 300MeV$, and hence being consistent with a change of the sign of the nuclear interaction of around 0.3 -0.4 fm, could be ascribed to the Pauli exclusion principle between quarks thus preventing identical fermions to overlap. However, the Pauli exclusion principle, by itself, turned out to produce a weak NN repulsion and thus, another physical agent needed to be at work to produce the observed strength. This extra ingredient was the dependence on a color hyperfine-like term $\vec{\lambda}_i \cdot \vec{\lambda}_j \; S_i \cdot S_j$, of the quark microscopic interaction. This dependence, together with the Pauli exchange, was enough to explain the observed repulsion.

To show this, we assume the nucleons to be in an antisymmetric combination of three valence quarks.

$$\Phi = \Psi_{space}(123) \; \chi_{su_{flavor}(4)}(123) \; \Sigma_{su_{color}(3)}(123) \qquad (18.7)$$

As usual, $\Psi_{space}(123)$ and $\chi_{su_{flavor}(4)}(123)$ are symmetric functions of the quark labels (123) and $\Sigma_{su_{color}(3)}(123)$, the color singlet wave function, is completely antisymmetric.

Another important ingredient is provided by the size of such bound states, which can be simulated, without loss of generality, by a gaussian 3-body wave function,

$$\Psi_{space}(123) = [\pi^3 3^{3/2} R_0^6]^{-1/2} \exp\{-\sum_{i=1}^{3} r_i^2/2R_0^2\}, \; \vec{r_i} = \vec{r_i} - \vec{R}_{C.M.} \quad (18.8)$$

with R_0 being a parameter that measures the nucleon size. Other spatial symmetric functions are possible without changing the physical effect. Being the set of harmonic oscillator wave functions $\psi_{n,l,m}(123)$, of which $\Psi_{space}(123)$ of Eq.(18.8) corresponds to the ground state, a complete set, it would merely have amounted to an ab-initio choice of another parameter R_0-i.e, a choice of a given h.o. basis and, hence, to an expansion of such wave functions on that basis. As it will be clear later, when describing the graphical-rules, it turns out that the Pauli exchange will rapidly filter out excitations higher than the ground state, so it is sufficient for the purpose of evaluating NN repulsive cores–an instance of what is called exotic scattering– to choose an R_0 which describes well the bulk size of the nucleon and go from there.

Nowadays we make use of the Wick contractions to take account of Pauli exchange between clusters of fermions. An equivalent and older form of implementing the Pauli principle was, as in the Resonating Group Method[13] to project out the antisymmetric combination of fermions by the use of an antisymmetrizer projector $\mathcal{A}$. We will use, with Wheeler, the stationary condition

$$\delta\langle\Psi^{R.G}|H - E|\Psi^{R.G.}\rangle = 0 \tag{18.9}$$

where E is the total energy (in this case of a six quark system) and H the microscopic many-body hamiltonian,

$$H = \sum_{i=1}^{6}(-\frac{\hbar^2}{2m_Q}\nabla_{x_i}) - T_{C.M.} + \sum_{i>j} V_{ij}. \tag{18.10}$$

$T_{C.M.}$ stands for the center of mass kinetic energy whereas V_{ij} is the two-body interaction between quarks i and j, sum of a color confining term plus a spin dependent term,[14]

$$V_{ij} = A\,\vec{\lambda}_i \cdot \vec{\lambda}_j; c(r_{ij}) - B\,\vec{\lambda}_i \cdot \vec{\lambda}_j\; S_i \cdot S_j\; f(r_{ij}) \tag{18.11}$$

The Resonating Group Method (RGM) wave function $\Psi^{R.G}$ is given by,

$$\Psi^{R.G}(\xi_1, \xi_2, \bar{R}) = \int d^3\bar{x} g(\bar{x})\mathcal{A}\{[\Phi^{(1)}(\xi_1)\Phi^{(2)}(\xi_2)]\delta^3(\bar{X} - \bar{R})\} \tag{18.12}$$

with the $\Phi^{\{1,2\}}$ being the completely antisymmetrized nucleon wave functions and $\mathcal{A}$ being given by,

$$\mathcal{A} = \frac{1}{N_A}[1 - \sum p^{ij} + \sum p^{ik}\, p^{jm} - \sum p^{il}\, p^{jm}\, p^{kn}], \tag{18.13}$$

Where p^{il} is responsible for the interchange of quark i, belonging to nucleon 1 with quark j belonging to nucleon 2. N_A is a suitable normalization coefficient. Because the results will not depend on it, we need not be concerned with its value. Finally, the symbol $[\Phi^{(1)}(\xi_1)\Phi^{(2)}(\xi_2)]$ appearing in Eq. (18.12) stands for a particular chosen symmetry for the combination of nucleons 1 and 2. Equation(18.9) leads to the following integral equation for $g(\bar{x})$

$$\int k_f(\bar{x}, \bar{x}')g(\bar{x}')d^3\bar{x}' = 0 \tag{18.14}$$

with $k_f(\bar{x}, \bar{x}')$-the full RG.M. kernel being given by,

$$k_f(\bar{x}, \bar{x}') = \langle \mathcal{A}(H - E) \rangle_{space \otimes SU(3)colour \otimes SU(4)flavor} =$$
$$\langle \Phi(\xi_1)\Phi(\xi_2)\delta^3(\bar{X} - \bar{R}) | \mathcal{A}(H - E) | [\Phi(\xi_1)\Phi(\xi_2)]\delta^3(\bar{X}' - \bar{R}) \rangle_{colour \otimes flavor} \tag{18.15}$$

In the expression above, an integration in the intrinsic coordinates $\{\xi_1, \xi_2\}$ is assumed. Now the Hamiltonian (18.10) can be cast as,

$$H_{6quarks} = H^{(1)}(3q) + H^{(2)}(3q) + H_{rel} \tag{18.16}$$

with H_rel standing for the relative Hamiltonian describing the movement between $\Phi^{(1)}$ and $\Phi^{(2)}$. Using for the intrinsic cluster (in this case, nucleons) wave functions $\Phi^{(1)}$ and $\Phi^{(2)}$ the eigenfunctions of $H_{3q}^{\{1,2\}}$ and writing the operator $\mathcal{A}$ as a sum of a direct and an exchange part $\mathcal{K}$,

$$\mathcal{A} = 1 - \mathcal{K} \tag{18.17}$$

we rewrite, Eq. (18.14) as follows,

$$[-\frac{\hbar^2}{M_N}\nabla_{\bar{R}} + V_D(\bar{R}) - E_{rel}]\ g(\bar{R}) + \int S_{L.R.}(\bar{R}, \bar{R}')g(\bar{R}')d^3\bar{R}' \tag{18.18}$$

with M_N being the nucleon mass and,

- $V_D(\bar{R}) = \int \langle \sum_{il} V_{il} \rangle d^3\bar{R}'$
- $S_{L.R.}(\bar{R}, \bar{R}') = -\langle \mathcal{K}(H_{rel} - E_{rel}) \rangle$

The kernel $\mathcal{K}$ contains, in principle, long range interaction due to the exchange of clusters $\Phi^{(1)}$ and $\Phi^{(2)}$,

$$p^{il}\ p^{jm}\ p^{kn} = p^{`}(\Phi^{(1)}, \Phi^{(2)}) \tag{18.19}$$

This is a trivial long range exchange, the net effect being the exchange of the two clusters as a whole and, provided we use the appropriate relative wave function $g(\bar{R})$, it can be discarded and $S_{L.R} \rightarrow S_{S.R.}$ i. e, the long range kernel $S_{L.R}$ becomes short ranged: $S_{S.R.}$. From here we could have gone to expand both $g(\bar{R})$ and $S_{S.R.}$ in spherical harmonics to arrive at a definite integro-differential equation to be solved numerically to obtain an effective NN repulsion of the right magnitude. Here we need not follow that path but merely extract the important ingredients that will be common to any such equation. Again, using the nucleon-nucleon scattering as a paradigm, we have altogether nine exchange operators $p^{\alpha\beta}$. It is not hard to understand that all these nine different exchange operators must lead to the same result: after all the numbering of quarks is a human mnemonic

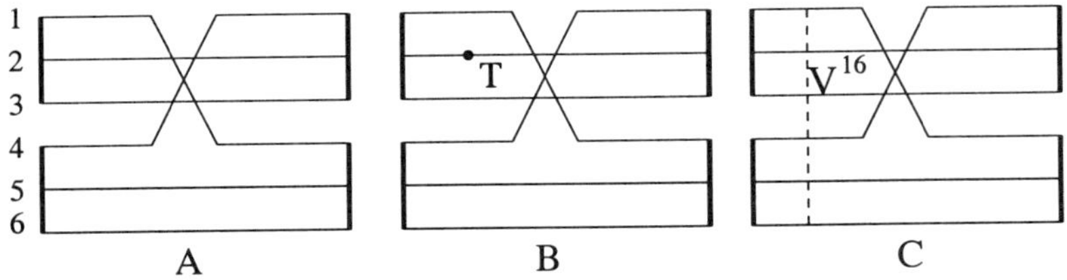

Fig. 18.2. Cluster Overlaps: Diagram A represents $\langle p^{\alpha\beta}\rangle$. Diagram B represents the kinetic insertion in quark 2 and contributes to $\langle p^{14}\nabla_{\bar{R}}\rangle$. Finally Diagram C depicts $\langle p^{14}V_{16}\rangle$.

and is devoid of physical significance. So we can forget all but one exchange, say p^{14}, and multiply the result by nine. So, for a given $P^{\alpha\beta}$, we have to contend with, $\mathcal{K}(\bar{R},\bar{R}') = +\langle p^{\alpha\beta}\rangle$, $T(\bar{R},\bar{R}') = -\langle p^{\alpha\beta}\nabla_{\bar{R}}\rangle$, and $V(\bar{R},\bar{R}') = -\langle p^{\alpha\beta}\sum_{ij}V_{ij}\rangle$, with i and j belonging to different nucleons.

The above three items can be respectively associated with the three diagrams of Fig.(18.2).

Although for the SU(n) groups it is easy to evaluate $\langle p^{\alpha\beta}\rangle_{SU(n)}$ by using the formula, $\lambda_i\cdot\lambda_j = \frac{1}{2}(p^{ij} + (n-1)\delta(i,j) - \frac{1}{n})$ (where $\delta(i,j)$ is the Kronecker delta), with the space overlaps it is quite a different story. For instance, diagram A of Fig.(18.2) yields, for ground state Harmonic Oscillator nucleon wave functions Φ with size R_0, a spatial overlap kernel which goes like,

$$\mathcal{K}(\bar{R},\bar{R}') \simeq \exp\{-15[(\bar{R}^2+\bar{R}'^2)+18\bar{R}\cdot\bar{R}']/16R_0^2]\}. \qquad (18.20)$$

This is a compact kernel and, therefore, possesses an eigenvalue decomposition,

$$\int \mathcal{K}(\bar{R},\bar{R}')\Phi_{n,l;m}(\bar{R}')d^3\bar{R}' = (1/3)^{(2n+l)}\Phi_{n,l;m}(\bar{R}) \qquad (18.21)$$

with $\Phi_{n,l;m}$ being the Harmonic oscillator w.f. in spherical coordinates with size R_0. This much was known in 1978. At this stage it is important to notice that this compactness is a direct consequence of the finite hadron size measure in units of R_0. In turn, this is a phenomenological consequence of confinement and, ultimately, this R_0 will be linked to the confining properties of QCD. For the moment, it is just a parameter. Finally this effective non-local potential only arises because of the Pauli principle (embodied by the antisymmetrizer A). In other words, the Pauli Principle produces forces.

So, any theory aiming at describing, under some unifying perspective, the diversity of hadronic phenomena, must, ab initio, be able to describe exotic scattering and to that extent one cannot do without having quarks interacting through a microscopic force homogeneous in $\vec{\lambda}\cdot\vec{\lambda}\,\vec{S}\cdot\vec{S}$.

Strong reactions do not deal solely with exotic scattering–where, as in the case of the NN soft core, we have no quarks annihilation–but also with reactions where quark-antiquark annihilation reactions, like the one depicted in Fig.(18.1), also take place.

Back in the OZI era,[2,15] it was understood that hadronic decay must be realized through the creation (and annihilation) of $q - \bar{q}$ pairs. The reason was that these pairs carried the vacuum quantum numbers. A quantitative description of this creation was not available and still evades us at present. The $^3P_0$ mechanism for hadronic decay is depicted in the figure (18.1). In the absence of any robust, first-principles, theoretical description of the $^3P_0$ mechanism, we would like to know at least the answer to three basic questions: 1-what is the strength of the $^3P_0$ creation.

2- where it is created.

3-and, finally, if the overlap between the 3_0^P pair, the incoming and outgoing clusters can be evaluated in the *same formalism* as the the exotic scattering kernels just discussed.

The first idea is to fit both parameters and hope for these fitted parameters to stay put when we hop from reaction to reaction. The shortcoming of such approach is that we do not have a model for these transition potentials which *do* (as it is clear by now) change from sector to sector. It is a matter of principle, and a crucially important one, that hadronic models, albeit deemed successful in one or another aspect, should be exportable to a variety of hadronic phenomena which were, at first glance and prior to the model application, unrelated with each other: *for instance, decay and exotic scattering*.

Therefore, we would like to contemplate a situation where we only have to deal at most with one scale: the size of hadrons. This scale can be thought to be the scale of the effective force among quarks. The rest should come from geometrical overlaps and group theory. Coming back to Eq. (18.20), we face the question whether we could substitute the cumbersome expression of that equation by a separable expansion and whether such expansion converged fast enough to be of any use. The answer to these question is affirmative and goes by the name of graphical rules.

In 1981, while staying at Nijmegen University, I developed an expedite way to evaluate kernels of the type of Eq.(18.20) so that I could unify, under the same formalism, such diverse kernels coming either from decay reactions or exotic scattering like the NN soft core just treated. This was published in 1982[16] for the non-trivial case of arbitrary angular momenta which was at the time the required formalism to deal with $^3P_0$ pairs. The case of radial excitations, as it was already stated in that paper, was a straightforward corollary of the proof. In any case, this corollary eventually got published one year later.[17]

This method is universal and can be used for any number of incoming and, not necessarily the same number, of outgoing clusters with or without quark-antiquark annihilation or creation. It can be applied to any compact operator $\hat{O}$ to obtain its cluster representation,

$$\mathcal{O}_{C_1,...,C'_m} = \langle \mathcal{C}_1...\mathcal{C}_n|\hat{O}|\mathcal{C}'_1...\mathcal{C}'_m\rangle, \tag{18.22}$$

where $\mathcal{C}_j(q_1,..ql)$ represents a given cluster of particles with Jacobean coordinates q_α. Relative coordinates between clusters can be thought of clusters of clusters. Think of equation (18.22) as the amplitude for having n particles simultaneously arranged in $\{\mathcal{C}_1...\mathcal{C}_n\}$ and in $\{\mathcal{C}'_1...\mathcal{C}'_m\}$ clusters.

An example of such compact operator is provided by the Pauli exchange operator $\hat{P}^{\alpha\beta}$. It is compact because $\hat{P}^{\alpha\beta}\hat{P}^{\alpha\beta} = 1$. Another example is given by the rearrangement operator $\mathcal{R}_i^{\alpha\beta}$–see Fig.(18.3). It is also compact $\mathcal{R}_i^{\alpha\beta}\mathcal{R}_i^{\beta\alpha} = 1$. In what follows, I will only use harmonic oscillator (H.O.) spectral decomposition of clusters, which is sufficient to study geometric overlaps. In principle, Hilbert spaces other than H.O. could be used too.

In figure (18.3) we give two examples of cluster representations for both $\hat{P}^{\alpha\beta}$ and $\hat{R}_i^{\alpha\beta}$. To the upper diagram we can associate the number $\mathcal{P}^{13}$,

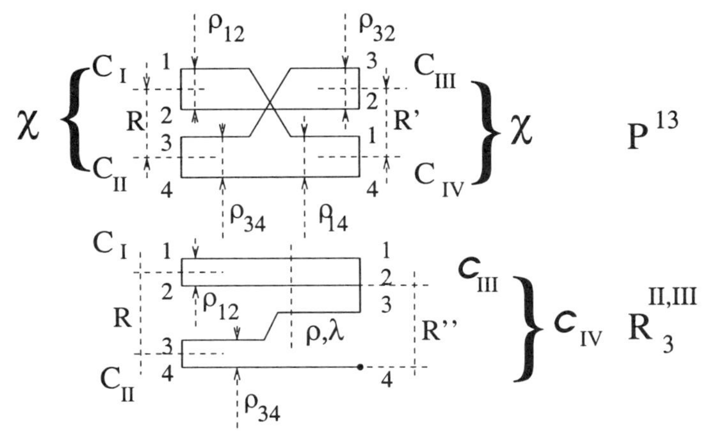

Fig. 18.3. Examples of cluster representations of compact operators. The upper diagram refers to Pauli exchange P^{13} between two clusters whereas the second diagram refers to the rearrangement operator $R_3^{I,II}$. The appropriate Jacobean coordinates are also presented.

$$\mathcal{P}^{13}_{C_1,{}^3P_0,C_{II},C_{III};\chi_1,\chi_2} = \int d^3\rho_{12} d^3\rho_{34} d^3R d^3R' \delta(R' - P^{13}R)$$
$$\chi_1(R) C_I(\rho_{12}) C_{II}(\rho_{34}) C^*_{III}(\rho_{32}) C^*_{IV}(\rho_{14}) \chi^*_2(R')$$
$$= \int d^3R d^3R' \chi_1(R) \mathcal{P}^{13}(R,R') \chi^*_2(R'). \quad (18.23)$$

We recognize $\mathcal{P}^{13}(R, R')$ as a typical Resonating Group Kernel.

It turns out that both kernels, the Pauli-exchange and the Rearrangement kernel, can be evaluated using the same graphical rules. We need *only* to know the Jacobean representation of these two operators. We have,

$$\langle P^{13}\rangle_{Jacobean} = \begin{bmatrix} \frac{1}{2} & \frac{1}{2} & -\frac{1}{\sqrt{2}} \\ \frac{1}{2} & \frac{1}{2} & \frac{1}{\sqrt{2}} \\ -\frac{1}{\sqrt{2}} & \frac{1}{\sqrt{2}} & 0 \end{bmatrix}, \ \langle R_3\rangle_{Jacobean} = \begin{bmatrix} 1 & 0 & 0 \\ 0 & -\frac{1}{\sqrt{3}} & \frac{\sqrt{2}}{\sqrt{3}} \\ 0 & \frac{\sqrt{2}}{\sqrt{3}} & \frac{1}{\sqrt{3}} \end{bmatrix} \quad (18.24)$$

Notice that $\mathcal{P}^{13} \cdot \mathcal{P}^{13} = 1$ and that $\mathcal{R}_3 \cdot \mathcal{R}_3 = 1$. In order to proceed let us enumerate the rules needed to construct a generic graphical-rules diagram associated with $\mathcal{O}_{C_1,\ldots,C'_m}$–see Fig.(18.4). They are:

1-Perform cluster decomposition in term of harmonic oscillator wave functions for both incoming and outgoing clusters: i.e
$C(1,...k) = \sum c_1^{n1,l1m1} ... c_k^{nk,lk,mk} \phi(n1, l1, m1)...\phi(nk, lk, mk)$.

2-Draw as many exterior legs as independent H.O. $\phi(ni, lj, mj)$, and label them with $\{nj, lj, mj\}$.

3-Connect all the exterior legs in all possible ways consistent with connecting one incoming leg with an outgoing leg. To these lines we call propagators.

4-Any given such propagator, connecting exterior lines i and j contributes with a number ${\mathcal{O}_{ij}}^{2n_{ij}+l_{ij}}$, being $\mathcal{O}_{ij}$ the ij th matrix element of the Jacobean representation of $\hat{O}$–see Eqs. (18.24)–and n_{ij}, l_{ij}, m_{ij} the quantum numbers flowing in that "propagator". Notice that $\mathcal{O}_{ij}$is always $\leq$1

5-Each Vertex contributes with a number as defined in Fig.(18.5)

6- Sum over all possible diagrams consistent with the conservation of quantum numbers of rule 5.

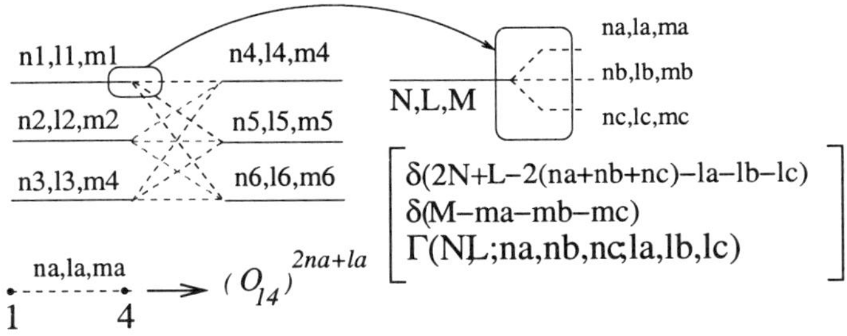

Fig. 18.4. Example of graphical rules for 3 clusters→3 clusters. The exterior legs correspond to incoming and outgoing clusters in a H.O. spectral decomposition. The slashed line $\{ij\}$-connecting exterior legs i and j-correspond to the ij th "propagator" $\hat{O}_{ij} = \mathcal{O}_{ij}{}^{2n_{ij}+l_{ij}}$, with $\mathcal{O}_{ij}$ the ij th matrix element of Jacobean representation of the operator $\hat{O}$ and $\{n_{ij}, l_{ij}, m_{ij}\}$, the H.O. quantum numbers . The vertices Γ are *universal for all operators* $\hat{O}$ we might be interested in, solely depending on the Hilbert space $\mathcal{H}$ we have chosen to for the cluster decomposition. Conservation of flowing quantum numbers $\{n, l, m\}$ is enforced at the vertices by delta functions.

to obtain a number,

$$\langle C_1^{out}(1,...)...C_M^{out}(1,...kM)|\hat{O}|C_1^{in}(1,...k1)...C_L^{in}(1,...kL)\rangle =$$
$$= \sum c_{1;in}^{n1in,l1in,m1in}...c_{N;in}^{nNin,lNin,mNin} c_{1;out}^{n1out,l1out,m1in}...c_{N;out}^{nNout,lNout,mNout}$$
$$\langle \phi(n1in, l1in, m1in)...\phi(nNin, lNin, mNin)|\hat{O}|$$
$$\phi(n1out, l1out, m1out)...\phi(nNout, lNout, mNout)\rangle =$$
$$= \sum \mathcal{O}(\{n1in, l1in, m1in\}, ..., \{nNout, lNout, mNout\})$$
$$c_{1;in}^{n1in,l1in,m1in}...c_{N;in}^{nNin,lNin,mNin} c_{1;out}^{n1out,l1out,m1in}...c_{N;out}^{nNout,lNout,mNout}. \tag{18.25}$$

Failure to integrate in any of the Jacobean coordinates, say Q_1 .... Q_J, will produce separable overlap kernels $\mathcal{O}(Q_1, ...Q_J)$ depending on those coordinates.

So far, this looks like a dreadfully complicated way to evaluate overlap kernels. Nevertheless, it is for practical cases quite powerful and allows, for most cases, a mental evaluation of overlap kernels. To see this, let us go back to the examples of figure (18.3). Consider the overlap kernel $\mathcal{P}^{13}(R, R')$, associated with the Pauli exchange P^{13},

$$\mathcal{P}^{13}(R, R') = \langle \phi_{0,0,0}(\rho_{12})\phi_{0,0,0}(\rho_{34})|P^{13}|\phi_{0,0,0}(\rho_{12})\phi_{0,0,0}(\rho_{34})\rangle \tag{18.26}$$

The graphical-rules diagram associated with Eq. (18.26), produces the overlap kernel–see Fig.(18.6),

$$\mathcal{P}^{13}(R,\ R') = \mathcal{P}^{13}_{\{[0]_{I,II,III,IV};[n,l,m],[n',l',m']\}}\phi_{0,0,0}(R)\ \phi_{0,0,0}(R'), \tag{18.27}$$

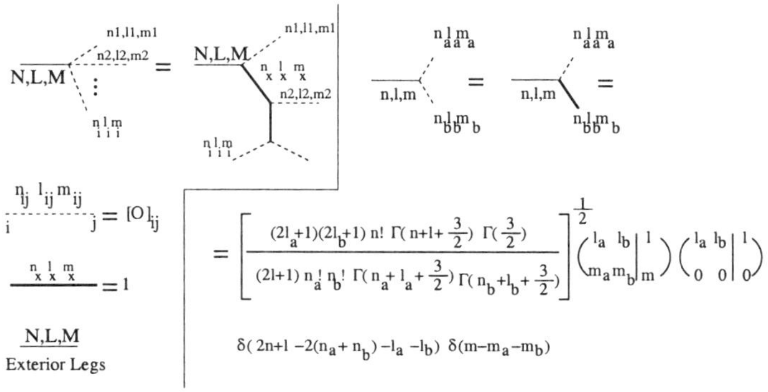

Fig. 18.5. Method to evaluate a generic vertex: Group all the "propagators" in pairs of neighbors by the introduction of fictitious carriers of quantum numbers n_x, l_x, m_x (thick lines). The function of these lines is just to bookkeep local conservation of the quantum numbers. Otherwise they play no role and are set to unity. Dashed lines correspond to the real "propagators". Exterior legs are represented with thin lines. The fundamental building-block vertices are depicted in the right of the figure together with their values.

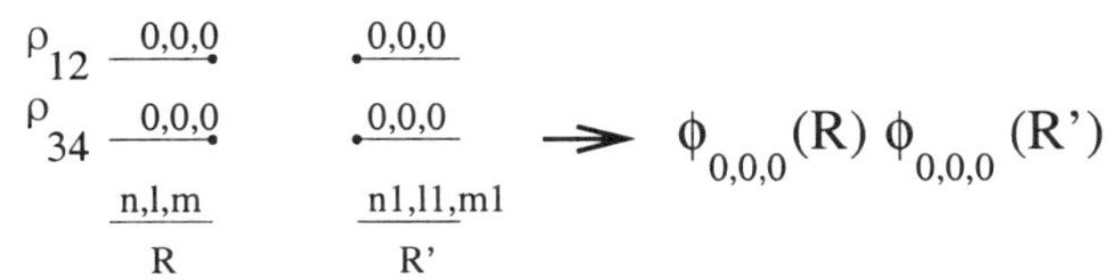

Fig. 18.6. A simple application of the graphical rules. No propagators are allowed connecting the ρ lines because of vertex conservation. Because we have no propagators the vertices are all equal to unity. Notice that because $P^{13}_{R,\ R'} = 0$-See Eq.18.24- we have the *same separable potential no matter what quantum numbers flow in the line RR'*.

with $\mathcal{P}^{13}_{\{[0]_{I,II,III}\cdots\}} = 1$ for the only non-zero case: $\{nlm\} = \{n1l1m1\} = \{000\}$. As an exercise, we leave for the reader to show Eq. (18.21). The rearrangement case is also quite simple. Now let us consider a typical amplitude involving a ${}^3P_0$ insertion as in figure (18.7). By now we understand this amplitude to be a familiar exchange kernel. We still have to answer two questions: one about the size of the ${}^3P_0$ pair and the other about the whereabouts of such creation. *It turns out that we can dump both these two parameters in one single parameter.* Let us see how this happens.

Consider the creation of the ${}^3P_0$ pair in one point of space. We have

$$\mathcal{P}(\rho_{12}, R') = \langle ...{}^3P_0(\rho_{34})\delta(R-a)|P^{13}|M_2(\rho_{12})M_3(\rho_{34})...\rangle. \qquad (18.28)$$

In Eq.(18.28) we only wrote, for the sake of simplicity, the relevant clusters and assume, from now on, the intervening clusters to be just given by single H.O wave functions of the same size (if this is not the case, then we must work with H.O. spectral decomposition of these clusters, an unessential complication). Using $\delta(R-a) = \phi_{000}(a)\phi_{000}(R) + \sum_{nlm\neq 000}\phi_{nlm}(a)\phi_{nlm}(R)$ we get for the overlap of Fig.(18.7),

$$\mathcal{P}^{a}_{\{a;...\}}(\rho_{12}, R') \simeq \phi_{000}(a)\ [\mathcal{P}^{13}]\ \phi_{000}(\rho_{12})\phi_{100}(R') + ... \qquad (18.29)$$

with $[\mathcal{P}^{13}]$ given by the graphical rules. The corrections due to the other terms in the HO spectral decomposition of $\delta(R-a)$ will be strongly filtered out because of the higher the nlm floating around, the much smaller the corrections–killed by the propagators $\mathcal{O}_{ij}{}^{2n_{ij}+l_{ij}}$ of rule 4–will be. Notice that the arbitrary distance **a** gets transformed in a multiplicative constant $\phi_{000}(\mathbf{a})$, which can be absorbed in the overall strength V_o of the transition potential. A similar conclusion would have been obtained for space distributions other than $\delta(\rho_{34}-a)$ with new multiplicative constants now different from $\phi_{000}(a)$. We can then obtain a simple transition

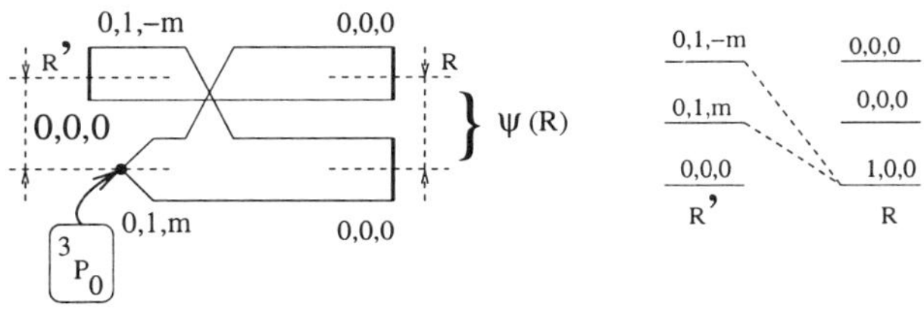

Fig. 18.7. Basic diagram for scalar decay. The bare scalar must contain a coordinate-space wave function with L=1. This is because quarks and antiquarks have opposite intrinsic parity. So we must have two internal propagators with L=1 which can only couple to n=1 or L=2. The decay process is dominated by the N=1,L=0 case.

potential for the decay of a scalar quark-antiquark bound state,

$$\mathcal{V}_T(\rho_{12}, R'') = V_o IS\ \mathcal{P}^{13}_{\{[0,1],[0,1],[0],[0];[0],[1,0]\}}\phi_{0,1}(\rho_{12}))\phi_{1,0}(R''). \qquad (18.30)$$

IS stands for the product of group 9-J symbols for trivial spin and isospin exchanges. V_o stands for the strength of $^3P_0$ creation, which is deemed universal. We see that we have *different* overlap kernels, depending on the particles involved, with *all related with one another* by a single unknown parameter V_o.

In 1986, we used this formalism to study the low lying scalar meson nonet.[18] We were surprised to see the appearance of new poles-hitherto hidden in such transition potentials-which, almost 20 years later, came to be experimentally discovered.[19]

The first thing to notice is that overlap kernels (transition potentials) *are non-local* and therefore any *local approximation to these potentials must be, not only energy dependent, but also process dependent*, i.e depending not only on the the sector we are studying (scalar, vectorial, and so on...) but also on the cluster sizes. The simplest way to construct such local approximations is to perform a delta shell fitting of these non-local potentials for phenomenological studies. Although numerically useful to verify, for diverse reactions, whether this new mechanism may come into effect, those parameters have no special physical significance and, as it will be clear when considering light quarks, they become mere fits.

And yet, despite the apparent phenomenological success of the formalism discussed so far, it **cannot be correct** because it grossly violates a symmetry that we know to be almost exact for reactions involving pions. In other words, we look for a model that *besides* possessing all the good properties discussed so far, should also, at one stroke (meaning: one parameter apart from the quark masses), simultaneously fulfill the well-known low energy theorems of Gell-Mann, Oakes, and Renner,[20] Goldberger and Treiman,[21] Adler self-consistency zero,[22] the Weinberg theorem,[23] and so on. To do this we must enter the realms of the Spontaneous Breakdown of Chiral Symmetry *S.B.C.S.*. In doing so we will be able to treat quark-antiquark annihilation and scattering as different aspects of the *same* interaction and hence get a microscopic grip on V_o. It is now appropriate to enter the physics of *S.B.C.S.*

18.4. The Physics of the Spontaneous Breakdown of Chiral Symmetry

The simplest class of Hamiltonians able to respond affirmatively to all the above requirements[24–27] is given by,

$$H = \int d^3x\, \psi^+(x) \left(-i\vec{\alpha}.\vec{\nabla}\right) \psi(x) + \int \frac{d^3x\, d^3y}{2} J_\mu^a(x) K_{\mu\nu}(x-y) J_\nu^a(y) \tag{18.31}$$

With, $J_\mu^a(x) = \overline{\psi}(x)\, \vec{\lambda}\, \gamma_\mu \frac{\lambda^a}{2}\, \psi(x)$ and $K_{\mu\nu}^{ab}(x-y) = \delta^{ab} K_{\mu\nu}(|\vec{x} - \vec{y}|)$. The ψ are Dirac spinors and represent the quarks. The $\vec{\lambda}$ are the color generators. Clearly the hyperfine term $\vec{\lambda}_i \cdot \vec{\lambda}_j\ S_i \cdot S_j$ must come from the Pauli matrices hidden in the small components of Dirac spinors.

18.4.1. *Building Fock Spaces*

To build a Fock space associated with the Hamiltonians (18.31) we must start by laying its foundations, i.e, the "vacuum" and then proceed to build the Fock space by applying to this "vacuum" strings of fermion creation operators. A general "vacuum" looks like,

$$|\widetilde{0}> = Exp\left\{\widehat{Q}_0^+ - \widehat{Q}_0\right\}|0> \tag{18.32}$$

with,

$$\widehat{Q}_0^+(\Phi) = -\sum_{cf} \int d^3p\, \phi(|\vec{p}|)\, M_{ss'}(\theta,\phi) \widehat{b}^+_{fcs}(\vec{p})\, \widehat{d}^+_{fcs'}(-\vec{p}). \tag{18.33}$$

The matrix $M_{ss'}(\theta,\phi)$ carries the $^3P_0$ pair coupling:

$$\begin{aligned} M_{ss'}(\theta,\phi) &= -4\pi\sqrt{2} \sum_{m_l m_s} \begin{bmatrix} 1 & 1 & |0 \\ m_l & m_s & |0 \end{bmatrix} \times \begin{bmatrix} 1/2 & 1/2 & |1 \\ s & s' & |m_s \end{bmatrix} y_{1m_l}(\theta,\phi) \\ &= i\vec{\sigma}\cdot\hat{k}\sigma_2. \end{aligned} \tag{18.34}$$

The functions $\phi(p)$ classify the set of infinite possible Fock spaces. The transformation of Eq.(18.32) is in fact a pseudo-unitary transformation because all the different Fock spaces corresponding to different profile functions $\phi(p)$ are orthogonal to each other for an infinite volume of space. As an illustration, let us evaluate $\langle 0|\widetilde{0}\rangle$. We have,

$$\langle 0|\widetilde{0}\rangle = \Pi_p \frac{1+\cos(2\phi(p))}{2} \rightarrow 0, \tag{18.35}$$

for a dense set of momenta and provided that $\phi(p)$ differs from being a constant equal to zero (which will be always the case for chiral condensation). The Fock space operators transform like, $\widetilde{\widehat{b}}_{cfs}(\overrightarrow{p}) = S\,\widehat{b}_{cfs}\,S^{-1}$, so that,

$$\begin{bmatrix} \widetilde{\widehat{b}} \\ \widetilde{\widehat{d}}^+ \end{bmatrix}_s = \begin{bmatrix} \cos\phi & \sin\phi M_{ss'} \\ -\sin\phi M^\star_{ss'} & \cos\phi \end{bmatrix} \begin{bmatrix} \widehat{b} \\ \widehat{d}^+ \end{bmatrix}_{s'}. \tag{18.36}$$

If we now consider the fermion field $\Psi_{fc}(\overrightarrow{x})$ as *an inner product between the Hilbert space spanned by the spinors {u,v} and the Fock space spanned by the operators {$\widehat{b}$,$\widehat{d}$}:*

$$\Psi_{fc}(\vec{x}) = \int \frac{d^3p}{(2\pi)^3} \left[u_s(p)\, b_{cfs}(\vec{p}) + v_s(-p)\, d^+_{cfs}(\vec{-p}) \right] e^{i\vec{p}.\vec{x}},$$

and require invariance of $\Psi_{fc}(\vec{x})$ under the Fock space rotations, we are led to a counter-rotation of the spinors u and v,

$$\begin{bmatrix} u_\phi \\ v_\phi \end{bmatrix}_s = \begin{bmatrix} \cos\phi & \sin\phi M^*_{ss'} \\ -\sin\phi M_{ss'} & \cos\phi \end{bmatrix} \begin{bmatrix} u \\ v \end{bmatrix}_{s'}. \tag{18.37}$$

The $\{u_\phi, v_\phi\}$ contain now the information on the angle $\phi(p)$. In what follows we will use the variable $\varphi(p)$ instead of $2\phi(p)$. φ is known as the chiral angle. For

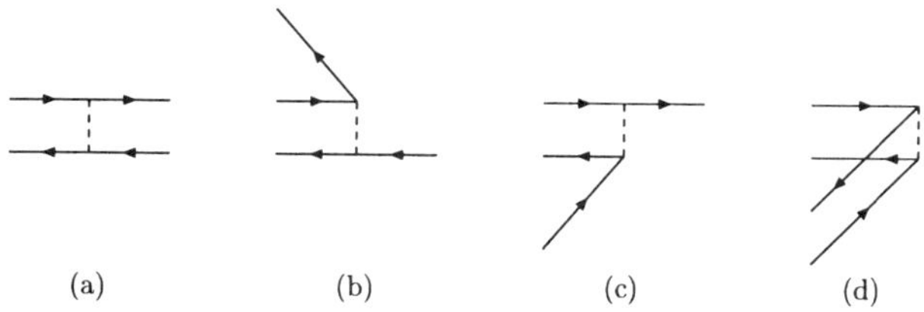

Fig. 18.8. We have four types of microscopic vertices. Diagram (a) is built with vertices $u_s^\dagger(p) \cdot \Gamma \cdot u_{s'}(p')$, (quark scattering) and $v_s(-p) \cdot \Gamma \cdot v_{s'}^\dagger(-p')$, (antiquark scattering). The remaining diagrams b,c,d, contain at least one vertex of either $u_s^\dagger(p) \cdot \Gamma \cdot v_s(p')$, (quark-antiquark creation) or $v_s^\dagger(-p) \cdot \Gamma \cdot u_s(p')$ (quark-antiquark annihilation). With those we can build all kind of ladder diagrams–having portions where we do not have annihilation (exotic scattering) alternating with quark-antiquark annihilation (creation) sectors.

instance, we have for $u(p),\ v(p)$,

$$\begin{cases} u(\vec{p}) = \frac{1}{\sqrt{2}}\left[\sqrt{1+\sin\varphi_p} + \sqrt{1-\sin\varphi_p}\ (\vec{\alpha}\hat{\vec{p}})\right] u(0), \\ v(-\vec{p}) = \frac{1}{\sqrt{2}}\left[\sqrt{1+\sin\varphi_p} - \sqrt{1-\sin\varphi_p}\ (\vec{\alpha}\hat{\vec{p}})\right] v(0), \end{cases} \tag{18.38}$$

denoting $\{\sin(1/2(\pi/2-\varphi(w))) = S_w,\ \cos(1/2(\pi/2-\varphi(w))) = C_w\}$, it is not difficult to see that we have for $\bar{u}(p)\gamma^0\, u(k)$,

$$\bar{u}(p)\gamma^0\, u(k) = C_p\, C_k + \sigma\cdot\hat{p}\,\sigma\cdot\hat{k}\, S_p\, S_k. \tag{18.39}$$

From here it is not difficult to find the quark Feynman projectors and, thus, the quark propagators. The Feynman projectors read ($\hat{p} = \vec{p}/|p|$),

$$\begin{aligned} \Lambda^{\pm}(p) &= \sum_s \begin{bmatrix} u_{[\phi]_s}(p) \otimes u^\dagger_{[\phi]_s}(p) \\ v_{[\phi]_s}(p) \otimes v^\dagger_{[\phi]_s}(p) \end{bmatrix} \\ &= \frac{1}{2}[1 \pm \sin(\varphi(p))\,\beta \pm \cos(\varphi(p))\,\vec{\alpha}\cdot\hat{p}\,]. \end{aligned} \tag{18.40}$$

Rather than proceeding with the usual Feynman rules, it is perhaps simpler to consider an equivalent set of Feynman rules whereby the Feynman projectors of Eq.(18.40) are disassembled, with the spinors u and v migrating towards the vertices. Among other simplifying features, this allows us to easily divide the microscopic quark interactions between those that produce quark-antiquark pair annihilation(creation) and those that do not. The former will be responsible for strong decay of hadrons while the latter will enter in the computation of bare hadronic masses, exotic scattering, like the NN soft core we studied, and alike– see Fig. (18.8).

To find the true vacuum of the Hamiltonians of Eq.(18.31), we must find the profile function $\phi(p)$. To do this, let us consider the transformation $\Psi \rightarrow \exp\left\{-i\alpha^a \frac{T^a}{2}\gamma_5\right\}$. Then,

$$H\left[m_q\right] \rightarrow H[m_q \cos\left(\frac{\alpha^2}{2}\right) - m_q \sin\left(\frac{\alpha^2}{2}\right) i\gamma_5]. \tag{18.41}$$

H is chirally symmetric iff $m_q = 0$. Assume that ϕ exists and construct,

$$Q_5^a = \int d^3x\, \overline{\Psi}\gamma_0\gamma_5\Psi. \tag{18.42}$$

We obtain,

$$\int d^3x \cos(\varphi)\left[\overrightarrow{p}.\overrightarrow{\sigma}\,\widehat{b}^+(p)\widehat{b}(p) + (\widehat{d}^+\widehat{d})\right] + \sin(\varphi)\left[\underbrace{\mu_{ss'}\widehat{b}^+(p)\widehat{d}^+(-p) + (\widehat{db})}\right]. \tag{18.43}$$

In the above expression, the under-braced term corresponds to two anomalous amplitudes: one for the creation of a $\mu_{ss'}\widehat{b}^+(p)\widehat{d}^+(-p)$ quark pair and its annihilation counterpart. Both having $\sin(\varphi)$ as, let's say, a wave function. It is true that for an arbitrary ϕ, $Q_5^a|0> \neq 0$, and because $[Q_5^a, H\,(m_q = 0)]$=0, we have that Q_5^a acting in the vacuum *creates a state*, which will turn out to be the pion ($\mu_{ss'}$ is the spin wave function for S=0, made out of two spin 1/2's). For massless quarks, this state is *degenerate with the vacuum.*

Notice that the Hamiltonians (18.31) can be decomposed as,

$$\begin{aligned}
\widehat{H} &= \widehat{H}_{normal}[\phi] + \widehat{H}_{anomalous}[\phi] \\
\widehat{H}_2\,[normal] &= \int d^3p\, E(p)\left[\widehat{b}^+_{fsc}(\overrightarrow{p})\widehat{b}_{fsc}(\overrightarrow{p}) + \widehat{d}^+_{fsc}(-\overrightarrow{p})\widehat{d}_{fsc}(\overrightarrow{-p})\right] \\
\widehat{H}_2[anomalous] &= \int d^3p\,[A[K]\sin(\varphi) - B[k]\cos(\varphi)] \\
&\quad \times\left[M_{ss'}\widehat{b}^+_{fsc}(\overrightarrow{p})\widehat{d}^+_{fsc}(\overrightarrow{-p}) + h.c.\right].
\end{aligned} \tag{18.44}$$

In general, $\widehat{H}|0> = \widehat{H}_{anomalous}[\phi]|0> \neq 0$. Therefore we must find φ such that,

$$[A[K]\sin(\varphi) - B[k]\cos(\varphi)] = 0. \tag{18.45}$$

This is *the Mass Gap Equation and solving it is tantamount to finding the true vacuum of a concrete Hamiltonian* .

We cannot *simultaneously* get rid of *both* the anomalous terms for $\widehat{H}$ and $\widehat{Q_5^a}$. $\widehat{Q_5^a}$ will remain anomalous: $\widehat{Q_5^a}|0> = |\pi>$. Then, $[H, Q_5] = 0$ implies *not only that π must be massless, but also that the two pion amplitudes–creation and annihilation of a pion must be the same.*

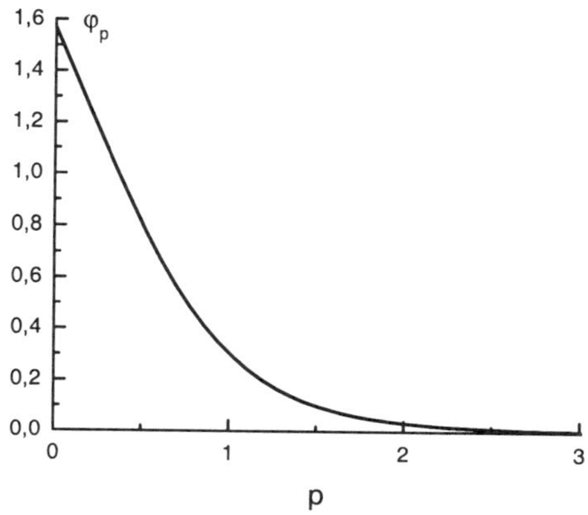

Fig. 18.9. A typical solution to the mass-gap equation: φ. The momentum p is measured in the units of the potential strength K_0.

So far we had no need to specify either the quark current $J_\mu^a(x) = \overline{\psi}(x)\,\vec{\lambda}\,\gamma_\mu \frac{\lambda^a}{2}\,\psi(x)$ or the effective microscopic kernel $K_{\mu\nu}^{ab}(x-y)$. A remarkable feature of the class of Hamiltonians of Eq.(18.31) is the robustness of the results against variations of the form and parameters of the confining potential $V_0(\vec{x})$. Usually a power-like form is adopted,

$$V(|\vec{x}|) = K_0^{\alpha+1}|\vec{x}|^\alpha, \tag{18.46}$$

with $0 \leqslant \alpha \leqslant 2$. The case of the linearly rising potential $\alpha = 1$ (see, for example,[28,29]) is strongly supported by hadronic phenomenology, whereas the marginal case of $\alpha = 2$ — the harmonic oscillator potential — leads to simpler, differential equations and is considered by many authors[24,25,27] because, despite its mathematical simplicity, it already yields a physical picture for the mechanism of dynamical chiral symmetry breaking not unlike the one given by linear confinement. Once the mass gap is solved for a definite microscopic quark interaction, we can look back to the Hamiltonian of Eq.(18.31) to identify several effective quark vertices-see Fig.(18.8)), with the spinors u and v carrying now the information on this interaction through the chiral angle $\varphi(p)$. In Fig.(18.9) we depict a typical profile for the chiral angle φ.

But there are a couple of features that every function $\varphi(p)$ must possess regardless of the actual form of the microscopic potential: it should start at $\pi/2$ for $p = 0$ and tail off to zero when $p \to 0$. The higher the quark current mass, the longer, in p, $\varphi(p)$ will take to approach zero. In fact, when the quark current mass $m_q \to \infty$, $\varphi \to \pi/2$.

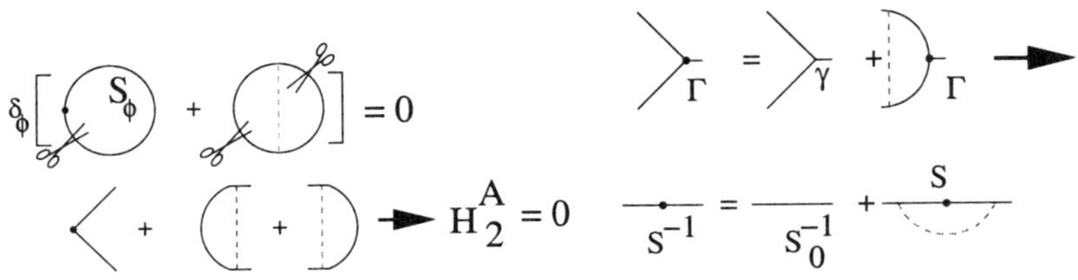

Fig. 18.10. Minimization of the vacuum energy is tantamount to writing the mass gap equation as the condition for the bilinear anomalous term of the Hamiltonian to be zero. The mass gap equation as a Dyson series for the fermion propagator is also depicted. This series can also be obtained using the Ward identity for the vectorial current-see Eq.(18.47).

We have several ways of arriving at the mass gap equation $\widehat{H}_2[anomalous] = 0$. Let us consider two different ways (albeit related) of obtaining the Mass Gap Equation. It can be done either by minimization of $H_0 : \frac{\delta H_0}{\delta \varphi} = 0$ or by the use of Ward identities.

Variation in ϕ is the same as cutting the fermion propagators S_ϕ–see Fig.(18.10), so as to obtain the mass gap equation as the condition for the bilinear anomalous term of the Hamiltonian to be zero. Another way to find the mass gap equation is to use the vectorial Ward identities.

$$
\begin{aligned}
&(p-p^{'})^\mu\, \Gamma_\mu(p,p^{'}) = (p-p^{'})^\mu\, \gamma_\mu + \\
&+i\int \frac{d^4q}{(2\pi)^4} K(q)\Omega\, S(p^{'}+q)\Gamma_\mu(p^{'}+q,p+q)\Omega S(p^{'}+q)i(p-p^{'})^\mu\, \Gamma_\mu(p,p^{'}) \\
&\quad = S^{-1}(p^{'}) - S^{-1}(p)
\end{aligned} \tag{18.47}
$$

18.4.2. *Salpeter Amplitudes*

With the Hamiltonians of Eq. (18.31) all we can do is to sum ladder diagrams–like the one depicted in Fig.(18.11). It turns out that *for any given physical process, summing the corresponding set of ladder diagrams is equivalent to solving the Dyson equation for that process S matrix. In turn, these Dyson equations are none other than the old RGM equations having, for incoming and outgoing clusters (hadrons), wave functions that coincide with the Salpeter amplitudes.*

We have come full circle and have placed the RGM equations–and the good phenomenology thereof–as a particular case of the field theory associated with the Hamiltonians (18.31).

In this chapter, we will only consider those physical processes that can be built by just using diagrams of type (a) and (d) of Fig.(18.8). We can proceed via two identical formalisms: either work in the Dirac Space or work in the Spin

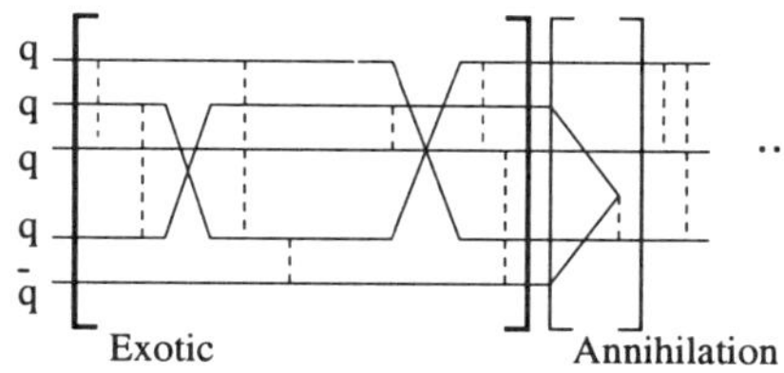

Fig. 18.11. A generic ladder diagram with sections without quark-antiquark annihilation (creation)– exotic scattering– alternating with sections where these annihilations (creations) take place.

Representation. To go from the Dirac representation to the spin representation, it is sufficient to construct the spin wave functions, as it is exemplified in Eq. (18.48)

$$\chi^{++}_{\alpha\beta}(k) = u_{s1;\alpha}(k)\Phi_{s1s2}(k)\overline{v}_{s2;\beta}(-k). \tag{18.48}$$

Reading from the Salpeter equation depicted in Fig.(18.12), we get, in general, for any given meson $\mathcal{M}$,

$$\begin{bmatrix} H^{++} & H^{+-} \\ H^{-+} & H^{--} \end{bmatrix} \begin{bmatrix} \Phi^{+} \\ \Phi^{-} \end{bmatrix} = m_{\mathcal{M}}\,\sigma_3 \begin{bmatrix} \Phi^{+} \\ \Phi^{-} \end{bmatrix} \tag{18.49}$$

with, $\Phi^+ \Leftarrow > \Phi^-$, $m_{\mathcal{M}} \Leftarrow > -m_{\mathcal{M}}$. ($m_{\mathcal{M}} = 0$, $\Phi^+ = \Phi^-$). Therefore, in the space of bare mesons we can rewrite the Hamiltonians of Eq. (18.31) as,

$$H = \sigma_3 \begin{bmatrix} \Phi^{+} \\ \Phi^{-} \end{bmatrix} m_{\mathcal{M}} \left[\Phi^{+},\, \Phi^{-} \right] \sigma_3 + \sigma_3 \begin{bmatrix} \Phi^{-} \\ \Phi^{+} \end{bmatrix} m_{\mathcal{M}} \left[\Phi^{-},\, \Phi^{+} \right] \sigma_3. \tag{18.50}$$

An important feature of the bound–state Eq. (18.49) is the fact that each mesonic state is described by means of a two–component wave function, $\Phi^{\pm}(p)$. The physical meaning of these two components is obvious: $\Phi^{+}(p)$ describes the forward motion, in time, of the mesonic quark–antiquark pair and $\Phi^{-}(p)$ its time backward motion. Strictly speaking, for a given set of quantum numbers, two pairs of amplitudes should be considered — two for the eigenvalue $m_{\mathcal{M}}$ and another two for $-m_{\mathcal{M}}$. As we have already seen, the case of $m_{\mathcal{M}} = 0$ corresponds to $\Phi^+ = \Phi^- = \sin(\varphi(p))$, i.e, the pion.

An interesting case arises when one of the quarks is very heavy. It is the case of the heavy-light mesons. Then the equation (18.49) becomes simpler: instead of two coupled amplitudes we will have just one (and the associated one for $\mathcal{M}_{hl} \rightarrow -\mathcal{M}_{hl}$. This is simple to understand: Due to the instantaneous interaction, both particles in the meson move forward and backward in time *in unison*. This leads us to an immediate conclusion that the heavy–light meson, with the static–quark Zitterbewegung suppressed by the heavy quark mass, can be described, when this heavy mass approaches infinity, with a one–component wave function ($H^{++} =$

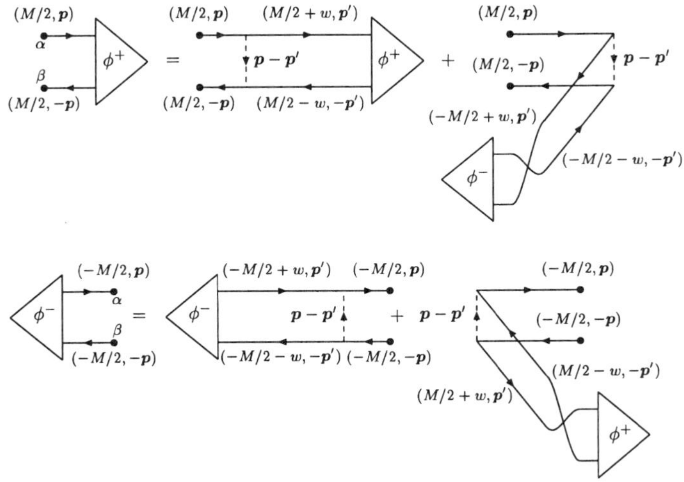

Fig. 18.12. Salpeter equation for mesons. Notice that the application of the microscopic quark interactions a and d of Fig.(18.8)-already yields two coupled channel equations. Further insertions of the microscopic interactions of Fig.(18.8)–b and c–will produce extra couplings of the meson state with two meson intermediate channels.

$-H^{--}$),

$$\begin{bmatrix} H & 0 \\ 0 & -H \end{bmatrix} \begin{bmatrix} \Phi \\ \Phi \end{bmatrix} = m_{\mathcal{M}}\, \sigma_3 \begin{bmatrix} \Phi \\ \Phi \end{bmatrix}. \tag{18.51}$$

In what follows we will choose the quark current to be $J^a_\mu(x) = \overline{\psi}(x)\ \vec{\lambda}\ \gamma_0 \frac{\lambda^a}{2}\ \psi(x)$. Other chirally symmetric currents are possible but the qualitative picture is not expected to change. Using Eq.(18.39) and the fact that for an infinitely heavy current quark mass M, we have, $\bar{u}_M \gamma_0\ u_M \to 1$, it is trivial to write the Salpeter equation (in fact a Schrödinger-like equation) for the Heavy-light mesonic wave function $\psi(\vec{p})$,

$$E_p \psi(\vec{p}) + \int \frac{d^3k}{(2\pi)^3} V(\vec{p}-\vec{k}) \left[C_p C_k + (\vec{\sigma}\hat{\vec{p}})(\vec{\sigma}\hat{\vec{k}}) S_p S_k \right] \psi(\vec{k}) = E \psi(\vec{p}), \tag{18.52}$$

with

$$C_w = \cos\frac{1}{2}\left(\frac{\pi}{2} - \varphi_w\right), \quad S_w = \sin\frac{1}{2}\left(\frac{\pi}{2} - \varphi_w\right). \tag{18.53}$$

Now consider $\psi'(\vec{p}) = (\vec{\sigma}\hat{\vec{p}})\psi(\vec{p})$ which, by construction, possesses an opposite parity to $\psi(\vec{p})$. The Schrödinger equation for $\psi'(\vec{p})$ can be easily derived by multiplying Eq. (18.52) by $(\vec{\sigma}\hat{\vec{p}})$ from the left,

$$E_p\psi'(\vec{p}) + \int \frac{d^3k}{(2\pi)^3} V(\vec{p}-\vec{k}) \left[S_pS_k + (\vec{\sigma}\hat{\vec{p}})(\vec{\sigma}\hat{\vec{k}})C_pC_k\right] \psi'(\vec{k}) = E\psi'(\vec{p}), \tag{18.54}$$

Notice that, for large momenta, we must have $C_p = S_p \to 1/\sqrt{2}$, $\varphi_p \to 0$. Therefore, for the cases where we have a large average of the light quark momentum, which is the case with highly excited states, the bound–state Eqs. (18.52) and (18.54) must go to

$$E_p\psi(\vec{p}) + \frac{1}{2}\int \frac{d^3k}{(2\pi)^3} V(\vec{p}-\vec{k}) \left[1 + (\vec{\sigma}\hat{\vec{p}})(\vec{\sigma}\hat{\vec{k}})\right] \psi(\vec{k}) = E\psi(\vec{p}), \tag{18.55}$$

so that the states $\psi(\vec{p})$ and $(\vec{\sigma}\hat{\vec{p}})\psi(\vec{p})$ become degenerate. In other words, the more highly excited a given parity heavy-light state is, the more degenerate it will be with its opposite parity partner. This is the scenario of progressive restoration of Chiral Symmetry.

And yet, for ground states, with moderate average momentum for the light quark, such a degeneracy does not exist, pointing out to the existence of an effective chiral breaking (scalar like) interaction. But we started with a full chiral symmetric interaction and therefore we face a puzzle. The answer to this puzzle was published in.[30] Here we merely restate the conclusions of that paper: "For small relative momenta the chiral regime takes place. Chiral symmetry breaking — spontaneous or explicit — plays a dominating role, the chiral angle being close to $\frac{\pi}{2}$. As a result, ... the effective inter-quark interaction *becomes purely scalar (even in the absence of any microscopic scalar force)*. For light quarks, the dressed–quark dispersive law E_p differs drastically from the free–particle form of $\sqrt{\vec{p}^2 + m^2}$, it may even become negative at small momenta — a necessary feature to have the light (massless) chiral pion. This chiral regime — where the effective interaction is purely scalar — holds below some effective dynamically generated low–energy scale, which we call the BCS scale $\Lambda_{\rm BCS}$, that is, holds for the mean relative inter-quark momentum $p \lesssim \Lambda_{\rm BCS}$. The BCS scale gives a measure of chiral symmetry breaking in the low–energy domain of the theory and it is closely related to the scale of confinement or, equivalently, to $\Lambda_{\rm QCD}$: $\Lambda_{\rm BCS} \simeq \Lambda_{\rm QCD}$. As soon as the mass of the quark–antiquark state grows, as does the mean relative inter-quark momentum, the vectorial part of the inter-quark interaction becomes more important. Finally, the chiral angle becomes negligibly small and the interaction acquires a purely vectorial nature. Thus, we arrive at the chiral symmetry restoration in the spectrum discussed in the literature".[31–33]

Now, for the Heavy-Light mesons, we can go a step further and allow for the existence of quark-antiquark annihilation processes. These processes will be

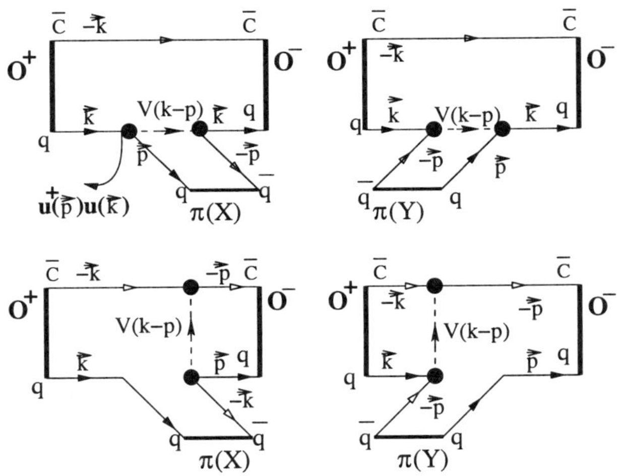

Fig. 18.13. The four diagrams contributing to $\langle \bar{D}'\pi|V|\bar{D}\rangle$.

responsible, for instance, for the Heavy-Light coupling to pions. Let us exemplify this with the case of $\langle \bar{D}'\pi|V|\bar{D}\rangle$.

From the diagrams of Fig.(18.13) and using the Salpeter equations (18.52) and (18.54), we can derive[34] the Goldberger-Treiman relation

$$\frac{1}{2}(E_{\bar{D}'} - E_{\bar{D}})G_A = f_\pi g_{nn'\pi}, \tag{18.56}$$

Notice that had we dropped the negative energy components of the pion wave function (as it is done in naive quark models) we would be:

1-violating the Goldberger–Treiman relation by 50%,

2-but also the two diagrams, A and B (with the interaction coupled solely to the light quarks), which must add to zero for pions at rest, would have now survived to give a result incompatible with G_A.

It should be also emphasized that the Goldberger–Treiman relation can be realized *only* by the use of the Salpeter solutions Ψ and Ψ' of (18.52, 18.54), which in turn contain the physics of progressive effective restoration of chiral symmetry for higher and higher excitations.

Finally, because $E_{\bar{M}'} \to E_{\bar{M}}$ and $G_A \to 1$ we conclude that the pion coupling to excited mesons decreases as the excitation number of the heavy–light mesons increases thus providing an explicit pattern of the Goldstone boson decoupling from excited hadrons.

18.4.3. *Conclusion*

The physics of Strong interactions, i.e the physics of quarks and their strong interactions, covers a wide range of apparently diverse reactions encompassing, among others, masses, scattering and decay. Many non-perturbative effects– equally strong–concur to obfuscate the dynamics of such processes. One must be able to disentangle such processes knowing what forces come from prima facie overlaps, from virtual channels–notarization–or directly from the microscopic quark forces.

In the absence of first principles derivation of such microscopic forces from QCD one is forced to simulate them through potentials.

Fortunately, Nature has come to our help and hadronic physics possesses an extra symmetry–the Chiral symmetry–which allows us, for the low energy domain, to gloss over the details of QCD.

It happens that a deceivably simple looking Nambu, Jonas-Lasinio class of Hamiltonians endowed with a confining potential, can, certainly qualitatively and, to a certain amazement, also many times quantitatively, describe the physics of low energy hadronic phenomena.

Soft core repulsions, Goldberger-Treiman relations, scalar decays, pion-ρ mass difference when compared with the N-Δ mass difference, Adler zeroes, the issue of scalar versus vectorial confinement, and many other phenomena, all seem to derive from an Hamiltonian of the type we have described in this chapter.

Any theory with ambitions to describe the realm of low energy hadronic physics must describe the whole set of hadronic phenomena as economically as possible. Models that, by construction, cannot be exported from one area of low energy hadronic physics to another area are of little use. This is specially true for any calculation that **does not take** into consideration the negative-energy component of the pion amplitude. Such a calculation cannot be made to respect the low energy theorems of hadronic physics and, thus, has a limited scope.

The mathematical methods used to treat these problems are identical to those used in the field of Condensed Matter Physics to treat similar non-perturbative interactions. We can say the the Physics of Chiral symmetry is the BCS treatment of $^3P_0$ quark-antiquark Cooper-like pairs.

A lot remains to be understood– mainly the physics of gluons–but I fear it will take quite some time before we have a sufficiently good machinery (as we have with quarks), able to deal with such elusive objects.

References

1. M. Gell-Mann, A schematic model of baryons and mesons, *Phys. Lett.* **8**, 214 (1964).
2. G. Zweig, *CERN Preprint TH* **401**, 412 (1964).
3. M. A. Tuve, N.P. Heydenburg and L. R.Hafstadt, The Scattering of Protons by Protons, *Phys. Rev.* **50**, 806 (1936).
4. N. Kemmer, The Charge-dependence of Nuclear Forces, *Proc. Camb. Phil. Soc.* **34**, 354 (1938).
5. B. Cassen and E. U. Condon, On Nuclear Forces , *Phys. Rev.* **50**, 846 (1936).
6. M. Gell-Mann, Isotopic spin and new unstable particles, *Phys. Rev.* **92**, 833 (1953).
7. T. Nakano and K. Nishijima, Charge Independence for V-particles, *Progr. Theor. Phys. (Kyoto)* **10**, 581 (1953).
8. in Fundamentals of Quark models, *Proc. of the seventh Scottish Universities Summer School in Physics 1976*, ed. I. Barbour and A. Davies.
9. M Y. Han and Y. Nambu, Three-Triplet Model with Double SU(3) Symmetry , *Phys. Rev.* **139**, B1006, 127 (1965).
10. H. J. Lipkin, Triality, exotics and the dynamical basis of the quark model, *Phys. Lett. B* **45**, 267 (1973).
11. J.E.F.T. Ribeiro, D. Phil. Thesis, Oxford (1978); *ibid.* Microscopic calculation of the repulsive core in the elastic nucleon-nucleon scattering, *Z. Phys. C.* **5**, 27 (1980).
12. V. Neudatchin, I.T. Obukhovsky, V. Kukulin, F. Golovanova, Attractive potential with forbidden states for the N-N interaction, *Phys. Rev. C* **11**, 128 (1975).
13. J. Wheeler, Molecular Viewpoints in Nuclear Structure, *Phys. Rev.* **52**, 1083 (1937); On the Mathematical Description of Light Nuclei by the Method of Resonating Group Structure, *Phys. Rev.* **52**, 1107 (1937).
14. A.Rujula, H. Georgi, S. Glashow, Hadron masses in a gauge theory, *Phys. Rev.* **D12**, 147 (1975).
15. S. Okubo, Phi-meson and unitary symmetry model, *Phys. Lett.* **5**, 105 (1963); J. Iizuka, K. Okada and O. Shito, Systematics and Phenomenology of Boson Mass Levels. III, *Prog. Theor. Phys.* **35**, 1061 (1966).
16. J.E.F.T. Ribeiro, Matrix elements of the exchange operator for arbitrary-angular-momentum two-meson states , *Phys. Rev. D* **25**, 2406 (1982).
17. E. van Beveren, Recoupling matrix elements and decay, *Z. Phys. C* **17**, 135 (1983).
18. E. van Beveren, T. A. Rijken, K. Metzeger, C. Dullemond, G. Rupp, J.E.F.T. Ribeiro, A low lying scalar meson nonet in a unitarized meson model, *Z. Phys. C* **30**, 615 (1986).
19. K. Hagiwara et al, Review of Particle Properties, *Phys. Rev. D* **66**, 010001 (2002); E. M. Aitala et al, E791 Collaboration, Dalitz Plot Analysis of the Decay $D^+ \to K^-\pi^+\pi^+$ and Indication of a Low-Mass Scalar Kπ Resonance, *Phys. Rev. Lett.* **89**, 121801 (2002).
20. M. Gell-Mann, R. J. Oakes, and B. Renner, Behavior of Current Divergences under $SU3 \times SU3$, *Phys. Rev.* **175**, 2195 (1968).
21. M. L. Goldberger and S. B. Treiman, Form Factors in beta Decay and muon Capture , *Phys. Rev.* **111**, 354 (1958).
22. S. L. Adler, Consistency Conditions on the Strong Interactions Implied by a Partially Conserved Axial-Vector Current , *Phys. Rev.* **137**, B1022 (1965).

23. S. Weinberg, Pion Scattering Lengths, *Phys. Rev. Lett.* **17**, 616 (1966).
24. A. Amer, A. Le Yaouanc, L. Oliver, O. Pene, and J.-C. Raynal, Instability of the Chiral-Invariant Vacuum for a Confining Potential , *Phys. Rev. Lett.* **50**, 87 (1983); A. Le Yaouanc, L. Oliver, O. Pene, and J.-C. Raynal, Chiral non-invariant solutions of the gap equation for a confining potential, *Phys. Lett. B* **134**, 249 (1984); *ibid.*, Spontaneous breaking of chiral symmetry for confining potentials, *Phys. Rev. D* **29**, 1233 (1984); *ibid.*, Quark model of light mesons with dynamically broken chiral symmetry, *Phys. Rev. D* **31**, 137 (1985).
25. P. Bicudo and J.E.F.T. Ribeiro, Current-quark model in a 3P0 condensed vacuum , *Phys. Rev. D* **42**, 1611 (1990); *ibid.*, π, K, ρ, ϕ as qq¯ bound states of the Salpeter equation in a 3P0 condensed vacuum , 1625 (1990); *ibid.*, Generalized resonating-group-method equations in a $^3P_0$ $q\bar{q}$ condensed vacuum. The ϕ and ρ resonances , 1635 (1990).
26. A. P. Szczepaniak, E. S. Swanson, Coulomb gauge QCD, confinement, and the constituent representation, *Phys. Rev. D* **65**, 025012 (2002).
27. P. Bicudo, N. Brambilla, J.E.F.T. Ribeiro, and A. Vairo, Confinement and chiral symmetry breaking in heavy-light quark systems, *Phys. Lett. B* **442**, 349 (1998).
28. S. L. Adler and A. C. Davis, Chiral symmetry breaking in Coulomb gauge QCD, *Nucl. Phys. B* **244**, 469 (1984); Y. L. Kalinovsky, L. Kaschluhn, and V. N. Pervushin, A new QCD inspired version of the Nambu-Jona-Lasinio model, *Phys. Lett. B* **231**, 288 (1989); P. Bicudo, J.E.F.T. Ribeiro, and J. Rodrigues, K-N S channel I=1,0 phase shifts as a direct measure of chiral condensation , *Phys. Rev. C* **52**, 2144 (1995); R. Horvat, D. Kekez, D. Palle, and D. Klabucar, Bilocal effective theory with the instantaneous funnel interaction and its renormalization, *Z. Phys. C* **68**, 303 (1995); N. Brambilla and A. Vairo, Non-perturbative dynamics of the heavy-light quark system in the non-recoil limit, *Phys. Lett. B* **407**, 167 (1997); Yu. A. Simonov and J. A. Tjon, String formation and chiral symmetry breaking in the heavy-light quark-antiquark system in QCD, *Phys. Rev. D* **62**, 014501 (2000); F. J. Llanes-Estrada and S. R. Cotanch, Meson Structure in a Relativistic Many-Body Approach , *Phys. Rev. Lett.* **84**, 1102 (2000); N. Ligterink, E. S. Swanson, Coulomb gauge model of mesons, *Phys. Rev. C* **69**, 025204 (2004).
29. Yu. A. Simonov, Theory of light quarks in a confining vacuum, *Yad. Fiz.* **60**, 2252 (1997) [*Phys. Atom. Nucl.* **60**, 2069 (1997)].
30. Yu. S. Kalashnikova, A. V. Nefediev, and J.E.F.T. Ribeiro, Confinement and parity doubling in heavy-light mesons, *Phys. Rev. D* **72**, 034020 (2005).
31. E. S. Swanson, Parity doubling in the meson spectrum, *Phys. Lett. B* **582**, 167 (2004).
32. L. Ya. Glozman, Parity doublets and chiral symmetry restoration in baryon spectrum, *Phys. Lett. B* **475**, 329 (2000); it ibidem,Chiral multiplets versus parity doublets in highly excited baryons, *Phys. Rev. D* **65**, 016006 (2002).
33. M. Nowak, M. Rho, and I. Zahed, Chiral Doubling of Heavy-Light Hadrons: BaBar 2317MeV and CLEO 2463 MeV Discoveries, *Acta Phys. Polon. B* **35**, 2377 (2004)
34. A. Szczepaniak, A. Nefediev and J.E.F.T.Ribeiro, Microscopic derivation of the pion coupling to heavy-light mesons., *hep-ph/0610430* .

PART 4

Quantum Entanglement

Chapter 19

Introduction to entanglement and applications to the simulation of many-body quantum systems

M. Almeida

ICFO-Institut de Ciencies Fotoniques, Mediterranean Technology Park, 08860 Castelldefels (Barcelona), Spain

Y. Omar

SQIG, Instituto de Telecomunicações, P-1049-001 Lisbon and CEMAPRE, ISEG, Technical University of Lisbon, P-1200-781 Lisbon, Portugal

V. Rocha Vieira

CFIF and Department of Physics, IST, Technical University of Lisbon, 1049-001 Lisbon, Portugal

We introduce and define the concept of quantum entanglement and its properties. We concentrate our discussion in the case of bipartite entanglement in a pure state, the Schmidt decomposition and the quantification of entanglement by the von Neumann entropy. We then review the Matrix Product State and Projected Entangled-Pair States representations of entangled states and their application to the classical simulation of many-body quantum systems, showing how these novel techniques allow us to extend the study of low-energy pure states to systems with periodic boundary conditions, as well as to obtain low-energy states and simulate the time evolution of d-dimensional systems at zero temperature and one-dimensional systems at finite temperature.

Contents

19.1. Introduction

The *entanglement* of states is a purely quantum feature associated to composite quantum systems (i.e. systems composed of several quantum objects which are denoted subsystems). We say there is entanglement when we have a superposition of different states of a composite system. Entangled states exhibit correlations that have no classical equivalent, and thus entanglement can also be thought of as a resource for quantum information processing. Moreover, entanglement is also related to quantum non-locality and the violation of Bell inequalities.

19.2. EPR vs. Bell

Let us start by analyzing an example, following David Bohm's version[1] of the Einstein-Podolsky-Rosen (EPR) problem.[2] Consider the creation of a electron-positron pair. Both particles have spin $1/2$ and we label them A and B respectively. Conservation laws impose that the previously inexistent total spin of the system remains zero, so the spin of the particles must be described by the (entangled) singlet state, the only one invariant under rotations (or basis transformations):

$$|\psi\rangle_{AB} = \frac{1}{\sqrt{2}} \left(|\uparrow\rangle_A |\downarrow\rangle_B - |\downarrow\rangle_A |\uparrow\rangle_B \right), \tag{19.1}$$

where $|\uparrow\rangle$ and $|\downarrow\rangle$ represent the two opposite spin components along some given direction. Recall that $\{|\uparrow\rangle_i, |\downarrow\rangle_i\}$ constitutes an orthonormal basis of the 2-dimensional Hilbert space.

Should we allow the two particles to move apart freely, say the electron towards Alice and the positron towards Bob, they will remain in the same spin singlet state. Imagine that then Alice performs a projective measurement on the spin of subsystem A along some direction θ and obtains $+\frac{1}{2}$, the eigenvalue associated to the state $|\uparrow\rangle_A$. Then, according to measurement postulate of quantum mechanics, the state $|\psi\rangle_{AB}$ is projected onto:

$$|\psi'\rangle_{AB} = |\uparrow\rangle_A |\downarrow\rangle_B \tag{19.2}$$

We can immediately see that a projective measurement of the spin of particle B along the same direction θ will yield $-\frac{1}{2}$, the eigenvalue associated to the state

$|\downarrow\rangle_B$, with probability one, and this is true regardless of when this second measurement takes place, inside or outside of the future light cone of the first. Moreover, note that there is an infinity of possible directions θ to choose from, and the above correlation will hold for each of them. This is impossible to simulate with any (finite) classical system: we say our entangled system exhibits perfect *quantum correlations*. But the most surprising is this non-local character they have, a feature difficult to understand, if not to accept (then again, just as is the projection due to the measurement). Note, though, that such correlations cannot be used for communication, so there is no superluminal transmission of information.

The presence of *non-locality* in quantum theory was first hinted at by Albert Einstein in the 1927 Solvay conference. In 1935, he raised this issue again, now together with Boris Podolsky and Nathan Rosen, using the original version of the above example (in terms of continuous position and momentum variables) to claim that the presence of this unacceptable[a] feature in quantum mechanics implied that the theory was an incomplete description of reality.[2] In the very same year, Erwin Schrödinger identified entanglement as "*the characteristic trait of quantum mechanics, the one that enforces its entire departure from classical lines of thought*"[3] (see also Ref. 4, and Ref. 5 for a bet about the true origin of the term *entanglement*). It was the beginning of a long and rather philosophical debate about the complete and non-local characters of quantum theory. Almost thirty years later, John Bell finally brought a quantitative tone to it and showed that theories based on EPR's assumptions of *locality* and *elements of physical reality* would have to obey some constraints on their statistical predictions, and proved that quantum mechanics violated such constraints,[6,7] thus confirming the theory's intrinsic non-local character. These constraints were later to be called *Bell inequalities* and generalized in several ways.[8] The first direct experimental evidence that quantum mechanics violates Bell inequalities was obtained in 1972,[9] despite the existence of possible loopholes. Since then, many other experiments have confirmed this result and solved some loopholes (e.g. see Ref. 10 and 11), although some rather conspirative ones could still be imagined.[8]

19.3. Definition and Properties

In the early nineties, it was realized that entanglement could be used as a *resource* for quite novel and unique information processing tasks, in particular for unconditionally secure quantum cryptographic key exchange,[12] dense coding of information[13] and quantum teleportation.[14] Moreover, entanglement plays an essential

[a]In their opinion...

role in the speed-up over classical computation by quantum algorithms operating on pure states[15] (although it should be noted that, in general, entanglement is not necessary for quantum computational speed-up[16–18]). Furthermore, entanglement is an important resource for quantum error correction[19] and quantum communication over long distances,[20] and can be used to improve frequency standards,[21] as well as to enhance positioning and clock synchronization.[22] Such results, including their respective implementations, contributed significantly to the initial development of quantum information. And only by then did entanglement start to be seen and studied as a full feature of quantum mechanics. Before that, entanglement was rarely explicitly taught or mentioned in textbooks.

Mathematically, for pure states, we can define entanglement as follows:

Definition 19.1 (Entanglement). *Given quantum systems $S_1, S_2, \ldots, S_n$ with underlying Hilbert spaces $\mathcal{H}_1, \mathcal{H}_2, \ldots, \mathcal{H}_n$ respectively, and the total composite system S with state space $\mathcal{H} = \mathcal{H}_1 \otimes \mathcal{H}_2 \otimes \ldots \otimes \mathcal{H}_n$, we say a state $|\chi\rangle \in \mathcal{H}$ is* entangled *if it cannot be described by a vector of the form:*

$$|\chi\rangle = |\chi_1\rangle \otimes |\chi_2\rangle \otimes \ldots \otimes |\chi_n\rangle \tag{19.3}$$

where each ket $|\chi_j\rangle \in \mathcal{H}_j$ ($j = 1, 2, \ldots, n$), i.e. a state is entangled if it cannot be written as a product state.

Product states are also called *separable*, whereas entangled states are called *non-separable*. Entanglement emerges when we consider superpositions of states in a composite system (technically, the tensor product plays here a key role). We will say there is entanglement between the respective subsystems if a measurement on the state of one of these subsystems is not independent from measurements on the states of the others. An entangled state contains more information than just the information of the states of each entangled party. It contains non-local correlations, such as the ones discussed in section 19.2 for state (19.1), specific to quantum physics, impossible to simulate in any classical system. These quantum correlations, until recently unexplored, offer us a rich and important resource for quantum information transmission and processing, as it was already mentioned. From a formal perspective, for pure states, we can identify non-separable states as the ones containing such correlations, as it was proven that they always violate Bell inequalities.[23] [b] A *caveat* is necessary, though: this is true for dis-

[b]Note that some authors use the term *non-locality* as a synonym of *entanglement*, although the former is not a purely quantum feature as the latter: in fact, one could conceive theories that are more non-local than quantum mechanics (see Ref. 24 and references therein). If those theories are physical or why is Nature not more non-local remain open questions!

tinguishable systems, but in indistinguishable ones these correlations are not accessible to us.[25] Just think of the example of two indistinguishable electrons. When the spin of the system is in the singlet state, it is described by the vector $|\kappa\rangle = \frac{1}{\sqrt{2}}(|\uparrow\rangle_1|\downarrow\rangle_2 - |\downarrow\rangle_1|\uparrow\rangle_2)$, formally identical to (19.1). Yet, there is an important difference: in the case of the indistinguishable electrons, the labels in the vector ($|\kappa\rangle$) describing their spin have no physical meaning. In an indistinguishable situation, we cannot address the particles and thus make use of the quantum correlations present in the system. For that, we need to be able to distinguish the particles, for instance when they are in separate locations.[c]

So far, we have only mentioned entanglement in bipartite pure states. This simplest case is where we best understand entanglement, and in particular how to quantify it. This case covers most of the situations that involve entanglement in the remainder of this chapter and will now be analyzed in more detail.

Given a vector $|\phi\rangle_{AB} \in \mathcal{H}_{AB} = \mathcal{H}_A \otimes \mathcal{H}_B$ describing a bipartite pure state, it can always be written in the following standard form, known as the *Schmidt decomposition* (e.g. see Ref. 27):

$$|\phi\rangle_{AB} = \sum_{i=1}^{\Delta} g_i |\zeta_i\rangle_A |\eta_i\rangle_B, \tag{19.4}$$

where $\{|\zeta_i\rangle_A\}$ and $\{|\eta_i\rangle_B\}$ are sets of orthonormal vectors, g_i are non-negative real numbers such that $\sum_{i=1}^{\Delta} g_i^2 = 1$ and $\Delta \equiv \min\{\dim \mathcal{H}_A, \dim \mathcal{H}_B\}$. Note that if two or more of the coefficients g_i are non-zero, then we can say $|\phi_{AB}\rangle$ is entangled, otherwise it is separable. Note also that it is in general not possible to find the Schmidt decomposition for systems involving more than two parties.[28]

If we now write our bipartite state in the form of a density operator:

$$\hat{\rho}_{AB} = |\phi\rangle_{AB}\, {}_{AB}\langle\phi|, \tag{19.5}$$

and trace it over all states of subsystem B, we obtain the reduced density operator for subsystem A:

$$\hat{\rho}_A = \mathrm{tr}_B\,(\hat{\rho}_{AB}) = \sum_{i=1}^{\Delta} g_i^2 |\zeta_i\rangle_A\, {}_A\langle\zeta_i|. \tag{19.6}$$

Similarly, we can get:

$$\hat{\rho}_B = \mathrm{tr}_A\,(\hat{\rho}_{AB}) = \sum_{i=1}^{\Delta} g_i^2 |\eta_i\rangle_B\, {}_B\langle\eta_i|. \tag{19.7}$$

We see that the reduced density operators of both subsystems have the same positive spectrum. Moreover, it is important to note that in the case of entanglement

[c]Still, the indistinguishability effects can be very useful for quantum information processing.[26]

$\hat{\rho}_A$ and $\hat{\rho}_B$ represent mixed states, whereas if the system is separable they represent pure states. We can now interpret the entanglement between parties A and B in the following way. Before we measure A (for instance), the uncertainty in subsystem B is given by the von Neumann entropy $S(\hat{\rho}_B) = -\mathrm{tr}(\hat{\rho}_B \log_2 \hat{\rho}_B)$ (this important quantity is defined in Appendix A). But after the measurement in the Schmidt basis there is no more uncertainty: if we obtain $|\zeta_i\rangle_A$ for A, then we can be sure that B is in state $|\eta_i\rangle_B$. Thus, the information gained is $S(\hat{\rho}_A) = S(\hat{\rho}_B)$. We have then two interpretations for the von Neumann entropy: it gives us the quantum uncertainty before a measurement or the information gained after a measurement.

The quantum correlations in entangled systems are not always the same, they actually depend on the particular non-local state the system happens to be in. In some cases these correlations are weaker, in some others they are stronger: some quantum systems are more entangled than others. How can we then quantify entanglement? This is in general not an easy problem, but for pure bipartite states we can use the degree of mixedness in the reduced density matrices of both parties — as given by the von Neumann entropy — as a measure of entanglement. When $\hat{\rho}_A$ and $\hat{\rho}_B$ are maximally mixed, we say the system is in a *maximally entangled state* and carries $S(\hat{\rho}_A) = S(\hat{\rho}_B)$ *e-bits* of entanglement. In particular, we can have at most one e-bit between two entangled quantum bits. [d] Should the system be in a separable state, it will have 0 e-bits of entanglement, i.e. no quantum correlations at all. Only for bipartite pure states do we have a unique measure of entanglement, the von Neumann entropy. Other measures can be defined for some specific tasks or situations, as we shall see in the following chapters. In fact, it was recently shown that there exists no unique measure of entanglement for mixed states.[29] Note also the interesting and recent attempts to define measures of entanglement for indistinguishable particles, taking into account the symmetrization terms (see Ref. 30, 31 and references therein).

To create entanglement between two parties A and B it is necessary to act on the total system. Quantum correlations cannot be created by local actions, i.e. operations acting only on subsystem A and on subsystem B. If the system is in a separable state, e.g. described by equation (19.4) with only one of the coefficients g_i different from zero, we see that local unitary operations $\hat{U} = \hat{U}_A \otimes \hat{U}_B$, as well as local measurements, cannot increase the number of terms in the Schmidt vector describing the state of the system, even if these local actions are supplemented with classical communication between A and B. In fact, even if the system already had some entanglement, such actions could not create new quantum

[d] A *quantum bit* or *qubit* is simply a quantum two-level system. It is the quantum analogue of a classical bit of information and can be defined as the fundamental unit of quantum information.

correlations in it. We usually say that entanglement cannot increase under *local operations and classical communication (LOCC)*.[32] In fact, we can even define an entangled state as a state that cannot be prepared out of a separable one using only LOCC. Note also that entanglement is preserved under unitary (reversible) operations, but can be decreased by local actions such as measurements, i.e. we can *consume* the resource. In particular, we can use maximally entangled states to increase the non-locality of other states, a scheme known as *entanglement concentration*.[33] Note that it is also possible to teleport entanglement from one set of particles to another.[34]

It should be noted that many questions about entanglement remain open. When we are considering entanglement between more than two parties, there are different senses in which some pure states can be more entangled than others, and it is not clear how this can be quantified.[35] In mixed states, entanglement is even less well understood. One way to define it is just to generalize definition 19.1 of non-separability to density operators. But then, even the simple task of deciding if a given state is entangled or not is in general not trivial, as a density operator has infinitely many decompositions and *a priori* all of them would have to be checked for non-separability. Necessary conditions for separability have been found, though, which are also sufficient for up to three parties.[36,37] Note also that there are some entangled (i.e. non-separable) mixed states — known as *Werner states* — that do not violate Bell inequalities (for single projective measurements).[38] For more details on some of these issues, including different measures of entanglement, see Ref. 39. The better understanding of entanglement, both from a qualitative and quantitative point of view, is currently an active and still developing research area, motivated not only by the novel and important applications entanglement can offer us, but also by the attempt to clarify a fundamental aspect of quantum mechanics. Furthermore, the role of entanglement in condensed matter is also a very promising research direction. In the remainder of this chapter, we will show how it can help us obtain more efficient simulations of many-body systems, and in the following chapters we will review the presence and role of quantum correlations in quantum phase transitions as well as in macroscopic systems at finite temperature.

Finally, a word about experiments, in particular about sources of entangled particles. The use of optical systems for quantum information tasks is nowadays very popular, as photons in fibres or in the air allow us to perform experiments over large distances (i.e. several kilometers). A common source for entangled pairs of photons uses the *parametric down-conversion* technique, where a single photon impinging on a non-linear crystal produces two photons entangled in polarization.[40] It is also possible to entangle pairs of atoms in cavities[41] and pairs of ions in electromagnetic traps[42] (for more details about all these techniques, see Ref. 43 or 44). Recently, sources for entangled pairs of electrons were also obtained.[45,46] Yet, despite the rapid progress of the experimental techniques, the best we can currently generate is four entangled photons[47] or eight entangled ions.[48]

19.4. Representations of Quantum States

So far, we have essentially discussed entanglement between two qubits. Let us now look at more general quantum systems and how their states can be represented.

19.4.1. *The Computational-Basis Representation*

Consider a pure quantum state $|\psi\rangle \in \mathcal{H}_d^{\otimes N}$ characterizing a system of N particles in a d-dimensional Hilbert space. If we want to represent this state, the most natural choice is perhaps to write it as the linear combination of the complete set of product states that compose the canonical basis (also called *computational basis*):

$$|\psi\rangle = \sum_{i_1,\cdots,i_N=1}^{d} c_{i_1\cdots i_N} |i_1, i_2, \cdots, i_N\rangle . \tag{19.8}$$

In general, d^N complex coefficients $c_{i_1\cdots i_N}$ are required for a complete description of the state.

19.4.2. *The Matrix Product State Representation*

Alternatively, we can use the *Matrix Product State* (MPS) representation,[49] where each coefficient $c_{i_1\cdots i_N}$ is replaced by the trace of the product of matrices $A_{i_k}^{[k]}$ with dimension $D_k \times D_{k+1}$:

$$|\psi\rangle = \sum_{i_1,\cdots,i_N=1}^{d} \mathrm{tr}\left[A_{i_1}^{[1]} A_{i_2}^{[2]} \cdots A_{i_N}^{[N]}\right] |i_1, i_2, \cdots, i_N\rangle . \tag{19.9}$$

Defining $D = \max_k D_k$, any pure quantum state has a (generally non-unique and unnormalized) MPS representation for sufficiently large D.[50] In order to give a more precise idea of MPS we present the following simple example.

Example 19.1 (GHZ state in the MPS representation). *Consider a system of 3 particles of spin $\frac{1}{2}$ in the Greenberger-Horne-Zeilinger (GHZ) tripartite-entangled state:*[51]

$$|GHZ\rangle = \frac{1}{\sqrt{2}} \left(|000\rangle + |111\rangle \right) . \tag{19.10}$$

This is a permutation invariant state and therefore we can choose a representation where the matrices A in (19.9) *do not depend on the particle label k but only on the spin state i_k: $A_{i_k}^{[k]} \equiv A_{i_k}$ (see Ref. 49). A possible (unnormalized) MPS representation of the GHZ state* (19.10) *is:*

$$|GHZ\rangle = \sum_{i_1,i_2,i_3=0}^{1} \mathrm{tr}\left(A_{i_1} A_{i_2} A_{i_3} \right) |i_1 i_2 i_3\rangle , \tag{19.11}$$

with

$$A_0 = \begin{pmatrix} 1 & 0 \\ 0 & 0 \end{pmatrix}, \qquad A_1 = \begin{pmatrix} 0 & 0 \\ 0 & 1 \end{pmatrix}.$$

At this point, it is important to ask what is the advantage of using this apparently more complicated representation. Although a complete answer will be given in the subsequent sections, it is possible to offer a hint right away. In the MPS representation, the weight of each product state is related to the product of the matrices $A_{i_k}^{[k]}$ associated to particle k. For this reason, MPS are particularly useful to describe 1-D quantum many-body systems with *local* interactions, and play an important role in the simulation of these systems.

19.5. Simulation of Many-Body Quantum Systems

Most physical systems have many particles or many degrees of freedom. Luckily, in many circumstances, a proper identification of their elementary excitations (usually taking into account the symmetries of the problem) enables their treatment by the use of mean field theories, for example. However, there are many situations, normally involving many scales or competing interactions, where this approach is not sufficient, at least not until the correct order parameters and correlations involved are correctly identified. These are very challenging topics, covering a large spectra of phenomena, including both classical and quantum phase transitions, for example, with extremely important applications.

Although there are some exact solvable models that give important insight in understanding the behavior of such systems, they are only valid under highly restrictive conditions, either on the dimension of the system or in the type of the interactions considered. In the next chapter we refer to some of these exactly solvable models. Not only their known correlation functions can be used to evaluate the bipartite entanglement present, as given by the concurrence or different entropies, but the analysis of the behavior of these quantities can be used to understand them better and to shed a new light on these physical systems and on the phenomena occurring in them, like phase transitions, for example. In the final chapter we make a similar analysis for free bosonic and fermionic systems, independently of the dimension, both at zero and finite temperatures.

In general, studying quantum many-body systems is a complex problem, mainly due to the exponential growth of the Hilbert space with the size of the system. Numerical techniques as renormalization algorithms or quantum Monte Carlo simulations have been frequently used to study systems with unknown analytical solution.

The Density Matrix Renormalization Group (DMRG) technique proposed by White[52] in 1992 was the first numerical method to successfully obtain ground states of one-dimensional many-body quantum systems, in general. It overcame the difficulties in generalizing Wilson's[53] real space renormalization group treatment of the Kondo problem to other spin and electron problems. Primarily applied to Heisenberg spin chains with $S = \frac{1}{2}$ and $S = 1$, the method produced accurate results for open boundary conditions (OBC) while for periodic boundary conditions (PBC) the results were inexplicably poor. In 1995, Ostlund and Rommer[54,55] observed that the DMRG technique is equivalent to a variational method over a ground state ansatz written in the MPS form.

Related to the DMRG are the Contractor Renormalization (CORE)[56,57] and the Effective Interactions in Real Space Renormalization Group (RSRG-EI) methods,[58] which is also presented and discussed in another chapter of this book.

Resonating valence bond (RVB) states are of special importance in the study of quantum states and phases in condensed matter physics. They go back to the work of Pauling and have been extensively used in quantum chemistry. In physics, they were promoted by Anderson, both before[59] and after[60] the discovery of high T_c superconductivity.

Majumdar and Ghosh[61] studied a spin $\frac{1}{2}$ chain with a special ratio of first and second-neighbor interactions and having two exact degenerate ground states given by nearest-neighbor valence bonds. Affleck, Kenneddy, Lieb and Tasaki (AKLT)[62,63] studied valence bond ground states for integer spin systems, and Arovas, Auerbach and Haldane[64] obtained the exact correlation functions of valence bond solids in one dimension. In Ref. 64, the Schwinger boson representation was used to replace the spins at each site of the lattice by a bilinear combination of boson operators, and the ground state was obtained from the vacuum by the creation of pairs of bosons, each from a different site of the bond linking the two sites. The MPS states explore and develop these type of ideas in a very systematic manner, in the study of the lowest states of physical systems.

19.5.1. *Entanglement and the Simulation of 1-D Systems*

In 2003, Vidal rediscovered the MPS representation and the essence of DMRG by studying the class of quantum computations that could be simulated in a classical computer.[50] Later, he extended the result to the simulation of 1-D quantum many-body systems.[65] The starting point of his work was to consider a pure quantum state $|\psi\rangle$ of N particles (19.8) and repeatedly apply the Schmidt decomposition to obtain a local representation of the state. In the first step, one bi-partitions the system into subsystems A (particle i_1) and B (particles $i_2, \cdots, i_N$) and uses (19.4) to obtain:

$$|\psi\rangle = \sum_{\alpha_1=1}^{D_1} \lambda_{\alpha_1}^{[1]} \left|\alpha_1^{[1]}\right\rangle \left|\alpha_1^{[2\cdots N]}\right\rangle \tag{19.12}$$

$$= \sum_{i_1=1}^{d} \sum_{\alpha_1=1}^{\chi_1} \Gamma_{i_1,\alpha_1}^{[1]} \lambda_{\alpha_1}^{[1]}\, keti_1 \left|\alpha_1^{[2\cdots N]}\right\rangle . \tag{19.13}$$

The matrix $\Gamma_{i_1,\alpha_1}^{[1]}$ represents a change of basis from the Schmidt to the computational basis. Now, one redefines subsystem B as the whole system and applies the same procedure. After $N-1$ steps, $|\psi\rangle$ is expressed as:

$$|\psi\rangle = \sum_{i_1,\cdots,i_N=1}^{d} \sum_{\alpha_1,\cdots,\alpha_{N-1}} \Gamma_{i_1,\alpha_1}^{[1]} \lambda_{\alpha_1}^{[1]} \Gamma_{i_2,\alpha_1,\alpha_2}^{[2]} \lambda_{\alpha_2}^{[2]} \cdots \Gamma_{i_N,\alpha_{N-1}}^{[N]} \left|i_1 i_2 \cdots i_N\right\rangle . \tag{19.14}$$

Comparing with (19.9), we verify that this expression is in fact a re-writing of the MPS, by setting $\Gamma_{i_k}^{[k]}\lambda^{[k]} \equiv A_{i_k}^{[k]}$ and defining $\lambda_{\alpha_N}^{[N]} \equiv 1$. But now there is the clear advantage that the dimensions of the matrices are interpreted as the Schmidt numbers[e] χ_k resulting from the bipartition. Although in general D grows exponentially with N, systems with small amount of entanglement also have MPS representations with small D. It is worth mentioning that states showing this property are simply called *MPS* (which should not be confused with the homonymous representation).

To simulate the dynamics of slightly entangled quantum systems does not require an exponential computational effort, as slightly entangled states can be

[e] The number of non-zero eigenvalues λ_i is called the Schmidt number and is a measure of entanglement.[43] A large amount of entanglement corresponds to a big χ and in the limit of $\chi = 1$, we have a separable state.

represented by matrices with a small D, that grows polynomially with N. The method proposed by Vidal[65] consists of truncating the MPS representation of the quantum state in study after each step of the simulation. The new dimension $\tilde{D} < D$ can be chosen according to the desired precision but such that the computational effort is polynomial. With this technique, it is possible to study the time evolution of ground states and low energy excitations of one-dimensional systems.[65] Shortly after Vidal's proposal, White *et al.*[66] proved that the DMRG can also be adapted to perform time evolution, and later Daley *et al.*[67] analyzed the strong resemblance between both methods.

19.5.2. *Simulation of 1-D Systems for $T = 0$*

One of the most important contributions of quantum information theory to the study of many-body quantum systems was given in 2004 by Verstraete *et al.*[68] Inspired by the AKLT model[62,63] and its extension to *finitely correlated states*,[69] a new conceptual framing for MPS was proposed, allowing a deeper understanding of the DMRG method and its further generalization. The idea is to associate two virtual D-dimensional systems a_k and b_k to each physical particle k of the system. Consider that virtual particle neighbors b_k and a_{k+1} are in the (unnormalized) *maximum entangled state*:

$$|\phi\rangle = \sum_{\alpha=1}^{D} |\alpha\rangle_{b_k} \otimes |\alpha\rangle_{a_{k+1}} . \tag{19.15}$$

The linear map $\hat{P}_k : \mathcal{H}_{a_{k+1}} \otimes \mathcal{H}_{b_k} \to \mathcal{H}_k$ projects the virtual particles onto the physical one:

$$\hat{P}_k = \sum_{i_k=1}^{d} \sum_{\alpha,\beta}^{D} A_{i_k,\alpha,\beta} |i_k\rangle \langle \alpha, \beta| , \tag{19.16}$$

and, by applying it to every particle of the system, it is easily seen that we obtain a MPS representation (19.9) of the state $|\psi\rangle$:

$$\begin{aligned} |\psi\rangle &= \hat{P}_1 \otimes \hat{P}_2 \otimes \cdots \otimes \hat{P}_N |\phi\rangle^{\otimes N} \\ &= \sum_{i_1,\cdots,i_N=1}^{d} \operatorname{tr}\left[A_{i_1}^{[1]} A_{i_2}^{[2]} \cdots A_{i_N}^{[N]}\right] |i_1, i_2, \cdots, i_N\rangle . \end{aligned} \tag{19.17}$$

Although the matrices $A_{i_k}^{[k]}$ in (19.17) have dimensions $D \times D$, it is possible to generalize it to the site-dependent case $D_k \times D_{k+1}$.[49] Notice that this strengthens the idea that the dimensions of matrices $A_{i_k}^{[k]}$ are related to the quantum correlations each particle k shares with its neighbors. Following this picture, states obtained by local projection of auxiliary entangled systems are named *Projected Entangled-Pair States* (PEPS) and, as it will be seen in 19.5.3, can be generalized to higher dimensions, as well as to mixed states.

Special attention must be payed to the extreme particles i_1 and i_N. In fact, the type of boundary conditions imposes restrictions on the matrices $A_{i_1}^{[1]}$ and $A_{i_N}^{[N]}$. For open boundary conditions, i_1 and i_N are only correlated to one neighbor and therefore matrices $A_{i_1}^{[1]}$ must be vectors. According to the above picture they have one virtual particle associated: b_1 for i_1 and a_N for i_N (see Fig. 19.1.a).

Verstraete *et al.*[68] proved that this kind of MPS is the ansatz used in the original DMRG method[52] and it justifies the good results obtained under OBC. An appropriate ansatz wave-function for systems with periodic boundary conditions is constructed by associating two auxiliary systems to all physical particles, such that the extreme virtual particles obey $a_1 = a_{N+1}$ and $b_N = b_0$ (see Fig. 19.1.b).

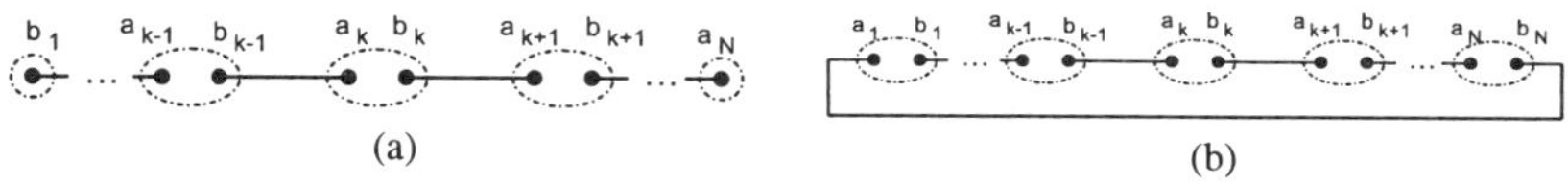

Fig. 19.1. Pictorial representation of MPS in 1-D. Each dot corresponds to a virtual particle and the surrounding ellipsis represent the physical particle. The maximum entangled state between the particles is represent by the bonds. (a) System with open boundary conditions. (b) System with periodic boundary conditions.

Variational Method Let us give a more precise idea of the proposed procedure. First, the system in study must be approximately *frustration free*, i.e. a minimum of its global energy $\langle\psi_0|\, H \,|\psi_0\rangle$ must be well approximated by minimizing local energies $\langle\psi_0|\, h_k \,|\psi_0\rangle$. Systems that exhibit such behavior have slightly entangled ground states,[70] which is in accordance with Vidal's derivation. Now, given that the expected value of the global energy is the product of expectation values of local operators h_k, we can write:

$$E = \frac{\langle\psi|\, H \,|\psi\rangle}{\langle\psi|\psi\rangle} = \frac{\langle\psi|\, h_1 \otimes \cdots \otimes h_N \,|\psi\rangle}{\langle\psi|\psi\rangle}. \tag{19.18}$$

Using (19.16) and (19.17), the previous expression is manipulated to give:

$$E = \frac{\left\langle\varphi^{[k]}\right|\, H_k \left|\varphi^{[k]}\right\rangle}{\left\langle\varphi^{[k]}\right|\, N_k \left|\varphi^{[k]}\right\rangle}. \tag{19.19}$$

On the one hand, (19.19) depends on the unknown local state $\left|\varphi^{[k]}\right\rangle = \oplus_i \left|\varphi_i^{[k]}\right\rangle$, where:

$$\left\langle\varphi_{i_k}^{[k]}\right| = \sum_{\alpha,\beta} A_{i_k,\alpha,\beta}^{[k]} \left\langle\alpha,\beta\right|, \tag{19.20}$$

and on the other hand, it is constructed with $d \times D^2$ matrices H_k and N_k which, besides being respectively related to the hamiltonian and the identity operators, also incorporate the information about the remaining states $\left|\varphi^{[i]}\right\rangle$, $i \neq k$.[68]

At each step of the variational method, the eigenstate $\left|\varphi^{[k]}\right\rangle$ with lower energy eigenvalue λ_0 is obtained by solving:

$$H_k \left|\varphi^{[k]}\right\rangle = \lambda_0 N_k \left|\varphi^{[k]}\right\rangle. \tag{19.21}$$

The procedure is repeated until the energy of the system reaches a stable point, which is guaranteed to be a minimum by the fact that the energy is a monotonically decreasing function of the step.[68,71] This method is only efficient in the case of translational invariant (TI) systems, since then the number of free parameters is significantly lower.

To test the validity of the model, Verstraete *et al.* apply the algorithm to a spin-$1/2$ Heisenberg chain with $N = 28$, imposing PBC. As expected, the results were considerably better than the ones obtained with DMRG and had accuracies comparable to those resulting from OBC simulations.

Convergence and Uniqueness of Solution The convergence of the method is profoundly related to the question of whether the MPS are a good approximation to exact ground states of one-dimensional quantum systems. In 2006, Verstraete and Cirac[72] proved that, for ground states of one-dimensional systems, there is a MPS representation where D scales polynomially with the number of particles N (see Ref. 49 and 72 for a discussion on the case of critical systems).

Although this is a promising argument, an efficient simulation also requires that such ground states are achievable with polynomial effort. In DMRG simulations, it is observed that the algorithm converges exponentially fast to the solution,[72] with relaxation time $\tau \sim \frac{1}{\Delta}$ where Δ is the energy gap between the ground state and the first excited state. It is interesting to mention that according to Hastings,[73] gapped systems imply exponentially decaying correlations, which intuitively agrees with the *locality* of the ground states.

For gapped non-critical systems and all known critical translational invariant systems, Δ decays at most with $\frac{1}{poly(N)}$. If in addition, the entanglement of any blocks of the system in the ground state grows at most logarithmically with N, then the computational effort to find the ground state is polynomial.[49]

It is important to stress that the minimum energy found using the variational method is not proven to be global, but in practice we rarely become trapped in a local minimum. To check that the obtained minimum energy is in fact global, the algorithm is usually tested with different initial states.

Quantum Phase Transitions A system experiencing a quantum phase transition (QPT) is known to exhibit special features, such as a diverging correlation length, a vanishing energy gap and the appearance of singularities in the thermodynamic limit (the role of entanglement in quantum phase transitions will be discussed in chapter 20). Studies on the amount of entanglement of blocks of such systems[74] concluded that it had a logarithmic diverging behavior in N and, for that reason, it was believed that a quantum phase transition could not be studied with the MPS variational method. Nevertheless, Wolf *et al.* discovered a new class of quantum phase transitions in spin chains, called *MPS-QPT*, where the entanglement remains finite (in fact, never exceeding $2\log_2 D$) and the ground state energy is analytic.[75] States displaying such QPT are MPS and can be efficiently studied with the DMRG technique. This result can be extended to higher dimensional systems.[75]

19.5.3. *Simulations of More General Systems*

Extension to Higher Dimensions The generalization of MPS to higher dimensions was proposed by Verstraete and Cirac,[76] following the rather intuitive picture offered by the *projected entangled-pair states*. Usually, this generalization to more than one dimension is simply called PEPS. For simplicity, let us start with a 2-D square lattice of $N_h \times N_v$ d-dimensional systems. A pure quantum state has a representation:[76]

$$|\psi\rangle = \sum_{i_{1,1},\cdots,i_{N_h,N_v}}^{d} F_2(\{A_{i_{h,v}}^{[h,v]}\})\,|i_{1,1},\cdots,i_{N_h,N_v}\rangle\,, \tag{19.22}$$

where the matrices $(A_i)_{\alpha,\beta}$ of the one-dimensional case (19.9) are now replaced by tensors $(A_{i_{h,v}})_{u,d,l,r}$ and F_2 is a suitable function that contracts the indexes (u,d,l,r). These represent the bonds that each non-extremal particle shares with its four nearest-neighbors; for the case of the particles at the boundaries, the indexes can be 2 or 3, according to their position in the lattice (see Fig. 19.2). This representation is also constructed with the projection of local maximally entangled virtual particles, and the number of auxiliary systems associated to each physical particle is given by the number of indexes of the corresponding tensor A.

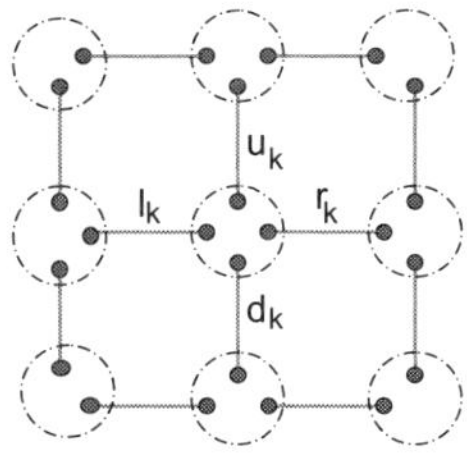

Fig. 19.2. Graphical representation of PEPS in a square lattice.

In a similar way to the 1-D case, any state can be written in the representation (19.22) and it is possible to obtain the ground state of Hamiltonians with short-range interactions by applying a variational method over an ansatz of PEPS.[76]

The above construction can be generalized to different geometries, given that the number of bounds and the contraction function F_d are properly chosen. To test the algorithm in the 2-D case, Verstraete and Cirac[76] performed an imaginary time evolution of a 4×4 spin-1/2 system with antiferromagnet Heisenberg interactions, that can be determined exactly. The results were considerably close to the expected values, even for a dimension as low as $D = 3$ for the tensors A.

Simulations with Mixed States The MPS representation was originally proposed for pure quantum states but it has a natural extension to mixed quantum states. Following Verstraete *et al.*,[77] this representation is obtained by transforming the linear maps $\hat{P}_k$ (19.16) in the most general operators allowed by quantum mechanics. Then, the *matrix product density operator* (MPDO) representation of a density matrix ρ of N d-dimensional systems is:

$$\hat{\rho} = \sum_{i_1,i'_1,\cdots,i_N,i'_N=1}^{d} \mathrm{tr}\left(M^{[1]}_{i_1,i'_1} \cdots M^{[N]}_{i_N,i'_N}\right) |i_1,\cdots,i_N\rangle \langle i'_1,\cdots,i'_N|, \quad (19.23)$$

where $M^{[k]}_{i_k,i'_k}$ are $D_k^2 \times D_{k'+1}^2$ matrices, related to the A matrices of the MPS representation (19.9):

$$M^{[k]}_{i,i'} = \sum_{a=1}^{d_k} A^{[k]}_{i,a} \otimes (A^{[k]}_{i',a})^*. \quad (19.24)$$

As before, the matrices $A_k^{s,k}$ have dimensions $D_k \times D_{k+1}$ and $d_k \leq dD_kD_{k+1}$. Once again, a sufficiently large D guarantees that any density operator can be expressed in this representation. The results derived for pure quantum states can be applied to mixed states by using the concept of *purification*[43]. The idea is to enlarge the Hilbert space such that $\hat{\rho}$ is a subsystem whose system is in a pure state $|\Psi\rangle$:

$$|\Psi\rangle = \sum_{i_1,\cdots,i_N} \sum_{a_1,\cdots,a_N} \mathrm{tr}\left(A^{[1]}_{i_1,a_1} \cdots A^{[N]}_{i_N,a_N}\right) |i_1 a_1, \cdots, i_N a_N\rangle . \quad (19.25)$$

The mixed state is obtained by the usual method: tracing out the ancilla system $\hat{\rho} = \mathrm{tr}_a |\Psi\rangle\langle\Psi|$. The maximum dimension of each auxiliary system a_k is d_k but states with $d \ll d_k$ are in the basis of many interesting phenomena. In this situation, the purification of $\hat{\rho}$ can be efficiently simulated and is used to study: (i) time evolution of mixed states, (ii) density operator for a system in thermal equilibrium and (iii) evolution of spin systems in the presence of decoherence. The results described here were almost simultaneously achieved by Zwolak and Vidal[78] and Verstraete *et al.*[77]

Finally, is is worth mentioning that recently new classes of ansatz wavefunctions have been proposed by Anders *et al.*[79] and Vidal.[80] Both works offer a more flexible structure in the ansatz, such that systems that possess interactions more general than the simple *nearest-neighbor* can also be efficiently simulated.

19.6. Summary

We have introduced and defined the concept of quantum entanglement and its properties. We discussed the case of bipartite entanglement in a pure state, the Schmidt decomposition and the quantification of entanglement by the von Neumann entropy. We then reviewed the novel matrix product state and projected entangled-pair states representations of entangled states and discussed how they are offering important advances in the simulation of many-body quantum systems. Taking a quantum information perspective of the problems, it was possible to identify that small amounts of entanglement are a necessary condition for efficient simulations, as well as to reinterpret the DMRG technique as a special case of a variational method over projected entangled-pair states. With this new conceptual framing, it was possible to extend the study of low-energy pure states to systems with periodic boundary conditions, as well as to obtain low-energy states and simulate the time evolution of d-dimensional systems at zero temperature and one-dimensional systems at finite temperature.

Acknowledgments

MA thanks Artur Garcia for so many clarifying discussions. The authors thank the support from Fundação para a Ciência e a Tecnologia (Portugal) through programs POCTI/POCI, partially funded by FEDER (EU), namely through projects POCI/MAT/55796/2004 QuantLog, POCTI-SFA-2-91 and POCI/FIS/58746/2004, and grant SFRH/BD/21915/2005, as well as the SQIG-IT EMSAQC initiative.

Appendix A. The von Neumann Entropy

The *von Neumann entropy* (also known as quantum entropy) plays a central role in quantum information theory. It is often (and not incorrectly) presented as the quantum version of the Shannon (classical) entropy, although historically the quantum entropy was introduced first:[81]

Definition 19.2 (von Neumann Entropy). *Given a quantum system lying in a Hilbert space $\mathcal{H}$ and described by a density operator $\hat{\rho}$, we define the* von Neumann entropy *of the system as*

$$S(\hat{\rho}) \equiv -\mathrm{tr}(\hat{\rho}\log_2\hat{\rho}) \tag{A.1}$$

The von Neumann entropy is a measure of *uncertainty* in a mixed state of a quantum system. Let $d \equiv \dim\mathcal{H}$, then we have:

$$0 \leq S(\hat{\rho}) \leq \log_2 d \tag{A.2}$$

The entropy is zero if and only if $\hat{\rho}$ describes a pure state, and is maximal if and only if $\hat{\rho}$ is the completely mixed density operator $\frac{1}{d}\hat{I}$. Finally, we introduce a formula useful for explicit calculations: if ρ has been diagonalized and we know its eigenvalues λ_i, then we can write:

$$S(\hat{\rho}) = -\sum_i \lambda_i \log_2 \lambda_i \tag{A.3}$$

(note that the eigenvalues of a density operator are non-negative real numbers and that we define $0\log 0 \equiv 0$). $S(\hat{\rho})$ has many other properties, useful in general in the context of quantum information.[82–84]

We can interpret the von Neumann entropy as the quantum generalization of the Shannon entropy. In fact, the former plays the same role in *Schumacher's Quantum Noiseless Coding Theorem*[85] that the latter plays in Shannon's original (i.e. classical) version of the theorem, giving us the minimum number of quantum bits per letter that are necessary to reliably encode the relevant information. Furthermore, in the particular case where we are considering orthogonal states, the quantum entropy is reduced to the classical one. But these features are far from exhausting the role of the von Neumann entropy in quantum information theory: for example, it is also used to quantify the amount of classical information that can be extracted from the optimal measurement of a quantum (pure) state, as well as to quantify the entanglement of a bipartite pure state.

References

1. D. Bohm, *Quantum Theory*. (Prentice Hall, 1951).
2. A. Einstein, B. Podolsky, and N. Rosen, Can quantum-mechanical description of physical reality be considered complete?, *Phys. Rev.* **47**, 777, (1935).
3. E. Schrödinger, Discussion of probability relations between separated systems, *Proc. Cambridge Philosophical Society*. **31**, 555, (1935).
4. E. Schrödinger, Die gegenwärtige Situation in der Quantenmechanik, *Naturwissenschaften*. **23**, 807–812, 823–823, 844–849, (1935). English translation: John D. Trimmer, *Proceedings of the American Philosophical Society*, **124**, 323–38 (1980); Reprinted in *Quantum Theory and Measurement*, 152 (1983).
5. The quest for the true origin of entanglement. URL `http://cam.qubit.org/users/matthias/Entanglement/Entanglement.php`.
6. J. S. Bell, On the Einstein-Podolsky-Rosen paradox, *Physics*. **1**, 195, (1964). Reprinted in: see Ref. 7.
7. J. S. Bell, *Speakable and Unspeakable in Quantum Mechanics*. (Cambridge University Press, 1987).
8. L. Hardy, Spooky action at a distance in Quantum Mechanics, *Contemp. Phys.* **39**, 419, (1998).
9. S. J. Freedman and J. F. Clauser, Experimental test of local hidden-variable theories, *Phys. Rev. Lett.* **28**, 938, (1972).
10. A. Aspect, P. Grangier, and G. Roger, Experimental tests of realistic local theories via Bell's theorem, *Phys. Rev. Lett.* **47**, 460, (1981).
11. H. Z. W. Tittel, J. Brendel and N. Gisin, Violation of Bell inequalities by photons more than 10 km apart, *Phys. Rev. Lett.* **81**, 3563, (1998).
12. A. K. Ekert, Quantum cryptography based on Bell's theorem, *Phys. Rev. Lett.* **67**, 661, (1991).
13. C. H. Bennett and S. J. Wiesner, Communication via one- and two-particle operators on Einstein-Podolsky-Rosen states, *Phys. Rev. Lett.* **69**, 2881–2884, (1992).
14. C. H. Bennett, G. Brassard, C. Crépeau, R. Jozsa, A. Peres, and W. K. Wootters, Teleporting an unknown quantum state via dual classical and Einstein-Podolsky-Rosen channels, *Phys. Rev. Lett.* **70**, 1895–1899, (1993).
15. R. Jozsa and N. Linden, On the role of entanglement in quantum computational speed-up, *Proc. R. Soc. A*. **459**, 2011, (2003).
16. E. Biham, G. Brassard, D. Kenigsberg, and T. Mor, Quantum computing without entanglement, *Theoretical Computer Science*. **320**, 15–33, (2004).
17. E. Knill and R. Laflamme, Power of one bit of quantum information, *Phys. Rev. Lett.* **81**, 5672–5675, (1998).
18. A. Datta and G. Vidal. On the role of entanglement and correlations in mixed-state quantum computation. quant-ph/0611157, (2006).
19. A. M. Steane, Error correcting codes in quantum theory, *Phys. Rev. Lett.* **77**, 793–797, (1996).

20. W. Dür, H.-J. Briegel, J. I. Cirac, and P. Zoller, Quantum repeaters based on entanglement purification, *Phys. Rev. A.* **59**, 169–181, (1999).
21. J. J. Bollinger, W. M. Itano, D. J. Wineland, and D. J. Heinzen, Optimal frequency measurements with maximally correlated states, *Phys. Rev. A.* **54**, R4649–R4652, (1996).
22. S. L. V. Giovannetti and L. Maccone, Quantum enhanced positioning and clock synchronization, *Nature.* **412**, 417, (2001).
23. N. Gisin, Bell's inequality holds for all non-product states, *Phys. Lett. A.* **154**, 201, (1991).
24. S. Popescu, Quantum mechanics: Why isn't nature more non-local?, *Nature Physics.* **2**, 507–508, (2006).
25. Y. Omar, Indistinguishable particles in quantum mechanics: An introduction, *Contemp. Phys.* **46**, 437, (2005).
26. Y. Omar, Particle statistics in quantum information processing, *Int. J. Quant. Inf.* **3**, 201, (2005).
27. A. Ekert and P. L. Knight, Entangled quantum systems and the Schmidt decomposition, *Am. J. Phys.* **63**, 415, (1995).
28. A. Peres, Higher order Schmidt decompositions, *Phys. Lett. A.* **202**, 16, (1995).
29. F. Morikoshi, M. F. Santos, and V. Vedral, Accessibility of physical states and non-uniqueness of entanglement measure, *J. Phys. A.* **37**, 5887, (2004).
30. J. Schliemann, J. I. Cirac, M. Kuś, M. Lewenstein, and D. Loss, Quantum correlations in two-fermion systems, *Phys. Rev. A.* **64**, 022303, (2001).
31. D. B. K. Eckert, J. Schliemann and M. Lewenstein, Quantum correlations in systems of indistinguishable particles, *Annals of Physics.* **299**, 88, (2002).
32. M. Plenio and V. Vedral, Entanglement in quantum information theory, *Contemp. Phys.* **39**, 431, (1998).
33. C. H. Bennett, H. J. Bernstein, S. Popescu, and B. Schumacher, Concentrating partial entanglement by local operations, *Phys. Rev. A.* **53**, 2046–2052, (1996).
34. S. Bose, V. Vedral, and P. L. Knight, Multiparticle generalization of entanglement swapping, *Phys. Rev. A.* **57**, 822–829, (1998).
35. G. V. W. Dür and J. I. Cirac, Three qubits can be entangled in two inequivalent ways, *Phys. Rev. A.* **62**, 062314, (2000).
36. P. H. M. Horodecki and R. Horodecki, Separability of mixed states: necessary and sufficient conditions, *Phys. Lett.A.* **223**, 1, (1996).
37. A. Peres, Separability criterion for density matrices, *Phys. Rev. Lett.* **77**, 1413–1415, (1996).
38. R. F. Werner, Quantum states with Einstein-Podolsky-Rosen correlations admitting a hidden-variable model, *Phys. Rev. A.* **40**, 4277–4281, (1989).
39. D. Bruss, Characterizing entanglement, *J. Math. Phys.* **43**, 4237, (2002).
40. Y. H. Shih and C. O. Alley, New type of Einstein-Podolsky-Rosen-Bohm experiment using pairs of light quanta produced by optical parametric down conversion, *Phys. Rev. Lett.* **61**, 2921–2924, (1988).
41. E. Hagley, X. Maître, G. Nogues, C. Wunderlich, M. Brune, J. M. Raimond, and S. Haroche, Generation of Einstein-Podolsky-Rosen pairs of atoms, *Phys. Rev. Lett.* **79**, 1–5, (1997).

42. B. E. K. C. Monroe, D. M. Meekhof and D. J. Wineland, A 'Schrödinger cat' superposition state of an atom, *Science*. **272**, 1131, (1996).
43. M. A. Nielsen and I. L. Chuang, *Quantum Computation and Quantum Information*. (Cambridge University Press, 2000).
44. D. Bouwmeester, A. Ekert, and A. Zeilinger, Eds., *The Physics of Quantum Information*. (Springer, 2000).
45. W. D. Oliver, F. Yamaguchi, and Y. Yamamoto, Electron entanglement via a quantum dot, *Phys. Rev. Lett.* **88**, 037901, (2002).
46. A. T. Costa, Jr., S. Bose, and Y. Omar, Entanglement of two impurities through electron scattering, *Phys. Rev. Lett.* **96**, 230501, (2006).
47. J.-W. Pan, M. Daniell, S. Gasparoni, G. Weihs, and A. Zeilinger, Experimental demonstration of four-photon entanglement and high-fidelity teleportation, *Phys. Rev. Lett.* **86**, 4435–4438, (2001).
48. H. Häffner and *et al.*, Scalable multiparticle entanglement of trapped ions, *Nature*. **438**, 643–646, (2005).
49. M. M. W. D. Péez-García, F. Verstraete and J. I. Cirac. Matrix product states representations. quant-ph/0608197, (2006).
50. G. Vidal, Efficient classical simulation of slightly entangled quantum computations, *Phys. Rev. Lett.* **91**, 147902, (2003).
51. D. M. Greenberger, M. Horne, and A. Zeilinger. Going beyond Bell's theorem. In ed. M. Kafatos, *Bell's Theorem, Quantum Mechanics and Conceptions of the Universe*, pp. 69–72. Kluwer Academic, (1989).
52. S. R. White, Density matrix formulation for quantum renormalization groups, *Phys. Rev. Lett.* **69**, 2863–2866, (1992).
53. K. G. Wilson, The renormalization group: Critical phenomena and the Kondo problem, *Rev. Mod. Phys.* **47**, 773–840, (1975).
54. S. Östlund and S. Rommer, Thermodynamic limit of density matrix renormalization, *Phys. Rev. Lett.* **75**, 3537–3540, (1995).
55. S. Rommer and S. Östlund, Class of ansatz wave functions for one-dimensional spin systems and their relation to the density matrix renormalization group, *Phys. Rev. B*. **55**, 2164–2181, (1997).
56. C. J. Morningstar and M. Weinstein, Contractor renormalization group technology and exact hamiltonian real-space renormalization group transformations, *Phys. Rev. D*. **54**, 4131–4151, (1996).
57. M. Weinstein, Contractor renormalization group and the Haldane conjecture, *Phys. Rev. B*. **63**, 174421, (2001).
58. J.-P. Malrieu and N. Guihéry, Real-space renormalization group with effective interactions, *Phys. Rev. B*. **63**, 85110, (2001).
59. P. W. Anderson, Resonating valence bonds – new kind of insulator, *Mater. Res. Bull.* **8**, 153–160, (1973).
60. P. W. Anderson, The resonating valence bond state in La_2CuO_4 and superconductivity, *Science*. **235**, 1196–1198, (1987).
61. C. K. Majumdar and D. K. Ghosh, On next-nearest-neighbor interaction in linear chain I, *J. Math. Phys.* **10**, 1388–1398, (1969).
62. I. Affleck, T. Kennedy, E. H. Lieb, and H. Tasaki, Rigorous results on valence-bond ground states in antiferromagnets, *Phys. Rev. Lett.* **59**, 799–802, (1987).

63. E. H. L. Ian Affleck, Tom Kennedy and H. Tasaki, Valence bond ground states in isotropic quantum antiferromagnets, *Comm. Mat. Phys.* **115**, 477–528, (1988).
64. D. P. Arovas and A. Auerbach, Extended Heisenberg models of antiferromagnetism: Analogies to the fractional quantum Hall effect, *Phys. Rev. Lett.* **60**, 531–534, (1988).
65. G. Vidal, Efficient simulation of one-dimensional quantum many-body systems, *Phys. Rev. Lett.* **93**, 040502, (2004).
66. S. R. White and A. E. Feiguin, Real-time evolution using the density matrix renormalization group, *Phys. Rev. Lett.* **93**, 076401, (2004).
67. A. J. Daley, C. Kollath, U. Schollwöck, and G. Vidal, Time-dependent density-matrix renormalization-group using adaptive effective Hilbert spaces, *J. Stat. Mech.* **2004**, P04005, (2004).
68. F. Verstraete, D. Porras, and J. I. Cirac, Density matrix renormalization group and periodic boundary conditions: A quantum information perspective, *Phys. Rev. Lett.* **93**, 227205, (2004).
69. B. N. M. Fannes and R. F. Werner, Finitely correlated states on quantum spin chains, *Comm. Mat. Phys.* **144**, 443, (1992).
70. C. M. Dawson and M. A. Nielsen, Frustration, interaction strength, and ground-state entanglement in complex quantum systems, *Phys. Rev. A.* **69**, 052316, (2004).
71. U. Schollwock, The density-matrix renormalization group, *Rev. Mod. Phys.* **77**, 259–316, (2005).
72. F. Verstraete and J. I. Cirac, Matrix product states represent ground states faithfully, *Phys. Rev. B.* **73**, 094423, (2006).
73. M. B. Hastings, Solving gapped hamiltonians locally, *Phys. Rev. B.* **73**, 085115, (2006).
74. G. Vidal, J. I. Latorre, E. Rico, and A. Kitaev, Entanglement in quantum critical phenomena, *Phys. Rev. Lett.* **90**, 227902, (2003).
75. M. M. Wolf, G. Ortiz, F. Verstraete, and J. I. Cirac, Quantum phase transitions in matrix product systems, *Phys. Rev. Lett.* **97**, 110403, (2006).
76. F. Verstraete and J. I. Cirac. Renormalization algorithms for quantum-many body systems in two and higher dimensions. cond-mat/0407066, (2004).
77. F. Verstraete, J. J. Garcia-Ripoll, and J. I. Cirac, Matrix product density operators: Simulation of finite-temperature and dissipative systems, *Phys. Rev. Lett.* **93**, 207204, (2004).
78. M. Zwolak and G. Vidal, Mixed-state dynamics in one-dimensional quantum lattice systems: A time-dependent superoperator renormalization algorithm, *Phys. Rev. Lett.* **93**, 207205, (2004).
79. S. Anders, M. B. Plenio, W. Dur, F. Verstraete, and H.-J. Briegel, Ground-state approximation for strongly interacting spin systems in arbitrary spatial dimension, *Phys. Rev. Lett.* **97**, 107206, (2006).
80. G. Vidal. Entanglement renormalization,. cond-mat/0512165, (2005).
81. J. von Neumann, *Mathematische Grundlagen der Quanten-Mechanik*. (Julius Springer-Verlag, 1932). English translation: *Mathematical Foundations of Quantum Mechanics*. (Princeton University Press, 1955).
82. A. Peres, *Quantum Theory: Concepts and Methods*. (Kluwer Academic Publishers, 1995).

83. A. Wehrl, General properties of entropy, *Rev. Mod. Phys.* **50**, 221, (1978).
84. V. Vedral, The role of relative entropy in quantum information theory, *Rev. Mod. Phys.* **74**, 197–234, (2002).
85. B. Schumacher, Quantum coding, *Phys. Rev. A.* **51**, 2738–2747, (1995).

Chapter 20

Entanglement in quantum phase transitions

P. Ribeiro

Laboratoire de Physique Théorique de la Matière Condensée, Université Pierre et Marie Curie, Place Jussieu, 75252 Paris Cedex 05, France

Y. Omar

SQIG, Instituto de Telecomunicações, P-1049-001 Lisbon and CEMAPRE, ISEG, Technical University of Lisbon, P-1200-781 Lisbon, Portugal

V. Rocha Vieira

CFIF and Department of Physics, IST, Technical University of Lisbon, 1049-001 Lisbon, Portugal

We review recent studies on the role of entanglement in many-body quantum systems at zero temperature, focusing on quantum phase transitions (QPT). We start by introducing the *entanglement entropy* and the *concurrence*, and then discuss how these measures of quantum correlations behave in one-, low- and infinite-dimensional spin and electron systems in (first- and second-order) phase transitions, giving us new insight into the correlations present in those systems, new order parameters and, in some cases, even new QPT points. Finally, we introduce the idea of the quantum computer, which in practice will be a many-body quantum system. We present the adiabatic model of quantum computation and discuss its connections to quantum phase transitions, as well as the role of entanglement in both phenomena.

Contents

20.1. Introduction

Classical phase transitions in a system are qualitative changes of the physical properties arising as the temperature approaches some particular value called the *critical temperature* T_c. Quantum phase transitions[1] (QPT) occur at zero temperature, when there are qualitative changes in the ground state of the system induced by a variation of the coupling constants in the Hamiltonian. So, there are fundamental differences between the phenomena of classical and quantum phase transitions: the former are driven by thermal fluctuations, whereas the latter are driven by quantum fluctuations.

As in the classical case, one defines the *order* of a quantum phase transition in relation to the non-analyticity of the *order parameter*. A discontinuous order parameter is obtained in a first order QPT, a continuous order parameter with a discontinuous first order derivative is the signature of a second order QPT, and so on. Phase transitions with order higher than two are rather rare in nature and so we will restrict our analysis to first and second order QPT.

In a general first order QPT point, the order parameter presents a discontinuity and the gap between the two lowest energy states goes to zero exponentially as the size of the system increases. For systems where a *correlation length* can be defined, it also presents a discontinuity and the spacial correlations decrease exponentially fast. As we shall see, bipartite entanglement measures such as the *entanglement entropy* and the *concurrence* (defined in section 20.1.1) are also, in general, discontinuous.

At second order QPT points, universal scaling behaviors are observed. The energy gap between the two lowest energy states vanishes polynomially as the size of the system increases. The correlation length diverges and the spacial correlations decrease algebraically. The behavior of these quantities is dictated by some non-trivial exponents that characterize the universality class of the QPT. The entanglement entropy also diverges and, near the QPT, the concurrence has a cusp-like form presenting a discontinuity (or a divergence) in its first derivative.

Away from the QPT points, it is possible to define a length scale giving the typical size within which the system variables have non-negligible correlations. In general, this length scale increases as we approach the critical point and, except in the case of first order phase transitions, it diverges at the point itself, where the system presents a long range order. Since the correlations vanish for separable states, the existence of strong correlations should be a hint for the existence of strong entanglement (see chapter 19) as well. It is thus natural to investigate the role of entanglement in quantum phase transitions, and try to use it as a new way of characterizing and even of detecting phase transitions.

20.1.1. *Enters Entanglement...*

The first analyses of entanglement in a quantum phase transition were done in 2002.[2,3] The authors pointed out that, for one-dimensional spin chains, the entanglement measures become non-analytic at the QPT points. Since then, other spin and electron models have also been studied.[4–7] The scaling properties of entanglement in spin chain systems, near and at a quantum critical point, were studied and a connection to conformal field theories made.[3,8]

The entanglement properties near a QPT were also studied in higher-dimensional systems, in particular in fully-connected spin systems, where it was shown that, as in the one-dimensional case, the entanglement measures become non-analytic at the QPT point.[9,10]

For some systems the entanglement entropy was also shown to play a role analogous to an order parameter[11] and it permits to study some QPT where the computation of more standard quantities was rather difficult.[12]

Before proceeding to review some of these important results, let us first introduce some relevant measures of entanglement. As it was discussed in chapter 19, the amount of entanglement in a bipartite pure state $\hat{\rho}_{AB}$ is given by the von Neumann entropy S of the reduced density matrix representing the state of each subsystem:

$$S(\hat{\rho}_A) = -tr(\hat{\rho}_A \log_2 \hat{\rho}_A) = S(\hat{\rho}_B), \tag{20.1}$$

where $\hat{\rho}_A = tr_B\left(\hat{\rho}_{AB}\right)$ and similarly $\hat{\rho}_B = tr_A\left(\hat{\rho}_{AB}\right)$. Throughout this chapter, we refer to the quantity given by Eq. (20.1) as the *entanglement entropy* of a quantum system in a bipartite pure state $\hat{\rho}_{AB}$ (in fact, for these states, all measures of entanglement coincide[13]).

In the case where $\hat{\rho}_{AB}$ is a mixed state, for instance given by $\hat{\rho}_{AB} = \sum_i p_i |\psi_i\rangle\langle\psi_i|$ (with probabilities p_i such that $\sum_i p_i = 1$), the amount of entanglement in the state can then be calculated by the *entanglement of formation*[14] E_F:

$$E_F(\hat{\rho}_{AB}) \equiv \min \sum_k p_k S(\hat{\rho}_A^k), \tag{20.2}$$

where the minimum is taken over all the possible realizations of the state $\hat{\rho}_{AB}$, with $\hat{\rho}_A^k = tr_B\left(|\psi_k\rangle\langle\psi_k|\right)$.

For the particular case of two quantum two-level systems (also know as quantum bits), Wootters proved there exists a closed formula[15] for Eq. (20.2). In this very useful — albeit very specific — situation, the entanglement of formation is called *concurrence*. It is a measure of the non-separability of two qubits density matrices, that goes from zero (separable) to one (maximally entangled states). To apply this measure to the study of many-body systems, one traces over all but two binary degrees of freedom obtaining a reduced density matrix $\hat{\eta}$. Then one computes the spin-flipped matrix $\hat{\eta}' = \hat{\sigma}_y \otimes \hat{\sigma}_y \, \hat{\eta}^* \, \hat{\sigma}_y \otimes \hat{\sigma}_y$, where $\hat{\eta}^*$ is the complex conjugate of $\hat{\eta}$ and $\hat{\sigma}_y$ is the second Pauli matrix. We can then define the concurrence C as:

$$C = \max\left\{0, \mu_1 - \mu_2 - \mu_3 - \mu_4\right\}, \tag{20.3}$$

where μ_j are the square roots of the four real eigenvalues of $\hat{\eta}\hat{\eta}'$, written in decreasing order. Note that the entanglement of formation varies monotonically with the concurrence.

Unfortunately, when the system is composed of more than two subsystems, the situation is much more complicated and a generalization is difficult to obtain.[13] Moreover, the use of bipartite entanglement has also some limitations, given a property known as the *entanglement monogamy*:[16] if two parties are already maximally entangled, none of them can share any bipartite entanglement with a third party. As a phase transition is approached,[2] the correlation length grows and each site starts developing entanglement with sites further away. Away from the QPT point, the correlations are exponentially damped, but at the critical point this is not the case. Although the overall entanglement of the system increases, its distribution between sites must respect the constraints of entanglement monogamy. This explains why the critical behaviors of concurrence in a number of spin models are not universal and shows the need for a deeper analysis.[17]

20.2. One- and Low-Dimensional Systems

The existence of non-trivial exactly solvable models is important since it allows to find or to verify, in a rigorous manner, new types of behavior or phenomena, departing from those already existing. This is the case of the classical two-dimensional Ising model, with non classical critical exponents, or the one-dimensional (1-D) Luttinger model, which is today the reference model for 1-D conductors, comparable to the Laudau liquid theory, in three dimensions. These models are also important to test new numerical techniques, before they are applied to more realistic models.

However, to find and solve exactly an interacting model is a very difficult task, normally only possible with important restrictions on the dimension of the system or in type of the interactions considered. In general, mean-field theories become valid in the limit of infinite dimensionality of the system or of the order parameter, and expansions in the inverse of these quantities are currently used. For 1-D systems special techniques exist, like the Jordan-Wigner transformation, the Bethe ansatz, the inverse scattering method and the bosonization and conformal field theory techniques.[18]

When some of these models were originally solved, those restrictions, in particular the one dimensionality, made them look unrealistic. However, with the impressive recent development in the study of low dimensional systems, e.g. surfaces, chains and dots, with important applications in the areas of the meso and nanosystems, for example, some of these models became the reference for some of those low dimensional physical systems.

It may also happen that the restrictions on the type of interactions, introduced to solve exactly the model with the mathematical technique used, makes them too special, i.e. with additional symmetries, and thus they belong to a different universality class, different from that of the models to which it was supposed to belong. This is what happens with some models[19–21] solved with the Bethe ansatz, normally gapless, independently of the value s of the spin.

In the study of 1-D spin chain models two results are important to take into consideration. One is the so-called Haldane[22,23] conjecture[a], which states that one-dimensional antiferromagnetic spin chains have an excitation gap for integer, but not half-integer, spin s. The other is the Lieb-Shultz-Mattis theorem,[24] initially proved for $\frac{1}{2}$, and then extended by Affleck and Lieb[25] to arbitrary half-integer s, which states that a half-integer-s spin chain, with a translationally and rotationally local Hamiltonian, is either gapless or has two degenerate ground sates and parity (i.e. reflection about a site) is spontaneously broken.

The known correlation functions of these models can be used to evaluate the two-site reduced density matrix and, from it, one evaluates the bipartite entanglement, as given by the concurrence and entropies. In turn, these quantities can be helpful in understanding and characterizing the phenomena involved, namely in the case of phase transitions.

There are several one-dimensional exactly solvable models which have been considered in this context.

The XY model in a transverse field:

$$\hat{H}_{XY} = -\frac{1}{2}\sum_j \left(\frac{1+\gamma}{2}\hat{\sigma}_j^x\hat{\sigma}_{j+1}^x + \frac{1-\gamma}{2}\hat{\sigma}_j^y\hat{\sigma}_{j+1}^y + \lambda\hat{\sigma}_j^z \right), \tag{20.4}$$

where $\hat{\sigma}_j^a$, $a = x, y, z$, are the Pauli matrices, can be solved by the Jordan-Wigner transformation (followed by a Bogoliubov-Valatin transformation), both for open

[a]Actually already confirmed, but still referred to as a conjecture.

and periodic boundary conditions.[24,26,27] When $\gamma = 1$, it is reduced to the Ising model in a transverse field, probably the simplest model with a quantum phase transition:

$$\hat{H}_{Ising} = -\frac{1}{2}\sum_j \left(\hat{\sigma}_j^x \hat{\sigma}_{j+1}^x + \lambda \hat{\sigma}_j^z\right), \tag{20.5}$$

and, for $\gamma = 0$, gives the isotropic XX model in a transverse field:

$$\hat{H}_{XX} = -\frac{1}{2}\sum_j \left(\frac{1}{2}\left(\hat{\sigma}_j^x \hat{\sigma}_{j+1}^x + \hat{\sigma}_j^y \hat{\sigma}_{j+1}^y\right) + \lambda \hat{\sigma}_j^z\right). \tag{20.6}$$

The XXZ model in a field:

$$\hat{H}_{XXZ} = \sum_j \left(\frac{1}{2}\left(\hat{\sigma}_j^x \hat{\sigma}_{j+1}^x + \hat{\sigma}_j^y \hat{\sigma}_{j+1}^y + \Delta \hat{\sigma}_j^z \hat{\sigma}_{j+1}^z\right) + \lambda \hat{\sigma}_j^z\right), \tag{20.7}$$

can be solved by the Bethe ansatz, for periodic boundary conditions.[28,29] For $\Delta = 1$, it gives the isotropic Heisenberg model, and for $\Delta = 0$, reduces again to the XX model in a transverse field.

These models have a very rich phase diagram, with different product states as ground states in some limits and quantum phase transitions, with strong entanglement, between them.[2] Understanding these phenomena is a rather difficult task where the concept of entanglement has been showing its relevance and importance, strengthening the connection between condensed matter physics and quantum information theory.

Entanglement in these models have been studied by several authors. In the Ising model in a transverse field[2–4] the first order derivative of the concurrence diverges at the transition point and obeys a scaling law in its vicinity. In the antiferromagnetic XXZ chain the concurrence is a continuous function of the anisotropy parameter and has a maximum at the transition point.[4,30–32]

The Hubbard model for electronic systems:

$$\hat{H} = -t \sum_{j,\delta=\pm 1,\sigma} \hat{c}_{j\sigma}^\dagger \hat{c}_{j+\delta\sigma} + U \sum_j \hat{n}_{j\uparrow}\hat{n}_{j\downarrow} - h \sum_j \hat{S}_j^z, \tag{20.8}$$

where $\hat{c}_{j\sigma}^\dagger$ and $\hat{c}_{j\sigma}$ are creation and destruction operators for electrons at site j and spin σ, $\hat{n}_{j\sigma} = \hat{c}_{j\sigma}^\dagger \hat{c}_{j\sigma}$ are the number operators and $\hat{S}_j^z = \frac{1}{2}(\hat{n}_{j\uparrow} - \hat{n}_{j\downarrow})$ is the z component of the magnetization, at site j, can also be solved by the Bethe ansatz, for periodic boundary conditions.[33]

The local entanglement entropy and its derivatives with respect to the magnetic field h and the chemical potential μ for this model have been obtained.[5] The exact solution has also been used to study the analyticity of the phase transition and to check the validity of the numerical exact diagonalization techniques in the study of the extended Hubbard model (having a nearest-neighbor Coulomb interaction term of the form $V \sum_{j,\sigma\sigma'} \hat{n}_{j\sigma}\hat{n}_{j\sigma'}$ added).[34]

The antiferromagnetic zig-zag ladder, i.e. a chain with nearest and next-nearest interactions, is defined by:

$$\hat{H}_{zig\text{-}zag} = \sum_j \left(J_1 \hat{\vec{S}}_j \cdot \hat{\vec{S}}_{j+1} + J_2 \hat{\vec{S}}_j \cdot \hat{\vec{S}}_{j+2} \right), \tag{20.9}$$

where $\hat{\vec{S}}_j = \frac{\hat{\vec{\sigma}}_j}{2}$ for spin $\frac{1}{2}$. The relative strength of the two interactions is given by $j = \frac{J_2}{J_1}$. In the case $j = 0$, one has the usual Heisenberg chain and, in the limit $j \to \infty$, one has two decoupled Heisenberg chains. For $j = \frac{1}{2}$ one has the Majumdar-Ghosh model,[35] with two perfectly dimerized ground states and a gap, in agreement with the Lieb-Shultz-Mattis theorem.[24,25] The gaped phase,[36] with degenerate valence bond ground states, starts at the critical point at $j_c \simeq 0.2411$, and extends beyond the Majumdar-Ghosh point. The anisotropic zig-zag chain has also been studied, using several methods, in particular the study of the crossing of the energy levels, together with their symmetries, i.e. quantum numbers.[37,38] This technique[39] is particularly useful, since it can give precise information on the occurrence of phase transitions and their locations, based on the study of finite systems, because it is less affected by the finite-size effects than the divergence of the correlations functions, only observed in the thermodynamic limit, for example. The concurrence is discontinuous at $j = \frac{1}{2}$, but does not show any special features at $j_c \simeq 0.2411$.[40]

The different types of behavior displayed by the concurrence at quantum critical points: a) discontinuous, b) continuous with a maximum at the QPT and c) continuous at the QPT with an extremum in higher order derivatives, have been related and analyzed in terms of the crossing of the ground states and first excited states.[17]

Looking at QPT from a quantum information perspective, namely identifying and studying the entanglement present in these macroscopic phenomena, has also allowed the discovery of new phase transitions. For instance, in gapped models exhibiting a finite correlation length, we can use entanglement to find a topological phase transition associated with non-local order parameters.[41]

The 2-D and 3-D XXZ models have also been investigated, namely numerically for the study of the concurrence.[32,42,43] In two- and three-dimensional systems, analytic results regarding the entanglement in ground states are difficult to obtain, and its study is much more challenging than in the one- and infinite-dimensional cases.

Given its analytic properties, as well as historical reasons,[44] the entanglement entropy of a connected subsystem is often considered. It is, in general, crudely proportional to the number of degrees of freedom in the boundary of the subsystem under consideration, which in turn is proportional to the area of the subsystem's boundary L^{d-1} (area law), with logarithmic corrections near the critical points, where L is the typical linear size of the subsystem.[45–48]

Other models or extensions of these models, but now without a know solution, can be studied by the usual methods in condensed matter physics, such as the exact diagonalization of finite systems, the Lanczos and modified Lanczos methods, the transfer matrix and the density matrix renormalization group. Furthermore, the technique of matrix product states (MPS), recently rediscovered in the context of quantum information, can offer us the possibility to efficiently simulate 1-D systems (see chapter 19 and references therein). Moreover, thanks to the MPS approach, it was possible to find (in fact, construct), in spin chains, a surprising and unconventional class of QPT points where the entanglement entropy remains finite and the ground state energy density is analytical (although every observable quantity is indeed non-analytical).[49] Using projected entangled-pair states (see chapter 19), this result can even be extended for higher-dimensional systems.[49]

Other methods used in physics and chemistry include the Contractor Renormalization (CORE) and the Effective Interactions in Real Space Renormalization Group (RSRG-EI), also discussed in previous chapters.

Finally, note that there exist other important macroscopic phenomena in low-dimensions, such as superconductivity and Bose-Einstein condensation. These will be reviewed in the next chapter, which deals with quantum many-body systems at finite temperatures.

20.3. Infinite-Dimensional Systems

Let us now briefly discuss the infinite-dimensional case and the role of entanglement in QPT of such systems. In the case of spin systems, this corresponds to the scenario where all the spins are symmetrically connected. In particular, in Ref. 6,7,50,51, this was studied for spin-$\frac{1}{2}$.

Other cases include two interacting bosonic modes or many spin-boson interacting models.[10] In these systems, the mean-field approach is exact in the thermodynamic limit. Furthermore, the Holstein-Primakov mapping between the $SU(2)$ and the boson-operators algebras allows us to compute the finite-size corrections were the entanglement is encoded. In Ref. 10, the authors describe the behavior of the entanglement entropy for a general quadratic Hamiltonian and show that this quantity diverges in every second order QPT point.

Another entanglement measure usually studied in infinite-dimensional systems is the *rescaled concurrence* C_R, defined as:

$$C_R \equiv (N-1)C, \tag{20.10}$$

where N is the number of binary degrees of freedom (such as spin-$\frac{1}{2}$) and C, given by Eq. (20.3), is the concurrence of the density matrix obtained by tracing out all-but-two spins of the system. Note that the pre-factor $(N-1)$ takes into account the coordination of each spin. Just as the entanglement entropy, the rescaled concurrence was shown to be non-analytical at the critical points for both first and second order QPT.[6,7,9]

One of the simplest models that exhibits a second order QPT is the Lipkin-Meshkov-Glick (LMG) model:

$$\hat{H}_{LMG} = h\hat{S}_z - \frac{\hat{S}_x^2}{N}, \tag{20.11}$$

where h is the global magnetic field (along the z-direction), $\hat{S}_k = \sum_{i=1}^{N} \hat{\sigma}_k^{(i)}$ and $\hat{\sigma}_k$ are the Pauli matrices (with $k = x, y, z$). This model has a second order QPT at $h = \pm 1$. Fig. 20.1 shows the behavior of the rescaled energy gap between the ground and the first excited states plotted as a function of the magnetic field h in the thermodynamic limit. As it can be observed, the gap vanishes at the second order QPT point. The scaling behavior with N and h is discussed in Ref. 51 and 52. The rescaled concurrence and the entanglement entropy are plotted in Fig. 20.2 and 20.3 respectively. These quantities are non-analytic at the QPT point and present a global maximum.

All these studies of entanglement in quantum phase transitions in infinite-dimensional systems show that entanglement can be seen as a good order parameter to detect critical points, as well as a new type of correlations that play a significant role in the QPT and that we can now identify and quantify.

20.4. Quantum Computation and Quantum Phase Transitions

The idea of the quantum computer, i.e. a quantum system that would process information, was proposed in the early 1980's independently by Benioff and Feynman. Benioff thought of a quantum Turing machine[53,54] that would use the reversible (unitary) time evolution of quantum mechanics as a way to do computation without dissipating any energy. Feynman had quite a different motivation and proposed a computer based on the laws of quantum physics as a natural and efficient way to simulate the dynamics of quantum systems,[55] whose encoding in a classical computer/memory grows exponentially with the number of degrees of freedom.

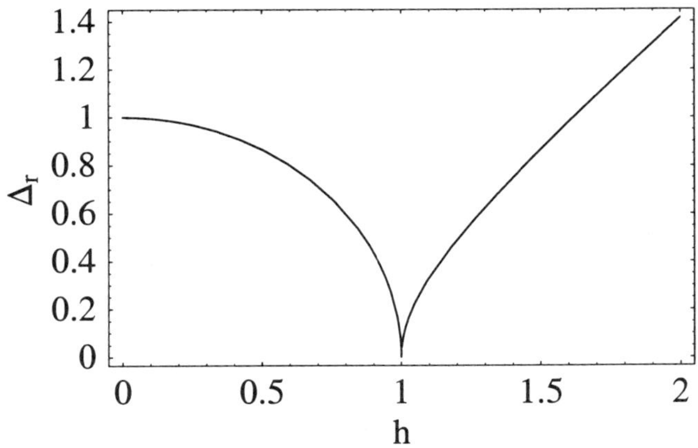

Fig. 20.1. Rescaled energy gap $\Delta_r = \Delta/N$ between the ground and the first excited states plotted as a function of the magnetic field h in the thermodynamic limit. The gap vanishes at the QPT ($h = 1$).

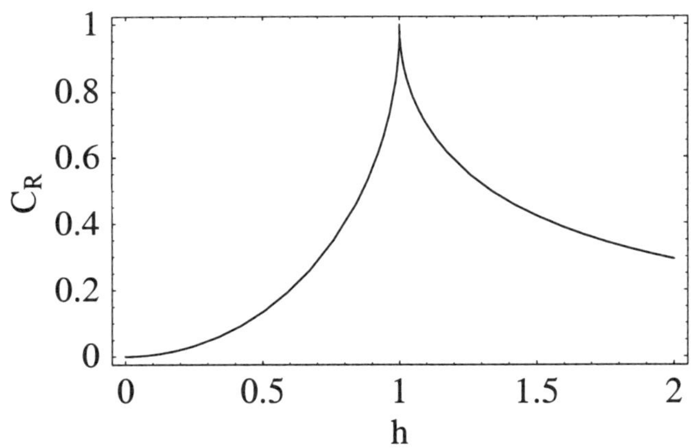

Fig. 20.2. Rescaled concurrence C_R of the ground state of the LMG model as a function of the magnetic field h. This quantity presents a non-analyticity seen by the cusp-like form at the phase transition where it reaches the maximal value $C_R = 1$.

Following the first proposal, by Deutsch,[56] in 1985, of a quantum algorithm that could outperform the efficiency of a (deterministic) classical one, Shor showed in 1994 that it was possible to factorize integers efficiently using a quantum algorithm.[57,58] This result has very important consequences, as it means that it is in theory possible to challenge the privacy of most current electronic communications. Shor's work sparked the attention of both theorists and experimentalists

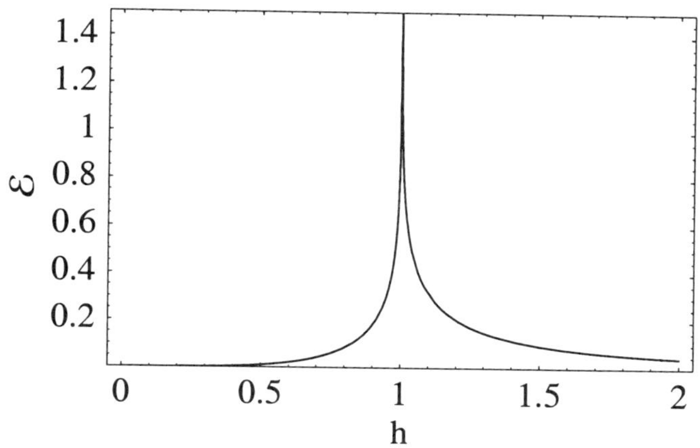

Fig. 20.3. Entanglement entropy $\mathcal{E}$ of the density matrix of the ground state of the LMG model after tracing out $\frac{N}{2}$ spins as a function of the magnetic field h. The entanglement entropy diverges logarithmically at the second order QPT point.

to the idea of the quantum computer and the race to build its respective hardware. Since then, a few other interesting quantum algorithms have emerged and some important experimental steps were achieved, but so far the quantum computer is still a machine that exists only in research laboratories, in a embryonary form (for more details on quantum algorithms and implementations of the quantum computer, see Ref. 59).

Entanglement is known to play an essential role in the speed-up over classical computation by quantum algorithms operating on pure states,[60] although it should be noted that in general, and namely for mixed states, entanglement is not a necessary condition for quantum computational speed-up.[61–63] Recent studies have investigated the role of entanglement in well known quantum algorithms,[64–68] but the exact mechanism by witch it intervenes is still unclear. In particular, Orús *et al.* proved that Shor's algorithm uses an exponentially large (in the size of the system) amount of entanglement,[67] which implies that it would be very hard to simulate it classically in an efficient way (see chapter 19 and Ref. 65).

In 2000, a new model of quantum computation was proposed by Farhi *et al.*,[69] based on the quantum adiabatic theorem,[70] and known as *adiabatic quantum computation* (AQC). This alternative model consists of using adiabatic evolution to go from an initial Hamiltonian H_0 — with a well known and easy-to-prepare ground state — to the ground state of a final Hamiltonian (or problem Hamiltonian) H_P which encodes the answer to a given classical problem. This approach has been proved to be equivalent to the standard model of quantum computation,[71,72] which is based on circuits of unitary gates. The time T necessary to perform the AQC is such that it satisfies the adiabatic theorem, limiting the probability of jumping to the first excited state. This roughly translates into the condition:

$$T >> \frac{|\partial_s H(s)|_{max}}{\Delta^2_{min}}, \tag{20.12}$$

where s is a parametrization of the time such that $t(s=0)=0$ and $t(s=1)=T$ (usually, we have $s=t/T$), and thus $H(0)=H_0$ and $H(1)=H_P$, and Δ is the value of the energy gap between the ground and the first excited states. The maximum and minimum in Eq. (20.12) are taken along the adiabatic evolution. Note that the term $|\partial_s H(s)|$ usually varies slowly and thus the time of the computation will be controlled essentially by the minimum value of the energy gap.

In the most interesting cases, the system will undergo a quantum phase transition, almost closing the gap along the Hamiltonian path. Therefore, the time scaling with the number N of quantum bits (i.e. binary degrees of freedom), and consequently the computational efficiency, will be dominated by the behavior of the energy gap. The time will then increase exponentially with N in first order QPT, and algebraically (i.e. polynomially) in second order ones. This makes all the difference in computational terms: in the former case the time of the computation quickly becomes very large for interesting N, whereas in the latter it is possible to perform efficient computations. Thus, when encoding computational problems in the adiabatic model, one should look for systems exhibiting second order quantum phase transitions, which will then offer an efficient algorithmic solution. Yet, in general, this is a highly non-trivial challenge, except in very simplistic toy models, such as the one proposed in Ref. 73.

Adiabatic quantum computation algorithms could also be used to prepare a quantum simulator in the ground state of a given Hamiltonian[74] and then study the respective system (e.g. many-body), say by measuring its correlation functions. Moreover, as it has been stated before, entanglement reaches maximal values at second order quantum phase transition points, the very same points where the power of an adiabatic algorithm seems to be maximized.[69,72,73,75] Adiabatic quantum computation thus appears as a new approach to quantum computation with direct connections to QPT, and offers interesting prospects and settings for studying both phenomena. A (useful) quantum computer is, after all, a many-body quantum system, and the understanding of the role of entanglement in the latter will certainly help us clarify the role of quantum correlations in the former, and vice-versa.

20.5. Summary

Quantum phase transitions represent the most significant part of the physics of macroscopic quantum phenomena at zero temperature. After having introduced the concept of quantum entanglement in the previous chapter, here we have reviewed recent studies on its role in quantum phase transitions. The progress in the understanding (and control) of this fundamental aspect of quantum mechanics came from the recent advent of quantum information theory. Yet, recently it has become clear that it also plays an important role in condensed matter as well.

After having introduced the *entanglement entropy* and the *concurrence*, we discussed how these measures of entanglement behave in one-, low- and infinite-dimensional spin and electron systems in both first- and second-order phase transitions. We saw how the study of these quantities can offer new insight into the (quantum) correlations present in those systems, new order parameters and, in some cases, even QPT points that were previously unknown. In general, systems undergoing a quantum phase transition are quite demanding in terms of entanglement. But, surprisingly, this is not always the case. We mentioned the recent discovery of an unconventional class of QPT where the entanglement entropy does not diverge and the ground-state energy remains analytical.

Finally, we introduced the idea of the quantum computer and presented the adiabatic model of quantum computation. We discussed the connections between quantum phase transitions and this non-standard model of computation, as well as the role of entanglement in both phenomena.

Acknowledgments

The authors thank the support from Fundação para a Ciência e a Tecnologia (Portugal) through programs POCTI/POCI, partially funded by FEDER (EU), namely through projects POCI/MAT/55796/2004 QuantLog, POCTI-SFA-2-91 and POCI/FIS/58746/2004, and grant SFRH/BD/16182/2004/2ZB5, as well as the SQIG-IT EMSAQC initiative.

References

1. S. Sachdev, *Quantum Phase Transitions*. (Cambridge University Press, 2001).
2. T. J. Osborne and M. A. Nielsen, Entanglement in a simple quantum phase transition, *Phys. Rev. A*. **66**, 032110, (2002).
3. A. Osterloh, L. Amico, G. Falci, and R. Fazio, Scaling of entanglement close to a quantum phase transitions, *Nature*. **416**, 608, (2002).

4. J. I. Latorre, E. Rico, and G. Vidal, Ground state entanglement in quantum spin chains, *Quantum Inf. Comput.* **4**, 48–92, (2004).
5. D. Larsson and H. Johannesson, Entanglement scaling in the one-dimensional Hubbard model at criticality, *Phys. Rev. Lett.* **95**, 196406, (2005).
6. J. Vidal, G. Palacios, and R. Mosseri, Entanglement in a second order quantum phase transition, *Phys. Rev. A.* **69**, 022107, (2004).
7. J. Vidal, R. Mosseri, and J. Dukelsky, Entanglement in a first order quantum phase transition, *Phys. Rev. A.* **69**, 054101, (2004).
8. G. Vidal, J. I. Latorre, E. Rico, and A. Kitaev, Entanglement in quantum critical phenomena, *Phys. Rev. Lett.* **90**, 227902, (2003).
9. J. Vidal, Concurrence in collective models, *Phys. Rev. A.* **73**, 062318, (2006).
10. J. Vidal, S. Dusuel, and T. Barthel. Entanglement entropy in collective models. cond-mat/0610833, (2006).
11. F. G. S. L. Brandão, Entanglement as a quantum order parameter, *New J. Phys.* **7**, 254, (2005).
12. A. Anfossi, C. D. E. Boschi, A. Montorsi, and F. Ortolani, Single-site entanglement at "superconductor"-insulator transition in the Hirsch model, *Phys. Rev. B.* **73**, 085113, (2006).
13. V. Vedral, The role of relative entropy in quantum information theory, *Rev. Mod. Phys.* **74**, 197–234, (2002).
14. C. H. Bennett, D. P. DiVincenzo, J. A. Smolin, and W. K. Wootters, Mixed-state entanglement and quantum error correction, *Phys. Rev. A.* **54**, 3824–3851, (1996).
15. W. K. Wootters, Entanglement of formation of an arbitrary state of two qubits, *Phys. Rev. Lett.* **80**, 2245, (1998).
16. V. Coffman, J. Kundu, and W. K. Wootters, Distributed entanglement, *Phys. Rev. A.* **61**, 052306, (2000).
17. S.-J. Gu, G.-S. Tian, and H.-Q. Lin, Entanglement and quantum phase transitions. (2005). quant-ph/0511243.
18. D. C. Mattis, *The Many-Body Problem–An Encyclopaedia of Exactly Solved Models in One Dimension.* (World Scientific, 1993).
19. B. Sutherland, Model for a multicomponent quantum system, *Phys. Rev. B.* **12**, 3795–3805, (1975).
20. H. M. Babujian, Exact solution of the one-dimensional isotropic Heisenberg chain with arbitrary spins S, *Phys. Lett. A.* **90**, 479–482, (1982).
21. L. A. Takhtajan, The picture of low-lying excitations in the isotropic Heisenberg chain of arbitrary spins, *Physics Letters A.* **87**, 479–482, (1982).
22. F. D. M. Haldane, Continuum dynamics of the 1-D Heisenberg anti-ferromagnet - identification with the O(3) non-linear sigma-model, *Phys. Lett. A.* **93**, 464–468, (1983).
23. F. D. M. Haldane, Nonlinear field theory of large-spin Heisenberg antiferromagnets: semiclassically quantized solitons of the one-dimensional easy-axis Néel state, *Phys. Rev. Lett.* **50**, 1153–1156, (1983).
24. E. Lieb, T. Schultz, and D. Mattis, 2 soluble models of an antiferromagnetic chain, *Ann. Phys. (NY).* **16**, 407–466, (1961).
25. I. Affleck and E. H. Lieb, A proof of part of Haldane conjecture on spin chains, *Lett. Math. Phys.* **12**, 57–69, (1986).

26. S. Katsura, Statistical mechanics of the anisotropic linear Heisenberg model, *Phys. Rev.* **127**, 1508 – 1518, (1962).
27. E. Barouch and B. M. McCoy, Statistical mechanics of the XY model. II. spin-correlation functions, *Phys. Rev. A.* **3**, 786–804, (1971).
28. H. Bethe, On the theory of metals, I: Eigenvalues and eignefunctions of a linear chain of atoms, *Zeits. Physik.* **74**, 205–226, (1931). Translated in *Selected Works of Hans A. Bethe with Commentary* (World Scientific, 1996).
29. R. Orbach, Linear antiferromagnetic chain with anisotropic coupling, *Phys. Rev.* **112**, 309–316, (1958).
30. S.-J. Gu, H.-Q. Lin, and Y.-Q. Li, Entanglement, quantum phase transition, and scaling in the XXZ chain, *Phys. Rev. A.* **68**, 042330, (2003).
31. O. F. Syljuåsen, Entanglement and spontaneous symmetry breaking in quantum spin models, *Phys. Rev. A.* **68**, 060301(R), (2003).
32. S.-J. Gu, G.-S. Tian, and H.-Q. Lin, Ground-state entanglement in the XXZ model, *Phys. Rev. A.* **71**, 052322, (2005).
33. C. Yang, A. N. Kocharian, and Y. L. Chiang, Phase transitions and exact ground-state properties of the one-dimensional Hubbard model in a magnetic field, *J. Phys.: Condens. Matter.* **12**, 7433–7454, (2000).
34. S.-J. Gu, S.-S. Deng, Y.-Q. Li, and H.-Q. Lin, Entanglement and quantum phase transition in the extended Hubbard model, *Phys. Rev. Lett.* **93**, 086402, (2004).
35. C. K. Majumdar and D. K. Ghosh, On next-nearest-neighbor interaction in linear chain. II, *J. Math. Phys.* **10**, 1399–1402, (1969).
36. S. R. White and I. Affleck, Dimerization and incommensurate spiral spin correlations in the zigzag spin chain: Analogies to the Kondo lattice, *Phys. Rev. B.* **54**, 9862–9869, (1996).
37. V. R. Vieira, N. Guihéry, J. P. Rodriguez, and P. D. Sacramento, Decoupling of the S=1/2 antiferromagnetic zig-zag ladder with anisotropy, *Phys. Rev. B.* **63**, 224417, (2001).
38. P. D. Sacramento and V. R. Vieira, Spin dynamics of the S = 1/2 antiferromagnetic zig-zag ladder with anisotropy, *J. Phys.: Condens. Matter.* **14**, 591–604, (2002).
39. K. Okamoto and K. Nomura, Fluid-dimer critical point in $s = \frac{1}{2}$ antiferromagnetic Heisenberg chain with next nearest neighbor interactions, *Phys. Lett. A.* **169**, 433–437, (1992).
40. S.-J. Gu, H. Li, Y.-Q. Li, and H.-Q. Lin, Entanglement of the Heisenberg chain with the next-nearest-neighbor interaction, *Phys. Rev. A.* **70**, 052302, (2004).
41. F. Verstraete, M. A. Martin-Delgado, and J. I. Cirac, Diverging entanglement length in gapped quantum spin systems, *Phys. Rev. Lett.* **92**, 087201, (2004).
42. H.-Q. Lin, J. S. Flynn, and D. D. Betts, Exact diagonalization and quantum Monte Carlo study of the spin-$\frac{1}{2}$ XXZ model on the square lattice, *Phys. Rev. B.* **64**, 214411, (2001).
43. O. F. Syljuåsen, Concurrence in the two-dimensional XXZ- and transverse field Ising-models, *Physics Letters A.* **322**, 25–30, (2004).
44. J. D. Bekenstein, Black holes and entropy, *Phys. Rev. D.* **7**, 2333–2346, (1973).
45. M. B. Plenio, J. Eisert, J. Dreissig, and M. Cramer, Entropy, entanglement, and area: analytical results for harmonic lattice systems, *Phys. Rev. Lett.* **94**, 060503, (2005).

46. M. M. Wolf, Violation of the entropic area law for fermions, *Phys. Rev. Lett.* **96**, 010404, (2006).
47. M. Cramer, J. Eisert, M. B. Plenio, and J. Dreissig, An entanglement-area law for general bosonic harmonic lattice systems, *Phys. Rev. A.* **73**, 012309, (2006).
48. T. Barthel, M.-C. Chung, and U. Schollwoeck, Entanglement scaling in critical two-dimensional fermionic and bosonic systems, *Phys. Rev. A.* **74**, 022329, (2006).
49. M. M. Wolf, G. Ortiz, F. Verstraete, and J. I. Cirac, Quantum phase transitions in matrix product systems, *Phys. Rev. Lett.* **97**, 110403, (2006).
50. J. I. Latorre, R. Orús, E. Rico, and J. Vidal, Entanglement entropy in the Lipkin-Meshkov-Glick model, *Phys. Rev. A.* **71**, 064101, (2005).
51. S. Dusuel and J. Vidal, Continuous unitary transformations and finite-size scaling exponents in the Lipkin-Meshkov-Glick model, *Phys. Rev. B.* **71**, 224420, (2005).
52. S. Dusuel and J. Vidal, Finite-size scaling exponents of the Lipkin-Meshkov-Glick model, *Phys. Rev. Lett.* **93**, 237204, (2004).
53. P. Benioff, The computer as a physical system: A microscopic quantum mechanical hamiltonian model of computers as represented by Turing machines, *J. Stat. Phys.* **22**, 563–591, (1980).
54. P. Benioff, Quantum mechanical models of Turing machines that dissipate no energy, *Phys. Rev. Lett.* **48**, 1581–1585, (1982).
55. R. P. Feynman, Simulating physics with computers, *Int. J. of Theor. Phys.* **21**, 467, (1982).
56. D. Deutsch, Quantum theory, the Church-Turing principle and the universal quantum computer, *Proc. R. Soc. Lond. A.* **400**, 97–117, (1985).
57. P. W. Shor. Algorithms for quantum computation: Discrete logarithms and factoring. In ed. S. Goldwasser, *Proceedings of the 35th Annual Symposium on the Foundations of Computer Science (FOCS'94)*, pp. 124–134, Los Alamitos, CA, (1994). IEEE Computer Society.
58. P. W. Shor, Polynomial-time algorithms for prime factorization and discrete logarithms on a quantum computer, *SIAM J. Comput.* **26**, 1484–1509, (1997). Presented at FOCS'94 (see Ref. 57).
59. M. Nielsen and N. Chuang, *Quantum computation and quantum information.* (Cambridge University Press, 2000).
60. R. Jozsa and N. Linden, On the role of entanglement in quantum computational speed-up, *Proc. R. Soc. A.* **459**, 2011, (2003).
61. E. Biham, G. Brassard, D. Kenigsberg, and T. Mor, Quantum computing without entanglement, *Theoretical Computer Science.* **320**, 15–33, (2004).
62. E. Knill and R. Laflamme, Power of one bit of quantum information, *Phys. Rev. Lett.* **81**, 5672–5675, (1998).
63. A. Datta and G. Vidal. On the role of entanglement and correlations in mixed-state quantum computation. quant-ph/0611157, (2006).
64. S. Parker and M. B. Plenio, Entanglement simulations of Shor's algorithm, *J. Mod. Opt.* **49**, 1325–1353, (2002).
65. G. Vidal, Efficient classical simulation of slightly entangled quantum computations, *Phys. Rev. Lett.* **91**, 147902, (2003).
66. R. Orús, J. I. Latorre, and M. A. Martin-Delgado, Systematic analysis of majorization in quantum algorithms, *Eur. Phys. Journal D.* **29**, 119, (2004).

67. R. Orús and J. I. Latorre, Universality of entanglement and quantum computation complexity, *Phys. Rev. A.* **69**, 052308, (2004).
68. R. Orús. *Entanglement, quantum phase transitions and quantum algorithms*. PhD thesis, Universitat de Barcelona, (2006). quant-ph/0608013.
69. E. Farhi, J. Goldstone, S. Gutmann, and M. Sipser. Quantum computation by adiabatic evolution. quant-ph/0001106, (2000).
70. A. Messiah, *Quantum Mechanics*. vol. II, (Amsterdam: North Holland, New York: Wiley, 1976).
71. W. van Dam, M. Mosca, and U. Vazirani. How powerful is adiabatic quantum computation? In *FOCS'01: Proceedings of the 42nd Annual IEEE Symposium on Foundations of Computer Science (FOCS'01)*, pp. 279–287, Washington, DC, USA, (2001). IEEE Computer Society.
72. D. Aharonov, W. van Dam, J. Kempe, Z. Landau, S. Lloyd, and O. Regev. Adiabatic quantum computation is equivalent to standard quantum computation. In *FOCS'04: Proceedings of the 45th Annual IEEE Symposium on Foundations of Computer Science (FOCS'04)*, pp. 42–51, Washington, DC, USA, (2004). IEEE Computer Society.
73. P. Ribeiro and R. Mosseri, Adiabatic computation: A toy model, *Phys. Rev. A.* **74**, 042333, (2006).
74. V. Murg and J. I. Cirac, Adiabatic time evolution in spin systems, *Phys. Rev. A.* **69**, 042320, (2004).
75. J. I. Latorre and R. Orús, Adiabatic quantum computation and quantum phase transitions, *Phys. Rev. A.* **69**, 062302, (2004).

Chapter 21

Macroscopic thermal entanglement

N. Paunković

SQIG, Instituto de Telecomunicações and IST, Technical University of Lisbon, P-1049-001 Lisbon, Portugal

Y. Omar

SQIG, Instituto de Telecomunicações, P-1049-001 Lisbon and CEMAPRE, ISEG, Technical University of Lisbon, P-1200-781 Lisbon, Portugal

V. Rocha Vieira

CFIF and Department of Physics, IST, Technical University of Lisbon, P-1049-001 Lisbon, Portugal

We review recent results on the existence and use of entanglement in macroscopic systems at finite temperatures. Entanglement represents a particular quantum form of correlations and we explore the relevance of these correlations on thermodynamic properties of macroscopic systems, as well as its survival at finite temperatures. Various lower bounds for *the entanglement temperature* (the temperature below which entanglement surely exists in a system) are discussed and it is shown that, in some cases, macroscopic entanglement can exist even at room temperatures. Furthermore, the relevant correlations determining the macroscopic behavior of some systems are present due to entanglement only, as we show in the example of high-temperature superconductivity. Entanglement can also affect various thermodynamic properties, such as magnetic susceptibility, internal energy and pressure. Motivated by the fact that entanglement is a crucial resource for quantum information processing, we also discuss how to extract it from macroscopic systems, namely from spin-chains. Finally, we illustrate the generation and manipulation of entanglement in a macroscopic system of non-interacting hopping bosons.

Contents

21.1. Introduction

In the previous chapter we have seen that at zero temperature, entanglement can exist in *macroscopic* systems. [a] Moreover, it plays a significant role, particularly in quantum phase transitions[1] (QPT), where ground states are pure [b] and entanglement is the only source of correlations among sub-systems of a composite system. [c] It is then natural to ask if entanglement can persist in more realistic situations, i.e. in macroscopic systems whose states are not given by pure, but by general mixed quantum states. In this chapter, we shall see not only that the answer to this question is affirmative, but that entanglement can affect a number of relevant thermodynamic properties of macroscopic systems in real physical experiments at temperatures different from zero, and maybe even in everyday life! Moreover, entanglement can be, at least in principle, extracted from various macroscopic systems and used as a resource in quantum information processing. We shall concentrate our discussion mainly on the case of systems that are in *thermal equilibrium* with the environment, but we will also briefly mention cases of open quantum systems and systems that are far from thermal equilibrium.

For a quantum system defined by the Hamiltonian $\hat{H}$, its thermal state is given by $\hat{\rho}(T) = \exp(-\beta\hat{H})/Z$, where $Z = \mathrm{Tr}(-\beta\hat{H})$ is the *partition function* and $\beta = 1/(k_B T)$, k_B being the Boltzmann constant and $T \in [0, \infty)$ the absolute temperature. It is a continuous function of the temperature T, therefore if the ground state of the system at $T = 0$ is an entangled pure state $|g\rangle$, then there always exists a finite *entanglement temperature* $T_e > 0$ below which entanglement will exist in the thermal state $\hat{\rho}(T)$ as well. [d] Further, as in the limit $T \to \infty$

[a] A composite system is said to be macroscopic if the number of its constituents N (or equivalently, the number of its degrees of freedom) is large. Mathematically, it means that formal calculations of all the relevant physical quantities are taken in the so-called thermodynamic limit, when $N \to \infty$.
[b] At $T = 0$, the ground state of a system is given by the pure (non-degenerate) quantum state with the lowest energy.
[c] Note that pure separable states are simple product states which contain no correlations whatsoever.
[d] Note that every reasonable measure of entanglement is a continuous function on the set of quantum states.

the thermal state of the physical system is proportional to the identity, [e] it follows that entanglement will always disappear at some finite *separability temperature* $T_s < \infty$, above which the thermal state of a system will be separable. [f]

A simple lower bound for T_e can be obtained using the notion of *global robustness of entanglement*.[6] For a given mixed state $\hat{\rho}$, its global robustness of entanglement $R(\hat{\rho})$ is the minimal value $t \geq 0$ for which there exists a mixed state $\hat{\sigma}$ such that $(\hat{\rho} + t\hat{\sigma})/(1+t)$ is separable. In other words, $R(\hat{\rho})$ is the minimal amount of "noise" (given by the state $\hat{\sigma}$) needed to disentangle the state $\hat{\rho}$. If $|i\rangle$ $(i = 1; \ldots d-1)$ are the excited states of a composite system given by the Hamiltonian $\hat{H}$ and E_i their corresponding energies (with E_g being the energy of the ground state $|g\rangle$), then the thermal state $\hat{\rho}(T)$ can be written in the following form:

$$\hat{\rho}(T) = \frac{e^{-E_g/k_B T}}{Z}|g\rangle\langle g| + \sum_{i=1}^{d-1} \frac{e^{-E_i/k_B T}}{Z}|i\rangle\langle i|. \tag{21.1}$$

The second term from the right-hand side of Eq. (21.1) can be interpreted as the "noise" term $t\hat{\sigma}/(1+t)$ from the above definition of the global robustness of entanglement; we refer to it as the "thermal noise", with respect to the ground state $|g\rangle$. In other words, if $e^{-\beta E_g}/Z = 1/(1+t) \geq 1/(1+R(|g\rangle))$, then $\hat{\rho}(T)$ is an entangled state and we arrive at a simple lower bound for the *entanglement temperature* T_e, the *transition temperature*[7] T_t given by:

$$\frac{e^{-E_g/k_B T_t}}{Z} = \frac{1}{1+R(|g\rangle)}. \tag{21.2}$$

Although very simple, the above condition for the existence of entanglement can be useful in certain cases. For example, it gives the exact entanglement temperature T_e in the case of two spin one-half particles coupled via the Heisenberg interaction (for more detailed study, see Ref. 7). On the other hand, it is based only on the properties (entanglement and energy) of the ground state, and it is obvious that, in general, the properties of thermal states are given by the whole spectrum of the system considered. That is to say, Eq. (21.2) is just a sufficient condition for the existence of entanglement and it may well be that entanglement can persist at temperatures much higher than T_t, depending on the statistical properties of the particular "thermal noise". Moreover, it can be shown that in some cases, contrary to what one might intuitively expect, entanglement can even *increase* with the rise

[e] $\lim_{T\to\infty} \hat{\rho}(T) = \frac{\hat{I}}{d}$, where d is the dimension of the system's Hilbert space $\mathcal{H}$ and $\hat{I}$ is the identity operator on $\mathcal{H}$.

[f] This is a consequence of the fact that $\hat{\rho}(\infty) = \frac{\hat{I}}{d}$ is an interior point of the set of separable states[2] and therefore there exists a neighborhood of $\hat{\rho}(\infty)$ consisting of separable states only (see the discussion given in section II. of Ref. 3; for a more formal treatment, see Refs. 4,5).

of temperature (in fact, that is precisely the reason for distinguishing between the *entanglement temperature* T_e and the *separability temperature* T_s).

Thermal entanglement was first discussed by M. A. Nielsen[8] on a system of two spins in an external magnetic field, coupled by the antiferromagnetic Heisenberg-like interaction. [g] To show its basic features, we will first present this simple two-spin case, and then following Ref. 9 the numerical results for up to $N = 10$ spins. The Hamiltonian of the system is $\hat{H} = J\hat{\vec{\sigma}}_1 \cdot \hat{\vec{\sigma}}_2 + B(\hat{\sigma}_1^z + \hat{\sigma}_2^z)$, where $J > 0$ is the coupling, B is the external magnetic field and $\hat{\vec{\sigma}}$ is the vector of Pauli spin operators. The eigenstates of this system are the triplet states $|\uparrow\uparrow\rangle$, $|\psi^+\rangle = \frac{1}{\sqrt{2}}(|\uparrow\downarrow\rangle + |\downarrow\uparrow\rangle)$, $|\downarrow\downarrow\rangle$, and the singlet state $|\psi^-\rangle = \frac{1}{\sqrt{2}}(|\uparrow\downarrow\rangle - |\downarrow\uparrow\rangle)$, with the corresponding energies $J+2B$, J, $J-2B$ and $-3J$ respectively. First, we see that in the absence of a magnetic field, the ground state is the maximally entangled singlet state $|\psi^-\rangle$. Therefore, following the above discussion we can conclude that there exists a finite temperature $T_e > 0$ below which entanglement surely persists in a thermal state. Moreover, one can explicitly calculate[7] the transition temperature $T_t = 4J/(k_B \ln 3)$, using the result for the global robustness of entanglement of the singlet state,[6] $R(|\psi^-\rangle) = 1$. On the other hand, it is also possible to calculate analytically[9] the entanglement of formation[11] $E_F(\hat{\rho}(T))$ for general B and T, by using the relation[11] between the entanglement of formation $E_F(\hat{\rho})$ of a two-qubit system and the concurrence $C(\hat{\varrho})$, where $\hat{\varrho} = \hat{\rho}(\hat{\sigma}_y \otimes \hat{\sigma}_y)\hat{\rho}^*(\hat{\sigma}_y \otimes \hat{\sigma}_y)$. It follows that, as noted above, in the case of $B = 0$, we have $T_e = T_t$. By increasing the magnetic field, the energy of the triplet state $|\downarrow\downarrow\rangle$ decreases until, for $B_c = 2J$, [h] it becomes equal to the ground state energy of the singlet state. Therefore, for $B > B_c$ the ground state becomes the triplet state $|\downarrow\downarrow\rangle$ and, still at $T = 0$, the system is disentangled. Yet, by increasing the temperature, the fraction of the singlet component from the "thermal noise" will start bringing back the entanglement so that within a certain range of temperatures (T^*, T_s), T_s being the separability temperature defined before, the thermal state $\hat{\rho}(T)$ will be entangled. The above analytical results have been also numerically confirmed for spin-chains of up to $N = 10$ spins, showing that a similar qualitative behavior for the nearest neighbor entanglement might persist in a macroscopic case as well. For more details, see Ref. 9. Especially interesting is the case of *magnetically induced entanglement* between next to nearest neighbors at finite temperatures.

---

[g] We will discuss only the antiferromagnetic case ($J > 0$), as it can be shown that no entanglement exists in the ferromagnetic case.[9] For a slightly different treatment of the two-spin entanglement in Heisenberg ferromagnets, see Ref. 10.

[h] At this point, the structure of the ground state of the system undergoes a dramatic change, which corresponds to the point of a quantum phase transition.[1] Note that this change in the ground state structure occurs with an equally sudden change of the ground state entanglement, from $E_F(|\psi^-\rangle) = 1$ to $E_F(|\downarrow\downarrow\rangle) = 0$.

The work along these lines was further conducted in Ref. 12, where an Ising chain of up to $N = 7$ spins in an external magnetic field was discussed. Also, similar investigations were taken for the cases of the one-dimensional Kondo necklace model[13] (up to two sites, $N = 4$ spins) and the XXZ model[14] (up to $N = 6$ spins). All these results are on thermal entanglement in composite systems consisting of up to $N = 10$ particles and at temperatures close to zero. Next, we will discuss survival of entanglement in genuine macroscopic systems, and at temperatures further away from the absolute zero.

21.2. Entanglement in High-Temperature Superconductivity

In this section, we shall examine a genuine case of a macroscopic system which contains entanglement even up to $T = 160$ K, which is the typical critical temperature for high-T_c superconductors. Following Ref. 15 we will present a basic analysis of the multipartite entanglement in high-temperature superconductivity.[16–19]

The existence of superconductivity (as well as superfluidity) is connected to the persistence of a special type of correlations called off-diagonal long-range order[20] (ODLRO). As well as in the "ordinary" low-temperature BCS model, where the existence of ODLRO is a consequence of the electron pairing in the ground state, so is the case in the high-temperature superconductor, [i] with two slight, yet quite relevant differences. First, unlike the BCS case, where the paired electrons are separated by a certain coherence length, in the so-called η-pairing mechanism[16–19] for high-temperature superconductivity, the paired electrons are situated on the same site of the lattice. The actual physical mechanism that leads to condensation of such on-site electron pairing is of no importance for our current study of entanglement, so we will skip it here. The important consequence of such a mechanism is that the *energy gap* between the ground and the first excited states is considerably bigger in this case than in the case of the low-temperature BCS superconductor. Therefore, the superconducting properties of the ground state will persist within a larger interval of temperatures in spite of the destructive influence of the "thermal noise". The crucial connection to entanglement comes from the following observation: as ODLRO is a form of correlation, its existence in the *pure* ground state can be only due to a certain type of multipartite entanglement. Further, as the superconducting properties of thermal states below the critical tem-

[i]Originally, Yang[16] introduced the so-called η-states as eigenstates of the Hubbard model having ODLRO, but the ground state of such simple model was shown *not* to be one of them. A slightly modified version of the Hubbard model whose ground state *is* one of the η-states was presented in Refs. 17–19. Strictly speaking, our analysis applies to such models only.

perature T_c are inherited *solely* from such properties (ODLRO) contained in the ground state, which are entanglement based, so entanglement must [j] persist at least up to T_c. In the following, we will formally explore the entanglement in η-states – the states that make up the ground state and the low-energy spectrum, and its relation to the existence of ODLRO. Then, we will give an upper bound to the multipartite entanglement contained in a thermal state made up out of such low-energy states only, as well as its relation to ODLRO and classical correlations in general.

Our model consists of spin one-half fermions (i.e. electrons) occupying a lattice with n sites labeled by $i \in \{1, \ldots n\}$. The electron modes are obtained by the anticommuting creation operators $\hat{c}^\dagger_{i,s}$, where $s \in \{\uparrow, \downarrow\}$ denote the spin index. The η-pair is an equal superposition of n orthogonal states each representing the electron pair residing at a different lattice site [k] and is obtained from the vacuum by the operator:

$$\hat{\eta}^\dagger = \sum_{i=1}^{n} \hat{c}^\dagger_{i,\uparrow}\hat{c}^\dagger_{i,\downarrow}. \tag{21.3}$$

If the system consists of $2k$ electrons, than the ground state of our model contains k identical η-pairs, and is given by $|k,n\rangle = (C^n_k)^{-1/2}(\hat{\eta}^\dagger)^k|\Omega\rangle$, where $C^n_k = \frac{n!}{k!(n-k)!}$ is the normalization constant and $|\Omega\rangle$ is the vacuum state. Therefore, the η-states $|k,n\rangle$ consist of equal superpositions of pure states (the number of them being equal to C^n_k), each having different k sites of the lattice occupied by one of the electron pairs. If by 0 and 1 we denote a vacant and occupied site respectively, we can represent η-states in the following form:

$$|k,n\rangle = (C^n_k)^{-1/2}(|\underbrace{00\ldots0}_{n-k}\underbrace{11\ldots1}_{k}\rangle + \ldots|\underbrace{11\ldots1}_{k}\underbrace{00\ldots0}_{n-k}\rangle). \tag{21.4}$$

In other words, each site of the lattice can be represented by a qubit with basis states $|0\rangle$ and $|1\rangle$, while η-states are symmetric n-qubit states. Representing η-states in such a way shows to be very convenient when calculating entanglement, due to their high degree of symmetry.

[j] Note that for the current argument the crucial assumption is that superconducting properties in a thermal state are inherited *only* from the ground state. This in turn means that the "thermal noise" can act only destructively and spoil such a property – ODLRO correlations, and therefore entanglement itself. But one can at least in principle construct a Hamiltonian for which the thermal mixing of the low-energy excitations would spoil the entanglement much below T_c, while preserving ODLRO correlations. In other words, the straightforward connection established between ODLRO and entanglement in pure states might be, at least in principle, much more subtle and complicated in the case of (thermal) mixed states.

[k] As they do not influence the overall entanglement,[15] we omit the phase factors originally introduced in Refs. 17–19.

As noted, the existence of ODLRO is given by the non-vanishing long-range correlations: $\langle \hat{c}^\dagger_{i,\uparrow}\hat{c}^\dagger_{i,\downarrow}\hat{c}_{j,\downarrow}\hat{c}_{j,\uparrow}\rangle \xrightarrow{|i-j|\to\infty} \alpha > 0$. The two-site partial state is (note that due to the symmetry of the η-states, all two-site states are isomorphic to each other):

$$\hat{\rho}_2(k,n) = \mathrm{Tr}_{3\ldots n}|k,n\rangle\langle k,n| = a|00\rangle\langle 00| + b|11\rangle\langle 11| + 2c|\psi^+\rangle\langle\psi^+|, \quad (21.5)$$

with $a = \frac{k(k-1)}{n(n-1)}$, $b = \frac{(n-k)(n-k-1)}{n(n-1)}$ and $c = \frac{k(n-k)}{n(n-1)}$. Using the Peres-Horodecki[21,22] partial transposition criterion for a two-qubit state separability one can easily find out that the bipartite entanglement will disappear in the thermodynamic limit ($n \to \infty$ *and* $\frac{k}{n} = \text{const}$). On the other hand, ODLRO survives, since we have that $\alpha = c \neq 0$.

Yet, although in the thermodynamic limit they do not contain the two-site entanglement, η-states have genuine multipartite entanglement as a consequence of the existence of ODLRO correlations. To evaluate this entanglement explicitly, we use the relative entropy of entanglement.[23] For a given mixed state $\hat{\rho}$, the relative entropy of entanglement $E_R(\hat{\rho}) = \inf_{\hat{\sigma}\in\mathcal{S}} S(\hat{\rho}||\hat{\sigma})$ represents the distance between $\hat{\rho}$ and the set $\mathcal{S}$ of separable states, measured in the sense of relative entropy $S(\hat{\rho}||\hat{\sigma}) = \mathrm{Tr}[\hat{\rho}(\log\hat{\rho} - \log\hat{\sigma})]$. The advantage of this measure is that it is quite general – it is defined for an arbitrary composite system and it captures the total multipartite entanglement, rather than being based on a particular division of a global system into its sub-systems. The drawback is that often it is quite difficult to explicitly evaluate it. But in our case of highly symmetric η-states, it can be proved (see Ref. 15 and the references therein) that

$$\hat{\sigma}(k,n) = \frac{1}{2\pi}\int_0^{2\pi} d\phi(|\phi\rangle\langle\phi|)^{\otimes n} = \sum_{l=0}^{n}\binom{n}{l}\left(\frac{n-k}{n}\right)^{n-l}\left(\frac{k}{n}\right)^{l}|l,n\rangle\langle l,n|, \quad (21.6)$$

where $|\phi\rangle = \sqrt{\frac{n-k}{n}}|0\rangle + \sqrt{\frac{k}{n}}e^{i\phi}|1\rangle$, is the closest separable state. Note that the probability $p_1 = \frac{k}{n}$ for finding an electron pair in each particular site is the same for $\hat{\sigma}(k,n)$ as for the η-state $|k,n\rangle$. It is now straightforward to calculate the relative entropy of entanglement for η-states

$$E_R(|k,n\rangle) = -\log\binom{n}{k} + k\log\frac{n}{k} + (n-k)\log\frac{n}{n-k}, \quad (21.7)$$

which is largest in the case of half-filling ($n = 2k$), with the logarithmic asymptotic behavior $E_R(|n/2,n\rangle) \approx 1/2(\log n + 2)$. In other words, the total entanglement does survive in the thermodynamic limit, as predicted (analogously, it can be shown[15] that a bipartite block-entanglement between l and $n - l$ qubits also survives in the thermodynamic limit – as one would expect from the persistence of ODLRO – and scales as $\log l$).

As already mentioned, the low-energy spectrum of our model consists of η-states as well, so that at low temperatures the thermal state is $\hat{\rho}_T = \sum_{k=o}^{n} w_k |k,n\rangle\langle k,n|$ with a Fermi-Dirac probability distribution $w_k = (\exp(E_k/k_B T) + 1)^{-1}$. In order to estimate the entanglement contained in the thermal state, we model the closest separable state $\hat{\sigma}_T$ as the thermal mixture of the separable states (21.6), $\hat{\sigma}_T = \sum_{k=0}^{n} w_k \hat{\sigma}(k,n)$ thus obtaining an upper bound for the entanglement $E_R(\hat{\rho}_T) \leq S(\hat{\rho}_T || \hat{\sigma}_T)$. For a detailed discussion of the low-energy spectrum and the exact analytic and asymptotic behavior of entanglement (and the classical correlations and mutual information as well [l]) see Ref. 15. Here we only note that, as expected, at high temperatures entanglement scales as $\log n/n$.

As noted, a similar mechanism for the existence of ODLRO correlations is present in BCS superconductivity and it can be shown that the same analysis holds for superfluidity as well.[25] This indicates that, at least in the cases of critical phenomena where the relevant correlation functions driving the phase transitions are induced by the ground state entanglement, entanglement itself can be taken as the order parameter (for more detailed discussion see Refs. 3,25,26).

Finally, we note that the above discussion on entanglement and ODLRO in η-states presents us with an alternative view of the relation between superconductivity, Meissner effect and flux quantization.[27] It is based on the observation that, from Eq. (21.5), the existence of ODLRO implies the non-vanishing component ($c \neq 0$) of the symmetric state $|\psi^+\rangle = \frac{1}{\sqrt{2}}(|0\rangle|1\rangle + |1\rangle|0\rangle)$ which, upon the adiabatic exchange of electron pairs on two sites, has to stay invariant. This is a consequence of the fact that before and after the exchange the system is in the same physical state. Thus, the quantum state $|\psi\rangle = \frac{1}{\sqrt{2}}(|0\rangle|1\rangle + e^{i\Phi}|1\rangle|0\rangle)$ after the exchange has to be the same as before ($|\psi\rangle \equiv |\psi^+\rangle$), and the Aharonov-Bohm phase[28] $\Phi = \frac{2e}{\hbar c}\int\int \vec{B}\cdot d\vec{S}$ acquired during the exchange process has to satisfy the condition $e^{i\Phi} = 1$. In other words, $\Phi = 2\pi n$, and in the case of connected regions we get the Meissner effect ($n = 0, \vec{B} = 0$), while for regions that are not simply connected, we have the flux quantization condition $\Phi = \int\int \vec{B}\cdot d\vec{S} = n\frac{\hbar c}{2e}$.

21.3. Macroscopic Witnesses of Entanglement

In the previous section, we have discussed a theoretical model for high-temperature superconductivity in which the existence of ODLRO correlations was

[l]For an interesting discussion on the relation between mutual information, classical correlations and entanglement in multipartite systems see Ref. 24 and references therein.

seen as a consequence of the entanglement contained in the η-states. Continuing our investigation of the role of entanglement in thermal states of macroscopic systems, in this section we will discuss the effects of genuine *quantum correlations* on macroscopic thermodynamic properties, such as magnetic susceptibility, internal energy and pressure. Here, by genuine quantum correlations we mean the following: we will consider thermodynamic properties which are given as functions of various sets of correlations, such that the value of each property is bounded for the set of separable states, in a sense similar to the case of the Bell inequalities.[29] Note the difference from the previous case of superconductivity: while before it was only important that ODLRO correlations did not vanish, here we require that the existence of entanglement is reflected in the overall amount of the *whole set* of correlations defining the particular thermodynamic property (this is precisely why entanglement is often, although somewhat misleading, referred to as the *excess correlation*). Because the presence of entanglement can be inferred from the value of such macroscopic quantities, we will call them *macroscopic witnesses of entanglement.* [m]

21.3.1. *Magnetic Susceptibility*

First, we will discuss the case of the magnetic susceptibility. In their 2003 paper[30] Gosh *et al.* have presented the experimental measurements of the z-component of the magnetic susceptibility for the insulating magnetic salt $LiHo_xY_{1-x}F_4$, with $x << 1$ being the probability of randomly substituting the Ho^{+3} magnetic ions with nonmagnetic Y^{+3} ions. It was shown that the experimental results below $T \approx 1K$ could be theoretically explained only after using the second instead of the first-order quantum mechanical perturbation expression for the magnetic susceptibility:[31]

$$\chi_z = \frac{N}{Z^{(0)}} \sum_n w_n \left\{ \frac{[E_n^{(1)}]^2}{k_B T} - 2E_n^{(2)} \right\}, \tag{21.8}$$

where $w_n = \exp[-E_n^0/k_BT]$ and $Z^{(0)} = \sum_n w_n$. Here, $E_n^{(0)}$ and $\{|n\rangle\}$ are the unperturbed energies and states of the system and $E_n^{(1)} = \langle n|\hat{M}_i^z|n\rangle$ and $E_n^{(2)} = \sum_m |\langle n|\hat{M}_i^z|m\rangle|^2/[E_n^{(0)} - E_m^{(0)}]$ are the first and the second order energy corrections due the application of the external magnetic field ($\hat{M}^z$ being the z-component magnetic moment operator). Note that the essential quantum-

[m]In quantum information science, the term *entanglement witness* by convention refers to an observable whose expectation value is non-negative for separable states. Here, we will consider any observable whose expectation values for separable states are within a certain domain $\mathcal{D} \subset \mathbb{R}$ to be an entanglement witness.

mechanical part of the above expression is given by the second-order term $E_n^{(2)}$ that contains the "off-diagonal" matrix elements $|\langle n|\hat{M}^z|m\rangle|^2$. In other words, for the proper theoretical fit of the experimental results, one has to take entangled states as the overall multi-particle energy eigenstates.

In general, the magnetic susceptibility $\chi_z = (\partial M_z/\partial B_z)$ is given in terms of the energy eigenvalues E_k only, through the expression for the magnetization $M_z = (1/\beta Z)(\partial Z/\partial B_z)$ (recall that the partition function is $Z = \sum_k e^{-\beta E_k}$). Yet, as we have just seen, it does not mean that the state structure has no role to play – the energy eigenstates determine the energy eigenvalues, which is best seen through the perturbation theory calculation, therefore affecting the overall thermal state as well. In order to examine the close relationship between the state structure of macroscopic quantum systems and the correlations determining the magnetic susceptibility, we will follow the approach by Wieśniak *et al.*[32]

We consider a lattice of N spins of length s, given by a generic Hamiltonian $\hat{H}_0$, in an external magnetic field $\vec{B}$. The interaction Hamiltonian is [n] $\hat{H}_1 = \vec{B}\cdot\sum_{i=1}^N \hat{\vec{s}}_i$, where $\hat{\vec{s}}_i = (\hat{s}_i^x, \hat{s}_i^y, \hat{s}_i^z)$ represents the vector operator for the i-th spin. Then, the expression for the magnetic susceptibility along an arbitrary (say, z) axis is (provided that $[\hat{H}_0, \hat{H}_1] = 0$):

$$\chi_z = \frac{1}{k_BT}\Delta^2(M_z) = \frac{1}{k_BT}(\langle M_z^2\rangle - \langle M_z\rangle^2) = \frac{1}{k_BT}\left(\sum_{i,j=1}^N \langle \hat{s}_i^z \hat{s}_j^z\rangle - \langle\sum_{i=1}^N \hat{s}_i^z\rangle^2\right) \tag{21.9}$$

We will now show that $\bar{\chi} = \chi_x + \chi_y + \chi_z$ is the witness of entanglement. Note that from Eq. (21.9) $\bar{\chi}$ can be written in terms of the one-site standard deviations $\Delta^2(\hat{s}_i^a) = \langle(\hat{s}_i^a)^2\rangle - \langle(\hat{s}_i^a)\rangle^2$ and the two-spin correlations $C_a(i,j) = \langle\hat{s}_i^a\hat{s}_j^a\rangle - \langle\hat{s}_i^a\rangle\langle\hat{s}_j^a\rangle$, with $a \in \{x,y,z\}$: $\bar{\chi} = \frac{1}{k_BT}\Big\{\sum_{i=1}^N [\Delta^2(\hat{s}_i^x) + \Delta^2(\hat{s}_i^y) + \Delta^2(\hat{s}_i^z)] + \sum_{i\neq j}^N [C_x(i,j) + C_y(i,j) + C_z(i,j)]\Big\}$. Using $\langle(\hat{s}_i^x)^2\rangle + \langle(\hat{s}_i^y)^2\rangle + \langle(\hat{s}_i^z)^2\rangle = s(s+1)$, $\langle\hat{s}_i^x\rangle^2 + \langle\hat{s}_i^y\rangle^2 + \langle\hat{s}_i^z\rangle^2 \le s^2$ and the translational invariance of the lattice, in the case of an uncorrelated *product* state ($C_a(i,j) = 0$), we obtain: $\bar{\chi} = (N/k_BT)[\Delta^2(\hat{s}_i^x)+\Delta^2(\hat{s}_i^y)+\Delta^2(\hat{s}_i^z)] \ge N/k_BT(s(s+1) - s^2) = Ns/k_BT$. The same holds for the case of a general mixed separable state $\hat{\rho} = \sum_n w_n\hat{\rho}_n^1 \otimes \ldots \hat{\rho}_n^N$, due to the convexity of the mixture (for a detailed proof, see Ref. 33). Therefore, if $\bar{\chi} \le Ns/k_BT$, then the thermal quantum state is surely non-separable and we have entanglement present

[n] Recall that, as usual, we use $\hbar = 1$ units. We also incorporate the g-factor and Bohr magneton μ_B into the Boltzmann constant in the expression (21.9) for the magnetic susceptibility ($g^2\mu_B^2 k_B^{-1} \to k_B^{-1}$ in $\chi_z = \frac{g^2\mu_B^2}{k_BT}\Delta^2(M_z)$).

in a macroscopic system!

Using the previously known theoretical results for the zero-field magnetic susceptibility in the infinite isotropic Heisenberg spin chains, it was possible to infer[32] the existence of entanglement below certain critical temperatures $T_c(s)$. It is interesting to note that, somewhat contrary to the intuitively expected disappearance of quantum features with the increasing spin value s, we have that $T_c(s=1/2) < T_c(s=1)$: the bigger the spins are, the more room for entanglement and quantum features to persist at higher temperatures.

Also, an interesting *macroscopic quantum complementarity relation* between the non-local $Q = 1 - k_B T\hat{\chi}/Ns$ and the local $P = \langle\hat{\vec{M}}\rangle^2/N^2s^2$ properties can be derived:[32] $Q + P \leq 1$ (for a similar relation between magnetization and susceptibility along *one* axis, see Ref. 34).

Subsequently,[35] experimental measurements of zero-field magnetic susceptibility in copper nitrate, performed in the 1960's, were shown to confirm the existence of entanglement in thermal macroscopic states. Neglecting correlations other than those between the nearest neighbors and using the system's isotropy ($\langle\hat{\vec{M}}\rangle = 0$ at zero field, and $\chi_x = \chi_y = \chi_z = \chi$), the expression (21.9) for the magnetic susceptibility along an arbitrary axis reads as $\chi = \frac{1}{3}(\chi_x + \chi_y + \chi_z) = (1/k_BT)[(1/4) + (\langle\hat{\vec{s}}_0 \cdot \hat{\vec{s}}_1\rangle/3)]$, for $s = 1/2$. Analogously to the previous case, using $|\langle\hat{\vec{s}}\rangle| \leq s$, one obtains the separable-state bound for the correlation $\langle\hat{\vec{s}}_0 \cdot \hat{\vec{s}}_1\rangle$. For product states, one has:

$$|\langle\hat{\vec{s}}_0 \cdot \hat{\vec{s}}_1\rangle| = |\langle\hat{s}_0^x\rangle\langle\hat{s}_1^x\rangle + \langle\hat{s}_0^y\rangle\langle\hat{s}_1^y\rangle + \langle\hat{s}_0^z\rangle\langle\hat{s}_1^z\rangle| \leq |\langle\hat{\vec{s}}_0\rangle| \cdot |\langle\hat{\vec{s}}_1\rangle| \leq 1/4. \quad (21.10)$$

Due to its convexity, the above expression holds for mixed separable states as well. Thus, one gets that for separable states, the magnetic susceptibility falls in the range $N/(6k_BT) \leq \chi \leq N/(3k_BT)$. In addition to this, from Eq. (21.10) it follows that the correlation [o] $\langle\hat{\vec{s}}_0 \cdot \hat{\vec{s}}_1\rangle$ can be used as an entanglement witness (it also determines Bell's parameter and the concurrence; for more details and experimental confirmation, see Ref. 35).

In both cases studied in Ref. 35 the critical temperature below which entanglement was witnessed was around $T_c \approx 5K$. In a related work[36] Vértesi *et al.* have subsequently shown that in the case of $Na_2V_3O_7$ the entanglement can survive even at room temperatures, as $T_c \approx 365K$! Also, see Ref. 37 where a theoretical discussion of a class of spin systems with $T_c \approx 25.4K$ was discussed.

[o] Often, for reasons of simplicity, we will refer to the mean product between the two subsystem-operators $\hat{A}$ and $\hat{B}$ as the correlation function $\langle\hat{A}\hat{B}\rangle$. Strictly speaking, the correlation function is given by $C(\hat{A}, \hat{B}) = \langle\hat{A}\hat{B}\rangle - \langle\hat{A}\rangle\langle\hat{B}\rangle$.

21.3.2. *Internal Energy*

Next, we discuss the relation between the internal energy U and entanglement in spin systems. Given an Hamiltonian $\hat{H}$, the internal energy is defined by $U = \langle \hat{H} \rangle = -(1/Z)(\partial Z/\partial\beta)$. In the simple case of a one-dimensional spin chain with N sites in an external magnetic field B along the z-axis and a nearest-neighbor Heisenberg interaction, [p] the Hamiltonian is $\hat{H} = J\sum_{i=1}^{N} \hat{\vec{s}}_i \cdot \hat{\vec{s}}_{i+1} + B\sum_{i=1}^{N} \hat{s}_i^z$. We can then construct[38] the following macroscopic thermodynamic entanglement witness (with $M = \langle \sum_{i=1}^{N} \hat{s}_i^z \rangle$):

$$\frac{U - BM}{NJ} = \frac{1}{N}\sum_{i=1}^{N} \langle \hat{\vec{s}}_i \cdot \hat{\vec{s}}_{i+1} \rangle, \tag{21.11}$$

whose absolute value for separable states is, according to Eq. (21.10), bounded by s^2. Therefore, for $\frac{|U-BM|}{N|J|} > s^2$ we have entanglement in the macroscopic thermal state. In the absence of an external magnetic field, this becomes solely the energy-based entanglement witness. In the case of the $s = 1/2$, XX Heisenberg model (with and without magnetic field), one obtains $T_c \approx 10K$ for the critical temperature.[38] It is interesting to note here that in this particular case of the Hamiltonian with nearest-neighbor interactions only, the magnetic susceptibility turns out to be a slightly better entanglement witness (see the discussion in Ref. 32). This is because the magnetic susceptibility depends on *all* two-site spin correlations and therefore is capable of capturing two-site entanglement beyond nearest-neighbors only. For further results regarding Heisenberg lattices, XY and Bose-Hubbard models, see Ref. 39. For a more general study of optimal bipartite entanglement witnesses, see also Ref. 40.

Using a slightly different approach Wang *et al.* have also discussed the connection between entanglement and internal energy in one-dimensional antiferromagnetic Heisenberg chains without an external magnetic field (see Ref. 41 and references therein). By explicitly calculating the partial state $\hat{\rho}_{12}$ of two nearest neighbor spins, they have obtained the expression for the concurrence:[11]

$$C = \frac{1}{2}\max[0, -\frac{U}{s^2 NJ} - 1], \tag{21.12}$$

which gives the same bound for separable states as the above Eq. (21.11) (note that in this case $J > 0$ and $M = 0$). The connection between entanglement (measured by the concurrence) and internal energy is established through the explicit

[p] We restrict here to one-dimensional chains and nearest-neighbor interactions with a single coupling constant J for reasons of simplicity only, but all the results are straightforward to generalize for multi-dimensional lattices, not necessarily with nearest-neighbor interactions only.

dependence of the two-site matrix elements on bipartite spin correlations[41] (which themselves define the internal energy U).

In the same paper, the connection between concurrence, internal energy and the measure of maximal Bell inequalities violation was also given, along the same lines as in the case of magnetic susceptibility.[35] Without loss of generality, we will consider neighboring sites 1 and 2, with partial state $\hat{\rho}_{12}$. The Bell measurement is given by four unit 3-D vectors corresponding to four axes of spin measurements – two for site 1 ($\vec{a}$ and $\vec{a}'$) and two for site 2 ($\vec{b}$ and $\vec{b}'$). The inequality itself reads

$$|\langle\hat{B}\rangle| = |\mathrm{Tr}[\hat{\rho}_{12}\hat{B}]| \leq 2s^2, \qquad (21.13)$$

where the Bell operator $\hat{B}$ is:

$$\hat{B} = \left[\vec{a}\cdot\hat{\vec{s}}_1\right] \otimes \left[(\vec{b}+\vec{b}')\cdot\hat{\vec{s}}_2\right] + \left[\vec{a}'\cdot\hat{\vec{s}}_1\right] \otimes \left[(\vec{b}-\vec{b}')\cdot\hat{\vec{s}}_2\right]. \qquad (21.14)$$

For a given state $\hat{\rho}_{12}$, the maximal value of $|\langle\hat{B}\rangle|$ is given by $\mathcal{B} = 2\sqrt{u+\tilde{u}}$, where u and $\tilde{u}$ are the highest two eigenvalues of the matrix $GG^\dagger$, where G is a 3×3 matrix whose elements are given by the spin-spin correlation functions $(G)_{nm} = \langle\hat{s}_1^n\hat{s}_2^m\rangle = \mathrm{Tr}[\hat{\rho}_{12}(\hat{s}_1^n\hat{s}_2^m)]$ (with $n,m\in\{x,y,z\}$). In the case of the isotropic Heisenberg Hamiltonian, we have[41] $G = \langle\hat{s}_1^x\hat{s}_2^x\rangle I_3$, where I_3 is the 3×3 identity matrix. Therefore, $\mathcal{B} = 2\sqrt{2}|\langle\hat{s}_1^x\hat{s}_2^x\rangle| \leq 2\sqrt{2}s^2$, which is precisely the maximal possible violation of any Bell inequality given by a set of four unit vectors $(\vec{a},\vec{a}',\vec{b},\vec{b}')$. On the other hand, the internal energy is then $U = 3NJ\langle\hat{s}_1^x\hat{s}_2^x\rangle$ and we finally have $\mathcal{B} = \frac{2\sqrt{2}}{3}\frac{|U|}{NJ}$. In other words, while the entanglement exists for $s^2 \leq \frac{|U|}{NJ} \leq 3s^2$, the Bell inequality is violated for $\frac{3}{\sqrt{2}}s^2 \leq \frac{|U|}{NJ} \leq 3s^2$.

Let us now define the *minimum separable energy* $E_s = \min_{\hat{\rho}\in\mathcal{S}}\mathrm{Tr}[\hat{\rho}\hat{H}]$, where $\mathcal{S}$ is the set of separable states. As already noted, thermal entanglement is generally a consequence of the ground state entanglement. Therefore, if thermal entanglement in a spin system can be witnessed by its internal energy, then the ground state energy E_g will be for sure strictly lower than the minimum separable energy E_s. In Refs. 39,42,43 the *entanglement gap* $G_e = (E_s - E_g)$ was studied as the witness of entanglement [q] for spin systems with 2-spin Hamiltonians. It showed to be a useful tool in detecting entanglement in such systems, as in the case of translational invariant bipartite lattices [r] the minimum separable energy can be determined by simply minimizing the 2-spin Hamiltonian $\hat{H}_{12}$ only:

$$E_s = \min_{\hat{\rho}\in\mathcal{S}}\mathrm{Tr}[\hat{\rho}\hat{H}] = \min_{|p\rangle_N\in\mathcal{P}_N}\langle p|\hat{H}|p\rangle_N = N\min_{|p\rangle_2\in\mathcal{P}_2}\langle p|\hat{H}_{12}|p\rangle_2 = NE_s^{(2)}, \qquad (21.15)$$

[q]Formally speaking, for systems with $G_e > 0$, the entanglement witness is defined by the observable $\hat{E}_W = \hat{H} - E_s\hat{I}$, whose negative expectation values signal out the entanglement.

[r]A lattice or graph is said to be bipartite if it can be divided into two sub-lattices A and B such that the sites from A are connected to sites of B only, and vice versa.

where $\mathcal{P}_N$ is the set of pure product states of N spins (N being the total number of lattice sites) and $E_s^{(2)}$ is the minimum separable energy for the two-site Hamiltonian $\hat{H}_{12}$. The second equality comes from the convexity of mixed states, while the third equality holds because the lattice is translational invariant *and* bipartite. Analogously, one gets for the ground state energy $E_g = N\mathrm{Tr}[\hat{\rho}_{12}^g \hat{H}_{12}] = NE_g^{(12)}$, with $\hat{\rho}_{12}^g$ being the unique two-site partial state for the ground state $|g\rangle$. This gives us the expression, as well as the upper bound for the entanglement gap ($E_g^{(2)}$ is the ground state energy for $\hat{H}_{12}$):

$$G_e = N(E_s^{(2)} - E_g^{(12)}) \leq N(E_s^{(2)} - E_g^{(2)}). \tag{21.16}$$

Also, we see that for bipartite lattices (not necessarily translational invariant ones), an entanglement gap based study can reveal only bipartite entanglement between interacting pairs of spins. On the other hand, for general n-partite lattices [s] the non-zero entanglement gap is in general associated with n-partite entanglement. It was also proved[42] that for bipartite lattices the entanglement gap per bond decreases with the coordination number, and the same property conjectured for general n-partite lattices.

The above study of the entanglement gap provides us yet another lower bound for the entanglement temperature T_e, the *entanglement-gap temperature*[42] T_g. It is defined as the temperature for which the thermal-state internal energy is equal to the minimum separable energy, $U(T_g) = E_s$. Obviously, below T_g the internal energy of a system in thermal equilibrium is smaller then E_s (due to the internal energy monotonicity), and we have entanglement present in the thermal state $\hat{\rho}(T)$. Intuitively, one would expect that the bigger the entanglement gap, the bigger the entanglement-gap temperature would be. Yet, that does not necessarily have to be the case! In order to make a "fair" comparison between systems given by different Hamiltonians, let us first define the scaled entanglement gap and the corresponding temperature. For a system with the maximal energy eigenvalue E_{max}, the *scaled entanglement gap* g_e is given by $g_e = G_e/(E_{\mathrm{max}} - E_g)$. Analogously, we define the *scaled entanglement-gap temperature* $t_e = k_B T_e/(E_{\mathrm{max}} - E_g)$. It can be proved[42] that the Hamiltonian with the biggest scaled entanglement gap is of the form $\hat{H}_g = \hat{I} - |\Psi_m\rangle\langle\Psi_m|$, where $|\Psi_m\rangle$ is the maximally entangled state according to the geometric measure of entanglement.[44] On the other hand, for bipartite systems the Hamiltonian [t] $\hat{H}_t = \hat{I} - \hat{P}_A = \hat{P}_S$ has the scaled entanglement-gap temperature $t_g = [\ln \frac{d+1}{d-1}]^{-1}$

[s] Often, these systems have competitive interactions and are referred to as frustrated systems, for which the ground state energy per bond is usually larger than that for two interacting spins alone.

[t] $\hat{P}_{S/A}$ represent the projectors onto the symmetric/antisymmetric subspaces. Note that for bipartite systems, $\hat{P}_S + \hat{P}_A = \hat{I}$, while all antisymmetric states are entangled.

which goes to infinity as the dimension of the subsystems d grows! On this simple example, we see the subtle interplay between the energy and entanglement content of the "thermal noise" – a lot of entangled states laying close (in terms of energy) to the ground state may cause slow energy change with respect to temperature rise, thus leaving the thermal state with internal energy below E_s for big (even infinite) range of temperatures.

21.3.3. *Entanglement in Systems of Non-Interacting Identical Particles*

In the last part of this section, we will discuss entanglement in systems of N non-interacting identical particles, confined in a given volume V. Although entanglement is commonly associated with the interaction between sub-systems of a composite system, it can be created solely due to the indistinguishability of identical particles[u]. We shall start with the analysis of the spin-entanglement in systems of identical electrons.[46] The electrons are described by the operators $\hat{c}_s^\dagger(\vec{k})$, where $\vec{k}$ is the momentum and s the spin index, which satisfy the standard fermionic [v] anti-commutation relations. Given the vacuum state $|0\rangle$, the ground state of the system is a simple product of uncorrelated momentum modes, $|g\rangle = \prod_{|\vec{k}|\leq k_F,s} \hat{c}_s^\dagger(\vec{k})|0\rangle$, with $k_F = \sqrt[3]{3\pi^2 n}$ being the Fermi momentum ($n = N/V$). Yet, it does not mean there is no entanglement present, only we have to look for it elsewhere. In this example, we will concentrate on the internal degrees of freedom and evaluate the entanglement in the spin between two electrons. In order to do that, we first evaluate the spin state of two electrons, one being in position $\vec{r}$ and the other in $\vec{r}'$. Its matrix elements are given by (s, s', t, t' are the four spin indexes), $\hat{\rho}_{ss',tt'} \propto \langle g|\hat{\psi}_t^\dagger(\vec{r})\hat{\psi}_{t'}^\dagger(\vec{r}')\hat{\psi}_s(\vec{r})\hat{\psi}_{s'}(\vec{r}')|g\rangle$, with the position operators $\hat{\psi}_t^\dagger(\vec{r}) = (2\pi)^{-3}\int d^3\vec{k}e^{-i\vec{k}\vec{r}}\hat{c}_t^\dagger(\vec{k})$ and the integral taken over all momenta $\vec{k} \in \mathbb{R}^3$. Upon straightforward calculation,[46] we obtain:

$$\hat{\rho}_{ss',tt'} \propto -\frac{n^2}{4}[\delta_{ts}\delta_{t's'} - f^2(|\vec{r}-\vec{r}'|)\delta_{ts'}\delta_{t's}], \tag{21.17}$$

where the "exchange interaction strength" is given by $f(|\vec{r} - \vec{r}'|) = \frac{2}{n}\int_{|\vec{k}|\leq k_F} d^3\vec{k}e^{i\vec{k}(\vec{r}-\vec{r}')} = \frac{3}{rk_F^3}\int_o^{k_F} k\sin(kr)dk = 3j_1(k_F r)/(k_F r)$, where j_1 is the spherical Bessel function. Thus, upon normalizing, one obtains the two-spin

[u]For a basic analysis of entanglement generation in a simple system of two identical particles, see Ref. 45.

[v]For reasons of simplicity, here we consider only spin one-half fermions (electrons), but all the results obtained can be easily generalized to fermions with higher spins.

density matrix for two electrons observed in positions $\vec{r}$ and $\vec{r}'$:

$$\hat{\rho}_{12} = \frac{1}{4-2f^2}\begin{bmatrix} 1-f^2 & 0 & 0 & 0 \\ 0 & 1 & -f^2 & 0 \\ 0 & -f^2 & 1 & 0 \\ 0 & 0 & 0 & 1-f^2 \end{bmatrix}. \qquad (21.18)$$

Note the implicit dependence of the state $\hat{\rho}_{12}$ on the relative distance r between the two electrons, through the exchange interaction strength $f = f(r)$. This can be written in a convenient way as a Werner state $\hat{\rho}_{12} = (1-p)\frac{\hat{I}}{4} + p|\Psi^-\rangle\langle\Psi^-|$, with $p = f^2/(2-f^2) \in [0,1]$, having the fidelity $F = \langle\Psi^-|\hat{\rho}_{12}|\Psi^-\rangle = (3p+1)/4 = (f^2+1)/(4-2f^2) \in [1/4,1]$ (see Ref. 47 and the references therein). We can now easily evaluate the entanglement in terms of various entanglement measures. The concurrence[11] C is given by

$$C = \max\{\frac{2f^2-1}{2-f^2}, 0\}, \qquad (21.19)$$

while the entanglement of formation[11] E_F and the relative entropy of entanglement[23] E_R are $E_F = h([1+\sqrt{1-C^2}]/2)$ and $E_R = 1 - h(F)$ respectively ($h(y) = -y\log y - (1-y)\log(1-y)$ is the Shannon entropy). For $p > 1/3$, i.e. for $f^2 > 1/2$, this state is entangled. Using the above expression for f in terms of the spherical Bessel function, one can find the "entanglement distance" r_e, above which the entanglement in spin disappears, solving $j_1(k_F r_e)/(k_f r_e) = 1/(3\sqrt{2})$.

Note that the above result is valid for the joint spin state of two electrons whose positions are given by points $\vec{r}$ and $\vec{r}'$. If we had used the realistic detectors with non-zero position measurement uncertainty, depending on the detector profiles, the entanglement distance r_e could be larger than in the case of ideal point-like measurements.[48] Also, we note that the two-particle entanglement analysis describes the overall entanglement contained in the system as it can be shown[49] that all the multipartite entanglement can be built up out of the above bipartite entanglement only (i.e. the system can not contain genuine multipartite entanglement, such as the tripartite GHZ entanglement, for example). Finally, we note that the same analysis could be conducted for systems of non-interacting bosons, but the resulting polarization state would have no entanglement (in fact, no correlations whatsoever) contained in it,[46] since at $T = 0$, all particles in the condensate have the same momentum.

Now, we turn to the case of non-zero temperatures. Using the Green's function approach, one obtains[47] the two-spin state $\hat{\rho}_{12}(r,T)$ (note the explicit dependence on temperature T). It is given by the same expression (21.18) for the two-spin state $\hat{\rho}_{12}$ with slightly different exchange interaction strength f which now depends on temperature as well: $f(r,T) = \frac{3}{rk_F^3}\int_o^\infty kn_k \sin(kr)dk$, where $n_k = [\exp(\frac{E_k-\mu}{k_BT}) + 1]^{-1}$ is the Fermi-Dirac distribution. The numerical and asymptotic analysis (for details, see Ref. 47) give: $\Delta f(r,T) = f(r,T) - f(r,T = 0) = \frac{\pi^2}{8}\left[\cos(k_Fr) - \frac{\sin(k_Fr)}{k_Fr}\right]\left(\frac{k_BT}{e_F}\right)^2$. The same quadratic dependence on the temperature can be observed for the differences of the measures of entanglement ΔC, ΔE_F and ΔE_R.

Note also the implicit dependence of all of the above formulas on the particle density n, through the Fermi momentum $k_F = \sqrt[3]{3\pi^2 n}$ featuring in the expression for f. This enables us to formulate the so-called "equation of state for entanglement" for the Fermi gas.[50] It is the relation between the macroscopic properties (temperature and pressure) and the spin entanglement of two electrons (separated by a relative distance r). The connection between the pressure of a nonrelativistic degenerate gas and its concentration n (a similar expression holds for a relativistic gas as well, see Ref. 50 and the references therein) is given by $P = \frac{(3\pi^2)^{2/3}\hbar^2}{5m}n^{5/3}$. Using $k_F = \sqrt[3]{3\pi^2 n}$, one obtains $k_F = (\frac{15\pi^2 m}{\hbar^2})^{1/5}P^{1/5}$. Thus, plugging the above result for $\Delta f(r,T)$ in formulas for different entanglement measures (Eq. (21.19) and below) we finally obtain the functions $C = C(r,T,P)$, $E_F = E_F(r,T,P)$ and $E_R = E_R(r,T,P)$ – the equation(s) of state for entanglement for the Fermi gas.

Next, we explore the existence of spatial entanglement in systems of identical particles and its connection with the internal energy. For reasons of simplicity, we start[51] with N non-interacting bosons confined in a one-dimensional box of length L. We are interested in the spatial correlations only, therefore we shall divide the box into M distinct consecutive segments A_j, $j = 1\ldots M$, of equal length $\Delta L = \frac{L}{M}$ and get the lower bound for the energy of any state separable with respect to such partition of the box, thus establishing the system's energy as an entanglement witness for the case of ideal Bose gas. The intuition behind our study relies on the fact that for non-interacting bosons all one-particle modes can occupy the lowest-energy one-particle states allowed by the constraints given. In our case, the constraint is the separability of the overall state, with respect to the given box partition defined by the number M of segments. Such a separability means that the overall state of the system can be expressed as a mixture of pure product states with respect to the box partition – states whose one-particle modes are each confined within only one segment A_j. Using the well known relation $E_0 \propto L^{-2}$ between the ground-state energy E_0 of a free boson and the length L of the box, one can immediately see that as long as we force each particle of the system to be confined within a region A_j smaller than the whole box, the overall energy would be bigger than if we let the particles freely occupy the entire box of length L (in which case the ground state could, and indeed will be entangled in space).

Thus, for a d-dimensional ideal Bose gas with the overall volume $V = L^d$, the formal mathematical description of the above analysis gives the energy bound for states separable with respect to a box partition defined by the number M of one-dimensional segments:

$$E(\hat{\rho}_s) \geq dN \frac{\hbar^2 \pi^2 M^2}{2mL^2}. \tag{21.20}$$

Using the result for the temperature dependence of the internal energy (see Ref. 51 and the references therein), we obtain the expression for the *entanglement-gap temperature* $T_g(M)$ below which equilibrium thermal states of free (ideal, non-interacting) d-dimensional Bose gas are spatially entangled with respect to a given one-dimensional partition (ζ is Reimann's zeta-function, m is the mass of the particles): $T_g(M) = \frac{2\pi\hbar^2}{k_B m L^2} \left(\frac{NM^2\pi}{2\zeta(1+d/2)} \right)^{2/(2+d)}$. Obviously, as M increases, the transition temperature increases as well, and thus we might conclude that for every temperature, entanglement will exist with respect to a partition defined by certain, big enough M. As the temperature and therefore the one-particle mean energy rise, the one-particle momentum rises as well which, due to the Heisenberg uncertainty principle, results in even smaller spatial spread of the one-particle modes. In other words, the bigger temperature is, the "smaller" particles are and we can fit them in boxes of smaller dimensions (determined by M). When the number of boxes equals the number of particles, $M^d = N$, we have that the entanglement-gap temperature with respect to such a partition results in a particle size small enough so that on average no two particles overlap. As the existence of multipartite entanglement in systems of non-interacting identical particles is precisely due to this overlap, we conclude that for the cases when $M^d > N$, the entanglement-gap temperature (and energy) witnesses the single-particle entanglement.[52] For $M^d = N$, we get (having fixed the particle density $n = N/V = N/L^d$): $T_g(M) = \frac{2\pi\hbar^2 n^{2/d}}{k_B m} \left(\frac{\pi}{2\zeta(1+d/2)} \right)^{2/(2+d)}$. On the other hand, the well known result for Bose-Einstein condensation (BEC) critical temperature T_c is: $T_c = \frac{2\pi\hbar^2 n^{2/d}}{k_B m} [\zeta(d/2)]^{2/d}$. We observe the same dependence on the particle density n and the mass m of both temperatures T_g and T_c. In fact, we have $T_g \approx 2T_c$ and the entanglement-gap temperature is the upper bound of the BEC critical temperature. [w] This has to be so because below T_g the one-particle states will start to overlap with each other and this is precisely the necessary condition for condensation to occur. Moreover, the same functional dependence of

[w]Note that in one and two dimensions, the critical BEC temperature tends to zero as $\zeta(d/2) \to \infty$. Thus, condensation can not occur in those dimensions, while entanglement can still exist.

both temperatures suggests[26,51] that in this case entanglement can be taken as an order parameter as well as in the previously discussed cases of superconductivity and superfluidity.[25] A further discussion of spatial entanglement with the formal introduction of operators defining spatial modes that allow for a detailed study of spatial (rather than momentum) entanglement was conducted in Ref. 53.

Finally, we note that an analogous analysis could be done for the case of non-interacting fermions,[51] but due to the Pauli exclusion principle, the entanglement-gap temperature obtained would be too small to be of any practical use – spatial entanglement in Fermi gases vanishes quickly with the rise of temperature.

21.4. Thermal Entanglement Generation and Manipulation

In the previous sections we have studied various theoretical models and verified the existence of thermal entanglement in macroscopic systems up to room temperatures. Moreover, we have shown that entanglement can affect various thermodynamic properties such as magnetic susceptibility, internal energy and pressure, and that in some cases, like in superconductivity and superfluidity, it can play a crucial role determining the features of macroscopic systems. Having these results in mind, one naturally wonders if macroscopic thermal entanglement could be used as a resource as well. This is of particular interest in the growing field of quantum information and computation, where entanglement is seen as one of the key resources in various quantum information tasks, such as quantum teleportation, quantum cryptography or quantum computation,[54] to name just a few. In other words, can we extract entanglement from macroscopic systems and use it for quantum teleportation, for example? Or, is it possible to generate and transfer useful entanglement in systems with many degrees of freedom? These are all questions addressing the issue of the manipulation of thermal entanglement, a topic of growing scientific interest. Here, we will only briefly mention two particular proposals, one for entanglement extraction, and the other for entanglement generation, while at the end of each subsection we will refer to some of the latest results done in the field, addressing these and similar questions.

Also, we note that the question of entanglement extraction is on its own interesting for purely theoretical reasons. All of the previous results confirming the existence of thermal entanglement are model-dependent in a sense that the entanglement-based correlations could be, in a different and probably more complex theory, reproduced by separable states. Thus, only if we manage to locally extract a pair of entangled particles which would then violate some Bell inequality and the requirements of local realism,[29,55] we can undoubtedly claim the existence of macroscopic thermal entanglement.

21.4.1. *Entanglement Extraction*

First, following Ref. 56 we present a theoretical proposal for extracting bipartite entanglement from a one-dimensional spin one-half chain to a pair of probe spin one-half particles labeled by L and R. We assume that either of the two probe particles interact with one spin from the chain only (the extraction is a local process), and that this bipartite interaction is of the same type as the one between the spins from the chain. We will consider two types of bipartite spin-spin interactions given by $\hat{H}_{ij}(\lambda) = J(\hat{s}^i_x \hat{s}^j_x + \hat{s}^i_y \hat{s}^j_y + \lambda \hat{s}^i_z \hat{s}^j_z)$: the Heisenberg interaction, for $\lambda = 1$ and the zero-field isotropic XY interaction, for $\lambda = 0$. Here, i and j label either the spin-chain vertices, $i, j \in \{1 \ldots N\}$, or one of the two probe particles, $i, j \in \{L, R\}$ (but not both of them – entanglement extraction is a local process and we do not allow for the two probes to interact between themselves, i.e. $\hat{H}_{LR} \equiv 0$), while $\hat{\vec{s}}$ is the vector of spin operators.

We are interested in the following problem. Having the probe particles initially prepared in the product state $|\phi\rangle_{LR} = |\downarrow\rangle_L |\downarrow\rangle_R$, is it possible to entangle them by coupling the particle L with spin n from the chain, and particle R with spin m? The interaction Hamiltonian between the probes and the spin-chain is given by $\hat{H}_I(\lambda) = \hat{H}_{Lm}(\lambda) + \hat{H}_{Rn}(\lambda)$. The two-spin partial state $\hat{\rho}_{mn}$ of the above Heisenberg and XY interactions $\hat{H}_{ij}(\lambda)$ is given by the Werner-like state:

$$\hat{\rho}_{mn} = \frac{\eta}{4}\hat{I}_{mn} + (1-\eta)|\psi^-\rangle\langle\psi^-|_{mn}, \tag{21.21}$$

where $\eta = 1 + 4G_{zz}$, with $G_{zz} = \langle \hat{s}^m_z \hat{s}^m_z \rangle$ being the spin correlation function. [x] This state is entangled for $0 \leq \eta < 2/3$, (for $-1/4 \geq G_{zz} \geq -1/12$), and we ask if it is possible to, at least partially, extract this entanglement by means of the above interaction Hamiltonian $\hat{H}_I(\lambda)$.

Although the bipartite entanglement in spin chains can not achieve its maximum (i.e. $G_{zz} > -1/4$ and $\eta > 0$), in order to gain some basic intuition of the process, let us first examine the extreme case of the initially maximally entangled two-spin state $|\psi^-\rangle$. Before the interaction starts, the initial state of the overall system consisting of two probe particles L and R and two corresponding spins m and n is $|\psi(0)\rangle_{LRmn} = |\downarrow\downarrow\rangle_{LR}|\psi^-\rangle_{mn}$. Upon evolving for a time τ under the Hamiltonian $\hat{H}_I(\lambda)$, the state of the system will be:

$$|\psi(t)\rangle_{LRmn} = \cos(2J\tau)|\downarrow\downarrow\rangle_{LR}|\psi^-\rangle_{mn} - i\sin(2J\tau)|\psi^-\rangle_{LR}|\downarrow\downarrow\rangle_{mn}. \tag{21.22}$$

[x]Note that when discussing internal energy and entanglement in spin chains in Sec. 21.3.2 we also encountered the problem of evaluating partial bipartite states and used the fact that for the Heisenberg and XY chains $G_{zz} = G_{xx} = G_{yy}$. It can be shown that this is a consequence of symmetries of the system: translational invariance, isotropy and null magnetization. For details, see Refs. 41,56 and the references therein.

Thus, the partial state of the probes is a mixture of the product and the maximally entangled singlet state, and its concurrence is $C_{LR} = \sin^2 2J\tau$. We see that no matter how long (or short!) the interaction is, for every τ (except for $\tau = k\pi, k \in \mathbb{N}$) the probes will be entangled, and the maximal value of 1 is achieved for $J\tau = (2k-1)\pi/4$. Numerical analysis for realistic mixed states $\hat{\rho}_{mn}$ given by $\eta > 0$ presents us with a similar qualitative picture[56] which means that, at least in principle, one can, by the means of local interactions only, extract entanglement from a solid modeled by the Heisenberg or XY interactions.

The above analysis was conducted for the ground state of a spin-chain. But as we have seen from previous sections, as long as the ground state is entangled, the thermal state of the spin-chain will be for a certain range of temperatures entangled as well, and therefore a similar qualitative picture of entanglement extraction would be obtained. Also, it was shown that in more realistic scenarios where each probe interacts with more than one spin from the chain, the above analysis would in general allow the extraction of multipartite entanglement as well.

In fact, one might argue that as long as the thermal state of a macroscopic system contains some entanglement (say, bipartite), leaving a pair of the same type of particles interact with the system as probes would eventually, upon thermalisation, evolve them to an entangled state. Indeed, environment-induced entanglement extraction from thermal baths was the subject of various papers, see Refs. 57–60 and the references therein. Particularly interesting is the recent paper by Yi *et al.* where the connection between the efficiency of the entanglement extraction from a spin chain and its regions of criticality was shown (see Ref. 61; for the connection between quantum decoherence rates, quantified by the Loschmidt echo, and quantum phase transitions quantified by the ground state fidelity, see Refs. 62,63). Finally, we mention the recent work on entanglement extraction from a system consisting out of indistinguishable identical bosons.[64] As seen in Sec. 21.3.3, systems of non-interacting identical particles can contain a great deal of entanglement, so it is not surprising that at least in principle it is possible to extract it. The interesting result of Ref. 64 is that it can be applied to the case of coherent states of Bose-Einstein condensates that are widely considered to be the quantum states closest to classical states. Thus, even the most "classical states" in quantum mechanics contain non-classical features and entanglement that can be extracted and used as a resource!

21.4.2. *Entanglement Generation*

We have seen that it is possible to extract the already existing entanglement contained in macroscopic systems in thermal equilibrium. In this subsection, we discuss the question of *engineering* macroscopic entanglement, i.e. creating entanglement in systems consisting of a large number of sub-systems and subsequently extracting it ready for use in quantum information processing protocols. We note that the current problem of entanglement creation is of relevance for other important problems regarding macroscopic entanglement manipulation, such as entanglement transfer[65] and entanglement dynamics in general.

Following Ref. 66, we will consider a system of M non-interacting bosons on a 1-D lattice consisting out of N sites, governed by the hopping Hamiltonian $\hat{H} = J\sum_{j=1}^{N}(\hat{a}_j^\dagger \hat{a}_{j+1} + \hat{a}_{j+1}^\dagger \hat{a}_j)$, where $\hat{a}_j^\dagger$ are creation operators representing bosons occupying sites labeled by $j = 1 \ldots N$. Initially, all the bosons are located in the central $\frac{N+1}{2}$ site (N is assumed to be odd), and we examine the state evolution and entanglement creation between the first site 1 and the last site N. Note that in this case we do not assume the cyclic boundary conditions – we aim to create entanglement between two *distant* sites.

Having only one boson in the central site, it is obvious that after some time t, the evolved one-particle state will be the coherent superposition of N one-particle states each representing the boson occupying one of $j = 1 \ldots N$ sites, with the corresponding amplitudes $f_j(t)$ (the vacuum state is given by $|0\rangle$): $|\Psi(0)\rangle = \hat{a}^\dagger_{\frac{N+1}{2}}|0\rangle \rightarrow |\Psi(t)\rangle = \sum_{j=1}^{N} f_j(t)\hat{a}_j^\dagger|0\rangle$. Using the correspondence between the hopping Hamiltonian $\hat{H}$ and the XY spin-chain Hamiltonian given by Jordan-Wigner transformation, one can evaluate the amplitudes $f_j(t)$ (for exact expressions, see Ref. 66 and the references therein). Thus, in the case of M bosons initially placed in the central lattice site, we obtain: $|\Psi(0)\rangle = \frac{(\hat{a}^\dagger_{\frac{N+1}{2}})^M}{\sqrt{M!}}|0\rangle \rightarrow |\Psi(t)\rangle = \frac{(\Sigma_{j=1}^{N} f_j(t)\hat{a}_j^\dagger)^M}{\sqrt{M!}}|0\rangle$. We are interested in the partial state between the sites 1 and N which, upon straightforward calculation (for details, see Ref. 66) reads as: $\hat{\rho}_{1N} = \sum_{r=0}^{M} P_r(t)|\psi_r\rangle\langle\psi_r|_{1N}$, with $\sqrt{P_r(t)} = \sqrt{C_r^M}(\sqrt{2}|f_1(t)|)^r(\sqrt{1-2|f_1(t)|^2})^{M-r}$ and $|\psi_r\rangle_{1N}$ representing states containing r bosons in sites 1 and N. In order to discuss and evaluate the amount of entanglement in the above state, it is convenient to express each of its pure components in the Schmidt biorthogonal form:

$$|\psi_r\rangle_{1N} = \frac{1}{2^{r/2}}\sum_{k=0}^{r}\sqrt{C_k^r}|k\rangle_1|r-k\rangle_N. \tag{21.23}$$

Note that the only dependence on time of the above state is given through $f_1(t)$. Numerical analysis of the amount of entanglement in terms of the logarithmic negativity[67] E_N, measured after some suitable time t_h that maximizes its value, presents very interesting results. First, already at $N = 21$ lattice sites, it is possible to generate the amount of entanglement $E_N > 1$, while in the corresponding case of the XY spin chain one can never create more entanglement than the amount contained in one singlet ($E_N(\text{singlet}) = 1$). Further, the amount of entanglement between the sites 1 and N created in this way is *bigger* than the corresponding entanglement contained in the ground state of this system.

Also, one can find interesting to discuss the average entanglement $\langle E_V \rangle = \sum_r P_r E_V(|\psi_r\rangle_{1N})$ in terms of von Neumann entropy of entanglement for pure states $E_V(|\psi_r\rangle_{1N}) = -\text{Tr}(\hat{\rho}_1 \log \hat{\rho}_1)$, where the partial state is given by $\hat{\rho}_1 = \text{Tr}_N(|\psi_r\rangle\langle\psi_r|_{1N})$. It is the average amount of entanglement obtained between the sites 1 and N, upon coherently measuring the overall number of bosons in the rest of the system $\{2, \ldots (N-1)\}$. In this way one obtains higher values of entanglement than originally contained in $\hat{\rho}_{1N}$. Further, as in the previous case, it shows to be more efficient to average the entanglement contained in $\hat{\rho}_{1N}$ than in the system's ground-state. Note that the number measurement is in fact a non-local operation, which is why one obtains more entanglement in this way than it is originally contained in the state $\hat{\rho}_{1N}$.

Finally, using the technique similar to that used in Ref. 56, it is possible, at least in principle, to extract this entanglement between the occupation number states of two sites to a system of two distant spin one-half particles, by means of local operations only.

In relation to this study, we mention Ref. 68 which discusses the creation of spatial entanglement between indistinguishable identical particles in the case of multiparticle quantum walks. There, the relation between the average distance $\langle|\hat{\Delta}|\rangle$ between two walkers (the "macroscopic feature" of the system) and entanglement was established and it was shown that the value of $\langle|\hat{\Delta}|\rangle$ is bounded on the set of separable initial states, thus proving to be an entanglement witness.

Finally, we mention other results regarding the existence of entanglement in open macroscopic systems in non-equilibrium steady states[69,70] that are not directly motivated by implementations in quantum information science.

21.5. Summary

Usually, the basic laws of physics, such as the three Newton laws of classical mechanics or the Schrödinger equation of quantum mechanics, are originally given in forms of various mathematical equations for single physical systems. Including the possible interaction terms, the composite systems are then defined by a set of such equations the number of which scales with the number of sub-systems, linearly in the case of Newton laws, exponentially in the case of quantum mechanics. Thus, at least in principle, every physical system, no matter how complex it is, can be described in terms of its subsystems (note that this is not the same as to say "in terms of subsystems' properties only", which in the case of entangled quantum states is indeed not true). The problems appear when dealing with large systems consisting of many particles, such as all macroscopic systems we encounter in everyday life are. Not only that we are unable to solve all those equations, neither analytically nor numerically, but also the initial conditions of such a system would be impossible to handle (neither we would be able to measure all the one- and multi- particle observables determining the system's state, nor we would be able to do it with enough precision – both in classical and quantum physics). Finally, even if one had all the equations solved for the right initial conditions, such a huge information would be of no practical use for us at all! In fact, when describing complex macroscopic systems, one is interested in determining only a few macroscopic features that persist on global level and are relevant for us, such as the system's magnetization and magnetic susceptibility, internal energy and heath capacity, etc. They reflect the *collective* behavior of large numbers of sub-systems, quantified in terms of various correlation functions, and one might say that statistical physics is in fact the science of correlations at the first place.

In this chapter, we discussed the particular type of such correlations that appear in quantum systems only – the entanglement – in genuine macroscopic systems on finite temperatures. We have showed, on the example of macroscopic high-temperature superconductivity, that the ODLRO correlations that determine the superconducting features of the system are entanglement-based. Thus, entanglement in such systems survives up to critical temperatures for the superconductor-insulator phase transition that can be as high as $160K$. Further, on the examples of magnetic susceptibility and internal energy, we showed that the genuinely non-classical part of entanglement-based correlations can affect the macroscopic thermodynamic properties which, in turn, can act as the witnesses of entanglement. This enabled us to derive various transition temperatures below which entanglement surely exists. In particular, one can show that in the case of $Na_2V_3O_7$ entanglement can exist up to $T = 365K$! Finally, we discussed the cases of spin and spatial entanglement in Fermi and Bose ideal gases and its connection with the pressure and the volume. In addition to this, we have shown that macroscopic thermal entanglement can also be generated and manipulated in order to extract it and use it as a resource in various quantum information tasks.

Acknowledgments

The authors thank the support from Fundação para a Ciência e a Tecnologia (Portugal) through programs POCTI/POCI, partially funded by FEDER (EU), namely through projects POCI/MAT/55796/2004 QuantLog, POCTI-SFA-2-91 and POCI/FIS/58746/2004, and grant SFRH/BPD/31807/2006, as well as the SQIG-IT EMSAQC initiative.

References

1. S. Sachdev, *Quantum Phase Transitions*. (Cambridge University Press, 2001).
2. K. Życzkowski, P. Horodecki, A. Sanpera, and M. Lewinstein, Volume of the set of separable states, *Phys. Rev. A*. **58**, 883, (1998).
3. D. Cavalcanti, F. G. S. L. Brandão, and M. O. Terra Cunha, Entanglement quantifiers and phase transitions, *New J. Phys.* **8**, 260, (2006).
4. B. V. Fine, F. Mintert, and A. Buchleitner, Equilibrium entanglement vanishes at finite temperature, *Phys. Rev. B*. **71**, 153105, (2005).
5. O. Osenda and G. A. Raggio, Entanglement in thermal equilibrium states, *Phys. Rev. A*. **72**, 064102, (2005).
6. G. Vidal and R. Tarrach, Robustness of entanglement, *Phys. Rev. A*. **59**, 141, (1999).
7. D. Markham, J. Anders, V. Vedral, and M. Murao. Survival of entanglement in thermal states. quant-ph/0606103, (2006).
8. M. A. Nielsen. *Quantum information theory*. PhD thesis, University of New Mexico, Albuquerque, New Mexico, USA (dec, 1998). quant-ph/0011036.
9. M. C. Arnesen, S. Bose, and V. Vedral, Natural thermal and magnetic entanglement in 1D Heisenberg model, *Phys. Rev. Lett.* **87**, 017901, (2001).
10. M. Asoudeh and V. Karimipour, Thermal entanglement of spins in the Heisenberg model at low temperatures, *Phys. Rev. A*. **70**, 052307, (2004).
11. W. K. Wootters, Entanglement of formation of an arbitrary state of two qubits, *Phys. Rev. Lett.* **80**, 2245, (1998).
12. D. Gunlycke, V. M. Kendon, V. Vedral, and S. Bose, Thermal concurrence mixing in one-dimensional Ising model, *Phys. Rev. A*. **64**, 042302, (2001).
13. A. Saguia and M. S. Sarandy, Entanglement in the one-dimensional Kondo necklace model, *Phys. Rev. A*. **67**, 012315, (2003).
14. U. Glaser, H. Büttner, and H. Feske, Entanglement and correlation in anisotropic quantum spin systems, *Phys. Rev. A*. **68**, 032318, (2003).
15. V. Vedral, High-temperature macroscopic entanglement, *New J. Phys.* **6**, 102, (2004).
16. C. N. Yang, η pairing and off-diagonal long-range order in a Hubbard model, *Phys. Rev. Lett.* **63**, 2144, (1989).
17. F. H. L. Essler, V. E. Korepin, and K. Schoutens, New exactly solvable model of strongly correlated electrons motivated by high-T_c superconductivity, *Phys. Rev. Lett.* **68**, 2960, (1992).
18. F. H. L. Essler, V. E. Korepin, and K. Schoutens, Electronic model for superconductivity, *Phys. Rev. Lett.* **70**, 73, (1993).
19. J. de Boer, V. E. Korepin, and A. Schadschneider, η pairing as a mechanism of superconductivity in models of strongly correlated electrons, *Phys. Rev. Lett.* **74**, 789, (1995).
20. C. N. Yang, Concept of off-diagonal long-range order and the quantum phases of liquid He and of superconductors, *Rev. Mod. Phys.* **34**, 694, (1962).
21. A. Peres, Higher-order Schmidt decompositions, *Phys. Lett. A*. **202**, 16, (1995).

22. M. Horodecki, P. Horodecki, and R. Horodecki, Separability of mixed states: necessary and sufficient conditions, *Phys. Lett. A.* **223**, 1, (1996).
23. V. Vedral, M. B. Plenio, M. A. Rippin, and P. L. Knight, Quantifying entanglement, *Phys. Rev. Lett.* **78**, 2275, (1997).
24. V. Vedral, Mean-field approximations and multipartite thermal correlations, *New. J. Phys.* **6**, 22, (2004).
25. F. G. S. L. Brandão, Entanglement as a quantum order parameter, *New J. Phys.* **7**, 254, (2005).
26. J. Anders and V. Vedral. Macroscopic entanglement and phase transitions. quant-ph/0610268, (2006).
27. V. Vedral. The Meissner effect and massive particles as witnesses of macroscopic entanglement. quant-ph/0410021, (2004).
28. Y. Aharonov and D. Bohm, Significance of electromagnetic potentials in the quantum theory, *Phys. Rev. (Ser. 2).* **115**, 485–491, (1959).
29. J. Bell, On the Einstein-Podolsky-Rosen paradox, *Physics.* **1**, 195, (1964).
30. S. Gosh, T. F. Rosenbaum, G. Aeppli, and S. N. Coppersmoth, Entangled quantum state of magnetic dipoles, *Nature.* **425**, 48, (2003).
31. J. H. V. Vleck, Quantum mechanics-The key to understanding magnetism, *Rev. Mod. Phys.* **50**, 181, (1978).
32. M. Wieśniak, V. Vedral, and Časlav Brukner, Magnetic susceptibility as a macroscopic entanglement witness, *New J. Phys.* **7**, 258, (2005).
33. H. F. Hofmann and S. Takeuchi, Violation of local uncertainty relations as a signature of entanglement, *Phys. Rev. A.* **68**, 032103, (2003).
34. B. C. Hiesmayr and V. Vedral. Thermodinamical versus optical complementarity. quant-ph/0501015, (2005).
35. Č. Brukner, V. Vedral, and A. Zeilinger, Crucial role of quantum entanglement in bulk properties of solids, *Phys. Rev. A.* **73**, 012110, (2006).
36. T. Vértesi and E. Bene, Thermal entanglement in the Nanotubular system $Na_2V_3O_7$, *Phys. Rev. B.* **73**, 134404, (2005).
37. I. Bose and A. Tribedi, Thermal entanglement properties of small spin clusters, *Phys. Rev. A.* **72**, 022314, (2005).
38. Č. Brukner and V. Vedral. Macroscopic thermodynamical witnesses of quantum entanglement. quant-ph/0406040, (2004).
39. G. Tóth, Entanglement witnesses in spin models, *Phys. Rev. A.* **71**, 01301(R), (2005).
40. L.-A. Wu, S. Bandyopadhyay, M. S. Sarandy, and D. A. Lidar, Entanglement observables and witnesses for interacting quantum spin systems, *Phys. Rev. A.* **72**, 032309, (2005).
41. X. Wang and P. Zanardi, Quantum entanglement and Bell inequalities in Heisenberg spin chains, *Phys. Lett. A.* **301**, 1, (2002).
42. M. R. Dowling, A. C. Doherty, and S. D. Bartlett, Energy as entanglement witness for quantum many-body systems, *Phys. Rev. A.* **70**, 062113, (2004).
43. O. Gühne and G. Tóth, Energy and multipartite entanglement in multidimensional and frustrated spin models, *Phys. Rev. A.* **73**, 052319, (2006).
44. T. C. Wei and P. M. Goldbart, Geometric measure of entanglement and applications to bipartite and multipartite quantum states, *Phys. Rev. A.* **68**, 042307, (2003).

45. N. Paunković. *The role of indistinguishability of identical particles in quantum information processing*. PhD thesis, University of Oxford, Oxford, UK (Trinity Term, 2004).
46. V. Vedral, Entanglement in the second quantization formalism, *Central Eur. J. Phys.* **1**, 289, (2003).
47. S. Oh and J. Kim, Entanglement of electron spins of noninteracting electron gases, *Phys. Rev. A.* **69**, 054305, (2004).
48. D. Cavalcanti, M. F. Santos, M. O. Terra Cunha, C. Lunkes, and V. Vedral, Increasing identical particle entanglement by fuzzy measurements, *Phys. Rev. A.* **72**, 062307, (2005).
49. C. Lunkes, Č. Brukner, and V. Vedral, Natural multiparticle entanglement in a Fermi gas, *Phys. Rev. Lett.* **95**, 030503, (2005).
50. C. Lunkes, Č. Brukner, and V. Vedral, Equation of state for entanglement in a Fermi gas, *Phys. Rev. A.* **71**, 034309, (2005).
51. J. Anders, D. Kaszlikowski, C. Lunkes, T. Ohshima, and V. Vedral, Detecting entanglement with a thermometer, *New J. Phys.* **8**, 140, (2005).
52. M. O. Terra Cunha, J. A. Dunningham, and V. Vedral. Entanglement in single particle systems. quant-ph/0606149, (2006).
53. L. Heaney, J. Anders, and V. Vedral. Spatial entanglement of a free Bosonic field. quant-ph/0607069, (2006).
54. M. A. Nielsen and I. L. Chuang, *Quantum computation and quantum information*. (Cambridge University Press, 2000).
55. A. Einstein, B. Podolsky, and N. Rosen, Can quantum-mechanical description of physical reality be considered complete?, *Phys. Rev.* **47**, 777, (1935).
56. G. D. Chiara, Č. Brukner, R. Fazio, G. M. Palma, and V. Vedral, A scheme for entanglement extraction from a solid, *New. J. Phys.* **8**, 95, (2006).
57. S. Scheel, J. Eisert, P. L. Knight, and M. B. Plenio, Hot entanglement in a simple dynamical model, *J. Mod. Opt.* **50**, 881, (2003).
58. F. Benatti, R. Floreanini, and M. Piani, Environment induced entanglement in Markovian dissipative dynamics, *Phys. Rev. Lett.* **91**, 070402, (2003).
59. M. Paternostro, W. Son, and M. S. Kim, Complete conditions for entanglement transfer, *Phys. Rev. Lett.* **92**, 197901, (2004).
60. J. Zhang and H. Yu. Entanglement generation in atoms immersed in a thermal bath of external quantum scalar field with a boundary. quant-ph/0611170, (2006).
61. X. X. Yi, H. T. Cui, and L. C. Wang. Entanglement induced in spin-$\frac{1}{2}$ particles by a spin-chain near its critical points. quant-ph/0511026, (2005).
62. H. T. Quan, Z. Song, X. F. Liu, P. Zanardi, and C. P. Sun, Decay of Loschmidt echo by quantum criticality, *Phys. Rev. Lett.* **96**, 140604, (2006).
63. P. Zanardi and N. Paunković, Ground state overlap and quantum phase transitions, *Phys. Rev. E.* **74**, 031123, (2006).
64. D. Kaszlikowski and V. Vedral. Local extraction of EPR entanglement from classical systems. quant-ph/0606238, (2006).
65. M. Christiandl, N. Datta, A. Ekert, and A. J. Landahl, Perfect state transfer in quantum spin networks, *Phys. Rev. Lett.* **92**, 187902, (2004).
66. S. Bose. Entanglement from the dynamics of an ideal Bose gas in a lattice. quant-ph/0610024, (2006).

67. G. Vidal and R. F. Werner, Computable measure of entanglement, *Phys. Rev. A*. **65**, 032314, (2002).
68. Y. Omar, N. Paunković, L. Sheridan, and S. Bose, Quantum walk on a line with two entangled particles, *Phys. Rev. A*. **74**, 042304, (2006).
69. L. Hartmann, W. Dür, and H. J. Briegel. Steady state entanglement in open and noisy quantum systems at high temperature. quant-ph/0512219, (2005).
70. L. Quiroga, F. J. Rodríguez, M. E. Ramírez, and R. París. Nonequilibrium thermal entanglement. quant-ph/0612046, (2006).

Subject Index